Mathematical Theory of Control Systems Design

Mathematics and Its Applications

Volume 341

Mathematical Theory of Control Systems Design

by

V. N. Afanas'ev,
V. B. Kolmanovskii

and

V. R. Nosov

Moscow University of Electronics and Mathematics,
Moscow, Russia

SPRINGER-SCIENCE+BUSINESS MEDIA, B.V.

Library of Congress Cataloging-in-Publication Data

Afanasiev, V. N.
Mathematical theory of control systems design / by V.N. Afanasiev, V.B. Kolmanovskii, and V.R. Nosov.
p. cm. -- (Mathematics and its applications ; v. 341)
Includes bibliographical references and index.
ISBN 978-90-481-4615-4 ISBN 978-94-017-2203-2 (eBook)
DOI 10.1007/978-94-017-2203-2
1. Control theory. I. Kolmanovskiĭ, Vladimir Borisovich. II. Nosov, V. R. III. Title. IV. Series: Mathematics and its applications (Kluwer Academic Publishers) ; v. 341.
QA402.3.A37 1995
629.8'312--dc20 95-20901

ISBN 978-90-481-4615-4

Printed on acid-free paper

This is a completely revised and
updated translation of the original Russian work
Mathematical Theory of Control Systems Design,
Vysshaya Shkola, Moscow, © 1989.

Dedicated to our wives
Natalia, Tatiana, and Irina

TABLE OF CONTENTS

Preface xix

Introduction xxi

PART ONE. STABILITY OF CONTROL SYSTEMS

Chapter I. Continuous and Discrete Deterministic Systems 3

§ 1. Basic Definitions of Stability Theory for Continuous systems 3

1. Stability 3
2. Asymptotic stability 8
3. Other stability definitions 10

§ 2. Lyapunov's Direct Method 11

§ 3. Examples of the Use of Lyapunov's Method 16

1. Stability of the motion of a shell 16
2. Motion of a rigid body fastened at one point 18

§ 4. Development of Lyapunov's Method 19

1. The Barbashin–Krasovskii theorem 20
2. The Matrosov criterion 22
3. The comparison principle 23
4. Stability with respect to part of the variables (partial stability) 24

§ 5. Stability of Linear Time-Invariant Systems 25

1. The Routh–Hurwitz criterion 25
2. Frequency criteria of stability 29
3. Automatic frequency control system of a heterodyne receiver 31
4. Linear single-circuit automatic control systems 33
5. Robust stability 34

§ 6. Stability of Linear Time-Varying Equations 36

1. On the method of "frozen" coefficients 36
2. Systems with an almost constant matrix 37
3. Linear systems with periodic coefficients 39
4. Equation of the second order with periodic coefficients 40
5. Parametric resonance in engineering applications 42

§ 7. Lyapunov Functions for Linear Time-Invariant Systems and First Approximation Stability 43
1. Lyapunov's matrix equation 43
2. Stability in the first approximation 46
3. Rotation of a shell 48
4. Time-Varying equations of the first approximation 48
§ 8. Synthesis of Control Systems for the Manipulator Robot 49
1. Single-link manipulator robot 49
2. The Cyclone-type robot 50
§ 9 Use of the Logarithmic Norm in Stability Theory 53
1. The definition and properties of the logarithmic norm 53
2. Stability of nonlinear systems 54
§ 10. Use of Degenerate Lyapunov Functions 59
§ 11. Stability of Discrete Systems 63
1. Lyapunov's direct method 64
2. Linear time-invariant equations 65
3. Stability in the first approximation 66
4. Stability with respect to specified variables 66
5. Unconditional minimization of functions 67

Main Results and Formulas of Chapter I 69

Chapter II. Stability of Stochastic Systems 73
§ 1. Introduction 73
§ 2. Some Preliminaries from Probability Theory and the Theory of Stochastic processes 73
1. Basic probability space 73
2. Random variables 74
3. Stochastic processes 76
§ 3. Stochastic Integrals and Stochastic Differential Equations 79
1. The stochastic integrals of Ito and Stratonovich 79
2. The Ito formula 82
3. Markov diffusion processes 86
4. Linear stochastic equations 88
§ 4. Definition of Stochastic Stability 90
§ 5. Application of Lyapunov's Direct Method 93
1. Sufficient stability conditions 93
2. Stability in mean square of linear systems 96
3. Scalar equations of the nth order 98
§ 6. Stability in Probability of Satellite Motion 99
1. The pitch stability of a symmetric satellite in a circular orbit 99
2. The yaw angle stability of a satellite in a circular equatorial orbit 100

Main Results and Formulas of Chapter II 101

Exercises of Part One 105

PART TWO. CONTROL OF DETERMINISTIC SYSTEMS

Chapter III. Description of Control Problems 127
§ 1. Introduction 127
§ 2. Statement of the Optimal Control Problem 128
1. The equations of evolution of a system 128
2. The functional to be minimized (cost functional) 129
3. Trajectory constraints 131
4. Control constraints 132
5. Joint constraints 135
§ 3. Examples of Optimal Control in Engineering 137
1. Optimal control of an electric motor 137
2. Optimization of the characteristics of nuclear reactors 139
3. Optimal control of spacecraft 145

Main Results and Formulas of Chapter III 150

Chapter IV. The Classical Calculus of Variations and Optimal Control 151
§ 1. Problems with a Variable End Point and Fixed Time 151
1. Main assumptions 151
2. Cauchy's formula 152
3. Necessary conditions for optimal control 153
4. The Boltz problem 156
§ 2. Optimal Control of Linear Systems with a Quadratic Functional 157
1. Necessary conditions for optimal control 157
2. Construction of an optimal control 158
3. Matrix Riccati equation 159
4. The scalar case 163
5. Optimal control of wire reeling 166
§ 3. Necessary Conditions for Optimal Control. The Method of Lagrange Multipliers 168
1. The method of Lagrange multipliers 168
2. Fixed initial and terminal moments, and fixed initial state 172
3. Fixed initial and terminal moments, and variable initial and terminal states 172
4. Problems with fixed values of some state variables at the initial and terminal moments 173
5. Problems with an unspecified terminal moment 174
6. The Chaplygin problem 175

7. Maximization of the rocket velocity immediately before putting the rocket into a rectilinear trajectory 177

Main Results and Formulas of Chapter IV 180

Chapter V. The Maximum Principle 183

§ 1. Problems with a Variable Terminal Point and Prescribed Transfer Time 183

1. The Mayer problem 183
2. The Boltz problem 186
3. About solutions of maximum principle equations 188
4. Rotation of a motor shaft through a maximum angle 189

§ 2. Problems with an Unspecified Terminal Moment 190

1. Reduction to a Mayer problem 190
2. Necessary conditions for the optimality of systems that are linear in a scalar control variable 191
3. Multivariable control 195
4. Time-invariant systems 196
5. Transfer of a system from one manifold to another manifold 196
6. Control problems with isoperimetric constraints 197
7. Sufficiency of the maximum principle 201
8. Connection between the maximum principle and the classical calculus of variations 201
9. The maximum principle for discrete systems 202

§ 3. Practical Applications of the Maximum Principle 205

1. Optimal configuration of a nuclear reactor 206
2. Control of a motion with regulated friction 208
3. Problem of the soft landing on the moon 214
4. Problem of the planar transfer of a space vehicle from one circular orbit to another 217

§ 4. Control of Ecological Systems 219

1. Equations describing the evolution of a single population 220
2. Communities of at least two species 222
3. Statements of typical control problems for ecological systems 225
4. Time-optimal control of the "predator–prey" system 227

Main Results and Formulas of Chapter V 234

Chapter VI. Linear Control Systems 239

§ 1. A Time-Optimal Problem 239

1. Evaluation of the number of switch points 239
2. The damping of the material point 242
3. The damping of a pendulum 244

4. Control of the rotation of an axially symmetric space vehicle 246
5. The controllability set 249
§ 2. Controllability of Linear Systems 252
1. Controllability of linear time-invariant systems 252
2. Controllability of linear time-varying systems 255
3. Canonical form of linear time-invariant control systems 258
4. Canonical form of linear time-varying control systems 260
5. The Hautus criterion for controllability 261
6. Controllability of a two-link manipulator 263
§ 3. Observation in Linear Systems. Observers 265
1. Statement of an observation problem. Duality of control and observation problems 265
2. On a method of determining the state vector 271
3. Observer of maximum order 272
4. Observer of reduced order (the Luenberger observer) 273
5. Observer of reduced order in the stabilization system of an aircraft 276
§ 4. Linear Feedback Control Systems 280
1. Various ways of describing feedback control systems. The realization problem 280
2. Criteria of performance for SISO-systems 283
3. Criteria of performance for MIMO-systems. The Hardy spaces H_2 and H_∞ 284
§ 5. Fundamentals of H_∞-Theory 286
1. Statement of the problem of constructing an H_∞-optimal controller 286
2. Estimates of the H_∞ and H_2 norms of the transfer matrix of an auxiliary system 287
3. The H_∞-problem for a static controller 288
4. The general case of a dynamic controller 290
5. Robust stability 291
§ 6. Zeros of a Linear Time-Invariant System and Their Use 292

Main Results and Formulas of Chapter VI 297

Chapter VII. Dynamic Programming Approach. Sufficient Conditions for Optimal Control 301
§ 1. The Bellman Equation and its Properties 301
1. The principle of dynamic programming. Heuristic derivation of The Bellman equation 301
2. Determining an F-control with the help of the dynamic programming approach 303

3. Connection between the dynamic programming approach and the maximum principle 305
4. Determining F-control in the problem of damping the motion of a rigid body 306
5. Optimal procedure for reducing the power of a nuclear reactor 307
6. The linear-quadratic problem 308

§ 2. Control on an Unbounded Time Interval. Stabilization of Dynamical Systems 309
1. Problem statements 309
2. Lyapunov's direct method for the optimal stabilization Problem 310
3. Exponential stabilization 312
4. Stabilization of the motion of a manipulator robot 315

§ 3. Stabilization of Linear Systems 320
1. Time-varying linear-quadratic problems 320
2. The use of the method of successive approximations for determining optimal control 321
3. Time-invariant linear-quadratic equation 322
4. The algebraic Riccati equation 323

§ 4. Stabilization of Quasilinear Systems 325
1. Quasioptimal stabilization and estimation of its error 325
2. Adaptive stabilization 331

§ 5. Sufficient Conditions for Optimal Control Using Auxiliary Functions 332
1. Conditions for optimal control 332
2. Sufficient conditions for the existence of a minimizing sequence 337
3. Problems with unspecified time 337

Main Results and Formulas of Chapter VII 339

Chapter VIII. Some Additional Topics of Optimal Control Theory 343

§ 1. Existence of Optimal Control 343
1. Problem statement and the main assumptions 343
2. Main theorem 344
3. Analysis of the conditions of the main theorem 349

§ 2. Singular Optimal Control 351
1. The definition and determination of singular controls 351
2. Optimality of singular control 355
3. Generalization of the Kelley and Kopp–Moyer conditions 358

§ 3. Chattering Mode 359

§ 4. Sliding Optimal Mode 365

Main Results and Formulas of Chapter VIII 372

Exercises of Part Two 375

PART THREE. OPTIMAL CONTROL OF DYNAMICAL SYSTEMS UNDER RANDOM DISTURBANCES

Chapter IX. Control of Stochastic Systems. Problem Statements and Investigation Techniques 393
§ 1. Statements of Control Problems for Stochastic Systems 393
1. Equations of motion of a system 393
2. Control constraints 394
3. Cost functional 396
§ 2. Dynamic Programming Approach 398
1. The Bellman function 398
2. The Bellman equation 399
3. Connection between the Bellman function and the Bellman equation 403
§ 3. Stochastic Linear-Quadratic Problem on a Finite Time Interval 406
1. Linear-quadratic problems in the case of an accurate measurement of phase coordinates 406
2. Linear-quadratic problems under incomplete information 409
3. Optimal program control in linear-quadratic problems 412
4. Linear-quadratic problem for Gaussian and Poisson disturbances 415
5. Control of wire reeling with regard to random disturbances 416
§ 4. Control on an Unbounded Time Interval. Stabilization of Stochastic Control Systems 417
1. Problem statement 417
2. Application of Lyapunov's direct method to optimal stabilization problems 417
3. Stabilization of linear stochastic systems 421
§ 5. Approximate Methods for Determining Optimal Control 422
1. Description of the algorithm of successive approximations 423
2. Zero approximation estimate 424
3. First approximation estimate 427
4. Higher-order approximation estimates 429
Main Results and Formulas of Chapter IX 431
Chapter X. Optimal Control on a Time Interval of Random Duration 435
§ 1. Time-Optimal Control 435

1. Statement of time-optimal problems in dynamical systems under random disturbances 435
2. Existence of an admissible control 436
3. An algorithm for constructing optimal control 438
4. Time-optimal control of the motion of a rigid body 439
5. Numerical construction of the time-optimal control of the motion of a material point 441

§ 2. Time-Optimality for a Gyrostat 444
1. Problem statement 444
2. The existence of an admissible control 446
3. Construction of an optimal control 448

§ 3. Control Problems with Stochastic Functional 450
1. Problem statement and method of solution 450
2. Optimal control of the motion of a material point involving a stochastic cost functional 452
3. Maximization of the probability that a point stays in a given region 455
4. Stochastic control of the motion of a simple pendulum 457
5. Control of a rigid body, given the stochastic cost functional 460
6. Maximization of the mean time for the system staying within a given region. Determination of optimal control for the motion of a rigid body 463

Main Results and Formulas of Chapter X 466

Chapter XI. Optimal Estimation of the State of the System 467

§ 1. Estimation Problems Involving Random Disturbances 467
1. Statement of an optimal estimation problem 467
2. Linear estimation 468
3. Optimal estimation of Gaussian variables 469
4. Linear estimation of stationary processes. The Wiener–Hopf equation 469

§ 2. The Kalman filter 472
1. Problem statement 472
2. The dual problem of optimal control 472
3. Equation for the estimation error 474
4. Equation for the optimal estimate 476
5. Stability of the filter 478
6. Filtering problem with constant parameters 480
7. Filtering problem with degenerate noise in the observation channel 481
8. Optimal extrapolation 484
9. Optimal interpolation 485
10. The discrete Kalman filter 486

§ 3. Some Relations of Nonlinear Filtering Theory 488

1. Statement of the general filtering problem 488
2. The Stratonovich differential equations for the conditional probability distribution 489
3. Conditionally Gaussian processes 490
4. Quasioptimal and quasilinear filtering (the scalar case) 491
5. Quasilinear filtering (the multidimensional case) 499

Main Results and Formulas of Chapter XI 503

Chapter XII. Optimal Control of the Observation Process 509
§ 1. Optimization of the Observation Process 509
1. Problem statement. Basic relations 509
2. Construction of optimal observation laws minimizing terminal variance 513
3. An example of observation control with integral cost functional 515
4. Optimal pulse observation laws 517
5. Pulsed observation of a material point 522
6. Optimal noise control in the observation channel 524
§ 2. Optimal Combination of Control and Observation 529
1. Linear-quadratic problem 529
2. A deterministic scalar system 532
3. A stochastic scalar system 534

Main Results and Formulas of Chapter XII 542

Exercises of Part Three 543

PART FOUR. NUMERICAL METHODS IN CONTROL SYSTEMS

Chapter XIII. Linear Time-Invariant Control Systems 555
§ 1. Stability of Linear Time-Invariant Systems 555
1. Precise methods for solving the complete eigenvalue problem 555
2. Iterative methods 559
3. The Routh–Hurwitz criterion 561
4. The Zubov method of a functional transformation of a matrix 562
§ 2. Methods of Solving the Lyapunov Equation 563
1. Preliminary remarks 563
2. The series method 564
3. Method of the matrix sign function 565
4. Application of the QR-algorithm 567
5. Construction of stabilizing control 568
6. Computation of the covariance matrix 570
§ 3. Controllability and Observability 570

§ 4. Linear-Quadratic Time-Invariant Problem of Optimal stabilization 572
1. Reduction to a sequence of Lyapunov equations 572
2. Use of the QR-algorithm 573
3. Landing control of a Boeing-747 airplane 576

Main Results and Formulas of Chapter XIII 579

Chapter XIV. Numerical Methods for the Investigation of Nonlinear Control Systems 581
§ 1. Analysis of Transients. The Runge–Kutta Methods 581
1. On numerical methods of investigating systems 581
2. One-step methods 581
3. Error estimates for Runge–Kutta methods 585
4. Estimation of solution errors for stable equations 588
5. Standard programs 589
§ 2. Analysis of Transients. Multistep Methods 591
1. General definitions 591
2. Particular multistep formulas 591
3. Computational scheme for multistep formulas 594
4. Error estimation 596
5. The Butcher formulas 599
§ 3. Stiff Systems of Equations 600
§ 4. Numerical Methods for the Design of Optimal Control in the Linear-Quadratic Problem 607
1. Determination of an optimal F-control for time-varying systems 607
2. Time-invariant linear-quadratic problem on a finite interval 608
3. Optimal stabilization problems and methods for solving the algebraic Riccati equation 612
4. The numerical design of the Kalman filter 613

Main Results and Formulas of Chapter XIV 617

Chapter XV. Numerical Design of Optimal Control Systems 619
§ 1. On Numerical Methods of Solving Optimal Control Problems 619
§ 2. Reduction to Nonlinear Programming 620
§ 3. Reduction to a Boundary-Value Problem 621
1. Boundary-value problem of the maximum principle 621
2. The Newton method of solving a boundary-value problem 622
3. An example of reducing an optimal control problem to a boundary-value problem 623
§ 4. Solution of Linear Boundary-Value Problems 625

1. Reduction of an optimal control problem to a linear boundary-value problem 625
2. Reduction of a linear boundary-value problem to Cauchy problems 626
3. Method of the transfer of boundary conditions 628
4. The Abramov method 630
§ 5. The Shatrovskii Method of Successive Improvements of Control 632
§ 6. The Fedorenko Method of Reducing to a Linear Programming Problem 635
Main Results and Formulas of Chapter XV 639
Exercises of Part Four 641
General References 657
Subject Index 663

PREFACE

Give, and it shall be given unto you.
ST. LUKE, VI, 38.

The book is based on several courses of lectures on control theory and applications which were delivered by the authors for a number of years at Moscow Electronics and Mathematics University. The book, originally written in Russian, was first published by Vysshaya Shkola (Higher School) Publishing House in Moscow in 1989. In preparing a new edition of the book we planned to make only minor changes in the text. However, we soon realized that we like many scholars working in control theory had learned many new things and had had many new insights into control theory and its applications since the book was first published. Therefore, we rewrote the book especially for the English edition. So, this is substantially a new book with many new topics.

The book consists of an introduction and four parts.

Part One deals with the fundamentals of modern stability theory: general results concerning stability and instability, sufficient conditions for the stability of linear systems, methods for determining the stability or instability of systems of various type, theorems on stability under random disturbances.

Part Two is devoted to the optimal control of deterministic systems. Here we present some principal statements of control problems based on specific applications, the connection of modern problems and methods of control theory with stability theory and the calculus of variations, the main results of control theory—Pontryagin's maximum principle and Bellman's dynamic programming approach, the theory of linear control systems, of singular, sliding, and chattering modes.

Part Three is concerned with problems of the control of systems under random disturbances of their parameters. Here we deal with the basic principles of estimation theory, with the Kalman filter and some relations of nonlinear filtering, including quasilinear filtering, and treat control problems related to the tracking of phase coordinates, and with the synthesis of optimal control in stochastic systems.

Part Four provides an outline of modern numerical methods of control theory, such as the numerical determination of stability conditions, methods of solving the Lyapunov and Riccati equations, the numerical treatment of the linear-quadratic problem, methods of analysis of control systems, including one-step and multistep methods, as well as methods of constructing optimal control in nonlinear systems.

The treatment of the problems presented in the book is rigorous but at the same time easy-to-grasp. This is due to a detailed discussion of specific control systems that are encountered in various branches of mechanics, astronautics, automatic control of manufacturing processes, and of ecology. A lot of attention is given to precise statements of mathematical problems and results, as well as to numerical and approximation procedures for constructing optimal (suboptimal) control, to the consideration of diverse examples of actual control systems, including computerized ones. These examples play a special role. They not only serve to illustrate the main assertions and algorithms but also assist the reader in acquiring the necessary skills needed for constructing the models of relevant phenomena, for designing nonlinear control systems, and for explaining the qualitative differences between various classes of control systems, as well as for using the results given in the book in the investigation of particular systems. In addition, almost 200 exercises designed to be solved by the reader are provided as well as diverse graphical material. The mastering of the material of the book will facilitate the assimilation of contemporary articles and monographs dealing with specialized problems of control theory.

The list of references is far from complete. Only the works that are directly related to the material of the book are included in the bibliography at the end of the book. In addition to this, numerous references to the original sources (more than 80 titles) are given in the text as footnotes. Special care is taken to highlight the most important results contained in the original sources which are unavailable in English.

During the preparation of the book we have benefited from discussions with many of our colleagues. We want to thank them all and apologize for not mentioning them by name. Furthermore, we would like to thank our students whose opinion was decisive in choosing the style of presentation of the material. The preparation of the English-language edition of the book was supported informally by prof. N. K. Kulman. Also support was given by the Soros International Science Foundation, by the Russian Foundation for Fundamental Research, and by INTAS-93-2622 "Optimal control and differential games: theory, numerical methods, and applications".

This book is a manual for the course of lectures "Control theory" and is intended for university and college students (both undergraduate and postgraduate) majoring in the subjects "Applied mathematics", "Mechanics", "Control problems in engineering", "Automation", "Cybernetics". It will also be useful to scientists and engineers working in these and neighboring areas of research and development.

As is usual in our practice, the names of the authors are given in strict alphabetical order. The authors realize that the material of the book is not devoid of shortcomings and will be grateful for any constructive criticism.

The authors

INTRODUCTION

The analysis and design of control systems which possess the required properties is the key problem of control theory. Foremost among these properties are stability, optimality, behavior under random disturbances, and so on.

The emphasis in this book is on the investigation of evolutionary systems. The general concept of an abstract system has evolved over the last 20–30 years; its very generality makes the rigorous definition of such a concept very complicated. We shall confine ourselves to some particular classes of evolutionary systems and provide for them all necessary definitions. On the descriptive level, by an evolutionary system we mean an engineering, physical, biological, ecological, or any other system whose changes evolving with time are studied. In mathematical terms, evolutionary systems can be described in various ways. The most commonly occurring classes of evolutionary systems are the following:

continuous systems described by ordinary differential equations;

discrete systems described by finite difference equations;

distributed parameter systems described by evolutionary partial differential equations, such as, for example, the heat conduction, the wave and the hydrodynamics equation, etc.;

systems with aftereffect for whose description functional-differential equations are used. Such systems appear in situations where the evolution of the process depends not only on the present state of the system, but also on the past history of the process;

stochastic systems; any of the above-mentioned systems can be stochastic if probabilistic concepts and methods are used for its description.

Let us cite a few examples of evolutionary systems:

(1) the solar system which is described with great accuracy by the system of ordinary differential equations defined by Newton's law of universal gravitation;

(2) fluid flow is usually described by the Navier–Stokes nonstationary partial differential equation; to describe turbulent flow, probabilistic methods are often used;

(3) the flight of an airplane; depending on the accuracy requirements, such a system is described by ordinary differential equations or, allowing for the elasticity of the structure members, by partial differential equations; the use of computers in the control loop gives rise to difference equations.

It should be noted that a physical system can be described by different mathematical models depending on the goal of the investigation and the accuracy requirements.

In the treatment of various practical control problems an essential role is played by stability theory. The term "stability" is very expressive and has a lot of connotations. By stability is usually meant the property of a system to preserve its state despite perturbations of the initial conditions, changes in the values of the parameters of the system, regular and random disturbances, and so on. Of course, mathematical rigor demands that the concept of stability be made more precise; however, the necessary formalization of the concept can be performed in a number of ways. This leads to different mathematical concepts of stability, some of which will be rigorously defined in the sequel.

Much attention is given in the book to the basic results concerning Lyapunov's direct method (the method of Lyapunov's functions). Lyapunov's method is not only one of the main methods for analysing the existing systems, but also an important method for the synthesis of new control systems. Therefore, it proves to be expedient to begin the presentation of the subject by first turning attention to stability theory and Lyapunov's direct method and subsequently to tie in all other material with them.

In the investigation of actual objects (plants) one often has to take into account various indeterminate factors connected with the system. These factors are often related to the uncertainty about forces acting on the system, to the retardation and errors of the control program, and so on. There are different ways to formalize such uncertainties in mathematical terms.

In the probabilistic approach used in this book various uncertainties are interpreted as random processes. Some basic notions from probability theory and the theory of stochastic processes are briefly sketched. Definitions of various types of stochastic stability are given, and the stochastic analog of Lyapunov's direct method is described.

A characteristic feature of contemporary science is a tremendous interest in extremal problems (i.e., in problems of determining maximum and minimum values, optimal conditions for a manufacturing process, and so on). It must be stressed that such problems have been encountered by man from the times of antiquity.

Extremal problems were the subject of investigation by J. Kepler, P. Fermat, C. Huygens, I. Newton, J. Bernoulli, J. Lagrange, L. Euler, K. Weierstrass among others.

However, it is at the present state of development that extremal problems have become especially important mainly due to the shortage of natural and material resources, the need for drastic cuts in energy and material consumption, the continuing population growth, and other reasons. Substantial progress in the solution and application of extremal problems is closely connected with the rapid development of computers. Contemporary extremal problems whose appearance was due to the needs of economics and flight mechanics now find application in such diverse areas of science and technology as physics, various manufacturing processes, nuclear power, ecology, medicine, etc.

One of the distinguishing features of the statements of modern control problems is that actions on the system can be divided into two classes: control actions and disturbing actions (disturbances). Typically, in these problems the constraints imposed on control actions are such that it is impossible to apply the methods of the classical calculus of variations. Within the framework of the control theory developed for such problems in the 50s two fundamental approaches were incorporated, namely Pontryagin's maximum principle and Bellman's dynamic programming approach. At present, scientists undertake intensive studies in different areas of the theory of optimal control.

Problems of the control of both deterministic and stochastic systems are treated in the book with a view to present an overall picture of modern control theory. In addition, considerable attention is payed to the study of approximation and numerical algorithms for solving particular classes of control problems.

PART ONE

STABILITY OF CONTROL SYSTEMS

CHAPTER I

CONTINUOUS AND DISCRETE DETERMINISTIC SYSTEMS

In this chapter basic problem statements of stability theory and methods for investigating such problems are outlined. Various examples are given, and modern results of stability theory are expounded.

§ 1. Basic Definitions of Stability Theory for Continuous Systems

1. Stability. Suppose that a continuous evolutionary system is described by a system of ordinary differential equations written in standard form:

$$\dot{x}_i(t) = f_i(t, x_1, ..., x_n), \qquad i = 1, \ldots, n; \qquad (1.1)$$
$$x_i(t_0) = x_{i0}.$$

Here t is an independent variable usually assumed to be time; $x_1(t), \ldots, x_n(t)$ are the desired functions, t_0 and x_{i0} are the given initial data, and a dot over a letter denotes a derivative of the corresponding function, such as $\dot{x}_i(t) = dx_i(t)/dt$. The functions $f_i(t, x_1, \ldots, x_n)$ are assumed to be real-valued. We shall write the system of differential equations (1.1) in vector form:

$$\dot{x}(t) = f(t, x), \qquad x(t_0) = x_0,$$
$$x = \begin{pmatrix} x_1 \\ \vdots \\ x_n \end{pmatrix} = (x_1, ..., x_n)' \in R^n.$$

Here and subsequently, a prime designates the transposition sign. Denote by S_H a sphere in Euclidean space R^n with norm

$$|x| = (x'x)^{1/2} = \left(\sum_{i=1}^{n} x_i^2\right)^{1/2}$$

of radius H with center at the origin. We denote by T an interval on the real axis of the form $T = \{a < t < \infty\}$, where a is $-\infty$ or some finite number. We assume that the function $f(t, x), f\colon T \times S_H \to R^n$ is continuous in both arguments and is Lipschitzian in the second argument, i.e.,

$$|f(t,x) - f(t,y)| \leqslant L|x - y|, \qquad L = \text{constant} > 0.$$

Under these conditions the theorems on local existence, uniqueness and continuous dependence on the initial data $t_0 \in T$ and $x_0 \in S_H$ are valid on a finite interval for the solution $x(t, t_0, x_0)$ to problem (1.1) [18].

Many fundamental results of stability theory were obtained by the great Russian mathematician A. M. Lyapunov (1857–1918) in his doctoral thesis "A general problem in the stability of motion" (1892) [15].

In Lyapunov's stability theory any solution may be tested for stability. Suppose that $z(t, t_0, z_0)$ is a solution of problem (1.1) continued on the right to infinity, i.e., $z(t)$ exists on the interval $[t_0, \infty)$ with $z(t) \in S_H$ for $t_0 \leqslant t < \infty$.

Definition 1.1. The solution $z(t, t_0, z_0)$ of problem (1.1) is said to be *stable in the sense of Lyapunov* (or *Lyapunov-stable*) if for any ε $(0 < \varepsilon < 0)$ there exists a $\delta = \delta(\varepsilon, t_0) > 0$ such that all solutions $x(t, t_0, x_0)$ of problem (1.1) are continued infinitely on the right whenever $|x_0 - z_0| \leqslant \delta(\varepsilon, t_0)$, and the following inequality holds for these solutions:

$$|x(t, t_0, x_0) - z(t, t_0, z_0)| \leqslant \varepsilon, \qquad t_0 \leqslant t \leqslant \infty.$$

Geometrically, Lyapunov stability means that any ε-tube, however narrow, of the solution $z(t)$ will hold all solutions $x(t)$ of problem (1.1) that at the initial time t_0 are displaced from z_0 by no more than $\delta(\varepsilon, t_0)$ (see Figure 1.1).

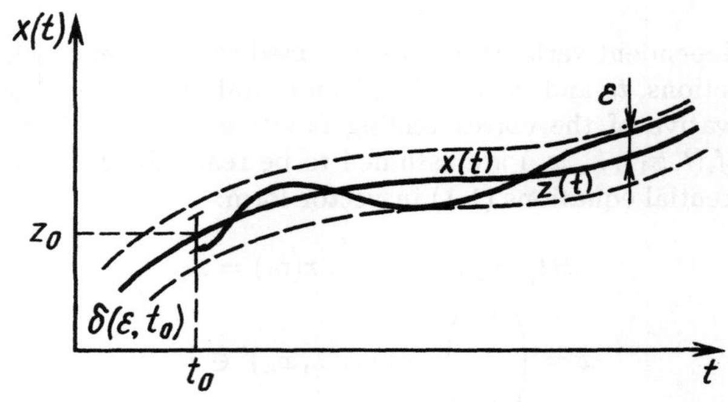

FIG. 1.1. Lyapunov stability of the unperturbed solution $z(t)$

Following Lyapunov, the solution $z(t)$ is called the *unperturbed* solution, and $x(t)$ is termed the *perturbed* solution.

It is important to note that the question is about the stability of one particular solution $z(t)$ and not of all solutions of equations (1.1).

Example 1.1 [2]. The general solution of the equation

$$\ddot{x}(t) = -\tfrac{1}{2}\left[x^2(t) + \left(x^4(t) + 4\dot{x}^2(t)\right)^{1/2}\right]x(t) \tag{1.2}$$

is of the form $x(t) = c\sin(ct + d)$, where c and d are arbitrary constants. The solution $z(t) \equiv 0$ of equation (1.2) is Lyapunov-stable, since all solutions of equation (1.2) can be continued infinitely, and $|x(t)| \leqslant |c| < \varepsilon$ if the initial conditions

$x_0(t_0) = c\sin(ct_0 + d)$ and $\dot{x}_0(t_0) = c^2\cos(ct_0 + d)$ are sufficiently small. Any other solution $x_1(t) = c_1\sin(c_1 t + d)$, $c_1 \neq 0$, is unstable. In fact, there always exists a solution $x_2(t) = c_2\sin(c_2 t + d)$ such that $c_2 \neq 0$ and the frequency ratio c_1/c_2 is irrational. For such solutions the equality

$$\overline{\lim_{t\to+\infty}} |x_1(t) - x_2(t)| = |c_1| + |c_2| \tag{1.3}$$

holds regardless of $x_2(t_0)$, $\dot{x}_2(t_0)$ and $x_1(t_0)$, $\dot{x}_1(t_0)$. Here $\overline{\lim}$ denotes the upper limit. Thus, the solutions $x_1(t)$ and $x_2(t)$ differ by the finite value $|c_1| + |c_2| \neq 0$ regardless of the proximity of their initial conditions.

For linear equations the stability of one solution implies the stability of any other solution; hence we may simply say that some linear system is stable (or unstable).

Theorem 1.1. *For the solution $z(t, t_0, z_0)$ to a linear system $\dot{y}(t) = A(t)y(t) + f(t)$ to be stable for any z_0 and any right-hand side $f(t)$, where $A(t)$ and $f(t)$ are continuous functions, it is necessary and sufficient that the trivial solution $x(t) \equiv 0$ to the homogeneous system $\dot{x}(t) = A(t)x(t)$ be stable.*

Proof. The necessity. As is known, all solutions of a linear equation can be continued infinitely on the right. Hence, the first requirement of Definition 1.1 is always met. Let $z(t, t_0, z_0)$ be some stable solution. This means that any other solution $y(t, t_0, y_0)$ satisfies the inequality $|y(t, t_0, y_0) - z(t, t_0, z_0)| \leqslant \varepsilon$, $t_0 \leqslant t < \infty$ if $|y_0 - z_0| \leqslant \delta(\varepsilon, t_0)$. The difference $x(t, t_0, x_0) = y(t, t_0, y_0) - z(t, t_0, z_0)$ is a solution of the homogeneous equation, and $|x(t, t_0, y_0 - z_0)| \leqslant \varepsilon$, $t_0 \leqslant t < \infty$. According to Definition 1.1, this implies that the solution $x(t) \equiv 0$ is Lyapunov-stable.

The sufficiency. Any solution $y(t)$ to the nonhomogeneous equation can be represented in the form $y(t) = z(t) + x(t)$, where $z(t)$ is a prescribed solution of the nonhomogeneous equation, and $x(t)$ is the solution of the corresponding homogeneous equation. Now the sufficiency immediately follows from Definition 1.1. □

The investigation of the stabiliy of any solution $z(t)$ of equations (1.1) can be reduced to the investigation of the stability of the trivial (zero) solution of some other equation. The function $y(t) = x(t) - z(t)$ satisfies the equation

$$\dot{y}(t) = f(t, y + z) - f(t, z) = g(t, y), \qquad g(t, 0) \equiv 0. \tag{1.4}$$

Henceforth, we shall investigate only the stability of the trivial solution $x(t) \equiv 0$ of equations (1.1) assuming that

$$f(t, 0) \equiv 0 \tag{1.5}$$

Definition 1.2. The trivial solution $x(t) \equiv 0$ of equation (1.4) is said to be *Lyapunov-stable* if for any $\varepsilon > 0$ there exists a $\delta(\varepsilon, t_0) > 0$ such that the inequality $|x(t, t_0, x_0)| \leqslant \varepsilon$ is satisfied for any $t \geqslant t_0$ whenever $|x_0| \leqslant \delta(\varepsilon, t_0)$ (see Figures 1.2 and 1.3).

A solution that is not stable is called *unstable*. To be more precise, we can supply the following definition.

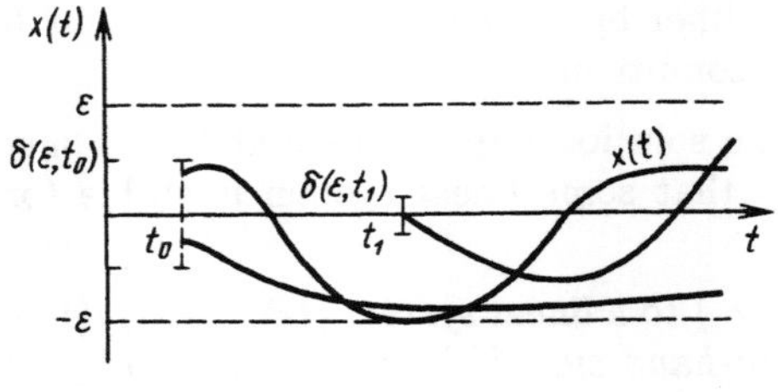

FIG. 1.2. Lyapunov stability of the trivial solution $x(t) \equiv 0$

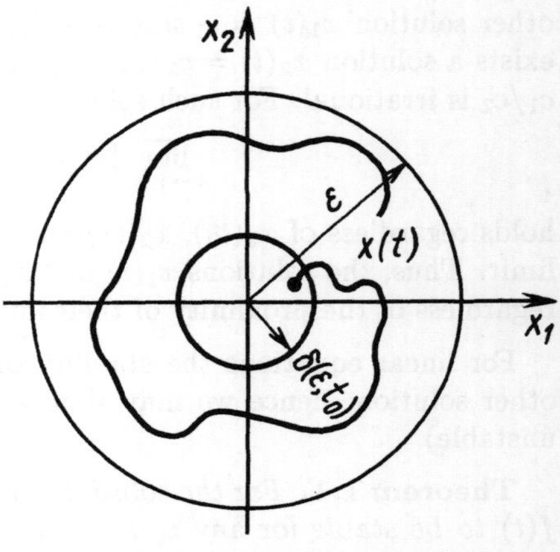

FIG. 1.3. Lyapunov-stable phase trajectory

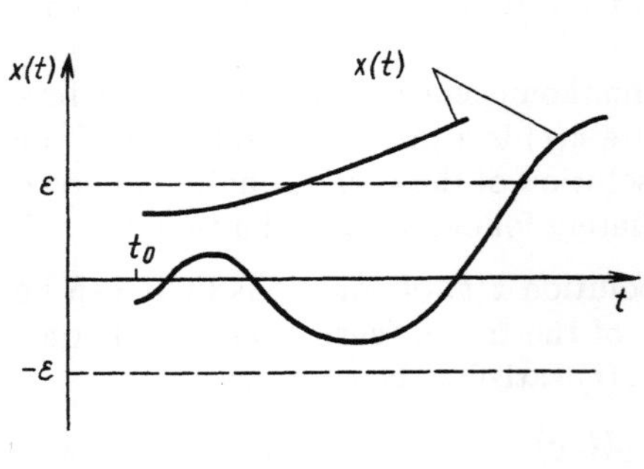

FIG. 1.4. Instability of the trivial solution $x(t) \equiv 0$

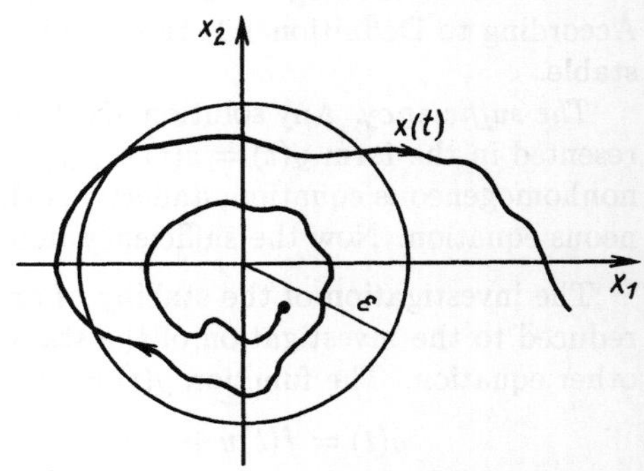

FIG. 1.5. Phase portrait of an unstable trajectory

Definition 1.3. The trivial solution $x(t) \equiv 0$ is said to be *unstable in the sense of Lyapunov* (or *Lyapunov-unstable*) if for some $\varepsilon > 0$ and any $\delta > 0$ there exists a solution $x(t, t_0, x_0)$ and an instant $t_1 > t_0$ such that $|x(t_1, t_0, x_0)| > \varepsilon$ although $|x_0| \leqslant \delta$ (see Figures 1.4 and 1.5).

Uniform stability with respect to the initial moment of time is stronger than the stability in the sense of Definition 1.2.

Definition 1.4. The solution $x(t) \equiv 0$ of equation (1.4) is called *uniformly stable* with respect to t_0 ($t_0 \in T$) if for any $\varepsilon > 0$ $(0 < \varepsilon < H)$ there exists a $\delta(\varepsilon) > 0$ independent of t_0 such that the inequality $|x(t, t_0, x_0)| \leqslant \varepsilon$ is satisfied for any $t \geqslant t_0$ ($t_0 \in T$) whenever $|x_0| \leqslant \delta(\varepsilon)$.

Example 1.2. The solution of the equation

$$\dot{x}(t) = t^{-2}x(t), \qquad x(t_0) = x_0, \ t_0 \in (0, \infty) \tag{1.6}$$

is of the form

$$x(t, t_0, x_0) = x_0 \exp(t_0^{-1} - t^{-1}).$$

Since $|x(t, t_0, x_0)| \leqslant |x_0| \exp(t_0^{-1})$, it follows that for any $\varepsilon > 0$, setting $\delta(\varepsilon, t_0) = \varepsilon \exp(-t_0^{-1})$, we infer that the solution $x(t) \equiv 0$ is stable according to Definition 1.2. However, there is no uniform stability, since $\delta(\varepsilon, t_0) \to 0$ for $t_0 \to 0$.

Theorem 1.2. *For an autonomous ($f(t,x) \equiv f(x)$) or an ω-periodic ($f(t + \omega, x) = f(t,x)$, $\omega > 0$) equation (1.5), Lyapunov stability implies stability uniform with respect to the initial time.*

Proof. For an arbitrary solution $x(t, t_0, x_0)$ to the periodic equations (1.1) the following identities hold:

$$\begin{gathered} x\big(t, t_1, x(t_1, t_0, x_0)\big) \equiv x(t, t_0, x_0), \\ x(t \pm m\omega, t_0 \pm m\omega, x_0) = x(t, t_0, x_0) \end{gathered} \tag{1.7}$$

for any integer m. The first of these identities follows immediately from the existence and uniqueness theorem, and the second arises from the ω-periodicity of equations (1.1). It follows from the second identity (1.7) that in proving the uniform stability it is sufficient to consider only t_0 on the segment $[0, \omega]$. We take an arbitrary ε $(0 < \varepsilon < H)$ and define $\eta(\varepsilon, \omega) > 0$ such that $|x(t, \omega, x_0)| \leqslant \varepsilon$, where $t \geqslant \omega$ and $|x_0| \leqslant \eta(\varepsilon, \omega)$. Due to the stability of the solution at time $t_0 = \omega$ there exists such an $\eta(\varepsilon, \omega)$. We now choose a $\delta(\varepsilon)$ so that

$$\max_{0 \leqslant t_0 \leqslant \omega} \ \max_{|x_0| \leqslant \delta(\varepsilon)} |x(\omega, t_0, x_0)| \leqslant \eta(\varepsilon, \omega).$$

In view of the fact that the solution on the finite segment $[0, \omega]$ depends continuously on the initial data, such a $\delta(\varepsilon)$ exists. Choosing such a $\delta(\varepsilon)$, we obtain for any $t_0 \in (-\infty, \infty)$

$$\begin{aligned} |x(t, t_0, x_0)| &= |x(t - m\omega, t_0 - m\omega, x_0)| = |x(t^1, t_0^1, x_0)| \\ &= \big|x\big(t^1, \omega, x(\omega, t_0^1, x_0)\big)\big| \leqslant \varepsilon, \qquad t \geqslant t_0, \ |x_0| \leqslant \delta(\varepsilon). \end{aligned}$$

Here the following notation is used: $t^1 = t - m\omega$, $t_0^1 = t_0 - m\omega$, and $t_0^1 \in [0, \omega]$; m is an integer. □

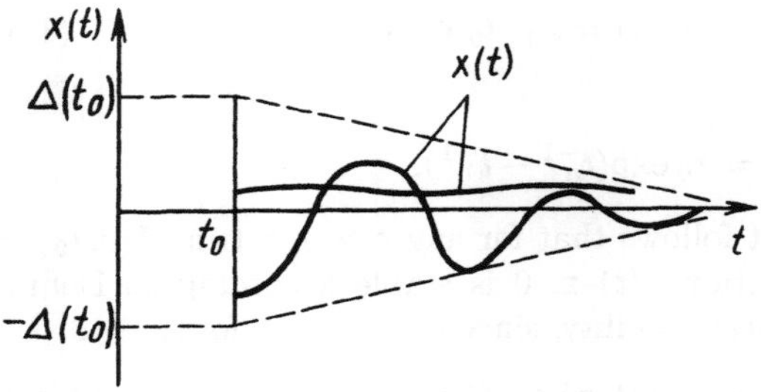

FIG. 1.6. Asymptotic stability of the trivial solution

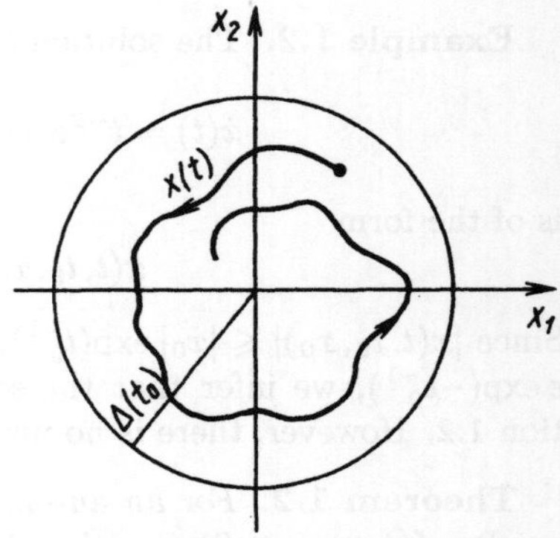

FIG. 1.7. Phase portrait of an asymptotically stable trajectory

2. Asymptotic stability. Asymptotic stability is stronger than the stability in the sense of Definition 1.2.

Definition 1.5. The trivial solution $x(t) \equiv 0$ of equations (1.1) is said to be *asymptotically stable* if

1°) it is Lyapunov-stable;

2°) for any $t_0 \in T$ there exists $\Delta = \Delta(t_0) > 0$ such that $x(t.t_0, x_0) \to 0, t \to \infty$, for $|x_0| \leqslant \Delta(t_0)$ (see Figures 1.6 and 1.7).

The set $\Omega(t_0) \subset R^n$ of all initial values x_0 such that $\lim_{t\to\infty} x(t, t_0, x_0) = 0$ is called the *attraction domain* of the trivial solution at the initial moment t_0. If $\Omega(t_0) = R^n$, then the solution $x(t) \equiv 0$ is said to be *globally asymptotically stable.*

If only condition 2° of Definition 1.5 is satisfied, then the trivial solution is said to be the *attractor.* Generally speaking, attraction does not imply stability. However, for a scalar ordinary differential equation, attraction always implies stability. Thus, one can construct examples where there is attraction but no stability only for $n \geqslant 2$ or for equations that are more general than ordinary differential equations. As an example, we cite the case of a functional-differential equation of the first order such that attraction does not lead to stability.

Example 1.3. [1] Consider the equation

$$\dot{x}(t) = t^{-1}x(t) - 27t(t+2)^{-3}x^3\left(\frac{t+2}{3}\right). \tag{1.8}$$

[1]Kolmanovskii, V. B., and Nosov, V. R., *Stability and Periodic Modes of Control Systems with Aftereffect*, Nauka, Moscow, (1981).

This equation involves the values of an unknown function $x(t)$ at different moments t and $(t+2)/3 = g(t)$. The function $g(t) \leqslant t$ for $t \geqslant 1$. Therefore, equation (1.8) describes a certain evolutionary process that depends on the past. For this equation the initial condition may be given in the form $x(1) = x_0$. Moreover, the definitions of stability and of asymptotic stability for equation (1.8) can be readily stated; they are quite similar to Definitions 1.2 and 1.5.

It can be demonstrated by a direct substitution that the solution $x(t,1,x_0)$ of equation (1.8) with initial condition $x(1) = x_0$ can be written as

$$x(t,1,x_0) = x_0 t \exp\bigl(-x_0^2(t-1)\bigr), \qquad t \geqslant 1$$

Obviously, for any x_0 we have

$$x(t,1,x_0) \to 0, \qquad t \to \infty.$$

Simultaneously, for $|x_0| = \delta$ we have at the moment $t = 1 + \delta^{-2}$

$$\begin{aligned} |x(1+\delta^2,1,x_0)| &= \delta(1+\delta^{-2})\exp\bigl(-\delta^2(1+\delta^{-2}-1)\bigr) \\ &= (\delta+\delta^{-1})e^{-1} \geqslant 2e^{-1}. \end{aligned}$$

Thus, all the solutions of equation (1.8) are characterized by attraction, but condition (1) of Definition 1.5 is not satisfied. The trivial solution of equation (1.8) is not stable and, a fortiori, is not asymptotically stable.

Definition 1.6. The trivial solution of equations (1.1) is said to be *uniformly asymptotically stable* if it is uniformly stable and for any number $\gamma > 0$ there exist numbers Δ_1 $(0 < \Delta_1 < H)$ and $T(\gamma)$ such that $|x(t,t_0,x_0)| \leqslant \gamma$ for $t \geqslant t_0 + T(\gamma)$ and $|x_0| \leqslant \Delta_1$ (see Figure 1.8).

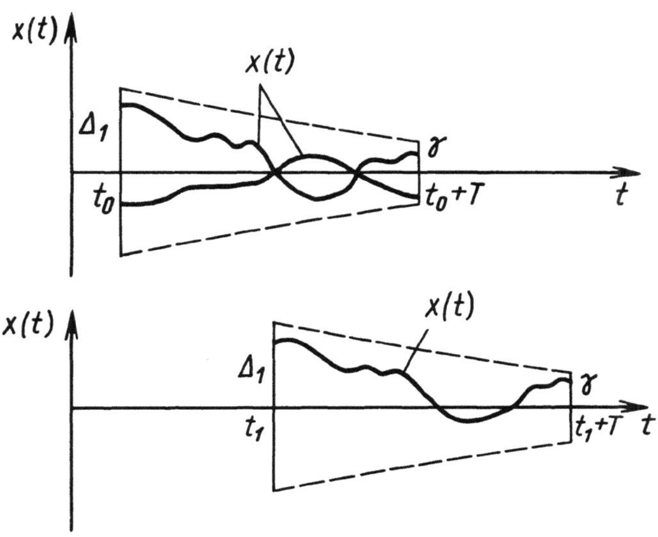

FIG. 1.8. Uniform asymptotic stability

In general, uniform asymptotic stability is stronger than asymptotic stability. However, similar to Theorem 1.2, one can prove that if the system (1.1) is autonomous or ω-periodic, and its trivial solution is asymptotically stable, then this solution is also uniformly asymptotically stable.

3. Other stability definitions.

Definition 1.7. The trivial solution of the system (1.1) is said to be *exponentially stable* if any solution $x(t, t_0, x_0)$ of this equation satisfies the estimate

$$|x(t, t_0, x_0)| \leqslant B|x_0| \exp(-\alpha(t - t_0)),$$
$$B > 0,\ \alpha > 0,\ t \geqslant t_0,\ |x_0| \leqslant H_1 < H.$$

(see Figure 1.9).

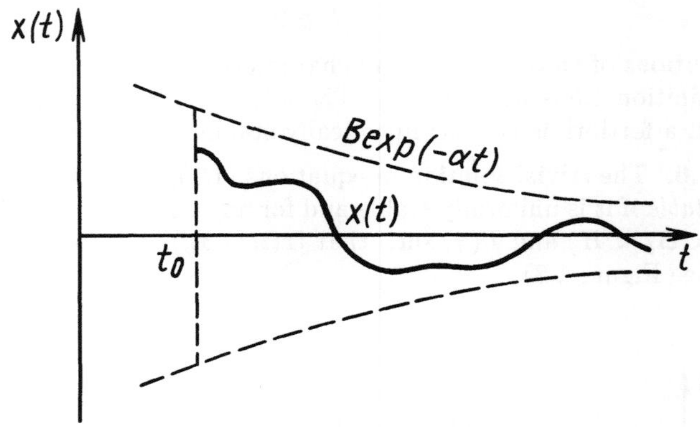

FIG. 1.9. Exponential stability

It should be noted that for linear equations $\dot{x}(t) = A(t)x(t)$ exponential stability is equivalent to uniform asymptotic stability.

In addition to the system (1.1) we consider the perturbed system

$$\dot{z}(t) = f(t, z) + \varphi(t, z), \qquad f(t, 0) = 0. \tag{1.9}$$

Definition 1.8. The trivial solution $x(t) \equiv 0$ of equations (1.1) is said to be *stable under steady-acting disturbances* if for any $\varepsilon > 0$ and $t_0 \in T$ there exists a $\delta(\varepsilon, t_0) > 0$ such that for $|z(t_0)| \leqslant \delta$ and for all disturbances $\varphi(t, z)$ such that $|\varphi(t, z)| \leqslant \delta$ all solutions of equation (1.9) satisfy the condition $|z(t)| \leqslant \varepsilon$.

In addition to the foregoing definitions of stability one sometimes uses other definitions, such as the definitions of partial stability (stability with respect to part of the variables), of stability on a finite time interval, of absolute stability, of orbital stability, and so on [1, 2, 4, 10, 14, 16, 20].

§ 2. Lyapunov's Direct Method

Lyapunov's direct method is a universal tool for investigating the stability of different types of systems. Certain special functions called *Lyapunov functions* are used as instruments of investigation in the direct method.

A real continuously differentiable (smooth) function $V(t,x)$, $V\colon T\times S_H\to R^1$, satisfying the condition $V(t,0)\equiv 0$ is called a *Lyapunov function.* A Lyapunov function of two variables x_1 and x_2 and its level curves $V(x_1,x_2)=C$ are shown in Figures 2.1 and 2.2.

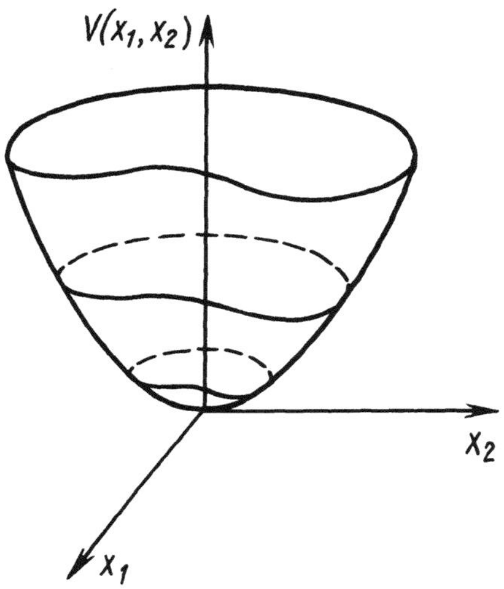

FIG. 2.1. Example of a Lyapunov function

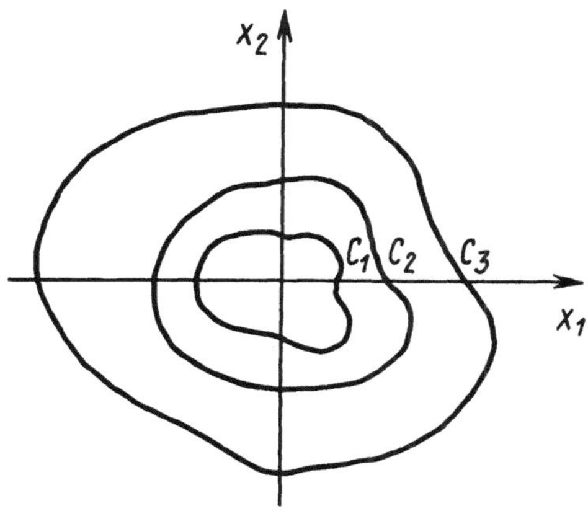

FIG. 2.2. Level curves for the Lyapunov function $V(x_1,x_2)=C_i$ $(i=1,2,3)$, $C_1<C_2<C_3$

We shall call the quantity

$$\dot V=\frac{\partial V}{\partial t}+\sum_{i=1}^{n}\frac{\partial V}{\partial x_i}f_i(t,x)=\frac{\partial V}{\partial t}+\big(\operatorname{grad}V,f(t,x)\big). \tag{2.1}$$

the *derivative* $\dot V$ *of the function* $V(t,x)$ *along the solution of equation* (1.1). If $x=x(t)$ is a solution of the system (1.1), then $\dot V$ is the total derivative of the composite function $V\big(t,x(t)\big)$ with respect to time. Note that it is not necessary to know the exact form of the solution $x(t)$ to compute $\dot V$. Henceforth, we denote by $\omega_i(u)$ $(u\geqslant 0,\ i=0,1,\dots)$ scalar continuous nondecreasing functions (also called wedges) such that $\omega_i(0)=0$ and $\omega(u)>0$ for $u>0$ (see Figure 2.3).

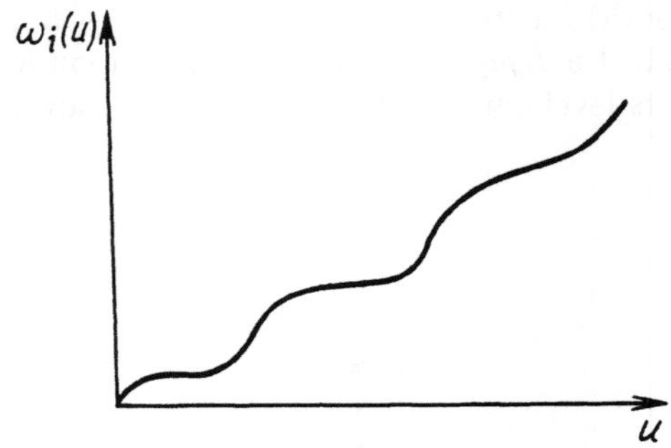

FIG. 2.3. A view of the function (wedge) $\omega_i(u)$

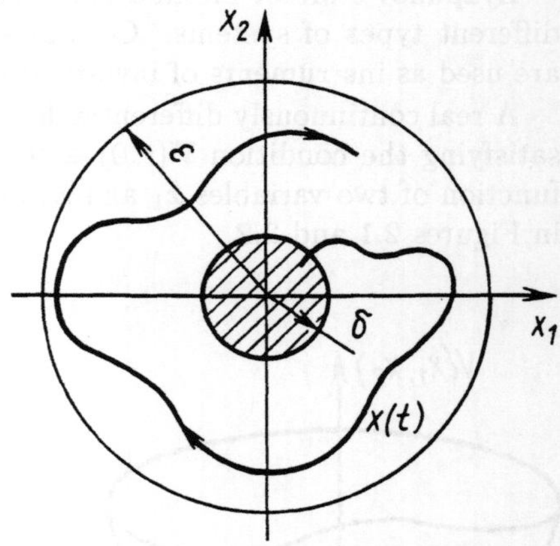

FIG. 2.4. Lyapunov's stability theorem illustrated

Theorem 2.1 (Lyapunov's first theorem). *Suppose there exists a Lyapunov function $V(t,x)$ such that*

$$\omega_1(|x|) < V(t,x), \tag{2.2}$$

$$\dot{V} \leqslant 0. \tag{2.3}$$

Then the trivial solution of the system (1.1) *is Lyapunov-stable.*

Proof. Take any ε $(0 < \varepsilon < H)$ and any $t_0 \in T$. A quantity $\delta(\varepsilon, t_0)$ will be taken as a number such that

$$\max_{|x_0| \leqslant \delta(\varepsilon,t_0)} V(t_0, x_0) \leqslant \omega_1(\varepsilon).$$

It follows from the continuity of the function $V(t,x)$ and the condition $V(t_0, 0) = 0$ that there exists such a $\delta(\varepsilon, t_0)$ (see Figure 2.4). Condition (2.3) indicates that the function $V\big(t, x(t)\big)$ does not increase along the solutions of the system (1.1). Using inequalities (2.2) and (2.3), we obtain the following inequality for $|x_0| \leqslant \delta(\varepsilon, t_0)$ and $t \geqslant t_0$:

$$\omega_1\big(|x(t,t_0,x_0)|\big) \leqslant V\big(t, x(t,t_0,x_0)\big) \leqslant V(t_0,x_0) \leqslant \max_{|x_0| \leqslant \delta(\varepsilon,t_0)} V(t_0,x_0) \leqslant \omega_1(\varepsilon).$$

By virtue of ω_1 being monotonic it follows that $|x(t,t_0,x_0)| \leqslant \varepsilon$. □

Functions $V(t,x)$ satisfying condition (2.2) are called *positive-definite.* Lyapunov himself used another definition for these functions, which is equivalent to (2.2) [15]. The derivative $\dot{V}$ satisfying condition (2.3) will be called *negative-semidefinite.*

Theorem 2.2 (K. P. Persidskii). *If in addition to the conditions of Theorem* 2.1 *the following inequality holds:*

$$V(t,x) \leqslant \omega_2(|x|), \tag{2.4}$$

then the trivial solution of equation (1.1) *is uniformly stable with respect to the initial moment* t_0.

Proof. For a given ε $(0 < \varepsilon < H)$ we define a number $\delta = \delta(\varepsilon)$ (that is independent of t_0) such that $\omega_1(\varepsilon) = \omega_2(\delta)$. Using (2.2–2.4) and the monotonicity of $\omega_2(u)$, we obtain the following inequality for $t \geqslant t_0$ and $|x_0| \leqslant \delta(\varepsilon)$:

$$\omega_1(|x(t)|) \leqslant V\big(t, x(t)\big) \leqslant V(t_0, x_0) \leqslant \omega_2(|x_0|) \leqslant \omega_2(\delta) = \omega_1(\varepsilon).$$

Because of the monotonicity of $\omega_1(u)$ it follows that $|x(t)| \leqslant \varepsilon$. □

Lyapunov functions satisfying condition (2.2) are said to possess an *infinitesimal upper limit* although another equivalent definition was used originally [15].

Theorem 2.3 (Lyapunov's second theorem). *Suppose there exists a Lyapunov function* $V(t,x)$ *such that*

$$\omega_1(|x|) \leqslant V(t,x) \leqslant \omega_2(|x|), \tag{2.5}$$

$$\dot V(t,x) \leqslant -\omega_3(|x|). \tag{2.6}$$

Then the trivial solution of the system (1.1) *is uniformly asymptotically stable.*

Proof. By Persidskii's theorem the solution $x(t) \equiv 0$ is uniformly stable. We shall show that the second condition in Definition 1.6 is also satisfied. We take $\gamma > 0$ and determine δ from the relation $\omega_1(\gamma) = \omega_2(\delta)$. $T(\gamma)$ will be taken as the number $T(\gamma) = 2\omega_2(\Delta_1)/\omega_1(\delta)$. On the interval $[t_0, t_0 + T(\gamma)]$ there exists a moment t_1 such that $t_0 < t_1 < t_0 + T(\gamma)$ and $|x(t_1, t_0, x_0)| \leqslant \delta$ for $|x_0| \leqslant \Delta_1$. If this is not so, then we shall have

$$\begin{aligned} &V\big(t_0 + T(\gamma), x(t_0 + T(\gamma), t_0, x_0)\big) - V(t_0, x_0) \\ &\quad = \int_{t_0}^{t_0+T(\gamma)} \dot V\big(s, x(s, t_0, x_0)\big)\, ds \leqslant -\int_{t_0}^{t_0+T(\gamma)} \omega_3(|x(s)|)\, ds \\ &\quad \leqslant -\omega_3(\delta) T(\gamma). \end{aligned}$$

Hence

$$\begin{aligned} 0 &\leqslant V\Big(t_0 + T(\gamma), x\big(t_0 + T(\gamma)\big)\Big) \leqslant V(t_0, x_0) - \omega_3(\delta)T(\gamma) \\ &\leqslant \omega_2(\Delta_1) - \omega_3(\delta) T(\gamma). \end{aligned} \tag{2.7}$$

If $T(\gamma) = 2\omega_2(\Delta_1)/\omega_3(\delta)$, then the inequality (2.7) cannot be true. Consequently, there exists a $t_1 \in [t_0, t_0 + T(\gamma)]$ such that $|x(t_1, t_0, x_0)| \leqslant \delta$. But then in view of uniform stability for all $t \geqslant t_0 + T(\gamma) \geqslant t_1 \geqslant t_0$ we have $|x(t, t_0, x_0)| \leqslant \varepsilon$. □

In Lyapunov's classical book [15] Theorem 2.3 was formulated as follows: *Suppose there exists a positive definite function* $V(t,x)$ *admitting an infinitesimal upper limit and having a negative definite derivative along the solution of equation* (1.1). *Then the trivial solution of equation* (1.1) *is asymptotically stable.*

If the inequality (2.4) is not assumed to hold, then the solution $x(t) \equiv 0$ may not be asymptotically stable.

Example 2.1 (J. L. Massera). Define the function $g(t) \in C^1[0,\infty)$, $g : [0,\infty) \to [0,1]$, coinciding with $\exp(-t)$ everywhere except for narrow peaks of width less than 2^{-n} with center at the point $t = n$; let $g(n) = 1$, $n = 1, 2, 3, \ldots$ (see Figure 2.5).

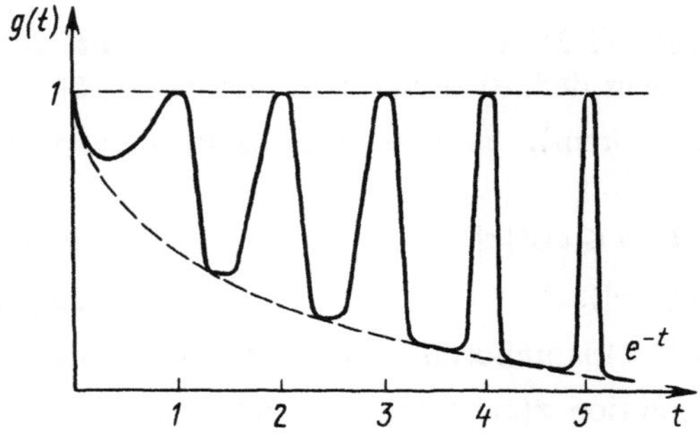

FIG. 2.5. A view of the function $g(t)$ of Example 2.1

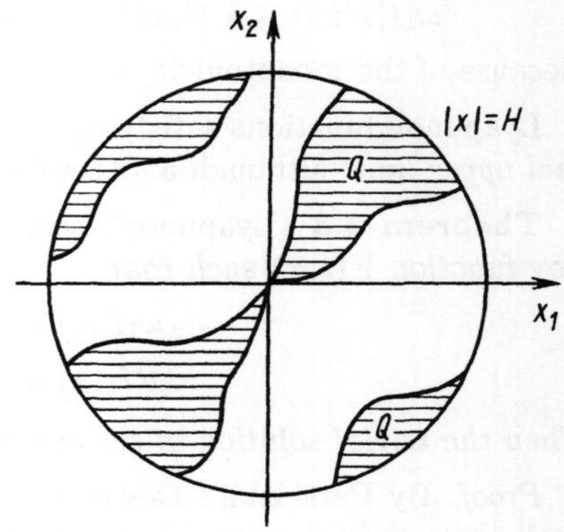

FIG. 2.6. Structure of the region $Q = \{V > 0\}$

Consider the equation $\dot{x} = \dot{g}(t)x/g(t)$. Its general solution is of the form $x(t) = g(t)x_0/g(t_0)$. It follows from this that the trivial solution $x(t) \equiv 0$ is not asymptotically stable.

However, if we take the Lyapunov function

$$V(t,x) = \frac{x^2(t)}{g^2(t)}\left[3 - \int_0^t g^2(s)\,ds\right], \tag{2.8}$$

then we obtain $\dot{V}(t,x) = -x^2$ and $V(t,x) \geqslant x^2$, since

$$\int_0^\infty g^2(s)\,ds < \int_0^\infty e^{-s}\,ds + \sum_{n=1}^{\infty} 2^{-n} = 2.$$

Moreover, the function (2.8) does not admit an infinitesimal upper limit, since there are points t, where $V(t,x) \geqslant e^{2t}x^2(t)$.

The following theorem due to V. P. Marachkov should be noted: *suppose that the righ-hand side of equation* (1.1) *is bounded in* $T \times S_H$: $|f(t,x)| \leqslant M$, $t \in T$, $x \in S_H$. *Then if there exists a function* $V(t,x)$ *satisfying the estimates* (2.2) *and* (2.6), *the trivial solution of equation* (1.1) *is asymptotically stable* [2].

Theorem 2.4 (E. A. Barbashin, N. N. Krasovskii). *Let all conditions of Theorem 2.3 be satisfied and, moreover,*

$$\omega_1(u) \to \infty, \qquad u \to \infty. \tag{2.9}$$

Then the solution $x(t) \equiv 0$ of the system (1.1) *is globally asymptotically stable*

Proof. We take an arbirary initial condition $x_0 \in R^n$ and define a number k such that $\omega_1(k) = \omega_2(|x_0|)$. Such a k always exists in view of (2.9). As in Theorem 2.2, it can be proved that the solution $x(t, t_0, x_0)$ remains inside the sphere S_k for $t \geqslant t_0$. Applying Theorem 2.3 to the sphere S_k, we prove the assertion of Theorem 2.4. □

When condition (2.9) is not satisfied, there may be no global stability. An appropriate example is given in [4].

We now prove a theorem on instability. Take a continuously differentiable function $V(t, x)$ and denote by Q a region of the form $Q = \{|x| < H,\ V(t, x) > 0\}$. Suppose that the region Q has the following properties:

1°) Q is composed of several connected open components;

2°) Q contains points x with an arbitrarily small norm $|x|$ (see Figure 2.6).

Theorem 2.5 (Chetaev's theorem). *Suppose there exists a function $V(t, x)$ such that the region $Q = \{|x| < H,\ V(t, x) > 0\}$ satisfies conditions 1° and 2°. Then if in the region Q the function $V(t, x)$ is bounded, and its derivative along the solution of the system* (1.1) *is positive definite (i.e., $\dot V(t, s) \geqslant \omega_4(|x|)$, $x \in Q$), the trivial solution of the system* (1.1) *is unstable.*

Proof. By hypothesis, in any neighborhood $|x| < \lambda$, however small, of the origin a point x_0 can be found such that $V(t_0, x_0) = V_0 > 0$. The function does not decrease along the solution $x(t, t_0, x_0)$, i.e., $V\big(t, x(t, t_0, x_0)\big) \geqslant V_0 > 0$. This means that the solution in question cannot reach the boundary $V = 0$ and leave the region Q. The solution $x(t, t_0, x_0)$ cannot always remain inside the region Q. In fact, for $V(t) \geqslant V_0$, in view of the hypothesis of the theorem, there exists an $\alpha > 0$ such that $\dot V(t) \geqslant \alpha$ for $t \geqslant t_0$. But then the following inequality would hold:

$$V(t) - V(t_0) = \int_{t_0}^{t} \dot V(s)\, ds \geqslant \alpha(t - t_0) \to \infty, \qquad t \to \infty.$$

But this is impossible, since V is bounded in the region Q. Consequently, the solution $x(t, t_0, x_0)$ will necessarily reach the boundary $|x| = H$ of the region Q in a finite time. □

Remarks. **1.** The conditions of Theorems 2.1–2.5 can be somewhat weakened. Specifically, we may require only that $V(t, x)$ be continuous in (t, x) and locally Lipschitzean in x. Then if in Theorems 2.1–2.4 we replace the derivative $\dot V$ associated with the system (1.1) by the right upper derivative D^+V also associated with the system (1.1), then the assertions of these theorems remain valid. In Theorem 2.5 instead of $\dot V$ we may consider the right lower derivative D_+V [20].

2. The method of Lyapunov functions is a universal method for investigating stability, and most theorems in the Lyapunov method admit inversion. More precisely, if the function $f(t, x)$ in the system (1.1) is continuously differentiable and the origin $x = 0$ is stable, then there exist a neighborhood Ω of the origin and a continuously differentiable function $V(t, x)$ defined on $T \times \Omega$ which satisfies all the hypotheses of Theorem 2.1. This assertion is due to K. P. Persidskii. The inversion of Theorem 2.2 on uniform stability was established by J. Kurzweil. J. L. Massera proved a theorem which is the converse of Theorem 2.3 on uniform asymptotic stability. The inversion of Chetaev's theorem on instability was obtained by I. Vrkoč. Proof of the inversion theorems and their dicussion can be found in N. N. Krasovskii's book [12].

§ 3. Examples of the Use of Lyapunov's Method

1. Stability of the motion of a shell. When considering grazing trajectories, we may assume that the center of gravity of a shell is in uniform and rectilinear motion. The motion of the shell relative to its center of mass can be described by the following quantities:

the angle α between the axis and the direction of motion of the shell,
the angle β between the axis of the shell and its projection on the vertical plane,
the projection n of the angular velocity of rotation of the shell on its axis.

It has been noted for a long time that when n is small, the shell begins "tumbling" in flight. Moreover, the angles α and β vary significantly. In this case the accuracy of shooting deteriorates appreciably. When the value of n is large, there is no "tumbling": the angles α and β change insignificantly during the flight; this results in a smaller dispersal of shells. Changing the parameters of grooves in a gun barrel, one can change n, thereby ensuring the stability of a shell in flight. Despite the major importance of this problem a complete theoretical investigation of the stability of a shell in flight was carried out only in the forties and the fifties of the 20th century.

The equations describing a change in time of the angles α and β were derived by A. N. Krylov and are of the form

$$\begin{aligned} A\ddot{\beta} + A\dot{\alpha}^2 \sin\beta\cos\beta - Cn\dot{\alpha}\cos\beta &= eR\sin\beta\cos\alpha, \\ A\ddot{\alpha}\cos\beta - 2A\dot{\alpha}\dot{\beta}\sin\beta + Cn\dot{\beta} &= eR\sin\alpha. \end{aligned} \tag{3.1}$$

In equations (3.1) the constants A, C, e, and R have the following meaning: C is the moment of inertia of the shell about its axis, A is the moment of inertia about the vertical axis passing through the center of gravity, e is the distance from the center of gravity to the center of pressure (the point at which the force of air resistance is applied), and R is the head resistance. (The derivation of Krylov's equations (3.1) is given in a number of books on theoretical mechanics and ballistics.)

Equations (3.1) have the solution $\alpha = \dot{\alpha} = \beta = \dot{\beta} = 0$. We shall determine the conditions of stability for this solution. This problem was solved by N. G. Chetaev by means of a sheaf of integrals, which he introduced. Chetaev started from the fact that energy and angular momentum are conserved during the flight of the shell. In other words, the equations of motion (3.1) have two first integrals:

the integral of energy

$$F_1(\alpha, \dot{\alpha}, \beta, \dot{\beta}) = \tfrac{1}{2}A(\dot{\beta}^2 + \dot{\alpha}^2 \cos^2 \beta) + eR(\cos\alpha\cos\beta - 1) = \text{constant};$$

the integral of angular momentum

$$\begin{aligned} F_2(\alpha, \dot{\alpha}, \beta, \dot{\beta}) = A(\dot{\beta}\sin\alpha - \dot{\alpha}\cos\beta\sin\beta\cos\alpha) \\ + Cn(\cos\alpha\cos\beta - 1) = \text{constant}. \end{aligned}$$

On differentiating these relations with regard to equations (3.1), we can easily verify that

$$\frac{dF_1}{dt} = \frac{dF_2}{dt} = 0. \tag{3.2}$$

However, neither F_1 nor F_2 is a positive definite function of α, $\dot{\alpha}$, β, and $\dot{\beta}$, i.e., neither function satisfies condition (2.2) for any function $\omega_1\left(\sqrt{\alpha^2 + \dot{\alpha}^2 + \beta^2 + \dot{\beta}^2}\right)$. Chetaev proposed seeking the Lyapunov function as a sheaf of integrals $V = F_1 - \lambda F_2$ (λ = constant), where λ must be chosen so that V becomes positive definite. We expand $V = F_1 - \lambda F_2$ in a series in α, $\dot{\alpha}$, β, and $\dot{\beta}$:

$$\begin{aligned} V = \tfrac{1}{2}\{A\dot{\alpha}^2 + (Cn\lambda - eR)\beta^2 + 2A\lambda\dot{\alpha}\beta\} + \dots \\ + \tfrac{1}{2}\{A\dot{\beta}^2 + (Cn\lambda - eR)\alpha^2 - 2A\alpha\dot{\beta}\} + \dots . \end{aligned} \tag{3.3}$$

The remaining terms in the expression (3.3) are of order not lower than the third in α, $\dot{\alpha}$, β, and $\dot{\beta}$. If λ is such that both quadratic forms in braces are positive definite, then the function V is also positive definite in a small neighborhood of the point $\alpha = \dot{\alpha} = \beta = \dot{\beta} = 0$. For both forms to be positive definite, it is necessary and sufficient that $A > 0$ and $A\lambda^2 - Cn\lambda + eR < 0$. The first inequality always holds, and whenever the condition $C^2n^2 - 4AeR > 0$ is satisfied, there exist values of λ such that the second inequality is also satisfied. Thus, whenever the so-called *Mayevskii-Krylov condition* is met:

$$n^2 > 4\frac{AeR}{C^2}, \tag{3.4}$$

there exist values of λ such that, in view of (3.2), the function $V = F_1 - \lambda F_2$ satisfies the hypotheses of Theorem 2.1. When condition (3.4) is met, the motion of the shell will be stable, without "tumbling". Using similar constructions and the instability theorem, one can prove that when condition (3.4) is not met, the shell will always "tumble" (see also Example 7.1 below).

2. Motion of a rigid body fastened at one point. We consider the motion of a rigid body fastened at a fixed point O which coincides with the center of inertia (see Figure 3.1). The principal moments of inertia of the body are A, B, and C. *Euler's equations* for the projections p, q, and r of the vector of angular velocity ω on the principal central axes of inertia are of the form

$$A\dot{p} = (B - C)qr, \qquad B\dot{q} = (C - A)rp, \qquad C\dot{r} = (A - B)pq. \tag{3.5}$$

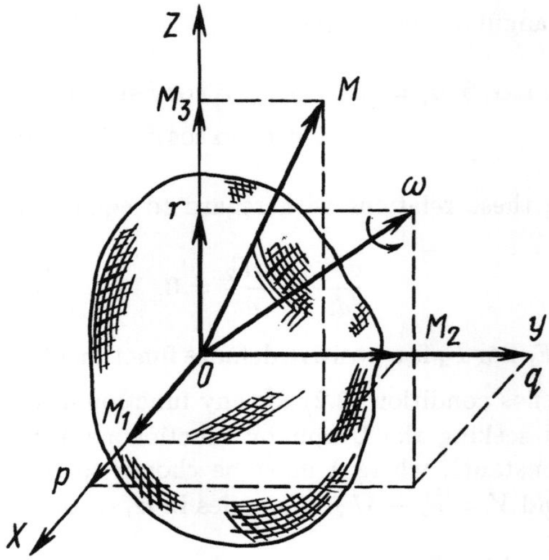

FIG. 3.1. Rotation of a rigid body about a fixed point O

We investigate the stability of steady-state rotation about the first axis $p = p_0$, $q = r = 0$. In the system (3.5) there are two first integrals

$$F_1 = Ap^2 + Bq^2 + Cr^2,$$
$$F_2 = B(B - A)q^2 + C(C - A)r^2.$$

For new variables $x = p - p_0$, $y = q$, $z = r$ we obtain the equations

$$A\dot{x} = (B - C)yz, \qquad B\dot{y} = (C - A)(p_0 + x)z, \qquad C\dot{z} = (A - B)(p_0 + x)y. \tag{3.6}$$

If the first axis is the major axis of the ellipsoid of inertia, then $A < B \leqslant C$, and we may take the Lyapunov function as the Chetaev sheaf of integrals for $\lambda = 1$

$$V_1 = B(B - A)y^2 + C(C - A)z^2 + [By^2 + Cz^2 + A(x^2 + 2xp_0)]^2. \tag{3.7}$$

For $A < B \leqslant C$ the function (3.7) is positive definite, and $\dot{V}_1 = 0$. Hence, according to Theorem 2.1, rotation about the major axis is stable.

If the first axis is the minor axis of the ellipsoid of inertia, then $A > B \geqslant C$, and for a Lyapunov function we may take a sheaf of integrals of the form

$$V_2 = B(A-B)y^2 + C(A-C)z^2 + [By^2 + Cz^2 + A(x^2 + 2xp_0)]^2.$$

Since $\dot{V}_2 = 0$, rotation about the minor axis is also stable.

In the case of a rotation about the mean axis ($C > A > B$), for an auxiliary function we take

$$V_3 = yz. \tag{3.8}$$

On the strength of equations (3.6) the derivative of the function (3.8) takes the form

$$\dot{V}_3 = (p_0 + x)\left[\frac{C-A}{B}z^2 + \frac{A-B}{C}y^2\right].$$

In the sphere S_H, where $H < p_0^2$, the hypotheses of Chetaev's theorem are satisfied. This means that rotation about the mean axis is unstable.

When using the foregoing results for the investigation of the motion of real objects, such as the angular motion of space vehicles, one must bear in mind the following. Real space vehicles are elastic, rather than rigid bodies. Internal dissipation of energy due to elastic effects in space vehicles does not allow to use assumptions made in the classical work by Euler and Poinsot for deriving equations (3.5). A more thorough investigation of the stability problem for the rotation of an elastic body about a fixed point shows that, as for a rigid body, rotation about the major axis of the ellipsoid of inertia remains stable and about the mean axis is also unstable. As to rotation about the minor axis, the effect of elastic forces is such that rotation becomes unstable. This fact was first discovered experimentally when observing first artificial satellites of the Earth. Radioastronomical observations showed that the first Soviet sputnik (1957) had a stable rotation, but the American satellite "Explorer 1" (1958) was unstable and passed from one extreme position into another every 90 minutes. The subsequent theoretical investigation of this phenomenon had, in fact, led to the conclusion that elastic forces affect the stability of rotation of the body about a fixed point.[2]

Numerous other results on the stability of mechanical systems can be found in the book by N. G. Chetaev [3] and also in [1, 2, 12, 19, 22], where one can find examples from other branches of science and engineering.

§ 4. Development of Lyapunov's Method

Although the theorems cited in § 2 provide a complete solution of most of the stability problems, it is sometimes difficult to use them, since it is not easy to construct Lyapunov functions satisfying the hypotheses of these theorems. There exist a number of other theorems which involve Lyapunov functions with other properties.

---

[2]Likins, P., Spacecraft Attitude Dynamics and Control—A Personal Perspective on Early Developments, *J. Guidance, Control, Dynamics*, (1986), vol. 9, no. 2.

1. The Barbashin–Krasovskii Theorem. Consider an autonomous system

$$\dot{x}(t) = f(x), \qquad f(0) = 0,\ x \in R^n, \qquad (4.1)$$
$$x(t_0) = x_0.$$

and denote by $x(t, x_0)$ a solution of the system (4.1).

Definition 4.1. A point $y \in R^n$ is called an *ω-limiting point* for the point x_0 if there exists a sequence t_n, $t_n \to \infty$, such that $y = \lim x(t_n, x_0)$.

Theorem 4.1. *The set* Ω *of all ω-limiting points of a given point* x_0 *is a closed set containing whole trajectories* (see Figure 4.1).

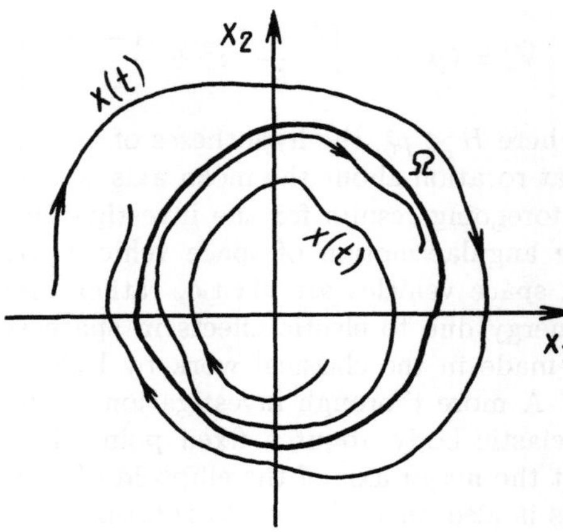

FIG. 4.1. Structure of the limiting set Ω

Proof. The assertion that the set Ω is closed can be verified in the following way. If $y^* = \lim_{n\to\infty} y_n$, $y_n \in \Omega$, then $y_n = \lim_{k\to\infty} x(t_{n_k}, x_0)$, and, consequently,

$$y^* = \lim_{n\to\infty} \lim_{k\to\infty} x(t_{n_k}, x_0) = \lim_{n_k\to\infty} x(t_{n_k}, x_0).$$

Further, if $y \in \Omega$, then necessarily $x(t, y) \in \Omega$ for all $t \geqslant t_0$. In fact, for all $t > t_0$ we have

$$x(t, y) = x\big(t, \lim_{n\to\infty} x(t_n, x_0)\big) = \lim_{n\to\infty} x(t_n + t, x_0).$$

Therefore, $x(t, y) \in \Omega$. $\square$

A set consisting of whole trajectories is called an *invariant* set. If $V(x) \geqslant 0$ is some Lyapunov function, and $\dot{V} \leqslant 0$, then all ω-limiting points for a given point x_0 lie on one level surface of the function $V(x)$, i.e., for $y_1, y_2 \in \Omega$ the equality $V(y_1) = V(y_2)$ holds. Indeed, if $y_1 = \lim_{n\to\infty} x(t_n, x_0)$ and $y_2 = \lim_{m\to\infty} x(t_m, x_0)$, then by continuity of $V(x)$ we obtain

$$V(y_1) = \lim_{n\to\infty} V\big(x(t_n, x_0)\big) = V_\infty = \lim_{m\to\infty} V\big(x(t_m, x_0)\big) = V(y_2).$$

Theorem 4.2 (E. A. Barbashin, N. N. Krasovskii). *Suppose there exists a Lyapunov function* $V(x)$ *such that* $V(x) \geqslant \omega_1(|x|)$, $\dot{V} \leqslant 0$. *Moreover, the set* $\{x \colon \dot{V} = 0\}$ *does not contain whole trajectories of the system* (4.1) *except for the point* $x = 0$. *Then the trivial solution of equation* (4.1) *is asymptotically stable.*

Proof. By Theorem 2.1 the solution $x(t) \equiv 0$ is stable, i.e., for a given $\varepsilon > 0$ there exists a $\delta > 0$ such that $|x_0| < \delta \Rightarrow |x(t, x_0)| \leqslant \varepsilon$. Let $\Omega \subset S_\varepsilon$ be the set of ω-limiting points for some $x_0 \in S_\delta$. Now we show that $\Omega = \{0\}$. Assume that this is not the case. Then there exists an $y \neq 0$, $y \in \Omega$. Since V is positive definite, we have $V(y) > 0$. According to Theorem 4.1, the trajectory $x(t, y) \in \Omega$; hence $V\big(x(t, y)\big) = V(y)$. Consequently, $\dot{V} = 0$ along the trajectory $x(t, y)$, i.e., $x(t, y)$ must belong to a set for which $\dot{V} = 0$. But according to the hypothesis $\dot{V} = 0$ does not contain whole trajectories. This contradiction provides the proof of the theorem. □

If the equation $\varphi(x_1, \ldots, x_n) = 0$ describes the set $\dot{V} = 0$, then the condition

$$\frac{d\varphi}{dt} = \sum_{i=1}^{n} \frac{\partial \varphi}{\partial x_i} f_i(x) \neq 0 \tag{4.2}$$

is sufficient for the absence of whole trajectories in the set $\dot{V} = 0$.

Note that Theorem 4.2 remains valid also for periodic systems (1.1).

Example 4.1. Consider the equation

$$\ddot{x} + \varphi(\dot{x}) + f(x) = 0, \qquad \varphi(0) = f(0) = 0. \tag{4.3}$$

This equation describes the oscillation of a mass point subjected to a nonlinear restoring force $f(x)$ in a medium whose resistance is a nonlinear function of the velocity $\dot{x}$. We write equation (4.3) in the form of a system

$$\dot{x} = y, \qquad \dot{y} = -f(x) - \varphi(y). \tag{4.4}$$

and introduce the Lyapunov function

$$V(x, y) = \frac{y^2}{2} + \int_0^x f(u)\, du. \tag{4.5}$$

Suppose the following conditions are satisfied:

$$x f(x) > 0, \qquad y\varphi(y) > 0, \qquad x \neq 0,\ y \neq 0. \tag{4.6}$$

Then the function (4.5) is positive definite, and its derivative is equal to $\dot{V} = -y\varphi(y)$. In this case the set $\dot{V} = 0$ is of the form $\{y = 0,\ x = \text{arbitrary}\}$. But if along some trajectory of the solution $y \equiv 0$, then also $\dot{y} \equiv 0$; hence $f(x) = 0$. In view of conditions (4.6) the point $x = 0$ is the sole zero of the function $f(x)$. It follows that in the set $\dot{V} = 0$ there are no whole trajectories except the point $x = 0$, $y = 0$. By Theorem 4.2 the trivial solution of equation (4.3) is asymptotically stable if inequalities (4.6) are satisfied. In addition, if

$$\lim_{|x| \to \infty} \int_0^x f(u)\, du = \infty,$$

then, according to Theorems 4.2 and 2.4, the trivial solution is globally asymptotically stable.

2. The Matrosov criterion. The Barbashin–Krasovskii theorem cannot be generalized to the nonautonomous case. Indeed, for the equation $\dot{x}(t) = -a(t)x$ ($x \in R^1$, $a(t) > 0$) the function $V(x) = x^2$ satisfies the hypotheses of Theorem 4.2. At the same time the solution $x(t) \equiv 0$ is asymptotically stable or unstable according as the integral $\int_{t_0}^{\infty} a(s)\,ds$ is convergent or divergent. For the nonautonomous equation (2.1) there exists a convenient stability criterion, which is derived by using two Lyapunov functions.

Theorem 4.3 (V. M. Matrosov). *For the trivial solution of equation (2.1) to be uniformly asymptotically stable, it is necessary and sufficient that there exist two functions $V(t,x)$ and $W(t,x)$ such that they possess the following properties:*

1°) $\omega_1(|x|) \leqslant V(t,x) \leqslant \omega_2(|x|)$; $x \in S_H$, $t_0 \leqslant t$;

2°) $\dot{V}(t,x) \leqslant a(|x|) \leqslant 0$;

3°) $|W(t,x)| \leqslant L =$ *constant*, $t_0 \leqslant t$, $x \in S_H$;

4°) $|\dot{W}(t,x)| \geqslant \omega_4(|x|)$, $t_0 \leqslant t$, $x \in E(\mu,\rho)$;

where $a(|x|)$ is an arbitrary continuous scalar function, $E(\mu,\rho)$ is a set of the form $\{x \in S_H,\ d(x, Q(a=0)) \leqslant \rho,\ \mu \leqslant |x| \leqslant H\}$ (see Figure 4.2). Here $Q(a=0)$ is a set of S_H such that $a(|x|) = 0$, and $d(x, Q(a=0))$ is the distance from the point x to the set $Q(a=0)$.

For the proof of this theorem, see [17].

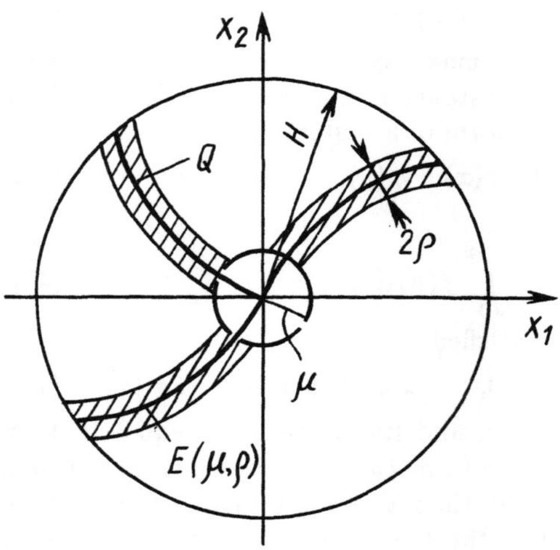

FIG. 4.2. An example of the set $E(\mu,\rho)$

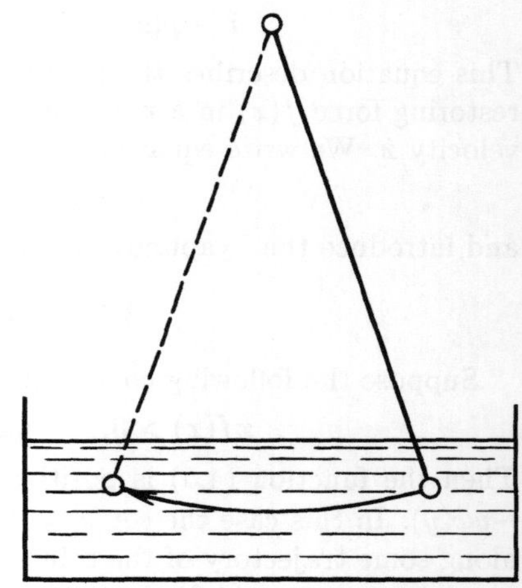

FIG. 4.3. A pendulum in a medium with viscous friction

Example 4.2. *Pendulum with time-dependent viscous friction.* The oscillations of this pendulum (see Figure 4.3) are described by the equation

$$\ddot{x} + h(t)\dot{x} + \sin x = 0, \qquad k_1 \leqslant h(t) \leqslant k_2, \tag{4.7}$$

or the equivalent system

$$\dot{x} = y, \qquad \dot{y} = -h(t)y - \sin x.$$

Take the function $V(x, y)$ in the form

$$V(x, y) = \frac{y^2}{2} + (1 - \cos x).$$

In the region $\{(x^2 + y^2)^{1/2} < \pi - \varepsilon,\ \varepsilon > 0\}$ we have

$$\frac{2}{\pi^2}(x^2 + y^2) \leqslant V(x, y) \leqslant \frac{1}{2}(x^2 + y^2).$$

The derivative $\dot{V} = -h(t)y^2 \leqslant -k_1 y^2 \leqslant 0$. The set $Q(\omega_3 = 0)$ is of the form $\{y = 0,\ x \text{ arbitrary}\}$. For the second function W we take $W(x, y) = -xy$. It is bounded inside any sphere S_H, and

$$\dot{W} = h(t)xy + x \sin x - y^2.$$

For $|y| \leqslant \rho$ and $|x| \geqslant \mu$, we have

$$\dot{W} \geqslant C\mu - k_2 \pi \rho - \rho^2 \geqslant \tfrac{1}{2} C\mu > 0,$$

where $C = \min\{\sin \mu, \sin \varepsilon\}$ and $k_2 \pi \rho + \rho^2 \leqslant C\mu/2$. By Theorem 4.3 the trivial solution of equation (4.7) is uniformly asymptotically stable.

3. The comparison principle. To solve the stability problem for complex or combined systems, it is convenient to use *Lyapunov vector functions*. Below we state some of the results [13].

In addition to equations (1.1) we consider the equation

$$\dot{u}(t) = F(t, u), \qquad u \in R^m. \tag{4.8}$$

The vector function $F(t, u) = (F_1, \dots, F_m)'$ is said to be *quasimonotonically increasing* if for each pair of points (t, u) and (t, v) such that

$$u_i = v_i, \qquad u_j \geqslant v_j, \qquad j \neq i,$$

the inequality $F_i(t, u) \geqslant F_i(t, v)$ holds for all $i = 1, \dots, m$.

Theorem 4.4. *Suppose there exists a vector continuously differentiable function $V(t, x) = \big(V_1(t, x), \dots, V_m(t, x)\big)'$ such that*

1°) $\max_i V_i(t, x) \geqslant \omega_1(|x|)$, $V(t, 0) = 0$;

2°) $\dot{V}_{(2.1)}(t, x) \leqslant F\big(t, V(t, x)\big)$.

Here $\dot{V}_{(2.1)}(t, x)$ is the derivative of the vector function $V(t, x)$ along the solution of the system (2.1) and defined by

$$\dot{V}_{(2.1)}(t, x) = \left(\frac{\partial V_1}{\partial t} + \sum_{j=1}^{n} \frac{\partial V_1}{\partial x_j} f_j, \dots, \frac{\partial V_m}{\partial t} + \sum_{j=1}^{n} \frac{\partial V_m}{\partial x_j} f_j\right).$$

The function $F(t,u)$ is assumed to be quasimonotonically increasing. Then

—the stability of the solution $u = 0$ of equation (4.8) implies the stability of the solution $x = 0$ of equation (2.1);

—the uniform asymptotic stability of the solution $u = 0$ implies the uniform asymptotic stability of the solution $x = 0$.

The proof of this assertion and also of a large number of other theorems associated with these problems can be found in, for example, [13, 20].

Example 4.3 [20]. Consider the equation

$$\dot{x} = \big(I \sin t + A(t,x)\big)x, \qquad x \in R^n, \tag{4.9}$$

where I is a unit $n \times n$ matrix, and the $n \times n$ matrix $A(t,x)$ with elements $a_{ij}(t,x)$, differentiable with respect to t and x, is skew-symmetric, i.e., such that

$$a_{ii}(t,x) \equiv 0, \qquad a_{ij}(t,x) = -a_{ji}(t,x).$$

Take the Lyapunov function in the form

$$V(x_1,\dots,x_n) = x_1^2 + \cdots + x_n^2 = x'x. \tag{4.10}$$

Since the matrix $A(t,x)$ is skew-symmetric, we have $x'A(t,x)x = 0$. Therefore, in view of the system (4.9) the derivative of the Lyapunov function (4.10) is

$$\dot{V} = 2\dot{x}'x = 2x'\big(I \sin t + A(t,x)\big)x = 2x'x \sin t = 2V \sin t.$$

The derivative $\dot{V}$ of the function (4.10) is not of constant signs, and thus we cannot use Lyapunov's theorem 2.1 immediately. However, the solution $u(t) \equiv 0$ of the scalar equation $\dot{u} = (2 \sin t)u$ is stable, since

$$u(t,t_0,u_0) = u_0 \exp\left(2 \int_{t_0}^{t} \sin s \, ds\right).$$

Consequently, by Theorem 4.4 we obtain that the trivial solution $x(t) \equiv 0$ of equation (4.9) is also stable.

4. Stability with respect to part of the variables (partial stability). In some cases one would expect from the behavior of the system that only some of the phase variables would remain stable during the evolution of the system. For example, when investigating the stability of the motion of a cyclist, it is only necessary to consider small displacements of his body from the vertical plane, but the translation velocity of the cyclist may vary over a wide range. Such a case can be conveniently investigated by means of the concept of *partial stability* [19].

Suppose that the system in question is described by the equations

$$\begin{gathered} \dot{x} = f(t,x,y), \qquad \dot{y} = g(t,x,y), \qquad x \in R^n, \qquad y \in R^m, \\ f(t,0,0) = g(t,0,0) = 0, \qquad z' = (x',y') \in R^{n+m}. \end{gathered} \tag{4.11}$$

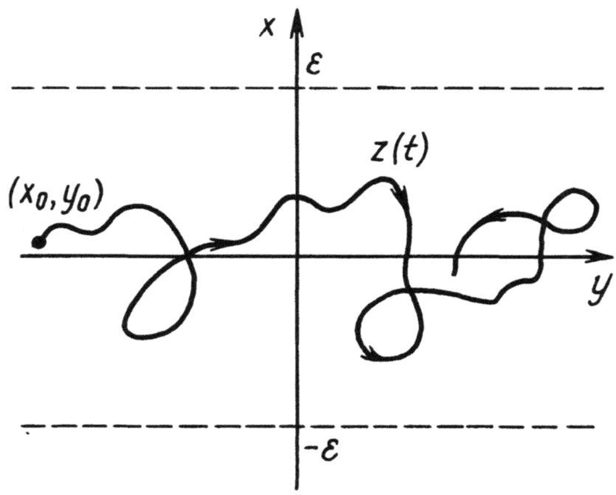

FIG. 4.4. A trajectory stable with respect to the variable x

Definition 4.2. The trivial solution $z(t) \equiv 0$ ($x(t) \equiv 0$, $y(t) \equiv 0$) of the system (4.11) is said to be *x-stable* (*stable with respect to x*) if for any $\varepsilon > 0$ and any $t_0 \in T$ there exists a $\delta(\varepsilon, t_0) > 0$ such that the inequality $|x(t, t_0, z_0)| \leqslant \varepsilon$, $t \geqslant t_0$, holds whenever $|z(t_0)| \leqslant \delta(\varepsilon, t_0)$. The solution $z(t) \equiv 0$ of the system (4.11) is said to be *asymptotically x-stable* if it is x-stable and for some $\Delta > 0$ the following relation is valid: $\lim_{t\to\infty} x(t, t_0, z_0) = 0$, $|z_0| \leqslant \Delta$, (see Figure 4.4).

Theorem 4.5 (V. V. Rumyantsev). *Suppose there exists a function $V(t, x, y)$ such that*

1°) $V(t, x, y) \geqslant \omega_1(|x|)$;

2°) $\dot V \leqslant 0$.

Then the solution $z(t) \equiv 0$ of the system (4.11) *is x-stable. But if the function $V(t, x, y)$ is such that*

3°) $\omega_1(|x|) \leqslant V(t, x, y) \leqslant \omega_2(|z|)$;

4°) $\dot V(t, x, y) \leqslant -\omega_3(|x|)$,

then the solution $z(t) \equiv 0$ is asymptotically x-stable.

§ 5. Stability of Linear Time-Invariant Systems

Even in the simplest cases, general theorems on stability (see § 2) do not allow one to pass judgement on the stability of solutions to equations solely on the basis of the knowledge of their coefficients. Therefore, obtaining the pertinent conditions for the coefficients is of great interest even for some particular classes of equations. Linear equations with constant coefficients will be studied first.

1. The Routh–Hurwitz criterion. Consider a system of linear differential equations with constant coefficients of the form

$$\dot x(t) = Ax(t), \qquad x \in R^n. \tag{5.1}$$

Recall several important facts concerning the structure of the general solution of equation (5.1).

Assume first that the eigenvalues (characteristic roots) λ_i and the corresponding eigenvectors (characteristic vectors) of the matrix A are all distinct, i.e., $Ah_i = \lambda_i h_i$, $i = 1, \ldots, n$. This case is often referred to as that of *simple* eigenvalues and eigenvectors. Then the general solution of equation (5.1) is of the form

$$x(t) = \sum_{i=1}^{n} C_i h_i e^{\lambda_i t},$$

where C_i are arbitrary constants.

To construct the general solution of equation (5.1) in the case of *multiple* eigenvalues, the matrix A must be reduced first to the *Jordan canonical form*. By the *Jordan canonical form* of a matrix we mean a block-diagonal matrix having Jordan boxes as the elements of its principal diagonal and all other elements zero. A *Jordan box* is a square matrix having the elements of its principal diagonal equal and not zero, the elements immediately above those in the diagonal unity, and all other elements zero (see Figure 5.1, where Jordan boxes are shown, and all other elements not belonging to the diagonal boxes are zero).

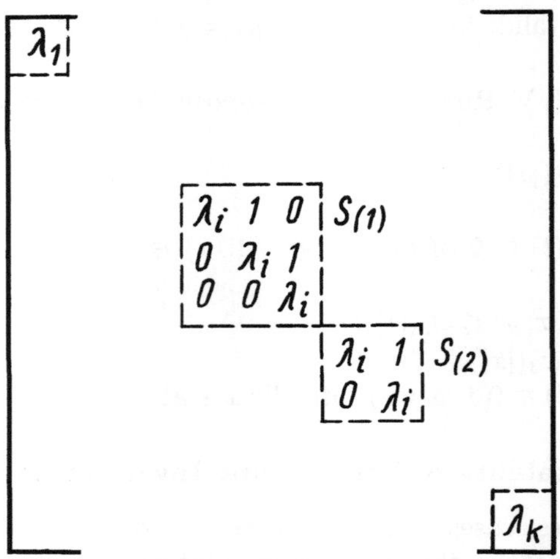

FIG. 5.1. The Jordan form of the matrix A

There are various ways of reducing a square matrix A of order n to Jordan form. We shall dwell on one of these[3] consisting of the following stages:

1°) For each k, $k = 0, \ldots, n$, find the greatest common divisor $D_k(\lambda)$ of all minors of order k of the characteristic matrix $\lambda I - A$. The leading coefficients of

[3]Gantmakher, F. R., *Theory of Matrices*, Nauka, Moscow, (1966).

all polynomials are taken to be unity. Thus, we have

$$D_n(\lambda) = \det(\lambda I - A), \qquad D_0(\lambda) = 1.$$

2°) Find invariant polynomials $\alpha_k(\lambda)$ (also called invariant multipliers) according to the formulas

$$\alpha_1(\lambda) = \frac{D_n(\lambda)}{D_{n-1}(\lambda)},\ \alpha_2(\lambda) = \frac{D_{n-1}(\lambda)}{D_{n-2}(\lambda)},\dots,\alpha_n(\lambda) = \frac{D_1(\lambda)}{D_0(\lambda)} = D_1(\lambda).$$

3°) Decompose each invariant polynomial into distinct irreducible polynomials called *elementary divisors* of the characteristic matrix $\lambda I - A$ or simply of the matrix A. An elementary divisor can be represented as a power of one of the differences $(\lambda - \lambda_i)$, where λ_i are eigenvalues of the matrix A. Thus, $\alpha_k(\lambda)$ is equal to

$$\alpha_k(\lambda) = (\lambda - \lambda_1)^{k_1}(\lambda - \lambda_2)^{k_2}\dots(\lambda - \lambda_j)^{k_j}.$$

Note that the product of all elementary divisors is equal to the product of all invariant multipliers, namely is equal to $D_n(\lambda)$.

4°) To each elementary divisor $(\lambda - \lambda_j)^{k_j}$ assign a Jordan box of order k_j with λ_j in its principal diagonal. A block-diagonal matrix having Jordan boxes so constructed as the elements of its principal diagonal is, in fact, the desired Jordan form of the matrix A. The Jordan canonical form is uniquely determined to within an order in which Jordan boxes are arranged in the principal diagonal.

In the case of multiple roots, the general solution of equation (5.1) can be obtained in the following way. Suppose that the matrix A has k different eigenvalues λ_i of multiplicities r_i, $i = 1,\dots,k$, $r_1 + r_2 + \dots + r_k = n$. With each eigenvalue several eigenvectors are associated. The degree of degeneracy of the eigenvalue λ_i is equal to the maximal number $m(i)$ of linearly independent eigenvectors associated with λ_i. The number $m(i)$ is also equal to the number of Jordan boxes associated with λ_i. Denote by B_j one of these boxes. If the order of the Jordan box B_j is $s(j)$, then

$$s(1) + s(2) + \dots + s\big(m(i)\big) = r_i.$$

In particular, it follows that the Jordan form of the matrix A is diagonal if the multiplicity of any of its eigenvalues is equal to the degree of degeneracy of this eigenvalue. In this case all eigenvalues possess simple elementary divisors. With each Jordan box B_j a sequence of vectors h_l, $l = 1,\dots,s(j)$, is associated such that

$$h_1 \neq 0,\ Ah_1 = \lambda_i h_1,\ Ah_2 = \lambda_i h_2 + h_1,\dots,Ah_{s(j)} = \lambda_s h_{s(j)} + h_{s(j)-1}.$$

The vectors h_l, $l = 1,\dots,s(j)$ belonging to a particular sequence are linearly independent. The subspace spanned by all the vectors h_l, $l = 1,\dots,s(j)$ associated with the box B_j is invariant, cyclic, and of dimension of the sequence, i.e., $s(j)$.

In the basis composed of vectors belonging to various series the matrix A is of Jordan form.

As to equation (5.1), its solutions are functions of the form

$$x_{iq}(t) = \omega_q(t)e^{\lambda_i t}, \tag{5.2}$$

$$\omega_q(t) = \frac{t^{q-1}}{(q-1)!}h_1 + \frac{t^{q-2}}{(q-2)!}h_2 + \cdots + h_q, \qquad q = 1, \ldots, s(j),$$

$$s(1) + s(2) + \cdots + s(m(i)) = r_i, \qquad r_1 + r_2 + \cdots + r_k = n.$$

The number of linearly independent solutions of the form (5.2) is equal to n. The general solution of equation (5.1) is a linear combination of particular solutions of the form (5.2).

This representation of the general solution of a linear time-invariant system immediately leads to the following theorem.

Theorem 5.1. *For the trivial solution of equation* (5.1) *to be stable, it is necessary and sufficient that all eigenvalues λ_i of the matrix A satisfy the condition* $\operatorname{Re}\lambda_i \leqslant 0$, *and eigenvalues λ_i such that* $\operatorname{Re}\lambda_i = 0$ *have simple elementary divisors (i.e., in formula* (5.2) *all* $s(j) = 1$). *For equation* (5.1) *to be asymptotically stable, it is necessary and sufficient that* $\operatorname{Re}\lambda_i < 0$.

Thus, whether equation (5.1) is stable, asymptotically stable, or unstable, is determined by the location of the roots of the characteristic equation of the matrix A:

$$\det(A - \lambda I) = 0. \tag{5.3}$$

relative to the imaginary axis. In this equation I is a unit $(n \times n)$-matrix. Developing the determinant, we can reduce equation (5.3) to the form

$$P_n(\lambda) = a_0 + a_1\lambda + \cdots + a_n\lambda^n = 0, \qquad a_n = 1. \tag{5.4}$$

The polynomial $P_n(\lambda)$ is said to be a *stable polynomial* or a *Hurwitz polynomial* if all of its roots λ_i have a negative real part, i.e.,

$$\operatorname{Re}\lambda_i < 0. \tag{5.5}$$

When condition (5.5) is satisfied, the matrix A is also said to be *stable.*

Theorem 5.2 (A. Stodola). *All coefficients of a stable polynomial are positive.*

Proof. Suppose that $\lambda_j = -\alpha_j \pm i\beta_j$ $(j = 1, \ldots, r,\ i^2 = -1)$ and $\lambda_k = -\gamma_k$ $(k = 1, \ldots, l,\ 2r + l = n)$ are the roots of a stable polynomial $P_n(\lambda)$. Then $\alpha_j > 0$ and $\gamma_k > 0$; factoring $P_n(\lambda)$, we obtain

$$P_n(\lambda) = \prod_{j=1}^{r}(\lambda^2 + 2\alpha_j\lambda + \alpha_j^2 + \beta_j^2)\prod_{k=1}^{l}(\lambda + \gamma_k).$$

Thus, $P_n(\lambda)$ is expressed as a product of polynomials with positive coefficients. Hence, all of its coefficients $a_j > 0$. □

The requirement that the coefficients be positive is a necessary but not a sufficient condition that inequalities (5.5) hold. For example, the polynomial $P_3(z) = 30 + 4z + z^2 + z^3$ has the roots $z_1 = -3$, $z_{2,3} = 1 \pm 3i$.

We now pass to a necessary and sufficient criterion for stability. By the *Hurwitz matrix* M_{P_n} of the polynomial (5.4) we mean a matrix of the form

$$M_{P_n} = \begin{pmatrix} a_1 & a_0 & 0 & 0 & \dots & 0 \\ a_3 & a_2 & a_1 & a_0 & \dots & 0 \\ \vdots & \vdots & \vdots & \vdots & \ddots & \vdots \\ a_{2n-1} & a_{2n-2} & \dots & \dots & \dots & a_n \end{pmatrix}. \tag{5.6}$$

In the matrix (5.6) all $a_s = 0$ whenever $s < 0$ and $s > n$. For example, the Hurwitz matrix for the polynomial of the sixth degree is of the form

$$M_{P_6} = \begin{pmatrix} a_1 & a_0 & 0 & 0 & 0 & 0 \\ a_3 & a_2 & a_1 & a_0 & 0 & 0 \\ a_5 & a_4 & a_3 & a_2 & a_1 & a_0 \\ 0 & a_6 & a_5 & a_4 & a_3 & a_2 \\ 0 & 0 & 0 & a_6 & a_5 & a_4 \\ 0 & 0 & 0 & 0 & 0 & a_6 \end{pmatrix}. \tag{5.7}$$

The Routh–Hurwitz Criterion. *For the polynomial $P_n(\lambda)$, with $a_0 > 0$ and $a_n \neq 0$, to be stable, it is necessary and sufficient that all principal diagonal minors*

$$\Delta_1 = a_1, \; \Delta_2 = \begin{vmatrix} a_1 & a_0 \\ a_3 & a_2 \end{vmatrix}, \dots, \Delta_n = a_n \Delta_{n-1}$$

of its Hurwitz matrix M_{P_n} be positive.

The Liénard–Chipart Criterion. *For a polynomial $P_n(\lambda)$ with positive coefficients to be stable, it is necessary and sufficient that one of the following two conditions are satisfied*: (a) $\Delta_1 > 0$, $\Delta_3 > 0, \dots$; (b) $\Delta_2 > 0$, $\Delta_4 > 0, \dots$.

For the proof of these criteria the reader is referred to [4].

Example 5.1. Consider the polynomial

$$P_3(\lambda) = a_0 + a_1\lambda + a_2\lambda^2 + \lambda^3 \qquad (a_i > 0).$$

Using the Liénard–Chipart criterion, we obtain from the inspection of the matrix (5.6) that the polynomial $P_3(\lambda)$ is stable whenever $0 < a_0 < a_1 a_2$.

2. Frequency criteria of stability. The application of the Routh–Hurwitz criterion to actual automatic control systems leads to cumbersome computations and does not allow to isolate the effects of individual parameters on the stability of the system. From an applications point of view the frequency criteria of A. V. Mikhailov and G. Nyquist are more convenient.

Consider the polynomial (5.4). Set $\lambda = i\omega$ $(i^2 = -1)$, and construct the *Mikhailov hodograph (gain–phase characteristic)* $P_n(i\omega)$ of this polynomial:

$$z = P_n(i\omega) = a_n(i\omega - \lambda_1)\dots(i\omega - \lambda_n).$$

Geometrically, $i\omega - \lambda_k$ is a vector in the complex plane z issuing from the point λ_k, the kth root of the polynomial $P_n(\lambda)$, and terminating at the point $\lambda = i\omega$.

Since $P_n(i\omega)$ is a complex number, we have

$$\arg P_n(i\omega) = \arg a_n + \arg(i\omega - \lambda_1) + \dots + \arg(i\omega - \lambda_n). \tag{5.8}$$

If the root λ_k lies in the left half-plane ($\operatorname{Re}\lambda_k < 0$), then, as ω changes from $-\infty$ to $+\infty$, the vector $i\omega - \lambda_k$ rotates in the positive direction (counterclockwise), and the change in the argument is equal to π:

$$\Delta \arg(i\omega - \lambda_k)|_{\omega=-\infty}^{\omega=\infty} = \pi. \tag{5.9}$$

Alternatively, if the root λ_l lies in the right half-plane, then the vector $i\omega - \lambda_l$ rotates in the negative direction (clockwise), and

$$\Delta \arg(i\omega - \lambda_l)|_{\omega=-\infty}^{\omega=\infty} = -\pi.$$

The following theorem can be deduced from relations (5.8) and (5.9).

Theorem 5.3. *A polynomial $P_n(\lambda)$ having no pure imaginary roots is stable if and only if*

$$\Delta \arg P_n(i\omega)|_{\omega=-\infty}^{\omega=\infty} = n\pi. \tag{5.10}$$

To simplify the conditions (5.10), we isolate the real and imaginary parts of $P_n(\lambda)$. Hence

$$\begin{aligned} P_n(i\omega) &= a_0 + a_1\, i\omega + a_2(i\omega)^2 + \dots + a_n(i\omega)^n \\ &= (a_0 - a_2\omega^2 + a_4\omega^4 + \dots) + i(a_1\omega - a_3\omega^3 + a_5\omega^5 + \dots). \end{aligned}$$

It is readily seen that $\operatorname{Re} P_n(\lambda)$ is an even function of ω, and $\operatorname{Im} P_n(\lambda)$ is an odd function. Therefore, the hodograph of the function $P_n(\lambda)$ consists of two parts: symmetric about the real axis, the first part corresponding to the change in ω from $-\infty$ to 0, and the second from 0 to ∞. Thus, we can deduce the following assertion.

Theorem 5.4(The Mikhailov criterion). *For a polynomial $P_n(\lambda)$ having no pure imaginary roots to be stable, it is necessary and sufficient that*

$$\Delta \arg P_n(i\omega)|_{\omega=0}^{\omega=\infty} = \frac{\pi}{2} n. \tag{5.11}$$

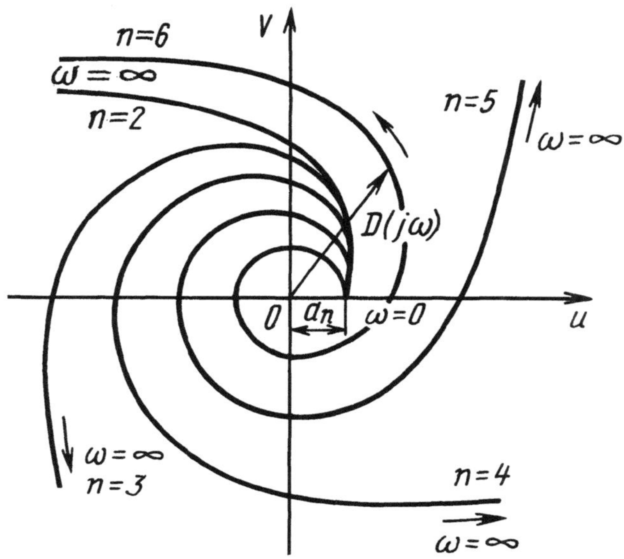

FIG. 5.2. Mikhailov hodographs for stable systems for $n = 2, 3, 4, 5, 6$

Various Mikhailov hodographs for $n = 2, \ldots, 6$ are shown in Figure 5.2. The closer the hodograph passes to the origin, the smaller the reserve of stability is. If the hodograph passes through the origin, then the system is at the boundary of the stability region. Thus, the critical values of the parameters a_{cr} and the frequency ω_{cr} for a linear system with characteristic equation (5.4) can be found from the conditions

$$\operatorname{Re} P_n(i\omega_{\text{cr}}, a_{\text{cr}}) = \operatorname{Im} P_n(i\omega_{\text{cr}}, a_{\text{cr}}) = 0$$

To stress the dependence of the polynomial (5.4) on the parameters $a = (a_0, \ldots, a_n)$, it is denoted by $P_n(\lambda, a)$. It is interesting to note that in practical work one can evaluate the gain–phase characteristic experimentally by using a signal of frequency ω and amplitude 1 as the input to the linear time-invariant system in question. The output signal has the same frequency ω, the amplitude $|P_n(i\omega)|$ and the phase-shift is $\arg P_n(i\omega)$.

3. Automatic frequency control system of a heterodyne receiver. In heterodyne receivers, to ensure good performance, the basic amplification is done, using a fixed, so-called intermediate, frequency (IF) ω_{IF}. Since the frequency ω_{sig} of the received signal is varying all the time due to various uncontrollable disturbances, an automatic frequency control (AFC) system is incorporated into the heterodyne receiver to boost its performance. The purpose of the AFC system is to maintain a prescribed value $\omega_{\text{IF}} = \omega_{\text{IF}}^{\circ}$. One of feasible AFC systems is shown in Figure 5.3.

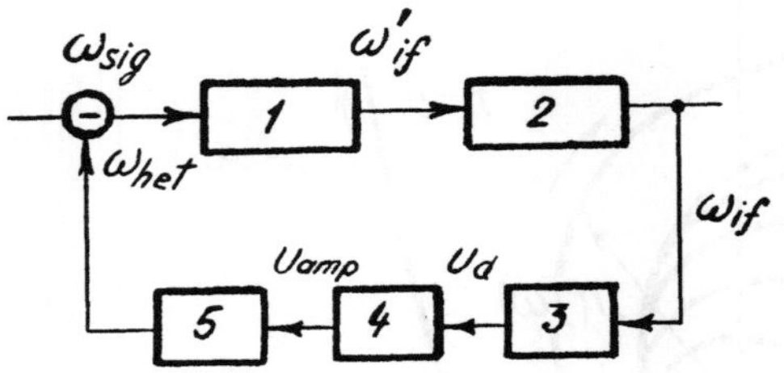

FIG. 5.3. Automatic frequency control system of a heterodyne receiver:
1,mixer; 2,IF amplifier; 3,discriminator; 4,amplifier; 5,heterodyne controller

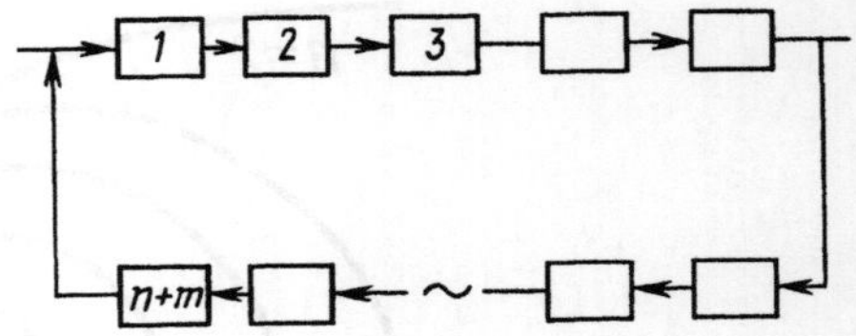

FIG. 5.4. An arbitrary single-loop automatic control system

The function of each of devices 1–5 can be described by the following relations involving the corresponding frequency deviations $\delta\omega$:

(1) Mixer: $\delta\omega_{\text{sig}} - \delta\omega_{\text{het}} = \delta\omega'_{\text{IF}}$.

(2) IF amplifier:

$$T_{\text{IF}}\frac{d}{dt}(\delta\omega_{\text{IF}}) + \delta\omega_{\text{IF}} = \delta\omega'_{\text{IF}}.$$

(3) Discriminator: $U_d = K_d\delta\omega_{\text{IF}}$.

(4) Amplifier: $T_{\text{d}}\frac{dU_{\text{amp}}}{dt} + U_{\text{amp}} = U_{\text{d}}$.

(5) Heterodyne controller:

$$T_{\text{het}}\frac{d}{dt}(\delta\omega_{\text{het}}) + \delta\omega_{\text{het}} = K_{\text{het}}U_{\text{amp}}.$$

We consider $\delta\omega_{\text{sig}}$ as the input to the AFC, and $\delta\omega_{\text{het}}$ as the output, i.e., the AFC is assumed to be a control system. Using the operator calculus, we obtain

$$\delta\omega_{\text{IF}} = \frac{1}{T_{\text{IF}}D+1}\delta\omega'_{\text{IF}},$$

$$U_{\text{d}} = \frac{K_{\text{d}}}{T_{\text{IF}}D+1}\delta\omega'_{\text{IF}},$$

$$U_{\text{amp}} = \frac{1}{T_{\text{d}}D+1}U_{\text{d}} = \frac{K_{\text{d}}}{(T_{\text{IF}}D+1)(T_{\text{d}}D+1)}\delta\omega'_{\text{IF}},$$

$$\delta\omega_{\text{het}} = \frac{K_{\text{het}}}{T_{\text{het}}D+1}U_{\text{amp}} = \frac{K_{\text{het}}K_{\text{d}}}{(T_{\text{het}}D+1)(T_{\text{IF}}D+1)(T_{\text{d}}D+1)}\delta\omega'_{\text{IF}}$$

$$= \frac{K(\delta\omega_{\text{sig}} - \delta\omega_{\text{het}})}{(T_{\text{het}}D+1)(T_{\text{IF}}D+1)(T_{\text{d}}D+1)}, \qquad K = K_{\text{het}}K_{\text{d}},\ D = \frac{d}{dt}.$$

Finally, we obtain the following AFC equation:

$$[(T_{\text{het}}D+1)(T_{\text{IF}}D+1)(T_{\text{d}}D+1) + K]\delta\omega_{\text{het}} = K\delta\omega_{\text{sig}}. \tag{5.12}$$

Next, we proceed to the stability conditions for the AFC described by equation (5.12). We use the Mikhailov criterion and construct the hodograph

$$G(i\omega) = \big(T_{\text{het}}(i\omega) + 1\big)\big(T_{\text{IF}}(i\omega) + 1\big)\big(T_{\text{d}}(i\omega) + 1\big) + K.$$

Now we separate the real and imaginary parts:

$$\begin{aligned}\operatorname{Re} G(i\omega) &= 1 + K - \omega^2(T_{\text{IF}}T_{\text{d}} + T_{\text{IF}}T_{\text{het}} + T_{\text{d}}T_{\text{het}}),\\ \operatorname{Im} G(i\omega) &= \omega(T_{\text{IF}} + T_{\text{d}} + T_{\text{het}} - \omega^2 T_{\text{IF}}T_{\text{d}}T_{\text{het}}).\end{aligned}$$

and determine the critical gain coefficient K_{cr} and the critical frequency ω_{cr} from the conditions $\operatorname{Re} G(i\omega) = \operatorname{Im} G(i\omega) = 0$. It follows from the condition $\operatorname{Im} G(i\omega) = 0$ that

$$\omega_{\text{cr}} = \sqrt{\frac{T_{\text{IF}} + T_{\text{d}} + T_{\text{het}}}{T_{\text{IF}}T_{\text{d}}T_{\text{het}}}}$$

and from the condition $\operatorname{Re} G(i\omega) = 0$ that

$$\begin{aligned}K_{\text{cr}} &= \omega_{\text{cr}}^2(T_{\text{IF}}T_{\text{d}} + T_{\text{IF}}T_{\text{het}} + T_{\text{d}}T_{\text{het}}) - 1\\ &= 2 + \frac{T_{\text{d}}}{T_{\text{het}}} + \frac{T_{\text{IF}}}{T_{\text{het}}} + \frac{T_{\text{d}}}{T_{\text{IF}}} + \frac{T_{\text{het}}}{T_{\text{IF}}} + \frac{T_{\text{het}}}{T_{\text{d}}} + \frac{T_{\text{IF}}}{T_{\text{d}}}.\end{aligned}$$

Example. For the values T_{IF}=0.003 sec, T_{d}=0.002 sec, and T_{het}=0.001 sec, we find $K_{\text{cr}} = 10$. Thus, the AFC system under consideration is stable for $K < 10$ and unstable for $K > 10$.

4. Linear single-circuit automatic control systems. As in the foregoing subsection, a similar assertion can be made about any single-loop automatic control system (ACS) which incorporates an arbitrary number of units (see Figure 5.4).

Suppose that the characteristic equation of a single-loop system is of the form

$$\prod_{j=1}^{m}(T_j\lambda + 1)\prod_{j=1}^{n}(a_j\lambda^2 + b_j\lambda + 1) + K = 0. \tag{5.13}$$

The first m factors in (5.13) are associated with aperiodic units, and the second n factors with oscillatory units.

Theorem 5.5. *For any single-loop ACS with characteristic function* (5.13) *there exists a critical value of the gain coefficient* K_{cr} *such that the system is stable whenever* $K < K_{\text{cr}}$ *and is unstable whenever* $K > K_{\text{cr}}$.[4]

[4]For the proof of this theorem, see Neimark, Yu. I., *Dynamical Systems and Controlled Processes*, Nauka, Moscow, (1978).

5. Robust stability[5]**.** The stability conditions given in the theorems of this section have been derived under the assumption that all coefficients of the system $\dot{x} = Bx$, i.e., all elements of the constant matrix B are known exactly *a priori*. This hypothesis is, however, not always valid. In a real situation only some region U containing the matrix B is known. It is clear that if stability occurs for any matrix B from the region U, then the real system is also stable. It turns out that in some situations one can form an opinion about the stability of all systems associated with different matrices $B \in U$ by considering some particular systems.

Now we proceed to some pertinent assertions.

Suppose that $P_n(\lambda)$ is the characteristic polynomial of the matrix B, and W is the set of all such polynomials associated with all matrices $B \in U$. An arbitrary polynomial $P_n(\lambda) \in W$ is taken to be of the form

$$P_n(\lambda) = \lambda^n + b_1\lambda^{n-1} + b_2\lambda^{n-2} + \cdots + b_n. \tag{5.14}$$

Further, suppose that the bounds of variation of the coefficients b_i, $i = 1, \dots, n$, are known as the polynomial $P_n(\lambda)$ ranges over the set W. In other words, we assume that certain numbers α_i, β_i, $i = 1, \dots, n$, are known such that

$$\begin{gathered} \alpha_i = \min_{P_n \in W} b_i, \qquad \beta_i = \max_{P_n \in W} b_i, \\ b_i \in [\alpha_i, \beta_i],\ 0 < \alpha_i \leqslant \beta_i,\ i = 1, \dots, n. \end{gathered} \tag{5.15}$$

It turns out that in order to test the stability of the whole infinite set of polynomials W, it is sufficient to verify the stability of only four polynomials $f_n^1(\lambda)$, $f_n^2(\lambda)$, $f_n^3(\lambda)$, and $f_n^4(\lambda)$.

The coefficients a_i^1 of the polynomial $f_n(\lambda)$ are equal to

$$a^1_{n-2j} = \begin{cases} \beta_{n-2j}, & \text{for } j \text{ even}, \\ \alpha_{n-2j}, & \text{for } j \text{ odd}; \end{cases}$$

$$a^1_{n-2j-1} = \begin{cases} \beta_{n-2j-1}, & \text{for } j \text{ even}, \\ \alpha_{n-2j-1}, & \text{for } j \text{ odd}. \end{cases}$$

The coefficients a_i^2 of the polynomial $f_n^2(\lambda)$ are equal to

$$a^2_{n-2j} = \begin{cases} \alpha_{n-2j}, & \text{for } j \text{ even}, \\ \beta_{n-2j}, & \text{for } j \text{ odd}; \end{cases}$$

$$a^2_{n-2j-1} = \begin{cases} \alpha_{n-2j-1}, & \text{for } j \text{ even}, \\ \beta_{n-2j-1}, & \text{for } j \text{ odd}. \end{cases}$$

The coefficients a_i^3 of the polynomial $f_n^3(\lambda)$ are equal to

$$a^3_{n-2j} = \begin{cases} \alpha_{n-2j}, & \text{for } j \text{ even}, \\ \beta_{n-2j}, & \text{for } j \text{ odd}; \end{cases}$$

$$a^3_{n-2j-1} = \begin{cases} \beta_{n-2j-1}, & \text{for } j \text{ even}, \\ \alpha_{n-2j-1}, & \text{for } j \text{ odd}. \end{cases}$$

[5] Kharitonov, V. L., On the Asymptotic Stability of the Equilibrium Position for a Family of Systems of Linear Differential Equations, *Differential Equations*, (1970), vol. 14, no. 11, 2086–2088.

The coefficients a_i^4 of the polynomial $f_n^4(\lambda)$ are equal to

$$a_{n-2j}^4 = \begin{cases} \beta_{n-2j}, & \text{for } j \text{ even}, \\ \alpha_{n-2j}, & \text{for } j \text{ odd}; \end{cases}$$
$$a_{n-2j-1}^4 = \begin{cases} \alpha_{n-2j-1}, & \text{for } j \text{ even}, \\ \beta_{n-2j-1}, & \text{for } j \text{ odd}. \end{cases}$$

Theorem 5.6 (V. L. Kharitonov). *For any polynomial $P_n(\lambda)$ of the form* (5.14), *with coefficients satisfying conditions* (5.15), *to be stable, it is necessary and sufficient that all four polynomials $f_n^1(\lambda)$, $f_n^2(\lambda)$, $f_n^3(\lambda)$, and $f_n^4(\lambda)$ be stable.*

Example 5.2. We turn now to deriving the necessary and sufficient conditions of asymptotic stability of the set of scalar systems of the third order

$$\dddot{x}(t) + b_1\ddot{x}(t) + b_2\dot{x}(t) + b_3 = 0 \qquad (5.16)$$

with coefficients satisfying the conditions

$$\alpha_i \leqslant b_i \leqslant \beta_i, \qquad i = 1, 2, 3. \qquad (5.17)$$

The characteristic polynomial of equation (5.16) is of the form

$$\lambda^3 + b_1\lambda^2 + b_2\lambda + b_3 = 0.$$

Further, construct the polynomials

$$\begin{aligned} f_3^1(\lambda) &= \lambda^3 + \alpha_1\lambda^2 + \beta_2\lambda + \beta_3, \\ f_3^2(\lambda) &= \lambda^3 + \beta_1\lambda^2 + \alpha_2\lambda + \alpha_3, \\ f_3^3(\lambda) &= \lambda^3 + \beta_1\lambda^2 + \beta_2\lambda + \alpha_3, \\ f_3^4(\lambda) &= \lambda^3 + \alpha_1\lambda^2 + \alpha_2\lambda + \beta_3. \end{aligned} \qquad (5.18)$$

Using the result of Example 5.1, we can obtain stability conditions for the polynomials (5.18) of the form

$$\begin{aligned} &0 < \beta_3 < \alpha_1\beta_2, \qquad 0 < \alpha_3 < \beta_1\alpha_2, \\ &0 < \alpha_3 < \beta_1\beta_2, \qquad 0 < \beta_3 < \alpha_1\alpha_2. \end{aligned} \qquad (5.19)$$

Since $\alpha_3 < \beta_3$ and $\alpha_1\alpha_2$ are less than any of the products $\alpha_1\beta_2$, $\beta_1\alpha_2$, and $\beta_1\beta_2$, then all of the conditions (5.19) will be satisfied simultaneously whenever one more condition is met:

$$0 < \beta_3 < \alpha_1\alpha_2. \qquad (5.20)$$

Thus, all of the equations in the set (5.16) with coefficients satisfying condition (5.17) are stable whenever the inequality (5.20) is met. Conditions (5.20) are quite natural. If they are not satisfied, it follows from (5.16), where

$$b_1 = \alpha_1, \qquad b_2 = \alpha_2, \qquad b_3 = \beta_3,$$

that the equation in question is unstable.

§ 6. Stability of Linear Time-Varying Equations

1. On the method of "frozen" coefficients. For the equation

$$\dot{x}(t) = A(t)x(t) \tag{6.1}$$

the determination of the stability conditions is a difficult problem, which has not been fully solved yet. In engineering applications one often resorts to the *method of "frozen" coefficients.* This method is basically as follows. For each t ($t_0 \leqslant t < \infty$) find all of the eigenvalues $\lambda_i(t)$ of the matrix $A(t)$. If they satisfy the condition $\operatorname{Re}\lambda_i(t) \leqslant -\sigma < 0$ ($i = 1, \dots, n$, $t_0 \leqslant t < \infty$), then the system (6.1) is considered to be asymptotically stable. But if for some s the condition $\operatorname{Re}\lambda_i(t) \geqslant \sigma > 0$ is satisfied, then the system (6.1) is unstable. These assertions are not valid without additional assumptions as the following examples illustrate.

Example 6.1[6]**.** The system

$$\begin{aligned} \dot{x}_1(t) &= (-1 - 9\cos^2 6t + 6\sin 12t)x_1 + (12\cos^2 6t + 4.5\sin 12t)x_2, \\ \dot{x}_2(t) &= (-12\sin^2 6t + 4.5\sin 12t)x_1 - (1 + 9\sin^2 6t + 6\sin 12t)x_2. \end{aligned} \tag{6.2}$$

has the solution

$$\begin{aligned} x_1(t) &= Ce^{2t}(\cos 6t + 2\sin 6t), \\ x_2(t) &= Ce^{2t}(2\cos 6t - \sin 6t), \end{aligned}$$

and hence is unstable. However, the eigenvalues of the matrix $A(t)$ are $\lambda_1(t) = -1$ and $\lambda_2(t) = -10$.

Example 6.2[7]**.** For the system

$$\begin{aligned} \dot{x}_1(t) &= (-5.5 + 7.5\sin 12t)x_1 + (7.5\cos 12t)x_2, \\ \dot{x}_2(t) &= (7.5\cos 12t)x_1 + (-5.5 - 7.5\sin 12t)x_2. \end{aligned} \tag{6.3}$$

the eigenvalues of the matrix A are as follows: $\lambda_1(t) \equiv 1$, $\lambda_2(t) = -13$. Moreover, the fundamental matrix $X(t)$ of solutions of the system (6.3) has elements of the form

$$\begin{aligned} x_{11}(t) &= \tfrac{1}{2}e^{-t}(\cos 6t + 3\sin 6t) + \tfrac{1}{2}e^{-10t}(\cos 6t - 3\sin 6t), \\ x_{12}(t) &= \tfrac{1}{6}e^{-t}(\cos 6t + 3\sin 6t) - \tfrac{1}{6}e^{-10t}(\cos 6t - 3\sin 6t), \\ x_{21}(t) &= \tfrac{1}{2}e^{-t}(3\cos 6t - \sin 6t) - \tfrac{1}{2}e^{-10t}(3\cos 6t + \sin 6t), \\ x_{22}(t) &= \tfrac{1}{6}e^{-t}(3\cos 6t - \sin 6t) + \tfrac{1}{6}e^{-10t}(3\cos 6t + \sin 6t). \end{aligned} \tag{6.4}$$

Hence, the trivial solution of the system (6.3) is asymptotically stable.

In the general case the following assertion holds.

---

[6]Vinograd, R. E., On a Lyapunov-Instability Criterion for the Solutions of Linear Systems of Ordinary Differential Equations, *Dokl. Akad. Nauk SSSR*, (1952), vol. 86, no. 2. (Translated into English as *Soviet Math. Dokl.*)

[7]Wu, M. U., A Note On Stability of Linear Time-Varying Systems, *IEEE Trans. Automat. Control*, (1974), vol. AC-19, no. 2.

Theorem 6.1. *Let $\mu_j(t) \leqslant -\sigma < 0$ ($j = 1, \dots, n$; $t_0 \leqslant t < \infty$), where $\mu_j(t)$ are the eigenvalues of the matrix $A^H(t) = [A(t) + A'(t)]/2$. Then the system (6.1) is asymptotically stable.*

Proof. The proof follows from the inequality

$$|x(t_0)| \exp\Big(\int_{t_0}^t \mu_{\min}(s)\, ds\Big) \leqslant |x(t)| \leqslant |x(t_0)| \exp\Big(\int_{t_0}^t \mu_{\max}(s)\, ds\Big) \tag{6.5}$$

called *Ważevski's inequality*. Indeed,

$$\frac{d}{dt}|x(t)|^2 = \frac{d}{dt}(x'x) = \dot{x}'x + x'\dot{x} = 2x'A^H(t)x.$$

Hence

$$2\mu_{\min}(t)|x(t)|^2 \leqslant \frac{d}{dt}|x(t)|^2 \leqslant 2\mu_{\max}(t)|x(t)|^2.$$

An integration of this inequality leads to the inequality (6.5) and the assertion of Theorem 6.1. □

2. Systems with an almost constant matrix. First, we cite a lemma which is important from an applications point of view.

Lemma 6.1 (the Gronwall–Bellman lemma). *Let scalar continuous functions $x(t) \geqslant 0$ and $g(t) \geqslant 0$ satisfy the inequality*

$$x(t) \leqslant \alpha(t) + \int_{t_0}^t g(s)x(s)\, ds, \qquad t \geqslant t_0, \tag{6.6}$$

where $\alpha(t)$ is a nondecreasing function. Then

$$x(t) \leqslant \alpha(t) \exp\left(\int_{t_0}^t g(s)\, ds\right). \tag{6.7}$$

Proof. Set $u(t) = \int_{t_0}^t g(s)x(s)\, ds$. It follows from the inequality (6.6) that

$$\dot{u}(t) = g(t)x(t) \leqslant g(t)[\alpha(t) + u(t)]. \tag{6.8}$$

Multiplying both sides of the inequality (6.8) by $\exp\left(-\int_{t_0}^t g(s)\, ds\right)$ gives

$$[\dot{u}(t) - g(t)u(t)] \exp\left(-\int_{t_0}^t g(s)\, ds\right) \leqslant g(t)\alpha(t) \exp\left(-\int_{t_0}^t g(s)\, ds\right),$$

or

$$\frac{d}{dt}\left[u(t) \exp\left(-\int_{t_0}^t g(s)\, ds\right)\right] \leqslant g(t)\alpha(t) \exp\left(-\int_{t_0}^t g(s)\, ds\right).$$

Hence, we have on integration

$$u(t) \leqslant \int_{t_0}^{t} g(s)\alpha(s) \exp\left(\int_{s}^{t} g(\xi)\, d\xi\right) ds, \qquad t \geqslant t_0.$$

Substituting this estimate in (6.6) and taking into account that the function $\alpha(t)$ is nondecreasing, we finally obtain

$$x(t) \leqslant \alpha(t) + \int_{t_0}^{t} g(s)\alpha(s) \exp\left(\int_{s}^{t} g(\xi)\, d\xi\right) ds$$
$$\leqslant \alpha(t)\left[1 + \int_{t_0}^{t} g(s) \exp\left(\int_{s}^{t} g(\xi)\, d\xi\right) ds\right] = \alpha(t) \exp\left(\int_{t_0}^{t} g(\xi)\, d\xi\right).$$

This completes the proof of the lemma. □

Consider the system with an almost constant matrix

$$\dot{x} = [A + B(t)]x, \qquad x \in R^n,\ t_0 \geqslant 0. \tag{6.9}$$

Suppose that the elements of the matrix $B(t)$ are continuous and

$$\int_0^\infty \|B(t)\|\, dt = \int_0^\infty \max_{1 \leqslant j \leqslant n} \sum_{i=1}^{n} |B_{ij}(t)|\, dt < \infty. \tag{6.10}$$

In this case the following assertion holds.

Theorem 6.2 (R. Bellman). *Let the solution $y = 0$ of the system $\dot{y} = Ay$ be stable (asymptotically stable) and the condition* (6.10) *be satisfied ($B(t) \to 0, t \to \infty$). Then the solution $x = 0$ of the system* (6.9) *is also stable (asymptotically stable).*

Proof. The fundamental matrix $Y(t)$ of the system $\dot{y} = Ay$ is bounded, i.e., $\|Y(t)\| \leqslant M$. Considering $B(t)x$ in (6.9) to be the right side of the equation and applying Cauchy's formula, we obtain

$$x(t) = Y(t)x(0) + \int_0^t Y(t-s)B(s)x(s)\, ds.$$

Hence

$$|x(t)| \leqslant M|x(0)| + \int_0^t M\|B(s)\|\, |x(s)|\, ds. \tag{6.11}$$

For $|x| = \sum_{i=1}^n |x_i|$, by virtue of Lemma 6.1, it follows from (6.11) that

$$|x(t)| \leqslant M|x(0)| \exp\left[M \int_0^\infty \|B(s)\|\, ds\right] \leqslant C|x(0)|.$$

Thus, the solution $x(t) \equiv 0$ is stable. In a similar way, it can be proved that the asymptotic stability of $y = 0$ implies the asymptotic stability of $x = 0$. □

In the case of a variable matrix A the assertion of Theorem 6.2, in general, does not hold. Occasionally, the conditions for the stability of the system (6.1) can be obtained by use of an appropriate change of variables. For more details, see problems 31–59.

3. Linear systems with periodic coefficients. Consider the system

$$\dot{x}(t) = A(t)x, \qquad A(t+\omega) = A(t) \in C^0, \ x \in R^n, \ \omega > 0. \tag{6.12}$$

Denote by $X(t)$ the fundamental matrix of solutions of equation (6.12), i.e., a matrix such that

$$\dot{X}(t) = A(t)X(t), \qquad X(0) = I. \tag{6.13}$$

The following theorem is given without proof.

Theorem 6.3 (A. Floquet). *The fundamental matrix* (6.13) *can be represented in the form*

$$X(t) = \Phi(t)e^{Lt}, \qquad \Phi(t+\omega) = \Phi(t) \in C^1, \ L = \text{constant matrix}. \tag{6.14}$$

The matrix $X(\omega)$ is called the *monodromy matrix*. It determines the *translation operator* by a period along the solutions of equation (6.12) according to the formula $x(\omega) = X(\omega)x(0)$. The eigenvalues ρ_j of the monodromy matrix $X(\omega)$ are called the *Floquet multipliers of the system* (6.12), i.e.,

$$\det[X(\omega) - \rho_j I] = 0. \tag{6.15}$$

It follows from equation (6.15), Viète's formula, and the Ostrogradsky-Liouville theorem that

$$\begin{gathered}\sum_{j=1}^{n} \rho_j = \operatorname{tr} X(\omega) = \sum_{i=1}^{n} x_{ii}(\omega), \\ \prod_{j=1}^{n} \rho_j = \det X(\omega) = \exp\left(\int_0^\omega \operatorname{tr} A(t)\,dt\right).\end{gathered} \tag{6.16}$$

Further, for any Floquet multiplier ρ_j there exists a solution $\xi_j(t) \neq 0$ of equation (6.12) such that $\xi_j(t+\omega) = \rho_j\xi_j(t)$. This solution can be constructed in the following way. Let ξ_j° be the eigenvector of the matrix $X(\omega)$ associated with ρ_j, i.e., $X(\omega)\xi_j^\circ = \rho_j\xi_j^\circ$. Then the desired solution is of the form

$$\xi_j(t) = X(t)\xi_j^0, \qquad \xi_j(t) = p(t)e^{\lambda_j t}, \qquad \lambda_j = \frac{1}{\omega}\ln\rho_j, \qquad p(t+\omega) = p(t). \tag{6.17}$$

If ρ_j is a multiple eigenvalue of the matrix $X(\omega)$ with Jordan boxes of order m, then it is associated with m Floquet solutions of the form

$$\begin{aligned}&\xi_{j1}(t) = p_{j1}(t)e^{\lambda_j t}, \\ &\xi_{j2}(t) = [tp_{j1}(t) + p_{j2}(t)]e^{\lambda_j t}, \\ &\xi_{jm}(t) = \left[\frac{t^{m-1}}{(m-1)!}p_{j1}(t) + \cdots + p_{jm}(t)\right]e^{\lambda_j t}, \\ &\quad p_{jk}(t+\omega) = p_{jk}(t), \qquad \lambda_j = \frac{1}{\omega}\ln\rho_j.\end{aligned} \tag{6.18}$$

Decomposing the initial value $x(0)$ in the basis composed of the eigenvectors and adjoined vectors of the matrix $X(\omega)$, we can obtain the representation of the general solution of equation (6.12) as a linear combination of solutions (6.17) and (6.18). It is not also difficult to verify that $\operatorname{Re}\lambda_j = (1/\omega)\operatorname{Re}\ln\rho_j < 0$ if $|\rho_j| < 1$, and $\operatorname{Re}\lambda_j > 0$ if $|\rho_j| > 1$. Thus, noting relations (6.17) and (6.18), one can state the following theorem.

Theorem 6.4. *Equation* (6.12) *is asymptotically stable if all of its Floquet multipliers ρ_j lie inside the unit circle $|\rho_j| < 1$. But if $|\rho_j| \leqslant 1$ and the multiplicators ρ_l lying on the unit circle $|\rho_l| = 1$ have simple elementary divisors, then equation* (6.12) *is stable. Otherwise, equation* (6.12) *is unstable.*

4. Equation of the second order with periodic coefficients. For a second-order equation it is possible to investigate the stability of its solutions in greater detail. Consider the equation

$$\ddot{x}(t) + p(t)x(t) = 0, \qquad p(t+\omega) = p(t),\ \omega > 0. \tag{6.19}$$

By means of the substitution $z(t) = x(t)\exp\left(-\frac{1}{2}\int_0^t a(s)\,ds\right)$ the more general equation $\ddot{z} + a(t)\dot{z} + b(t)z = 0$ can be reduced to the foregoing equation.

We write equation (6.19) in the form of the system (6.12):

$$\begin{aligned} \dot{x} &= y, \\ \dot{y} &= -p(t)x. \end{aligned} \tag{6.20}$$

It is readily seen that

$$A(t) = \begin{bmatrix} 0 & 1 \\ -p(t) & 0 \end{bmatrix}, \qquad \operatorname{tr} A(t) = 0.$$

The fundamental matrix $X(t)$ is the following:

$$X(t) = \begin{vmatrix} \varphi(t) & \psi(t) \\ \dot{\varphi}(t) & \dot{\psi}(t) \end{vmatrix}, \qquad \begin{aligned} \varphi(0) &= 1, & \psi(0) &= 0, \\ \dot{\varphi}(0) &= 0, & \dot{\psi}(0) &= 1. \end{aligned}$$

Equation (6.15) in this case becomes

$$\rho^2 - a\rho + \det X(\omega) = 0. \tag{6.21}$$

By virtue of (6.16), (6.20), and (6.21) we have

$$\det X(\omega) = 1, \qquad a = \varphi(\omega) + \dot{\psi}(\omega), \qquad \rho_{1,2} = \tfrac{1}{2}(a \pm \sqrt{a^2 - 4}).$$

The quantity $a = \varphi(\omega) + \dot{\psi}(\omega)$ is called the *Lyapunov constant*; it determines the stability of equation (6.19). Three cases are possible:

(a) $|a| > 2$; in this case both multiplicators ρ_1 and ρ_2 are real, and $|\rho_1| > 1$ and $|\rho_2| < 1$; by Theorem 6.4 equation (6.19) is *unstable*;

(b) $|a| < 2$; in this case the multiplicators ρ_1 and ρ_2 are distinct complex conjugates, and $|\rho_1| = |\rho_2| = 1$; by Theorem 6.4 equation (6.19) is *stable*;

(c) $|a| = 2$ is a double root $\rho_1 = \rho_2$; this is a *critical case*, which requires a more detailed study.

Thus, if

$$|\varphi(\omega) + \dot{\psi}(\omega)| < 2, \tag{6.22}$$

then equation (6.19) is stable. For each particular equation, the relation (6.22) makes it possible to find $\varphi(\omega)$ and $\dot{\psi}(\omega)$ by numerically solving two Cauchy problems, and hence to determine whether equation (6.19) is stable.

One of the equations that are often met in applications is the *Mathieu equation*

$$\ddot{x} + (\omega^2 + \varepsilon \cos t)x = 0. \tag{6.23}$$

For $\varepsilon = 0$ equation (6.23) has the form $\ddot{x} + \omega^2 x = 0$; its solutions are expressed by the formulas $\varphi(t) = \cos \omega t$ and $\psi(t) = (1/\omega) \sin \omega t$. Hence, $a = 2 \cos 2\pi\omega$. Clearly, $|a| < 2$ for $\omega \neq 1/2, 1, 3/2, 2, \ldots$. The points $\omega = 1/2, 1, 3/2, 2, \ldots$ may be regarded as the boundaries of the regions of stability. For $\varepsilon \neq 0$ the regions of stability in the parameter space (ω, ε) can be found numerically. They are shown in Figure 6.1.

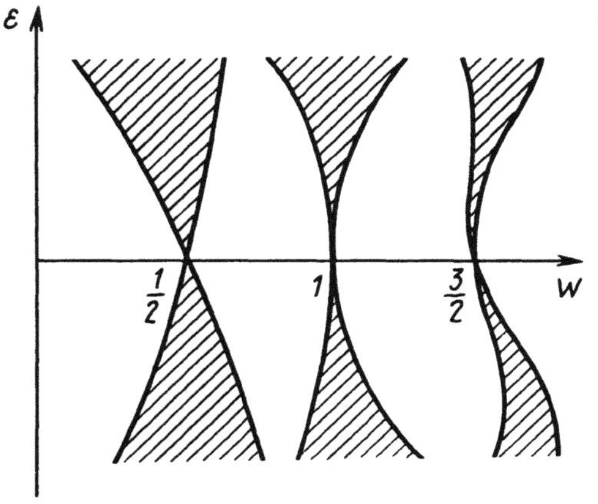

FIG. 6.1. Instability regions for Mathieu's equation

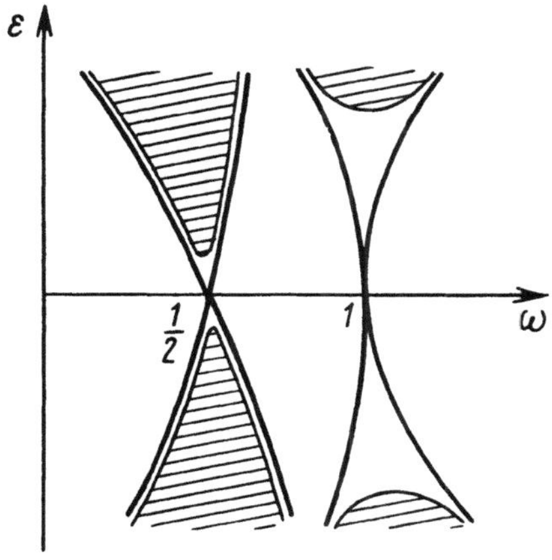

FIG. 6.2. Instability regions for Mathieu's equation with friction

In Figure 6.1 the parameter values inside the hatched regions correspond to the instability of the equation, while those outside correspond to its stability. For Mathieu's equation with friction

$$\ddot{x} + \alpha\dot{x} + (\omega^2 + \varepsilon \cos t)x = 0, \qquad \alpha > 0, \tag{6.24}$$

instability regions are shown in Figure 6.2.

Instability regions are getting wider near the point $\omega = 1/2$. This region of instability accounts for the oscillations of a swing building up. The swing is described by equation (6.24). Periodically bending and unbending the knees, i.e., periodically changing the moment of inertia, one can make the swing become unstable, and the amplitude of oscillations starts to increase. This phenomenon is called *parametric resonance.* A parametric resonance is easiest to obtain if the knee-bending frequency is equal 1/2, i.e., if two sit-ups are made over each period.

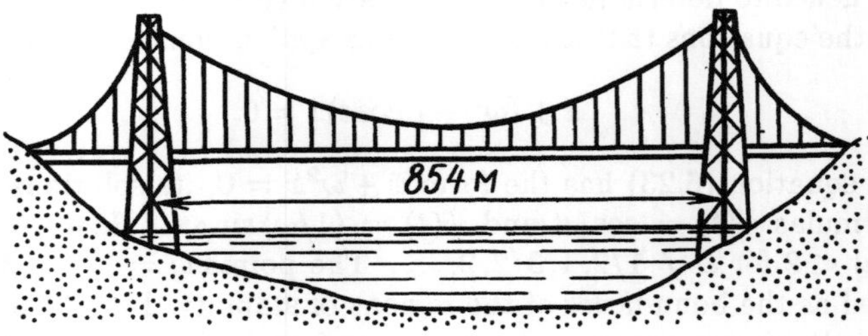

FIG. 6.3. The Takoma bridge

5. Parametric resonance in engineering applications. In practice one often encounters the phenomenon of parametric resonance. Sometimes it plays a positive role as in parametric amplifiers in radioengineering. In other situations parametric resonance may lead to disasters, among which the destruction of suspension bridges figures prominently. It has been noted for a long time that suspension bridges, despite their technical, esthetical, and economical advantages, are rather unstable with respect to disturbances. To avoid resonance, the French emperor Napoleon introduced a section in the infantry regulations that explicitly forbade to march on bridges.

Wind-induced resonance caused the collapse of bridges across the Niagara in 1864 and 1889 (in the former case the bridge was 320 meters long, and in the latter the bridge had a length of 386 m). In the summer of 1940 a suspension bridge over the Takoma River in the Far West of the USA was built (see Figure 6.3). The bridge had a single span 854 m long and 11.9 m wide and was supported by two steel cables of diameter 44 cm and of sag 70.7 m each. An appreciable wind sensitivity of the bridge was immediately discovered. Attempts to eliminate oscillations by introducing additional constraints were made. However, on November 7, 1940, about 10 a.m., when the velocity of the wind increased to 18.7 m/sec, bending-rotational oscillations of the bridge were set up with angles of inclination of about 45°. The bridge withstood such oscillations for about an hour, then collapsed and fell into the river (see Figure 6.4). It is worth noting that the bridge was designed to withstand a statical wind load of 50 m/sec, but under a dynamical oscillatory load it fell when the wind velocity was considerably smaller. For greater detail on the stability of equations with periodic coefficients and on parametric resonance the reader is referred to [19, 23].

FIG. 6.4. The Takoma bridge collapsing

§ 7. Lyapunov Functions for Linear Time-Invariant Systems and First Approximation Stability

1. Lyapunov's matrix equation. For the linear time-invariant system

$$\dot{x} = Ax, \qquad x \in R^n \tag{7.1}$$

we wish to find the Lyapunov function as a quadratic form

$$V(x) = x'Bx, \qquad B' = B. \tag{7.2}$$

We require that the derivative $\dot{V}$ be equal to a given arbitrary quadratic form $x'Cx$ associated with the system (7.1), i.e.,

$$\dot{V}(x) = x'Cx, \qquad C' = C. \tag{7.3}$$

Substituting expression (7.2) in (7.3) gives

$$\dot{V}(x) = \dot{x}'Bx + x'B\dot{x} = x'(A'B + BA)x = x'Cx. \tag{7.4}$$

The identity (7.4) for all x holds if and only if

$$A'B + BA = C. \tag{7.5}$$

The matrix equation (7.5) is called *Lyapunov's equation.* In this equation C is a given symmetric matrix, and B is the desired symmetric matrix.

Denote by $\lambda_i(A)$ $(1 \leqslant i \leqslant n)$ the eigenvalues of the matrix A.

Theorem 7.1. *Let*

$$\lambda_i(A) + \lambda_k(A) \neq 0 \qquad (1 \leqslant i \leqslant k \leqslant n). \tag{7.6}$$

Then Lyapunov's equation (7.5) *can be solved uniquely for any symmetric matrix* C.

Proof. In the space R^N, $N = n(n+1)/2$, of symmetric $n \times n$ matrices we introduce a linear operator F according to the formula $F(B) = A'B + BA$, $B \in R^N$. It is necessary and sufficient for the existence of a bounded inverse operator F^{-1} (or, equivalently, for the existence of a unique solution of equation (7.5)) that the operator F have no zero eigenvalues. In other words, if $F(B) = \mu B$ $(B \neq 0)$, then necessarily $\mu \neq 0$. Suppose that all $\lambda_i(A)$ are distinct, and all sums $\lambda_i + \lambda_k$ are also distinct. Denote by x' the (column) eigenvectors of the matrix A', i.e., $A'x^i = \lambda^i x^i$. Note that the eigenvalues of the matrices A and A' coincide, but they have, in general, distinct eigenvalues. Define the matrix $B^{ik} \in R^N$ as

$$B^{ik} = x^i(x^k)' + x^k(x^i)'.$$

Then

$$\begin{aligned} F(B^{ik}) &= A'[x^i(x^k)' + x^k(x^i)'] + [x^i(x^k)' + x^k(x^i)']A \\ &= [\lambda_i x^i(x^k)' + \lambda_k x^k(x^i)' + \lambda_k x^i(x^k)' + \lambda_i x^k(x^i)'] \\ &= (\lambda_i + \lambda_k)[x^i(x^k)' + x^k(x^i)'] = (\lambda_i + \lambda_k)B^{ik}. \end{aligned}$$

Consequently, B^{ik} is the characteristic matrix of the operator F associated with the eigenvalue $\mu_{ik} = \lambda_i + \lambda_k$. Therefore, when the foregoing hypotheses and the condition (7.6) are satisfied, the operator F has exactly $n(n+1)/2$ eigenvalues μ_{ik}, none of which is equal to zero. There can be no other eigenvalues of F. Thus, in the case being considered the proof is complete. For the proof in the case of multiple roots, see [1]. □

Note that henceforth the notation $C > 0$ $(C < 0)$ indicates that the symmetric matrix C is positive definite (negative definite).

Theorem 7.2 (A. M. Lyapunov). *Suppose that*

$$\operatorname{Re} \lambda_j(A) < 0, \qquad 1 \leqslant j \leqslant n. \tag{7.7}$$

Then for any matrix $C < 0$ there exists a unique solution $B > 0$ to Lyapunov's equation (7.5). Moreover, the following formula holds:

$$B = -\int_0^\infty e^{A't} C e^{At}\, dt. \tag{7.8}$$

Conversely, if for at least one matrix $C < 0$ there exists a solution $B > 0$ of equation (7.5), then the matrix A is stable (i.e., the condition (7.7) holds).

Proof. Condition (7.7) implies the condition (7.6). Hence Lyapunov's equation has a solution for any matrix $C < 0$. Show that simultaneously the matrix $B > 0$. We introduce the form $V(x) = x'Bx$ and assume that it is negative semidefinite. Then there exists a point $x_0 \neq 0$ such that either $V(x_0) < 0$ or $V(x_0) = 0$. In the first case $V(kx_0) = k^2 V(x_0) < 0$ for all $k > 0$. Therefore, in any neighborhood of the origin there exist points of the form kx_0 at which $V(kx_0) < 0$ and $\dot V = k^2 x_0' C x_0 < 0$. Hence, in this case Theorem 2.5 (in which V is replaced by $-V$) would imply that the trivial solution of the system (7.1) is unstable, which contradicts the condition (7.7).

In the case when $V(x_0) = 0$ ($x_0 \neq 0$) we have $\dot V(x_0) < 0$, and there always exists a point $x_1 \neq 0$ such that $V(x_1) < 0$, i.e., we obtain the previous case and thus a contradiction.

To prove formula (7.8), we note that

$$\begin{aligned} V(x(t)) - V(x_0) &= x'(t)Bx(t) - x_0'Bx_0 = \int_0^t \dot V(x(s))\, ds \\ &= \int_0^t x'(s)Cx(s)\, ds = x_0'\left[\int_0^t e^{A's} C e^{As}\, ds\right] x_0. \end{aligned} \tag{7.9}$$

Passing to the limit in the equality (7.9) as $t \to \infty$ and noting that $x(t) \to 0$ as $t \to \infty$, we immediately obtain the following identity:

$$x_0'Bx_0 = -x_0'\left[\int_0^\infty e^{A's} C e^{As}\, ds\right] x_0,$$

which gives formula (7.8).

But if for some matrix $C < 0$ there exists a solution $B > 0$ of equation (7.5), then the function $V(x) = x'Bx$ satisfies Theorem 2.3 on asymptotic stability. Hence, the matrix A satisfies the condition (7.7). □

Remarks. **1.** Theorem 7.2 furnishes a method of calculating integrals of the form

$$\int_0^\infty x'(t)Dx(t)\, dt, \qquad D > 0. \tag{7.10}$$

The integral (7.10) describes the mean-square deviation of the solution $x(t)$ of the system (7.1) from its equilibrium state and serves as an integral estimate of transient processes in the system (7.1). Suppose that the matrix A is stable, and denote by $P > 0$ the unique solution of the equation $A'P + PA = -D$. Writing the relation (7.9) for the matrices D and P gives

$$\int_0^\infty x'(t)Dx(t)\,dt = V(x_0) = x_0'Px_0. \tag{7.11}$$

2. Along with the representation (7.8), the solution of equation (7.5) can be written in the form[8]

$$\mathrm{vec}[B] = -(A' \oplus A)^{-1}\,\mathrm{vec}[C],$$

where $\oplus$ is a Kronecker sum of matrices, and vec$[B]$ is a vector whose elements are columns of B written one below another. Recall that for two matrices $A = (a_{ij})$ and $B = (b_{ij})$, $i, j = 1, \ldots, n$, the block-partitioned matrix $A \otimes B = (a_{ij}B)$ is the Kronecker product and $A \oplus B = A \otimes I + I \otimes B$, where I is an $n \times n$ identity matrix.[9]

The next assertion is given without proof.

Theorem 7.3 (A. M. Lyapunov). *If for at least one j ($1 \leqslant j \leqslant n$) the inequality* $\mathrm{Re}\,\lambda_j(A) > 0$ *is satisfied, then for any matrix $C > 0$ there exist a matrix B and a number $\alpha > 0$ such that $A'B + BA = C + \alpha B$, with matrix B positive semidefinite.*

Corollaries to Theorems 7.2 and 7.3.

1°) *If the trivial solution of equation* (7.1) *is asymptotically stable, then there exists a Lyapunov function $V(x) = x'Bx$ that satisfies Lyapunov's theorem on asymptotic stability.*

2°) *If the trivial solution of equation* (7.1) *is unstable, then there exists a Lyapunov function $V(x) = x'Bx$ that satisfies Chetaev's theorem on instability.*

Specific numerical algorithms for solving Lyapunov's matrix equation will be dealt with in Chapter XIII.

2. Stability in the first approximation. In applications stability of solutions is often investigated using the so-called *equations of the first approximation.* Suppose it is required to test the stability of the solution $x(t) \equiv 0$ of the system

$$\dot{x}(t) = f(t,x), \qquad f(t,0) = 0, \tag{7.12}$$

where $f(t,x)$ is a differentiable function. We linearize equation (7.12) in the neighborhood of solution $x(t) \equiv 0$. To this end, we write (7.12) in the form

$$\dot{x}(t) = A(t)x + R(t,x), \qquad |R(t,x)| \leqslant C_1|x|^{1+\alpha},\ \alpha > 0,\ C_1 > 0. \tag{7.13}$$

Along with (7.13), we consider the linear system

$$\dot{x}(t) = A(t)x, \qquad A(t) = \left.\frac{\partial f}{\partial x}\right|_{x=0}. \tag{7.14}$$

---

[8]Bellman, R., Kronecker Products and the Second Method of Lyapunov, *Math. Nachrichten*, (1959), vol. 20, no. 1–2.

[9]Graham, A., *Kronecker Products and Matrix Calculus with Applications*, Ellis Horwood, Chichester, (1981).

The system (7.14) is called the *equation of the first approximation* for (7.13). Further, the stability of equation (7.14) is tested and the result is then extended to cover the nonlinear system (7.13). This procedure was used long before Lyapunov's work. Lyapunov was the first to investigate the validity of such procedure.

Suppose now that the equation of the first approximation (7.14) is *time-invariant*, i.e., it coincides with (7.1): $A(t) \equiv A$, $\dot{x}(t) = Ax$.

Theorem 7.4 (A. M. Lyapunov). *Suppose the trivial solution of equation (7.1) is asymptotically stable. Then the trivial solution of the nonlinear equation (7.13) is also asymptotically stable.*

Proof. Denote by B the solution of the equation $A'B + BA = -I$, where I is a unit matrix. Consider the Lyapunov function $V(x) = x'Bx$, and find its derivative $\dot{V}$ according to the system (7.13). We have

$$\begin{aligned}\dot{V} &= \dot{x}'Bx + x'B\dot{x} = (x'A' + R'(t,x))Bx + x'B(Ax + R(t,x)) \\ &= x'(A'B + BA)x + R'Bx + x'BR \leqslant -|x|^2 + 2C_1\|B\||x|^{2+\alpha} \\ &\leqslant |x|^2(-1 + 2C_1\|B\||x|^\alpha \leqslant -\frac{1}{2}|x|^2\end{aligned}$$

for x small such that

$$2C_1\|B\||x|^\alpha < \frac{1}{2}. \tag{7.15}$$

Thus, the function $V(x) = x'Bx$ satisfies Theorem 2.3 on asymptotic stability, and hence the trivial solution of equation (7.13) is asymptotically stable. □

Remark. Note that the attraction domains of equations (7.13) and (7.14) do not coincide. For equation (7.14) the attraction domain is the total space R^n, whereas for (7.13) one can guarantee that $x(t, t_0, x_0) \to 0$ only for x_0 complying with the estimate (7.15).

Theorem 7.5 (A. M. Lyapunov). *If the trivial solution of the equation of the first approximation (7.1) is unstable, then the trivial solution of the nonlinear equation (7.13) is also unstable.*

The proof of this theorem is similar to that of the foregoing one except that the Lyapunov function is constructed using Theorem 7.3, and then Chetaev's theorem on instability is used.

The cases treated in Theorems 7.4 and 7.5 are said to be *noncritical.* In noncritical cases the matrix A has no eigenvalues on the imaginary axis. In such cases the type of stability of the original nonlinear system coincides with that of the system of the first approximation regardless of the form of nonlinear terms.

The cases in which the eigenvalues $\lambda_j(A)$ of the matrix A are located on the imaginary axis, i.e., $\operatorname{Re}\lambda_j(A) = 0$, are called *critical.* In critical cases stability (or lack of it) depends on the form of nonlinear terms. The investigation of critical cases is complicated, but it is important for a lot of applications. For example, the stability of the Solar System reduces to the investigation of a complicated critical case. At the moment the cases of one zero root and of a pair of pure imaginary roots are well studied. A few results have also been obtained for the cases of one zero root and two pure imaginary roots and of two pairs of pure imaginary roots. For more details on critical cases, see [16].

3. Rotation of a shell. This type of motion is described by equations (3.1). Substituting in these equations $\sin\alpha \approx \alpha$, $\cos\alpha \approx 1$ and $\sin\beta \approx \beta$, $\cos\beta \approx 1$ and discarding terms of order higher than one yields equations of the first approximation for (3.1) of the form

$$\begin{aligned} A\ddot{\beta} - Cn\dot{\alpha} &= eR\beta, \\ A\ddot{\alpha} + Cn\dot{\beta} &= eR\alpha. \end{aligned} \tag{7.16}$$

The characteristic equation of the linear system (7.16) is of the form

$$\begin{vmatrix} A\lambda^2 - eR & -Cn\lambda \\ Cn\lambda & A\lambda^2 - eR \end{vmatrix} = 0,$$

which yields

$$(A\lambda^2 - eR)^2 + C^2n^2\lambda^2 = 0. \tag{7.17}$$

The roots of equation (7.17) are the following:

$$\lambda_{1,2,3,4} = (2A)^{-1}[\pm Cni \pm \sqrt{4AeR - C^2n^2}]. \tag{7.18}$$

If $4AeR - C^2n^2 > 0$ (i.e., the Mayevskii–Krylov conditions (3.4) are not satisfied), then two of the roots (7.18) have a positive real part, and the solution of the linear system (7.16) is unstable. Moreover, on the strength of Theorem 7.5 the solution of the nonlinear system (3.1) is also unstable.

But if $4AeR - C^2n^2 < 0$, then equation (7.17) has two pairs of pure imaginary roots. In this critical case the equations of the first approximation (7.16) are insufficient to test the stability of the original system (3.1). This demonstrates yet again the importance of Chetaev's result discussed in detail in § 3.

Although the linearization method is the main method for solving various engineering problems, one must bear in mind that it leads to correct results only in noncritical cases. The unjustified use of the linearization method may lead to results that are not even superficially correct.[10]

4. Time-varying equations of the first approximation. In the case of time-varying equations of the first approximation (7.14) Theorems 7.4 and 7.5 do not hold in general.

---

[10] In this connection it is worth quoting what prof. P. Likins had to say on the subject in his article (referred to in the footnote at the end of § 3): "I think it's fair to say that in the early 1960s very few people with doctorates in the mechanically oriented disciplines of engineering understood the limitations of linearization as a technique in motion stability analysis; the fundamental work on stability theory by Lyapunov and others was familiar in this country only to those working in certain areas of mathematics and control theory. (The Russians seemed better prepared in this respect.) The flood of conceptually new problems introduced by space exploration caught many of us with educational deficiencies that we scrambled to repair."

Example 7.1. The solution of equation $\dot{x}(t) = -xt^{-1}$, $t \geqslant 2$, has the form

$$x(t, t_0, x_0) = x_0 t_0 t^{-1}, t_0 \geqslant 2. \tag{7.19}$$

It is readily seen that the trivial solution (7.19) is asymptotically stable. On the other hand, the nonlinear equation

$$\dot{x}(t) = -xt^{-1} + b(t)x^2, \tag{7.20}$$
$$b(t) = \frac{1 + \ln t}{t(\ln t)^2},$$

whose coefficient $b(t)$ is bounded for $t \geqslant 2$, has the solution $x(t) = \ln t$. The trivial solution of equation (7.20) is unstable.

However, the following theorem holds.

Theorem 7.6. *Suppose that the trivial solution of the time-varying equation of the first approximation* (7.14) *is exponentially stable. Then the trivial solution of equation* (7.13) *is asymptotically stable.*

§ 8. Synthesis of Control Systems for the Manipulator Robot

1. Single-link manipulator robot. Consider a single-link manipulator made of an ideally rigid homogeneous rectilinear rod of length L and mass M (see Figure 8.1). At one end the rod is connected with a fixed base via an ideal cylindrical hinge O, while at the other end a weight (to be moved) of mass m is made fast. The control torque u is applied to the axis of the hinge O. The motion of the manipulator occurs in the vertical plane in the field of gravity.

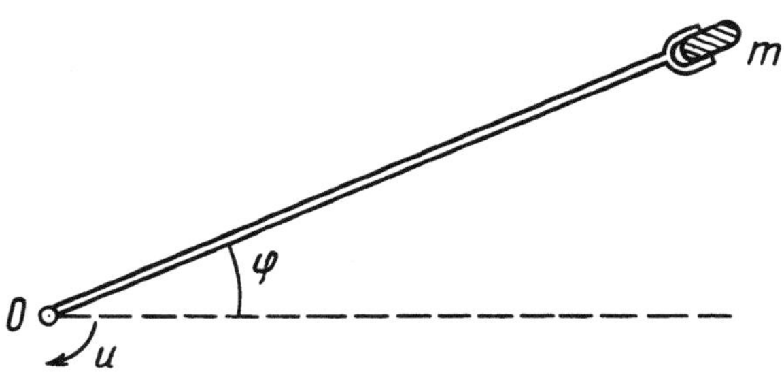

FIG. 8.1. A single-link manipulator

The equations of motion are of the form

$$L^2 m_1 \ddot{\varphi} + \alpha\dot{\varphi} + gL\left(m + \frac{M}{2}\right)\sin\varphi = u, \tag{8.1}$$
$$\varphi(0) = \varphi_0,\ \dot{\varphi}(0) = \dot{\varphi}_0,\ m_1 = m + \frac{M}{3},$$

where φ is the angle between the axis of the rod and the vertical line passing through the point O, g denotes the gravity acceleration, and α, $\alpha \geqslant \alpha_0 > 0$, is the coefficient of viscous friction. The value of α is usually not known precisely: it is assumed that only α_0 is known, and the specific value of α is unknown. The problem is to choose a control torque u so that the weight of known mass m is moved from an arbitrary initial position $(\varphi_0, \dot\varphi_0)$ to the equilibrium position $(\varphi = 0, \dot\varphi = 0)$. Since the coefficient of friction α is not known, the time of attaining the equilibrium position is, naturally, not prescribed *a priori*. Therefore, one of the ways to formalize the problem is to choose a control u so that

$$\lim \dot\varphi(t) = 0, \qquad \lim \varphi(t) = 0, \qquad t \to \infty. \tag{8.2}$$

We assume the control torque to be of the form $u = -b_1\varphi(t)$, $b_1 > 0$; this type of control can easily be implemented with the help of standard proportional regulators. For this control function, the equations of motion (8.1) can be written thus:

$$\dot\varphi_1 = \varphi_2, \qquad \dot\varphi_2 = -a_1\varphi_2 - a_2 \sin\varphi_1 - b\varphi_1, \tag{8.3}$$
$$a_1 = \alpha L^{-2} m_1^{-1},\ a_2 = gL^{-1}(m + \tfrac{1}{2}M)m_1^{-1},\ b = b_1 L^{-2} m_1^{-1}.$$

The Lyapunov function V for the system (8.3) will be taken as

$$V = 2b\varphi_1^2 + \varphi_2^2 + (\varphi_2 + a_1\varphi_1)^2.$$

The derivative $\dot V$ of the function V along the trajectories of the system (8.3) for any $\delta > 0$ satisfies the relation

$$\begin{aligned} \dot V &= 4b\varphi_1\varphi_2 + 2\varphi_2(-a_1\varphi_2 - a_2\sin\varphi_1 - b\varphi_1) + 2(\varphi_2 + a_1\varphi_1)(-b\varphi_1 - a_2\sin\varphi_1) \\ &= 2a_1\varphi_2^2 - 4a_2\varphi_2\sin\varphi_1 - 2a_1 b\varphi_1^2 - 2a_1a_2\varphi_1\sin\varphi_1 \\ &\leqslant -2\varphi_2^2(a_1 - \delta a_2) - 2a_1\varphi_1^2(b - a_2 - a_2\delta^{-1}). \end{aligned}$$

Choose and fix a δ so that $a_1 - \delta a_2 = \alpha L^{-2} m_1^{-1} - \delta a_2 > 0$, and after that choose a b_1 so that $b - a_2 - a_2\delta^{-1} > 0$. Then $\dot V$ will be negative definite. This fact together with Theorem 2.3 yields equalities (8.2). Thus, in our case a sufficiently strong coordinate feedback stabilizes the motion of the manipulator.

2. The Cyclone-type robot. We shall investigate the stabilization problem for a Cyclone-type robot (see Figure 8.2). The robot arm of length L is activated by the double-action pneumatic cylinders via the transfer mechanism with arm l. The weight of unknown mass m is placed in the gripper of the robot arm.

The equations of plane motion are of the form

$$\ddot\varphi = 2PFl(m_1L^2)^{-1} + f(\dot\varphi),$$
$$\dot P = -4P_{\text{ave}}Fl(V)_{\dot\varphi}^{-1} + 2RTgV^{-1}.$$

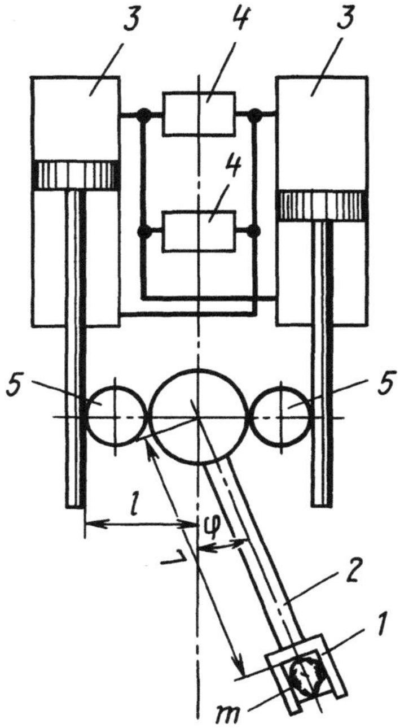

FIG. 8.2. The Cyclone-type manipulator robot:
1, gripper of the robot with the weight of mass m; 2, manipulator arm of length L and mass M; 3, double-action pneumatic cylinders; 4, distribution mechanism of the pneumatic cylinders; 5, transfer mechanism with arm l

Here φ is the angular displacement of the manipulator arm; $m_1 = m + M/3$, where M is the mass of the manipulator arm; P is the current pressure in pneumatic cylinders; F is the piston cross-section area; R is the gas constant; T is the absolute temperature of the gas; V is the average volume of a pneumatic cylinder; P_{ave} is the average pressure inside the pneumatic cylinder; g is the mass flow rate in the space of the cylinders; $f(\dot\varphi) = -\alpha\dot\varphi$ is the viscous friction force in the final-control element with unknown coefficient of friction $\alpha > 0$. Assume

$$x_1 = \dot\varphi, \qquad x_2 = \varphi, \qquad x_3 = P$$
$$a_1 = 2Fl(m_1L^2)^{-1}, \qquad a_3 = 4P_{\text{ave}}FlV^{-1},$$
$$b = 2RTV^{-1}, \qquad u = bg.$$

Then the equations of motion take the form

$$\dot x_1 = a_1x_3 - \alpha x_1, \qquad \dot x_2 = x_1, \qquad \dot x_3 = -a_3x_1 + u. \tag{8.4}$$

The problem is in choosing a control u such that

$$\lim_{t\to\infty}\big(|x_1(t)| + |x_2(t)| + |x_3(t)|\big) = 0 \tag{8.5}$$

for any initial values of the variables x_i. Note that this problem is equivalent to the problem of moving an unknown weight of mass m from an arbitrary position in the phase space (x_1, x_2, x_3) to the origin, i.e., to a position such that its angular displacement, its velocity, and its pressure are equal to zero. Put $u = -b_1x_2 - b_2x_3$. Choose the constants $b_i > 0$ so that the relation (8.5) is satisfied. Consider a Lyapunov function

$$\begin{aligned} V = {} & (x_3 + b_2a_1^{-1}x_1)^2 + (x_3 + b_2a_1^{-1}x_1 + a_3x_2)^2 + x_3^2 \\ & + 2a_1^{-1}x_1^2(a_3 + \alpha b_2a_1^{-1}) + b_2a_1^{-1}x_2^2(2b_1 + \alpha a_3). \end{aligned}$$

The total derivative $\dot{V}$ of this function along the trajectories of the system (8.4) for a chosen control is of the form

$$\begin{aligned} \dot{V} = {} & 2(x_3 + b_2a_1^{-1}x_1)(\dot{x}_3 - b_2a_1^{-1}\alpha x_1 + b_2x_3) \\ & + 2(x_3 + x_1b_2a_1^{-1} + a_3x_2)(\dot{x}_3 + b_2x_3 + a_3x_1 - b_1x_2 - b_2x_3) \\ & + 4a_3x_1x_3 - 4a_3a_1^{-1}\alpha x_1^2 + 2x_1x_2a_1^{-1}b_2(2b_1 + \alpha a_3) \\ & + 4x_1x_3(a_3 + a_1^{-1}\alpha b_2) - 4x_1^2a_1^{-1}\alpha(a_3 + a_1^{-1}\alpha b_2). \end{aligned}$$

Thus, noting the equations of motion (8.4), we find

$$\begin{aligned} \dot{V} = {} & -x_1^2(2a_3b_2a_1^{-1} + 4b_2^2a_1^2\alpha + 4a_3a_1^{-1}\alpha) \\ & - 2a_3b_1x_2^2 - 2b_2x_3^2 - 6b_1x_2x_3. \end{aligned} \tag{8.6}$$

But for any $\delta > 0$ the following inequality is satisfied:

$$6b_1x_2x_3 \leqslant 3b_1(x_2^2\delta + x_3^2\delta^{-1}).$$

Choose and fix a $\delta > 0$ so that

$$2a_3 - 3\delta > 0. \tag{8.7}$$

Then choose any $b_1 > 0$ and b_2 satisfying the condition

$$2b_2 > 3b_1\delta^{-1}. \tag{8.8}$$

It follows from the conditions (8.6)–(8.8) that $\dot{V}$ is a negative definite function. Hence, and from the fact that V is positive definite we obtain the relation (8.5). So, for chosen b_1 and b_2 the proposed control furnishes the solution to the problem under consideration. Note that if it is required to transfer the system (8.4) from an arbitrary initial position to a final one $(0, \bar{x}_2, 0)$, with $\bar{x}_2$ arbitrary and final velocity x_1 zero, then we can use the control

$$u = -b_1(x_2 - \bar{x}_2) - b_2x_3$$

with the previous values of the feedback coefficients b_1 and b_2.

§ 9 Use of the Logarithmic Norm in Stability Theory

1. The definition and properties of the logarithmic norm. To characterize the degree of stability of nonlinear systems, it is often convenient to use the logarithmic norm of the Jacobi matrices of the right sides of the equations under consideration.

First, we recall several properties of the logarithmic norm.[11] Let R^n be a real linear space of column vectors $x = (x_i)$, $i = 1, \ldots, n$, endowed with a norm $\|\cdot\|$. In the linear space of square matrices $A = (a_{ij})$, $i, j = 1, \ldots, n$, we define a norm $\|A\|_1$ by

$$\|A\|_1 = \max_{x, \|x\|=1} \|Ax\| = \max_{x \neq 0} \frac{\|Ax\|}{\|x\|}.$$

It is readily seen that $\|Ax\| \leqslant \|A\|_1 \|x\|$ for any vector $x \in R^n$. The norm of the matrix $\|A\|_1$ introduced above is called the *operator norm* of the matrix A subordinate to the norm $\|x\|$ of the vector x. It has the property that for any matrix A there exists a vector x not equal to zero such that $\|Ax\| = \|A\|_1 \|x\|$. In addition, $\|I\|_1 = 1$, where I is a unit matrix of order $n \times n$.

The *logarithmic norm* $\gamma(A)$ of the matrix A is defined by

$$\gamma(A) = \lim_{h \to +0} \tfrac{1}{h}[\|I + hA\|_1 - 1].$$

The logarithmic norm may assume negative values, i.e., it is not a norm in the ordinary sense. Denote by $\lambda(A)$ any eigenvalue of the matrix A. Then the following relations hold:[12]

$$\begin{gathered} -\|A\|_1 \leqslant -\gamma(-A) \leqslant \operatorname{Re}\big(\lambda(A)\big) \leqslant \gamma(A) \leqslant \|A\|_1; \\ \gamma(cA) = c\gamma(A), \qquad c \geqslant 0; \\ \gamma(A + cI) = \gamma(A) + c, \qquad -\infty < c < \infty; \\ \gamma\big(cA + (1-c)B\big) \leqslant c\gamma(A) + (1-c)\gamma(B), \qquad 0 \leqslant c \leqslant 1; \end{gathered} \tag{9.1}$$

where A and B are arbitrary square $n \times n$ matrices, $\operatorname{Re}\lambda$ denotes the real part of the number λ, and c is a constant. In Table 9.1 expressions for some norms of the vector $x \in R^n$ together with subordinate operators and logarithmic norms of the matrix A are given.

In Table 9.1 the prime denotes transposition, $\rho(A)$ is the spectral norm of the matrix A equal to $\rho(A) = \big(\lambda_{\max}(A'A)\big)^{1/2}$, the number $\lambda_{\max}(A'A)$ is the maximum eigenvalue of the matrix $A'A$, the prescribed constants $\alpha_i > 0$. The spectral norm $\rho(A)$ is equal to the maximum singular value of the matrix A. The

---

[11] Lozinskii, S. M., Error Evaluation for Numerical Integration of Ordinary Differential Equations, *Izv. Vyssh. Uchebn. Zaved. Mat.*, (1958), no. 5, 52-90, (Translated into English as *Soviet Math. (Iz. VUZ)*); Dahlquist, G., Stability and Error Bounds in the Numerical Integration of Ordinary Differential Equations, *Trans. Royal Inst. of Techn.*, Stockholm, (1959), no. 130.

[12] The reader is also referred to Brunner, H., and van der Houwen, P. J., *The Numerical Solution of Volterra Equations*, North Holland, New York, (1986).

TABLE 9.1

	Norm of the vector $\|x\|$	Operator norm of the matrix $\|A\|_1$	Logarithmic norm $\gamma(A)$
1.	$\sum_{i=1}^{n} \|x_i\|$	$\max_{1\leqslant j\leqslant n} \sum_{i=1}^{n} \|a_{ij}\|$	$\max_{1\leqslant j\leqslant n} (a_{jj} + \sum_{i\neq j} \|a_{ij}\|)$
2.	$\max_{1\leqslant i\leqslant n} \|x_i\|$	$\max_{1\leqslant i\leqslant n} \sum_{j=1}^{n} \|a_{ij}\|$	$\max_{1\leqslant i\leqslant n} (a_{ii} + \sum_{j\neq i} \|a_{ij}\|)$
3.	$(x'x)^{1/2}$ $= \left(\sum_{i=1}^{n} x_i^2\right)^{1/2}$	$\rho(A)$ $= \left(\lambda_{\max}(A'A)\right)^{1/2}$	$\frac{1}{2}\lambda_{\max}(A + A')$ $= \lambda_{\max}\left(\frac{A+A'}{2}\right)$
4.	$\sum_{i=1}^{n} \frac{\|x_i\|}{\alpha_i}$	$\max_{1\leqslant j\leqslant n} \sum_{i=1}^{n} \frac{\alpha_i}{\alpha_j}\|a_{ij}\|$	$\max_{1\leqslant j\leqslant n} \left(\alpha_j a_{jj} + \sum_{i\neq j} \alpha_i\|a_{ij}\|\right)$
5.	$\max_{1\leqslant i\leqslant n} \frac{\|x_i\|}{\alpha_i}$	$\max_{1\leqslant i\leqslant n} \sum_{j=1}^{n} \frac{\alpha_j}{\alpha_i}\|a_{ij}\|$	$\max_{1\leqslant i\leqslant n} \left(\alpha_i a_{ii} + \sum_{j\neq i} \alpha_j\|a_{ij}\|\right)$

first of the norms in Table 9.1 is called *octahedral*, the second *cubic*, and the third *Euclidean* or *spheric*, since level surfaces of the form $\|x\| = 1$ for these norms are an octahedron, a cube, and a sphere, respectively.

Further, consider a scalar function $Q(x, y)$ of two independent variables $x \in R^n$ and $y \in R^n$ given by

$$Q(x,y) = \lim_{h\to+0} \tfrac{1}{h}\left[\|x + hy\| - \|x\|\right]. \tag{9.2}$$

The function Q can also be used to evaluate the logarithmic norm of the matrix A. Indeed,[13]

$$\gamma(A) = \sup_{x,\|x\|\neq 0} \frac{1}{\|x\|} Q(x, Ax), \qquad x \in R^n. \tag{9.3}$$

Recall that for any continuously differentiable function $x(t) \in R^n$ and for any norm $\|\cdot\|$ in R^n we have, noting formula (9.2),

$$\frac{d^+\|x(t)\|}{dt} = \lim_{h\to+0} \tfrac{1}{h}\left[\|x(t+h)\| - \|x(t)\|\right] = Q\big(x(t), \dot{x}(t)\big). \tag{9.4}$$

2. Stability of nonlinear systems. Consider now the problem of the stability of the zero solution of the system

$$\dot{x}(t) = F\big(t, x(t)\big), \qquad F(t,0) \equiv 0,\ t \geqslant t_0,\ x(t_0) = x_0 \tag{9.5}$$

with respect to the perturbations of the initial position x_0. Here $F(t, x)$ is a continuous function of all its arguments and continuously differentiable with respect

[13]Bylov, B. F., Vinograd, R. E., Grobman, D. N., and Nemytskii, V. V., *Theory of Lyapunov Exponents*, Nauka, Moscow, (1966).

to x. Denote by $f(t,x)$ the Jacobi matrix: $f(t,x) = \partial F(t,x)/\partial x$ and by $\mathcal{D}$ the domain $(t \geqslant t_0) \times (\|x\| \leqslant r)$.

Theorem 9.1. *Suppose there exist numbers $r > 0$, $a > 0$, and a norm $\|\cdot\|$ in R^n such that the associated logarithmic norm $\gamma(f(t,x))$ satisfies the relation*

$$\sup_{(t,x)\in\mathcal{D}} \gamma(f(t,x)) = -a < 0. \tag{9.6}$$

Then the zero solution of the system (9.5) *is asymptotically stable.*

Proof. Let us show that under the hypotheses of Theorem 9.1 there exists a Lyapunov function V satisfying the requirements of Remark 1 in § 2 in the domain $\mathcal{D}$. Take an arbitrary point $(t,x) \in \mathcal{D}$ and set $V(t,x) = \|x\|$. Clearly, this function is positive definite and has an infinitesimal upper limit in the sense of definitions given in § 2, since all norms in R^n are equivalent. It should be recalled that two norms $\|\cdot\|_a$ and $\|\cdot\|_b$ are said to be *equivalent* if there exist constants $C_1 > 0$ and $C_2 > 0$ such that for any vector $x \in R^n$ the following inequalities hold:

$$C_1\|x\|_a \leqslant \|x\|_b \leqslant C_2\|x\|_a.$$

Further, note that for Theorem 2.3 to be valid it is sufficient by virtue of Remark 1 on this theorem, that the right total derivative of the function V satisfy the inequality (2.6). We check that the right total derivative $\dot V$ is negative definite in $\mathcal{D}$. On the strength of (9.4) and (9.5) we have

$$\dot V(t,x) = Q(x, \dot x(t)) = Q(x, F(t,x)). \tag{9.7}$$

Let us transform the right side of (9.7). First, note that

$$F(t,x) = \left[\int_0^1 f(t,sx)\,ds\right]x. \tag{9.8}$$

It follows that

$$Q(x, F(t,x)) = Q\left(x, \left[\int_0^1 f(t,sx)\,ds\right]x\right). \tag{9.9}$$

However, because of (9.3) the following estimate is valid:

$$Q\left(x, \left[\int_0^1 f(t,sx)\,ds\right]x\right) \leqslant \gamma\left(\int_0^1 f(t,sx)\,ds\right)\|x\|. \tag{9.10}$$

Moreover, due to the convexity of the logarithmic norm (see relations (9.1)) we have

$$\gamma\left(\int_0^1 f(t,sx)\,ds\right) \leqslant \int_0^1 \gamma(f(t,sx))\,ds. \tag{9.11}$$

Relations (9.9)–(9.11) imply that

$$Q(x, F(t,x)) \leqslant \left(\int_0^1 \gamma(f(t,sx))\, ds \right) \|x\|.$$

Therefore, for any point $(t,x) \in \mathcal{D}$ we have

$$Q(x, F(t,x)) \leqslant -a\|x\|.$$

Thus, noting relation (9.7), one can write

$$\dot{V} \leqslant -a\|x\|, \qquad (t,x) \in \mathcal{D}.$$

Theorem 9.1 is proved. □

Remark. The choice of a particular norm depends merely on the convenience of calculating the logarithmic norm and the degree of consistency between the stability conditions obtained and desired.

As an example, consider the two-dimensional linear system

$$\dot{x}_1(t) = -11x_1(t) + 10x_2(t),$$
$$\dot{x}_2(t) = 2x_1(t) - 3x_2(t).$$

The Jacobi matrix f for this system is constant and equal to

$$f = \begin{pmatrix} -11 & 10 \\ 2 & -3 \end{pmatrix}.$$

Therefore, $\gamma(f) = -1$ for the cubic norm, $\gamma(f) = 7$ for the octahedral norm, and $\gamma(f) = 0.211$ for the Euclidean norm. Hence, we can verify whether the system in question is stable or not by means of the cubic norm.

Consider the system

$$\dot{x}_1(t) = -11x_1(t) + 2x_2(t),$$
$$\dot{x}_2(t) = 10x_1(t) - 3x_2(t).$$

This time the Jacobi matrix f is of the form

$$f = \begin{pmatrix} -11 & 2 \\ 10 & -3 \end{pmatrix}.$$

Therefore, $\gamma(f) = 7$ for the cubic norm, $\gamma(f) = -1$ for the octahedral norm, and $\gamma(f) = 0.211$ for the Euclidean norm. Hence, we can verify whether the system in question is stable or not by means of the octahedral norm.

Finally, consider the system

$$\dot{x}_1(t) = -x_1(t) + 3x_2(t),$$
$$\dot{x}_2(t) = -3x_1(t) - 2x_2(t).$$

This time the Jacobi matrix f is of the form

$$f = \begin{pmatrix} -1 & 3 \\ -3 & -2 \end{pmatrix}.$$

Therefore, $\gamma(f) = 2$ for the cubic norm, $\gamma(f) = 2$ for the octahedral norm, and $\gamma(f) = -1$ for the Euclidean norm. Hence, we can verify whether the system in question is stable or not by means of the Euclidean norm.

Corollary 1. *Suppose that the system (9.5) is a linear one and has the form*

$$\dot{x}(t) = A(t)x(t), \qquad t \geqslant t_0. \tag{9.12}$$

Then if

$$\lambda_{\max}\big(A(t) + A'(t)\big) \leqslant -2\sigma < 0, \qquad t_0 \leqslant t < \infty,$$

the system (9.12) is asymptotically stable.

Proof. For a norm in R^n, choose the Euclidean norm $\|x\| = (x'x)^{1/2}$. Since the matrix $f = A(t)$ and the logarithmic norm $\gamma(A)$ is specified by Table 9.1, it follows from Theorem 9.1 that if $\lambda_{\max}\big(A(t)+A'(t)\big) \leqslant -2\sigma < 0$, $t_0 \leqslant t < \infty$, then the system (9.12) is asymptotically stable. This result has already been obtained in Theorem 6.1. □

Corollary 2. *Let the matrix $f(t,x)$ be uniformly negative definite in $\mathcal{D}$, i.e., for any $y \in R^n$*

$$y'f(t,x)y \leqslant -Cy'y, \qquad (t,x) \in \mathcal{D}, \tag{9.13}$$

where $C > 0$ is a constant. Then the system (9.5) is asymptotically stable.

Proof. Indeed, for $\|x\|$ take the Euclidean norm in R^n. From the equality (9.2) it follows that

$$\gamma(f) = \tfrac{1}{2}\lambda_{\max}(f + f'). \tag{9.14}$$

However, since the matrix $f + f'$ is symmetric, we have

$$\lambda_{\max}(f + f') = \max_{y\neq 0} y'(f + f')y\|y\|^{-2}.$$

Thus, noting relation (9.13), we can write

$$\lambda_{\max}(f + f') = 2\max_{y\neq 0} y'fy\|y\|^{-2} \leqslant -2C.$$

Hence, by virtue of (9.2) the logarithmic norm $\gamma(f) \leqslant -C$, i.e., the requirement (9.6) of Theorem 9.1 is met. □

Corollary 3. *Suppose that the system (9.5) is of the form*

$$\dot{x}(t) = F_1\big(t, x(t)\big)x(t), \tag{9.15}$$

where $F_1\big(t,x(t)\big)$ is an $n \times n$ matrix with continuous bounded elements meeting requirements of the (9.6) type, i.e.,

$$\sup_{(t,x)\in\mathcal{D}} \gamma\big(F_1(t,x)\big) = -a < 0. \tag{9.16}$$

Then the zero solution of the system (9.15) is asymptotically stable.

Proof. The proof of this assertion is similar to that of Theorem 9.1. Indeed, take $V = \|x\|$. Then by virtue of (9.3) we obtain in a way similar to (9.7)

$$\dot{V} = Q\big(x, \dot{x}(t)\big) = Q\big(x, F_1(t,x)x\big) \leqslant \gamma\big(F_1(t,x)\big)\|x\| \leqslant -a\|x\|.$$

Thus, the function $V = \|x\|$ satisfies the hypotheses of Theorem 2.3. It follows that the system (9.15) is asymptotically stable whenever the condition (9.15) is satisfied. □

Example 9.1. Test the stability of the zero solution of the two-dimensional system

$$\dot{x}(t) = Ax(t), \qquad x = \begin{pmatrix} x_1 \\ x_2 \end{pmatrix}, \quad A = \begin{pmatrix} -a & \alpha(t, x(t)) \\ -\alpha(t, x(t)) & -a \end{pmatrix}, \quad t \geqslant t_0, \quad (9.17)$$

where a is a positive constant, $\alpha(t, x)$, $t \geqslant t_0$, $x \in R^2$, is a continuous bounded function. For $\|\cdot\|$ take the Euclidean norm in R^2. Then in view of (9.2) the logarithmic norm $\gamma(A)$ of the matrix A is equal to $-a$, i.e., on the strength of Corollary 1 the zero solution of the system (9.17) is asymptotically stable for $a > 0$ and any function α.

Example 9.2. Suppose that the system (9.5) is of the form

$$\dot{x}(t) = \begin{pmatrix} F_{11}(t, x(t)) & F_{12}(t, x(t)) \\ F_{21}(t, x(t)) & F_{22}(t, x(t)) \end{pmatrix} x(t), \qquad x \in R^2, \ t \geqslant t_0, \qquad (9.18)$$

where $F_{ij}(t, x)$ are continuous bounded functions. For the norm $\|x\|$ we take the octahedral norm. Then the asymptotic stability condition for the zero solution of the system (9.18) is the existence of a domain $\mathcal{D}$ of the form $(t \geqslant t_0) \times (\|x\| \leqslant r)$ such that

$$\sup_{(t,x)\in\mathcal{D}} \max\big[F_{11}(t, x) + |F_{21}(t, x)|, F_{22}(t, x) + |F_{12}(t, x)|\big] < 0.$$

But if the cubic norm is taken, then the following stability condition holds:

$$\sup_{(t,x)\in\mathcal{D}} \max\big[F_{11}(t, x) + |F_{12}(t, x)|, F_{22}(t, x) + |F_{21}(t, x)|\big] < 0.$$

Using the 4th and the 5th norms of Table 9.1, we obtain the following respective stability conditions:

$$\sup_{(t,x)\in\mathcal{D}} \max\big[\alpha_1 F_{11}(t, x) + \alpha_2 |F_{21}(t, x)|, \alpha_2 F_{22}(t, x) + \alpha_1 |F_{12}(t, x)|\big] = -a < 0,$$

$$\sup_{(t,x)\in\mathcal{D}} \max\big[\alpha_1 F_{11}(t, x) + \alpha_2 |F_{12}(t, x)|, \alpha_2 F_{22}(t, x) + \alpha_1 |F_{21}(t, x)|\big] = -a < 0,$$

where $\alpha_1 > 0$ and $\alpha_2 > 0$ are positive constants that are chosen so that the most convenient stability conditions are obtained.

Example 9.3. Consider the nonlinear two-dimensional system

$$\begin{aligned} \dot{x}_1 &= -\sin x_1 + \cos x_2 - 1, \\ \dot{x}_2 &= \cos x_1 - \sin x_2 - 1. \end{aligned} \qquad (9.19)$$

The Jacobi matrix $f(t, x)$ for this system can be written as

$$f(t, x) = \begin{pmatrix} -\cos x_1 & -\sin x_2 \\ -\sin x_1 & -\cos x_2 \end{pmatrix} \qquad (9.20)$$

Since for $|x_1| \leqslant \delta < \pi/4$ and $|x_2| \leqslant \delta < \pi/4$ the following inequality holds:

$$\begin{aligned}\gamma(f(t,x)) &= \sup_{\substack{|x_1|\leqslant\delta,\\ |x_2|\leqslant\delta}} \max\bigl[-\cos x_1 + |-\sin x_2|, -\cos x_2 + |-\sin x_1|\bigr]\\ &\leqslant -\cos\delta + \sin\delta = -a < 0,\end{aligned}$$

it follows that the logarithmic norm of the matrix (9.20) subordinate to the cubic norm in R^n satisfies the estimate

$$\gamma(f(t,x)) \leqslant -\cos\delta + \sin\delta = -a < 0.$$

Therefore, under this condition the trivial solution of the system (9.19) is asymptotically stable with the attraction domain $|x_1| \leqslant \delta < \pi/4$, $|x_2| \leqslant \delta < \pi/4$.

The asymptotic stability of the system (9.19) can also be established, using Theorem 7.4 on stability in the first approximation, since in our case the system of the first approximation is of the form

$$\begin{aligned}\dot{x}_1 &= -x_1,\\ \dot{x}_2 &= -x_2.\end{aligned}$$

However, no estimate of the attraction domain can be derived from Theorem 7.4. Thus, the use of the logarithmic norm sometimes makes it possible to obtain more information about the system compared with other methods.

§ 10. Use of Degenerate Lyapunov Functions

The basic method to test the stability of nonlinear systems is Lyapunov's direct method. This method enables reduction of the problem of stability analysis to that of constructing a positive definite Lyapunov function, whose derivative, in accord with the system in question, is either negative definite or of constant signs. Despite the validity of inversion theorems the use of Lyapunov's classical theorems for the investigation of particular systems is wrought with considerable difficulties caused by the above-mentioned constraints on the Lyapunov function. Due to this, a lot of work has been done to obtain sufficient stability conditions that are valid under less limiting assumptions concerning the Lyapunov function.

In this section we give several assertions on the stability of systems of ordinary differential equations based on the use of a scalar degenerate Lyapunov function.

Consider the equation

$$x^{(n+1)}(t) = F\bigl(t, x(t), \dots, x^{(n)}(t)\bigr), \qquad t \geqslant t_0. \tag{10.1}$$

Here the vector $x(t) \in R^m$, $n \geqslant 1$ is an integer; the continuous function $F \in R^m$ satisfies the condition

$$F(t, 0, \dots, 0) \equiv 0.$$

We introduce the notation

$$S_k\big(x(t)\big) = \sum_{i=1}^{k} |x^{(i)}(t)|.$$

The initial conditions for equation (10.1) are of the form

$$x(t_0) = x_0^0, \dots, x^{(n)}(t_0) = x_0^{(n)}. \tag{10.2}$$

It is assumed that for the problem (10.1), (10.2) the conditions of the local existence and uniqueness theorem are satisfied. Definition 1.2 in this case takes the following form.

Definition 10.1. The trivial solution $x(t) \equiv 0$ of equation (10.1) is said to be *stable* if for any $\varepsilon > 0$ there exists a $\delta = \delta(\varepsilon, t_0) > 0$ such that, whenever $S_n\big(x(t_0)\big) \leqslant \delta(\varepsilon, t_0)$, the solution associated with these initial data satisfies the inequality $S_n\big(x(t)\big) \leqslant \varepsilon$ for $t \geqslant t_0$. If the solution $x(t) \equiv 0$ is stable, and, moreover, $S_n\big(x(t)\big) \to 0$ as $t \to \infty$ for the initial data $S_n\big(x(t_0)\big) \leqslant \Delta(t_0)$, then the trivial solution is called *asymptotically stable.*

In a similar way, some other definitions, namely those of uniform stability, uniform asymptotic stability, etc., can be introduced.

For an analysis of stability one can use continuous Lyapunov functions of the form

$$V(t) = V\Big(t, x(t), \dots, x^{(n)}(t), g\big(t, x(t), \dots, x^{(n)}(t)\big)\Big), \tag{10.3}$$

where g is a continuously differentiable function such that $g(t, 0, \dots, 0) = 0$. The only constraint on the function V is $V \geqslant 0$, but, in general, V is not positive definite with respect to the arguments $x(t), \dots, x^{(n)}(t)$. Such functions V are called *degenerate Lyapunov functions.*

Henceforth, the concepts connected with the stability of inequalities of the form

$$|g\big(t, y(t), \dots, y^{(n)}(t)\big)| \leqslant f(t), \qquad t \geqslant t_0 \tag{10.4}$$

are of considerable importance.

Denote by $y(t, t_0, y_0^0, \dots, y_0^{(n-1)})$ any solution of the differential the inequality (10.4) with initial condition

$$y(t_0) = y_0^0, \dots, y^{(n-1)}(t_0) = y_0^{n-1}. \tag{10.5}$$

Definition 10.2. The trivial solution $y(t) \equiv 0$ of the inequality (10.4) is said to be *f-stable* if for any $\varepsilon > 0$ there exists a quantity $\delta(\varepsilon, t_0) > 0$ such that for all initial conditions and right sides satisfying the inequalities $S_{n-1}\big(y(t_0)\big) \leqslant \delta(\varepsilon, t_0)$ and $f(t) \leqslant \delta(\varepsilon, t_0)$, $t \geqslant t_0$, the inequality $S_n\big(y(t, t_0, y_0^0, \dots, y_0^{(n-1)})\big) \leqslant \varepsilon$ is satisfied for all $t \geqslant t_0$.

The solution $y(t) \equiv 0$ is said to be *asymptotically f-stable* if it is f-stable and, moreover,

$$\lim_{t\to\infty} S_n\big(y(t, t_0, y_0^0, \dots, y_0^{(n-1)})\big) = 0$$

for all initial conditions (10.5) satisfying the inequality $S_{n-1}\big(y(t_0)\big) \leqslant \Delta(t_0)$ and for any right side $f(t)$ such that $f(t) \to 0$ as $t \to \infty$.

Theorem 10.1. *Suppose there exists a function V of the form* (10.3) *which satisfies the following conditions:*

1°) $\omega_1\Big(\big|g\big(t, x(t), x^{(n)}(t)\big)\big|\Big) \leqslant V \leqslant \omega_2\Big(S_n\big(x(t)\big)\Big)$;

2°) $\dot V \leqslant 0$;

3°) *the trivial solution of equation* (10.4) *is f-stable.*

Then the trivial solution of equation (10.1) *is uniformly stable.*

Proof. Take an arbitrary $\varepsilon > 0$. On account of the f-stability of the inequality (10.4), for this ε there exists a $\delta_1 > 0$ such that any solution of the inequality (10.4) with initial condition (10.2) satisfies the relation $S_n\big(y(t)\big) < \varepsilon$ whenever $f(t) \leqslant \delta_1$ and $S_{n-1}\big(y(t_0)\big) \leqslant \delta_1$. Further, choose a δ_2 so that $0 < \delta_2 \leqslant \delta_1$ and $\omega_2(\delta_2) = \omega_1(\delta_1)$. Then for $S_n\big(x(t_0)\big) \leqslant \delta_2$ we have according to conditions 1° and 2°

$$\begin{aligned}\omega_1\Big(\big|g\big(t, x(t), \dots, x^{(n)}(t)\big)\big|\Big) &\leqslant V(t)\\ &\leqslant V\Big(t_0, x(t_0), \dots, x^{(n)}(t_0), g\big(x(t), \dots, x^{(n)}(t_0)\big)\Big)\\ &\leqslant \omega_2\Big(S_n\big(x(t_0)\big)\Big) \leqslant \omega_2(\delta_2) = \omega_1(\delta_1).\end{aligned}$$

Thus, noting the monotonicity of the function ω_1, we find that $\big|g\big(t, x(t), \dots, x^{(n)}(t)\big)\big| = f(t) \leqslant \delta_1$. Taking into account the f-stability of the inequality (10.4), we obtain $S_n\big(x(t)\big) \leqslant \varepsilon$ $(t \geqslant t_0)$ for all initial conditions such that $S_n\big(x(t_0)\big) \leqslant \delta_2 \leqslant \delta_1$. □

Remark. If under the conditions of Theorem 10.1 it can be proved that $f(t)$ is not only bounded, but tends to zero as $t \to \infty$ according to some law, then it is sufficient to test the f-stability of the inequality (10.4) only for such right sides $f(t)$.

Example 10.1. Consider the equation

$$\ddot x = -\frac{(1 + x^2)\dot x + 2x\dot x^2}{1 + x^2} \tag{10.6}$$

and the function $g(x, \dot x) = (1 + x^2)\dot x$. Then equation (10.6) can be written in the form

$$\dot g = -g. \tag{10.7}$$

Consider $V\big(g(x, \dot x)\big) = g^2(x, \dot x)$. It follows from (10.7) that

$$\big|g\big(x(t), \dot x(t)\big)\big| \leqslant Ce^{-(t-t_0)}.$$

Thus, we must establish the f-stability of the solution $x(t) \equiv 0$ of the inequality

$$|(1 + x^2)\dot x| \leqslant Ce^{-(t-t_0)}. \tag{10.8}$$

It follows from the inequality (10.8) that $|x(t) + x^3(t)/3| \leqslant C_1$ or $|x(t)| \leqslant C_2$. Further, the inequality (10.8) implies that

$$|\dot{x}| \leqslant \frac{C}{1+C_2^2} e^{-(t-t_0)} \leqslant C_3.$$

On the basis of Theorem 10.1 and the remark associated with it we conclude that the solution $x(t) \equiv 0$ of equation (10.6) is stable.

To test the asymptotic stability, assume that

$$\left|\frac{\partial g}{\partial t}\right| + \left|\frac{\partial g}{\partial x}\right| + \cdots + \left|\frac{\partial g}{\partial x^{(n)}}\right| + |F(t, x, \ldots, x^{(n)})| \leqslant M \tag{10.9}$$

for $S_n(x(t)) \leqslant H$.

Theorem 10.2. *Suppose there exists a function V of the form* (10.3) *such that:*

1°) *Condition* 1° *of Theorem* 10.1 *is satisfied;*

2°) $\dot{V}(t) \leqslant -\omega_3\big(|g(t, x(t), \ldots, x^{(n)}(t))|\big)$;

3°) *the trivial solution of the inequality* (10.4) *is asymptotically f-stable.*

Then the trivial solution of equation (10.1) *is asymptotically stable.*

Proof. According to Theorem 10.1, the solution $x(t) \equiv 0$ of equation (10.1) is stable. Therefore, $S_n(x(t)) \leqslant H$, $t \geqslant t_0$, for all initial conditions such that $S_n(x(t_0)) \leqslant \delta(H)$. We shall show that in this case $|g(t, x(t), \ldots, x^{(n)}(t))| \to 0$ as $t \to \infty$. Assume that this is not so. Then there exist a number γ $(0 < \gamma < H)$ and a sequence $t_i \to \infty$ such that $|g(t_i, x(t_i), \ldots, x^{(n)}(t_i))| > \gamma$. For $S_n(x(t)) \leqslant H$, we have by virtue of the inequality (10.9) and equation (10.1)

$$\left|\frac{dg(t, x(t), \ldots, x^{(n)}(t))}{dt}\right| = \left|\frac{\partial g}{\partial t} + \sum_{i=0}^{n} \frac{\partial g}{\partial x_i} x^{(i+1)}(t)\right| \leqslant MH + M^2 = L.$$

Moreover, for $\tau \in [t_i - \gamma/(2L), t_i + \gamma/(2L)]$ we find $|g(\tau, x(\tau), \ldots, x^{(n)}(\tau))| \geqslant \gamma/2$. Denote by $n(t)$ the number of points $t_i \in [t_0, t]$; we obtain

$$\begin{aligned} V(t) - V(t_0) &= \int_{t_0}^{t} \dot{V}(s)\, ds \leqslant -\int_{t_0}^{t} \omega_3\big(|g(s, x(s), \ldots, x^{(n)}(s))|\big)\, ds \\ &\leqslant -\gamma L^{-1}\omega_3(\gamma/2)n(t) \to -\infty, \qquad t \to \infty. \end{aligned}$$

This contradicts the condition 1°. Therefore, we have proved that $|g(t, x(t), \ldots, x^{(n)}(t))| \to 0$ as $t \to \infty$ and $S_n(x(t_0)) \leqslant \delta(H)$. Hence, due to the asymptotic f-stability of the inequality (10.4) it follows that the solution $x(t) \equiv 0$ of equation (10.1) is asymptotically stable. □

Corollary. *Suppose that we have the equation*

$$\dot{u}(t) = F_1\big(t, u(t)\big), \qquad u \in R^k,\ F_1(t,0) \equiv 0, \tag{10.10}$$

and let the function $g\big(t, x(t), \dots, x^{(n)}(t)\big) \in R^k$ *for* $t \geqslant t_0$. *Then, if the trivial solution of equation* (10.10) *is uniformly asymptotically stable, and the trivial solution of the inequality* (10.4) *is asymptotically* f*-stable, then the trivial solution of equation*

$$\frac{d}{dt} g\big(t, x(t), \dots, x^{(n)}(t)\big) = F_1\big(t, g(t, x(t), \dots, x^{(n)}(t)\big)$$

is asymptotically stable.

Let us make a few remarks about the notions of the f-stability and of the asymptotic f-stability of solutions of the differential inequality (10.4). It is readily seen that whenever the condition (10.9) is satisfied, the f-stability of the trivial solution of the inequality (10.4) follows from the stability under steady-acting disturbances of the trivial solution of the equation

$$g\big(t, x(t), \dots, x^{(n)}(t)\big) = 0.$$

The notion of asymptotic f-stability cannot be reduced to the notions introduced above. However, this notion has a simple mechanical meaning. Suppose that the inequality (10.4) describes a mechanical system, and $f(t)$ is an estimate of the norm of all external forces acting on this system. In that case the requirement of asymptotic f-stability means that if the external forces vanish for $t \to 0$, then the system must tend to its equilibrium state.

On the basis of other concepts it was established [1] that for the linear time-varying inequality

$$|y^{(n)}(t) + A_1(t) y^{(n-1)}(t) + \cdots + A_n(t) y(t)| \leqslant f(t) \tag{10.11}$$

the exponential stability of the trivial solution of the equation

$$y^{(n)}(t) + A_1(t) y^{(n-1)}(t) + \cdots + A_n(t) y(t) = 0$$

implies the asymptotic f-stability of the trivial solution of the inequality (10.11).

§ 11. Stability of Discrete Systems

When describing dynamical systems, we often come up against the situation in which the system state $x(t)$ is known (is measured) only at discrete moments $t_k = kh$ ($k = 0, 1, 2, \dots$; $h > 0$). The dynamics of such a system can be described by an equation of the form

$$x(t_{k+1}) = f\big(t_k, x(t_k), x(t_{k-1}), x(t_{k-2}), \dots, x(t_{k-p})\big). \tag{11.1}$$

Denoting $x(t_k) = x_k$ and $t_k = k$, we write (11.1) in the form

$$x_k = f(k, x_{k-1}, \dots, x_{k-p}). \tag{11.2}$$

Equation (11.2) is called a *difference equation of order p*. Introduce the vector $X_n = (x_n, x_{n+1}, \dots, x_{n+p-1})$, we can write equation (11.2) in the form of the system

$$X_{n+1} = F(n, X_n). \tag{11.3}$$

The initial condition for the system (11.3) is of the form $X_{n_0} = X^0$. Systems of the form (11.3) are often used when describing pulse systems. We are also concerned with discrete systems in cases when a computer is included in the control loop.

1. Lyapunov's direct method. We shall give basic facts about the stability of discrete systems. Consider the equation

$$x_{k+1} = f(k, x_k), \qquad x_k \in R^n, \ f(k, 0) \equiv 0. \tag{11.4}$$

The solution of equation (11.4) with initial condition $x_{k_0} = x^0$ will be denoted by $x(k, k_0, x^0)$.

Definition 11.1. The solution of equation (11.4) is said to be *uniformly stable* if for any $\varepsilon > 0$ there exists a $\delta(\varepsilon) > 0$ such that the inequality $|x_{k_0}| \leqslant \delta(\varepsilon)$ leads to $|x_k| \leqslant \varepsilon$ for all $k \geqslant k_0$. The solution is said to be *uniformly asymptotically stable* if it is uniformly stable, and for any $\gamma > 0$ there exist an integer $N(\gamma) > 0$ and a number $H_1 > 0$ such that $|x(k, k_0, x^0)| \leqslant \gamma$ for $k \geqslant k_0 + N(\gamma)$ and $|x^0| \leqslant H_1$.

Other definitions of stability for the difference equation (11.4) are similar to corresponding definitions for the differential equation (1.5).

For the analysis of the stability of the solution $x_k \equiv 0$ of equation (11.4), Lyapunov's direct method can also be used. We shall cite several assertions without proof.

Theorem 11.1. *For the solution $x_k \equiv 0$ of equation* (11.4) *to be uniformly stable, it is necessary and sufficient that a function $V(k, x)$ exist, satisfying the conditions*

$$\omega_1(|x|) \leqslant V(k, x) \leqslant \omega_2(|x|), \qquad V\big(k+1, f(k, x)\big) - V(k, x) \leqslant 0.$$

Theorem 11.2. *Suppose there exists a function $V(k, x)$ such that*

$$\omega_1(|x|) \leqslant V(k, x) \leqslant \omega_2(|x|),$$
$$V\big(k+1, f(k, x)\big) - V(k, x) \leqslant -\omega_3(|x|).$$

Then the solution $x_k \equiv 0$ is uniformly asymptotically stable.

The other theorems of § 2 and 4 can be readily extended to difference equations of the form (11.4).[14]

[14] See, for example, Agarwal, R. P., *Difference Equations and Inequalities*, Marcel Dekker, New York, (1992); Lakshmikantham, V., and Trigante, D., *Theory of Difference Equations: Numerical Methods and Applications*, Academic Press, New York, (1988).

2. Linear time-invariant equations. Consider the case of a linear time-invariant difference equation

$$x_{k+1} = Ax_k, \qquad x_0 = x^0, \ x \in R^n. \tag{11.5}$$

As in subsection 1 of § 5, the general solution of equation (11.5) can be obtained as follows. If $\lambda_1(A)$ is a simple eigenvalue of the matrix A with eigenvector h_1, then it is associated with the solution of equation (11.5) having the form $x_k^1 = \lambda_1^k h_1$. But if $\lambda_2(A)$ is a multiple eigenvalue with Jordan box of order m, then it is associated with m solutions of the form

$$x_k^2 = \lambda_2^k h_2;\ x_k^3 = k\lambda_2^{k-1} h_2 + \lambda_2^k h_3; \dots ;$$
$$x_k^{m+1} = C_k^{m-1} \lambda_2^{k-m+1} h_2 + \dots + \lambda_2^k h_{m+1},$$

where $h_2, \dots, h_{m+1}$ is a sequence of vectors satisfying the conditions

$$h_2 \neq 0,\ Ah_2 = \lambda_2 h_2, \dots, Ah_{m+1} = \lambda_2 h_{m+1} + h_m.$$

In a similar way, solutions associated with other Jordan boxes and eigenvalues $\lambda_i(A)$ can be obtained. A linear combination of these solutions is the general solution of equation (11.5). This implies the following theorem.

Theorem 11.3. *If $|\lambda_i(A)| < 1$ for all i, then the solution $x_k \equiv 0$ of equation (11.5) is uniformly asymptotically stable. But if $|\lambda_i(A)| \leqslant 1$, and all eigenvalues λ_k such that $|\lambda_k(A)| = 1$ are associated with Jordan boxes of order one, then the solution $x_k \equiv 0$ is uniformly stable. In all other cases the solution $x_k \equiv 0$ is unstable.*

Let us state the Schur–Kohn criterion which represents a necessary and sufficient condition that all roots of the polynomial (5.4):

$$P_n(\lambda) = a_0 + a_1\lambda + \dots + a_n\lambda^n = 0, \qquad a_n = 1,$$

lie inside a unit circle. The proof of this criterion invokes the Hurwitz criterion (see § 5) that there are no roots of the polynomial (5.4) in the right half-plane and also uses the linear fractional transformation that maps the left half-plane into the interior of the unit circle. For each $m = 1, 2, \dots, n$ we define the matrices A_{0m} and A_m of order $m \times m$ by the relations

$$A_{0m} = \begin{pmatrix} a_n & 0 & 0 & \dots & 0 \\ a_{n-1} & a_n & 0 & \dots & 0 \\ a_{n-2} & a_{n-1} & a_n & \dots & 0 \\ \vdots & \vdots & \vdots & \ddots & \vdots \\ a_{n-m+1} & a_{n-m+2} & a_{n-m+3} & \dots & a_n \end{pmatrix},$$

$$A_m = \begin{pmatrix} a_0 & a_1 & a_2 & \dots & a_{m-1} \\ 0 & a_0 & a_1 & \dots & a_{m-2} \\ 0 & 0 & a_0 & \dots & a_{m-3} \\ \vdots & \vdots & \vdots & \ddots & \vdots \\ 0 & 0 & 0 & \dots & a_0 \end{pmatrix}.$$

Next, consider the sequence of the determinants

$$\Delta_m = \begin{vmatrix} A_{0m} & A_m \\ A'_m & A'_{0m} \end{vmatrix}, \qquad m = 1, 2, \ldots, n.$$

The Schur–Kohn criterion. *In order that all roots of the polynomial* (5.4) *lie inside a unit circle, it is necessary and sufficient that* $\Delta_m > 0$ *for all* $m = 1, 2, \ldots, n$. *Here* $\Delta_m = |A_{0m}A'_{0m} - A_m A'_m|$.

Let us now state the analog of Theorem 7.2 for discrete systems of the form (11.5). We take a Lyapunov function for the system (11.5) as $V = x'Bx$. The matrix B satisfies the equation called the *discrete-time Lyapunov equation*:

$$A'BA - B = C, \tag{11.6}$$

provided the chosen Lyapunov function meets the conditions of Theorem 11.2.

Theorem 11.4. *If* $|\lambda_i(A)| < 1$ *for all* i, *then for any matrix* $C < 0$ *there exists a unique solution* $B > 0$ *of equation* (11.6) *of the form*[15]

$$\text{vec}[B] = -(A' \otimes A' - I)^{-1} \text{vec}[C],$$

where $\otimes$ *is a Kronecker product of matrices, and* vec$[B]$ *is a vector whose elements are columns of* B *written one below another.*

Conversely, if for at least one matrix $C < 0$ *there exists a solution* $B > 0$ *of equation* (11.6), *then* $|\lambda_i(A)| < 1$ *for all* i.

3. Stability in the first approximation. Consider the equation

$$x_{k+1} = Ax_k + R(k, x_k), \qquad |R(k, x_k)| \leqslant C|x_k|^{1+\alpha}, \ \alpha > 0. \tag{11.7}$$

Theorem 11.5. *If the zero solution of the linear equation of the first approximation* $x_{k+1} = Ax_k$ *is asymptotically stable, then the solution* $x_k \equiv 0$ *of equation* (11.7) *is also asymptotically stable. If there exists an eigenvalue* $\lambda_j(A)$ *such that* $|\lambda_j(A)| > 1$, *then the solution* $x_k \equiv 0$ *of equation* (11.7) *is unstable regardless of the form of the nonlinearity* $R(k, x)$.

Cases in which there are eigenvalues $\lambda_i(A)$, with $|\lambda_i(A)| = 1$, are critical. In such cases the stability of the nonlinear system (11.7) depends on the form of the nonlinearity $R(k, x_k)$.

4. Stability with respect to specified variables[16]**.** We shall give a theorem on the stability of the difference equation (11.4). In addition to equation (11.4) we consider a vector of variables y_k having the form

$$y_k = q(k, x_k), \qquad y \in R^l.$$

[15]See the second of the remarks following the proof of Theorem 7.2.

[16]Nosov, V. R., and Furasov, V. D., Stability of Discrete Processes with respect to Specified Variables and Convergence of Certain Optimization Algorithms, *Zh. Vychisl. Mat. i Mat. Fiz.*, (1979), no. 2, 316–328. (Translated into English as *U.S.S.R. Comput. Math. and Math. Phys.*)

The function $q(k, x_k)$ is assumed to be continuous with respect to the second argument.

Definition 11.2. The solution $x_k \equiv 0$ of equation (11.4) is said to be *stable with respect to the variables* $y_k = (y_k^1, \dots, y_k^l)$ if for any $\varepsilon > 0$ there exists a $\delta = \delta(\varepsilon, k_0)$ such that for any initial conditions x^0 satisfying the inequality $|y_{k_0}| = |q(k_0, x_{k_0})| \leqslant \delta$ the condition $|y_k| = |q(k, x_k)| \leqslant \varepsilon$ is satisfied for all $k \geqslant k_0$. But if, in addition,

$$|y_k| = |q(k, x_k)| = \left|q\left(k, x(k, k_0, x^0)\right)\right| \to 0, \qquad k \to \infty,$$

for all x^0 being inside the attraction domain of the solution $x_k \equiv 0$, having the form $|q(k_0, x_{k_0})| \leqslant \delta$, then the solution $x_k \equiv 0$ of equation (11.4) is said to be *asymptotically stable with respect to the variables* $y_k = q(k, x_k)$.

Definition 11.2 generalizes the definition of stability and of partial stability. Other definitions of stability with respect to specified variables can be introduced in a similar way.

Theorem 11.6. *Suppose there exists a Lyapunov function* $V(k, x)$ *such that*

$$\omega_1\left(|q(k,x)|\right) \leqslant V(k,x) \leqslant \omega_2\left(|q(k,x)|\right),$$
$$V\left(k+1, f(k,x)\right) - V(k,x) \leqslant 0.$$

Then the solution $x_k \equiv 0$ *is uniformly stable with respect to the variables* $y_k = q(k, x_k)$.

Theorem 11.7. *Suppose there exists a Lyapunov function* $V(k, x)$ *such that*

$$0 \leqslant V(k,x), \qquad V(k_0, x) \leqslant \omega_2\left(|q(k_0,x)|\right),$$
$$V\left(k+1, f(k,x)\right) - V(k,x) \leqslant -\omega_3\left(|q(k,x)|\right).$$

Then the solution $x_k \equiv 0$ *is asymptotically stable with respect to the variables* $y_k = q(k, x_k)$.

5. Unconditional minimization of functions. When looking for minima of functions $f(x)$, where $x \in R^n$, $f(x) \in C^1(R^n)$, one often uses *gradient methods with adjusted step.*[17] The algorithm of the gradient method consists in the following:

1°) A point $x_0 \in R^n$ is chosen.

2°) The sequence x_k of the form

$$x_{k+1} = x_k - \lambda_k f_x(x_k), \qquad f_x = \frac{df(x)}{dx} = \operatorname{grad} f(x) \tag{11.8}$$

[17] For the details of the proofs of the assertions cited in this subsection, see Polak, E., *Computational Methods in Optimization. A Unified Approach*, Academic Press, New York–London, (1971).

is constructed.

3°) A step λ_k is chosen so that $0 \leqslant \lambda_k \leqslant A$ and

$$f(x_k - \lambda_k f_x(x_k)) \leqslant f(x_k) - \frac{1}{2}\lambda_k |f_x(x_k)|^2. \tag{11.9}$$

Suppose that the function $f(x) \in C^2(R^n)$ and is strongly convex, i.e., the matrix of second derivatives $f_{xx}(x)$ is uniformly positive definite:

$$(f_{xx}(x)p, p) \geqslant \alpha(p, p), \qquad \alpha > 0,\ \forall x, p \in R^n,\ (x, p) = x'p.$$

In addition, let the matrix f_{xx} be bounded above: $|f_{xx}(x)p| \leqslant M|p|$. It is known that such functions $f(x)$ have a unique minimum point.

Under these conditions the sequence (11.8) converges to the minimum point of the function $f(x)$.

To prove this, consider the nonnegative function

$$V(x) = f(x) - \min_{x \in R^n} f(x). \tag{11.10}$$

The strong convexity of $f(x)$ implies the following estimates:

$$\frac{m}{8M^2}|f_x(x)|^2 \leqslant V(x) \leqslant \frac{2M}{m^2}|f_x(x)|^2, \qquad \frac{1}{2M} \leqslant \lambda_k \leqslant \frac{1}{M}.$$

It follows from (11.9) that

$$V(x_{k+1}) \leqslant V(x_k) - \frac{1}{2}\lambda_k |f_x(x_k)|^2 \leqslant V(x_k) - \frac{1}{4M}|f_x(x_k)|^2.$$

Thus, the function (11.10) satisfies the conditions of Theorem 11.7 on the asymptotic stability of the solutions of equation (11.8) with respect to the variable $y_k = f_x(x_k)$. In other words, the gradient method (11.8), (11.9) yields a sequence converging to a point x^* such that $f_x(x^*) = 0$, i.e., to the unique minimum point x^* of the function $f(x)$.

Main Results and Formulas of Chapter I

Lyapunov's first theorem. Suppose there exists a function $V(t,x)$ such that $\omega_1(|x|) \leqslant V(t,x)$, $\dot{V}(t,x) \leqslant 0$. Then the trivial solution is Lyapunov-stable.

Lyapunov's second theorem. If $\omega_1(|x|) \leqslant V(t,x) \leqslant \omega_2(|x|)$, $\dot{V}(t,x) \leqslant -\omega_3(|x|)$. Then the trivial solution is uniformly asymptotically stable.

Global stability. If, in addition to the conditions of Lyapunov's second theorem, $\omega_1(u) \to \infty$, $u \to \infty$, then the trivial solution is globally uniformly asymptotically stable.

Chetaev's theorem on instability. If in the region $V(t,x) > 0$ the inequality $\dot{V}(t,x) \geqslant \omega_4(|x|)$ holds, then the trivial solution is unstable.

The Barbashin–Krasovskii theorem. If the system is autonomous, $V(x) \geqslant \omega_1(|x|)$, $\dot{V}(x) \leqslant 0$, and, moreover, the set $\{x : \dot{V}(x) = 0\}$ does not contain whole trajectories, then the trivial solution is asymptotically stable.

The Matrosov criterion. If there exist two functions $V(t,x)$ and $W(t,x)$ such that $\omega_1(|x|) \leqslant V(t,x) \leqslant \omega_2(|x|)$, $\dot{V}(t,x) \leqslant a(|x|) \leqslant 0$, $|W(t,x)| \leqslant L =$ constant, and $|\dot{W}(t,x)| \geqslant \omega_4(|x|)$ in $E(\mu,\rho)$, then the trivial solution is uniformly asymptotically stable.

Comparison principle. Suppose there exists a vector function $V(t,x) = \big(V_1(t,x),\dots,V_m(t,x)\big)$ such that $\max V_i(t,x) \geqslant \omega_1(|x|)$, $V(t,0) = 0$, and $\dot{V}(t,x) \leqslant F\big(t,V(t,x)\big)$, where $F(t,u)$ is a quasimonotonic increasing function. Then the stability (asymptotic stability) of the solution $u \equiv 0$ of the equation $\dot{u} = F(t,u)$ implies the stability (asymptotic stability) of the solution $x \equiv 0$.

Rumyantsev's theorem on partial stability. If there exists a function $V(t,x,y)$ such that $\omega_1(|x|) \leqslant V(t,x,y) \leqslant \omega_2(|z|)$, $\dot{V}(t,x,y) \leqslant -\omega_3(|x|)$, then the solution $z \equiv 0$ is asymptotically x-stable.

The Routh–Hurwitz criterion. For the polynomial $P_n(\lambda)$ to be stable, it is necessary and sufficient that all principal diagonal minors of its Hurwitz matrix M_{P_n} be positive.

Robust stability. For the polynomial $P_n(\lambda)$ to be stable, it is necessary and sufficient that all four polynomials $f_n^1(\lambda)$, $f_n^2(\lambda)$, $f_n^3(\lambda)$, and $f_n^4(\lambda)$ be stable.

The Mikhailov criterion. For the polynomial $P_n(\lambda)$ having no pure imaginary roots to be stable, it is necessary and sufficient that

$$\Delta \arg P_n(i\omega)\big|_{\omega=0}^{\omega=\infty} = \frac{\pi}{2}n.$$

The Gronwall–Bellman lemma. If the scalar functions $x(t), g(t) \geqslant 0$ and the nondecreasing function $\alpha(t)$ satisfy the inequality

$$x(t) \leqslant \alpha(t) + \int_{t_0}^{t} g(s)x(s)\,ds, \qquad t \geqslant t_0,$$

then

$$x(t) \leqslant \alpha(t) \exp\left[\int_{t_0}^{t} g(s)\,ds\right].$$

The Bellman theorem. Let the solution $y = 0$ of the system $\dot{y} = Ay$ be stable (asymptotically stable), and $\int_0^\infty \|B(t)\|\,dt < \infty$. Then the solution $x = 0$ of the system $\dot{x} = [A + B(t)]x$ is stable (asymptotically stable).

The Floquet theorem. The fundamental matrix $X(t)$ of a linear ω-periodic system can be represented in the form

$$X(t) = \Phi(t)e^{Lt}, \qquad \Phi(t+\omega) = \Phi(t),\ L = \text{constant matrix}.$$

If all Floquet multipliers of a linear ω-periodic system lie inside a unit circle, then the system is asymptotically stable.

The Lyapunov theorem. Suppose that $\operatorname{Re}\lambda_j(A) < 0$, $1 \leqslant j \leqslant n$. Then for any matrix $C < 0$ there exists a unique solution $B > 0$ of Lyapunov's matrix equation $A'B + BA = C$.

Lyapunov theorem on stability in the first approximation. Suppose that $\operatorname{Re}\lambda_j(A) < 0$, $1 \leqslant j \leqslant n$. Then the trivial solution of the nonlinear equation

$$\dot{x}(t) = Ax + R(t,x), \qquad |R(t,x)| \leqslant C|x|^{1+\alpha},\ \alpha > 0,$$

is asymptotically stable. But if for at least one k $(1 \leqslant k \leqslant n)$ $\operatorname{Re}\lambda_k(A) > 0$, then the trivial solution of the nonlinear equation is unstable.

Application of the logarithmic norm. The trivial solution of the nonlinear equation $\dot{x} = F(t,x)$ is asymptotically stable if the logarithmic norm of its Jacobi matrix is negative, i.e.,

$$\gamma \frac{\partial F(t,x)}{\partial x} \leqslant -a < 0.$$

Degenerate Lyapunov functions. Suppose there exists a function

$$V(t) = V\Big(t, x(t), \ldots, x^{(n)}(t),\ g\big(t, x(t), \ldots, x^{(n)}(t)\big)\Big),$$

such that

$$\omega_1\Big(\big|g\big(t, x(t), \ldots, x^{(n)}(t)\big)\big|\Big) \leqslant V \leqslant \omega_2\Big(S_n\big(x(t)\big)\Big), \qquad \dot{V} \leqslant 0,$$
$$\dot{V}(t) \leqslant -\omega_3\Big(\big|g\big(t, x(t), \ldots, x^{(n)}(t)\big)\big|\Big),$$

and the trivial solution of the inequality $\big|g\big(t, x(t), \ldots, x^{(n)}(t)\big)\big| \leqslant f(t)$ is f-stable (asymptotically f-stable). Then the trivial solution of equation $x^{n+1}(t) = F\big(t, x(t), \ldots, x^{(n)}(t)\big)$ is uniformly stable (asymptotically stable).

Stability of discrete systems. If

$$\omega_1(|x|) \leqslant V(k, x) \leqslant \omega_2(|x|),$$
$$V\big(k+1, f(k, x)\big) - V(k, x) \leqslant -\omega_3(|x|),$$

then the trivial solution of the discrete system $x_{k+1} = f(k, x_k)$ is uniformly asymptotically stable. If

$$0 \leqslant V(k, x), \qquad V(k_0, x) \leqslant \omega_2\big(|q(k_0, x)|\big),$$
$$V\big(k+1, f(k, x)\big) - V(k, x) \leqslant -\omega_3\big(|q(k, x)|\big),$$

then the solution $x_k \equiv 0$ is asymptotically stable with respect to the variables $y_k = q(k, x_k)$.

CHAPTER II

STABILITY OF STOCHASTIC SYSTEMS

In this chapter we present basic statements of the stability theory of dynamic systems under random perturbations. Further, we expound necessary details pertaining to the theory of stochastic differential equations.

§ 1. Introduction

We encounter stability problems under incomplete information in different areas of engineering, mechanics, economics, biology, etc. In these problems the evolution of a system occurs under certain partially known circumstances. Consider, for instance, the motion of an aircraft in the atmosphere. In actual practice this motion is affected by various weakly controllable or unpredictable factors which are interwoven in a complicated manner and are of random nature. Among these factors are external not fully known forces, disturbances, and errors in the measurements of the initial conditions, incomplete knowledge of aerodynamic properties and of actual parameters of an aircraft, the occurrence of wind gusts, the variations of atmospheric density, of magnetic or gravitational fields, and so on.

Lack of information may be due to noise in the observation channel during the motion of the system. It may also be caused by the delay due to the amount of time required for carrying out the observations and processing the obtained data.

Stability problems for dynamical systems under incomplete information can be formalized in different ways. In this book we adopt the stochastic approach, where the incompleteness of information is treated as a result of random perturbations, with known stochastic parameters, acting upon the system.

The stability problems under random perturbations are related, in fact, to the questions about the conditions ensuring that certain stochastic characteristics of motion (e.g., the expectation of coordinates and velocities, their covariance matrix, and so on) undergo a slight change, given a small variation of the disturbing factors. In this chapter we present general statements of stochastic stability problems, some methods for their solutions, and relevant examples.

§ 2. Some Preliminaries from Probability Theory and the Theory of Stochastic Processes

1. Basic probability space. Let a space Ω consisting of elements ω be given. Suppose that in the space Ω there is a chosen family $\mathcal{U}$ of subsets forming a σ-*algebra.* This means that for any finite or countable collection of subsets $A_i \in \mathcal{U}$ their intersection $\cap_i A_i$, sum $\cup_i A_i$, and complements $\overline{A_i}$ belong to $\mathcal{U}$.

Denote by P a *probability measure* on $\mathcal{U}$, i.e., a nonnegative countably additive function P with domain $\mathcal{U}$ such that $P(\Omega) = 1$.

The sets belonging to $\mathcal{U}$ are called *random events*, while elements $\omega \in \Omega$ are called *elementary events*. The triplet $(\Omega, \mathcal{U}, P)$ is called the *basic probability space*.

Example 2.1. Suppose that a coin is thrown up once. By rejecting the possibility for the coin to fall on the edge or to disappear, we infer that $\Omega = \{0, 1\}$, where $\omega = 0$ ($\omega = 1$) corresponds to heads (tails). If the coin is thrown up n times, then the number of possible events is equal to 2^n; the space Ω takes the form $\Omega = \{\omega : \omega = (\omega_1, \dots, \omega_n)\}$, and the probability of any elementary event is equal to 2^{-n}. Here ω_i is either 0 or 1.

Example 2.2. Suppose that two dice are thrown, and their faces are numbered from 1 to 6. The set Ω containing 36 elements is just the set of pairs (n, m), i.e., $\Omega = \{(n, m) : n, m = 1, 2, \dots, 6\}$. The probability of an elementary event is equal to 1/36.

2. Random variables. A $\mathcal{U}$-measurable almost everywhere finite function $\xi(\omega)$ with values in (R^n, B), i.e., a map $\xi : (\Omega, \mathcal{U}) \longrightarrow (R^n, B)$ is called a *random vector*. Recall that the Borel *σ-algebra* of the space R^n is exactly the minimal σ-algebra containing all open subsets $\mathcal{O}$ (a σ-algebra containing all open subsets $\mathcal{O}$ is said to be minimal if it coincides with the intersections of all σ-algebras containing these subsets).

A random vector $\xi(\omega)$ can be defined by means of its *distribution function* $F(x_1, \dots, x_n)$ which is equal to the probability of the event $(\xi_1 < x_1, \dots, \xi_n < x_n)$, where ξ_i are the components of the vector ξ. In other words

$$F(x_1, \dots, x_n) = P(\xi_1 < x_1, \dots, \xi_n < x_n).$$

The function $F(x_1, \dots, x_n)$ is nondecreasing with respect to any of its arguments $x_1, \dots, x_n$, with all the remaining ones fixed. Further, it follows from the definition of F that

$$\lim_{x_i \to -\infty} F(x_1, \dots, x_{i-1}, x_i, \dots x_n) = 0 \qquad (i = 1, \dots, n),$$

$$\lim_{x_i \to \infty} F(x_1, \dots, x_{i-1}, x_i, x_{i+1}, \dots, x_n)$$
$$= P(\xi_1 < x_1, \dots, \xi_{i-1} < x_{i-1}, \xi_{i+1} < x_{i+1}, \dots, \xi_n < x_n).$$

In a number of cases the function F can be characterized by means of a probability density. A Lebesgue measurable nonnegative function $f(x_1, \dots, x_n)$ such that

$$P(\xi_1 < x_1, \dots, \xi_n < x_n) = \int_{-\infty}^{x_1} \cdots \int_{-\infty}^{x_n} f(y_1, \dots, y_n)\, dy_1 \dots dy_n$$
$$= F(x_1, \dots, x_n)$$

is called a *probability density*. This equality implies the following relation between probability densities and distribution functions:

$$f(x_1, \dots, x_n) = \frac{\partial^n}{\partial x_1 \dots \partial x_n} F(x_1, \dots, x_n).$$

It is clear that the integral of a probability density f over the entire space R^n is equal to unity, i.e.,

$$\int_{-\infty}^{\infty} \cdots \int_{-\infty}^{\infty} f(x_1, \ldots, x_n)\, dx_1 \ldots dx_n = 1.$$

A random vector $\xi(\omega)$ is said to possess a *nondegenerate Gaussian distribution* if there exist a vector $m \in R^n$ and a symmetric positive definite matrix D such that the probability density f of the vector ξ takes the form

$$f(x) = [(2\pi)^n (\det D)]^{-1/2} \exp\left[-\tfrac{1}{2}(x-m)' D^{-1}(x-m)\right], \tag{2.1}$$

where $x = (x_1, \ldots, x_n)$.

The *mathematical expectation* (also called the *expectation* or the *mean*) $E\xi$ of a random vector $\xi(\omega)$ is defined by

$$E\xi = \int_\Omega \xi(\omega) P(d\omega). \tag{2.2}$$

Here we assume that the function $|\xi(\omega)|$ is P-integrable (i.e., integrable with respect to the measure P). By evaluating the corresponding integrals (2.2), we derive from equality (2.1) that the expectation $E\xi$ of a Gaussian vector ξ is equal to m, and the covariance matrix is

$$E(\xi - m)(\xi - m)' = D.$$

The *conditional mathematical expectation* (also called the *conditional expectation* or the *conditional mean*) $E(\xi \mid \sigma_0)$ of a random vector ξ with respect to a σ-algebra $\sigma_0 \subset \mathcal{U}$ is by definition a random vector $\eta(\omega) = E(\xi \mid \sigma_0)$ such that: a) it is σ_0-measurable; b) the equality

$$\int_A \eta(\omega) P(d\omega) = \int_A \xi(\omega) P(d\omega) \tag{2.3}$$

holds for any set $A \in \sigma_0$. These two requirements determine the conditional expectation η uniquely up to measure 0. The conditional expectation always exists provided that $E|\xi| < \infty$.

We point out some properties of the conditional expectation to be used below. First of all, by virtue of (2.3) we have

$$E\xi = E[E(\xi \mid \sigma_0)].$$

Moreover, if a random variable ζ is measurable with respect to σ_0, then

$$E(\zeta\xi \mid \sigma_0) = \zeta E(\xi \mid \sigma_0).$$

If a random variable ξ does not depend on σ-algebra σ_0, then

$$E(\xi \mid \sigma_0) = E\xi.$$

Recall that two random variables ξ_1 and ξ_2 are called *independent* if for any x_1 and x_2 the equality

$$P(\xi_1 < x_1, \xi_2 < x_2) = P(\xi_1 < x_1)P(\xi_2 < x_2)$$

holds.

3. Stochastic processes. Let B_1 be the Borel σ-algebra of an interval $T = [0, \tau]$; denote by $B_1 \times \mathcal{U}$ the minimal σ-algebra of subsets of $T \times \Omega$ containing all subsets of the form $\Delta \times A$, where $\Delta \in B$, $A \in \mathcal{U}$.

A function $\zeta(t, \omega)$ of two arguments $t \in T$ and $\omega \in \mathcal{U}$ with values in (R^n, B) is called a *measurable stochastic process* if it is a $B_1 \times \mathcal{U}$ −measurable one and for any fixed $t \in T$ the vector $\zeta(t, \omega)$ is a random variable. The function $\zeta(t, \omega)$ with ω fixed is called a *trajectory* (or *sample path* of the stochastic process).

One usually considers separable stochastic processes. A stochastic process is said to be *separable* if its behavior for all $t \in T$ is determined uniquely up to events of probability zero by its restriction to a dense countable subset $\Delta \subset T$. It is known that for any process $\zeta(t, \omega)$ which is stochastically continuous (that is continuous with respect to convergence in probability) outside some countable subset of T there exists a separable process $\zeta_1(t, \omega)$ such that

$$P\big(\zeta(t, \omega) = \zeta_1(t, \omega)\big) = 1, \qquad \forall t \in T.$$

One can use finite-dimensional distributions to describe a stochastic process $\zeta(t, \omega)$. Denote by $A_i \subset R^n$ arbitrary Borel subsets, and by t_i $(i = 1, \ldots, m)$ arbitrary points of T.

The probabilities

$$F(t_1, \ldots, t_m, A_1, \ldots A_m) = P\big(\zeta(t_1, \omega) \in A_1, \ldots, \zeta(t_m, \omega) \in A_m\big) \tag{2.4}$$

are said to be the *finite-dimensional distributions* of the stochastic process $\zeta(t, \omega)$. According to the Kolmogorov theorem [11], for any compatible set of functions $F(t_1, \ldots, t_m, A_1, \ldots A_m)$ there exists a stochastic process such that these functions are its finite-dimensional distributions.

The expectation of a stochastic process ζ is defined by the formula which is similar to (2.2)

$$E\zeta(t, \omega) = \int_{\Omega} \zeta(t, \omega) P(d\omega).$$

An important characteristic of a stochastic process is the *covariance matrix*, which is equal to

$$E[\zeta(t, \omega) - E\zeta(t, \omega)][\zeta(t, \omega) - E\zeta(t, \omega)]'.$$

A stochastic process is said to be *Gaussian* if all of its finite-dimensional distributions are Gaussian. These distributions depend only on the expectation and the covariance matrix.

A process $\zeta(t,\omega)$ for all $t \in T$ is said to be a *process with independent increments* if the random variables $\zeta(t_1,\omega)-\zeta(0,\omega),\ldots,\zeta(t_m,\omega)-\zeta(t_{m-1},\omega)$ are independent for any $t_1,\ldots,t_m$ such that $0 < t_1 < t_2 < \cdots < t_m \leqslant \tau$. It is worth observing that any process with independent increments and continuous sample paths is Gaussian [5].

A Gaussian process with independent increments is called a *Brownian motion* (or a *Wiener process*).

A Wiener process $\xi(t,\omega)$ can serve as a mathematical model for the description of the Brownian motion, i.e., the motion of microparticles in a liquid (see Figures 2.1 and 2.2).

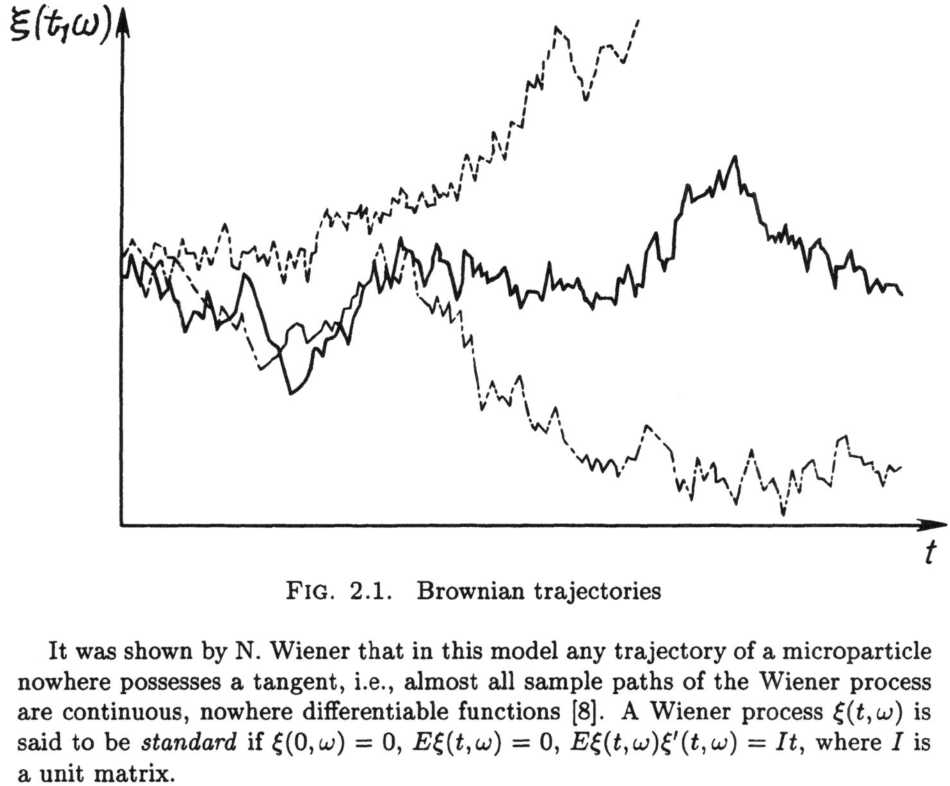

FIG. 2.1. Brownian trajectories

It was shown by N. Wiener that in this model any trajectory of a microparticle nowhere possesses a tangent, i.e., almost all sample paths of the Wiener process are continuous, nowhere differentiable functions [8]. A Wiener process $\xi(t,\omega)$ is said to be *standard* if $\xi(0,\omega)=0$, $E\xi(t,\omega)=0$, $E\xi(t,\omega)\xi'(t,\omega)=It$, where I is a unit matrix.

Further, we present a formula for the density of finite-dimensional distributions of the scalar standard Wiener process $\xi(t,\omega)$. Consider an arbitrary natural integer m and moments $0 < t_1 < t_2 < \cdots < t_{m-1} < t_m$. Then the probability density of the random vector $(\xi(t_1,\omega),\ldots,\xi(t_m,\omega))$ takes the form

$$f(x_1,\ldots,x_m) = [(2\pi)^m t_1(t_2-t_1)\ldots(t_m-t_{m-1})]^{-1/2} \times \exp\left[-\frac{x_1^2}{2t_1} - \frac{1}{2}\sum_{j=1}^{m-1}\frac{(x_{j+1}-x_j)^2}{t_{j+1}-t_j}\right].$$

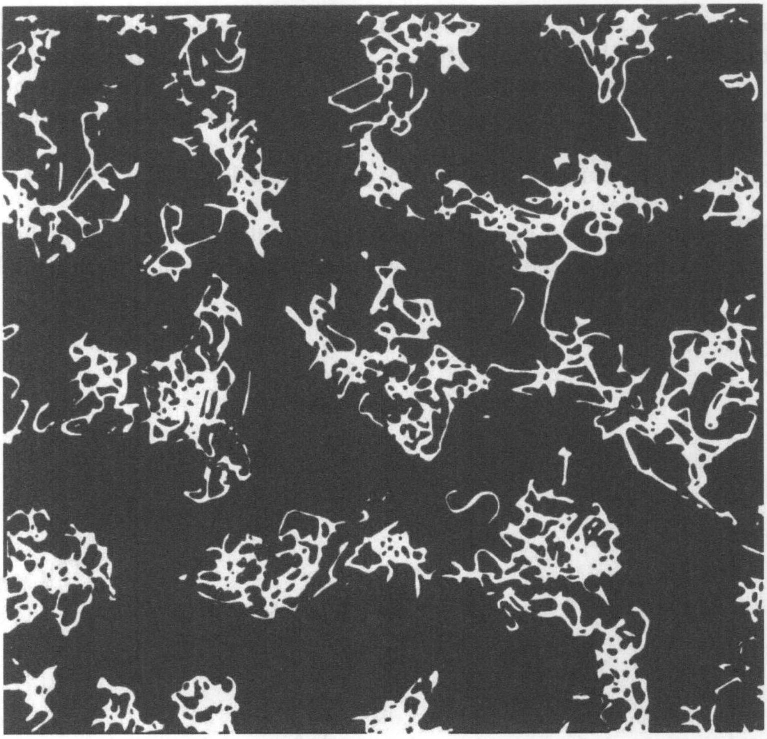

FIG. 2.2. Computer simulated molecular trajectories illustrating two-dimensional Brownian motion

The relation named the *iterated logarithm law*

$$\overline{\lim_{t\to\infty}}\, \xi(t,\omega)(2t\ln\ln t)^{-1/2} = 1, \tag{2.5}$$

where $\overline{\lim}$ stands for the upper limit, holds for the Wiener process $\xi(t,\omega)$ with probability 1.

Moreover, the trajectories of a Wiener process have unbounded variation on any interval $[0,T]$, $T > 0$. In other words, for any positive δ the relation

$$P\left[\sum_{j=0}^{m-1} |\xi(t_{j+1},\omega) - \xi(t_j,\omega)| > \delta\right] \to 1 \tag{2.6}$$

holds as $\max(t_{j+1} - t_j) \to 0$, $0 = t_0 < t_1 < \cdots < t_m = T$. The Wiener process is a fundamental mathematical tool for describing a number of natural phenomena: nonequilibrium processes of statistical thermodynamics, heat fluctuations, measurement noises, disturbing forces, and torques acting upon a moving body, measurement errors, and so on (see [8]).

It is usual in the theory of stochastic processes to omit the variable ω and denote a stochastic process simply by $\zeta(t)$. Henceforth, we follow this agreement.

The detailed proofs of the above assertions can be found in [5, 6, 17].

§ 3. Stochastic Integrals and Stochastic Differential Equations

1. The stochastic integrals of Ito and Stratonovich. There are a lot of different classes of stochastic processes (for instance, stationary ones, processes with independent increments, martingales etc.). Among them an important place is occupied by Markov processes. Heuristically, a *Markov process* is a process whose value at a given time t_0 determines uniquely its future behavior independent of the past. A Wiener process provides an example of a Markov process. A solution of an ordinary deterministic differential equation is also a Markov process. By means of a Wiener process $\xi(t)$ one can construct a wide class of Markov processes with continuous sample paths defined as solutions of *stochastic differential equations* of the form

$$dx(t) = f\big(t, x(t)\big)\,dt + b\big(t, x(t)\big)\,d\xi(t), \qquad t \geqslant 0, \tag{3.1}$$
$$x(0) = x_0,\ x \in R^n,\ \xi \in R^m.$$

Equation (3.1) is a symbolic representation for the following integral identity

$$x(t) = x_0 + \int_0^t f\big(s, x(s)\big)\,ds + \int_0^t b\big(s, x(s)\big)\,d\xi(s). \tag{3.2}$$

The latter integral on the right-hand side of equality (3.2) is called the *stochastic integral.* Different schemes for constructing stochastic integrals are possible. We present two of them due to K. Ito and to R. L. Stratonovich [8, 17, 21]. Consider an arbitrary partition of the interval $[0, t]$ by points t_i such that $0 = t_0 < t_1 < \cdots < t_N = t$, and examine the Riemann sum

$$\sum_{i=1}^{N-1} b\big(t_i, x(t_i)\big)\big(\xi(t_{i+1}) - \xi(t_i)\big) = J. \tag{3.3}$$

Suppose that the sum (3.3) converges in mean square to a limit as the maximal length of intervals $[t_i, t_{i+1}]$ tends to zero. Then this limit is called the *Ito stochastic integral* and is denoted by $\int_0^t b\big(s, x(s)\big)\,d\xi(s)$.

Here convergence in mean square means the following property:

$$E\bigg[\int_0^t b\big(s, x(s)\big)\,d\xi(s) - J\bigg]^2 \to 0$$

as $\max(t_{i+1} - t_i) \to 0$.

We note that, because of (2.6), the Ito stochastic integral can fail to exist in the sense of Stieltjes. The convergence in mean square of sums (3.3) implies their convergence in probability but not almost everywhere. However, it follows from the convergence of J in mean square that for some sequence $\{t_i\}$ of partitions the corresponding J also converges almost everywhere.

Remark. Sometimes the stochastic equation (3.1) is written in the form

$$\dot{x}(t) = f\big(t, x(t)\big) + b\big(t, x(t)\big)\dot{\xi}(t), \qquad t \geqslant 0,$$

where $\dot{\xi}(t)$ is labeled as "white noise". It should be stressed that irrespective of the particular form of the stochastic equation it is always understood in the form of the stochastic integral identity (3.2).

Example 3.1. Let us evaluate the integral $\int_0^T \xi(t)\,d\xi(t)$, where $\xi(t)$ is a standard Wiener process. The corresponding integral sum takes the form

$$\sum_{j=0}^{m-1} \xi(t_j)\big(\xi(t_{j+1}) - \xi(t_j)\big), \qquad 0 = t_0 < t_1 < \cdots < t_m = T.$$

Note that

$$\xi(t_j)\big(\xi(t_{j+1}) - \xi(t_j)\big) = \tfrac{1}{2}\big[\big(\xi^2(t_{j+1}) - \xi^2(t_j)\big) - \big(\xi(t_{j+1}) - \xi(t_j)\big)^2\big].$$

Hence,

$$\sum_{j=0}^{m-1} \xi(t_j)\big(\xi(t_{j+1}) - \xi(t_j)\big) = \frac{1}{2}\sum_{j=0}^{m-1}[\xi^2(t_{j+1}) - \xi^2(t_j)] - \frac{1}{2}\sum_{j=0}^{m-1}\big(\xi(t_{j+1}) - \xi(t_j)\big)^2.$$

Here the first sum on the right-hand side is equal to $\xi^2(T)/2$, while the second one, by virtue of the properties of the Wiener process, tends to $T/2$ as $\max(t_{j+1} - t_j) \to 0$. Thus,

$$\int_0^T \xi(t)\,d\xi(t) = \tfrac{1}{2}[\xi^2(T) - T].$$

To construct the Stratonovich integral, we consider the Riemann sum

$$\sum_{i=1}^{N-1} b\Big(t_i, \tfrac{1}{2}\big(x(t_i) + x(t_{i+1})\big)\Big)\big(\xi(t_{i+1}) - \xi(t_i)\big). \tag{3.4}$$

The limit in mean square of this sum is called the *Stratonovich stochastic integral* and is denoted by

$$\int_0^t b\big(s, x(s)\big)\,\overline{d\xi(s)}.$$

Stochastic equation (3.2) must be regarded in the same way as the stochastic integral in (3.2). It turns out that if a process $\xi(t)$ satisfies the *Stratonovich equation*

$$dx(t) = f\big(t, x(t)\big)\, dt + b\big(t, x(t)\big)\, \overline{d\xi(t)}, \tag{3.5}$$

then it also satisfies the *Ito equation*

$$dx(t) = f\big(t, x(t)\big)\, dt + b\big(t, x(t)\big)\, d\xi(t) + \frac{1}{2}\frac{\partial b\big(t, x(t)\big)}{\partial x} b\big(t, x(t)\big)\, dt. \tag{3.6}$$

Here $(\partial b/\partial x)b$ denotes the vector whose ith component is equal to

$$\sum_{j=1}^{n}\sum_{k=1}^{m}\left(\frac{\partial b_{ik}(t,x)}{\partial x_j}\right) b_{jk}(t,x), \qquad i = 1, \dots, n.$$

Let us present a heuristic explanation of the equivalence of equations (3.5) and (3.6) for the scalar case. We write the difference equation whose solutions converge in mean to the solution of (3.5):

$$\begin{aligned} x(t_{i+1}) - x(t_i) &= f\big(t_i, x(t_i)\big)(t_{i+1} - t_i) \\ &\quad + b\Big(t_i, \tfrac{1}{2}\big(x(t_{i+1}) + x(t_i)\big)\Big)\big(\xi(t_{i+1}) - \xi(t_i)\big). \end{aligned} \tag{3.7}$$

By using the Taylor formula, one can write, up to higher order infinitesimals,

$$\begin{aligned} b\Big(t_i, \tfrac{1}{2}\big(x(t_{i+1}) + x(t_i)\big)\Big) &= b\big(t_i, (x(t_i)\big) \\ &\quad + \frac{1}{2}\frac{\partial b\big(t_i, x(t_i)\big)}{\partial x}\big(x(t_{i+1}) - x(t_i)\big). \end{aligned} \tag{3.8}$$

By inserting (3.8) in (3.7), we obtain

$$\begin{aligned} &\left[I - \frac{1}{2}\frac{\partial b\big(t_i, x(t_i)\big)}{\partial x}\big(\xi(t_{i+1}) - \xi(t_i)\big)\right]\big(x(t_{i+1}) - x(t_i)\big) \\ &\qquad = f\big(t_i, x(t_i)\big)(t_{i+1} - t_i) + b\big(t_i, (x(t_i)\big)\big(\xi(t_{i+1}) - \xi(t_i)\big), \end{aligned}$$

where I stands for a unit matrix. This implies that

$$\begin{aligned} x(t_{i+1}) - x(t_i) &= \left[I - \frac{1}{2}\frac{\partial b\big(t_i, x(t_i)\big)}{\partial x}\big(\xi(t_{i+1}) - \xi(t_i)\big)\right]^{-1} \\ &\quad \times f\big(t_i, x(t_i)\big)(t_{i+1} - t_i) + b\big(t_i, x(t_i)\big)\big(\xi(t_{i+1}) - \xi(t_i)\big) \\ &= f\big(t_i, x(t_i)\big)(t_{i+1} - t_i) + b\big(t_i, x(t_i)\big)\big(\xi(t_{i+1}) - \xi(t_i)\big) \\ &\qquad\qquad + \frac{1}{2}\left(\frac{\partial b\big(t_i, x(t_i)\big)}{\partial x}\right) b\big(t_i, x(t_i)\big). \end{aligned}$$

The latter relation is a difference approximation scheme for the Ito equation (3.6).

In view of the equivalence of equations (3.5) and (3.6) we confine ourselves to the consideration of Ito equations.

Let us present some properties of Ito stochastic integrals. Define a family F_t of σ-algebras on the basic probability space $(\Omega, \mathcal{U}, P)$ of a Wiener process $\xi(t)$ such that

1°. For any t and s, $t < s$, the inclusion $F_t \subset F_s$ is valid.

2°. The Wiener process $\xi(t)$ is F_t-measurable. The latter means that for any Borel set $A \in R^n$ the event $\{\omega : \xi(t) \in A\}$ belongs to F_t.

3°. For any nonnegative t and s the process $\xi(t+s) - \xi(t)$ does not depend on the σ-algebra F_t.

The Ito stochastic integral

$$\zeta(t) = \int_0^t \alpha(s)\, d\xi(s) \tag{3.9}$$

is defined for any stochastic process $\alpha(t)$ obeying the following requirements:

1) the process $\alpha(t)$ is F_t-measurable for any t;

2) the integral

$$J(\alpha) = \int_0^T E(\,|\alpha(t)|^2 \mid F_0\,)\, dt \tag{3.10}$$

is finite with probability one.

When regarded as a function of its upper limit, the stochastic integral (3.9) defines a stochastic process $\zeta(t)$ with zero expectation and covariance matrix:

$$E\zeta(t) = E\left(\int_0^t \alpha(s)\, d\xi(s) \mid F_0 \right) = 0$$

$$E\zeta(t_1)\zeta'(t_2) = E \int_0^{\min(t_1,t_2)} E(\,\alpha(s)\alpha'(s) \mid F_0\,)\, ds.$$

The values of the stochastic integral (3.9) for all possible upper limits t can be adjusted in such a way that the process $\zeta(t)$ is continuous (and, in particular, separable) one.

2. The Ito formula. Consider a process $\gamma(t)$ such that it is F_t-measurable for any t and

$$\gamma(t_2) - \gamma(t_1) = \int_{t_1}^{t_2} f(t)\, dt + \int_{t_1}^{t_2} \sigma(t)\, d\xi(t), \qquad 0 \leqslant t_1 \leqslant t_2 \leqslant T.$$

Here the vector $f(t)$ and the matrix $\sigma(t)$ with random elements are F_t-measurable for any t and $J(t) < \infty$, $J(\sigma) < \infty$, where the functional J is defined by formula (3.10). The sum $f(t)\, dt + \sigma(t)\, d\xi(t)$ is said to be the *stochastic differential* $d\gamma(t)$ of the process $\gamma(t)$ and is written as

$$d\gamma(t) = f(t)\, dt + \sigma(t)\, d\xi(t). \tag{3.11}$$

Suppose that for a scalar function $V(t,x)$ there exist partial derivatives $V_t = \partial V/\partial t$, $V_x = \partial V/\partial x$, and $V_{xx} = \partial^2 V/\partial x^2$. Then, if the process $\gamma(t)$ possesses the stochastic differential (3.11), the process $\eta(t) = V(t,\gamma(t))$ also possesses a stochastic differential $d\eta(t)$. This differential is described by the *Ito formula*:

$$\begin{aligned} d\eta(t) = \big[&V_t(t,\gamma(t)) + V_x'(t,\gamma(t))f(t) \\ &+ \tfrac{1}{2}\operatorname{tr}\sigma(t)\sigma'(t)V_{xx}(t,\gamma(t))\big]\,dt + V_x'(t,\gamma(t))\sigma(t)\,d\xi(t). \end{aligned} \tag{3.12}$$

Here tr is the trace of a matrix; $V_x \in R^n$ is the vector with components $\partial V/\partial x_i$; V_{xx} is the square matrix with elements $\partial^2 V/\partial x_i \partial x_j$ $(i,j = 1,\dots,n)$. The Ito formula is an analog of the chain rule for the differential of a composite function.

We present a heuristic derivation of this formula. Let τ_1 and τ_2 $(\tau_1 < \tau_2)$ be arbitrary moments. We divide the interval $[\tau_1,\tau_2]$ into subintervals by the points t_i: $\tau_1 = t_0 < t_1 < \cdots < t_m = \tau_2$. It is clear that

$$\eta(\tau_2) - \eta(\tau_1) = \sum_{j=0}^{m-1}\big[V(t_{j+1},\gamma(t_{j+1})) - V(t_j,\gamma(t_j))\big].$$

By using the Taylor formula, we rearrange the right-hand side as follows:

$$\begin{aligned} &V(t_{j+1},\gamma(t_{j+1})) - V(t_j,\gamma(t_j)) \\ &\quad = V_t\big(t_j + \theta_j(t_{j+1}-t_j),\gamma(t_j)\big)(t_{j+1}-t_j) \\ &\qquad + V_x'(t_j,\gamma(t_j))\big(\gamma(t_{j+1}) - \gamma(t_j)\big) \\ &\qquad + \tfrac{1}{2}\operatorname{tr} V_{xx}\Big(t_j,\gamma(t_j) + \lambda_j\big(\gamma(t_{j+1}) - \gamma(t_j)\big)\Big) \\ &\qquad \times \big(\gamma(t_{j+1}) - \gamma(t_j)\big)\big(\gamma(t_{j+1}) - \gamma(t_j)\big). \end{aligned}$$

Here θ_j and λ_j are some numbers from the interval $[0,1]$. According to identity (3.11) one can represent the increment of the process $\gamma(t)$ in the form

$$\gamma(t_{j+1}) - \gamma(t_j) = f(t_j)(t_{j+1}-t_j) + \sigma(t_j)\big(\xi(t_{j+1}) - \xi(t_j)\big)$$

up to some random variables which tend to zero with probability 1 as $\max(t_{i+1} - t_i) \to 0$.

It follows from the latter two formulas that with the same accuracy

$$\begin{aligned} &V(t_{j+1},\gamma(t_{j+1})) - V(t_j,\gamma(t_j)) \\ &\qquad = \big[V_t(t_j,\gamma(t_j)) + V_x'(t_j,\gamma(t_j))f(t_j)\big](t_{j+1}-t_j) \\ &\qquad\quad + V_x'(t_j,\gamma(t_j))\sigma(t_j)\big(\xi(t_{j+1}) - \xi(t_j)\big) \\ &\qquad\qquad + \tfrac{1}{2}\operatorname{tr} V_{xx}(t_j,\gamma(t_j))\sigma(t_j)\sigma'(t_j)(t_{j+1}-t_j). \end{aligned}$$

In the derivation of the latter equality use was made of the fact that the random variables

$$\sum_{j=0}^{m-1}\operatorname{tr} V_{xx}(t_j,\gamma(t_j))f(t_j)f'(t_j)(t_{j+1}-t_j)^2,$$

$$\sum_{j=0}^{m-1} \operatorname{tr} V_{xx}\big(t_j, \gamma(t_j)\big) f(t_j)(t_{j+1} - t_j)\big(\xi(t_{j+1}) - \xi(t_j)\big)'\sigma'(t_j),$$

$$\sum_{j=0}^{m-1} \operatorname{tr} V_{xx}\big(t_j, \gamma(t_j)\big)\sigma(t_j)\big[\big(\xi(t_{j+1}) - \xi(t_j)\big)\big(\xi(t_{j+1}) - \xi(t_j)\big)' - I(t_{j+1} - t_j)\big]\sigma'(t_j)$$

converge in probability to zero as $\max(t_{i+1} - t_i) \to 0$.

Thus, finally, we have

$$\eta(t_2) - \eta(t_1) = \sum_{j=0}^{m-1} \Big[\Big(V_t\big(t_j, \gamma(t_j)\big) + V_x'\big(t_j, \gamma(t_j)\big) f(t_j) + \tfrac{1}{2} \operatorname{tr} V_{xx}\big(t_j, \gamma(t_j)\big)\sigma(t_j)\sigma'(t_j)\Big)(t_{j+1} - t_j) + V_x'\big(t_j, \gamma(t_j)\big)\sigma(t_j)\big(\xi(t_{j+1}) - \xi(t_j)\big)\Big].$$

By observing that the limits of all sums in this expression are equal to the corresponding integrals, we arrive at the Ito formula (3.12).

Now we consider some examples of the computation of stochastic integrals.

Example 3.2. Let $\xi(t)$ be a standard scalar Wiener process. We evaluate the Ito integral

$$\int_0^T \xi(t)\, d\xi(t). \tag{3.13}$$

The integral was already found in Example 3.1. Here we present another way of its computation. Denote by $x(t)$ a stochastic process satisfying the equation

$$dx(t) = d\xi(t), \qquad x(0) = 0. \tag{3.14}$$

It is clear that $x(t) = \xi(t)$. Now, we apply the Ito formula to the function $V\big(t, x(t)\big) = x^2(t)$. Taking relation (3.12) into account, we have

$$dx^2(t) = 2x(t)\, d\xi(t) + dt = 2\xi(t)\, d\xi(t) + dt.$$

Integrating both sides of this equality gives

$$\int_0^T dx^2(t) = x^2(T) - x^2(0) = \xi^2(T) = 2\int_0^T \xi(t)\, d\xi(t) + T.$$

Thus, we finally find that

$$\int_0^T \xi(t)\, d\xi(t) = \tfrac{1}{2}[\xi^2(T) - T]. \tag{3.15}$$

Example 3.3. Let us find a similar Stratonovich integral : $\int_0^T \xi(t)\,\overline{d\xi(t)}$. One can evaluate this integral by the same method as above. However, it is possible to deal with it in another way by using its very definition. By virtue of (3.4) this integral is the limit in mean square of Riemann sums

$$\frac{1}{2}\sum_{j=0}^{N-1}\big(\xi(t_{j+1})+\xi(t_j)\big)\big(\xi(t_{j+1})-\xi(t_j)\big)$$

$$=\sum_{j=0}^{N-1}\xi(t_j)\big(\xi(t_{j+1})-\xi(t_j)\big)+\frac{1}{2}\sum_{j=0}^{N-1}\big(\xi(t_{j+1})-\xi(t_j)\big)^2. \tag{3.16}$$

The first term on the right-hand side is a Riemann sum corresponding to the Ito integral (3.13). Hence, in view of (3.15) it converges to $(\xi^2(T)-T)/2$. In accordance with a basic property of the Wiener process, the second sum converges in probability to $T/2$, since

$$\sum_{j=0}^{N-1}\Big[\big(\xi(t_{j+1})-\xi(t_j)\big)^2-(t_{j+1}-t_j)\Big]\to 0.$$

Thus,

$$\int_0^T \xi(t)\,\overline{d\xi(t)}=\tfrac{1}{2}\xi^2(T).$$

This example shows that, by its formal properties, the Stratonovich integral resembles the usual definite integral to a greater extent than its Ito counterpart.

Example 3.4. Consider the system of two stochastic Ito equations

$$\begin{aligned} dx(t) &= -\tfrac{1}{2}x(t)\,dt+y(t)\,d\xi(t),\\ dy(t) &= -\tfrac{1}{2}y(t)\,dt-x(t)\,d\xi(t), \end{aligned} \tag{3.17}$$

where $\xi(t)$ is a standard scalar Wiener process. Let us show that processes $x(t)=\sin\xi(t)$, $y(t)=\cos\xi(t)$ provide a solution of the system (3.17). Using the Ito formula we have

$$dx(t)=\big(\cos\xi(t)\big)\,d\xi(t)-\tfrac{1}{2}\big(\sin\xi(t)\big)\,dt=y(t)\,d\xi(t)-\tfrac{1}{2}x(t)\,dt,$$
$$dy(t)=-\big(\sin\xi(t)\big)\,d\xi(t)-\tfrac{1}{2}\big(\cos\xi(t)\big)\,dt=-x(t)\,d\xi(t)-\tfrac{1}{2}y(t)\,dt.$$

Example 3.5. Let $F(x)$ be a scalar twice continuously differentiable function. Let us evaluate the Ito stochastic integral:

$$\int_0^T f\big(\xi(t)\big)\,d\xi(t),\qquad f(x)=\frac{\partial F(x)}{\partial x}.$$

We apply the Ito formula to the function $F\big(x(t)\big)$, where the process $x(t)$ is described by relations (3.14). We have

$$dF\big(x(t)\big)=dF\big(\xi(t)\big)=f\big(\xi(t)\big)\,d\xi(t)+\tfrac{1}{2}f_x\big(x(t)\big)\,dt.$$

Hence, we obtain by integration

$$\int_0^T f\big(\xi(t)\big)\,d\xi(t)=F\big(\xi(t)\big)-F(0)-\frac{1}{2}\int_0^T f_x\big(\xi(t)\big)\,dt.$$

3. Markov diffusion processes. A solution of the stochastic differential equation (3.1) is defined as a stochastic process which is F_t-measurable for any t and satisfies the integral identity (3.2) for any t with probability 1. We present here some existence conditions for solutions of equation (3.1).

Theorem 3.1 [6]. *Let functions $f(t,x)$ and $b(t,x)$ be measurable with respect to the arguments $t \geqslant 0$, $x \in R^n$ and satisfy the conditions*

$$|f(t,x)|^2 + |b(t,x)|^2 \leqslant C(1+|x|^2), \qquad C \geqslant 0, \tag{3.18}$$

$$|f(t,x) - f(t,y)| + |b(t,x) - b(t,y)| \leqslant C|x-y| \tag{3.19}$$

for any $x, y \in R^n$, $|x| = (x'x)^{1/2}$. Then there exists a unique solution of equation (3.1) *on any interval* $[0,T]$. *Here the uniqueness is understood as follows: if $x_1(t)$ and $x_2(t)$ are two continuous solutions of equation* (3.1), *then for any $T > 0$ the following relation holds:*

$$P(\sup_t |x_1(t) - x_2(t)| > 0) = 0, \qquad 0 \leqslant t \leqslant T.$$

A solution of (3.1) is a Markov process, and its transition probability is defined by

$$P(t,x,t_1,A) = P\big(X_{t,x}(t_1) \in A\big). \tag{3.20}$$

The process $X_{t,x}(s)$ involved in equality (3.20) is the solution of equation (3.1) for $s > t$ with initial condition $X_{t,x}(t) = x$. If x is a random vector then, under the assumptions (3.18), the inequality

$$E|X_{t,x}(t_1)|^4 \leqslant C_1(1 + E|x|^4) \tag{3.21}$$

holds. Here the constant C_1 depends on t, t_1, and on the constant C in (3.18).

In many problems it is necessary to evaluate some functionals of solutions to equation (3.1). In a number of cases one can reduce this problem to the solution of a boundary-value problem for a partial differential equation. Suppose, for example, that we need to compute the expectation

$$EF\big(X_{t,x}(s)\big), \qquad s \geqslant t. \tag{3.22}$$

The function $F(x)$ is a given bounded continuous scalar function. The value of functional (3.22) depends on the initial time t and on the initial vector x. Set

$$u(t,x) = EF\big(X_{t,x}(s)\big), \qquad s \geqslant t, \tag{3.23}$$

where s is fixed. Suppose that the coefficients $f(t,x)$ and $b(t,x)$ in equation (3.1) and the function F in (3.22) are defined for $0 \leqslant t \leqslant T$, $x \in R^n$ and in this domain have bounded continuous derivatives with respect to x up to order two. Then the function $u(t,x)$ in (3.23) assumes continuous derivatives with respect to x up to order two, is differentiable with respect to t, and satisfies the equation

$$Lu(t,x) = 0, \qquad t \leqslant s,\ x \in R^n. \tag{3.24}$$

The differential operator L, called the *generator of the Markov process* (3.1) is given by the formula

$$L = \frac{\partial}{\partial t} + f' \frac{\partial}{\partial x} + \frac{1}{2} \operatorname{tr} \sigma\sigma' \frac{\partial^2}{\partial x^2}. \tag{3.25}$$

The initial condition for the function $u(t, x)$

$$\lim_{t \to s} u(t, x) = F(x), \qquad x \in R^n \tag{3.26}$$

follows from (3.23).

By solving the Cauchy problem (3.24)–(3.26), we obtain the value of the functional (3.22) for any (t, x). Equation (3.24) is called the *backward Kolmogorov equation.*

We now explain the probabilistic meaning of the operator (3.25). Consider a bounded domain $Q \in R^n$. Suppose that at time t_0 a solution is located in Q, i.e., $P\{x(t_0) \in Q\} = 1$. Denote by τ_Q the first exit time for the domain Q of the process $X_{t_0,x(t_0)}(t)$ $(t \geqslant t_0)$. Set $\tau_Q(t) = \min(\tau_Q, t)$. Then, for any function $V(t, x)$ which is continuously differentiable with respect to t and twice continuously differentiable with respect to x the following formula holds:

$$\begin{aligned} &E\Big[V\Big(\tau_Q(t), X_{t_0,x(t_0)}\big(\tau_Q(t)\big)\Big) - V\big(t_0, x(t_0)\big)\Big] \\ &\qquad = E \int_{t_0}^{\tau_Q(t)} LV\big(s, X_{t_0,x(t_0)}(s)\big)\, ds. \end{aligned} \tag{3.27}$$

Example 3.6. Consider a scalar process $x(t)$ obeying the equation

$$dx(t) = ax(t)\,dt + \sigma x(t)\,d\xi(t), \qquad x(0) = x_0,\ T \geqslant t \geqslant 0.$$

Here a, x_0, σ are given constants and $\xi(t)$ is a standard scalar Wiener process. Let us find the second moment $Ex^2(T)$. In the example under consideration it can be found in two ways: by a direct application of the Ito formula or by use of the backward Kolmogorov equation. By applying the Ito formula to the function $x^2(t)$, we obtain

$$dx^2(t) = 2x(t)\big(ax(t)\,dt + \sigma x(t)\,d\xi(t)\big) + \sigma^2 x^2(t)\,dt.$$

Integrating both sides of this relation from 0 to t and evaluating expectations of both sides of the obtained equality leads to the identity

$$Ex^2(t) - x_0^2 = (2a + \sigma^2) \int_0^t Ex^2(s)\,ds.$$

Hence, the function $Ex^2(t)$ is a solution of the differential equation

$$\frac{d}{dt} Ex^2(t) = (2a + \sigma^2) Ex^2(t), \qquad Ex_0^2(0) = x_0^2.$$

Solving this equation, we infer that

$$Ex^2(T) = x_0^2 \exp[(2a + \sigma^2)T].$$

Now we present another method of determining $Ex^2(T)$. Set

$$V(t, x) = E[x^2(T) \mid x(t) = x], \qquad 0 \leqslant t \leqslant T.$$

We write the backward Kolmogorov equation for the function $V(t, x)$:

$$LV = V_t + axV_x + \frac{1}{2}\sigma^2 x^2 V_{xx} = 0, \tag{3.28}$$
$$V(T, x) = x^2.$$

We look for solutions of the Cauchy problem (3.28) of the form

$$V(t, x) = P(t)x^2, \tag{3.29}$$

where the function $P(t)$ is be found. By inserting expression (3.29) into (3.28), we obtain the equation

$$\dot{P}(t) + (2a + \sigma^2)P(t) = 0, \qquad P(T) = 1,$$

fom which $P(t)$ can be determined. Hence,

$$P(t) = \exp[(2a + \sigma^2)(T - t)].$$

Therefore,

$$Ex^2(T) = V(0, x_0) = P(0)x_0^2 = x_0^2 \exp[(2a + \sigma^2)T].$$

4. Linear stochastic equations. As in Example 3.1., one can obtain expressions for moments of solutions of a linear system of the form

$$dx(t) = A(t)x(t)\, dt + \sigma\, d\xi(t) + f(t)\, dt, \qquad t \geqslant 0, \tag{3.30}$$
$$x(0) = x_0, \qquad x \in R^n,\ \xi \in R^n,\ f \in R^n.$$

Here A and σ are given matrices of dimensions $n \times n$ and $n \times m$ with continuous elements; f is a given continuous function. The random variable x_0 has the following characteristics:

$$Ex_0 = m_0, \qquad E(x_0 - m_0)(x_0 - m_0)' = D_0.$$

Denote by $m(t)$ the expectation (the mean) and by $D(t)$ the covariance matrix of the process $x(t)$:

$$m(t) = Ex(t), \qquad D(t) = E\big[\big(x(t) - m(t)\big)\big(x(t) - m(t)\big)'\big].$$

Using the Ito formula, we infer that the following equations are valid:

$$\dot{m}(t) = A(t)m(t) + f(t), \qquad m(0) = m_0, \tag{3.31}$$
$$\dot{D}(t) = A(t)D(t) + D(t)A'(t) + \sigma(t)\sigma'(t), \qquad D(0) = D_0 \tag{3.32}$$

Denote by $Z(t, s)$ the Cauchy matrix of the linear equation $\dot{x}(t) = A(t)x(t)$. Then, by virtue of (3.31), (3.32) the expectation of process (3.30) and the covariance matrix can be represented as

$$\begin{aligned} m(t) &= Z(t,0)m_0 + \int_0^t Z(t,s)f(s)\,ds, \\ D(t) &= Z(t,0)D_0Z'(t,0) + \int_0^t Z(t,s)\sigma(s)\sigma'(s)Z'(t,s)\,ds. \end{aligned} \tag{3.33}$$

Similarly, one can express moments of solutions to linear equations in which the diffusion term depends on the phase coordinates:

$$dx(t) = A(t)x(t)\,dt + \sigma x(t)\,d\xi(t) + f(t)\,dt, \qquad t \geqslant 0,$$
$$x(0) = x_0, \ \xi \in R^1,$$

where A and σ are $n \times n$ matrices with continuous elements. In this case the equation for $m(t)$ can be written in the form (3.31), while the equation for $D(t)$ in the form

$$\dot{D}(t) = A(t)D(t) + D(t)A'(t) + \sigma(t)D(t)\sigma'(t), \qquad D(0) = D_0.$$

Example 3.7. Consider the following second-order system (see [5] of Part Two):

$$\begin{aligned} \dot{x}_1 &= x_2, \\ \dot{x}_2 &= -\omega^2 x_1 - 2\alpha\omega x_2 + \omega^2\sqrt{q}\dot{\xi}(t), \end{aligned} \tag{3.34}$$

where $\alpha > 0$, $q > 0$, and $\omega > 0$ are given constants. The vector $(x_1(0), x_2(0))$ has zero expectation and covariance matrix $D(0)$ with elements $d_{ij}(0)$, $i, j = 1, 2$. The matrix $D(0)$ can be represented in the form (3.33), where the elements $z_{ij}(t,s) = z_{ij}(t-s)$ and $\sigma(s)$ are as follows:

$$\begin{aligned} z_{11}(t) &= e^{-\alpha\omega t}(\cos\beta t + \frac{\alpha\omega}{\beta}\sin\beta t), \\ z_{12}(t) &= \tfrac{1}{\beta}\sin\beta t \\ z_{21}(t) &= -\omega^2\beta^{-1}\sin\beta t, \\ z_{22}(t) &= \cos\beta t - \alpha\omega\beta^{-1}\sin\beta t, \qquad \beta = \omega(1-\alpha^2)^{1/2}, \\ \sigma(t) &= \begin{pmatrix} 0 \\ \omega^2\sqrt{q} \end{pmatrix}. \end{aligned}$$

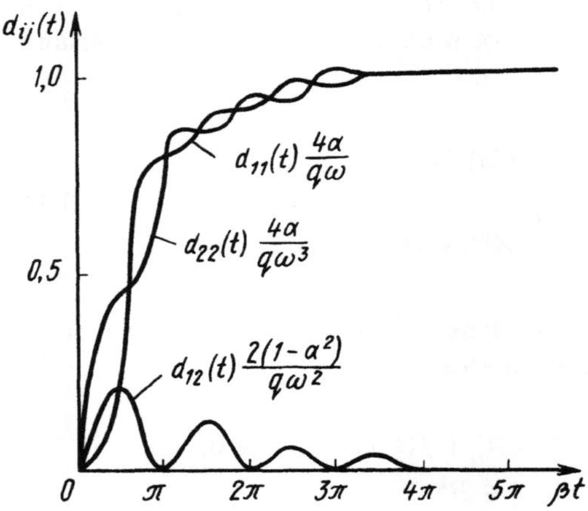

FIG. 3.1. Graphs of covariance matrix elements of the second-order Gaussian Markov process with small damping coefficient $\alpha = 0.2$

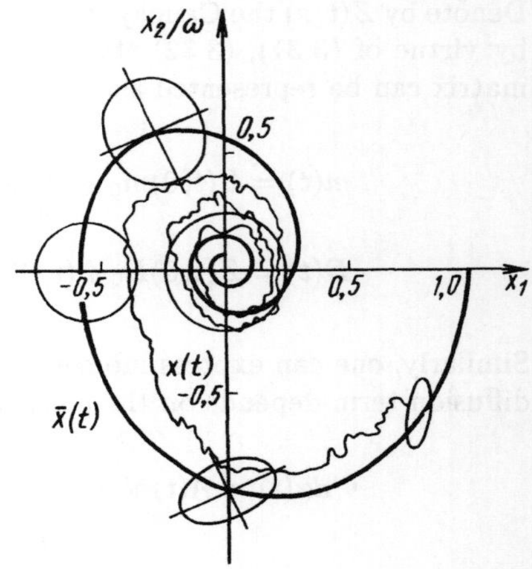

FIG. 3.2. Phase trajectory $x(t)$ and expectation $\bar{x}(t)$ of a second-order Gaussian Markov process with small damping

It follows from the representation of the elements $d_{ij}(t)$ of the matrix $D(t)$ that

$$\lim_{t\to\infty} d_{11}(t) = q\omega(4\alpha)^{-1}, \qquad \lim_{t\to\infty} d_{12}(t) = 0, \qquad \lim_{t\to\infty} d_{22}(t) = q\omega^3(4\alpha)^{-1}.$$

The (scaled) graphs of the functions $d_{11}(t)$, $d_{12}(t)$, $d_{22}(t)$ corresponding to $\alpha = 0.2$ and $D(0) = 0$ are shown in Figure 3.1. In Figure 3.2 the function $(Ex_1(t), Ex_2(t))$ for trajectories of the system, starting at point $x_1(0) = 1$, $x_2(0) = 0$, is displayed by the bold curve, while a sample path by the broken line. In this figure for some instants t ellipses are drawn such that the probability for locating the solution of the system (3.34) at time t inside this ellipse is 0.865.

§ 4. Definition of Stochastic Stability

It is common in stability theory to reduce, by a change of variables, the problem of stability for a solution of equation (3.1) to the investigation of the stability problem for the trivial solution. Therefore, one can assume that

$$f(t,0) \equiv 0, \qquad \sigma(t,0) \equiv 0, \qquad t \geqslant 0. \tag{4.1}$$

Under the assumptions (4.1) equation (3.1) admits the trivial solution $x(t) \equiv 0$. The stability of the trivial solution of equation (3.1) is understood as the property to change slightly under a slight variation of the initial conditions. Different definitions of stability are possible depending on the particular concept of "slight variation of the solution". Some of them are presented below.

Definition 4.1. The trivial solution of equation (3.3) is called:

weakly stable in probability if for all $\varepsilon > 0$ and $\delta > 0$ there exists an $\eta(\varepsilon, \delta) > 0$ such that for $t \geqslant t_0$ and $|x_0| < \eta(\varepsilon, \delta)$ the inequality

$$P\{|X_{t_0,x_0}(t)| > \varepsilon\} < \delta \tag{4.2}$$

holds;

asymptotically weakly stable in probability if it is weakly stable in probability and, further, for all $\varepsilon > 0$ there exists an $\eta(\varepsilon) > 0$ such that for $|x_0| < \eta(\varepsilon)$ the relation

$$P\{|X_{t_0,x_0}(t)| > \varepsilon\} \to 0, \qquad t \to \infty$$

is valid;

globally asymptotically weakly stable in probability if it is asymptotically weakly stable in probability and, further, for all $x_0 > 0$ and $\varepsilon_1 > 0$ the relation

$$\lim_{t\to\infty} P\{|X_{t_0,x_0}(t)| > \varepsilon_1\} = 0$$

holds true;

uniformly globally asymptotically weakly stable in probability with respect to the initial conditions if in the previous definition the convergence to zero is uniform for all $t_0 \geqslant 0$ and $|x_0| < \eta$.

Definition 4.2. The trivial solution of equation (3.1) is called :

p-stable ($p > 0$) if for all $\varepsilon > 0$ there exists an $\eta(\varepsilon) > 0$, such that for $|x_0| < \eta(\varepsilon)$ the inequality

$$E|X_{t_0,x_0}(t)|^p \leqslant \varepsilon, \qquad \forall t \geqslant t_0$$

holds;

asymptotically p-stable if it is p-stable and, moreover, for $|x_0| < \eta$ with some fixed η the identity

$$\lim_{t\to\infty} E|X_{t_0,x_0}(t)|^p = 0$$

holds;

exponentially p-stable if there exist positive constants C_0 and C_1 such that

$$E|X_{t_0,x_0}(t)|^p \leqslant C_0|x_0|^p \exp\bigl(-C_1(t-t_0)\bigr);$$

globally asymptotically p-stable if it is asymptotically p-stable and for all x_0 the identity

$$\lim_{t\to\infty} E|X_{t_0,x_0}(t)|^p = 0$$

holds true.

In addition to the above definitions one also uses the concept of *stability with probability one.* This is understood as a kind of stability such that all trajectories of the system (possibly, except for some zero probability subset of the set of trajectories) are stable in the corresponding sense.

Suppose that the zero solution of equation (3.1) is p-stable ($p > 0$). Then, for any $p_1 \in (0, p)$ we have, by virtue of the Hölder inequality, that

$$E|X_{t_0,x_0}(t)|^{p_1} \leqslant [E|X_{t_0,x_0}(t)|^p]^{p_1/p}.$$

Hence, in this case the trivial solution is also p_1-stable for any $p_1 \in (0, p)$. Similarly, the asymptotic p-stability implies the asymptotic p_1-stability for any $p_1 \in (0, p)$.

Now we invoke the *Chebyshev inequality:*

$$P\{|X_{t_0,x_0}(t)| > \varepsilon\} \leqslant EV(t, X_{t_0,x_0}(t)) \left[\inf_{\substack{t \geqslant t_0, \\ |x_0| > \varepsilon}} V(t, x) \right]^{-1}.$$

Here $V(t,x)$ is an arbitrary nonnegative function such that there exists $EV(t, X_{t_0,x_0}($ By applying the Chebyshev inequality to $V(t,x) = |x|^p$, we have for any $p > 0$

$$P\{|X_{t_0,x_0}(t)| > \varepsilon\} \leqslant \varepsilon^{-p} E|X_{t_0,x_0}(t)|^p.$$

Therefore, the p-stability of the trivial solution of equation (3.1) implies its weak stability in probability, while the asymptotic p-stability implies its asymptotic weak stability in probability. Note also that p-stability for $p = 2$ is also called *stability in mean square.*

Definition 4.3. The trivial solution of equation (3.1) is called:
stable in probability if for all $s \geqslant t_0$ and $\varepsilon > 0$ the relation

$$\lim_{x_0 \to 0} P\{\sup_{t \geqslant s} |X_{s,x_0}(t)| > \varepsilon\} = 0 \tag{4.3}$$

holds;

asymptotically stable in probability if it is stable in probability and

$$\lim_{x_0 \to 0} P\{\lim_{t \to \infty} X_{t_0,x_0}(t) = 0\} = 1;$$

uniformly stable in probability if $P\{|X_{t_0,x_0}(t)| > \varepsilon\}$ tends to zero uniformly in x_0 and $s \geqslant 0$ as $x_0 \to 0$;

globally asymptotically stable in probability if it is stable in probability and for all t_0 and x_0 the relationship

$$P\{\lim_{t \to \infty} X_{t_0,x_0}(t) = 0\} = 1$$

holds true.

Stability in probability is much stronger than weak stability and signifies that all trajectories of the process starting at time t_0 from the point x_0 remain inside any preassigned neighborhood of the trivial solution with probability that tends to 1 as $x_0 \to 0$. It seems that in stochastic stability problems the most interesting questions are related to the study of stability in probability and of p-stability.

§ 5. Application of Lyapunov's Direct Method

1. Sufficient stability conditions. We state some sufficient conditions for stability by using the concept of Lyapunov functions. Theorems 5.1–5.3 to be stated below are due to R. Z. Hasminskii [7].

Recall that a function $V(t,x)$ is said to be positive definite if $V(t,0) \equiv 0$ and $V(t,x) \geqslant \omega_1(|x|)$. As in Chapter I, we denote by $\omega_i(x)$ a scalar continuous nondecreasing function such that $\omega_i(u) > 0$ for $u > 0$ and $\omega_i(0) = 0$. We say that the function $V(t,x) \in C_2^1(U)$, where $U = \{t \geqslant t_0\} \times \{|x| \leqslant r\}$, if this function is twice continuously differentiable with respect to x and once with respect to t everywhere in the domain U except, possibly, the set $\{x = 0\}$ and is continuous in the closed set $U_\varepsilon = \{t \geqslant t_0\} \times \{\varepsilon \leqslant |x| \leqslant r\}$ for any $\varepsilon > 0$.

Theorem 5.1. *Suppose that in the domain U there exists a continuous function $V(t,x) \in C_2^1(U)$ such that for any $x \neq 0$ the inequalities*

$$\omega_1(|x|) \leqslant V(t,x), \qquad LV \leqslant 0 \tag{5.1}$$

hold, where the operator L is defined by formula (3.25). *Then the trivial solution of equation* (3.1) *is stable in probability.*

Proof. Consider an arbitrary $\varepsilon \in (0,r)$, an arbitrary moment s, and an arbitrary initial point x_0 such that $|x_0| < \varepsilon$. We verify that under the conditions of Theorem 5.1 the process $X_{s,x_0}(t)$ satisfies the relation (4.3). Denote by τ_ε the hitting time of the surface $|x| = \varepsilon$ by the process $X_{s,x_0}(t)$. If some trajectory of the process X does not attain the surface $|x| = \varepsilon$ in any finite amount of time, then we take $\tau_\varepsilon = \infty$ for this trajectory. Put $\tau_\varepsilon(t) = \min(\tau_\varepsilon, t)$. Then, by virtue of (3.27), we have

$$\begin{aligned} E\Big[V\Big(\tau_\varepsilon(t), X_{s,x_0}\big(\tau_\varepsilon(t)\big)\Big) - V(s,x_0)\Big] \\ = E\int_s^{\tau_\varepsilon(t)} LV\big(s_1, X_{s,x_0}(s_1)\big)\, ds_1. \end{aligned}$$

This identity, together with the assumption (5.1), implies that

$$EV\Big(\tau_\varepsilon(t), X_{s,x_0}\big(\tau_\varepsilon(t)\big)\Big) \leqslant V(s,x_0). \tag{5.2}$$

Set

$$V_\varepsilon = \inf_{\substack{t \geqslant t_0 \\ \varepsilon \leqslant |x| \leqslant r}} V(t,x).$$

Then, by virtue of the Chebyshev inequality, the estimate (5.2), and the definition of V_ε we find that

$$\begin{aligned} P\Big\{\sup_{s \leqslant s_1 \leqslant t} |X_{s,x_0}(s_1)| > \varepsilon\Big\} &\leqslant V_\varepsilon^{-1} V\Big(\tau_\varepsilon(t), X_{s,x_0}\big(\tau_\varepsilon(t)\big)\Big) \\ &\leqslant V_\varepsilon^{-1} V(s,x_0). \end{aligned}$$

If we put $t \to \infty$ in the latter relation, then we obtain

$$P\Big\{\sup_{t \geqslant s} |X_{s,x_0}(t)| > \varepsilon\Big\} \leqslant V_\varepsilon^{-1} V(s, x_0).$$

But, since the function $V(s, x_0)$ is continuous and satisfies the first inequality (5.1), we conclude that the relation (4.3) must hold. This completes the proof of the theorem. □

Similarly to Theorem 5.1., one can use Lyapunov functions to state conditions ensuring stability in another sense. We present some of them.

Theorem 5.2. *Suppose there exists a Lyapunov function $V(t, x)$ satisfying the conditions of Theorem 5.1 in the domain $(t \geqslant t_0) \times R^n$ such that*

$$\omega_1(|x|) \leqslant V(t, x) \leqslant \omega_2(|x|), \qquad \omega_1(u) \to \infty,\ u \to \infty,$$
$$LV \leqslant -\omega_3(|x|).$$

Then, the trivial solution of equation (3.1) is globally asymptotically stable in probability.

Theorem 5.3. *Suppose there exists a function $V(t, x) \in C_2^0\ ((t \geqslant t_0) \times R^n)$ satisfying the inequalities*

$$k_1|x|^p \leqslant V(t, x) \leqslant k_2|x|^p, \tag{5.3}$$
$$LV(t, x) \leqslant -k_3|x|^p, \qquad k_i > 0,\ p > 0. \tag{5.4}$$

Then the trivial solution of equation (3.1) is exponentially p-stable for $t \geqslant 0$.

Proof. The existence of $EV\big(t, X_{s,x}(t)\big)$ for all $t \geqslant s$, as well as the equality,

$$EV\big(t, X_{s,x}(t)\big) - V(s, x) = \int_s^t ELV\big(\tau, X_{s,x}(\tau)\big)\, d\tau.$$

follows from the conditions of the theorem and from (3.22). By differentiating both sides of this equality with respect to t and taking (5.3) and (5.4) into account, we obtain

$$\frac{d}{dt} EV\big(t, X_{s,x}(t)\big) = ELV\big(t, X_{s,x}(t)\big) \leqslant -k_3 E|X_{s,x}(t)|^p$$
$$\leqslant -k_3 k_2^{-1} EV\big(t, X_{s,x}(t)\big).$$

Hence,

$$EV\big(t, X_{s,x}(t)\big) \leqslant V(s, x) \exp\{-k_3 k_2^{-1}(t - s)\}.$$

This inequality and (5.3) imply the estimate

$$k_1 E|X_{s,x}(t)|^p \leqslant k_2 |x|^p \exp\{-k_3 k_2^{-1}(t - s)\},$$

which means that the trivial solution of equation (3.1) is exponentially p-stable. □

Example 5.1. Consider a scalar equation

$$dx(t) = bx(t) + \sigma x(t)\, d\xi(t), \qquad t \geqslant 0,\ x(0) = x_0. \tag{5.5}$$

Take the function $V(x) = |x|^p$, $p > 0$, for the Lyapunov function. Then,

$$L = bx\frac{\partial}{\partial x} + \frac{1}{2}\sigma^2 x^2 \frac{\partial^2}{\partial x^2},$$
$$LV(x) = p|x|^p\left[b + \frac{\sigma^2}{2}(p-1)\right].$$

Hence, by virtue of Theorem 5.3 for $2b + \sigma^2(p-1) < 0$ the trivial solution of (5.5) is exponentially p-stable. We observe that since the function $V = |x|^p$ satisfies all conditions of Theorem 5.2, the trivial solution of (5.5) is also globally asymptotically stable in probability.

Example 5.2. In examining actual systems under uncertain conditions, one can use different classes of random variables and stochastic processes to simulate different causes that lead to incompleteness of information. The question of a proper choice of the model is highly substantial and extremely complicated. It should deserve a special study for any particular case. Suppose, for example, that a scalar system under random forces is described by equation (5.5). Then, according to the Ito formula (3.12) solutions of equation (5.5) take the form

$$x(t) = x_0 \exp\left(\left(b - \frac{\sigma^2}{2}\right)t + \sigma\xi(t)\right).$$

Hence, by using the iterated logarithm law, we infer that for sufficiently high intensity of disturbances (that is, for sufficiently great σ) the system (5.5) is asymptotically stable in probability. In particular, the system can be stabilized by uncertain factors. However, if equation (5.5) is regarded in the Stratonovich sense this is no longer true. In fact, by virtue of (3.5) and (3.6) the Stratonovich equation (5.5) is equivalent to the following Ito equation:

$$dx(t) = \left(b - \frac{\sigma^2}{2}\right)x(t)\, dt + \sigma x(t)\, d\xi(t).$$

According to the Ito formula, any solution of this equation takes the form

$$x(t) = x_0 \exp\bigl(bt + \sigma\xi(t)\bigr).$$

Hence, the scalar system treated in this way can not be stabilized by random forces. Thus, this example shows that for modelling concrete physical systems the Stratonovich integral is preferable in a number of cases.

2. Stability in mean square of linear systems. Now we study conditions for the exponential stability in mean square of systems of the form

$$dx(t) = A(t)x(t)\,dt + \sum_{i=1}^{m} b_i(t)x(t)\,d\xi_i(t), \qquad t \geqslant t_0, \tag{5.6}$$

$$x \in R^n,\ \xi_i(t) \in R^1$$

Here $A(t)$ and b_i are given $n \times n$ matrices with bounded continuous elements, and the standard Wiener processes are independent.

Theorem 5.4. *It is necessary and sufficient for exponential stability in mean square of the system* (5.6) *that there exist a function* $V(t,x) \in C_2^0\ ((t \geqslant t_0) \times R^n)$ *satisfying the estimates* (5.3), (5.4), *where* $p = 2$.

Proof. The sufficiency follows from Theorem 5.3. To prove the necessity, we define function $V(t,x)$ by

$$V(t,x) = \int_t^{t+\tau} E|X_{t,x}(s)|^2\,ds. \tag{5.7}$$

Here the constant parameter $\tau > 0$ is to be chosen below, and $X_{t,x}(s)$ stands for the solution of equation (5.6) for $s \geqslant t$ with initial value $X_{t,x}(t) = x$, $x \in R^n$. By the definition of exponential stability in mean square, we have

$$E|X_{t,x}(s)|^2 \leqslant C_0|x|^2 e^{C_1(t-s)}, \qquad t \leqslant s. \tag{5.8}$$

It follows from (5.7), (5.8) that the inequality

$$V(t,x) \leqslant C_0|x|^2 \int_t^{t+\tau} e^{-C_1(t-s)}\,ds = k_2|x|^2 \tag{5.9}$$

holds. To verify that the function $V(t,x)$ is positive definite, we note that, by virtue of the Ito formula, the stochastic differential with respect to s of the process $|X_{t,x}(s)|^2$ is equal to

$$d|X_{t,x}(s)|^2 = \Big[y'A(s)\frac{\partial}{\partial y}|y|^2 + \frac{1}{2}\sum_{i=1}^{m} \operatorname{tr}\Big(\sigma_i(s)yy'\sigma_i'(s)\frac{\partial^2}{\partial y^2}|y|^2\Big)\Big]\,ds + \sum_{i=1}^{m} y'\sigma_i'(s)\frac{\partial}{\partial y}|y|^2\,d\xi_i(s), \tag{5.10}$$

where the prime stands for transposition and $y = X_{t,x}(s)$. The expression in the square brackets is, in view of (3.25), the result of applying the generator L of the process (5.6) to the function $|y|^2$. In other words,

$$d|X_{t,x}(s)|^2 = \big(L(|y|^2)\big)\,ds + \sum_{i=1}^{m} y'\sigma_i'(s)\frac{\partial}{\partial y}|y|^2\,d\xi_i(s) \tag{5.11}$$

$$y = X_{t,x}(s).$$

Recall that

$$|y|^2 = \left(\sum_{i=1}^{n} y_i^2\right).$$

Therefore,

$$\frac{\partial}{\partial y}|y|^2 = 2y, \qquad \frac{\partial^2}{\partial y^2}|y|^2 = 2.$$

This and the boundedness of coefficients A and σ_i of equation (5.6) imply the existence of a constant $k_1 > 0$ such that

$$|L(|y|^2)| \leqslant \frac{1}{2k_1}|y|^2. \tag{5.12}$$

Furthermore, in view of (3.21) the fourth moment of the process $X_{t,x}(s)$ is bounded in any bounded interval of the argument $s \geqslant t$. Hence and from properties of Ito integrals it follows that

$$E\int_t^{t+\tau} X'_{t,x}(s)\sigma'_i(s)X_{t,x}(s)\, d\xi_i(s) = 0. \tag{5.13}$$

Now we integrate both sides of equality (5.11) with respect to s from $s = t$ to $s = t + \tau$, and then evaluate the expectation. Taking (5.11)–(5.13) into account, we conclude that

$$E|X_{t,x}(t+\tau)|^2 - |x|^2 \geqslant -\frac{1}{2k_1}\int_t^{t+\tau} E|X_{t,x}(s)|^2\, ds = -\frac{1}{2k_1}V(t,x). \tag{5.14}$$

Because of (5.8), the parameter $\tau > 0$ can be chosen so that for all t

$$E|X_{t,x}(t+\tau)|^2 \leqslant \tfrac{1}{2}|x|^2.$$

This and (5.14) imply the estimate

$$V(t,x) \geqslant k_1|x|^2.$$

Thus, we verified that the function (5.7) obeys the relations (5.3) with $p = 2$. Let us verify that for the function (5.7) the inequality (5.4) with $p = 2$ holds true. In the case under consideration this inequality has the form

$$LV(t,x) \leqslant -k_3|x|^2, \qquad k_3 > 0,$$

where the operator L is given by

$$\begin{aligned} LV &= \frac{\partial V(t,x)}{\partial t} + x'A(t)\frac{\partial V(t,x)}{\partial x} \\ &+ \frac{1}{2}\sum_{i=1}^{m} \operatorname{tr}\left(\sigma_i(t)xx'\sigma'_i(t)\frac{\partial^2 V(t,x)}{\partial x^2}\right). \end{aligned}$$

By applying the operator L to both sides of formula (5.7), we obtain

$$LV(t,x) = E|X_{t,x}(t+\tau)|^2 - |x|^2 + \int_t^{t+\tau} LE|X_{t,x}(s)|^2\,ds. \qquad (5.15)$$

But, by virtue of (3.24),

$$LE|X_{t,x}(s)|^2 = 0, \qquad s \geqslant t.$$

Therefore, by virtue of (5.15),

$$LV(t,x) = E|X_{t,x}(t+\tau)|^2 - |x|^2 \leqslant -\frac{1}{2}|x|^2.$$

The proof is complete. □

3. Scalar equations of the nth order. We now present conditions for the asymptotic stability in mean square of linear equations whose coefficients a_i are disturbed by independent Wiener processes $\xi_i(t)$. The relevant equations take the form

$$x^{(n)}(t) + (a_1 + \sigma_1\dot{\xi}_1(t))x^{(n-1)}(t) + \cdots + (a_n + \sigma_n\dot{\xi}_n(t))x(t) = 0, \qquad t > 0, \quad (5.16)$$

where a_i and σ_i are given constants and the notation

$$x^{(n)}(t) = \frac{d^n x(t)}{dt^n}, \qquad x(t) \in R^1$$

is used.

A solution of equation (5.16) for $t > 0$ is determined by assigning the values of $x(0), \dots, x^{(n-1)}(0)$ at the initial time $t = 0$. The conditions for the asymptotic stability in mean square of the system (5.16) with respect to disturbances of the initial state can be stated as follows.

Theorem 5.5. *It is necessary and sufficient for the asymptotic stability in mean square of the system* (5.16) *that the deterministic system derived from* (5.16) *by putting* $\sigma_1 = \sigma_2 = \cdots = \sigma_n = 0$ *be asymptotically stable and that the determinant*

$$\Delta = \det \begin{pmatrix} \sigma_1^2 & -\sigma_2^2 & \sigma_3^2 & \dots & (-1)^{n-2}\sigma_{n-1}^2 & (-1)^{n-1}\sigma_n^2 \\ 1 & a_2 & a_4 & \dots & 0 & 0 \\ 0 & a_1 & a_3 & \dots & 0 & 0 \\ 0 & 1 & a_2 & \dots & 0 & 0 \\ \vdots & \vdots & \vdots & \ddots & \vdots & \vdots \\ 0 & 0 & 0 & \dots & 0 & a_n \end{pmatrix} \qquad (5.17)$$

be positive.

§ 6. Stability in Probability of Satellite Motion

1. The pitch stability of a symmetric satellite in a circular orbit[1]. The motion of an actual earth satellite is determined by various factors. Among them only the gradient of the gravitational force can be regarded as a deterministic one. All the remaining ones (for instance, aerodynamical and magnetic torques, sun radiation, the earth's electric field, meteor flow, and so on) possess random components. Moreover, the values of moments of inertia of the satellite, being affected by thermo-elastic oscillations of antennas and sun batteries, the motion of liquid in tanks, are also random.

Lyapunov's direct method can be applied to consideration of the satellite stability under random perturbations. Consider a plane motion of a symmetric satellite along a circular orbit. We assume that the pitch angle x and the velocity $y = \dot{x}$ are described by the system of two Ito equations

$$\begin{aligned} &\dot{x} = y, \\ &dy = (-\sin x + l \sin 2x - \eta_0 y)\, dt - \sqrt{\sigma}(\sin x + \eta y)\, d\xi. \end{aligned} \tag{6.1}$$

Here $\xi(t)$ is a standard Wiener process, the constants η and $l < \frac{1}{2}$ are defined by the parameters of the satellite motion, σ is the intensity of random atmospheric disturbances. Moreover, η is the dynamic coefficient of the aerodynamic torque, l is the torque of gravitational forces. The problem is to determine conditions for the stability in probability of the zero solution of the system (6.1). To this end, we apply Theorem 5.1 and set

$$V(x, y) = \tfrac{1}{2} b\eta x^2 + bxy + y^2 + 4\left(1 - \eta\sigma - 2l \cos^2 \frac{x}{2}\right) \sin^2 \frac{x}{2}, \tag{6.2}$$

where the parameter b is to be chosen below.

By using (6.1), (6.2), and (3.25), we evaluate LV:

$$\begin{aligned} LV &= -(2\eta - b - \sigma\eta^2)y^2 - \left(b - 2bl \cos x - \sigma \frac{\sin x}{x}\right) x \sin x \\ &\leqslant -(2\eta - b - \sigma\eta^2)y^2 - (b - 2bl - \sigma) x \sin x. \end{aligned} \tag{6.3}$$

It follows from relations (6.2), (6.3) that the conditions (5.1) of Theorem 5.1 are met if

$$0 < b < 2\eta, \qquad \sigma < \min\left[b(1 - 2l), \frac{2\eta - b}{\eta^2}\right].$$

By examining the dependence of the interval of admissible values of σ on b, we conclude that the maximal admissible domain corresponds to

$$b = \frac{2\eta}{1 + \eta^2(1 - 2l)}.$$

[1] Sagirow, P., Stochastic Methods in Dynamics of Satellites, *Courses and Lectures*, International Center for Mechanical Sciences, Udine, (1970), no. 57.

Thus, pitch stability in probability occurs provided that

$$\sigma < \frac{2\eta(1-2l)}{1+\eta^2(1-2l)}. \tag{6.4}$$

One can see from this condition that, if there are no disturbances ($\sigma = 0$), then the pitch motion is stable provided that the torque l of gravitational forces satisfy the inequality $l < \frac{1}{2}$. If disturbances are present, then condition (6.4) implies that

$$l < \frac{1}{2}\left(1 - \frac{\sigma}{2\eta - \sigma\eta^2}\right).$$

Hence, taking account of disturbances diminishes the range of possible values of gravitational torques such that the pitch stability of the satellite is preserved.

2. The yaw angle stability of a satellite in a circular equatorial orbit. Consider plane oscillations of a symmetric satellite in a circular equatorial orbit caused by the random fluctuations of the earth's magnetic field. The equations for the angle of yaw x take the form

$$\begin{aligned} \dot{x} &= y, \\ dy &= (-\alpha \sin x + \beta y)\,dt - (2\sin x + \beta y)\sqrt{\sigma}\,d\xi, \end{aligned} \tag{6.5}$$

where σ is the intensity of the random fluctuations of the magnetic field, the positive constants α and β depend on the deterministic components of the magnetic field and of the satellite characteristics.

We analyze, by means of Theorem 5.1, the stability in probability of the zero equilibrium point of the system (6.5). Set

$$V(x,y) = \frac{1}{2}\left(\sqrt{\beta}x + \frac{y}{\sqrt{\beta}}\right)^2 + \frac{2\beta}{\alpha}y^2 + 4(\alpha+4\beta^2)\left(\frac{1}{2\beta} - \sigma\right)\sin\frac{x}{2},$$

Evaluating LV, we obtain

$$LV \leqslant -\frac{\alpha+4\beta^2}{2\alpha\beta}\left(\frac{2\beta}{\alpha+4\beta^2} - \sigma\right)(\alpha^2 x\sin x + 4\beta^2 y^2).$$

Hence, it is seen that the conditions (5.1) of Theorem 5.1 are met, provided that

$$\sigma < \frac{2\beta}{\alpha+4\beta^2}, \qquad \alpha > 0,\ \beta > 0. \tag{6.6}$$

Thus, the angle of the yaw oscillations caused by the random fluctuations of the earth's magnetic field, are stable in probability if the intensity of these fluctuations satisfies (6.6).

Main Results and Formulas of Chapter II

Probability density for a Gaussian vector:

$$f(x) = [(2\pi)^n[\det D]]^{-1/2}\exp\left[-\tfrac{1}{2}(x-m)'D^{-1}(x-m)\right].$$

Expectation (mean) of a random vector:

$$E\xi = \int_\Omega \xi(\omega)P(d\omega).$$

Conditional expectation (conditional mean) $E(\xi \mid \sigma_0) = \eta(\omega)$ **of a random vector** ξ with respect to a σ-algebra σ_0 satisfies the equality

$$\int_A \eta(\omega)P(d\omega) = \int_A \xi(\omega)P(d\omega), \qquad \forall A \in \sigma_0.$$

Finite-dimensional distributions of the stochastic process $\zeta(t,\omega)$:

$$F(t_1,\dots,t_m,A_1,\dots A_m) = P(\zeta(t_1,\omega)\in A_1,\dots,\zeta(t_m,\omega)\in A_m)$$

Covariance matrix of a stochastic process:

$$E[\zeta(t,\omega)-E\zeta(t,\omega)][\zeta(t,\omega)-E\zeta(t,\omega)]'.$$

Density of a finite-dimensional distribution for a standard Wiener process $\xi(t,\omega)$:

$$f(x_1,\dots,x_n) = [(2\pi)^m t_1(t_2-t_1)\dots(t_m-t_{m-1})]^{-1/2} \times\exp\left[-\frac{x_1^2}{2t_1}-\frac{1}{2}\sum_{j=1}^{m-1}\frac{(x_{j+1}-x_j)^2}{t_{j+1}-t_j}\right].$$

The iterated logarithm law for a standard Wiener process:

$$\overline{\lim_{t\to\infty}}\,\xi(t,\omega)(2t\ln\ln t)^{-1/2} = 1,$$

The Ito stochastic integral $\int_0^t b(s,x(s))\,d\xi(s)$ is the limit in mean square of Riemann sums

$$\sum_{i=1}^{N-1} b(t_i,x(t_i))\left(\xi(t_{i+1})-\xi(t_i)\right).$$

The Stratonovich stochastic integral $\int_0^t b(s,x(s))\,\overline{d\xi(s)}$ is the limit in mean square of Riemann sums

$$\sum_{i=0}^{N-1} b\Big(t_i, \tfrac{1}{2}\big(x(t_{i+1}) + x(t_i)\big)\Big)\big(\xi(t_{i+1}) - \xi(t_i)\big).$$

The Ito stochastic differential equation:

$$dx(t) = f\big(t,x(t)\big)\,dt + b\big(t,x(t)\big)\,d\xi(t), \qquad x(0) = x_0.$$

The Stratonovich stochastic differential equation:

$$dx(t) = f\big(t,x(t)\big)\,dt + b\big(t,x(t)\big)\,\overline{d\xi(t)}, \qquad x(0) = x_0$$

Any solution of the Stratonovich stochastic differential equation also satisfies the Ito equation

$$dx(t) = \left[f\big(t,x(t)\big)\,dt + \frac{1}{2}\frac{\partial b\big(t,x(t)\big)}{\partial x} b\big(t,x(t)\big)\right] dt + b\big(t,x(t)\big)\,d\xi(t)$$

The Ito formula. If $V = V(t,x)$ and $d\gamma(t) = f(t)\,dt + \sigma(t)\,d\xi(t)$, then

$$\begin{aligned} dV\big(t,\gamma(t)\big) = {} & \big[V_t\big(t,\gamma(t)\big) + V_x'\big(t,\gamma(t)\big)f(t) \\ & + \tfrac{1}{2}\operatorname{tr}\sigma(t)\sigma'(t)V_{xx}\big(t,\gamma(t)\big)\big]\,dt + V_x'\big(t,\gamma(t)\big)\sigma(t)\,d\xi(t). \end{aligned}$$

Existence and uniqueness theorem for solutions of the Ito stochastic differential equation. Suppose that

$$\begin{gathered} |f(t,x)|^2 + |b(t,x)|^2 \leqslant C(1 + |x|^2), \\ |f(t,x) - f(t,y)| + |b(t,x) - b(t,y)| \leqslant C|x-y|. \end{gathered}$$

Then there exists a unique solution of the Ito equation on any interval $[0,T]$.

Generator for the Ito stochastic differential equation:

$$L = \frac{\partial}{\partial t} + f'\frac{\partial}{\partial x} + \frac{1}{2}\operatorname{tr}\sigma\sigma'\frac{\partial^2}{\partial x^2}.$$

Hasminskii's theorem. Suppose that there exists a function $V(t,x)$ such that

$$\omega_1(|x|) \leqslant V(t,x), \qquad LV(t,x) \leqslant 0.$$

Then the trivial solution of the Ito equation is stable in probability. But if

$$\begin{gathered} \omega_1(|x|) \leqslant V(t,x) \leqslant \omega_2(|x|), \qquad \omega_1(u) \to \infty,\ u \to \infty, \\ LV \leqslant -\omega_3(|x|), \end{gathered}$$

then the trivial solution of the Ito equation is globally asymptotically stable in probability.

Stability in mean square of linear systems. It is necessary and sufficient for exponential stability in mean square of the system

$$dx(t) = A(t)x(t)\,dt + \sum_{i=1}^{m} b_i(t)x(t)\,d\xi_i(t), \qquad t \geqslant t_0,$$
$$x \in R^n,\ \xi_i(t) \in R^1,\ i = 1, \dots, m,$$

that there exist a function $V(t,x) \in C_2^0\ \big((t \geqslant t_0) \times R^n\big)$ satisfying the estimates

$$k_1|x|^2 \leqslant V(t,x) \leqslant k_2|x|^2,$$
$$LV(t,x) \leqslant -k_3|x|^2, \qquad k_1, k_2, k_3 > 0.$$

It is necessary and sufficient for the asymptotic stability in mean square of the equation

$$x^{(n)}(t) + \big(a_1 + \sigma_1\dot{\xi}_1(t)\big)x^{(n-1)}(t) + \cdots + \big(a_n + \sigma_n\dot{\xi}_n(t)\big)x(t) = 0$$
$$a_i, \sigma_i = \text{constant},\ x \in R^1,\ \xi_i \in R^1,\ i = 1, \dots, n,\ t \geqslant 0$$

that the deterministic equation

$$x^{(n)}(t) + a_1x^{(n-1)}(t) + \cdots + a_nx(t) = 0$$

be asymptotically stable and that the determinant

$$\Delta = \det\begin{pmatrix} \sigma_1^2 & -\sigma_2^2 & \sigma_3^2 & \dots & (-1)^{n-2}\sigma_{n-1}^2 & (-1)^{n-1}\sigma_n^2 \\ 1 & a_2 & a_4 & \dots & 0 & 0 \\ 0 & a_1 & a_3 & \dots & 0 & 0 \\ 0 & 1 & a_2 & \dots & 0 & 0 \\ \vdots & \vdots & \vdots & \ddots & \vdots & \vdots \\ 0 & 0 & 0 & \dots & 0 & a_n \end{pmatrix}$$

be positive.

EXERCISES OF PART ONE

1. Test the stability of the linear system $\dot{x}(t) = Ax(t)$ if:

$$\text{(a)}\quad A = \begin{pmatrix} 1 & 2 & 0 \\ 0 & 1 & 1 \\ 1 & 3 & 1 \end{pmatrix}; \qquad \text{(b)}\quad A = \begin{pmatrix} -2 & -2 & 1 \\ 0 & -1 & 4 \\ 1 & 0 & -10 \end{pmatrix}.$$

2. Determine the values of the parameters α or (α, β) for the stability of the following equations:

$$\text{(a)}\quad \ddot{x} + (2 - \alpha^2)\dot{x} + \alpha x = 0;$$

$$\text{(b)}\quad \dddot{x} + \ddot{x} + \alpha^2 \dot{x} + 5\alpha x = 0;$$

$$\text{(c)}\quad \dot{x} = -x + \alpha y, \quad \dot{y} = -\beta x - y + \alpha z, \quad \dot{z} = \beta y - z.$$

3. Using the Routh-Hurwitz criterion, test the stability of the following polynomials:

$$\text{(a)}\quad z^5 + 7z^4 + 33z^3 + 88z^2 + 122z + 60 = 0;$$

$$\text{(b)}\quad z^4 + 7z^3 + 19z^2 + 23z + 10 = 0;$$

$$\text{(c)}\quad z^6 + 3z^5 + 8z^4 + 15z^3 + 61z^2 + 12z + 3 = 0;$$

$$\text{(d)}\quad \frac{5\cdot 4\cdot 3\cdot 2\cdot 1}{10\cdot 9\cdot 8\cdot 7\cdot 6}\frac{z^5}{5!} + \frac{5\cdot 4\cdot 3\cdot 2}{10\cdot 9\cdot 8\cdot 7}\frac{z^4}{4!} + \frac{5\cdot 4\cdot 3}{10\cdot 9\cdot 8}\frac{z^3}{3!} + \frac{5\cdot 4}{10\cdot 9}\frac{z^2}{2!} + \frac{5}{10}z + 1 = 0.$$

(This polynomial occurs in Padé approximations of the exponent.)

4. Using the Mikhailov criterion, test the stability of the following equations:

$$\text{(a)}\quad \dddot{x} + 2\ddot{x} + \dot{x} + x = 0;$$

$$\text{(b)}\quad \dddot{x} + x = 0;$$

$$\text{(c)}\quad x^{(\mathrm{iv})} + 13\dddot{x} + 28\ddot{x} + 23\dot{x} + 6x = 0;$$

$$\text{(d)}\quad x^{(\mathrm{iv})} + \dddot{x} + \dot{x} + x = 0;$$

$$\text{(e)}\quad x^{(\mathrm{v})} + 2x^{(\mathrm{iv})} + \dddot{x} + 2\ddot{x} + \dot{x} + 2x = 0;$$

$$\text{(f)}\quad x^{(\mathrm{iv})} + \dddot{x} + 4\ddot{x} + \dot{x} + x = 0.$$

5. *Problem of the automatic stabilization of the ship's heading* The simplest model of a neutrally stable ship is described by the equation

$$I\ddot{\varphi} + h\dot{\varphi} = -k\psi, \qquad I > 0,\ h > 0,\ k > 0,$$

where φ is the angle of deviation of the ship from the prescribed heading, ψ is the angle of deviation of the rudder from the ship's axis, I is the moment of inertia of the ship about the vertical axis passing through the ship's center, h is the coefficient of fluid friction, and k is the coefficient of rudder efficiency (see Figure E.1).

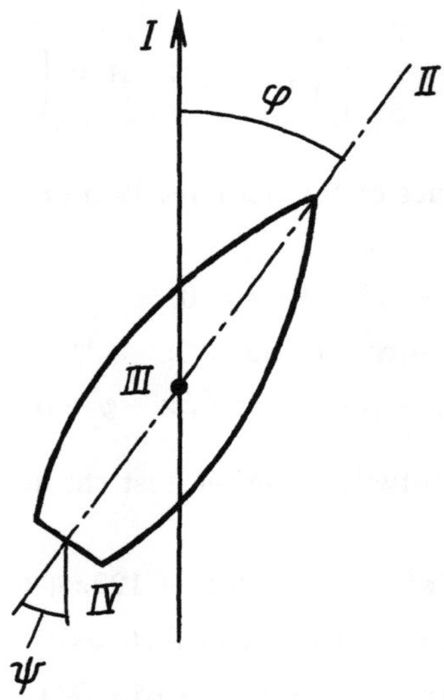

FIG. E.1. The simplest model of a ship:
I, ship's heading; *II*, ship's axis; *III*, ship's center of inertia; *IV*, rudder

Suppose that the equations of the ship's autopilot are of the form

$$T\dot{\psi} + \psi = \alpha\xi + \beta\dot{\xi}, \qquad T > 0.$$

Here, T is the time constant of the autopilot, α and β are the autopilot parameters whose values can be chosen, ξ is the measured angle of deviation φ from the prescribed heading. A change in the angle of deviation does not occur instantaneously but rather is described by the equation

$$\tau\dot{\xi} + \xi = \varphi, \qquad \tau > 0,$$

where τ is the time constant of the measuring device.

In the parameter space (α, β) associated with the autopilot, find the region of asymptotic stability for the closed system composed of the ship and the autopilot for the following values of the remaining parameters: $I = 10^3$, $h = 10^{-2}$, $k = 1$, $T = 10^2$, $\tau = 10^{-1}$.

6. Find a periodic solution of the equation

$$\dddot{x} + x = \cos t,$$

and test its stability.

7. Using Lyapunov's theorems, in each case determine the stability or instability of the trivial solution of the following equations:

$$\begin{aligned}
&\text{(a)} \quad \dot{x} = -2x - 3y + x^5, \quad \dot{y} = x + y - y^2, \quad V = 4x^2 + 14xy + 19y^2;\\
&\text{(b)} \quad \dot{x} = -xy^4, \quad \dot{y} = yx^4, \quad V = x^4 + y^4;\\
&\text{(c)} \quad \dot{x} = -2y - x^5, \quad \dot{y} = 5x - y^5, \quad V = 5x^2 + y^2;\\
&\text{(d)} \quad \dot{x} = -x - \frac{y}{\ln(x^2+y^2)}, \quad \dot{y} = -y + \frac{x}{\ln(x^2+y^2)}.
\end{aligned}$$

8. Determine the values of the parameters (α, β) which ensure:
(a) the stability of the trivial solution of the system

$$\dot{x} = \alpha x + y - x^5, \qquad \dot{y} = -x - \beta y^5;$$

(b) the instability of the trivial solution of the above system.

9. Using Chetaev's theorem, test the stability of the trivial solution of the system:

$$\begin{aligned}
&\text{(a)} \quad \dot{x} = y^3 + x^5, \quad \dot{y} = x^3 + y^5, \quad V = x^4 - y^4;\\
&\text{(b)} \quad \dot{x} = y + x^2, \quad \dot{y} = 0, \quad V = (\alpha x - y)y, \quad \alpha > 0.
\end{aligned}$$

10. Test the stability of the trivial solution of the system:

$$\begin{aligned}
&\text{(a)} \quad \dot{x} = -y - x^3, \quad \dot{y} = x - y^3;\\
&\text{(b)} \quad \dot{x} = x - y + x^2 + y^2 \sin t, \quad \dot{y} = x + y - y^2;\\
&\text{(c)} \quad \dot{x} = 2x + 8\sin y, \quad \dot{y} = 2 - e^x - 3y - \cos y;\\
&\text{(d)} \quad \dot{x} = y, \quad \dot{y} = -x^n, \quad n > 1;\\
&\text{(e)} \quad \dot{x} = x^2 y, \quad \dot{y} = -x^n, \quad n > 1.
\end{aligned}$$

11. Suppose that in a one-loop system shown in Figure 5.4 each section is aperiodic, i.e., let the system in question be described by the equations

$$\begin{aligned}
\dot{x}_1 &= -a_1 x_1 + b_1 x_m,\\
\dot{x}_i &= -a_i x_i + b_i x_{i-1}, \qquad i = 2, \ldots, m,\\
&a_i > 0,\ b_i > 0,\ i = 1, \ldots, m.
\end{aligned}$$

Show that this one-loop system is asymptotically stable whenever

$$\prod_{i=1}^{m} \frac{a_i}{b_i} > 1.$$

12. Test the stability of the system

$$\dot{x} = y, \qquad \dot{y} = -2x - 3y,$$

using Lyapunov functions

$$V_1(x, y) = x^2 + y^2, \qquad V_2(x, y) = 12x^2 + 4xy + 3y^2.$$

Construct level curves for the functions V_1 and V_2 and explain in geometric terms why the function V_1 cannot be used to test the stability, whereas the function V_2 can.

13. Can the function

$$V(x, y) = x^2 + 1 - e^{-y^2}$$

be used to test the stability of the system

$$\dot{x} = (-x + yx)e^{-y^2}, \qquad \dot{y} = -y - x^2.$$

14. Test the stability of Van der Pol's equation

$$\ddot{x} + \varepsilon(x^2 - 1)\dot{x} + x = 0, \qquad \varepsilon > 0; \qquad V = x^2 + y^2.$$

15. Show for Van der Pol's equation that the attraction domain for the trivial solution contains the disk $x^2 + y^2 < 3$.

16. Find stability conditions for the trivial solution of Liénard's equation

$$\ddot{x} + f(x)\dot{x} + g(x) = 0.$$

Hint. Write this equation as the system

$$\dot{x} = y,$$
$$\dot{y} = -g(x) - f(x)y,$$

and use the Lyapunov function

$$V = \frac{1}{2}y^2 + \int_0^x g(s)\,ds.$$

17. Using the Lyapunov function

$$V = (dx - by)^2 + 2d\int_0^x f(s)\,ds - bcx^2 = (dx - by)^2 + 2\int_0^x (df(s) - bcs)\,ds,$$

find stability conditions for the trivial solution of the system

$$\dot{x} = f(x) + by,$$
$$\dot{y} = cx + dy.$$

18. Using the Lyapunov function

$$V = (dx - by)^2 + 2\int_0^x [df_1(s) - bf_2(s)]\,ds,$$

find stability conditions for the trivial solution of the system

$$\dot{x} = f_1(x) + by,$$
$$\dot{y} = f_2(x) + dy.$$

19. Consider a pendulum with friction defined by

$$\ddot{x} + \beta\dot{x} + \omega^2 \sin x = 0, \qquad \beta > 0,\ \omega^2 > 0.$$

Physically, it is clear that the friction leads to the asymptotic stability of the equilibrium position $x = \dot{x} = 0$.

Can we use the total energy of the pendulum

$$E(x, \dot{x}) = \frac{\dot{x}^2}{2} + (1 - \omega^2 \cos x)$$

as a Lyapunov function?

Investigate the stability of the equilibrium position of the inverted pendulum ($x = \pi$, $\dot{x} = 0$).

20. Prove that the equilibrium position of a pendulum with friction is asymptotically stable (see Exercise 19)

(a) by using the theorem of stability in the first approximation

(b) by using the Lyapunov function

$$V(x, \dot{x}) = \dot{x}^2 + (\dot{x} + x)^2 + 4(1 - \omega^2 \cos x).$$

21. Test the stability of the equilibrium position of a pendulum with friction (see Exercise 19) that is acted upon by a constant moment L. The pendulum is assumed to be described by the equation

$$\ddot{x} + \beta\dot{x} + \omega^2 \sin x = L, \qquad |L| \leqslant \omega^2.$$

22. Consider a pendulum with a friction coefficient rapidly increasing with time. Suppose that its linearized equation is of the form

$$\ddot{x} + (2 + e^t)\dot{x} + x = 0.$$

Verify that the function

$$x(t) = a(1 + e^{-t}),$$

with a arbitrary, is a solution of the equation and thus the equilibrium position of the pendulum is not asymptotically stable. Explain this in physical terms.

23. Using the Lyapunov vector function

$$V'(x,y) = \left((x+y)^2, (x-y)^2\right)$$

and Theorem 4.4, test the stability of the trivial solution of the system

$$\dot{x} = e^{-t}x + y\sin t - (x^3 + xy^2)\sin^2 t,$$
$$\dot{y} = x\sin t + e^{-t}y - (x^2y + x^3)\sin^2 t.$$

24. The growth of an isolated biological population can be described by the logistic equation

$$\dot{N} = N(\alpha - \gamma N), \qquad \alpha, \gamma > 0,$$

where $N(t)$ is the population size at time t.

Find the stationary states $N(t) = N_i$ $(i = 0, 1)$ of the logistic equation. Verify that the stationary state $N_1 > 0$ is asymptotically stable.

26. The growth of an isolated population can be described by the Alley law

$$\dot{N}(t) = \varepsilon(N)N,$$

where $\varepsilon(N)$ is a nonlinear reproduction coefficient. Experimentally derived Alley relationships are shown in Figure E.2. For a linear relation $\varepsilon(N) = \alpha - \gamma N$, the Alley law leads to the logistic equation.

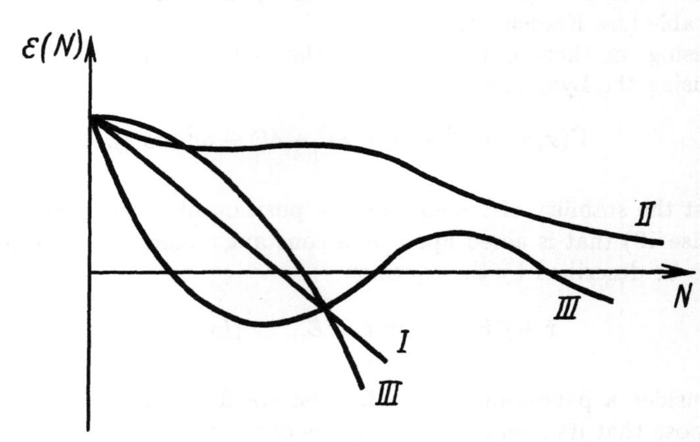

FIG. E.2. Experimentally derived Alley relationships for $\varepsilon(N) = \dot{N}/N$ and certain types of population growth:
I, the logistic equation; *II*, monotonic laws for population growth; *III*, nonmonotonic laws

Find the equilibrium states N_i of an Alley-type population, and test them for stability. Use the Lyapunov function $V(N) = \varepsilon^2(N)$.

27. A biological community consisting of only two species: the predator type (such as the lynx) and the prey type (such as the hare) can be described by the Volterra model

$$\dot{x}(t) = \alpha x - \beta xy, \qquad \dot{y}(t) = k\beta xy - my,$$
$$\alpha, \beta, k, m > 0.$$

Here, $x(t)$ is the number of prey, $y(t)$ is the number of predators, α is the natural reproduction coefficient for the prey, m is the natural mortality coefficient for the predators, β is the coefficient that characterizes the probability of a predator eating its prey during their encounter, the probability of which is proportional to the product xy. It is required:

(a) to find the stationary states (x_0, y_0) and (x_1, y_1) in the model "predator–prey";

(b) to test the stability of the state (x_1, y_1), using the Lyapunov function

$$V = m\left(\frac{x}{x_1} - \ln\frac{x}{x_1} - 1\right) + \alpha\left(\frac{y}{y_1} - \ln\frac{y}{y_1} - 1\right).$$

28. The model "predator–prey" with consideration for intraspecific competition of victims is described by the following equations:

$$\dot{x} = \alpha x - \beta xy - \gamma x^2, \qquad \dot{y} = k\beta xy - my,$$
$$\alpha, \beta, \gamma, k, m > 0.$$

Find the equilibrium states of this model, and test them for stability.

29. In the theory of the genetic control of bacteria the Goodwin equations are used:

$$\dot{x} = \frac{\gamma}{y} - \alpha, \qquad \dot{y} = Ax - \beta,$$

where x is the concentration of the initial genetic material, y is the concentration of eximatic protein, and the constants γ, α, A, β are positive.

Test the stability of the nonzero equilibrium position $\bar{x} = \beta/A$, $\bar{y} = \gamma/\alpha$. Use the Lyapunov function

$$V(x, y) = W(x, y) - W(\bar{x}, \bar{y}),$$

where

$$W(x, y) = -\gamma \ln y + \alpha y + \tfrac{1}{2}Ax^2 - \beta x.$$

Is the function $V(x, y)$ positive definite?

30. In another model for the genetic control of bacteria the following equations are used:

$$\dot{x} = \frac{\gamma}{y} - \alpha, \qquad \dot{y} = Ax - \beta y,$$
$$\gamma, \alpha, A, \beta > 0,$$

where the meaning of the variables is the same as in the preceding exercise.

Test the stability of the nonzero equilibrium position $\bar{x} = \beta\gamma/A\alpha$, $\bar{y} = \gamma/\alpha$. Use the Lyapunov function

$$V(x,y) = W(x,y) - W(\bar{x},\bar{y}),$$

where

$$W(x,y) = -\gamma \ln y + \alpha y + \frac{1}{2A}(Ax - \beta y)^2.$$

Is the function $V(x,y)$ positive definite?

31. Prove the following variants of the *Bellman theorem.* Let the system $\dot{y} = Ay$, $y \in R^n$, be asymptotically stable and

(a) the matrix $B(t)$ be continuous and $B(t) \to 0$ as $t \to \infty$;

(b) $\|B(t)\| + \max_i \operatorname{Re} \lambda_i(A) < 0$, $i = 1, \dots, n$, where $\lambda_i(A)$ are the eigenvalues of the matrix A;

(c) $\overline{\lim}_{t\to\infty} \frac{1}{t} \int_0^t \|B(s)\|\, ds + \max_i \operatorname{Re} \lambda_i(A) < 0$.

Then the system $\dot{x}(t) = [A + B(t)]x$ is also asymptotically stable.

Hint. Use the estimate of the Cauchy matrix $Z(t,t_0)$ of the equation $\dot{y} = Ay$ of the form

$$|Z(t,t_0)| \leqslant e^{\alpha(t-t_0)},$$

where $\alpha = \max_i \operatorname{Re} \lambda_i(A)$, $i = 1, \dots, n$.

32. Prove that the linear system

$$\dot{x} = (A_0 t^m + A_1 t^{m-1} + \cdots + A_m)x, \qquad x \in R^n,$$

is asymptotically stable if A_0 is a Hurwitz matrix.

33. Prove that the system

$$\dot{x} = \left(\frac{A_1}{t} + \frac{A_2}{t^2} + \cdots + \frac{A_m}{t^m}\right)x, \qquad x \in R^n,$$

is asymptotically stable if A_1 is a Hurwitz matrix. *Hint.* Make the change of the time variable $\tau = \ln t$ and use the Bellman theorem.

34. Suppose that for the system

$$\dot{x}(t) = [A + B(t)]x(t) + f(t), \qquad x \in R^n,$$

the following conditions are satisfied:

(a) A is a constant Hurwitz matrix;

(b) the matrix $B(t)$ is continuous, and $\int_{t_0}^{\infty} \|B(t)\|\, dt < \infty$;

(c) the vector $f(t)$ is continuous, and $f(t) \to 0$ as $t \to \infty$.

Then all solutions of this system satisfy the condition

$$\lim_{t\to\infty} x(t) = 0.$$

Hint. Use the Gronwall–Bellman lemma.

35. Verify that the trivial solution of the system $\dot{x} = A(t)x$ with matrix

$$A(t) = \begin{pmatrix} -1 + ae^{-t}\sin t & 2 - b\sin t \\ -2 + b\cos t & -1 - ae^{-t}\cos t \end{pmatrix}$$

is asymptotically stable if $b < \sqrt{2}$.

Hint. UseWażevski's inequality.

36. A linear system of the form

$$\dot{x}(t) = A(t)x(t), \qquad t \geqslant t_0 > 0, \tag{E.1}$$

is called *reducible* if there exists a smooth change of variables (the Lyapunov transformation) of the form

$$y = L(t)x, \qquad \|L(t)\| \leqslant \text{constant}, \ \|L^{-1}(t)\| \leqslant \text{constant}, \tag{E.2}$$

which transforms the system (E.1) into the system with constant matrix [2, 4, 15].

Prove the *Yerugin theorem*:[1] For the system (E.1) to be reducible, it is necessary and sufficient that the fundamental matrix $X(t)$ of the system (E.1) can be expressed as

$$X(t) = L(t)e^{Bt},$$

where $L(t)$ satisfies the condition (E.2), and B is a constant matrix.

Hint. To prove the necessity, use the Cauchy matrix of the system $\dot{y} = By$, $B =$ constant, to find the matrix $X(t)$ of the system (E.1); to prove the sufficiency, make the change of variables $x(t) = L(t)y(t)$, $L(t) = X(t)e^{-Bt}$.

---

[1] Yerugin, N. P., *Linear Systems of Ordinary Differential Equations with Periodic and Quasiperiodic Coefficients*, AN BSSR Publishing House, Minsk, (1963).

37. Two matrices $A(t)$ and $B(t)$ continuous and bounded on $[t_0,\infty)$ are called *kinematically similar* [15] if there exists a smooth matrix $L(t)$ satisfying the relation (E.2) such that

$$B(t) = L^{-1}(t)A(t)L(t) - L^{-1}(t)\dot{L}(t), \qquad \|L(t)\| < \infty, \ \|L^{-1}(t)\| < \infty.$$

Verify that systems of the form (E.1) with kinematically similar matrices are either simultaneously reducible (to the same system) or not.

38. Prove the *Lyapunov theorem*: the linear system (E.1) with a continuous periodic matrix $A(t)$ is reducible [2, 15].

Hint. Use the theorems of Floquet and Yerugin.

39. Verify that the linear system (E.1) with matrix $A(t)$ of the form

$$A(t) = \begin{pmatrix} 0 & 1 \\ 2t^{-2} & 0 \end{pmatrix}$$

is irreducible.

40. Verify that the system (E.1) with matrix

$$A(t) = \sum_{k=2}^{m} \frac{A_k}{t^k}, \qquad A_k = \text{constant}, \ t \geqslant t_0 > 0$$

is reducible.

41. The matrix $A(t)$ is said to satisfy the *Lappo-Danilevskii condition for commutativity* if

$$A(t)\left(\int_0^t A(s)\,ds\right) = \left(\int_0^t A(s)\,ds\right)A(t) \tag{E.3}$$

Verify that

(a) the condition (E.3) is equivalent to the condition

$$A(t)\dot{A}(t) = \dot{A}(t)A(t);$$

(b) the condition (E.3) is equivalent to the condition

$$A(t)A(\tau) = A(\tau)A(t)$$

for all $t,\tau \geqslant t_0$;

(c) the fundamental matrix of the system (E.1), given the condition (E.3), is of the form

$$X(t) = e^{\int_0^t A(s)\,ds};$$

(d) the system (E.1) with matrix $A(t)$ satisfying the condition (E.3) is reducible.

42. Verify that the matrix

$$A(t) = \begin{pmatrix} \alpha_1(t) + b_1\alpha_2(t) & \alpha_2(t) \\ b_2\alpha_2(t) & \alpha_1(t) \end{pmatrix},$$

where $\alpha_1(t)$, $\alpha_2(t)$ are arbitrary continuous functions, and b_1, b_2 are arbitrary constants, satisfy the Lappo-Danilevskii condition (E.3).

43. Suppose that the matrix $A(t)$ can be expressed as

$$A(t) = \sum_{k=0}^{m} \alpha_k(t) A_k, \tag{E.4}$$

where $\alpha_k(t)$ are linearly independent continuous scalar functions, and the matrices A_j commute pairwise: $A_j A_k = A_k A_j$, $j, k = 1, \ldots, m$.

Verify that the matrix $A(t)$ of the form (E.4) satisfies the condition (E.3).

44. Verify that the system (E.1) with matrix of rotation $A(t)$ of the form

$$A(t) = \begin{pmatrix} \cos\omega t & \sin\omega t \\ -\sin\omega t & \cos\omega t \end{pmatrix}$$

is stable for any angular velocity ω.

Hint. Express the matrix of rotation $A(t)$ in the form (E.4)

45. Show that the Cauchy matrix $Z(t, t_0)$ of the system (E.1) satisfying the condition (E.4) is of the form

$$Z(t, t_0) = \prod_{k=0}^{m} \exp[A_k \beta_k(t, t_0)],$$

$$\beta_k(t, t_0) = \int_{t_0}^{t} \alpha_k(s)\, ds, \qquad k = 0, 1, \ldots, m.$$

46. Show that the change of variables and time $x = T(t)y$, $T(t) = \prod_{k=0,\, k\neq j}^{m} \exp[A_k\beta_k(t, t_0)]$, $\tau = \beta_j(t, t_0) = \int_{t_0}^{t} \alpha_j(s)\, ds$ transforms the system (E.1), (E.4) into the system with a constant matrix $dy/d\tau = A_j y$.

47. Let the matrix $A(t)$ satisfy the condition (E.4), let there exist the limit

$$P_0 = \lim_{t\to\infty} \frac{1}{t} \int_{t_0}^{t} A(s)\, ds,$$

and let P_0 be a Hurwitz matrix. Prove that under these conditions the system (E.1) with matrix $A(t)$ is asymptotically stable [4].

Hint. Show that the Cauchy matrix in this case can be expressed as

$$Z(t, t_0) = \exp[P_0(t - t_0) + (t - t_0)P_1(t, t_0)],$$
$$P_1(t, t_0) \to 0, \qquad t \to \infty.$$

48. Test the stability of the system (E.1) with matrix[2]

$$A(t) = \begin{pmatrix} \sin t & \cos t + \sin t & 0 \\ 0 & \sin t & \cos t + \sin t \\ -6\cos t - 6\sin t & -11\cos t - 11\sin t & -6\cos t - 5\sin t \end{pmatrix}.$$

Hint. Express the matrix $A(t)$ in the form (E.4), where $\alpha_0(t) = \cos t$, $\alpha_1(t) = \sin t$, and

$$A_0 = \begin{pmatrix} 0 & 1 & 0 \\ 0 & 0 & 1 \\ -6 & -11 & -6 \end{pmatrix}, \quad A_1 = \begin{pmatrix} 1 & 1 & 0 \\ 0 & 1 & 1 \\ -6 & -11 & -5 \end{pmatrix}.$$

49. The time-varying matrix $A(t)$ and the constant matrix D form the *Lax pair*[3] if

$$\dot{A}(t) = DA(t) - A(t)D. \tag{E.5}$$

Suppose that a matrix D has been chosen so that it forms the Lax pair with the matrix $A(t)$ of the system (E.1). Show that by the change of variables $x = e^{Dt}$ the system (E.1) can be reduced to the system $\dot{y} = By$ with constant matrix B equal to $B = A(0) - D$.

50. Show that the trivial solution of the system (E.1) with matrix

$$A(t) = \begin{pmatrix} -\cos\omega t & \sin\omega t \\ \sin\omega t & \cos\omega t \end{pmatrix}$$

is stable for $\omega > 2$ and unstable for $\omega \leqslant 2$. It should be noted that the corresponding time-invariant system is unstable for $\omega = 0$. Hence, in this system parametric stabilization occurs for sufficiently high frequencies ω of parameter changes.

Hint. The matrices $A(t)$ and $D = \begin{pmatrix} 0 & \omega/2 \\ -\omega/2 & 0 \end{pmatrix}$ form a Lax pair.

51. Verify that the system (E.1) with matrix

$$A(t) = \begin{pmatrix} a + b\cos\omega t & d - b\sin\omega t \\ -d - b\sin\omega t & a - b\cos\omega t \end{pmatrix}$$

is stable if

$$a < 0, \qquad a^2 - b^2 + \left(d - \frac{\omega}{2}\right)^2 > 0.$$

Hint. The matrices $A(t)$ and $D = \begin{pmatrix} 0 & \omega/2 \\ -\omega/2 & 0 \end{pmatrix}$ form a Lax pair.

[2]Wu, M.-Y., and Sherif, A., On the Commutative Class of Linear Time-Varying Systems, *Internat. J. Control*, (1976), vol. 23, no. 3.

[3]Lax, P. D., Integrals of Nonlinear Equations of Evolution and Solitary Waves, *Comm. Pure Appl. Math.*, (1968), vol. 21, 467–490.

52. Reduce the systems in Examples 6.1 and 6.2 of Chapter I to systems with constant matrices.

53. The linearized equations of the perturbed motion of a spacecraft along a circular orbit in a central force field, referred to a fixed coordinate system, coincide with the system of equations (E.1) with matrix

$$A(t) = \begin{pmatrix} 0 & 0 & 1 & 0 \\ 0 & 0 & 0 & 1 \\ \frac{1}{2}(1+3\cos 2\omega t) & \frac{3}{2}\sin 2\omega t & 0 & 0 \\ \frac{3}{2}\sin 2\omega t & \frac{1}{2}(1-3\cos 2\omega t) & 0 & 0 \end{pmatrix},$$

where ω is the angular velocity of the spacecraft.

Reduce this system to a time-invariant one and test it for stability.

Hint. The matrices $A(t)$ and $D = \begin{pmatrix} 0 & -1 & 0 & 0 \\ 1 & 0 & 0 & 0 \\ 0 & 0 & 0 & -1 \\ 0 & 0 & 1 & 0 \end{pmatrix}$ form a Lax pair.

54. Consider [2] the system (E.1) with matrix [2]

$$A(t) = \begin{pmatrix} 0 & -1 & \varepsilon\cos\omega t & \varepsilon\sigma_2^{-1}\sin\omega t \\ \sigma_1^2 & 0 & \varepsilon\sigma_1\sin\omega t & -\varepsilon\sigma_1\sigma_2^{-1}\cos\omega t \\ -\varepsilon\sin\omega t & \varepsilon\sigma_1^{-1}\cos\omega t & 0 & -1 \\ \varepsilon\sigma_2\cos\omega t & \varepsilon\sigma_2\sigma_1^{-1}\sin\omega t & \sigma_2^2 & 0 \end{pmatrix}.$$

Show that for any values of the parameters $\sigma_1, \sigma_2, \varepsilon, \omega$ such that $\varepsilon \neq 0$, $\sigma_1^2+\sigma_2^2 \neq \omega$ this system is unstable.[4]

Hint. The given matrix $A(t)$ and the matrix D given in the form

$$D = \begin{pmatrix} 0 & \alpha & 0 & 0 \\ -\sigma_1^2\alpha & 0 & 0 & 0 \\ 0 & 0 & 0 & \beta \\ 0 & 0 & -\sigma_2^2\beta & 0 \end{pmatrix}, \qquad \sigma_1\alpha + \sigma_2\beta = -\omega,$$

form a Lax pair. Further, use the Routh–Hurwitz criterion.

55. Suppose there exist a smooth scalar function $\varphi(t)$, $\varphi(t) > 0$, $t \geqslant t_0$, and a constant matrix D such that for $t \geqslant t_0$

$$\frac{d}{dt}\left(\frac{A(t)}{\varphi(t)}\right) = DA(t) - A(t)D. \tag{E.6}$$

Verify that by the simultaneous change of coordinates and time

$$x = \exp[D\psi(t, t_0)]y,$$

$$\tau = \psi(t, t_0) = \int_{t_0}^{t} \psi(s)\, ds$$

[4] Morozov, V. M., and Kalyonova, V. I., *Estimation and Control in Time-Varying Linear Systems*, Moscow University Press, Moscow, (1988). See also [2]

the system (E.1) with matrix $A(t)$ satisfying the relation (E.6) reduces to the following system with constant matrix:

$$\frac{dy}{d\tau} = By, \qquad B = \lim_{t\to\infty} \frac{A(t)}{\varphi(t)} - D.$$

56. Reduce the system (E.1) with matrix

$$A(t) = \begin{pmatrix} 0 & 1 & 0 \\ 0 & 0 & 1 \\ -a_3 t^{-3} & -a_2 t^{-2} & -a_1 t^{-1} \end{pmatrix}$$

to the system with constant coefficient.

Hint. Make the change of variables

$$x = T(t)y, \qquad T(t) = \begin{pmatrix} 1 & 0 & 0 \\ 0 & t^{-1} & 0 \\ 0 & 0 & t^{-2} \end{pmatrix},$$

and then set $\tau = \ln t$.

57. Determine the stability or instability of the system (E.1) with matrix $A(t)$ of the form:

$$\text{(a)} \quad A(t) = \begin{pmatrix} -4 & 3e^{-8t} \\ -e^{8t} & 0 \end{pmatrix};$$

$$\text{(b)} \quad A(t) = \begin{pmatrix} 2 & -e^{-t} \\ e^{-t} & 1 \end{pmatrix};$$

$$\text{(c)} \quad A(t) = \begin{pmatrix} \frac{\sin t}{t} & 1 \\ -2 + e^{-t} & -3 - \frac{\sin t}{t} \end{pmatrix};$$

$$\text{(d)} \quad A(t) = \begin{pmatrix} 0 & t^{-1} \\ -2t^{-1} & -(t + t^2)^{-1} \end{pmatrix}.$$

58. *Small oscillations of a stabilizer gyrocompass.*[5] The equations of small oscillations of a stabilizer gyrocompass, subject to the condition $V^2 \ll gR^{-1}$, where V is the velocity of the point of suspension of the gyrocompass, g is the acceleration of gravity, and R is the radius of the earth, can be written in the form (E.1) with matrix

$$A(t) = \begin{pmatrix} 0 & \omega_0 & 0 & \Omega(t) \\ -\omega_0 & 0 & \Omega(t) & 0 \\ 0 & -\Omega(t) & 0 & -\omega_0 \\ -\Omega(t) & 0 & -\omega_0 & 0 \end{pmatrix}$$

[5] Ishlinskii, A. Yu., *Attitude Control, Gyroscopes, and Inertial Navigation*, Nauka, Moscow, (1976).

Here the phase coordinates x_i denote the following dimensionless variables:

$$x_1 = \frac{V(t)\delta_1}{\sqrt{gR}}, \qquad x_2 = \delta_2, \qquad x_3 = \delta_3, \qquad x_4 = \frac{2B\sin\varepsilon_0}{ml\sqrt{gR}}\delta_4,$$

$V(t)$ is the absolute velocity of the point of suspension, $\Omega(t)$ is the projection of the absolute angular velocity of the gyrocompass sensor on the geocentric vertical reference, $\omega_0 = \sqrt{g/R}$, $2B\cos\varepsilon_0 = mlV$; B, m, l, ε_0 are the parameters of the gyrocompass; δ_1, δ_2, δ_3, δ_4 are the angles specifying the position of the axes of the sensor relative to some fixed coordinate system (see Figure E.3).

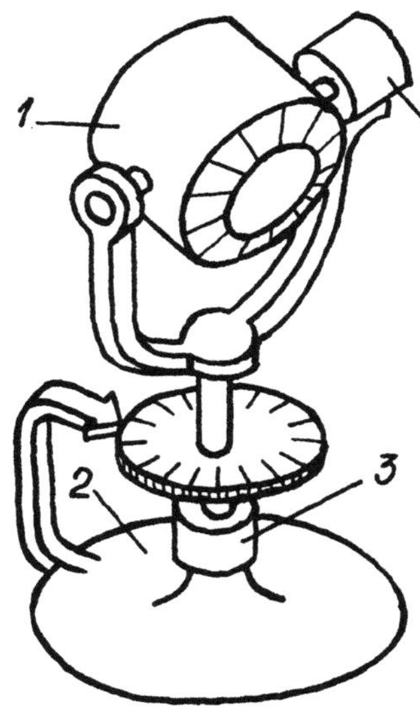

FIG. E.3. Stabilizer gyrocompass:
I, gimbal-suspended astatic gyroscope; II, platform stabilized in the horizontal; III, gyro-rotor drive; IV, platform drive

Verify that the system of small oscillations of the stabilizer gyrocompass is stable for any velocity $\Omega(t)$ (see the reference to Exercise 54).

Hint. Show that the matrix $A(t)$ can be expressed in the form (E.4), where $\alpha_1(t) \equiv 1$, $\alpha_2(t) = \Omega(t)$, and

$$A_1 = \begin{pmatrix} 0 & \omega_0 & 0 & 0 \\ -\omega_0 & 0 & 0 & 0 \\ 0 & 0 & 0 & -\omega_0 \\ 0 & 0 & \omega_0 & 0 \end{pmatrix}, \qquad A_2 = \begin{pmatrix} 0 & 0 & 0 & 1 \\ 0 & 0 & 1 & 0 \\ 0 & -1 & 0 & 0 \\ -1 & 0 & 0 & 0 \end{pmatrix},$$

$$A_1A_2 = A_2A_1.$$

Next, by the change of variables

$$x = \exp\left[A_2 \int_0^t \Omega(s)\,ds\right] y$$

reduce the system of small oscillations of the stabilizer gyrocompass to the form $\dot{y} = A_1 y$.

59. *Vertical gyroscope with spinning vessels.* This represents a gimbal-suspended gyroscope with its spin axis aligned with the vertical. On the gyro housing a platform is located, which spins with a constant angular velocity ω about the axis that coincides geometrically with the rotor axis. Two communicating vessels with mercury connected by means of a capillary tube are mounted on the platform. When the mercury is equally distributed in both vessels, the center of gravity of the whole system coincides with the point of support of the gyroscope. The platform is spinning in the same way as the gyro rotor (see Figure E.4).

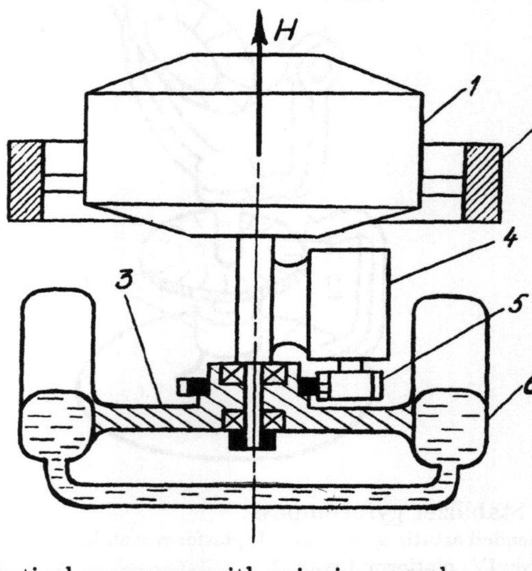

FIG. E.4. Vertical gyroscope with spinning vessels:
I, gyroscope with spin axis; II, gimbal suspension of the gyroscope; III, platform; IV, platform motor; V, platform drive; VI, communicating vessels with mercury

Discarding the nutation of the gyroscope, we have equations of motion of the gyroscope of the form (E.1) with matrix

$$A(t) = \begin{pmatrix} 0 & 0 & c\sin\omega t \\ 0 & 0 & -c\cos\omega t \\ -b\cos\omega t & -b\sin\omega t & -b \end{pmatrix}, \qquad x = \begin{pmatrix} \alpha \\ \beta \\ \varphi \end{pmatrix}.$$

Here α, β are the angles of the rotor axis, and φ is the angle of inclination of the liquid table; b, c are the gyroscope parameters.

Determine the conditions for the stability of the vertical gyroscope.

Hint. Show that the matrix $A(t)$ satisfies the condition (E.5) with a matrix D of the form

$$D = \begin{pmatrix} 0 & -\omega & 0 \\ \omega & 0 & 0 \\ 0 & 0 & 0 \end{pmatrix}$$

On reducing the system to the form $\dot{y} = By$, $B = A(0) - D$, use the Routh–Hurwitz criterion (see the reference to Exercise 54).

60. For the matrices of Exercise 1 calculate the norms and the logarithmic norms.

61. Construct matrices whose logarithmic norms are negative.

62. For the norms cited in Table 9.1 and for any vector $x \in R^n$ prove the inequality $\|Ax\| \leqslant \|A\|\|x\|$. Then indicate vectors x such that $\|Ax\| = \|A\|\|x\|$.

63. Prove the expressions for logarithmic norms cited in Table 9.1.

64. Prove the properties of the logarithmic norm as described by the relations in Table 9.1.

65. The norm of a vector $x \in R^n$ is described by

$$\|x\| = \left(\sum_{i=1}^{n} \alpha_i x_i^2\right)^{1/2}, \qquad \alpha_i > 0.$$

Determine the operator norm and the logarithmic norm of a matrix that are subordinate to the above definition of the norm.

66. Using the logarithmic norm, find the stability conditions for the systems described in Exercises 13 and 15.

67. Evaluate[6]

$$M\left(\int_0^T \xi^n(t)\, d\xi(t)\right)^2, \quad n = 0, 1, \dots; \qquad M\left(\int_0^T e^{\xi(t)}\, d\xi(t)\right)^2.$$

68. Evaluate the stochastic integral

$$J = \int_0^T \cos \xi(t)\, d\xi(t).$$

Show that

$$J = \sin \xi(T) + \frac{1}{2} \int_0^T \sin \xi(s)\, ds.$$

[6]In Exercises 67–80 $\xi(t)$ denotes a standard Wiener process.

69. Suppose that a random process $(x(t), y(t))$ has the stochastic differential

$$dx(t) = -\big(\sin \alpha(t)\big)\, d\xi(t), \qquad dy(t) = -\big(\cos \alpha(t)\big)\, d\xi(t),$$
$$t \geqslant 0,\ x(0) \neq 0,\ \alpha(t) = \arctan \frac{y(t)}{x(t)}.$$

Applying the Ito formula, prove that

$$x^2(t) + y^2(t) = x^2(0) + y^2(0) + t, \qquad t \geqslant 0.$$

70. Suppose that a function $F(x)$, $x \in (-\infty, \infty)$, is twice continuously differentiable. Using the Ito formula, prove that

$$F(0) + \frac{1}{2} \int_0^t F''\big(\xi(s)\big)\, ds = F\big(\xi(t)\big) - \int_0^t F'\big(\xi(s)\big)\, d\xi(s).$$

71. Verify that the functions $x(t) = \operatorname{sh} \xi(t)$, $y(t) = \operatorname{ch} \xi(t)$ are a solution of the Ito stochastic equations

$$dx(t) = \tfrac{1}{2} x(t)\, dt + y(t)\, d\xi(t),$$
$$dy(t) = \tfrac{1}{2} y(t)\, dt + x(t)\, d\xi(t).$$

72. Consider the oscillations of a weather vane subject to random gusts of wind. The equations of motion are of the form

$$\ddot{x} + a_1 \dot{x} + a_2 x = \sigma \dot{\xi}, \qquad t > 0,\ x(0) = 0,\ \dot{x}(0) = 0,$$

where $a_1 > 0$, $a_2 > 0$, σ are given constants. Find the covariance matrix of the process $\big(x(t), \dot{x}(t)\big)$.

73. Find a solution of the Ito stochastic differential equation

$$dx(t) = \tfrac{1}{2} x(t)\, dt + x(t)\, d\xi(t).$$

Hint. Transform this equation to the corresponding Stratonovich equation.

74. Show that for a Gauss–Markov process $x(t)$ satisfying the linear stochastic equation

$$dx(t) = -\gamma x(t)\, dt + g\, d\xi(t), \qquad t \geqslant 0,\ x(0) = x_0,$$

where $\gamma > 0$, g are given constants, the covariance function is

$$Ex(t)x(t+\tau) = \frac{g^2}{2\gamma}(1 - e^{-2\gamma t}) e^{-\gamma |\tau|}, \qquad -t < \tau < \infty.$$

75. Show that for Gauss–Markov processes $x_1(t)$ and $x_2(t)$ satisfying the system of linear stochastic equations

$$dx_1(t) = -\gamma x_1(t)\,dt + g\,d\xi_1(t),$$
$$dx_2(t) = -\gamma x_2(t)\,dt + Rg\,d\xi_1(t) + \sqrt{1-R^2}g\,d\xi_2(t),$$

where $\xi_1(t)$ and $\xi_2(t)$ are independent Wiener processes, and $\gamma > 0$, g, R are given constants, the cross-covariance function for $t \gg \gamma^{-1}$ is

$$Ex_1(t)x_2(t+\tau) = R\frac{g^2}{2\gamma}e^{-\gamma\tau}, \qquad 0 \leqslant \tau < \infty.$$

76. An amplitude- and phase-modulated harmonic signal

$$s(t) = A(t)\cos\bigl(\omega_0 t + \varphi(t)\bigr), \qquad t > 0,\ A > 0,$$

is often expressed in terms of its quadratures as

$$s(t) = \lambda_1(t)\cos\omega_0 t + \lambda_2(t)\sin\omega_0 t. \tag{E.7}$$

Suppose that the amplitude, A, and phase, φ, satisfy the stochastic equations

$$\begin{aligned} dA(t) &= \left(-\gamma A(t) + \frac{g^2}{2A(t)}\right)dt + g\,d\xi_1(t),\\ d\varphi(t) &= \frac{g}{A(t)}\,d\xi_2(t), \end{aligned} \tag{E.8}$$

where $\xi_1(t)$ and $\xi_2(t)$ are independent Wiener processes, and $\gamma > 0$, g are given constants.

Verify that the processes $\lambda_1(t)$ and $\lambda_2(t)$ satisfy the following stochastic equations:

$$\begin{aligned} d\lambda_1(t) &= -\gamma\lambda_1(t)\,dt + g\,d\tilde{\xi}_1(t),\\ d\lambda_2(t) &= -\gamma\lambda_2(t)\,dt + g\,d\tilde{\xi}_2(t), \end{aligned} \tag{E.9}$$

where $\tilde{\xi}_1(t) = (\cos\varphi)\xi_1(t) - (\sin\varphi)\xi_2(t)$ and $\tilde{\xi}_2(t) = -(\sin\varphi)\xi_1(t) - (\cos\varphi)\xi_2(t)$ are also independent Wiener processes.

Hint. Noting the relations $\lambda_1 = A\cos\varphi$ and $\lambda_2 = -A\sin\varphi$, use the Ito formula.

77. Suppose that, instead of the system (E.8), the amplitude, A, and the phase, φ, of the signal (E.7) satisfy the following system of stochastic differential equations:

$$\begin{aligned} dA(t) &= \left(-\gamma A(t) + \frac{g^2}{2A(t)}\right)dt + g\,d\xi_1(t), \qquad \gamma > 0,\\ d\varphi(t) &= \Omega(t)\,dt + \frac{g}{A(t)}\,d\xi_2(t),\\ d\Omega(t) &= -\beta\Omega(t)\,dt + h\,d\xi_3(t), \end{aligned} \tag{E.10}$$

where $\xi_1(t)$, $\xi_2(t)$, $\xi_3(t)$ are independent Wiener processes, and $\gamma > 0$, g, h are given constants. The additional term $\Omega(t)\,dt$ in the equation for $\varphi(t)$ describes, obviously, frequency modulation.

Verify that the processes $\lambda_1(t)$ and $\lambda_2(t)$ satisfy the following stochastic equations:

$$\begin{aligned} d\lambda_1(t) &= -\gamma\lambda_1(t)\,dt + \Omega(t)\lambda_2(t)\,dt + g\,d\tilde{\xi}_1(t), \\ d\lambda_2(t) &= -\gamma\lambda_2(t)\,dt - \Omega(t)\lambda_1(t)\,dt + g\,d\tilde{\xi}_2(t), \end{aligned} \tag{E.11}$$

where $\tilde{\xi}_1(t)$ and $\tilde{\xi}_2(t)$ are the same as in the preceding exercise.

Hint. The same as in the preceding exercise.

78. Test the mean square stability of the scalar system

$$dx(t) = ax(t)\,dt + \sigma \sin x(t)\,d\xi(t).$$

Using the Lyapunov function $V = x^2$, check that the stability condition is of the form

$$a + \frac{\sigma^2}{2} < 0.$$

79. The following system of Stratonovich stochastic equations is given:

$$\begin{aligned} dx(t) &= \omega y(t)\,dt, \\ dy(t) &= -\big(ky(t) + \omega x(t)\big)\,dt - \sigma(t)y(t)\,d\xi(t). \end{aligned}$$

These equations model the motion of an oscillatory system with natural frequency ω and coefficient of viscous friction $k + \sigma\dot{\xi}(t)$.

By using the Lyapunov function

$$W = \left(\frac{1}{k-\sigma^2} + \frac{1}{2\omega^2}\left(k - \frac{\sigma^2}{2}\right)\right)x^2 + \frac{1}{\omega}xy + \frac{1}{k-\sigma^2}y^2,$$

show that the mean square stability condition is of the form $\sigma^2 < k$.

80. Given the scalar equation

$$dx(t) = -ax(t)\,dt + \sigma(t)\,d\xi(t), \qquad t > 0,$$

show that any solution of this equation determined by the deterministic initial condition $x(0)$ tends to zero as $t \to \infty$ whenever the constant $a > 0$ and $\int_0^\infty \sigma^2(t)\,dt < \infty$. To this end, use the relations

$$x(t) = x_0 e^{-at} + e^{-at}\int_0^t e^{as}\sigma(s)\,d\xi(s),$$

$$\begin{aligned} M[x(t) - x_0 e^{-at}]^2 &= e^{-2at}\int_0^t e^{2as}\sigma^2(s)\,ds \\ &\leqslant e^{-at}\int_0^{t/2}\sigma^2(s)\,ds + \int_{t/2}^t \sigma^2(s)\,ds. \end{aligned}$$

PART TWO

CONTROL OF DETERMINISTIC SYSTEMS

CHAPTER III

DESCRIPTION OF CONTROL PROBLEMS

This chapter is intended to describe various mathematical problem statements in the area of optimal control of continuous deterministic systems. Some particular engineering problems are considered, such as nuclear reactor design, the soft landing on the moon, etc.

§ 1. Introduction

Various real processes taking place around us are *controllable* in the sense that the processes in question follow some definite course depending on the action generated by the controller. Moreover, it seems natural to choose an optimal control action, i.e., an action that is best among all possible control actions. Optimization problems date back to ancient times; however, they have become of special interest in modern times due to several reasons, among which the following can be noted: the scarcity of natural resources and the development of technology and computers that has facilitated the calculation and implementation of complicated laws of control.

Early results of optimization theory have become part of the classical calculus and the calculus of variations; the latter is concerned with the investigation of smooth functions and functionals defined on a total space or on a smooth manifold, necessary conditions for an extremum being written as conditions for stationarity (the vanishing of the gradient of a function, Euler's equation etc.).

However, recently some new control problems have arisen, in which the control parameters may belong to some closed set.

An example of problems of this type is to determine the control program for the motion of the rocket in order to minimize fuel consumption, given the height or range of flight. Note that despite the fact that such problems became especially important in the forties of this century, in as early as 1919 P. Goddard, one of the pioneers of astronautics, investigated the problem of minimizing fuel consumption for a rocket attaining a given height. He draw the conclusion that it is not possible to solve this problem within the framework of the classical calculus of variations.

Necessary conditions for optimal control in the problems with control constraints (essentially extending the main results of the classical calculus of variations) can be written as Pontryagin's maximum principle. Another approach is based on the dynamic programming method of R. Bellman.

§ 2. Statement of the Optimal Control Problem

A particular problem of optimal control is specified by the type of equations describing the evolution of a system, by the form of a functional to be minimized called a *cost functional* or a *performance (criterion) index*, and by trajectory and control constraints. Let us concentrate on these notions.

1. The equations of evolution of a system. Depending on the character of the process under consideration, in order to describe it, we can use ordinary differential equations, equations with aftereffect, difference equations, partial differential equations, stochastic equations, and so on. Consider an object whose state at time t is determined by the *phase vector* $x(t) \in R^n$, where R^n is an n-dimensional Euclidean space. Sometimes the phase vector $x(t)$ is also called the *state vector*. Henceforth, all vectors $x \in R^n$ are considered as column vectors with components $x_1, \ldots, x_n$, i.e.,

$$x = \begin{pmatrix} x_1 \\ \vdots \\ x_n \end{pmatrix}.$$

Assume that the motion of an object for $t \geqslant t_0$ is described by the system of ordinary differential equations

$$\dot{x}(t) = f(t, x(t), u), \quad \dot{x}(t) = \frac{dx(t)}{dt}. \tag{2.1}$$

where $u \in R^m$ is a control, and $f \in R^n$ is a prescribed function. By making the control u assume various admissable values, we traverse the state space of the object, whose "best" (optimal) state is to be chosen.

Example 2.1. *Equations of motion of a rigid body.* The motion of a rigid body is composed of the motion of its mass center and the rotation about the mass center considered as a fixed point. If $r(t)$ is a three-dimensional state vector of the mass center, then in most cases the equation of motion of the mass center is of the form

$$\ddot{r}(t) = u. \tag{2.2}$$

where u is the resultant vector of forces, including control forces, acting on the body.

The equations of motion of a rigid body relative to its mass center have been derived by L. Euler and named after him. Let A, B, C be the principal central moments of inertia, and p, q, r be the projections of the angular velocity vector on the principal central axes of inertia Ox, Oy, Oz. Then Euler's equations can be written in the form

$$\begin{aligned} A\dot{p} + (C - B)qr &= M_1, \\ B\dot{q} + (A - C)rp &= M_2, \\ C\dot{r} + (B - A)pq &= M_3. \end{aligned} \tag{2.3}$$

where M_1, M_2, M_3 are the projections on Ox, Oy, Oz of the control torque.

The stability of rotation of a rigid body was investigated in § 3 of Chapter 1.

An important class of control systems is composed of *linear systems*. In this case the equations of motion (2.1) are linear in the phase coordinates and the control, i.e.,

$$\dot{x}(t) = A(t)x(t) + B(t)u.$$

Linear equations can be used to approximate the behavior of real systems in the neighborhood of the nominal (unperturbed) trajectory.

Another class of systems that is used for describing real objects is called *bilinear*; such systems are characterized by bilinear equations whose right sides are linear in coordinates, given fixed values of control variables, and linear in control variables, given fixed values of coordinates:

$$\dot{x}_i(t) = \sum_{j=1}^{n} a_{ij}(t)x_j + \sum_{j=1}^{n}\sum_{k=1}^{m} u_k(t)B_{ijk}(t)x_j + Bu_0.$$

2. The functional to be minimized (cost functional). System (2.1) is controlled by u in order to achieve a certain goal, which can be expressed mathematically in terms of the minimization of a certain functional depending on the system's trajectory and the control function. Subject to the manner in which the *functional to be minimized*, also called the *cost functional* or *performance criterion*, is prescribed, it is customary to distinguish between the problems of Lagrange, Mayer, and Boltz.

In the *Lagrange problem* the cost functional J_0 is of the form

$$J_0 = \int_{t_0}^{T} F_0(t, x(t), u)\, dt, \qquad x \in R^n,\ u \in R^m, \tag{2.4}$$

where F_0 is a prescribed scalar function, and T is the moment of the end of motion, or the *terminal moment.* The moment T can be either prescribed beforehand or determined by a particular trajectory of motion. In the latter case T may be considered as an additional optimization parameter.

In the *Mayer problem* the cost functional J_0 depends on the trajectory of the system only at the terminal moment T:

$$J_0 = \varphi_0(T, x(T)). \tag{2.5}$$

Finally, in the *Boltz problem* it is necessary to minimize the functional J of mixed type:

$$J_0 = \int_{t_0}^{T} F_0(t, x(t), u)\, dt + \varphi_0(T, x(T)). \tag{2.6}$$

Note that the foregoing classification of control problems by the form of the functional to be minimized is somewhat arbitrary. In fact, the Boltz problem (and hence the Lagrange problem) can easily be reduced to the Mayer problem.

To demonstrate this, we introduce another scalar variable $x_{n+1}(t)$ defined by the equations

$$\dot{x}_{n+1}(t) = F_0(t, x(t), u), \qquad t \geqslant t_0, \ x_{n+1}(t_0) = 0. \tag{2.7}$$

In that case

$$x_{n+1}(T) = \int_{t_0}^{T} F_0\big(t, x(t), u\big)\, dt.$$

Therefore, the functional (2.6) can be written as (2.5) in the following way:

$$J_0 = x_{n+1}(T) + \varphi_0\big(T, x(T)\big). \tag{2.8}$$

Thus, for the controlled process (2.1), (2.7) we have obtained a Mayer problem with cost functional (2.8), which is equivalent to the original problem (2.1), (2.6). In a similar way, in the case of a differentiable function $\varphi_0(t, x)$ a Mayer problem can be reduced to a Lagrange problem.

If in the functional (2.4) the function $F_0(t, x, u) = 1$, then the problem of minimizing the control time is called the *time-optimal control problem*.

In a number of cases the control objective is *to stabilize a prescribed programmed motion*, i.e., to hold the actual trajectory of motion within some neighborhood of the desired trajectory. In that case the functions F_0 and φ_0 in the functional (2.6) must characterize the deviation of the actual trajectory from the programmed one.

To this end, one often uses *quadratic functionals* of the form

$$J_0 = x'(T)N_1x(T) + \int_{t_0}^{T} \big(x'(t)N_2(t)x(t) + u'(t)N_3(t)u(t)\big)\, dt. \tag{2.9}$$

In the functional (2.9) the matrices $N_i \geqslant 0$ (i.e., N_i are positive-semidefinite); the prime denotes a transposition. The latter summand in (2.9) describing the cost of control fits naturally into a number of problems.

Example 2.2. Consider the problem of controlling a rocket in flight by means of the acceleration $u(t)$ caused by the thrust. Denote by m the mass of the vehicle. One of the requirements on control is that the consumption of fuel be minimal, i.e., the following relation be valid:

$$\frac{m(T)}{m(t_0)} \to \max_{u(t)}. \tag{2.10}$$

We transform the left side of relation (2.10). It follows from the definition of the reaction force that

$$u(t) = -\frac{v}{m} \cdot \dot{m}. \tag{2.11}$$

where $\dot{m}(t)$ is the mass flow rate, v is the velocity of particles in the propulsive jet relative to the vehicle. For a number of low-thrust jet engines, such as ion and plasma engines, the power γ in the jet remains constant and equal to

$$\gamma = -\frac{v^2}{2} \cdot \dot{m}. \tag{2.12}$$

Eliminating the exhaust velocity v from equations (2.11), (2.12) gives $2\gamma m^{-2}\dot{m} = -u^2(t)$.

Solving this equation, we find that the ratio on the left side of (2.10) is of the form

$$\frac{m(T)}{m(t_0)} = \left[1 + \frac{m(t_0)}{2\gamma}\int_{t_0}^{T} u^2(t)\,dt\right]^{-1}.$$

Therefore, the control objective (2.10) is reached once the integral $\int_{t_0}^{T} u^2(t)\,dt$ is minimized.

If the initial data (i.e., the moment t_0 and the vector $x(t_0)$) of the system can also be varied, then the cost functional may be written as

$$\begin{aligned} J_0 &= J_0\big(t_0, x(t_0), u, T, x(T)\big) \\ &= \int_{t_0}^{T} F_0\big(t, x(t), u\big)\,dt + \varphi_0\big(t_0, x(t_0), T, x(T)\big). \end{aligned} \tag{2.13}$$

Sometimes problems with prescribed t_0 and T are called *fixed-time problems*, and those in which either t_0 or T is not determined are called *problems with unspecified time.*

Henceforth, we shall write only those arguments of the cost functional J_0 that can be varied independently of one another. In particular, in fixed-time problems the cost functional (2.13) is of the form $J_0\big(x(t_0), u, x(T)\big)$, and in problems with prescribed initial data $\big(t_0, x(t_0)\big)$ and terminal data $\big(T, x(T)\big)$ the functional (2.13) depends on the control function only, i.e., $J_0 = J_0(u)$.

3. Trajectory constraints. In a number of practical situations the system is barred from entering various regions of the phase space. This is reflected by corresponding constraints on the trajectory of motion $x(t)$ of the system: at each instant t a region $G(t) \in R^n$ is prescribed, to which the state vector $x(t)$ belongs.

Depending on the form of the constraints, different types of control problems are distinguished. In *problems with fixed end points* the initial $x(t_0)$ and terminal $x(T)$ states are given. But if $x(t_0)$ (or $x(T)$) is not given, then the corresponding problem is called a *problem with a variable left (right) end point.* A *problem with variable end points* is a problem in which the moments t_0 and T are prescribed but the initial $x(t_0)$ and terminal $x(T)$ states may vary within the respective regions $G(t_0)$ and $G(T)$.

In *problems with isoperimetric constraints* the values of the integrals

$$J_j = \int_{t_0}^{T} F_j\big(t, x(t)\big)\,dt \leqslant 0 \qquad (j = 1, \dots, m_1), \tag{2.14}$$

F_j being given scalar functions, are prescribed.

Problems with constraints of the form (2.14) represent a generalization of a problem in the calculus of variations, namely that of determining the shape of a curve of given length (given perimeter) maximizing the area enclosed between the curve and the prescribed Cartesian axis.[1] Incidently, the term "isoperimetric constraints" is due to this problem.

In addition to trajectory constraints of the equality type (2.14) one can also meet inequality-type constraints:

$$J_i = \int_{t_0}^{T} F_i\big(t, x(t)\big)\, dt = 0 \qquad (i = m_1 + 1, \dots, m_2). \tag{2.15}$$

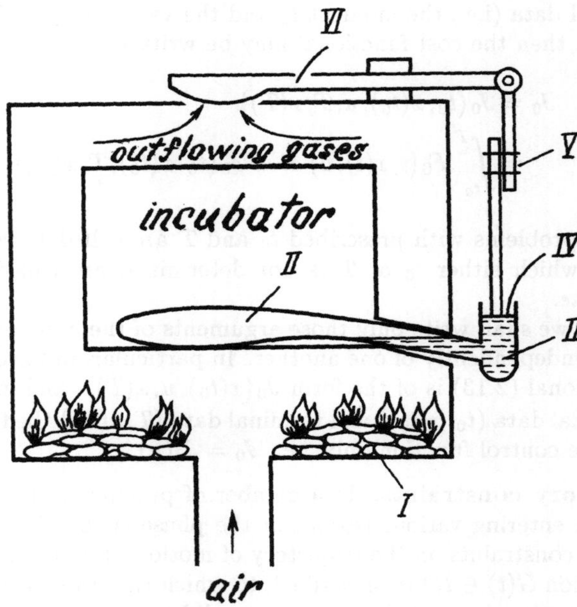

FIG. 2.1. The temperature regulator by Cornelius Drebbel:
I, burning coal; II, alcohol; III, mercury; IV, float; V, adjustment of the arm length; VI, damper valve

4. Control constraints. When designing an optimal control for the system (2.1), *a priori* information about the system is of great importance. If the phase vector $x(t)$ cannot be measured, then the optimal control is called the *program control* or *P-control*; it is sought as a function of t only, i.e., $u(t)$. Alternatively, if at each instant t the phase vector $x(t)$ is known exactly, then the optimal control is sought in the class of functions depending on time and the phase coordinates, and is of the form $u = u\big(t, x(t)\big)$. In this case the control

[1]This classic problem seems to be the first isoperimetric problem whose solution was known in ancient times; it is connected with the Tyrian princess Dido who thereby maximized the land upon which she was to found Carthage. This problem is also known as Dido's problem.

$u = u(t, x(t))$ is called the *feedback control* or *F-control*. Clearly, F-control is more flexible compared to P-control, i.e., it renders a value of the cost functional that is not greater than the corresponding value for P-control. The simplest example of feedback is proportional feedback, in which the control function varies proportionally with the phase vector.

The notion of feedback originated nearly two thousand years ago in connection with the design of a water clock, in which the water level was maintained by means of a float valve. Float-valve level regulators have found numerous engineering applications, as in car carburetors. In such regulators the weight of the float is balanced by the force of Archimedes. If the liquid level deviates from its nominal value, then the float moves either up or down. This, in turn, activates the feed or run-off mechanisms.

In 1620 the Dutchman Cornelius Drebbel constructed a stove with automatically controlled temperature (see Figure 2.1). In this stove a slight increase in the temperature inside the incubator leads to the expansion of the alcohol, and this, in turn, leads to the rising of the float, which produces the covering of the hole by the damper valve and, consequently, turns down heat for the incubator. By changing the length of the arm of the damper valve, one can regulate the nominal value of the temperature inside the incubator. Modern investigations of this stove have shown that the temperature can be stabilized to within 0.6° C with the help of such a simple regulator.

Other classical examples of the use of feedback are the centrifugal governors of I. Polzunov (1766) and J. Watt (1784). The latter one is shown in Figure 2.2.

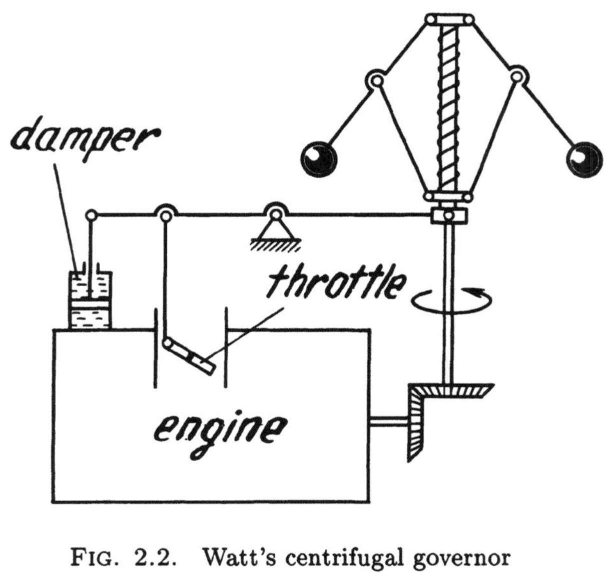

FIG. 2.2. Watt's centrifugal governor

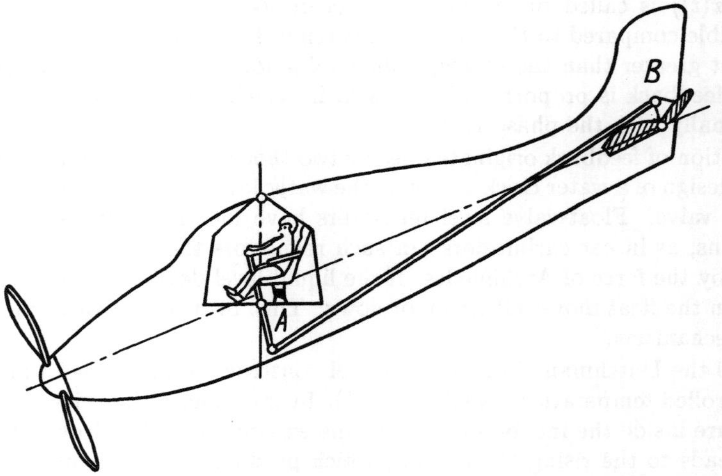

FIG. 2.3. Olkhovskii's pendulum autopilot for pitch hold

The principle of feedback was also used by Olkhovskii (1912) and the Sperry brothers (1914) for constructing the first aviation autopilot (see Figure 2.3). All these regulators belong to the class of *direct-action regulators*, in which a measuring device acts directly on the regulator. Such control is feasible only for low-power machines. However, an indirect control is more often used; it involves amplifiers of various kind and other final control elements in the control loop. A hydraulic amplifier in the control loop was used for the first time in 1873.

If the coordinates of the system can be measured precisely and their values on the time interval from the initial moment t_0 to the current moment t stored, then an optimal F-control can be sought as a functional $u(t, x_t)$, where $x_t = \{x_\tau, \tau \in (t_0, t)\} = \{x(t+\theta), \theta \in (t_0 - t, 0)\}$ is the corresponding segment of the trajectory of the system (2.1).

In this case substitution of the control $u(t, x) = u(t, x_t)$ in equation (2.1) produces a change in the type of the equation: it now becomes an equation with aftereffect.

However, for deterministic systems treated in this chapter the optimal values of the cost functional attained in the class of F-controls and those for P-controls are the same. Indeed, if an F-control for the system (2.1) has been constructed, then solving (2.1) for $x(t)$, given a control $u(t, x)$ and an initial condition, yields a P-control $v(t) = u\big(t, x(t)\big)$. Conversely, once a P-control $v_{x,t_0}(t)$ for an arbitrary initial condition $x(t_0) = x$ is obtained, the corresponding F-control $u(t, x)$ is determined by the formula $u(t, x) = v_{x,t}(t)$. Henceforth, in view of the above, it suffices to consider P-controls only.

In addition to the information constraints discussed above which are related to the amount of knowledge available to the controller, constraints of another type are possible, namely those that involve restrictions on control resources, *control constraints*.

These constraints can be written in the form

$$u(t) \in U(t), \tag{2.16}$$

where $U(t) \subset R^m$ is a given set.

An example of constraints of the form (2.16) is the following: $\big(u'(t), u(t)\big)^{1/2} = |u(t)| \leqslant u_0$. Such constraints arise, for example, in controlling the rotation of a spacecraft if the torque M is due to a pair of steering jet engines.

The relation (2.16) means that the control $u(t)$ is subject to constraints at each instant. Sometimes control constraints are those of integral type similar to trajectory constraints (2.14) and (2.15):

$$J_j = \int_{t_0}^{T} F_j\big(t, u(t)\big)\, dt \leqslant 0 \qquad (j = 1, \ldots, m_1), \tag{2.17}$$

$$J_k = \int_{t_0}^{T} F_k\big(t, u(t)\big)\, dt = 0 \qquad (k = m_1 + 1, \ldots, m_2). \tag{2.18}$$

In some cases relations (2.17), (2.18) can be interpreted as constraints on the cost of the control process.

Note that constraints of the type (2.14) or (2.17) can be reduced to trajectory constraints at the moments t_0 and T by augmenting the phase vector. In fact, in the case of, say, constraints of the type (2.14) it suffices to introduce a vector $y(t) \in R^m$ satisfying the equations

$$\dot{y}_j(t) = F_j\big(t, x(t)\big), \qquad y(t_0) = y(T) = 0,\ j = 1, \ldots, m_1.$$

Then the original problem of the control of the vector $x(t)$ subject to the constraints (2.14) is equivalent to that of the control of the vector $\big(x(t), y(t)\big)$ without the constraints (2.14).

5. Joint constraints. We have considered above some problem statements in optimal control. They are specified by the equations of motion (2.1), by the cost functional (2.13), by the trajectory constraints (2.14) and/or (2.15) and by the control constraints (2.16), (2.17), (2.18). One can obtain new problem statements in the case of joint constraints, i.e., in cases where the constraints in question depend on both the trajectory and the control.

Let us summarize the relations that specify the problem statement in the case under consideration:

(1) the equations of motion (2.1)

$$\dot{x}(t) = f\big(t, x(t)\big), \qquad x \in R^n,\ u \in R^m,\ t_0 \leqslant t \leqslant T; \tag{2.19}$$

(2) the scalar cost functional

$$J_0\big(t_0, x(t_0), u, T, x(T)\big) \to inf. \tag{2.20}$$

(the cost functional is often designated by J without the subscript "0");

(3) the constraints

$$J_j\big(t_0, x(t_0), u, T, x(T)\big) \leqslant 0 \qquad (j = 1, \dots, m_1), \tag{2.21}$$

$$J_i\big(t_0, x(t_0), u, T, x(T)\big) = 0 \qquad (i = m_1 + 1, \dots, m_2), \tag{2.22}$$

$$u(t) \in U. \tag{2.23}$$

In these relations the scalar functionals J_k are of the form

$$J_k\big(t_0, x(t_0), u, T, x(T)\big) = \int_{t_0}^{T} F_k\big(t, x(t), u(t)\big)\, dt + \varphi_k\big(t_0, x(t_0), T, x(T)\big), \qquad k = 0, \dots, m_2. \tag{2.24}$$

The problem of evaluating the minimum of the functional (2.20) with constraints of the form (2.19), (2.21), and (2.22) is called the *Lagrange problem.* In the relations (2.19)–(2.23) the functions f, $F = (F_k)$, and $\varphi = (\varphi_k)$ are prescribed:

$$f : R^1 \times R^n \times R^m \to R^n; \qquad F_k : R^1 \times R^n \times R^m \to R^1;$$
$$\varphi_k : R^1 \times R^n \times R^1 \times R^n \to R^1, \quad k = 0, \dots, m_2.$$

Note that in the relations (2.20)–(2.22) some of the functions φ_k or F_k may vanish. The moments t_0 and T that are associated with the beginning of the motion and its termination, respectively, belong to some prescribed finite segment D of the number line.

The control $u(t)$ is sought in the class of piecewise continuous functions, and the trajectory $x(t)$ in the class of piecewise continuous differentiable functions of $t \in D$.

A set of variable parameters $q = \big(t_0, x(t_0), u, T, x(T)\big)$ is said to be *admissible* if it satisfies all the constraints (2.19), (2.21)–(2.23). An admissible parameter set q is said to be *optimal* if for any near (in a certain sense) admissible set $\bar{q}$ the value of the functional (2.20) corresponding to q is not greater than that for $\bar{q}$. Depending on the definition of proximity of q and $\bar{q}$, there exist different types of extremum (e.g., a strong extremum, a weak extremum, etc.). It is convenient to express the proximity of the parameter sets q and $\bar{q}$ in terms of the proximity of the trajectories $x(t)$ and $\bar{x}(t)$ that correspond to q and $\bar{q}$. Everywhere in Part Two necessary conditions for the existence of a locally optimal strong extremum are considered.

Definition 2.1. An admissible parameter set q is said to be *locally optimal in the strong sense* if there exists an $\varepsilon > 0$ such that for any admissible set $\bar{q}$ satisfying the estimate

$$|t_0 - \bar{t}_0| + |T - \bar{T}| + \max_t |x(t) - \bar{x}(t)| < \varepsilon, \qquad t \in [t_0, T] \cap [\bar{t}_0, \bar{T}],$$

the following inequality holds:

$$J_0(\bar{t}_0, \bar{x}(t_0), \bar{u}, \bar{T}, \bar{x}(\bar{T})) \geqslant J(t_0, x(t_0), u, T.x(T)).$$

Remark. Suppose that the moments t_0 and T are given, but the differential relation (2.19) is not given, and the functions F_k in (2.24) do not depend on the trajectory $x(t)$. Then the problem of the minimization of the functional (2.20) under the constraints (2.21), (2.22) is called the *Lyapunov problem* [2].

§ 3. Examples of Optimal Control in Engineering

1. Optimal control of an electric motor. Consider the optimal control problem for the rotation of the shaft of a direct-current motor controlled by the exciting current. The excitation circuit is shown in Figure 3.1. The control voltage $u(t)$ is applied to the excitation winding. Changes in the current $i(t)$ through the winding and in the magnetic flux $\varphi(t)$ are described by the equations

$$L\frac{di}{dt} + Ri = u(t), \qquad i(0) = i_0,\ \varphi(t) = k_1 i(t), \tag{3.1}$$

where R and L are the resistance and inductance of the excitation winding, respectively.

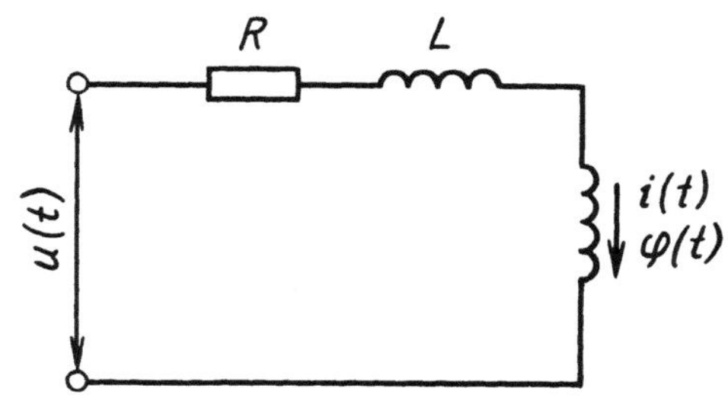

FIG. 3.1. The excitation circuit of an electric motor

The torque $M(t)$ of the armature, through which a direct current flows, is equal to $M(t) = k_2\varphi(t) = k_1 k_2 i(t)$, where $k_1, k_2 > 0$ are factors of proportionality. The output angular displacement $x(t)$ of the motor shaft satisfies the equation

$$J\ddot{x}(t) = M(t) = ki(t), \qquad k > 0, \tag{3.2}$$

where J is the moment of inertia of all rotating elements. Substituting expression (3.2) in (3.1) gives

$$LJ\dddot{x}(t) + JR\ddot{x}(t) = ku(t). \tag{3.3}$$

For the cost functional associated with the control $u(t)$, it is natural to take the energy E_{in} that is transferred to the input of the motor in the control interval from $t = 0$ to $t = T$. Obviously,

$$E_{\text{in}} = E_R + E_L, \tag{3.4}$$

where E_R is the energy liberated in the resistance R, and E_L is the energy stored in the inductance L during the time from $t = 0$ till $t = T$. As is known,

$$\begin{gathered} E_R = R \int_0^T i^2(t)\, dt, \\ E_L = L \int_0^T i(t) \frac{di}{dt}\, dt = L[i^2(T) - i^2(0)]. \end{gathered} \tag{3.5}$$

It follows from equation (3.1) that

$$i(t) = i(0) e^{-RL^{-I}t} + \int_0^t \exp\bigl(-RL^{-1}(t-s)\bigr) u(s)\, ds = g(u), \tag{3.6}$$

where $g\bigl(u(t+\theta)\bigr)$, $-t \leqslant \theta \leqslant 0$, is a functional of the form (3.6) depending on the values of the control $u(s)$ for $0 \leqslant s \leqslant t$.

Using the relations (3.5) and (3.6), we obtain the final form of the cost functional E_{in}:

$$J_0(u) = R \int_0^T g^2\bigl(u(t+\theta)\bigr)\, dt + L\bigl[g^2\bigl(u(T+\theta)\bigr) - i^2(0)\bigr]. \tag{3.7}$$

Two constraints are imposed on the control $u(t)$:

$$\begin{gathered} 0 \leqslant u(t) \leqslant U_{\max}, \\ L\left|\frac{di}{dt}\right| = \left|\frac{Ldg\bigl(u(t+\theta)\bigr)}{dt}\right| \leqslant U_{\text{br}}. \end{gathered} \tag{3.8}$$

The physical meaning of these constraints is as follows. The first constraint indicates that the input voltage cannot exceed the voltage of the supply line the motor is connected to. Further, abrupt changes in the input voltage $u(t)$ may lead to abrupt changes in the voltage across the inductance, and this may result in the break-through of the excitation winding. To prevent this, it is necessary to impose the second of the constraints (3.8), where the upper limit is identified with the break-through voltage U_{br} for the excitation winding.

Thus, the optimal control problem for the rotation of the shaft of an electric motor can be stated in the following way: transfer the shaft of the motor from the initial state $x(0) = x_0$ to the terminal state $x(T) = x_1$ so that the functional (3.7) has a minimum value and the rotation is constrained by the relations (3.8).

This problem is not an easy one. Therefore, a simplified model with $L = 0$ is often used. Consequently, the order of equation (3.3) is reduced and the functional (3.7) is simplified, in which the second summand vanishes and $i(t)g(u(t+\theta))$ in the first summand can be replaced by $u(t)$, since $Ri(t) = u(t)$. In addition, the second constraint (3.8) becomes superfluous. The simplified problem is of the form

$$\begin{gathered} \ddot{x}(t) = k_3 u(t), \qquad k_3 = kJ^{-1}R^{-1}, \\ x(0) = x_0,\ x(T) = x_1, \\ J_1(u) = \int_0^T u^2(t)\,dt, \qquad 0 \leqslant u(t) \leqslant U_{\max}. \end{gathered} \tag{3.9}$$

The investigation of the relationship between the solutions of the problem (3.3), (3.7), (3.8) and those of the problem (3.9) is of interest from a theoretical point of view. Can we assume that for sufficiently small $L > 0$ the solutions of these problems are close to one another? Suppose that by solving the problem (3.9) we have determined the optimal control of the original system (3.3). What error in the value of the functional (3.7) is then made? This and other similar problems are dealt with in the perturbation theory for optimal control problems; some of these problems will be considered in the sequel.

In the case of the problems (3.3), (3.7), (3.8) and (3.9) the situation is made more complicated because of the small parameter L being the multiplier of the highest-order derivative in equation (3.3). Such problems are called *singular perturbation problems.* It turns out that solutions of singular perturbation (for $L > 0$) and generative (for $L = 0$) problems are close. The determination of conditions for such proximity is of special interest. Investigation of singularly perturbed equations was carried out by A. N. Tikhonov. Various control problems for singularly perturbed systems were considered by many authors.[2]

2. Optimization of the characteristics of nuclear reactors[3]. In the process of designing and using nuclear reactors, several optimization problems arise. They can be divided into two large groups.

The first group comprises dynamical problems, in which certain time characteristics of nuclear reactors need to be adjusted. Such, for example, is the problem of the optimal burning-out of xenon-135, etc.

The second group comprises space optimization reactor problems, such as the maximization of reactor power with a constrained density of neutron flux, the optimal configuration of the absorber in the reactor, and so on.

Finally, there exist problems of mixed time-space type, such as the optimization problem for the burning-out of nuclear fuel.

---

[2]Dontchev, A. L., *Perturbations, Approximations, and Sensitivity Analysis of Optimal Control Systems*, Springer-Verlag, Berlin, (1983); Vasil'eva, A. B., Dmitriev, M. G., Singular Perturbations in Optimal Control Problems, *Mathematical Analysis (Itogi Nauki i Tekhniki, vol. 20)*, VINITI, Moscow, (1982).

[3]Rudik, A. P., *Optimization of the Physical Characteristics of Nuclear Reactors*, Atomizdat, Moscow, (1979); Rudik, A. P., *Nuclear Reactors and Pontryagin's Maximum Principle*, Atomizdat, Moscow, (1970).

We proceed to more detailed problem statements for the cases of optimization of reactor power and of xenon transient processes.

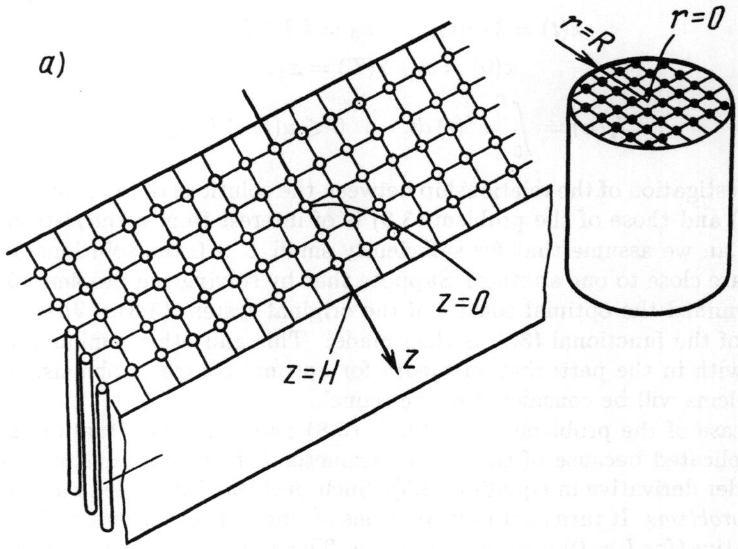

FIG. 3.2. Simplified models of nuclear reactors:
a, plane reactor; *b*, cylidrical reactor

Problem of the optimization of reactor power. Consider a simplified model of a reactor which gives a fairly good description of the properties of actual gas-cooled energy reactors (see Figure 3.2). A regular grid occupies all the available space of the reactor. At the vertices of the grid, tubes with heat releasing elements are located. By changing the shape of the tubes, we can change the amount of nuclear fuel in the tube, i.e., the concentration of nuclear fuel per unit volume. We shall confine ourselves with one-dimensional problems concerning plane, cylindrical, and spherical reactors. In addition, we consider the so-called one-group approximation that takes into account neutrons of one type only. From a physical point of view this approximation is valid for thermal neutron reactors with a good moderator or for fast neutron reactors with practically no thermal neutrons.

We denote by z the space variable and assume that $z = 0$ corresponds to the center of the reactor and $z = H$ to its boundary. Under these assumptions for a plane reactor, the neutron desity $N(z)$ satisfies the equation

$$\frac{d^2N(z)}{dz^2} + \alpha^2(u)N(z) = 0. \tag{3.10}$$

A typical empirical plot of the coefficient $\alpha^2(u) \geqslant 0$ is shown in Figure 3.3, its particular shape depending on the reactor design; however, in the neighborhood of points I and II the relationship in question may be taken to be linear, i.e.,

$$\alpha^2(u) = a + bu. \tag{3.11}$$

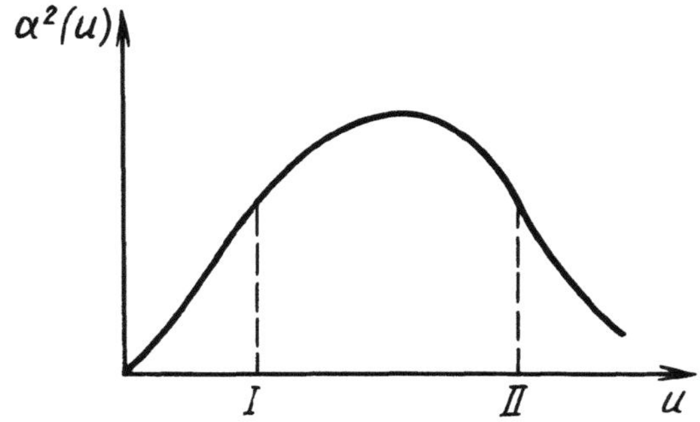

FIG. 3.3. An empirical plot of the coefficient $\alpha^2(u)$

The boundary conditions for equation (3.10) are of the form

$$\begin{aligned} N(0) &= N_{\max}, & \left.\frac{dN}{dz}\right|_{z=0} &= 0, \\ N(H) &= p, & \left.\frac{dN}{dz}\right|_{z=H} &= q. \end{aligned} \tag{3.12}$$

We normalize in a way such that $N_{\max} = 1$. The control must be constrained by

$$0 \leqslant u_{\min} \leqslant u \leqslant u_{\max}, \tag{3.13}$$

where $u_{\min}$ and $u_{\max}$ are specified by the reactor design, in particular, by the number of uranium rods in the tube.

The meaning of the optimized functional W is the power released in unit reactor volume. To within an arbitrary constant, this functional can be evaluated by

$$W = \int_0^H u(z)N(z)\,dz. \tag{3.14}$$

Indeed, the probability of capture of a neutron by the nuclear fuel is proportional to the product of the neutron concentration $N(z)$ by the fuel concentration u. Each neutron capture induces the fission of uranium accompanied by the liberation of a fixed amount of heat. Therefore, the overall heat power of the reactor is proportional to the integral in expression (3.14).

One often has to take into account technological restrictions that lead to temperature constraints or to constraints on the amount of heat liberated at all points of the reactor. These constraints can be written in the form

$$u(z)N(z) \leqslant p_1 \tag{3.15}$$

or in the following form:

$$u(z)N(z) + \int_0^z u(s)N(s)\,ds \leqslant p_2. \tag{3.16}$$

Thus, the statement of the problem of maximizing the reactor power is as follows: find a distribution $u(z)$ of nuclear fuel satisfying the constraints (3.13), (3.15) such that the power W of the reactor (3.14), where $N(z)$ is the solution of the problem (3.10)–(3.12), is maximized.

For a cylindrical reactor, the problem of maximizing reactor power is of the form

$$\begin{gathered}
\frac{1}{r}\frac{d}{dr}\left[r\frac{dN(r)}{dr}\right] + \alpha^2(u)N(r) = 0,\\
N(0) = N_{\max}, \frac{dN}{dr} = 0, \qquad N(R) = 0,\\
W = \int_0^R N(r)u(r)r\,dr \to \max,\\
0 \leqslant u_{\min} \leqslant u \leqslant u_{\max}.
\end{gathered}$$

We may also consider the problem of maximizing reactor power in the two-group approximation. Denote by $N(z)$ the neutron density in the thermal group, and by $n(z)$ in the moderation group. Then, if neutron breeding occurs only in the thermal group and there is no neutron capture in the moderation group, two-group equations for the plane reactor can be written as

$$\begin{aligned}
&\frac{d^2N(z)}{dz^2} + \alpha^2(u)N(z) = n(z),\\
&\frac{d^2n(z)}{dz^2} - \frac{n(z)}{\tau} = k(u)\frac{\alpha^2(u)}{\tau}N(z).
\end{aligned} \tag{3.17}$$

where τ is the moderation length squared, which is independent of control, $k(u)$ is the breeding coefficient depending on control. For a control we take either the quantity $u_1(z)$, which is proportional to the concentration of the nuclear fuel or the quantity $u_2(z)$, which is proportional to the concentration of the moderator that is put expressly inside the reactor. A general assumption is that

$$\begin{gathered}
\alpha^2(u_1) = a + bu_1 \qquad k(u_1)\alpha^2(u_1) = \eta u_1,\\
u_{1\,\min} \leqslant u_1 \leqslant u_{1\,\max}
\end{gathered}$$

or

$$\alpha^2(u_2) = a + bu_2, \qquad k(u_2)\alpha^2(u_2) = \eta,$$
$$u_{2\,\min} \leqslant u_2 \leqslant u_{2\,\max}.$$

It is required to choose a control so that it maximizes the overall reactor power, which is proportional to the integral

$$\int_0^H N(z)u(z)\,dz \to \max,$$

under the technological constraint $N(z)u(z) \leqslant p_1$.

Problem of the optimization of xenon transient processes. Among the various products of fission of uranium-235 (^{235}U) we find the unstable isotope, xenon-135 (^{135}Xe). The significance of ^{135}Xe for the performance of a nuclear reactor is due to the fact that it has a large cross-section of capture of thermal neutrons, and, consequently, plays the role of a neutron absorber. In the process of exploiting a nuclear reactor one of the problems that often arises is the switching of the reactor from one level of power to another, for example, its shut-down in order to carry out some work. At the beginning of the process there is a sharp increase in the amount of ^{135}Xe in the reactor, but in the subsequent 1.5–2 days it decays spontaneously. All this time the reactor is out of work. Thus, a problem arises, which is called the *problem of the xenon poisoning of the reactor.* Therefore, we must consider the problem of finding a way of shutting down the reactor so that there is no xenon poisoning, which is equivalent to the problem of the optimal control of transient processes during the formation of ^{135}Xe.

We now state this problem in greater detail. The formation and decay scheme for ^{135}Xe is the following:

$$^{235}U + n \longrightarrow \begin{cases} ^{135}Te \xrightarrow[\beta]{2\,\text{min}} {}^{135}I \xrightarrow[\beta]{9.7\,\text{h}} {}^{135}Xe \xrightarrow[\beta]{13.4\,\text{h}} {}^{135}Cs \xrightarrow[\beta]{2\cdot 10^6\,\text{yr}} {}^{135}Ba, \\ ^{135}Xe \xrightarrow[\beta]{13.4\,\text{h}} {}^{135}Cs \xrightarrow[\beta]{2\cdot 10^6\,\text{yr}} {}^{135}Ba. \end{cases} \tag{3.18}$$

The figures over the arrows indicate the lifetimes of the corresponding isotopes. The probability of the upper chain of reactions is quite large: $\gamma_1 = 0.06$, and that of the lower one is $\gamma_2 = 0.003$. Thus, ^{135}Xe is formed mostly from ^{135}I. Denote by $x_1(t)$ and $x_2(t)$ the concentrations of ^{135}I and ^{135}Xe at the instant t, and by $u(t)$ the control taken to be the mean density of thermal neutrons in the reactor. It is assumed that the concentration of nuclear fuel does not change during the control time. In this case the following equations hold:

$$\begin{aligned} \dot{x}_1(t) &= -\lambda_1 x_1(t) + \gamma_1\sigma_1 u(t); \\ \dot{x}_2(t) &= \lambda_1 x_1(t) - \lambda_2 x_2(t) + \gamma_2\sigma_1 u(t) - \sigma_2 u(t) x_2(t). \end{aligned} \tag{3.19}$$

Here λ_1, λ_2, γ_1, γ_2, σ_1, σ_2 are some quantities which may be considered as constants under the assumptions made.

Equations (3.19) have a simple physical meaning. The first equation describes the variation of the number of ^{135}I atoms with time; the term $\lambda_1 x_1$ is the number of iodine atoms that disappear per unit time due to β-decay (λ_1 is the inverse of the lifetime of ^{135}I equal to $T = 9.7\,\text{h}$); the term $\gamma_1\sigma_1 u(t)$ is the amount of ^{135}I produced due to nuclear fission ($\gamma_1 = 0.06$ is the probability of the upper reaction chain in the scheme (3.18), and σ_1 is the macroscopic cross-section of nuclear fission that remains unchanged due to the assumption of the concentration of nuclear fuel being constant). In a similar way, the term $\lambda_1 x_1$ in the second equation indicates the amount of ^{135}Xe produced due to the β-decay of ^{135}I; the term $\lambda_2 x_2$ describes the β-decay of ^{135}Xe; the term $\gamma_2\sigma_1 u(t)$ is the amount of ^{135}Xe produced due to the fission of ^{235}U according to the lower reaction chain in the scheme (3.18); and the term $-\sigma_2 u x_2$ describes the burning-out of ^{135}Xe due to the reaction that is associated with the capture of thermal neutrons by ^{135}Xe.

The initial conditions for equations (3.19) are usually prescribed according to the physical problem statement. Suppose that it is required to effect a switch of the reactor from a given level of power W_{in} to the terminal level $W_{\text{term}} = 0$. The level of power W_{in} for a fixed concentration of ^{235}U determines uniquely the neutron flux $u = u\text{in}$. Further, substituting this value in equation (3.19) and assuming $\dot{x}_1(t) = \dot{x}_2(t) = 0$ gives $x_1(0)$ and $x_2(0)$. The curve ξ in Figure 3.4 shows the dependence of $x_1(0)$ and $x_2(0)$ on the concentration of ^{235}U, i.e., on σ_1 (the measurement unit being taken as $10^16\,\text{cm}^{-2}$).

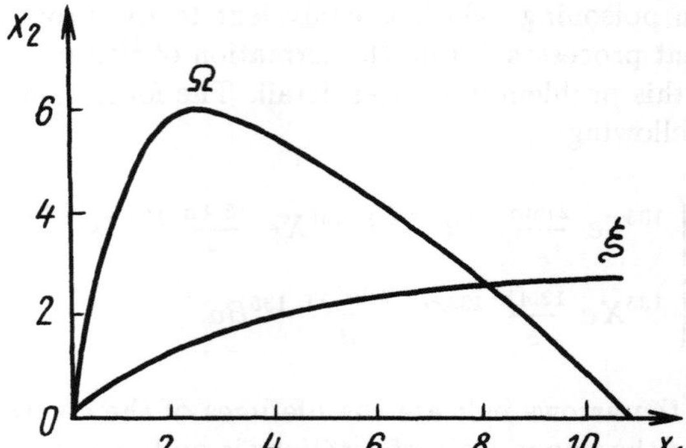

FIG. 3.4. Initial and terminal conditions in the optimization problem for xenon transient processes

The terminal values $x_1(T)$ and $x_2(T)$ are evaluated in a similar way from condition $W_{\text{term}} = 0$. The curve Ω in Figure 3.4 shows the dependence of $x_1(T)$ and $x_2(T)$ on the initial concentration of ^{135}I.

We assume that the cost functional is of the form

$$J(u) = \int_0^T [a + bu(t)]\, dt. \tag{3.20}$$

It should be stressed again that in view of the fact that the concentration of ^{235}U is constant the released energy $W(t)$ is proportional to $u(t)$. For different values of a and b we obtain the following problems:

—time-optimality problem ($a = 1$, $b = 0$);

—minimum energy-release problem ($a = 0$, $b = 1$);

—maximum energy-release problem ($a = 0$, $b = -1$);

—problem of minimum energy deviation from the maximum energy ($a = U_{\max}$, $b = -1$).

Other optimal control problems for nuclear reactors are discussed in the books by A. P. Rudik cited above.

3. Optimal control of spacecraft. Various interesting problems arise in rocket dynamics and spacecraft navigation. The importance of optimal control in astronautics is largely due to strict weight limitations. As is known, placing 1 kg in orbit requires from 30 to 50 kg of fuel; hence maneuvering in space must require a minimum amount of fuel. Therefore, the solution of various optimization problems in astronautics is justified from both an engineering and an economic point of view.

Numerous problems of the optimal control of space vehicles can be divided into several groups. One group is composed of problems in which a space vehicle can be treated as a material point, and hence equations of motion of the type (2.2) can be used. To this group belong problems of maximizing the height attained by a rocket, of moving a rocket to a given point, of landing on the moon, and so on. Another group comprises problems in which a space vehicle is considered as a rigid body described by Euler's equations (2.3). In the main, these are stabilization and rotation problems for space vehicles. In a number of problems one has to take account of the elastic properties of space vehicles or of some of their parts, such as antennas, etc. These problems comprise yet another group. Finally, it is convenient to consider a special group consisting of navigation problems, say, connected with space flights from the earth to the moon, from the earth to Mars, and so on.

Maximization of the height attained by a rocket in vertical flight (Goddard's problem). Suppose that a rocket with initial mass m_0 is launched along the vertical Oy (see Figure 3.5). Owing to the burning-out of the fuel, the rocket mass $m(t)$ varies according to the law $\dot{m}(t) = -u(t)$, where $u(t)$ is the fuel flow (fuel consumption per second). The rocket is acted on by the force of gravity $P = m(t)g$, a drag force Q and a thrust force $F = \beta u(t)$. Let the force Q be of the form $Q = Cv^2(t)$, where $v(t) = \dot{y}$. We assume the parameters C and β to be constants depending on the rocket design and the rocket to be a material point. Then the second law of Newton leads to the following equations of motion of the rocket (the

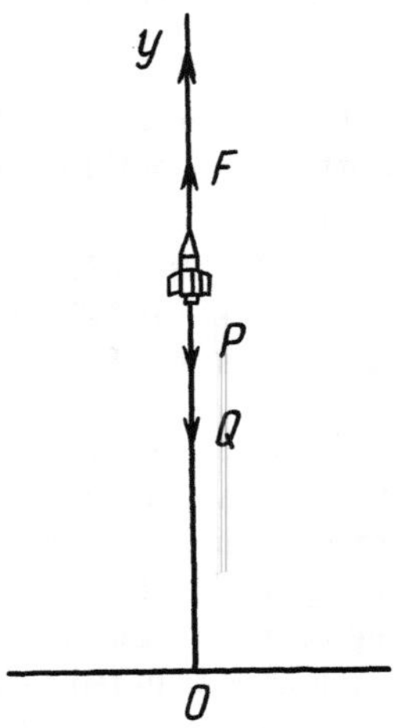

FIG. 3.5. The vertical flight of a rocket

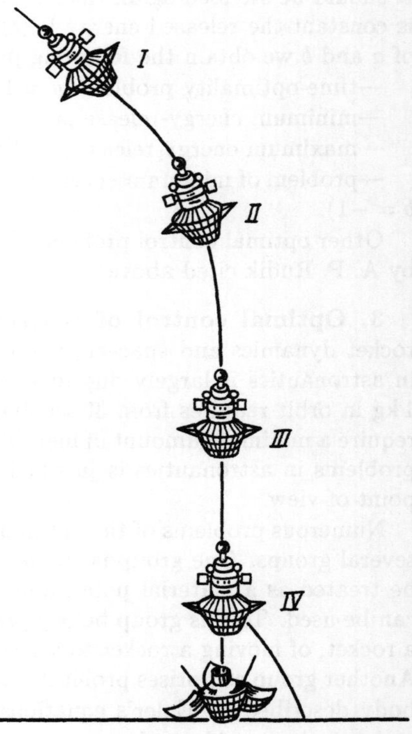

FIG. 3.6. The soft landing on the moon of the space station "Luna-9": *I*, the height of 8300 km, orientating by the lunar vertical; *II*, the height of 75 km, the firing of the main brake engine; *III*, the height of 150 m, the firing of the low-thrust engines; *IV*, the height of 5 m, the pin gauge comes into contact with the lunar surface

Meshchersky equations)

$$\dot{v} = \frac{\beta u(t) - Q}{m} - g, \qquad \dot{y} = v, \ \dot{m} = -u(t). \tag{3.21}$$

The initial conditions for the problem (3.21) are of the form

$$y(0) = 0, \qquad v(0) = 0, \ m(0) = m_0. \tag{3.22}$$

The terminal conditions are the following: the moment T is not given, but

$$v(T) = 0, \qquad m(T) = m_1. \tag{3.23}$$

Goddard's problem is in choosing a control $u(t)$ so that it maximizes the height attained by the rocket, i.e.,

$$J(u) = y(T) \to \max. \tag{3.24}$$

Thus, the problem (3.21)–(3.24) is a Mayer problem with an unspecified terminal moment.

Problem of the soft landing on the moon. Consider the problem of the lunar landing of a space vehicle. Suppose that the landing occurs along the vertical. Since there is no atmosphere on the moon, a soft landing can be ensured by jet engines only; this requires a greater accuracy of the control of the jet engines.

Using the same notation as above, we write the equations of soft landing in the form

$$\dot{y} = v, \qquad \dot{v} = \frac{\beta u(t)}{m} - g_{\mathrm{m}}, \qquad \dot{m} = -u(t) \tag{3.25}$$

with the initial conditions $y(0) = y_0$, $v(0) = v_0$, $m(0) = m_0$.

At some unspecified terminal moment T the following conditions must hold: $y(T) = 0$, $v(T) = 0$. In equations (3.25) g_{m} is the gravity acceleration on the moon, which is equal to $1.62\,\mathrm{m/sec}^2$. Choose a control $u(t)$, which has the meaning of fuel flow, so that it minimizes the fuel consumption during the lunar landing, i.e.,

$$J(u) = m_0 - m(T) \to \inf \tag{3.26}$$

subject to the constraint

$$0 \leqslant u(t) \leqslant u_{\max}. \tag{3.27}$$

The problem (3.25)–(3.27) of the lunar landing looks very much like Goddard's problem (3.21)–(3.24). However, the model under consideration leads to a control that is difficult to implement. Indeed, suppose the optimal P-control $u_{\mathrm{opt}}(t)$ for the problem (3.25)–(3.27) has been found. The control process can be split into two stages:

—a free fall from the instant $t = 0$ till the instant $t = t_1$ for $u_{\mathrm{opt}}(t) = 0$;

—braking in the interval $[t_1, T]$ for $u_{\mathrm{opt}}(t) = u_{\max}$.

The implementation of such a P-control leads to poor results. Suppose that the firing time t_{br} for the brake engine does not coincide with t_1, which always happens. If $t_{\mathrm{br}} < t_1$, the space vehicle will stop some distance from the moon and then will fly away. But if $t_{\mathrm{br}} > t_1$, the landing velocity will not be zero, and, consequently, there will be no soft landing. The instant t_1 depends on the initial height $y(0)$ and velocity $v(0)$; its precise evaluation requires a very precise evaluation of the initial data $y(0)$, $v(0)$, which is practically impossible.

In view of the foregoing, the actual soft landing system is more complicated. Besides the main brake engine it incorporates low-thrust engines. In this case the purpose of the main brake engine is to assure the programmed motion as specified by the problem

$$\begin{gathered} \dot z_1 = z_2, \qquad \dot z_2 = \frac{\beta\omega(t)}{z_3} - g_m, \qquad \dot z_3 = -\omega(t), \\ z(0) = z_{10}, \qquad z_2(0) = z_{20},\ z_3(0) = z_{30}, \\ z_1(T) = z_{11} > 0, \qquad z_2(T) = z_{21} > 0, \\ J(\omega) = z_3(0) - z_3(T) \to \inf, \\ 0 \leqslant \omega(t) \leqslant \omega_{\max}. \end{gathered} \tag{3.28}$$

where z_{11} and z_{21} are some small, *a priori* given, quantities. The control $\omega(t)$ is assured by the main brake engine.

The second control loop involves the low-thrust engines. We introduce perturbed variables y_1, y_2, y_3, ξ such that $y = z_1 + y_1$, $v = z_2 + y_2$, $m = z_3 + y_3$, $u = \omega + \xi$. In the first approximation we obtain the following equations for the perturbed variables:

$$\dot y_1 = y_2; \qquad \dot y_2 = -\frac{C\omega}{z_3^2} y_3 + \frac{C}{z_3}\xi; \qquad \dot y_3 = -\xi. \tag{3.29}$$

The initial conditions for equations (3.29) are evaluated with the accuracy of the execution of the programmed motion (3.28). It can be assumed that the initial conditions are given by arbitrary quantities y_{10}, y_{20}, y_{30} subject to the constraint

$$\alpha_1 y_{10}^2 + \alpha_2 y_{20}^2 + \alpha_3 y_{30}^2 \leqslant R^2. \tag{3.30}$$

In the problem (3.29), (3.30) it is natural to take the cost functional as a functional characterizing the accuracy of the execution of the soft landing:

$$\begin{gathered} J_1(\xi) = \bigl(z_{11} + y_1(T_1)\bigr)^2 + \alpha_4\bigl(z_{21} + y_2(T_1)\bigr)^2 + \alpha_5 y_3^2(T_1) \to \inf, \\ 0 \leqslant \xi(t) \leqslant \xi_{\max}. \end{gathered} \tag{3.31}$$

where T_1 is an unspecified (beforehand) moment of soft landing, the α_i are relative weight coefficients, R is a quantity that determines the accuracy of execution of the programmed motion.

It should be noted that in the problem (3.29)–(3.31) it is necessary to find the F-control $\xi = \xi(y_1, y_2, y_3)$. It is also important that problems (3.29)–(3.31), and (3.28) are not independent of each other and must be solved jointly. As a rule, a joint fuel resource Q is allocated for solving both problems, i.e., the following condition must be satisfied:

$$\omega_{\max}(T - t_1) + \xi_{\max}(T_1 - T) \leqslant Q.$$

Further, the accuracy of the execution of the programmed motion determines the value of R in the inequality (3.30) and the minimal value of $J_1(\xi)$, i.e., the value of the touch-down velocity.

The soft lunar landing, the first in history, of the Soviet automatic space station "Luna-9" took place on February 3, 1966.[4] It was controlled according to the foregoing scheme. The impact velocity that had to be canceled was 2600 m/sec. The soft-landing scheme for the station "Luna-9" is shown in Figure 3.6. At the height of 8300 km, an hour before the landing, the trajectory of the station was adjusted so that it began to fall along the lunar vertical down to the height of 75 km. At this height, 48 sec before landing, the main brake engine was fired by the signal from the on-board radio altimeter. At the height of about 150 m the main engine was cut off and the low-thrust engines were fired, which insured the touch-down velocity of 4–7 m/sec. At the height of 5 m the special pin gauge came into contact with the lunar surface, which led to the inflation by compressed gas of an elastic shell that enclosed the station. On making a few jumps, the station stopped moving, the shell broke into two parts that were thrown aside, and the "Luna-9" was standing on the lunar soil in the Ocean of Tempests.

[4]Levantovskii, B. I., *The Mechanics of Space Flights Made Easy*, Nauka, Moscow, (1980).

Main Results and Formulas of Chapter III

General statement of the optimal control problem consists of:
the equations of evolution of the system

$$\dot{x}(t) = f(t, x, u);$$

the cost functional

$$J_0\big(t_0, x(t_0), u, T, x(T)\big) \to \inf_u;$$

trajectory constraints

$$J_j\big(t_0, x(t_0), u, T, x(T)\big) = 0, \qquad j = 1, \dots, m_1;$$
$$J_j\big(t_0, x(t_0), u, T, x(T)\big) \leqslant 0, \qquad j = m_1 + 1, \dots, m_2;$$

control constraints

$$u(t) \in U(t)$$

or

$$\int_{t_0}^{T} F_j\big(t, u(t)\big)\, dt = 0, \qquad j = 1, \dots, m_3,$$
$$\int_{t_0}^{T} F_j\big(t, u(t)\big)\, dt \leqslant 0, \qquad j = m_3 + 1, \dots, m_4.$$

The Lagrange problem:

$$J_0 = \int_{t_0}^{T} F_0\big(t, x(t) u(t)\big)\, dt.$$

The Mayer problem:

$$J_0 = \varphi_0\big(T, x(T)\big).$$

The Boltz problem:

$$J_0 = \int_{t_0}^{T} F_0\big(t, x(t) u(t)\big)\, dt + \varphi_0\big(T, x(T)\big).$$

Time-optimality problem:

$$F_0\big(t, x(t), u(t)\big) = 1.$$

CHAPTER IV

THE CLASSICAL CALCULUS OF VARIATIONS AND OPTIMAL CONTROL

In this chapter we consider necessary conditions for a locally optimal (in the strong sense) extremum in problems with no control constraints. First, necessary conditions for problems with a variable trajectory end point are established. Next, we give a general solution procedure for linear-quadratic problems; this can be used as a basis for solving particular problems. The method of Lagrange multipliers is also treated.

§ 1. Problems with a Variable End Point and Fixed Time

1. Main assumptions. Consider necessary conditions for the optimal control of the system

$$\dot{x}(t) = f\big(t, x(t), u(t)\big), \qquad x(t_0) = x_0,\ t_0 \leqslant t \leqslant T, \qquad (1.1)$$
$$x \in R^n,\ u \in R^m,$$

with the cost functional

$$J(u) = \varphi\big(x(T)\big) \to \inf. \qquad (1.2)$$

The initial (t_0) and terminal (T) moments and also the initial state x_0 are prescribed. Suppose that the continuous functions

$$f(t, x, u)\colon R^1 \times R^n \times R^m \to R^n, \qquad \varphi(x)\colon R^n \to R^1$$

are continuously differentiable with respect to x and u. The derivatives of the functions $f(t, x, u)$ will be denoted by

$$f_x(t, x, u) = \frac{\partial f(t, x, u)}{\partial x}, \qquad f_u(t, x, u) = \frac{\partial f(t, x, u)}{\partial u},$$
$$\varphi_x(x) = \frac{\partial \varphi(x)}{\partial x}.$$

Recall that f_x is an $n \times n$ matrix with elements $\partial f_j(t, x, u)/\partial x_i$, where $i, j = 1, \dots, n$ (i being the row number, and j the column number), and $\varphi_x(x)$ is a column vector with components $\partial\varphi(x)/\partial x_i$, where $i = 1, \dots, n$.

We assume that the derivatives $f_x(t, x, u)$ and $\varphi_x(x)$, as well as the function $f(t, x, u)$, satisfy the Lipschitz condition with respect to (x, y), i.e., there exists a constant $C > 0$ such that for any $t \in R^1$, $x, y \in R^n$, and $u, v \in R^m$ the following conditions are satisfied:

$$\begin{aligned} |f(t,x,u) - f(t,y,v)| &\leqslant C(|x-y| + |u-v|), \\ \|f_x(t,x,u) - f_x(t,y,v)\| &\leqslant C(|x-y| + |u-v|), \\ |\varphi_x(x) - \varphi_x(y)| &\leqslant C|x-y|. \end{aligned} \tag{1.3}$$

In the problem (1.1), (1.2) the control $u(t)$ is sought in the class of continuous functions.

2. Cauchy's formula. Consider the system of linear equations

$$\dot{x}(t) = A(t)x(t), \qquad x \in R^n. \tag{1.4}$$

The square $n \times n$ matrix $A(t)$ has piecewise continuous bounded elements. Denote by $X(t)$ the fundamental matrix of the system (1.4), i.e., the matrix whose columns are any n linearly independent solutions of equation (1.4). The Cauchy matrix $Z(t,s)$ for the system (1.4) is defined by

$$Z(t,s) = X(t)X^{-1}(s) \tag{1.5}$$

The solution of the linear nonhomogeneous equation

$$\dot{x}(t) = A(t)x(t) + f(t), \qquad x(t_0) = x_0, \tag{1.6}$$

where $f(t)$ is a piecewise continuous bounded function, is expressed by the following *Cauchy's formula*:

$$x(t) = Z(t,t_0)x_0 + \int_{t_0}^{t} Z(t,s)f(s)\,ds. \tag{1.7}$$

Moreover, the following properties of the matrix $Z(t,s)$ can be put to use:

$$\begin{aligned} &\frac{\partial Z(t,s)}{\partial t} = A(t)Z(t,s), \qquad \frac{\partial Z(t,s)}{\partial s} = -Z(t,s)A(s), \\ &Z(s,t) = Z^{-1}(t,s), \qquad \frac{\partial Z^{-1}(t,s)}{\partial t} = -Z^{-1}(t,s)A(t), \\ &Z(t,s)Z(s,t_1) = Z(t,t_1). \end{aligned} \tag{1.8}$$

Note that for a constant matrix $A(t) \equiv B$ the Cauchy matrix $Z(t,s)$ is of the form

$$Z(t,s) = exp[B(t-s)]. \tag{1.9}$$

Below we use the Cauchy-Bunyakovskii inequality in the following form. For any two scalar functions $x(t)$ and $y(t)$ that are square-integrable on the segment $[t_0, T]$ the following *Cauchy-Bunyakovskii inequality* holds:

$$\left[\int_{t_0}^{T} |x(t)y(t)|\,dt\right]^2 \leqslant \left(\int_{t_0}^{T} x^2(t)\,dt\right)\left(\int_{t_0}^{T} y^2(s)\,ds\right). \tag{1.10}$$

3. Necessary conditions for optimal control. We introduce the following notation:

$$H(t,x,u,\psi) = \psi' f(t,x,u), \qquad \psi \in R^n. \tag{1.11}$$

Theorem 1.1 (necessary conditions for optimal control). *Let $u_0(t)$ be an optimal P-control in the problem (1.1), (1.2) and $x_0(t)$ be the corresponding trajectory satisfying the equation similar to (1.1):*

$$\dot{x}_0(t) = f\big(t, x_0(t), u_0(t)\big), \qquad x_0(t_0) = x_0.$$

Then

$$\frac{\partial H}{\partial u}\bigg|_{\substack{x_0(t)\\ u_0(t)}} = H_u\big(t, x_0(t), u_0(t), \psi(t)\big) \equiv 0, \qquad t_0 \leqslant t \leqslant T, \tag{1.12}$$

where $\psi(t) \in R^n$ is a solution of the Cauchy problem

$$\dot{\psi}(t) = -H_x\big(t, x_0(t), u_0(t), \psi(t)\big), \qquad \psi(T) = -\varphi_x\big(x_0(T)\big). \tag{1.13}$$

Proof. Consider an arbitrary continuous control $v(t)$, $t_0 \leqslant t \leqslant T$. We try to find the trajectory $y(t)$ of the system (1.1) that corresponds to the control $u_0(t) + v(t)$:

$$\dot{y}(t) = f\big(t, y(t), u_0(t) + v(t)\big), \qquad y(t_0) = x_0,\ t_0 \leqslant t \leqslant T.$$

Let us evaluate the deviation $\Delta(t) = y(t) - x_0(t)$ of the optimal trajectory due to a change in the optimal control by a value of $v(t)$. It is readily seen that

$$\Delta(t) = \int_{t_0}^{t} \big[f\big(s, y(s), u_0(s) + v(s)\big) - f\big(s, x_0(s), u_0(s)\big)\big]\, ds.$$

Hence, noting the Lipschitz condition (1.3), we find that

$$|\Delta(t)| \leqslant C \int_{t_0}^{t} \big(|\Delta(s)| + |v(s)|\big)\, ds.$$

It follows that by virtue of the Gronwall-Bellman lemma (see § 6 of Chapter I) we have

$$|\Delta(t)| \leqslant e^{C(t-t_0)} \int_{t_0}^{t} |v(s)|\, ds \leqslant C_1 \int_{t_0}^{t} |v(s)|\, ds. \tag{1.14}$$

Here and subsequently, we denote by C_i some nonnegative constants.

Let us now evaluate the change in the cost functional (1.2), i.e., we evaluate the difference $\Delta J = J(u_0 + v) - J(u_0)$. Taking account of (1.2), we can write

$$\Delta J = \varphi_x'\big(x_0(T)\big)\Delta(T) + N_1, \tag{1.15}$$

$$N_1 = \big[\varphi_x'\big(x_0(T) + \theta\Delta(T)\big) - \varphi_x'\big(x_0(T)\big)\big]\Delta(T). \tag{1.16}$$

Here and subsequently, we denote by θ some numbers belonging to the segment $(0,1)$. It follows from the relations (1.14), (1.16) and the Lipschitz condition (1.3) that

$$|N_1| \leqslant C_2\left(\int_{t_0}^{T} |v(s)|\,ds\right)^2. \tag{1.17}$$

Next, we transform the first summand on the right side of (1.15). Using the expression (1.13), we obtain

$$\begin{aligned}
\varphi_x'(x(T))\Delta(T) &= -\psi'(T)\Delta(T) = -\int_{t_0}^{T} \frac{d}{dt}(\psi'(t)\Delta(t))\,dt \\
&= -\int_{t_0}^{T} (\dot\psi'(t)\Delta(t) + \psi'(t)\dot\Delta(t))\,dt \\
&= \int_{t_0}^{T} H_x'(t, x_0(t), u_0(t), \psi(t))\,\Delta(t)\,dt \\
&\quad - \int_{t_0}^{T} \psi'(t)\left[f(t, y(t), u_0(t) + v(t)) - f(t, x_0(t), u_0(t))\right] dt.
\end{aligned}$$

Hence and from the relations (1.11), (1.15) it follows that

$$\begin{aligned}
\Delta J = &-\int_{t_0}^{T} \left[H(t, y(t), u_0(t) + v(t), \psi) - H(t, x_0(t), u_0(t), \psi)\right] dt \\
&+ \int_{t_0}^{T} H_x'(t, x_0(t), u_0(t), \psi)\,\Delta(t)\,dt + N_1.
\end{aligned} \tag{1.18}$$

Further, note that

$$\begin{aligned}
H(t, y(t), u_0(t) + v(t), \psi) &= H(t, x_0(t), u_0(t) + v(t), \psi) \\
&\quad + H_x'(t, x_0(t) + \theta\Delta(t), u_0(t) + v(t), \psi)\,\Delta(t).
\end{aligned}$$

Therefore, the equality (1.18) can be written as

$$\begin{aligned}
\Delta J = -\int_{t_0}^{T} &\left[H(t, x_0(t), u_0(t) + v(t), \psi)\right. \\
&\left. - H(t, x_0(t), u_0(t), \psi)\right] dt + N_1 + N_2,
\end{aligned} \tag{1.19}$$

where

$$\begin{aligned}
N_2 = \int_{t_0}^{T} &\left[H_x'(t, x_0(t) + \theta\Delta(t), u_0(t) + v(t), \psi)\right. \\
&\left. - H_x'(t, x_0(t), u_0(t), \psi)\right]\Delta(t)\,dt.
\end{aligned}$$

The equality (1.13) and the conditions of the theorem imply that the function $|\psi(t)|$ is uniformly bounded for $0 \leqslant t \leqslant T$. Hence, noting the definition of N_2 and the conditions (1.3), we find that

$$|N_2| \leqslant C_3 \int_{t_0}^{T} (|\Delta(t)|^2 + |\Delta(t)||v(t)|)\, dt. \tag{1.20}$$

Using relation (1.14), we obtain

$$\begin{aligned} \int_{t_0}^{T} |\Delta(t)||v(t)|\, dt &\leqslant C_1 \int_{t_0}^{t} |v(t)| \left(\int_{t_0}^{T} |v(s)|\, ds \right) dt \\ &\leqslant C_1 \int_{t_0}^{T} |v(t)|\, dt \int_{t_0}^{T} |v(s)|\, ds = C_1 \left[\int_{t_0}^{T} |v(t)|\, dt \right]^2. \end{aligned} \tag{1.21}$$

In a similar way, by virtue of (1.14) we have

$$\int_{t_0}^{T} |\Delta(t)|^2\, dt \leqslant (T - t_0) C_1^2 \left[\int_{t_0}^{T} |v(s)|\, ds \right]^2. \tag{1.22}$$

Formulas (1.20)–(1.22) imply that

$$|N_2| \leqslant C_4 \left[\int_{t_0}^{T} |v(t)|\, dt \right]^2. \tag{1.23}$$

Thus, for any arbitrary admissable variation $v(t)$ of the optimal control $u_0(t)$ we have proved the main relation (1.19) for an increment ΔJ of the cost functional.

Next, we assume that $v(t) = \varepsilon\omega(t)$, where $\omega(t)$ is an arbitrary continuous control, and the parameter $\varepsilon > 0$. Then it follows from (1.19) that

$$0 \leqslant \Delta J = - \int_{t_0}^{T} H'_u(t, x_0(t), u_0(t) + \theta\varepsilon\omega(t), \psi)\varepsilon\omega(t)\, dt + N_1 + N_2. \tag{1.24}$$

It should be noted that for $v = \varepsilon\omega$, on the strength of inequalities(1.17), (1.23), we have $\lim_{\varepsilon\to 0} N_1\varepsilon^{-1} = 0$ and $\lim_{\varepsilon\to 0} N_2\varepsilon^{-1} = 0$. Hence, dividing both sides of (1.24) by ε and passing to the limit as $\varepsilon \to 0$ gives

$$0 \leqslant - \int_{t_0}^{T} H'_u(t, x_0(t), u_0(t), \psi)\omega(t)\, dt.$$

Next, modifying the main lemma of the calculus of variations leads to the conclusion that the last inequality holds if and only if the equality (1.12) is valid. □

4. The Boltz problem. Consider the problem

$$\dot{x}(t) = f(t, x(t), u(t)), \qquad t_0 \leqslant t \leqslant T,\ x(t_0) = x_0, \tag{1.25}$$

$$J(u) = \int_{t_0}^{T} F(t, x(t), u(t))\, dt + \varphi(x(T)) \to \inf, \tag{1.26}$$

where t_0 and T are given moments, and $x_0 \in R^n$ is a given initial state vector. The terminal state $x(T)$ is variable. Necessary conditions for optimal control in the problem (1.25), (1.26) can be deduced by means of Theorem 1.1, using the method (discussed in § 2 of Chapter III) of reducing a Boltz problem to a Mayer problem. According to this method, the problem (1.25), (1.26) is equivalent to the following problem:

$$\begin{gathered} \dot{x}_{n+1}(t) = F(t, x(t), u(t)), \qquad x_{n+1}(t_0) = 0, \\ \dot{x}(t) = f(t, x(t), u(t)), \qquad x(t_0) = x_0, \\ x_{n+1}(T) + \varphi(x(T)) \to \inf \end{gathered} \tag{1.27}$$

The necessary conditions for optimal control in the problem (1.27) are stated in Theorem 1.1. In the case under consideration, by virtue of the relations (1.27), (1.11), the function H becomes

$$H = \psi'(t) f(t, x, u) + \psi_{n+1}(t) F(t, x, u).$$

The equation for the vector $\psi(t) \in R^n$ preserves the same form (1.12) as before, and the scalar function $\psi_{n+1}(t)$ satisfies the relations

$$\dot{\psi}_{n+1}(t) = -\frac{\partial H}{\partial x_{n+1}} = 0, \qquad \psi_{n+1}(T) = -1.$$

This implies that $\psi_{n+1}(t) \equiv -1$. Hence,

$$H(t, x, u, \psi) = -F(t, x, u) + \psi'(t) f(t, x, u).$$

In summary, the necessary conditions for optimal control in the problem (1.25), (1.26) that are satisfied by the optimal trajectory $x_0(t)$ and by the optimal control $u_0(t)$ are expressed by the equations

$$\dot{x}_0(t) = f(t, x_0(t), u_0(t)), \qquad t_0 \leqslant t \leqslant T, \qquad x_0(t_0) = x_0, \tag{1.28}$$

$$\dot{\psi}(t) = -f_x(t, x_0(t), u_0(t))\psi(t) + F_x(t, x_0(t), u_0(t)), \tag{1.29}$$

$$\psi(T) = -\varphi_x(x_0(T)),$$

$$f_u(t, x_0(t), u_0(t))\psi(t) - F_u(t, x_0(t), u_0(t)) = 0. \tag{1.30}$$

If there exists a solution of the optimal control problem (1.25), (1.26) and it is unique, then it is fully determined by equations (1.28), (1.29), the optimal control being constructed step by step as follows:

1°. Equation (1.30) is solved for u_0, and the function $u_0 = u_0(t, \psi(t), x_0(t))$ determined.

2°. The expression obtained for u_0 is substituted in equations (1.28), (1.29). As a result, we come up against a two-point boundary-value problem for the unknowns $x_0(t)$ and $\psi(t)$.

3°. The resulting boundary-value problem is solved for the functions of time $x_0(t)$ and $\psi(t)$. Substituting the obtained functions $x_0(t)$, $\psi(t)$ in the formula specifying the optimal control gives the optimal program control $u_0(t)$.

The implementation of this procedure in specific situations is fraught with numerous difficulties that are encountered, in particular, when solving the boundary-value problem (1.28), (1.29) in the case of a large dimension of the system in question, of the nonlinear character of the system, etc. [6]. Furthermore, one must take account of the fact that in order to prove the optimality of the control thus obtained, it is necessary to establish the existence of a solution of the optimal control problem and the uniqueness of the solution of the boundary-value problem (1.28), (1.29), in which u is replaced by $u_0(t, \psi(t), x_0(t))$. A numerical realisation of this method will be considered in Chapter XV.

Nevertheless, the method discussed above can be effectively implemented in some important problems. One of such problems, namely the control of a linear system with a quadratic functional, will be discussed in the next section.

§ 2. Optimal Control of Linear Systems with a Quadratic Functional

1. Necessary conditions for optimal control. The problem of minimizing the quadratic functional

$$J(u) = x'(T)N_1x(T) + \int_{t_0}^{T} \big(x'(t)N_2(t)x(t) + u'(t)N_3(t)u(t)\big)\,dt \tag{2.1}$$

on the trajectories of a linear system

$$\dot{x}(t) = A(t)x(t) + B(t)u(t), \qquad t \in [t_0, T],\ x \in R^n,\ u \in R^m;$$
$$x(t_0) = x_0 \tag{2.2}$$

is called the *linear-quadratic problem*, or the *LQ-problem* for short. The moments of time t_0 and T and also the initial state x_0 are given. The matrices $A(t)$, $B(t)$, $N_2(t)$, and $N_3(t)$ of corresponding dimensions are assumed to have piecewise continuous elements. It is also assumed that

$$N_1 \geqslant 0, \qquad N_2(t) \geqslant 0, \qquad N_3(t) > 0. \tag{2.3}$$

It should be recalled that the matrix $N \geqslant 0$ if $x'Nx \geqslant 0$ for any vector $x \in R^n$, and $N > 0$ if $x'Nx > 0$ for any $x \in R^n$, $x \neq 0$.

To deduce an optimal control in the LQ-problem (2.1), (2.2), we invoke the necessary conditions (1.28)–(1.30) for optimal control that were obtained in § 1.

In the case under consideration they are determined by equations (2.2) and by the following relations:

$$\begin{aligned} H\big(t, x(t), u(t), \psi(t)\big) &= -x'(t)N_2(t)x(t) \\ &-u'(t)N_3(t)u(t) + \psi'(t)[A(t)x(t) + B(t)u(t)], \\ -\frac{\partial H}{\partial x} = \dot\psi(t) &= -A'(t)\psi(t) + 2N_2(t)x(t), \qquad t_0 \leqslant t \leqslant T, \\ \psi(T) &= -2N_1x(T), \\ \frac{\partial H}{\partial u} &= B'(t)\psi(t) - 2N_3(t)u(t) = 0. \end{aligned} \tag{2.4}$$

By virtue of the conditions (2.3) the functional (2.1) is convex with respect to u. As in the finite-dimensional case, we can use this property to show that the relations (2.2), (2.4) express not only necessary but also sufficient conditions for optimal control.

2. Construction of an optimal control. To deduce an optimal control by means of the necessary conditions (2.2), (2.4), we use the method discussed in § 1:

1°. It follows from equation (2.4) that an optimal control u_0 is determined by the formula

$$u_0 = \tfrac{1}{2}N_3^{-1}(t)B'(t)\psi(t). \tag{2.5}$$

2°. Insert the expression (2.5) in (2.2), (2.4). As a result, we obtain a boundary-value problem for the optimal trajectory $x_0(t)$ and for the vector of adjoint variables $\psi(t)$:

$$\begin{aligned} \dot x_0(t) &= A(t)x_0(t) + \tfrac{1}{2}B_1(t)\psi(t), \\ \dot\psi(t) &= -A'(t)\psi(t) + 2N_2(t)x_0(t), \\ x_0(t_0) &= x_0, \ \psi(T) = -2N_1x_0(T), \\ B_1(t) &= B(t)N_3^{-1}(t)B'(t). \end{aligned} \tag{2.6}$$

3°. To solve the boundary-value problem (2.6), we shall seek the function $\psi(t)$ in the form

$$\psi(t) = -2P(t)x_0(t), \tag{2.7}$$

where a symmetric matrix $P(t)$ is still to be defined. Differentiating both sides of (2.7) with respect to t gives

$$\dot\psi(t) = -2[\dot P(t)x_0(t) + P(t)\dot x_0(t)].$$

Here we replace the derivatives $\dot\psi$ and $\dot x_0$ using the corresponding equalities (2.6), and then substitute the expression (2.7) for $\psi(t)$. Thus, we obtain

$$\begin{aligned} -A'(t)\psi(t) + 2N_2(t)x_0(t) &= -2[\dot P(t)x_0(t) \\ &+P(t)A(t)x_0(t) + \tfrac{1}{2}P(t)B_1(t)\psi(t)], \\ [\dot P(t) + A'(t)P(t) + P(t)A(t) - P(t)B_1(t)P(t) + N_2(t)]x_0(t) &= 0, \\ P(T)x_0(T) &= N_1x_0(T). \end{aligned}$$

These equalities are satisfied for any vector $x_0(t)$ if

$$\dot{P}(t) + A'(t)P(t) + P(t)A(t) - P(t)B_1(t)P(t) + N_2(t) = 0, \qquad (2.8)$$
$$P(T) = N_1.$$

Equation (2.8) that serves to determine the matrix $P(t)$ is called a *matrix Riccati equation.* It plays an important role in the theory of LQ-control problems.

Once the matrix $P(t)$ is found, the optimal control u_0, in view of the equalities (2.5) and (2.7), is expressed by

$$u_0(t) = -N_3^{-1}(t)B'(t)P(t)x_0(t). \qquad (2.9)$$

Moreover, as will be shown in Chapter VI, the minimal value of the cost functional (2.1) that corresponds to the control (2.9) is given by

$$\min_u J(u) = J(u_0) = x_0' P(t_0)x_0. \qquad (2.10)$$

Thus, the construction of the matrix $P(t)$ leads to a complete solution of the control problem (2.1), (2.2).

We note further that the control (2.9) has been obtained as an F-control, i.e., as a function of time and phase coordinates. Indeed, denoting by x the realised value of the optimal trajectory $x_0(t)$, we deduce that the optimal control for the system of state x at time t becomes

$$u_0(t,x) = -N_3^{-1}(t)B'(t)P(t)x. \qquad (2.11)$$

3. Matrix Riccati equation. Consider some properties of the matrix Riccati equation (2.8).

1°. *There exists a unique solution of equation (2.8) for all $t < T$.*

The local (i.e., belonging to a neighborhood of the point T) existence and uniqueness properties of the solution of the problem (2.8) are an immediate consequence of the general theorems on the existence of a solution for the Cauchy problem in the theory of ordinary differential equations. Considerably more complicated is to prove that there exists a solution of the problem (2.8) for all $t < T$. This difficulty is related to the fact that equation (2.8) involves the quadratic nonlinearity $P(t)B_1(t)P(t)$. Not any equation with quadratic nonlinearity has solutions that can be continued to the whole axis. For example, the scalar equation

$$\dot{y}(t) = 1 + y^2(t), \qquad y(0) = 0$$

has the solution $y(t) = \tan t$ that tends to infinity as $t \to \pm\pi/2$, and hence it cannot be continued beyond the point $t \to \pi/2$ to the right and beyond the point $t \to -\pi/2$ to the left. However, the special structure of equation (2.8) makes it possible to establish the fact that solutions can be continued to the semiaxis $t \leqslant T$.

Consider a scalar equation of the type (2.8):

$$\dot{p}(t) - b_1 p^2(t) + n_2 = 0, \qquad p(T) = n_1; \qquad (2.12)$$
$$b_1, n_1, n_2 > 0.$$

The solution of equation (2.12) cannot intersect the straight line $p(t) = 0$. In fact, if at the point $t = t_1 < T$ the solution intersects the line $p(t) = 0$, then we must have $p(t_1) = 0$ and $\dot{p}(t_1) \geqslant 0$. But then equation (2.12) is not valid at the point t_1. The solution $p(t)$ cannot also cut the straight line $p(t) = \sqrt{n_2/b_1}$, since $-b_1 p^2(t) + n_2 = 0$ on this line, and $-b_1 p^2(t) + n_2 < 0$ everywhere above this line. Thus, the solution of equation (2.12) is always included between the following bounds:

$$0 \leqslant p(t) \leqslant \sqrt{n_2/b_1}, \qquad t \leqslant T.$$

Consequently, $p(t)$ cannot tend to infinity for finite values of t as distinct from solutions of equation (2.12).

Using a similar argument, R. Kalman proved that solutions of equation (2.8) exist for all $t \leqslant T$.

2°. *The solution of equation (2.8) is nonnegative, i.e.,* $P(t) \geqslant 0$. Relation (2.10) implies that

$$x_0' P(t_0) x_0 = J(u_0) \geqslant 0, \qquad (2.13)$$

since always the functional $J(u) \geqslant 0$. In the above relation x_0 as well as the moment $t_0 < T$ can be chosen in an arbitrary way. This means that $P(t) \geqslant 0$.

3°. *Let $N_2(t) \equiv 0$ and $N_1 > 0$. Then equation* (2.8) *can be integrated in explicit form.*

Equation (2.8) for $N_2(t) \equiv 0$ is called *Bernoulli's equation* and is of the form

$$\dot{P}(t) + A'(t)P(t) + P(t)A(t) - P(t)B_1(t)P(t) = 0, \qquad (2.14)$$
$$P(T) = N_1, \; N_1 > 0.$$

Consider the matrix $D(t) = P^{-1}(t)$. Since

$$\frac{d}{dt}(P^{-1}(t)P(t)) = \dot{P}^{-1}(t)P(t) + P^{-1}(t)\dot{P}(t) = 0,$$

it follows that $\dot{D}(t) = -P^{-1}(t)\dot{P}(t)P^{-1}(t)$. Multiplying equation (2.14) on the left and on the right by $P^{-1}(t)$, we obtain that the matrix $D(t)$ is a solution of the linear equation

$$\dot{D}(t) - D(t)A'(t) - A(t)D(t) + B_1(t) = 0, \qquad (2.15)$$
$$D(T) = N_1^{-1}.$$

The solution of equation (2.15) can be written as

$$D(t) = Z(t,T)N_1^{-1}Z'(t,T) + \int_t^T Z(t,s)B_1(s)Z'(t,s)\,ds. \qquad (2.16)$$

(This can be verified by a direct substitution in the equation.) Here the Cauchy matrix $Z(t,s)$ of the linear equation (2.15) is determined from the relations (1.5) and (1.8). It follows from (2.16) and (1.8) that

$$\begin{aligned} P(t) &= \Big[Z(t,T)(N_1^{-1} + Z^{-1}(t,T) \\ &\quad\times \int_t^T Z(t,s)B_1(s)Z'(t,s)\,ds Z'(t,T)^{-1})Z'(t,T)\Big]^{-1} \\ &= Z'(T,t)\Big[I + N_1 \int_t^T Z(T,s)B_1(s)Z'(T,s)\,ds\Big]^{-1} N_1 Z(T,t). \end{aligned} \tag{2.17}$$

As is seen from formula (2.17), the condition $N_1 > 0$ is not necessary for it to be valid. It is only sufficient to impose the condition that the matrix in square brackets in (2.17) be nondegenerate.

4°. *Formula for the solution of the problem* (2.8), *given an arbitrary matrix* $N_2(t)$. Let us now deduce a formula for the solution of the problem (2.8), using a Cauchy matrix $G(t,s)$ of dimensions $2n \times 2n$. We consider the system of $2n$ linear equations

$$\begin{aligned} \dot{x}_1(t) &= A(t)x_1(t) - B(t)x_2(t), \qquad x_1 \in R^n; \\ \dot{x}_2(t) &= -N_2(t)x_1(t) - A'(t)x_2(t), \qquad x_2 \in R^n; \\ x_2(T) &= N_1 x_1(T). \end{aligned} \tag{2.18}$$

Find a matrix $K(t)$ such that $K(T) = N_1$ and $x_2(t) = K(t)x_1(t)$ for all $t \leqslant T$. In view of equations (2.18), we obtain

$$\begin{aligned} \dot{x}_2(t) &= -N_2(t)x_1(t) - A'(t)K(t)x_1(t) \\ &= \dot{K}(t)x_1(t) + K(t)[A(t)x_1(t) - B_1(t)K(t)x_1(t)]. \end{aligned}$$

Hence

$$[\dot{K}(t) + A'(t)K(t) + K(t)A(t) - K(t)B_1(t)K(t) + N_2(t)]x_1(t) = 0.$$

The last equality holds for all $x_1(t)$ if the expression in square brackets vanishes for all t. Thus, $K(t)$ satisfies equation (2.8) and the same initial condition $K(T) = N_1$ as for $P(t)$. In view of the uniqueness property (see item 1°) we find that $K(t) \equiv P(t)$.

We represent the matrix $G(t,s)$ as a block-diagonal matrix composed of four $n \times n$ blocks:

$$G(t,s) = \begin{pmatrix} G_{11}(t,s) & G_{12}(t,s) \\ G_{21}(t,s) & G_{22}(t,s) \end{pmatrix}. \tag{2.19}$$

By virtue of Cauchy's formula the solution of the system (2.18) can be written as

$$\begin{aligned} x_1(t) &= G_{11}(t,T)x_1(T) + G_{12}(t,T)x_2(T), \\ x_2(t) &= G_{21}(t,T)x_1(T) + G_{22}(t,T)x_2(T). \end{aligned}$$

Therefore,

$$x_2(t) = \big(G_{21}(t,T) + G_{22}(t,T)N_1\big)x_1(T),$$
$$x_2(t) = P(t)x_1(t) = P(t)\big(G_{11}(t,T) + G_{12}(t,T)N_1\big)x_1(T).$$

Comparing these two formulas and noting that $x_1(T)$ is arbitrary, we obtain

$$P(t) = [G_{21}(t,T) + G_{22}(t,T)N_1][G_{11}(t,T) + G_{12}(t,T)N_1]^{-1} \tag{2.20}$$

The represention (2.20) of the solution of the problem (2.8) is valid under the condition that the matrix in the second brackets in (2.20) be nondegenerate.

We shall now discuss another method to obtain the relations (2.20). Set $P(t) = \beta(t)\alpha(t)^{-1}$, where $\alpha(t)$ and $\beta(t)$ are the two square matrices to be determined. Differentiating the above relation and taking account of (2.8), we can write

$$\begin{aligned}\dot\beta(t)\alpha^{-1}(t) - \beta\alpha^{-1}\dot\alpha\alpha^{-1} = \dot P(t) &= -A'P - PA + PB_1P - N_2\\ &= -A'\beta\alpha^{-1} - \beta\alpha^{-1}A + \beta\alpha^{-1}B_1\beta\alpha^{-1} - N_2.\end{aligned}$$

Multiply both sides of this expression by α on the right and by $\alpha\beta^{-1} = (\beta\alpha^{-1})^{-1}$ on the left. Then we find that equation (2.8) can be satisfied, provided the matrices α and β satisfy the following equations that are similar to (2.18):

$$\dot\alpha = A\alpha - B_1\beta, \qquad \dot\beta = -A'\beta - N_2\alpha,$$
$$\alpha(0) = I,\ \beta(0) = N_1.$$

Writing the general solution of these equations with the help of Cauchy's formula yields the representation (2.20).

5°. *Method of successive approximations.* Consider a sequence of symmetric matrices $P_i(t)$ $(i \geqslant 1)$ introduced with the help of the recurrence relations

$$\begin{aligned}&\dot P_0(t) + A'(t)P_0(t) + P_0(t)A(t) + N_2(t) = 0, \qquad P_0(T) = N_1,\\ &\dot P_i(t) + A_i'(t)P_i(t) + P_i(t)A_i(t) + P_{i-1}(t)B_1(t)P_{i-1}(t) + N_2(t) = 0,\\ &P_i(T) = N_1, \qquad A_i(t) = A(t) - B_1(t)P_{i-1}(t), \qquad i = 1,2,\dots .\end{aligned} \tag{2.21}$$

Using Cauchy's formula and equations (2.21), we obtain

$$\begin{aligned}P_i(t) = \int_t^T &W_i'(t,s)[P_{i-1}(s)B_1(s)P_{i-1}(s) + N_2(s)]W_i(t,s)\,ds\\ &+ W_i'(t,T)N_1W_i(t,T), \qquad i \geqslant 1,\end{aligned} \tag{2.22}$$

where W_i is the Cauchy matrix of solutions of the homogeneous equation $\dot y(t) = -A_i(t)y(t)$. The relation (2.22) and the similar representation for the matrix $P_0(t)$

imply that all $P_i(t) \geqslant 0$. We set $Q_i = P_{i+1} - P_i$. Then by virtue of (2.21) taken for $i \geqslant 2$ we obtain the equations

$$\dot{Q}_i(t) + A_i'(t)Q_i(t) + Q_i(t)A_i(t) - Q_{i-1}(t)B_1(t)Q_{i-1}(t) = 0$$
$$Q_i(T) = 0.$$

Thus, noting Cauchy's formula, we can write

$$Q_i(t) = -\int_t^T W_i'(t,s)Q_{i-1}(s)B_1(s)Q_{i-1}(s)W_i(s,T)\,ds.$$

The last equalities imply that all matrices $Q_i(t) \leqslant 0$, i.e., $0 \leqslant P_i(t) \leqslant P_{i-1}(t)$ for $i \geqslant 2$. Therefore, $\{P_i(t)\}$ is a nonincreasing sequence of positive semidefinite matrices, i.e., $P_{i+1} - P_i \leqslant 0$. It can be proved that the sequence $\{P_i(t)\}$ has the limit $P(t) \geqslant 0$ as $t \to \infty$, the convergence $P_i(t) \to P(t)$ being uniform on any bounded time interval. Consequently, passing to the limit, as $i \to \infty$, in the integral identity that corresponds to equation (2.21), we conclude that the matrix $P(t)$ is a solution of the problem (2.8). Thus, for an approximate solution of Riccati's equation (2.8) it is sufficient to solve the linear equation (2.21) whose solution is expressed by (2.22). Denote by $\|P\|$ the norm of the matrix P which is equal to the square root of the sum of the squares of all the elements. Then the following estimate holds:

$$\max_{t_0 \leqslant t \leqslant T} \|P(t) - P_i(t)\| \leqslant \frac{C}{i!}.$$

4. The scalar case [3]. As an example, consider a control problem of the form (2.1), (2.2), with constant coefficients a, b, $N_2 \geqslant 0$, $N_3 > 0$, and $N_1 \geqslant 0$:

$$\dot{x}(t) = ax(t) + bu(t), \qquad 0 \leqslant t \leqslant T,\ x(0) = x_0, \tag{2.23}$$
$$J(u) = N_1 x^2(T) + \int_0^T \big(N_2 x^2(t) + N_3 u^2(t)\big)\,dt.$$

In this problem, according to the equality (2.9) the optimal control $u_0(t)$ is expressed by

$$u_0(t) = -bN_3^{-1}P(t)x(t),$$

where $x(t)$ is a solution of the Cauchy problem:

$$\dot{x}(t) = \big(a - r^{-1}P(t)\big)x(t), \qquad x(0) = x_0,\ r = b^{-2}N_3.$$

On the strength of Cauchy's formula (1.14) and relation (1.16) we obtain that the optimal trajectory $x(t)$ is of the form

$$x(t) = x_0 \exp\left[at - r^{-1}\int_0^t P(s)\,ds\right].$$

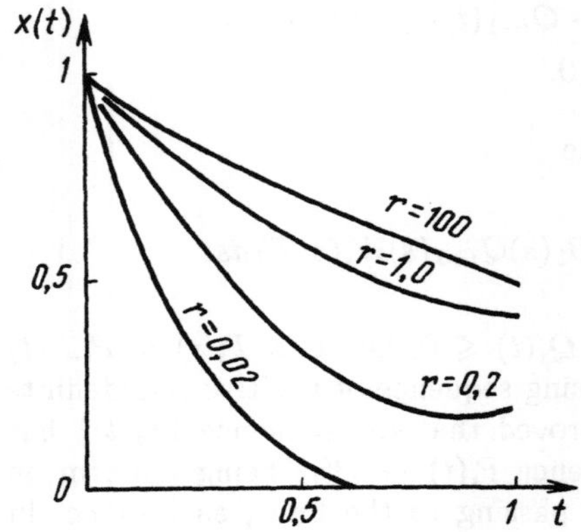

FIG. 2.1. The dependence on the parameter r of the transient processes in the problem (2.23)

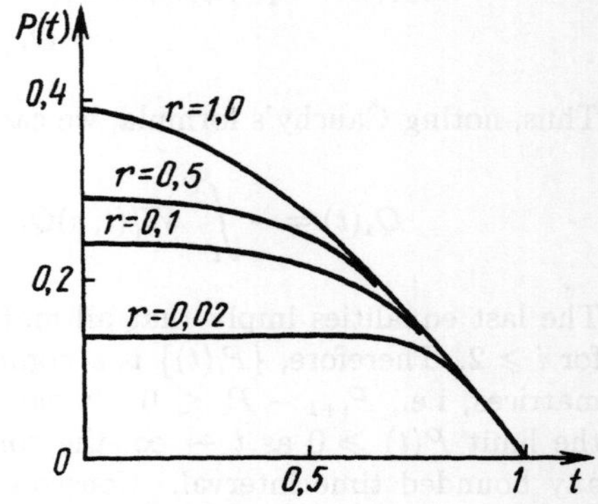

FIG. 2.2. The r-dependence of the solution of Riccati's equation (2.24)

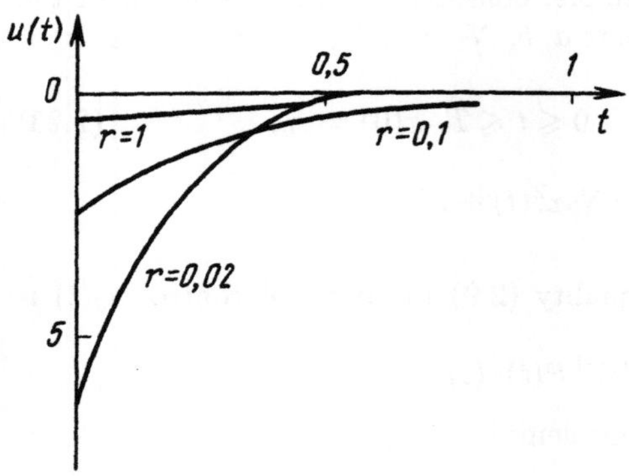

FIG. 2.3. Optimal control in the problem (2.23)

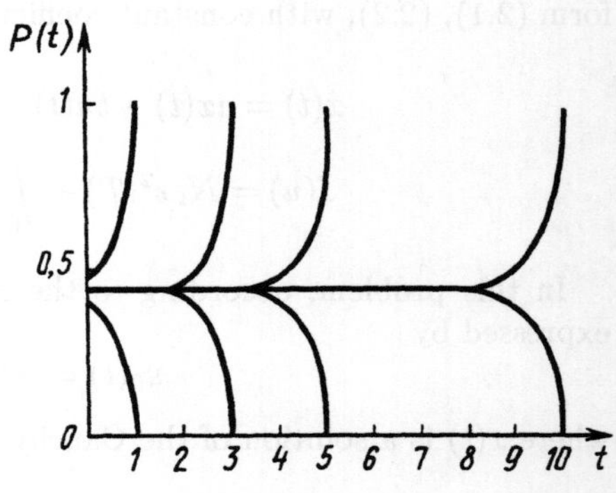

FIG. 2.4. The plot of the solution of Riccati's equation (2.24) versus control time:
$N_1 = 0$ corresponds to lower branches;
$N_1 = 1$ corresponds to upper branches

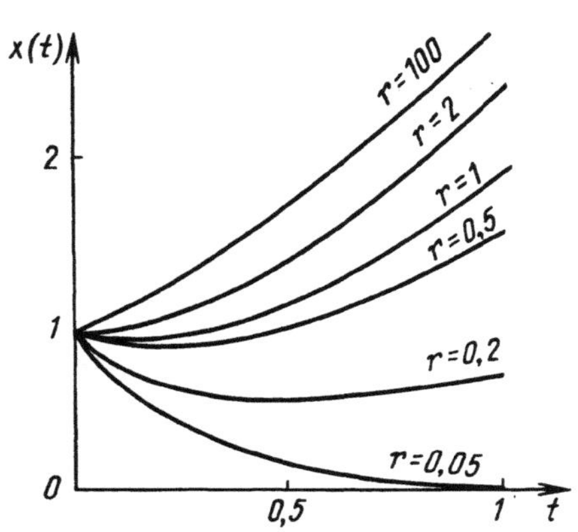

FIG. 2.5. The r-dependence of the optimal trajectories

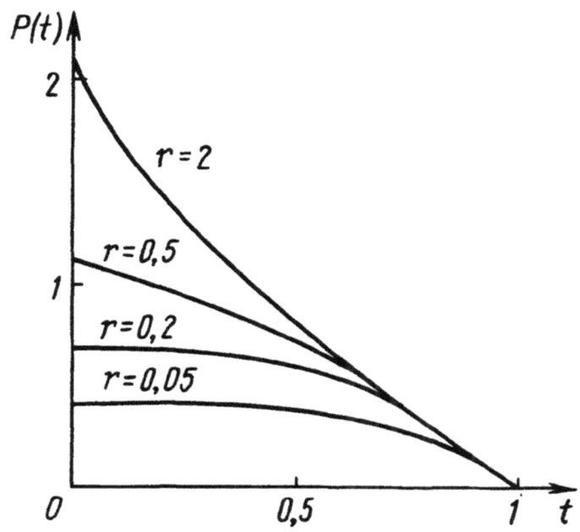

FIG. 2.6. The r-dependence of the solution of Riccati's equation

The scalar function $P(t)$ is determined by solving the problem (2.8), which in our case is of the form

$$\dot{P}(t) = -N_2 - 2aP(t) + r^{-1}P^2(t), \qquad P(T) = N_1. \tag{2.24}$$

Hence, separating the variables and integrating gives

$$\begin{aligned} P(t) = r[N_1 r - a + \beta - (N_1 r - a - \beta)\alpha(t)]^{-1}[(\beta + a)(N_1 r - a + \beta) \\ + (\beta - a) - (N_1 r - a - \beta)\alpha(t)], \\ \alpha(t) = \exp[-2\beta(T - t)], \qquad \beta = \sqrt{a^2 + N_2 r^{-1}}. \end{aligned}$$

The obtained expressions for $P(t)$ and the optimal trajectory $x(t)$ make it possible to investigate their form for different values of the parameters.

Figure 2.1 depicts optimal trajectories $x(t)$ for $a = -1$, $N_1 = 0$, $N_2 = 1$, $T = 1$, $x(0) = 1$ for values of $r = b^{-2}N_3$ that are equal to 100, 1, 0.2, and 0.02. As shown, the decay rate of $x(t)$ depends essentially on the value of r, which characterizes the control cost (the multiplier N_3) and the effectiveness of the control (the multiplier b^{-2}).

In Figure 2.2 we see the graphs of the solutions of equation (2.24) for $r = 1$, 0.5, 0.1, and 0.02. For $t \leqslant T - 3\beta/2$ the function $P(t)$ is close to a constant equal to $N_2(\beta - a)$. This means that for $t \leqslant T - 3\beta/2$ the control of the system (2.23) is effected with an almost constant feedback gain coefficient.

In Figure 2.3 the graphs of the optimal control are shown for $r = 1$, 0.1, and 0.02. As is seen, with increasing control cost (increasing N_3) or with increasing control effectiveness (increasing b) the optimal control tends to a delta-function, i.e., it takes an impulse form.

In Figure 2.4 we see the graphs of the solutions of the problem for $a = -1$, $r = 1$, $T = 1$, 3, 5, 10, and $N_1 = 0$, 1. For $t \leqslant T - 3\beta/2$, as above, the function $P(t)$ is close to a constant equal to $N_2/(\beta - a)$.

Figure 2.5 depicts optimal trajectories for $a = 1$, $N_1 = 0$, $N_2 = 1$, $T = 1$, $x(0) = 1$ for the values of $r = 100$, 2, 1, 0.5, 0.2, 0.05. The original system (2.23) is unstable at $b = 0$. As shown, for small control resources (i.e., for large values of r) the effect of the control on the system's trajectories is weak.

Finally, in Figure 2.6 the graphs of $P(t)$ are depicted for $a = 1$, $N_1 = 0$, $T = 1$ and $r = 2$, 0.5, 0.2, 0.05.

5. Optimal control of wire reeling [12]. The device for wire reeling is shown in Figure 2.7. The electric motor rotates the reel that takes up wire. To avoid wire breaks or sagging, the take-up velocity must be maintained at a constant level. During the take-up process the diameter of the coil increases, a factor that leads to an increase in the moment of inertia of the coil $g(t)$. To maintain a constant linear take-up velocity, $v(t) = v_0$, it is necessary to decrease the angular velocity $\omega(t)$, so that

$$r(t)\omega(t) = v(t) = v_0. \tag{2.25}$$

The equation of rotation of the coil is of the form

$$\frac{d}{dt}[g(t)\omega(t)] = \varkappa U(t) - \psi\omega(t), \tag{2.26}$$

where $U(t)$ is the voltage across the input terminals of the motor, $\varkappa$ is the coefficient of proportionality between the torque of the motor and its input voltage, and ψ is the coefficient of pivoting friction. The control quantity is $U(t)$, and the controlled quantity is $\omega(t)$, which is to be varied in a way such that the condition (2.25) is satisfied. We seek a dependence of $r(t)$ and $g(t)$ on time, given the constant take-up velocity v_0.

Over the time of one revolution t_{rev} which is necessary for taking up one layer of wire the radius changes from r_0 to $r(t_{\text{rev}})$, so that

$$r^2(t_{\text{rev}}) - r_0^2 = kv_0 t_{\text{rev}}, \tag{2.27}$$

where $k > 0$ is a coefficient characterizing the drive. Formula (2.27) for $t_{\text{rev}} \ll 1$ yields

$$r(t) = \sqrt{r_0^2 + Ct}. \tag{2.28}$$

Since the moment of inertia for the reel with wire considered as a variable-mass disk of radius R is proportional to R^4, we have

$$g(t) = g(0) + C_1[r^4(t) - r^4(0)].$$

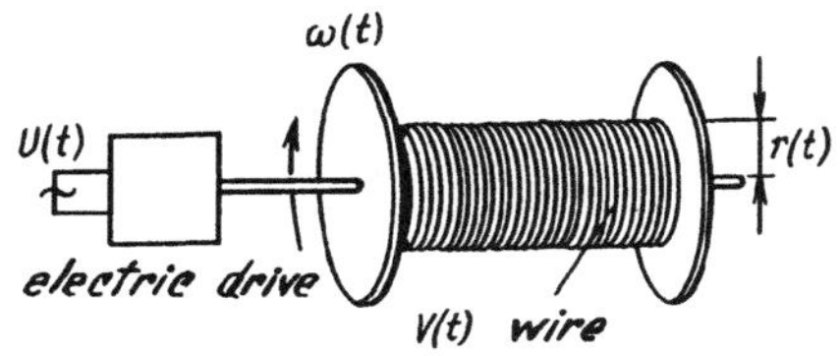

FIG. 2.7. Device for wire reeling: $U(t)$, input voltage to the motor; $\omega(t)$, angular velocity of the reel; $r(t)$, radius of the reel; $v(t)$, linear take-up velocity

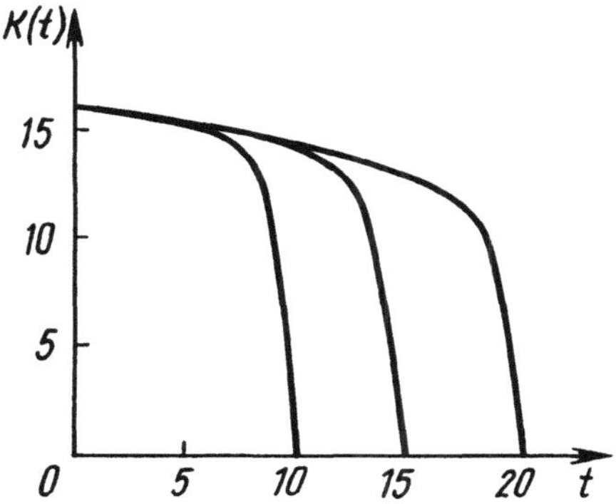

FIG. 2.8. Optimal gain coefficient $K(t)$ in the wire reeling problem

The nominal value $\omega_{\text{nom}}(t)$ for angular velocity is given by

$$\omega_{\text{nom}}(t) = v_0 r^{-1}(t),$$

and the nominal control voltage $U_{\text{nom}}(t)$ is

$$U_{\text{nom}}(t) = \frac{1}{\varkappa}\left[\frac{d}{dt}\big(g(t)\omega_{\text{nom}}(t)\big) + \psi\omega_{\text{nom}}(t)\right].$$

In stability theory $\omega_{\text{nom}}(t)$ and $U_{\text{nom}}(t)$ are called *unperturbed solutions.* For the perturbed values of the variables

$$\begin{aligned} x(t) &= g(t)[\omega(t) - \omega_{\text{nom}}(t)], \\ u(t) &= U(t) - U_{\text{nom}}(t) \end{aligned}$$

we obtain

$$\dot{x}(t) = -\psi g^{-1}(t)x(t) + \varkappa u(t). \tag{2.29}$$

The cost functional, which is to be minimized, is appropriately taken to be

$$J(u) = \int_0^T \left[\big(r(t)g^{-1}(t)x(t)\big)^2 + \rho u^2(t)\right] dt, \qquad \rho > 0. \tag{2.30}$$

The first summand in the integral (2.30) is proportional to the kinetic energy of the revolving reel, and the second to the electrical energy consumed by the motor (see § 3 of Chapter III).

Thus, the optimal control problem for wire reeling reduces to the LQ-problem (2.29), (2.30) with variable coefficients. The F-control being the solution of the problem (2.29), (2.30) is of the form

$$u(t) = -K(t)x(t), \qquad K(t) = \varkappa\rho^{-1}P(t),$$

where $P(t)$ is a solution of Riccati's equation:

$$\dot{P}(t) - 2\psi g^{-1}(t)P(t) - \varkappa^2\rho^{-1}P^2(t) = -r^2(t)g^{-2}(t), \qquad P(T) = 0. \tag{2.31}$$

Equation (2.31) can be integrated numerically. Take, for example, the following numerical values of the parameters of the problem (2.31):

$$g(t) = 0,02 + 66,67[r^4(t) - r^4(0)]\,(\mathrm{kg}\cdot\mathrm{m}^2),$$
$$r(t) = \sqrt{0,01 + 0,0005t}\,(\mathrm{m}),$$
$$\psi = 0,01\,\frac{\mathrm{kg}\cdot\mathrm{m}^2}{\mathrm{sec}}, \qquad \varkappa = 0,1\,\frac{\mathrm{kg}\cdot\mathrm{m}^2\cdot\mathrm{rad}}{\mathrm{V}\cdot\mathrm{sec}}, \qquad \rho = 0,06\,\frac{\mathrm{m}^2}{\mathrm{V}^2\cdot\mathrm{sec}^2}.$$

Figure 2.8 shows graphs of the optimal gain coefficient $K(t)$ against time t for $T = 10$, 15, 20 sec. It is seen that the optimal gain coefficient $K(t)$ behaves in the same way for different T everywhere except for the terminal time interval.

§ 3. Necessary Conditions for Optimal Control. The Method of Lagrange Multipliers

1. The method of Lagrange multipliers. To obtain necessary conditions for optimal control in finite-dimensional optimization problems and in the calculus of variations, it is both convenient and efficatious to use the *method of Lagrange multipliers.* Using the concept of Lagrange multipliers, we shall present the necessary conditions for optimal control in the following problem:

$$\begin{gathered}
\dot{x}(t) = f\big(t, x(t), u(t)\big), \qquad x \in R^n,\ u \in R^m,\ t_0 \leqslant t \leqslant T,\\
J_0\big(t_0, x(t_0), u, T, x(T)\big) \to \inf_u,\\
J_j\big(t_0, x(t_0), u, T, x(T)\big) \leqslant 0, \qquad j = 1, \dots, m_1,\\
J_i\big(t_0, x(t_0), u, T, x(T)\big) = 0, \qquad i = m_1 + 1, \dots, m_2,\\
J_k = \int_{t_0}^{T} F_k\big(t, x(t), u(t)\big)\,dt + \varphi_k\big(t_0, x(t_0), T, x(T)\big), \qquad k = 0, \dots, m_2.
\end{gathered} \tag{3.1}$$

Suppose that there exists a solution of the problem (3.1), and it is of the form $\{t_0, T, u_0(t), x_0(t)\}$. Here $u_0(t)$ is an optimal control in the problem (3.1), $x_0(t)$ is an optimal trajectory t_0 and T are unspecified initial and terminal moments.

We assume that the functions f, F, f_x, f_u, F_x, φ, φ_x are continuous in some neighborhood of $\big(t, x_0(t), u_0(t)\big)$, $t \in [t_0, T]$, where $F = (F_0, \dots, F_{m_2})$, $\varphi = (\varphi_0, \dots, \varphi_{m_2})$.

Theorem 3.1. *Suppose that the problem (3.1) has a solution of the form $\{t_0, T, u_0(t), x_0(t)\}$. Then there exist Lagrange multipliers $\lambda = (\lambda_0, \dots, \lambda_{m_2})$, $\psi(t) \in R^n$, not both zero, where $\psi(t)$ is continuously differentiable on $[t_0, T]$, such that the following relations hold:*

the evolution equations

$$\dot{x}_0(t) = f\big(t, x_0(t), u_0(t)\big); \tag{3.2}$$

the Euler equations

$$\dot{\psi}(t) = -f_x\big(t, x_0(t), u_0(t)\big)\psi(t) - F_x\big(t, x_0(t), u_0(t)\big)\lambda; \tag{3.3}$$

the transversality conditions

$$\psi(t_0) = -\varphi_x(t_0, x, T, y)\lambda, \qquad \psi(T) = \varphi_y(t_0, x, T, y)\lambda \tag{3.4}$$

for $x = x_0(t_0)$, $y = x_0(T)$;

the stationarity conditions

$$-F_u\big(t, x_0(t), u_0(t)\big)\lambda + f_u\big(t, x_0(t), u_0(t)\big)\psi(t) = 0; \tag{3.5}$$

the conditions for variable end points t_0 and T (taken account of in problems with variable end points)

$$\begin{aligned} F'\big(t_0, x_0(t_0), u_0(t_0)\big)\lambda - [\varphi_{t_0}\big(t_0, x_0(t_0), T, x_0(T)\big) \\ + \varphi'_x(t_0, x, T, y)\dot{x}_0(t_0)]'\lambda = 0, \end{aligned} \tag{3.6}$$

$$\begin{aligned} F'\big(T, x_0(T), u_0(T)\big)\lambda + [\varphi_T\big(t_0, x_0(t_0), T, x_0(T)\big) \\ + \varphi'_y(t_0, x, T, y)\dot{x}_0(t_0)]'\lambda = 0 \end{aligned} \tag{3.7}$$

for $x = x_0(t_0)$ and $y = x_0(T)$;

the conditions for complementary slackness

$$\lambda_j\left[\int_{t_0}^{T} F_j\big(t, x_0(t), u_0(t)\big)\,dt + \varphi_j\big(t_0, x_0(t_0), T, x_0(T)\big)\right] = 0, \qquad j = 1, \dots, m_1; \tag{3.8}$$

the nonpositivity condition

$$\lambda_i \leqslant 0, \qquad i = 0, 1, \dots, m_1. \tag{3.9}$$

For the proof of this theorem, the reader is referred to [2].

Note that the condition (3.6) should be used if the initial moment t_0 is variable, and the condition (3.7) in the case of the variable moment T. But if one of the moments, t_0 or T, is fixed, then the respective condition(3.6) or (3.7) is not used.

Moreover, if λ_0 is different from zero, i.e., if $\lambda_0 < 0$, then λ_0 may be taken as any negative constant.

Remark. The conditions (3.6), (3.7) are conveniently written in terms of the function $H(t, x, u, \psi)$ equal to

$$H(t, x, u, \psi) = \psi'(t) f(t, x, u) + \lambda' F(t, x, u).$$

The condition (3.6) may be represented as

$$H(t, x, u, \psi)\big|_{t=t_0} = \lambda' \varphi_{t_0}\big(t_0, x_0(t_0), T, x_0(T)\big).$$

The condition (3.7) may be represented in the form

$$H(t, x, u, \psi)\big|_{t=T} = -\lambda' \varphi_T\big(t_0, x_0(t_0), T, x_0(T)\big).$$

Here $H\big|_{t=T}$ is the value of H for $t = T$.

In summary, the necessary conditions for optimality in the control problem (3.1) are expressed by the relations (3.1)–(3.9).

The method for determining an optimal control from the necessary conditions for optimality consists in the following :

1°. Equation (3.5) is solved for the control u_0 as a function of the time t, of the trajectory $x_0(t)$, and of the Lagrange multipliers λ, ψ, i.e.,

$$u_0 = u_0\big(t, x_0(t), \psi(t), \lambda\big).$$

2°. The control obtained is inserted in equations (3.2), (3.3), (3.6)–(3.9). Next, we determine the trajectory $x_0(t)$ and the vector $\psi(t)$ by solving the boundary-value problem composed of the system of equations (1.1), (1.3), of the boundary conditions (3.4), (3.6), (3.7), and of the relations (3.8), (3.9).

3°. A general solution $x_0(t), \psi(t)$ of the system of $2n$ equations (1.1), (3.3) is constructed. This general solution depends on $2n$ arbitrary constants, and also on Lagrange multipliers λ, and two moments t_0 and T.

Observe that the relations (3.1)–(3.9) preserve their form under the substitution of $C\psi$ for ψ and $C\lambda$ for λ, where $C > 0$ is an arbitrary constant. Hence, the number of independent components of the vector $\lambda \in R^{m_2+1}$ is at most m_2. Therefore, the general solution of system (3.2), (3.3) depends on $2n + m_2 + 2$ constants; they can be determined from the $2n$ conditions (3.4), from the two conditions (3.6), (3.7) for variable ends t_0, T, and, finally, from the m_1 conditions (3.8) and from the $m_2 - m_1$ equalities of the problem (3.1). Hence, the number of conditions that are used for constructing an optimal solution coincides with the number of desired constants.

In what follows, we shall present conditions for optimal control in typical situations.

Example 3.1. Consider a system consisting of a driver cell connected via a resistor R with a capacitor C. It is required to determine a control $u(t)$ for the driver voltage such that the voltage $x(t)$ across the capacitor is changed from $x_0 = x(0)$ to $x_1 = x(T)$ and the energy J dissipated by the resistor minimized. The equations of the problem are the following:

$$RC\dot{x}(t) = -x(t) + u(t), \qquad 0 \leqslant t \leqslant T,\ x(0) = x_0,\ x(T) = x_1,$$
$$J = \frac{1}{R}\int_0^T \big(u(t) - x(t)\big)^2\, dt \to \min_u .$$

It immediately follows that the optimal control $u_0(t)$ and the associated voltage $x_0(t)$ are given by

$$u_0(t) = x_0 + \frac{4}{T}(x_1 - x_0)(RC + t),$$
$$x_0(t) = x_0(1 - \frac{t}{T}) + x_1\frac{t}{T}.$$

Let us now obtain the same solution using Theorem 3.1. We have

$$H(x, u, \psi) = \lambda_0\frac{(u-x)^2}{R} + \frac{\psi}{RC}(u - x).$$

If $\lambda_0 = 0$, the condition $\partial H/\partial u = 0$ would imply that $\psi \equiv 0$, i.e., the vector of Lagrange multipliers (λ_0, ψ) is zero, which cannot be true. Hence, $\lambda_0 < 0$; then we can put $\lambda_0 = -1$. Further, the condition $\partial H/\partial u = 0$ implies that

$$u_0(t) = \frac{1}{2C}\psi(t) + x(t).$$

Substituting this expression in the equation for $\psi(t)$ gives

$$\dot{\psi}(t) = -\frac{\partial H}{\partial x} = -2\frac{u-x}{R} + \frac{\psi}{RC} = 0.$$

Therefore, $\psi(t)$ is a constant, i.e., $\psi(t) \equiv \psi_0$. To determine ψ_0, we substitute u_0 in the equation for $x(t)$. This gives

$$RC\dot{x}(t) = -x(t) + \frac{1}{2C}\psi_0 + x(t) = \frac{1}{2C}\psi_0.$$

Integrating this equation leads to the relation $x_0(T) = x_1 = x_0 + \psi_0 T(2RC^2)^{-1}$. It follows that $\psi_0 = 2RC^2T^{-1}(x_1 - x_0)$. Hence, using the equation for x and the expression for u_0, we obtain the required solution.

2. Fixed initial and terminal moments, and fixed initial state. Let us consider the control problem for the system (3.2) with prescribed end moments t_0 and T, initial state $x(t_0)$, and cost functional

$$J_0(u, x(T)) = \int_{t_0}^{T} F_0(t, x(t), u(t))\, dt + \varphi_0(x(T)) \to \inf .$$

There are no other constraints of (3.1). In the case under consideration the necessary conditions for optimal control are expressed by Theorem 1.1. The statement of Theorem 1.1 can be obtained as a consequence of the general necessary conditions for optimal control that were stated in Theorem 3.1. The vector λ of Lagrange multipliers is reduced in our case to the scalar λ_0, which is nonpositive by virtue of the conditions (3.9). Therefore, the equations (3.3), (3.4) take the form

$$\dot{\psi}(t) = -f_x(t, x_0(t), u_0(t))\psi(t) - \lambda_0 F_{0x}(t, x_0(t), u_0(t)), \tag{3.10}$$

$$\psi(T) = \lambda_0 \varphi_{0x}(x_0(T)) \tag{3.11}$$

The condition (3.4) leads to the equation

$$f_u(t, x_0(t), u_0(t))\psi(t) + \lambda_0 F_{0u}(t, x_0(t), u_0(t)) = 0. \tag{3.12}$$

Since the moments t_0 and T and the initial state $x_0(t)$ are prescribed and fixed, and there are no constraints of the type (3.1), the remaining necessary conditions (3.6)–(3.8) for optimal control are not considered. Thus, for the problem (1.1), (1.2) the necessary conditions for optimal control are reduced to the existence of a number $\lambda_0 \leqslant 0$ and of a vector $\psi(t) \in R^n$, not both zero, which satisfy equations (1.1), (3.10), (3.11).

Observe further that if $\lambda_0 = 0$, then it follows from (3.10), (3.11) that the function $\psi(t) \equiv 0$, which cannot be true. Hence, $\lambda_0 < 0$, i.e., λ_0 may be taken as any negative constant. Put $\lambda_0 = -1$. In that case the relations (3.10)–(3.12) transform into the conditions for optimal control stated in Theorem 1.1 and proved above.

3. Fixed initial and terminal moments, and variable initial and terminal states. Consider the control problem for the system (3.2) with prescribed moments t_0, T and the cost functional

$$J = \int_{t_0}^{T} F_0(t, x(t), u(t))\, dt + \varphi_0(x(t_0), x(T)) \to \inf .$$

There are no other constraints of (3.1). The initial $x(t_0)$ and the terminal $x(T)$ states are variable. In our case the necessary conditions (3.3)–(3.9) for optimal control take the form of equations (3.10), (3.12) and of boundary conditions

$$\psi(t_0) = -\lambda_0 \varphi_x(x, y), \qquad \psi(T) = \lambda_0 \varphi_y(x, y),$$
$$x = x_0(t_0),\ y = x_0(T).$$

Using the same argument as in subsection 2, we may set $\lambda_0 = -1$. Thus, in the case under consideration necessary conditions for optimal control are of the form

$$\begin{aligned} \dot{x}_0(t) &= f(t, x_0(t), u_0(t)), \qquad t_0 \leqslant t \leqslant T, \\ \dot{\psi}(t) &= -f_x(t, x_0(t), u_0(t))\psi(t) + F_{0x}(t, x_0(t), u_0(t)), \\ \psi(t_0) &= \varphi_{0x}(x, y), \qquad x = x_0(t_0), \\ \psi(T) &= -\varphi_{0y}(x, y), \qquad y = x_0(T), \\ f_y(t, x_0(t), &u_0(t))\psi(t) - F_{0u}(t, x_0(t), u_0(t)) = 0 \end{aligned} \tag{3.13}$$

4. Problems with fixed values of some state variables at the initial and terminal moments. Consider the same control problem as in the previous subsection but with prescribed values of some of the components of the vectors $x(t_0)$ and $x(T)$. The equations of motion have the form (3.2), with cost functional J, the moments t_0 and T being prescribed. In addition, q_1 components $x_1^0, x_2^0, \ldots, x_{q_1}^0$ of the vector $x_0(t)$ and q_2 components $x_1^T, x_2^T, \ldots, x_{q_2}^T$ of the vector $x(T)$, with the integers q_1, q_2 at most n, are also given.

We render the last constraints in the form (3.1):

$$\begin{aligned} x_i(t_0) - x_i^0 = 0, \qquad i = 1, \ldots, q_1, \\ x_j(t) - x_j^T = 0, \qquad j = 1, \ldots, q_2. \end{aligned} \tag{3.14}$$

Due to the constraints (3.14) it is natural to assume that the function φ in the functional J depends only on the remaining values of the phase coordinates, i.e.,

$$\varphi(x(t_0, x(T))) = \varphi(x_{q_1+1}(t_0), \ldots, x_n(t_0), x_{q_2+1}(T), \ldots, x_n(T)).$$

The vector of Lagrange multipliers is of the form $\lambda = (\lambda_0, \lambda_1, \ldots, \lambda_{m_2})$, where $m_2 = q_1 + q_2$, $\lambda_0 \leqslant 0$. Therefore, the necessary conditions (3.3)–(3.9) for optimal control can be written as

$$\begin{aligned} \dot{x}_0(t) &= f(t, x_0(t), u_0(t)), \qquad t_0 \leqslant t \leqslant T, \\ \dot{\psi}(t) &= f_x(t, x_0(t), u_0(t))\psi(t) - \lambda_0 F_{0x}(t, x_0(t), u_0(t)), \\ f_u(t, x_0(t), &u_0(t))\psi(t) + \lambda_0 F_{0u}(t, x_0(t), u_0(t)) = 0. \end{aligned} \tag{3.15}$$

Simultaneously the transversality conditions hold:

$$\begin{aligned} \psi_i(t_0) &= -\lambda_i, \qquad i = 1, \ldots, q_1, \\ \psi_j(t_0) &= -\lambda_0 \varphi_{0x_j(t_0)}(x(t_0), x(T)), \qquad j = q_1 + 1, \ldots, n, \\ \psi_k(T) &= \lambda_{q_1+k}, \qquad k = 1, \ldots, q_2, \\ \psi_l(T) &= \lambda_0 \varphi_{0x_l(T)}(x(t_0), x(T)), \qquad l = q_2 + 1, \ldots, n \end{aligned} \tag{3.16}$$

Example 3.2. Consider the following one-dimensional control problem:

$$\dot{x}(t) = ax(t) + u(t), \qquad 0 \leqslant t \leqslant T, \tag{3.17}$$

$$x(0) = 0,\ x(T) = 1,\ \int_0^T u^2(t)\,dt \to \min,$$

where a and T are given constants.

The necessary conditions (3.15), (3.16) for optimal control yield the relations

$$\dot{\psi} = -a\psi(t), \qquad -2\lambda_0 u(t) - \psi(t) = 0, \tag{3.18}$$

$$\psi(0) = -\lambda_1, \qquad \psi(T) = \lambda_2. \tag{3.19}$$

If $\lambda_0 = 0$, then by virtue of (3.18) the function $\psi(t) \equiv 0$. Therefore, on the strength of (3.19) the constants $\lambda_1 = 0$, $\lambda_2 = 0$, i.e., the Lagrange multipliers $(\lambda_0, \lambda_1, \lambda_2, \psi(t))$ are all zero, which is not possible. Hence, λ_0 may be taken as any negative constant. Set $\lambda_0 = -1/2$. Then $u(t) = \psi(t)$ on the strength of (3.18). Hence, noting Cauchy's formula, we can write $\psi(t) = -\lambda_1 \exp(-at) = u(t)$. Substituting this expression for the control $u(t)$ in (3.17) gives

$$x(t) = -\lambda_1 \int_0^t e^{a(t-2s)}\, ds.$$

Setting $t = T$ in this equation and using the boundary condition $x(T) = 1$, we obtain

$$\lambda_1 = -2a\big[e^{aT} - e^{-aT}\big]^{-1}.$$

Consequently, the optimal control $u_0(t)$ is of the form

$$u_0(t) = 2ae^{-at}\big[e^{aT} - e^{-aT}\big]^{-1}, \tag{3.20}$$

and the trajectory associated with this control has the form

$$x(t) = \big[e^{at} - e^{-at}\big]\big[e^{aT} - e^{-aT}\big]^{-1}. \tag{3.21}$$

Formulas (3.20), (3.21) hold for any a. In particular, if $a = 0$, then, noting the L'Hospital rule, we have $u_0(t) = 1/T$ and $x(t) = t_0/T$.

5. Problems with an unspecified terminal moment. Consider the control problem (3.1) with a prescribed initial moment t_0 and state $x(t_0)$, with variable terminal moment T and state $x(T)$, with a scalar cost functional

$$J_0\big(u, T, x(T)\big) \to \inf, \tag{3.22}$$

and with constraints of the form

$$J_i\big(u, T, x(T)\big) = 0, \qquad i = 1, \dots, m_2. \tag{3.23}$$

Here the scalar functionals J_k are determined by

$$J_k\big(u, T, x(T)\big) = \int_{t_0}^{T} F_k\big(t, x(t), u(t)\big)\, dt + \varphi_k\big(T, x(T)\big),$$
$$k = 0, \dots, m_2. \tag{3.24}$$

The necessary conditions for optimal control in the problem (3.1), (3.22)–(3.24) correspond to the existence of a vector $\lambda = (\lambda_0, \ldots, \lambda_{m_2})$, $\lambda_0 \leqslant 0$, and a function $\psi(t) \in R^n$, not both zero, such that equations (3.2), (3.3)–(3.5), (3.7) hold (in these equations $F = (F_0, \ldots, F_{m_2})$ and $\varphi = (\varphi_0, \ldots, \varphi_{m_2})$).

6. The Chaplygin problem. Suppose that an aircraft flies in the horizontal xy-plane with constant velocity v that makes an angle u with the x-axis. The aircraft motion is affected by the wind whose velocity ω is constant and directed along the x-axis (see Figure 3.1). The *Chaplygin problem* is to determine a control u (a relative bearing of the aircraft) so that the aircraft, starting from the point $A(x_0, y_0)$, will fly around a figure of maximum area S in a given time T.

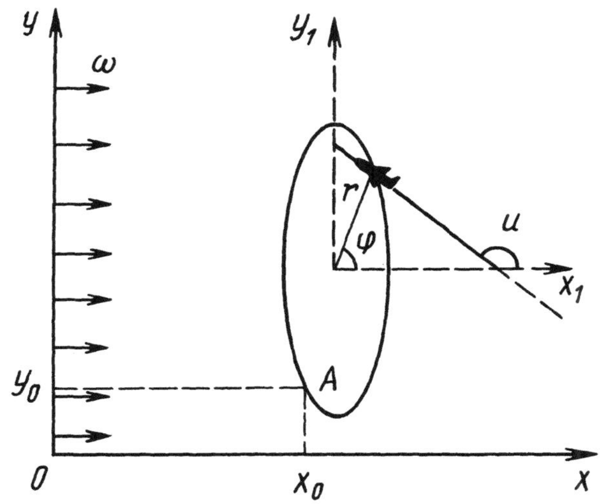

FIG. 3.1. The Chaplygin problem

Denote by $x(t)$ and $y(t)$ the moving coordinates of the aircraft. Then the equations of motion of the aircraft, the initial and terminal conditions are of the form

$$\dot{x} = v\cos u - \omega, \qquad \dot{y} = v\sin u, \tag{3.25}$$

$$x(0) = x(T) = x_0, \; y(0) = y(T) = y_0. \tag{3.26}$$

The flown-around area S is determined by

$$S = \frac{1}{2}\int_0^T \big(x(t)\dot{y}(t) - y(t)\dot{x}(t)\big)\,dt \to \max.$$

Hence, the cost functional $J(u)$ (to be minimized) may be taken as

$$J(u) = -\int_0^T [xv\sin u - y(v\cos u - \omega)]\,dt. \tag{3.27}$$

Thus, the Chaplygin problem is a classical variational problem of the Lagrange type.

To solve the Chaplygin problem, we rely on the procedure outlined above and construct the function

$$H = [xv \sin u - y(v \cos u - \omega)] + \psi_1 (v \cos u - \omega) + \psi_2 v \sin u.$$

Then equations (3.3) for adjoint variables ψ_1, ψ_2 and the stationarity conditions (3.5) yield the following equations:

$$\begin{aligned} -\dot{\psi}_1 &= \frac{\partial H}{\partial x} = v \sin u = \dot{y}, \\ -\dot{\psi}_2 &= \frac{\partial H}{\partial y} = -(v \cos u - \omega) = -\dot{x}; \\ \frac{\partial H}{\partial u} &= \lambda_1 v \sin u - \lambda_2 v \cos u = 0. \end{aligned} \tag{3.28}$$

The first two equations (3.28) imply that $\psi_1 = -y + C_2$, $\psi_2 = x + C_1$. Then it follows from the last equation of (3.8) that

$$(y + C_2) \sin u + (x + C_1) \cos u = 0. \tag{3.29}$$

Equations (3.25), (3.29) define a curve in parametric form. We can show that this curve for $\omega < v$ is an ellipse. We now introduce new coordinates

$$x_1 = x + C_1, \qquad y_1 = y + C_2; \qquad \varphi = \frac{\pi}{2} + u.$$

Then equations (3.25), (3.29) become

$$\dot{x}_1 + v \sin \varphi - \omega = 0, \qquad \dot{y}_1 - v \cos \varphi = 0, \qquad y_1 \cos \varphi - x_1 \sin \varphi = 0. \tag{3.30}$$

It follows from the last equation of (3.30) that

$$x_1 = r \cos \varphi, \qquad y_1 = r \sin \varphi, \qquad r^2 = x_1^2 + y_1^2. \tag{3.31}$$

Differentiating (3.31) and using (3.30) yields

$$\begin{aligned} \dot{r} = \frac{d}{dt}\left(\sqrt{x_1^2 + y_1^2}\right) &= \frac{2(x_1 \dot{x}_1 + y_1 \dot{y}_1)}{2\sqrt{x_1^2 + y_1^2}} \\ &= \dot{x}_1 \cos \varphi + \dot{y}_1 \sin \varphi = \omega \cos \varphi = \frac{\omega}{v} \dot{y}_1. \end{aligned}$$

An integration of this relation gives

$$r = \frac{\omega}{v} y_1 + C = r \frac{\omega}{v} \sin \varphi + C,$$

or

$$r = \frac{C}{1 + \frac{\omega}{v}\cos\left(\frac{\pi}{2} + \varphi\right)}. \tag{3.32}$$

As is known,[1] equation (3.32) for $\omega < v$ specifies an ellipse in polar coordinates with eccentricity ω/v, focal parameter C, and major axis parallel to the y-axis. On determining the constant of integration C from the conditions (3.26), we find that the maximum area S that is flown around by the aircraft in time T is as follows:

$$S_{\max} = \frac{v^2T^2}{4\pi}\left(1 - \frac{\omega^2}{v^2}\right)^{3/2}.$$

7. Maximization of the rocket velocity immediately before putting the rocket into a rectilinear trajectory [5]. Suppose that a rocket regarded as a material point of constant mass m moves in the xy-plane by action of thrust $F_{\text{th}} = ma(t)$, $0 \leqslant t \leqslant T$, (see Figure 3.2). The thrust acceleration $a(t)$ is taken to be a known function of time, such as a constant $a(t) = a$ (in the simplest case). Such a model of motion for a real rocket is justified if there is no appreciable change in the mass of the rocket ($(m_T - m_0) \ll m_0$) and the mass of the rocket can be disregarded ($g \ll a(t)$). Denote by $v_x(t)$ and $v_y(t)$ the horizontal and vertical components of the rocket velocity, i.e., $v_x = \dot{x}$, $v_y = \dot{y}$. For a control we take the thrust direction angle, i.e., the angle $\vartheta = u$ between the thrust force F_{th} and the x-axis. The motion of the rocket is then described by the equations

$$\dot{x} = v_x, \qquad \dot{y} = v_y, \qquad \dot{v}(x) = a\cos u, \qquad \dot{v}_y = a\sin u. \tag{3.33}$$

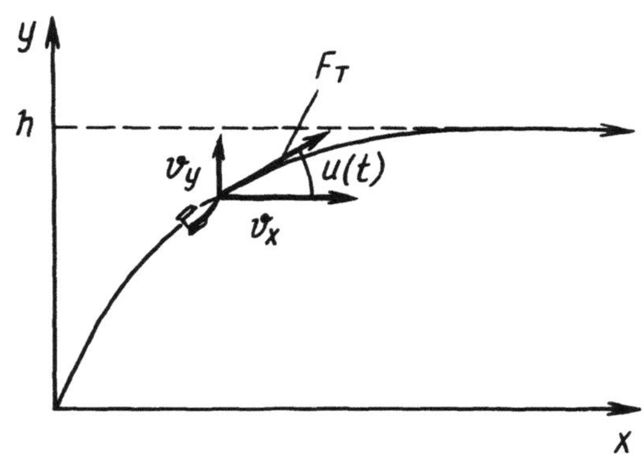

FIG. 3.2. Putting the rocket into a horizontal trajectory

[1] Korn, G. A., and Korn, T. M, *Mathematical Handbook for Scientists and Engineers*, McGraw–Hill, New York, (1968).

It is desired to transfer the rocket in a prescribed time T to a trajectory parallel to the x-axis and a distance h apart such that a maximum value of the velocity is attained. In other words, the initial conditions are of the form

$$x(0) = y(0) = v_x(0) = v_y(0) = 0. \tag{3.34}$$

The terminal conditions for a given T have the form

$$y(T) = h, \qquad v_y(T) = 0, \qquad x(T) \quad \text{is arbitrary.} \tag{3.35}$$

We must deduce a control $\theta(t)$ such that the following functional is maximized:

$$J(u) = v_x(T) \to \max. \tag{3.36}$$

No constraints are imposed on the angle $\theta = u$. Therefore, the problem under consideration is a classical variational Mayer problem.

The Hamiltonian H for our problem can be written as

$$H = \psi_1 a \cos\vartheta + \psi_2 a \sin\vartheta + \psi_3 v_x + \psi_4 v_y.$$

For the adjoint variables ψ_i the following equations are valid:

$$\dot\psi_1 = 0, \qquad \dot\psi_2 = 0, \qquad \dot\psi_3 = -\psi_1, \qquad \dot\psi_4 = -\psi_2;$$

hence,

$$\psi_1 = C_1, \qquad \psi_2 = C_2, \qquad \psi_3 = -C_1 t + C_3, \qquad \psi_4 = -C_2 t + C_4.$$

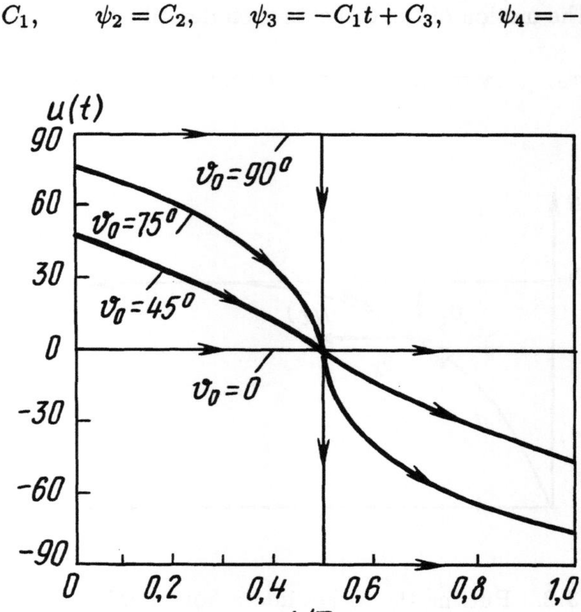

FIG. 3.3. Optimal P-control of the thrust angle u_{opt}

It follows from the conditions for optimal control

$$\frac{\partial H}{\partial u} = -\psi_3 \sin u + \psi_4 \cos u = 0$$

that the optimal direction of the thrust force is expressed by

$$\tan \vartheta_{\text{opt}} = \frac{\psi_4}{\psi_3} = \frac{-C_2 t + C_4}{-C_1 t + C_3}. \tag{3.37}$$

The law of change for the thrust direction angle (3.37) is called the *law of linear-fractional tangent.*

It follows from the boundary conditions (3.34), (3.35) that $\psi_1 \equiv 0$ and $\psi_3 \equiv 1$. Therefore, the law (3.37) takes the form

$$\tan u = \tan \vartheta_0 - Ct.$$

The values of ϑ_0 and C are determined from two boundary conditions $v_y(T) = 0$, $y(T) = h$. Figure 3.3 shows the programs of the control of the thrust direction angle u_{opt} for various parameter values ϑ_0. Other problems of guidance system design are discussed in the numerous literature devoted to the subject.

Main Results and Formulas of Chapter IV

The Mayer problem with unconstrained control:

$$\dot{x}(t) = f\big(t, x(t), u(t)\big), \qquad x(t_0) = x_0,\ t_0 \leqslant t \leqslant T,$$
$$\varphi\big(x(T)\big) \to \inf_u;$$

necessary conditions for optimal control:

$$\frac{\partial H}{\partial u} = 0, \qquad H(t, x, u, \psi) = \psi'(t) f(t, x, u),$$
$$\dot{\psi}(t) = -H_x(t, x, u, \psi), \qquad \psi(T) = -\varphi_x\big(x(T)\big).$$

The Boltz problem with unconstrained control:

$$\dot{x}(t) = f\big(x, x(t), u(t)\big), \qquad t_0 \leqslant t \leqslant T,\ x(t_0) = x_0;$$
$$\int_{t_0}^{T} F\big(t, x(t), u(t)\big)\, dt + \varphi\big(x(T)\big) \to \inf_u;$$

necessary conditions for optimal control:

$$\frac{\partial H}{\partial u} = 0, \quad \text{where } H(t, x, u, \psi) = -F(t, x, u) + \psi'(t) f(t, x, u);$$
$$\dot{\psi}(t) = -H_x(t, x, u, \psi), \qquad \psi(T) = -\varphi_x\big(x(T)\big).$$

Linear-quadratic problem:

$$\dot{x}(t) = A(t)x(t) + B(t)u(t), \qquad t_0 \leqslant t \leqslant T,\ x(t_0) = x_0,$$
$$J(u) = x'(T)N_1 x(T) + \int_{t_0}^{T} \big(x'(t)N_2(t)x(t) + u'(t)N_3(t)u(t)\big)\, dt,$$
$$N_1 \geqslant 0,\ N_2(t) \geqslant 0,\ N_3(t) > 0;$$

optimal control:

$$u_0(t) = -N_3^{-1}(t)B'(t)P(t)x_0(t);$$

matrix Riccati equation:

$$\dot{P}(t) + A'(t)P(t) + P(t)A(t) - P(t)B(t)N_3^{-1}(t)B'(t)P(t) + N_2(t) = 0,$$
$$0 \leqslant t \leqslant T,\ P(T) = N_1.$$

The method of Lagrange multipliers. Let $(t_0, T, u_0(t), x_0(t))$ be a solution of the general optimal control problem (3.1) with phase constraints and unspecified initial and terminal moments. Then there exist Lagrange multipliers $\lambda = (\lambda_0, \dots, \lambda_{m_2})$, $\psi(t) \in R^n$, not both zero, where $\psi(t)$ is continuously differentiable on $[t_0, T]$, such that the following relations hold:

the evolution equations

$$\dot{x}_0(t) = f\big(t, x_0(t), u_0(t)\big);$$

the Euler equations

$$\dot{\psi}(t) = -f_x\big(t, x_0(t), u_0(t)\big)\psi(t) - F_x\big(t, x_0(t), u_0(t)\big)\lambda;$$

the transversality conditions

$$\psi(t_0) = -\varphi_x(t_0, x, T, y)\lambda, \qquad \psi(T) = \varphi_y(t_0, x, T, y)\lambda,$$
$$x = x_0(t_0), \ y = x_0(T);$$

the stationarity conditions

$$-F_u\big(t, x_0(t), u_0(t)\big)\lambda + f_u\big(t, x_0(t), u_0(t)\big)\psi(t) = 0;$$

the conditions for variable end points t_0 and T (taken account of in problems with variable end points)

$$\begin{gathered} F'\big(t_0, x_0(t_0), u_0(t_0)\big)\lambda - [\varphi_{t_0}\big(t_0, x_0(t_0), T, x_0(T)\big) \\ + \varphi'_x(t_0, x, T, y)\dot{x}_0(t_0)]'\lambda = 0, \\ F'\big(T, x_0(T), u_0(T)\big)\lambda + [\varphi_T\big(t_0, x_0(t_0), T, x_0(T)\big) \\ + \varphi'_y(t_0, x, T, y)\dot{x}_0(T)]'\lambda = 0, \\ x = x_0(t_0), \ y = x_0(T) \end{gathered}$$

or

$$H(t, x, u, \psi)\big|_{t=t_0} = \lambda'\varphi_{t_0}\big(t_0, x_0(t_0), T, x_0(T)\big),$$
$$H(t, x, u, \psi)\big|_{t=T} = -\lambda'\varphi_T\big(t_0, x_0(t_0), T, x_0(T)\big);$$

the conditions for complementary slackness

$$\lambda_j\left[\int_{t_0}^T F_j\big(t, x_0(t), u_0(t)\big)\,dt + \varphi_j\big(t_0, x_0(t_0), T, x_0(T)\big)\right] = 0,$$
$$j = 1, \dots, m_1;$$

the nonpositivity condition

$$\lambda_j \leqslant 0, \qquad j = 0, 1, \dots, m_1.$$

CHAPTER V

THE MAXIMUM PRINCIPLE

In practical problems certain essential constraints are usually imposed on the control set. For such problems the necessary conditions for optimal control stated in the preceding chapter are, in general, not suitable. Necessary conditions for optimality in such problems are furnished by Pontryagin's maximum principle, which is the subject of this chapter. It turns out that only the stationarity condition needs to be modified, whereas all other necessary conditions for optimal control preserve the form [1, 2, 7, 8, 15].

§ 1. Problems with a Variable Terminal Point and a Prescribed Transfer Time

1. The Mayer problem. Consider the following control problem:

$$\dot{x}(t) = f\big(t, x(t), u(t)\big), \qquad x \in R^n,\ u \in R^m,\ t_0 \leqslant t \leqslant T,\ x(t_0) = x_0, \quad (1.1)$$

$$J(u) = \varphi\big(x(T)\big) \to \inf, \quad (1.2)$$

$$u(t) \in U. \quad (1.3)$$

As compared to the preceding chapter, the constraint (1.3) is a new element in the problem statement. If the set U is closed and bounded, then the optimal control may, in general, belong to the set boundary U. In that case the control cannot be varied arbitrarily; hence the methods of the classical calculus of variations are not applicable. We assume that f and φ satisfy the constraints (1.3)–(1.5) of Chapter IV, and we look for an optimal control in the class of piecewise-continuous functions $u(t)$. Since the values of $u(t)$ at the points of discontinuity do not affect the cost functional (1.2), we can choose the control values at the points of discontinuity in an arbitrary way. For definiteness, control functions $u(t)$ are taken to be piecewise continuous on the right, i.e.,

$$u(t) = \lim_{s \to t+0} u(s), \qquad t \in [t_0, T),\ u(T) = u(T-0).$$

If there are no constraints (1.3), the necessary conditions for optimal control are given in Theorem 1.1 of Chapter IV. The constraint (1.3), however, requires a modification of the condition (1.7) in the above-mentioned theorem of Chapter IV because the optimal control $u(t)$ may belong to the boundary of the set U; hence the derivative $\partial H/\partial u$ may be different from zero. The necessary conditions for optimal control, called *Pontryagin's maximum principle*, are stated as follows.

Theorem 1.1 (Pontryagin's maximum principle). *Let $u_0(t)$ be the optimal control in the problem (1.1)–(1.3), and $x_0(t)$ be the corresponding optimal trajectory. Then $u_0(t)$ satisfies the maximum condition*

$$\max_{u \in U} H\big(t, x_0(t), u, \psi(t)\big) = H\big(t, x_0(t), u_0(t), \psi(t)\big). \tag{1.4}$$

Here the Hamiltonian H is determined by the expression

$$H\big(t, x(t), u(t), \psi(t)\big) = \psi'(t) f\big(t, x(t), u(t)\big),$$

and the vector of adjoint variables $\psi(t)$ is a solution of the problem

$$\dot{\psi}(t) = -H_x\big(t, x_0(t), u_0(t), \psi(t)\big), \qquad \psi(T) = -\varphi_x\big(x_0(T)\big).$$

Remark. As compared to the necessary conditions for optimal control in the classical variational control problem (Theorem 1.1 of Chapter IV), in Pontryagin's maximum principle only the stationarity condition $\partial H/\partial u = 0$ is replaced by the condition (1.4) for a maximum. All other conditions remain unaltered.

Proof. The proof of this theorem up to formula (1.26) of Chapter IV is a verbatim repetition of the proof of Theorem 1.1 of Chapter IV. The change in the proof caused by the control constraint (1.3) affects the subsequent argument. We turn now to that part of the proof.

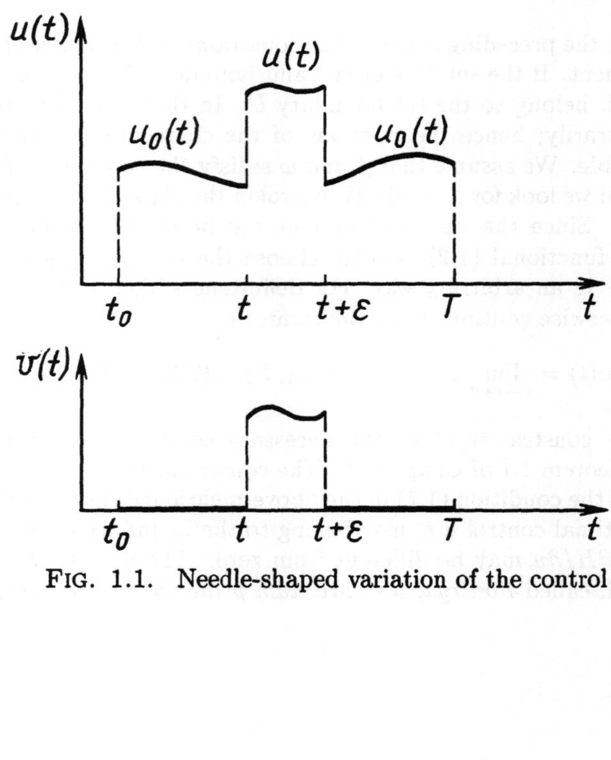

FIG. 1.1. Needle-shaped variation of the control

Take an arbitrary element $u \in U$ and the moments t and $t+\varepsilon$ on the segment $[t_0, T]$, where $\varepsilon > 0$. Denote by $v(s)$ the *needle-shaped variation of the control* (see Figure 1.1) defined by

$$v(s) = -u_0(s) + u, \qquad t \leqslant s \leqslant t+\varepsilon,$$
$$v(s) = 0, \qquad s \in [t_0, T] \setminus [t, t+\varepsilon].$$

Note that $u_0(t) + v(t) \in U$, $t_0 \leqslant t \leqslant T$. Then it follows from (1.19) of Chapter IV (as in the deduction of (1.24) of Chapter IV) that

$$0 \leqslant \Delta J = -\int_t^{t+\varepsilon} \gamma(s)\,ds + N_1 + N_2,$$
$$\gamma(t) = H\big(t, x_0(t), u, \psi(t)\big) - H\big(t, x_0(t), u_0(t), \psi(t)\big),$$

where the summands N_1 and N_2 are expressed by formulas (1.16) and (1.19) of Chapter IV.

Using the Cauchy-Bunyakovski inequality, the estimates (1.17) and (1.23) of Chapter IV for the summands N_1 and N_2, and also the expressions for the variation v, we conclude that for some constant $C > 0$ the following inequality holds:

$$|N_1| + |N_2| \leqslant C\left[\int_t^{t+\varepsilon} |v(s)|\,ds\right]^2 \leqslant \varepsilon C \int_t^{t+\varepsilon} |v(s)|^2\,ds.$$

On the strength of the assumptions made the function H is a continuous function of its arguments. Therefore, in view of the control $u_0(t)$ being piecewise continuous, there exists a small $\varepsilon > 0$ such that the function $\gamma(t)$ is continuous for $s \in [t, t+\varepsilon]$. Consequently, for some $\theta \in [0,1]$ we obtain

$$\int_t^{t+\varepsilon} \gamma(s)\,ds = \varepsilon\gamma(t+\theta\varepsilon).$$

Thus, noting the expression for ΔJ, we can write

$$0 \leqslant \Delta J \leqslant -\varepsilon\gamma(t+\theta\varepsilon) + \varepsilon C \int_t^{t+\varepsilon} |v(s)|^2\,ds, \qquad \varepsilon > 0.$$

Dividing both sides of the obtained inequality by ε and passing to the limit as $\varepsilon \to 0$ yields $\gamma(t) \leqslant 0$. Since the element $u \in U$ is an arbitrary one, it follows that relation (1.4) has been established for $t \in [t_0, T]$. In a similar way, we can prove that the equality (1.4) also holds for $t = T$ by putting $v(s) = u - u_0(s)$, $T - \varepsilon \leqslant s \leqslant T$ and $v(s) = 0$, $s \leqslant T - \varepsilon$. □

It should be noted that, given the conditions of this theorem, the function $H\big(t, x_0(t), u_0(t), \psi(t)\big)$ is continuous with respect to t, since the value of the function H at any instant is equal to the left side of the expression (1.4) irrespective of whether the control is continuous or has a jump.

2. The Boltz problem. As a consequence of Theorem 1.1, we can obtain necessary conditions in the problem of the control of the system (1.1) with the constraints (1.3) and the cost functional

$$J(u) = \int_{t_0}^{T} F_0\big(t, x(t), u(t)\big)\, dt + \varphi(x(T)) \to \inf. \tag{1.5}$$

The instants t_0 and T, and also the initial condition $x(t_0) = x_0 \in R^n$ are prescribed.

By analogy with the deduction of the necessary conditions in Theorem 1.1 of Chapter IV, we can establish the following theorem.

Theorem 1.2. *Let $u_0(t)$ be the optimal control in the problem* (1.1), (1.3), (1.5), *and $x_0(t)$ be the optimal trajectory. Then there exists a vector $\psi(t)$ satisfying the equations*

$$\dot\psi(t) = -H_x\big(t, x_0(t), u_0(t), \psi(t)\big), \qquad t_0 \leqslant t \leqslant T,$$
$$\psi(T) = -\varphi_x\big(x_0(T)\big), \tag{1.6}$$

such that the condition for a maximum

$$\max_{u\in U} H\big(t, x_0(t), u, \psi(t)\big) = H\big(t, x_0(t), u_0(t), \psi(t)\big) \tag{1.7}$$

is satisfied, where H is determined by the relation

$$H\big(t, x(t), u, \psi(t)\big) = \psi'(t) f\big(t, x(t), u\big) - F_0\big(t, x(t), u\big). \tag{1.8}$$

It should be noted that the maximum of the function $H\big(t, x(t), u, \psi(t)\big)$ on the left side of the condition (1.7) is calculated with respect to the parameter $u \in U$ for fixed values of the other arguments t, $x_0(t)$, and $\psi(t)$.

The optimal control in the problem (1.1), (1.3), (1.5) is constructed in the same way as in the classical variational control problem, the only difference being the exclusion of the control due to the use of relation (1.7). Thus, the method consists of the following steps:

1°. The control $u_0\big(t, \psi(t), x_0(t)\big)$ is determined from equation (1.7).

2°. The obtained control u_0 is substituted in (1.1), (1.6), and the resulting boundary-value problem is solved for $x_0(t)$ and $\psi(t)$.

3°. The obtained values for $x_0(t)$ and $\psi(t)$ are substituted in the expression for u_0. As a result, we obtain the P-control $u_0(t)$. As in problems without constraints (1.3), the resulting P-control is optimal if the solution of the original control problem (1.1), (1.3), (1.5) exists, and the solution of the boundary-value problem in 2° is unique.

Example 1.1. Consider the problem of optimal control for the scalar system with constant coefficients a and b

$$\dot x(t) = ax(t) + bu(t), \qquad 0 \leqslant t \leqslant T,\ x(0) = 0, \tag{1.9}$$

and the cost functional

$$J = \int_0^T [x^2(t) + u^2(t)]\, dt \to \min. \tag{1.10}$$

Suppose first that the control $u(t)$ is unconstrained, i.e., $-\infty < u(t) < +\infty$. It is readily seen that the solution of the problem (1.9), (1.10) is $u_0(t) = 0$ and $x_0(t) = 0$. We shall obtain it also with the help of the maximum principle. By virtue of the equality (1.8) the function H becomes

$$H = \psi(t)[ax(t) + bu(t)] - x^2(t) - u^2(t). \tag{1.11}$$

Noting (1.11), the condition (1.7), which becomes

$$\left.\frac{\partial H}{\partial u}\right|_{u_0} = 0,$$

gives $u_0 = b\psi(t)/2$. Then equation (1.6) for the adjoint variable $\psi(t)$ takes the form

$$\dot{\psi}(t) = 2x(t) - a\psi(t), \qquad \psi(T) = 0. \tag{1.12}$$

Now, substitute for u in (1.9) the expression $\psi(t)$. Then, to determine the trajectory $x_0(t)$ and the function $\psi(t)$, we obtain the boundary-value problem formed by equation (1.12) and the equation

$$\dot{x}(t) = ax(t) + b^2\psi(t)/2, \qquad x(0) = 0. \tag{1.13}$$

The solution of the boundary-value problem (1.12), (1.13) is furnished by $x_0(t) \equiv 0$, $\psi(t) \equiv 0$, i.e., the optimal control is $u_0(t) = b\psi(t)/2 \equiv 0$.

Let now $u(t)$ be constrained by $|u(t)| \leqslant 1$. Then it follows from the condition (1.7) that

$$u_0 = \beta(\psi) = \begin{cases} 1, & \text{for } \frac{b\psi(t)}{2} \geqslant 1, \\ \frac{b\psi(t)}{2}, & \text{for } -1 \leqslant \frac{b\psi(t)}{2} \leqslant 1, \\ -1, & \text{for } -1 \leqslant \frac{b\psi(t)}{2}. \end{cases}$$

The boundary-value problem (1.12), (1.13) then becomes

$$\dot{x}(t) = ax(t) + b\beta(\psi), \qquad x(0) = 0,$$
$$\dot{\psi}(t) = 2x(t) - a\psi(t), \qquad \psi(T) = 0.$$

This problem also has the solution $x_0(t) \equiv 0$, $\psi(t) \equiv 0$, i.e., the optimal control is $u_0(t) = \beta(0) \equiv 0$.

3. About solutions of maximum principle equations. Suppose there exists a solution of the Mayer optimal control problem (1.1)-(1.3) (or of the Boltz problem (1.1), (1.3), (1.5)) and, in addition, only one admissible control satisfies the maximum principle. Then this control is optimal. However, it should be noted that in the general case the maximum principle is only a necessary condition for optimal control but not a sufficient one. In other words, the trajectory $x(t)$ and the control $u(t)$ can satisfy all the conditions of the maximum principle and still be not optimal. Consider the following illustrative examples.

Example 1.2. Let $x(t) \in R^1$ satisfy the equation

$$\dot{x}(t) = u(t), \qquad 0 \leqslant t \leqslant 1,\ x(0) = 0,\ x(1) = 0. \tag{1.14}$$

It is required to determine a control $u(t)$ such that the functional

$$J(u) = \int_0^1 [u^2(t) - 4x(t)u^3(t) + 2tu^4(t)]\, dt \to \inf_u . \tag{1.15}$$

be minimized.

The Hamiltonian for the problem (1.14), (1.15) is of the form

$$H = H(u^2 - 4xu^3 + 2tu^4) + \psi(t)u,$$

Therefore,

$$\dot{\psi}(t) = -\frac{\partial H}{\partial x} = -4u^3(t). \tag{1.16}$$

Note that the functions

$$x_0(t) = 0, \qquad \psi_0(t) = 0, \qquad u_0(t) = 0. \tag{1.17}$$

satisfy equations (1.14), (1.16) and also the following condition for a maximum:

$$H(t, x_0, \psi_0, u_0) \equiv 0 \geqslant H(t, x_0, \psi_0, u) = -u^2(1 + 2tu^2),$$

which holds for any control u. In other words, the functions (1.17) satisfy all the conditions of the maximum principle. We shall show, however, that the trajectory $x_0(t) = 0$ and the control $u_0(t) = 0$ are not a solution of the optimal control problem (1.14), (1.15). The value of the functional (1.15) associated with the control $u_0(t)$ is $J(u_0) = 0$. We now construct a control sequence characterized by a negative cost functional (even tending to $-\infty$). Set

$$u_\varepsilon(t) = \begin{cases} \varepsilon^{-1}, & \text{for } 0 \leqslant t \leqslant \varepsilon, \\ (\varepsilon - 1)^{-1}, & \text{for } \varepsilon < t \leqslant 1, \end{cases} \tag{1.18}$$

where $\varepsilon \in (0,1)$. The corresponding trajectory $x_\varepsilon(t)$ has the form

$$x_\varepsilon(t) = \begin{cases} t\varepsilon^{-1}, & \text{for } 0 \leqslant t \leqslant \varepsilon, \\ (1-t)(1-\varepsilon)^{-1}, & \text{for } \varepsilon \leqslant t \leqslant 1. \end{cases} \tag{1.19}$$

Let us evaluate the cost functional (1.15) that is associated with equations (1.18), (1.19). We have

$$J(u_\varepsilon) = [\varepsilon(1-\varepsilon)]^{-1} - \varepsilon^{-2} + 2(1-\varepsilon)^{-2} + (1+\varepsilon)(1-\varepsilon)^{-3}. \tag{1.20}$$

Hence,

$$J(u_\varepsilon) \to -\infty, \qquad \varepsilon \to 0. \tag{1.21}$$

It follows from (1.20) and (1.21) that the control $u_0(t) \equiv 0$ is not optimal.

4. Rotation of a motor shaft through a maximum angle. We shall treat this problem using the simplified model for an electric motor, the one we have already discussed in § 3 of Chapter III. In the notation of that section we have

$$\dot{x}_1 = x_2, \qquad \dot{x}_2 = u, \qquad |u| \leqslant b, \tag{1.22}$$

$$x_1(0) = x_2(0) = 0, \tag{1.23}$$

$$x_2(T) = 0, \ x_1(T) \to \max. \tag{1.24}$$

Here x_1 is the angle of rotation of the motor, u is the normalized torque, T is the prescribed terminal time (the time of the end of the rotation). It is required to determine a control $u(t)$ such that at the instant T the rotation velocity is zero, and the angle of rotation is a maximum.

In order to use the methods discussed in this section, we reduce the problem (1.22)–(1.24) to a Mayer problem by introducing a *penalty function.* It should be noted that penalty functions are often used to take account of various constraints.

We introduce the functional

$$J_k(u) = -x_1(T) + \frac{k}{2} x_2^2(T) \to \min. \tag{1.25}$$

Here $k \gg 1$ is called the *penalty coefficient.* It can be proved that for $k \to \infty$ the optimal control in the problem (1.22), (1.23), (1.25) must tend to the optimal control in the problem (1.22)–(1.24).

On the strength of Theorem 1.1 we obtain for the problem (1.22), (1.23), (1.25)

$$\begin{gathered} H = \psi_1 x_2 + \psi_2 u, \qquad \dot{\psi}_1 = -\frac{\partial H}{\partial x_1} = 0, \qquad \psi_1(T) = 1, \\ \dot{\psi}_2 = -\frac{\partial H}{\partial x_2} = -\psi_1, \qquad \psi_2(T) = -kx_2(T), \qquad \max_{|u| \leqslant b} H = \max_{|u| \leqslant b} \psi_2 u. \end{gathered} \tag{1.26}$$

This implies that the optimal control is of the form

$$\begin{gathered} \psi_1(t) \equiv 1, \qquad \psi_2(t) = C - t, \qquad u_\theta(t) = b \operatorname{sgn} \psi_2(t), \\ u_0^k(t) = \begin{cases} b, & \text{for } 0 \leqslant t \leqslant \tau, \\ -b, & \text{for } \tau < t \leqslant T. \end{cases} \end{gathered} \tag{1.27}$$

Here τ is the switch time for which $\psi_2(\tau) = 0$; it will be determined below. Solving equations (1.22) for $x_2(t)$ and $x_1(t)$ in the case of the control satisfying equation (1.27) yields

$$\begin{gathered} x_2(t) = \begin{cases} bt, & \text{for } 0 \leqslant t \leqslant \tau, \\ 2b\tau - bt, & \text{for } \tau \leqslant t \leqslant T; \end{cases} \\ x_1(t) = \begin{cases} b\frac{t^2}{2}, & \text{for } 0 \leqslant t \leqslant \tau, \\ b\frac{t^2}{2} + 2b\tau(t-\tau) - \frac{b}{2}(t^2 - \tau^2) & \text{for } \tau \leqslant t \leqslant T. \end{cases} \end{gathered} \tag{1.28}$$

To determine the switch time τ we use the boundary condition $\psi_2(T) = -kx_2(T)$. Hence, we can write

$$\psi_2(T) = C - T = -k(2b\tau - bT), \qquad C = T - k(2b\tau - bT).$$

Since $\psi_2(\tau) = 0$, we have

$$T - k(2b\tau - bT) - \tau = 0.$$

Thus, we obtain the following formula to determine the switch time:

$$\tau = T(1 + bk)(1 + 2bk)^{-1}. \tag{1.29}$$

There is another possible procedure to determine the switch time τ. Let us evaluate the functional $J_k(u)$ along the trajectory (1.28):

$$J_k(u) = \frac{bT^2}{2} + b\tau^2 + \frac{kb^2}{2}(2\tau - T)^2 - 2bT\tau. \tag{1.30}$$

Then the switch time τ can be determined from the condition for a minimum of the expression (1.30) with respect to τ:

$$\frac{dJ}{d\tau} = 2b\tau + 2kb^2(2\tau - T) - 2bT = 0,$$

where

$$\tau = T(1 + bk)(1 + 2bk)^{-1},$$

i.e., we obtained the same result as before.

It is readily seen from (1.29) that $\tau \to T/2$ as $k \to \infty$. Further, since

$$x_2(T) = bT(1 + 2bk)^{-1},$$

it follows that $x_2(T) \to 0$ as $k \to \infty$. Therefore, the optimal control $u_0^k(t)$ has been deduced; for sufficiently large k it is close to the optimal control $u_0(t)$ in the problem (1.22)–(1.24):

$$u_0(t) = \begin{cases} b, & \text{for } 0 \leqslant t \leqslant T/2, \\ -b, & \text{for } T/2 \leqslant t \leqslant T. \end{cases}$$

§ 2. Problems with an Unspecified Terminal Moment

1. Reduction to a Mayer problem. Consider the control problem for the system (1.1) with constraint (1.3) and the cost functional

$$J(u) = \int_{t_0}^{T} F_0\big(t, x(t), u(t)\big)\, dt \to \inf_u, \qquad x \in R^n. \tag{2.1}$$

It will be assumed that the initial state $x(t_0) = x_0$ and the terminal state $x(T) = x_1$ are prescribed, the moment t_0 is specified, and the moment T is not. Now we state the problem somewhat differently by using the same procedure as in transforming the Lagrange problem into the Mayer problem (see § 1 of Chapter IV).

Consider the vectors

$$y(t) = \big(x_1(t), \dots, x_n(t), x_{n+1}(t)\big), \qquad y(t) \in R^{n+1},$$
$$F = (f_1, \dots, f_n, F_0), \qquad F \in R^{n+1}.$$

and the equation

$$\dot{x}_{n+1}(t) = F_0\big(t, x(t), u(t)\big), \qquad x_{n+1}(0) = 0.$$

Then the problem (1.1), (1.3), (2.1) can be transformed into the following problem:

$$\begin{aligned} \dot{y}(t) = F\big(t, y(t), u(t)\big), &\qquad t \geqslant t_0,\ y(t_0) = (x_0, 0), \\ y_1(T) = x_1(T), \dots, y_n(T) = x_n(T), &\qquad y_{n+1}(T) \to \inf_u,\ u(t) \in U(t). \end{aligned} \tag{2.2}$$

Note that the right side F of equation (2.2) does not depend on the variable $y_{n+1}(t)$.

The terminal moment T is not specified but is deduced from the condition of transfer of the system (2.2) to a specified terminal state $x(T)$. If the function $F_0(t, x, u) \equiv 1$, then the condition (2.1) takes the form $T \to \inf_u$, and the corresponding problem is then referred to as a time-optimality problem.

2. Necessary conditions for the optimality of systems that are linear in a scalar control variable. We turn to the deduction of necessary conditions for the optimal control of a system of the form (2.2) that is linear in a control variable:

$$\begin{gathered} \dot{y}(t) = F\big(t, y(t)\big) + b(t)u(t), \qquad t_0 \leqslant t \leqslant T \\ y \in R^{n+1},\ u \in R^1,\ y(t_0) = y_0. \end{gathered} \tag{2.3}$$

The function F is assumed to be independent of y_{n+1} and to satisfy the Lipschitz conditions (1.3) of Chapter IV, and $b(t) \in R^{n+1}$ is a vector with bounded and piecewise continuous components. The scalar control $u(t)$ is subject to the constraint

$$|u(t)| \leqslant a, \qquad a = \text{constant}. \tag{2.4}$$

The initial moment t_0, the initial state $y(t_0)$, and the first n components $y(T)$ are prescribed. The moment T and the control $u(T)$ can be determined from the condition of transfer to the terminal state $y(T)$ with minimal value $y_{n+1}(T)$:

$$y_{n+1}(T) \to \inf_u. \tag{2.5}$$

Suppose there exists a solution of the problem (2.3)–(2.5). Denote by $u_0(t)$ an optimal control, and by $y_0(t)$ an optimal trajectory with components $y_{0,1}(t), \dots, y_{0,n+1}(t)$.

Let $v(t)$ be an arbitrary piecewise continuous function such that $u_0(t) + v(t)$ satisfies the constraint (2.4), i.e.,

$$|u_0(t) + v(T)| \leqslant a. \tag{2.6}$$

Such a function, $v(t)$, will be called *admissible.*

Denote by $\alpha(t) \in R^{n+1}$ a solution of the equationin variations:

$$\begin{gathered}\dot{\alpha}(t) = A(t)\alpha(t) + b(t)v(t), \qquad \alpha(t_0) = 0,\\ A(t) = F_y(t, y_0(t)).\end{gathered} \tag{2.7}$$

The solution of the problem (2.7) associated with a given $v(t)$ is denoted by $\alpha(t, v)$.

In the space R^{n+1} we introduce the set $Q(T)$ (see Figure 2.1) defined thus: $Q(T) = \{q \in Q(T) : q = y(T) + \alpha(T, v),$ where v is any admissible function$\}$.

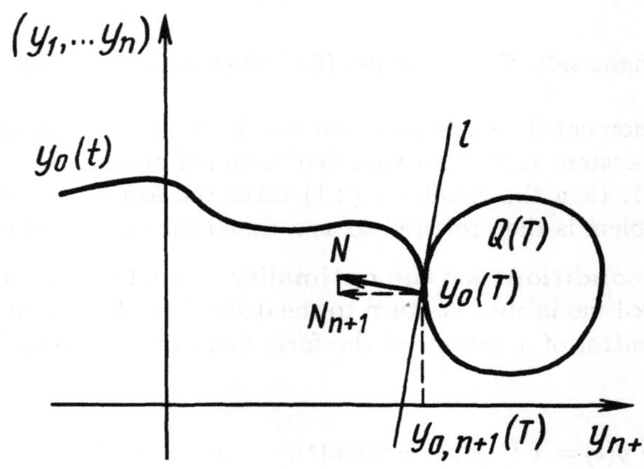

FIG. 2.1. A view of the set $Q(T)$, support hyperplane l, and the normal N

Let us establish some properties of the set $Q(T)$ that are essential in obtaining necessary conditions for optimal control:

1°. *The set $Q(T)$ is bounded.*

Since the set of admissible functions $v(t)$ is bounded ($|v(t)| \leqslant 2a$) and the matrix $A(t)$ is also bounded on $[t_0, T]$, all the $\alpha(t, v)$ are bounded on $[t_0, T]$.

2°. *The set $Q(T)$ is convex.*

It will be recalled that the set $Q \in R^n$ is called *convex* if for any two points $q_1 \in Q$, $q_2 \in Q$ and for any number ν, $\nu \in [0, 1]$, the point $q_\nu = \nu q_1 + (1 - \nu)q_2$ also belongs to Q.

Let q_1 and q_2 be two points belonging to $Q(T)$. Show that all points q_ν of the form $q_\nu = \nu q_1 + (1-\nu)q_2$ also belong to $Q(T)$. If $q_1, q_2 \in Q(T)$, then there exist admissible $v_1(t)$ and $v_2(t)$ such that

$$q_1 = y_0(T) + \alpha(T, v_1), \qquad q_2 = y_0(T) + \alpha(T, v_2).$$

Consequently,

$$q_\nu = y_0(T) + \nu\alpha(T, v_1) + (1-\nu)\alpha(T, v_2).$$

For the linear equation (2.7) the superposition principle holds, whereby

$$\nu\alpha(T, v_1) + (1-\nu)\alpha(T, v_2) = \alpha(T, \nu v_1 + (1-\nu)v_2).$$

The function $v_\nu = \nu v_1 + (1-\nu)v_2$ for all $v \in [0,1]$ and $t \in [t_0, T]$ is admissible, since

$$\begin{aligned} |u_0(t) + v_\nu(t)| &= |u_0(t) + \nu v_1(t) + (1-\nu)v_2(t)| = \\ &= \big|\nu\big(u_0(t) + v_1(t)\big) + (1-\nu)\big(u_0(t) + v_2(t)\big)\big| \\ &\leqslant \nu|u_0(t) + v_1(t)| + (1-\nu)|u_0(t) + v_2(t)| \leqslant a. \end{aligned}$$

Hence, $a(T, v_\nu)$ for all $\nu \in [0,1]$ corresponds to the admissible function $v_\nu(t)$. Therefore, by definition,

$$q_\nu = y_0(T) + \alpha(T, v_\nu) \in Q(T),$$

which proves that the set $Q(T)$ is convex.

3°. *The point $y_0(T)$ is a boundary point of the set $Q(T)$.*

It will be recalled that the point $q \in Q \subset R^n$ is called an *interior point* of the set Q if it belongs to Q together with some sphere centered at the point q. The aggregate of all interior points of the set Q is denoted by Q_0, and the closure of the set Q by $\bar{Q}$. Any point of the set $(\bar{Q} \setminus Q_0)$ is called a *boundary point* of the set Q. In other words, in any neighborhood of a boundary point there exist points belonging to Q and other points not belonging to Q.

Prove now that $y_0(T)$ is a boundary point of the set $Q(T)$. Assume the opposite, i.e., that $y_0(T)$ is an interior point of the set $Q(T)$. Then there exists a number $\beta < y_{0,n+1}(T)$ such that a line segment parallel to the axis y_{n+1} and joining the points $y_0(T)$ and $q_\beta = \big(y_{0,1}(T), \ldots, y_{0,n}(T), \beta\big)$ lies entirely in the set $Q(T)$. Since $q_\beta \in Q(T)$, there exists an admissible function $v_\beta(t)$ such that $q_\beta = y_0(T) + \alpha(T, v_\beta)$.

Denote by y_β the solution of equation (2.3) generated by the admissible control function $u_0(t) + v_\beta(t)$. By virtue of formulas (1.17) and (1.23) of Chapter IV we deduce that for some constants C_1 and C_2 the following inequalities are valid:

$$|y_0(T) - y_\beta(T)| \leqslant C_1\left(\int_{t_0}^{T} |v_\beta(t)|\,dt\right),$$

$$|q_\beta - y_\beta(T)| \leqslant C_2\left(\left[\int_{t_0}^{T} |v_\beta(t)|\,dt\right]^2\right).$$

Hence, for β sufficiently close to $y_{0,n+1}(T)$ the inequality $y_{\beta,n+1}(T) < y_{0,n+1}(T)$ is valid. However, this inequality cannot hold, since it contradicts the optimality of $y_0(T)$. This contradiction proves the assertion that $y_0(T)$ is a boundary point of the set $Q(T)$.

4°. *There exists a vector $N \neq 0$ such that for any $\alpha(T,v)$ satisfying the condition $\big(y_0(T) + \alpha(T,v)\big) \in Q(T)$ the following inequality holds:*

$$N'\alpha(T,v) \leqslant 0. \tag{2.8}$$

It will be recalled [8] that through a boundary point $y_0(T)$ of a convex set $Q(T)$ we can construct a support hyperplane such that all points of $Q(T)$ are located to one side of this hyperplane. For N we take a unit vector that is normal to the hyperplane l at the point $y_0(T)$ and points into the side where there are no points of the set $Q(T)$ (see Figure 2.1). Clearly, the vector N satisfies the inequality (2.8).

Now we continue with the deduction of necessary conditions. Let $Z(t,s)$ be the Cauchy matrix for the system $\dot{\alpha}(t) = A(t)\alpha(t)$. Denote by $\psi(t) = (\psi_1, \ldots, \psi_n, \psi_{n+1})$ the vector of adjoint variables. We have

$$\psi(t) = Z'(T,t)N,$$

where N is the vector whose existence was established in 4°. It is readily seen that $\psi(t) \not\equiv 0$, since $N \neq 0$ and the matrix $Z'(T,t)$ is nondegenerate. By virtue of the property (1.8) (Chapter IV) of the Cauchy matrix $Z(t,s)$ we can write

$$\dot{\psi}(t) = \frac{\partial Z'(T,t)}{\partial t} N = -A'(t)Z'(T,t) = -A'(t)\psi(t). \tag{2.9}$$

Further, we introduce the Hamiltonian $H\big(t, y(t,u,\psi(t))\big)$ for the problem (2.3)–(2.5) by the equality

$$H\big(t, y(t), u, \psi(t)\big) = \psi'(t)\big[F\big(t, y(t) + b(t)u\big)\big]. \tag{2.10}$$

Then equation (2.9) for the adjoint variables $\psi(t)$ can be written in the form

$$\dot{\psi}(t) = -H_y\big(t, y_0(t), u_0(t), \psi(t)\big), \qquad \psi(T) = N. \tag{2.11}$$

We now prove that the optimal control $u_0(t)$ furnishes a maximum of the Hamiltonian H, i.e., the following condition for a maximum holds

$$\max_{u, |u(t)| \leqslant a} H\big(t, y_0(t), u, \psi(t)\big) = H\big(t, y_0(t), u_0(t), \psi(t)\big). \tag{2.12}$$

It follows from formula (2.10) for the Hamiltonian and from the relation (2.12) that

$$u_0(t) = a \operatorname{sgn}\big(\psi'(t)b(t)\big). \tag{2.13}$$

Suppose that the equality (2.13) does not hold. Consider first those points $t \in [t_0, T]$ where $\psi'(t)b(t) = 0$. At these points the condition (2.12) for a maximum, obviously, holds. Therefore, the condition (2.12) may not hold only at those points $t \in [t_0, T]$ where $\psi'(t)b(t) \neq 0$. Since $\psi'(t)b(t)$ is piecewise continuous, a segment $[t_1, t_2] \subset [t_0, T]$, $t_1 < t_2$, can be found such that either

$$\psi'(t)b(t) < 0, \qquad u_0(t) \geqslant -a_1 > -a, \ t \in [t_1, t_2],$$

or

$$\psi'(t)b(t) > 0, \qquad u_0(t) \leqslant a_1 < a, \ t \in [t_1, t_2], \ t_1 < t_2. \tag{2.14}$$

To be definite, we consider only the case (2.14). It follows from the relations (2.7), (2.8) and Cauchy's formula that

$$N'a(T, v) = N' \int_{t_0}^{T} Z(T, t)b(t)v(t)\,dt \leqslant 0.$$

This relation and the definition of $\psi(t)$ lead to the inequality

$$\int_{t_0}^{T} \psi'(t)b(t)v(t)\,dt \leqslant 0. \tag{2.15}$$

We set $v(t) = C$ for $t \in [t_1, t_2]$ and $0 < C \leqslant a - a_1$, and $v(t) = 0$ for $t \notin [t_1, t_2]$. Then, according to (2.14), formula (2.9) leads to the inequality

$$N'a(T) = \int_{t_0}^{T} b'(t)\psi(t)v(t)\,dt = C \int_{t_1}^{t_2} b'(t)\psi(t)\,dt > 0,$$

which contradicts the inequality (2.15). Thus, we have proved that the equality (2.13) is valid. The relations (2.10), (2.11) imply that $\dot{\psi}_{n+1}(t) \equiv 0$, i.e., $\psi_{n+1}(t)$ is constant, and $\psi_{n+1}(T) = N_{n+1}$. Moreover, it can be proved that $N_{n+1} \leqslant 0$ (see Figure 2.1). Therefore, $\psi_{n+1}(t) = N_{n+1} \leqslant 0$.

Thus, the following theorem has been proved.

Theorem 2.1. *Suppose that $u_0(t)$ and $y_0(t)$ are the optimal control and trajectory in the problem (2.3)–(2.5). Then there exists a nonzero vector $\psi(t)$ satisfying equation (2.11) such that the condition (2.12) for a maximum holds (ψ_{n+1} being constant and nonpositive).*

3. Multivariable control. Let us state necessary conditions for optimal control in the problem

$$\begin{gathered} \dot{x}(t) = f\big(t, x(t), u(t)\big), \qquad t \geqslant t_0, \ x \in R^n, \ u \in R^m, \\ x(t_0) = x_0, \ x(T) = x_1, \ u(t) \in U(t), \\ \int_{t_0}^{T} F_0\big(t, x(t), u(t)\big)\,dt \to \inf_u . \end{gathered} \tag{2.16}$$

Here t_0, x_0, x_1 are prescribed, and T is unspecified.

Theorem 2.2. *If $u_0(t)$, $x_0(t)$ are the optimal control and the trajectory in the problem (2.16), then there exists a nonzero vector $(\psi(t), \lambda_0)$, $\psi(t) \in R^n$, such that*

$$\psi_{n+1}(t) = constant \leqslant 0, \qquad \dot{\psi}(t) = -H_x\big(t, x_0(t), u_0(t), \psi(t)\big). \tag{2.17}$$

The Hamiltonian H for the problem (2.16) is of the form

$$H\big(t, x, u, \psi(t)\big) = \psi'(t) f\big(t, x(t), u(t)\big) + \lambda_0 F_0(t, x, u). \tag{2.18}$$

The following condition for a maximum holds:

$$\max_{u \in U(t)} H\big(t, x_0(t), u, \psi(t)\big) = H\big(t, x_0(t), u_0(t), \psi(t)\big). \tag{2.19}$$

Moreover,

$$H\big(t, x_0(t), u_0(t), \psi(t)\big) = \int_T^t H_s(s, x_0(s), u_0(s), \psi(s))\, ds, \tag{2.20}$$
$$H_s = \frac{\partial H}{\partial s}.$$

4. Time-invariant systems. Let us discuss necessary conditions for optimal control in the problem (2.16) using the additional assumption of stationarity. The stationarity of the problem (2.16) means that f, F_0, and U do not depend explicitly on time, i.e.,

$$f(t, x, u) \equiv f(x, u), \qquad F_0(t, x, u) = F_0(x, u), \qquad U(t) \equiv U. \tag{2.21}$$

If the time-invariant problem (2.16), (2.21) possesses an optimal control $u_0(t)$ and an optimal trajectory $x_0(t)$, then there exists a nonzero vector of adjoint variables $\big(\psi(t), \psi_{n+1}\big)$, $\psi(t) \in R^n$, satisfying the conditions (2.17), and the condition (2.19) for a maximum is valid. Moreover, in the time-invariant case the following identity holds:

$$H\big(t, x_0(t), u_0(t), \psi(t)\big) \equiv 0. \tag{2.22}$$

5. Transfer of a system from one manifold to another manifold. We shall give necessary conditions for optimal control in the time-invariant problem (2.16) where it is assumed that the initial and terminal states, x_0 and x_1, belong to prescribed smooth manifolds, M_0 and M_T whose dimension is less than n, i.e., $x_0 \in M_0$, $x_1 \in M_T$. The necessary conditions for optimal control in this problem coincide with those given in the preceding subsection, and, moreover, certain transversality conditions must be met; these conditions mean that $\psi(t_0)$ is perpendicular to all tangent vectors of the manifold M_0 at the point x_0 and that $\psi(T)$ is perpendicular to all tangent vectors of the manifold M_T at the point x_1.

For particular problems the explicit analytical form of transversality conditions is given below.

6. Control problems with isoperimetric constraints. We give necessary conditions for optimal control in the problem (2.19)–(2.23) of Chapter III:

$$\dot{x}(t) = f\big(t, x(t), u(t)\big), \qquad x \in R^n,\ u \in R^m,\ t_0 \leqslant t \leqslant T, \tag{2.23}$$

$$\begin{aligned} &J_0\big(t_0, x(t_0), u, T, x(T)\big) \to \inf, \\ &J_j\big(t_0, x(t_0), u, T, x(T)\big) \leqslant 0, \qquad j = 1, \ldots, m_1, \\ &J_i\big(t_0, x(t_0), u, T, x(T)\big) = 0, \qquad i = m_1 + 1, \ldots, m_2,\ u(t) \in U. \end{aligned} \tag{2.24}$$

Here the scalar functionals J_k are of the form

$$\begin{aligned} J_k\big(t_0, x(t_0), u, T, x(T)\big) = &\int_{t_0}^{T} F_k\big(t, x(t), u(t)\big)\, dt \\ &+ \varphi_k\big(t_0, x(t_0), T, x(T)\big), \qquad k = 0, 1, \ldots, m_2. \end{aligned}$$

For an unconstrained control $u(t)$, necessary conditions for optimal control are given in Chapter IV. The constraint on the control $u(t) \in U$ only leads to the necessity of modifying the stationarity condition (3.5) of Chapter IV.

Denote by $\lambda = (\lambda_0, \ldots, \lambda_{m_2})$, $\psi(t) \in R^n$ the vectors of Lagrange multipliers. Suppose there exists a solution $\big(t_0, T, u_0(t), x_0(t)\big)$ of the problem (2.23), (2.24). It is assumed that the functions $f(t, x, u)$, $F = \big(F_0(t, x, u), \ldots, F_{m_2}(t, x, u)\big)$ and their partial derivatives with respect to x and u are continuous in some neighborhood of $\big(t, x_0(t), u_0(t)\big)$, $t \in [t_0, T]$, and the function $\varphi = \big(\varphi_0(t_0, \xi, T, \zeta), \ldots, \varphi_{m_2}(t_0, \xi, T, \zeta)\big)$ is continuously differentiable in the neighborhood of the point $\big(t_0, x_0(t_0), T, x_0(T)\big)$.

Assertion. *Under the assumptions made there exist Lagrange multipliers λ and $\psi(t)$, not both zero, such that relation* (2.23) *and also relations* (3.2), (3.3), (3.5)–(3.9) *of Chapter IV hold:*

the evolution equations

$$\dot{x}_0(t) = f\big(t, x_0(t), u_0(t)\big); \tag{2.25}$$

the Euler equations

$$\dot{\psi}(t) = -f_x\big(t, x_0(t), u_0(t)\big)\psi(t) - F_x\big(t, x_0(t), u_0(t)\big)\lambda; \tag{2.26}$$

the transversality conditions

$$\psi(t_0) = -\varphi_x(t_0, x, T, y)\lambda, \qquad \psi(T) = \varphi_y(t_0, x, T, y)\lambda \tag{2.27}$$

for $x = x_0(t_0)$, $y = x_0(T)$;

the maximality condition (replacing the stationarity conditions (3.4) of Chapter IV)

$$\max_{u\in U} H(t, x_0(t), u, \psi(t)) = H(t, x_0(t), u_0(t), \psi(t)), \tag{2.28}$$

where the Hamiltonian is of the form

$$H(t, x, u, \psi) = \psi'(t)f(t, x, u) + \lambda' F(t, x, u);$$

the conditions for variable end points t_0 and T (taken account of in problems with variable end points)

$$\begin{aligned} &F'(t_0, x_0(t_0), u_0(t_0))\lambda - [\varphi_{t_0}(t_0, x_0(t_0), T, x_0(T)) \\ &\qquad + \varphi_x'(t_0, x, T, y)x_0(t_0)]'\lambda = 0, \end{aligned} \tag{2.29}$$

$$\begin{aligned} &F'(T, x_0(T), u_0(T))\lambda + [\varphi_T(t_0, x_0(t_0), T, x_0(T)) \\ &\qquad + \varphi_y'(t_0, x, T, y)x_0(T)]'\lambda = 0 \end{aligned} \tag{2.30}$$

for $x = x_0(t_0)$ and $y = x_0(T)$;

the conditions for complementary slackness

$$\lambda_j\left[\int_{t_0}^{T} F_j(t, x_0(t), u_0(t))\,dt + \varphi_j(t_0, x_0(t_0), T, x_0(T))\right] = 0, \tag{2.31}$$

$$j = 1, \ldots, m_1;$$

the nonpositivity condition

$$\lambda_i \leqslant 0, \qquad i = 0, 1, \ldots, m_1. \tag{2.32}$$

Remark. In terms of the function H the conditions (2.29) and (2.30) can be written as

$$\begin{aligned} H(t, x, u, \psi)\big|_{t=t_0} &= \lambda'\varphi_{t_0}(t_0, x_0(t_0), T, x_0(T)), \\ H(t, x, u, \psi)\big|_{t=T} &= -\lambda'\varphi_T(t_0, x_0(t_0), T, x_0(T)). \end{aligned}$$

Here $H\big|_{t=\tau}$ is the value of the function H for $t = \tau$.

Corollary 2.1. *The function $H(t, x_0(t), u_0(t), \psi(t))$ is continuous in t for all $t \in [t_0, T]$.*

Proof. This assertion is, in general, not obvious, since the control $u_0(t)$ may have discontinuities. To prove that the function H is continuous, we take two arbitrary moments t and τ, on the interval $[t_0, T]$ and consider the difference

$$\Delta H = H(\tau, x_0(\tau), u_0(\tau), \psi(\tau)) - H(t, x_0(t), u_0(t), \psi(t)).$$

It will be shown that $\Delta H \to 0$ as $\tau \to t$. Observe that by virtue of (2.28) the following inequalities hold:

$$H\big(t, x_0(t), u_0(t), \psi(t)\big) \geqslant H\big(t, x_0(t), u_0(\tau), \psi(t)\big),$$
$$H\big(\tau, x_0(\tau), u_0(\tau), \psi(\tau)\big) \geqslant H\big(\tau, x_0(\tau), u_0(t), \psi(\tau)\big).$$

Hence,

$$H\big(\tau, x_0(\tau), u_0(t), \psi(\tau)\big) - H\big(t, x_0(t), u_0(t), \psi(t)\big) \leqslant \Delta H$$
$$\leqslant H\big(\tau, x_0(\tau), u_0(\tau), \psi(\tau)\big) - H\big(t, x_0(t), u_0(\tau), \psi(t)\big).$$

However, the left and right sides of this relation tend to zero as $\tau \to t$ in view of the continuity of the functions $x_0(t)$, $\psi(t)$, f, F and the fact that the control u_0 enters these functions with the same value of the argument. Therefore, $\Delta H \to 0$ as $\tau \to t$. □

Corollary 2.2. *Suppose that in addition to the foregoing assumptions the functions $f(t,x,u)$ and $F(t,x,u)$ are continuously differentiable with respect to t. Then for those $t \in [t_0, T]$, where the optimal control $u_0(t)$ is continuous, the following equality holds:*

$$\dot H\big(t, x_0(t), u_0(t), \psi(t)\big) = \frac{\partial H\big(t, x_0(t), u_0(t), \psi(t)\big)}{\partial t}.$$

Here $\dot H$ is a total derivative of the function H with respect to t, and $\partial H/\partial t$ is a partial derivative of H with respect to the first argument.

Proof. The proof of this assertion is analogous to that of Corollary 2.1. Take any point t, where the optimal control $u_0(t)$ is continuous. We have

$$\frac{\Delta H}{\tau - t} \geqslant \alpha(t,\tau), \qquad \tau > t,$$

where the function $\alpha(t,\tau)$ is equal to

$$\alpha(t,\tau) = \frac{1}{\tau - t}\Big[\Big(H\big(\tau, x_0(\tau), u_0(t), \psi(\tau)\big) - H\big(\tau, x_0(t), u_0(t), \psi(\tau)\big)\Big)$$
$$+ \Big(H\big(\tau, x_0(t), u_0(t), \psi(\tau)\big) - H\big(\tau, x_0(t), u_0(t), \psi(t)\big)\Big)$$
$$+ \Big(H\big(\tau, x_0(t), u_0(t), \psi(t)\big) - H\big(t, x_0(t), u_0(t), \psi(t)\big)\Big)\Big], \qquad \tau > t.$$

Hence, noting equations (2.23), (2.24), we have

$$\lim_{\tau \to t} \alpha(t,\tau) = \frac{\partial H'(t, x_0(t), u_0(t), \psi(t))}{\partial x}\dot x_0(t) + \frac{\partial H'(t, x_0(t), u_0(t), \psi(t))}{\partial \psi}\dot\psi(t)$$
$$+ \frac{\partial H\big(t, x_0(t), u_0(t), \psi(t)\big)}{\partial t} = \frac{\partial H\big(t, x_0(t), u_0(t), \psi(t)\big)}{\partial t}.$$

Therefore, denoting by $\underline{\lim}$ the lower limit, we can write

$$\underline{\lim}_{\tau\to t} \frac{\Delta H}{\tau - t} \geqslant \lim_{\tau\to t} \alpha(t,\tau) = \frac{\partial H\big(t, x_0(t), u_0(t), \psi(t)\big)}{\partial t}.$$

In a similar way, we obtain

$$\alpha(t,\tau) \geqslant \frac{\Delta H}{\tau - t},$$

$$\overline{\lim}_{\tau\to t} \frac{\Delta H}{\tau - t} \leqslant \lim_{\tau\to t} \alpha(t,\tau) = \frac{\partial H\big(t, x_0(t), u_0(t), \psi(t)\big)}{\partial t},$$

where $\overline{\lim}$ denotes the upper limit. The case $\tau < t$ can be considered in a similar way. This implies the assertion of Corollary 2.2. □

Example 2.1. Consider a scalar control system

$$\dot{x}(t) = ax(t) + u(t), \qquad 0 \leqslant t \leqslant T,\ x(0) = x_0,\ x(T) \to \min. \tag{2.33}$$

The control $u(t)$ at each instant can assume two values: $u(t) = 1$ and $u(t) = 0$ (no control). The following isoperimetric constraint on the control time is prescribed:

$$\int_0^T u(t)\,dt = T_0, \qquad 0 < T_0 < T, \tag{2.34}$$

where a, T, T_0, x_0 are given constants.

To solve the problem (2.33), (2.34), we use the necessary conditions (2.25)–(2.32) for optimal control. For the problem under consideration the function H is of the form

$$H = \psi(t)\big(ax(t) + u(t)\big) + \lambda_1 u(t), \tag{2.35}$$

and the equation for the function $\psi(t)$ has the form

$$\dot{\psi}(t) = -a\psi(t) = -H_x, \qquad \psi(T) = \lambda_0. \tag{2.36}$$

In the expressions (2.35), (2.36) λ_0 and λ_1 denote constant Lagrange multipliers.

If $u_0(t)$ is an optimal control, then according to (2.28) the following relation must be valid:

$$\max_{u;u=0,1} \big(\psi(t) + \lambda_1\big)u = \big(\psi(t) + \lambda_1\big)u_0. \tag{2.37}$$

The vector $\big(\psi(t), \lambda_0, \lambda_1\big)$ must not be zero. It follows from this that $\lambda_0 \neq 0$. Indeed, if $\lambda_0 = 0$, then according to (2.36) the function $\psi(t) \equiv 0$, i.e., $\lambda_1 \neq 0$. However, by virtue of (2.37) the control $u_0(t)$ must be constant, which cannot be true, since this would contradict the isoperimetric constraint (2.34).

Therefore, $\lambda_0 \neq 0$, which means that λ_0 may be taken as any negative constant. We set $\lambda_0 = -1$.

Thus, noting equation (2.36), we can write that $\psi(t) = -\exp[a(T-t)]$. This expression shows that $\psi(t)$ monotonically increases with t for $a > 0$ and monotonically decreases for $a < 0$. Hence, using the equality (2.37), we obtain that if $a > 0$, then $u_0(t) = 0$ for $0 \leqslant t < \tau$ and $u_0(t) = 1$ for $\tau \leqslant t \leqslant T$; but if $a < 0$, then $u_0(t) = 1$ for $0 \leqslant t < \tau$ and $u_0(t) = 0$ for $\tau \leqslant t \leqslant T$. Here τ is the control switching time equal to the unique zero $t = \tau$ of the monotonic function

$$\lambda_1 + \psi(t) = \lambda_1 - \exp[a(T-t)].$$

Choosing the constant λ_1 so that the constraint (2.36) is satisfied, we conclude that $\tau = T - T_0$ for $a > 0$ and $\tau = T_0$ for $a < 0$. But if $a = 0$, then any control equal to zero or unity at each instant t and satisfying the condition (2.34) is optimal.

7. Sufficiency of the maximum principle. In a number of cases the maximum principle is not only a necessary but also a sufficient condition for optimal control. Consider, for example, the problem of control for a linear system [8]

$$\begin{gathered}\dot{x}(t) = A(t)x(t) + B(t)u(t), \qquad 0 \leqslant t \leqslant T,\\ x(t_0) = x_0,\ x(T) = x_1,\ u \in U,\\ \int_{t_0}^{T} F_0\big(t, x(t), u(t)\big)\,dt \to \min, \qquad x \in R^n,\ u \in R^m.\end{gathered} \tag{2.38}$$

Here the moments t_0, T and the vectors x_0, x_1 are prescribed. The function $F_0(t,x,u)$ is convex and continuously differentiable with respect to x, u and continuous in t; the matrices $A(t)$, $B(t)$ are continuous. The set $U \in R^m$ is closed and convex, and, moreover, the aggregate of all interior points of the set U is not empty. Finally, $B'(t)\psi(t) \not\equiv 0$ for any nonzero solution $\psi(t)$ of the system $\dot{\psi}(t) = -A'(t)\psi(t)$. Then under these assumptions necessary conditions for optimal control in the problem (2.38) having the form (2.26), (2.28) are sufficient.

8. Connection between the maximum principle and the classical calculus of variations. In problems without constraints, using the maximum principle, we can obtain conditions for an extremum that are known in the calculus of variations.

As an example, consider the problem of finding a minimum of the functional

$$\int_{t_0}^{T} F\big(x(t), \dot{x}(t)\big)\,dt, \qquad x \in R^n,\ x(t_0) = x_0,\ x(T) = x_1. \tag{2.39}$$

Here t_0, T, x_0, x_1 and the scalar function F are prescribed. The continuously differentiable function $x(t)$ must be chosen so that the prescribed boundary conditions are satisfied and the minimum of the functional (2.39) is attained.

Necessary conditions for the optimality of the function $x(t)$ are described by *Euler's equation*

$$\frac{d}{dt}F_{\dot{x}}\big(x(t), \dot{x}(t)\big) - F_x\big(x(t), \dot{x}(t)\big) = 0. \tag{2.40}$$

We shall show that equation (2.40) can be obtained using the maximum principle. Set $u = \dot{x}$. Then the problem (2.39) is equivalent to the following problem of optimal control:

$$\begin{gathered}\dot{x}(t) = u, \qquad x(t_0) = x_0,\ x(T) = x_1, \\ \int_{t_0}^{T} F\big(x(t), u(t)\big)\, dt \to \min. \end{gathered} \tag{2.41}$$

Necessary conditions for optimal control in the problem (2.41) are expressed by Theorem 1.2. In order to write them, consider the Hamiltonian $H = \psi' u - F(x,u)$. Since no constraints on the control u have been imposed, the condition for a maximum of H with respect to u leads to the relation

$$\frac{\partial H}{\partial u} = \psi(t) - \frac{\partial F}{\partial u} = 0.$$

The equation for the vector of adjoint variables $\psi(t)$ gives

$$\dot{\psi}(t) = -\frac{\partial H}{\partial x}.$$

The last two equations imply that

$$\dot{\psi}(t) = F_x\big(x(t), u(t)\big), \qquad \psi(t) = F_u\big(x(t), u(t)\big).$$

Hence, noting the designation $u = \dot{x}$, we obtain

$$\frac{d}{dt} F_u\big(x(t), u(t)\big) = F_x\big(x(t), u(t)\big) = \frac{d}{dt} F_{\dot{x}}\big(x(t), \dot{x}(t)\big);$$

this coincides with equation (2.40).

Note that under certain assumptions other necessary conditions for a minimum (the conditions of Weierstrass, Legendre, and Jacobi) can also be derived from the maximum principle.

9. The maximum principle for discrete systems[1]. Consider a discrete control system of the form

$$\begin{gathered} x(j+1) = f\big(x(j), u(j)\big), \qquad j = 0, 1, \ldots, N-1,\ x(0) = x_0, \\ x \in R^n,\ u \in R^m,\ f : R^n \times R^m \to R^n. \end{gathered} \tag{2.42}$$

Suppose that

$$u(j) \in U \subset R^m. \tag{2.43}$$

[1]Propoy, A. I., On the Maximum Principle for Discrete Control Systems, *Avtomat. i Telemekh.*, (1965), vol. 26, no. 7, 1177–1187. (Translated into English as *Automat. Remote Control*)

It is required to deduce the control $u(j)$ such that for the prescribed integer N and the vector x_N the condition

$$x(N) = x_N \tag{2.44}$$

is met, and the functional

$$J(u) = \sum_{j=0}^{N-1} F_0\big(x(j), u(j)\big) \to \inf_u, \qquad F_0\colon R^n \times R^m \to R^1 \tag{2.45}$$

minimized. Denote by $H(x, u, \psi)$ a function of the form

$$H(x, u, \psi) = \lambda_0 F_0(x, u) + \psi' f(x, u),$$

where ψ_{n+1} is a constant.

The vector of adjoint variables $\psi(j) \in R^n$ satisfies the equation

$$\psi(j-1) = \frac{\partial H\big(x(j), u(j), \psi(j)\big)}{\partial x}, \qquad j = N-1, \dots, 1. \tag{2.46}$$

In the space R^{n+1} consider for each $x \in R^n$ the set

$$Q(x) = \bigcup_{u \in U} \big(F_0(x, u), f(x, u)\big),$$

which is the aggregate of all points $\big(F_0(x, u), f(x, u)\big)$ for all $u \in U$.

Theorem 2.3. *Let the set $Q(x)$ be convex for any $x \in R^n$. Then, for the control $u_0(j)$ and the corresponding trajectory $x_0(j)$ of the system* (2.42)–(2.45) *to be optimal, it is necessary that there exist a nonzero vector $\big(\lambda_0, \psi(j)\big)$, where the constant $\lambda_0 \leqslant 0$ and $\psi(j)$ satisfies equation* (2.46), *such that for any $j = 0, 1, 2, \dots, N-1$ the following relation is valid*:

$$\max_{u \in U} H\big(x(j), u, \psi(j)\big) = H\big(x_0(j), u_0(j), \psi(j)\big). \tag{2.47}$$

The comparison of the necessary conditions for optimal control for the discrete problem (2.42)–(2.45) and for the corresponding continuous problem shows that in the discrete case it is additionally required that the set $Q(x)$ be convex. Without this condition the discrete maximum principle does not, in general, hold true. Now we cite an appropriate example.

Example 2.2. Consider a two-dimensional control system, with a scalar control, of the form

$$\begin{aligned} x_1(j+1) &= f\big(x_1(j)\big) + f_1\big(u(j)\big), \qquad j = 0, 1, \\ x_2(j+1) &= f\big(x_2(j)\big) + f_2\big(u(j)\big), \qquad 0 \leqslant u \leqslant 1, \end{aligned} \tag{2.48}$$

$$x_1(0) = x_2(0) = 0. \tag{2.49}$$

The right side of the system (2.48) is determined by the relations

$$f(x) = -(x-2)^2 + 4,$$
$$f_1(u) = \begin{cases} 1-2u, & \text{for } 0 \leqslant u \leqslant \frac{1}{2}, \\ 0, & \text{for } \frac{1}{2} \leqslant u \leqslant 1; \end{cases} \tag{2.50}$$
$$f_2(u) = \begin{cases} 0, & \text{for } 0 \leqslant u \leqslant \frac{1}{2}, \\ -2+4u, & \text{for } \frac{1}{2} \leqslant u \leqslant 1. \end{cases}$$

The cost functional is of the form

$$-x_1(2) - x_2(2) \to \min. \tag{2.51}$$

Clearly, the set $Q(x)$ associated with the problem under consideration is not convex. If in the problem (2.48)–(2.51) the discrete maximum principle were valid, then the optimal control $u_0(j)$ and trajectory $x_0(j)$ would satisfy relation (2.47), in which the functions ψ_1 and ψ_2 are determined from the equations

$$\begin{aligned} \psi_1(j-1) &= \psi_1(j)\frac{\partial f(x_1(j))}{\partial x_1}, \qquad \psi_1(1) = 1, \\ \psi_2(j-1) &= \psi_2(j)\frac{\partial f(x_2(j))}{\partial x_2}, \qquad \psi_2(1) = 1. \end{aligned} \tag{2.52}$$

However, we shall show that the optimal control $u_0(0)$ does not satisfy relation (2.47).

To prove the optimality of the control in question, we must evaluate the cost functional (2.51) for some control, and then find its minimum. According to (2.48), (2.58) we have

$$\begin{aligned} x_1(1) = f_1(u(0)), \qquad & x_2(1) = f_2(u(0)), \\ x_1(2) &= f\big(f_1(u(0))\big) + f_1(u(1)), \\ x_2(2) &= f\big(f_2(u(0))\big) + f_2(u(1)). \end{aligned} \tag{2.53}$$

Substitute (2.53) in (2.51) and select the summands in the cost functional that depend on the chosen control $u(0)$. As a result, we obtain that the optimal control $u_0(0)$ must satisfy the condition

$$\max_{0 \leqslant u \leqslant 1} \varphi(u) = \varphi\big(u_0(0)\big),$$

where

$$\varphi(u) = f(f_1(u)) + f(f_2(u)).$$

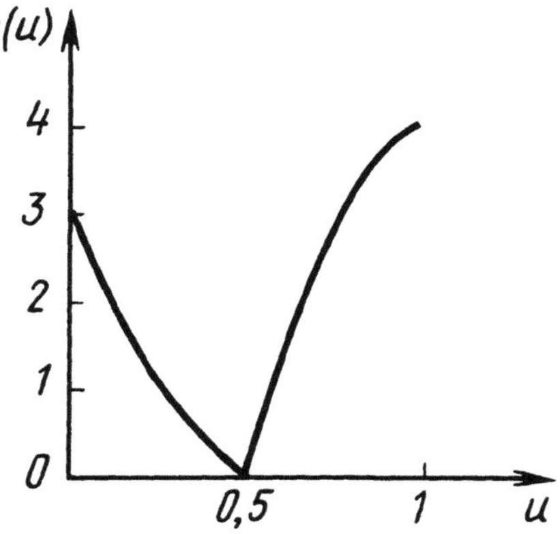

FIG. 2.2. Graph of the cost functionial $\varphi(u)$ as a function of the initial value of the control u

The graph of the function $\varphi(u)$ (see Figure 2.2) demonstrates that the optimal control $u_0(0) = 1$. Therefore, according to (2.47), (2.49) the optimal trajectories are the following:

$$x_1(1) = 0, \qquad x_2(1) = 2. \tag{2.54}$$

Thus, noting (2.52), we can write

$$\begin{aligned} \psi_1(0) &= \psi_1(1)\frac{\partial f\big(x_1(1)\big)}{\partial x_1} = -2\big(x_1(1) - 2\big) = 4, \\ \psi_2(0) &= \psi_2(1)\frac{\partial f\big(x_2(1)\big)}{\partial x_2} = -2\big(x_2(1) - 2\big) = 0. \end{aligned} \tag{2.55}$$

In our case the function $H(x, u, \psi)$ is of the form

$$\begin{aligned} H\big(x(j), u(j), \psi(j)\big) =& \psi_1(j)[f\big(x_1(j)\big) + f_1\big(u(j)\big)] \\ &+ \psi_2(j)[f\big(x_2(j)\big) + f_2\big(u(j)\big)]. \end{aligned} \tag{2.56}$$

Consequently, according to (2.54)–(2.56) we have for $j = 0$

$$H\big(x(0), u, \psi(0)\big) = 4f_1(u). \tag{2.57}$$

As is seen from (2.57), the maximum of the function $H\big(x(0), u, \psi(0)\big)$ with respect to u is attained, according to (2.49), at $u = 0$, and it is equal to $H\big(x(0), 0, \psi(0)\big) = 4$. Simultaneously, for the optimal control $u_0(0) = 1$, the function $H\big(x_0(0), u_0(0), \psi(0)\big) = 0$, i.e., it does not attain its maximum value.

§ 3. Practical Applications of the Maximum Principle

Pontryagin's maximum principle is widely used for investigating various engineering control systems. With its help, the task of determining an optimal control reduces to that of investigating certain boundary-value problems. Though an analytical determination of optimal control is, in general, fraught with difficulties, nevertheless, it can be brought to a satisfactory result in a number of interesting and important cases. These cases are illustrated in the examples that are treated in this section. To determine and investigate optimal controls in more complicated cases, approximation and numerical methods are used; they are treated in Part Four of the book.

1. Optimal configuration of a nuclear reactor. Consider the problem of the optimal configuration of a nuclear reactor without a reflector in the one-group approximation (see § 3 (subsection 2) of Chapter III). The size of the reactor L is not specified. Denote by $x_1(z)$ the neutron density $N(z)$ and set $x_2(z) = dN(z)/dz$. For the control u we take the concentration of nuclear fuel per reactor unit volume. Then the problem under consideration is of the form

$$\begin{gathered} \frac{dx_1(z)}{dz} = x_2(z), \qquad \frac{dx_2(z)}{dz} = -(a+bu)x_1(z), \\ x_1(0)=1,\ x_2(0)=0,\ x_1(L)=0, \\ -\int_0^L u(z)x_1(z)\,dz \to \min, \qquad 0 \leqslant u_1 \leqslant u \leqslant u_2, \\ a+bu(z)>0, \qquad 0 \leqslant z \leqslant L, \end{gathered} \tag{3.1}$$

where u_1, u_2, a, $b \neq 0$ are prescribed constants.

To solve the problem, we use the necessary conditions for optimal control (see Theorem 2.2 and the transversality conditions).

The Hamiltonian H for the problem (3.1) is of the form

$$\begin{gathered} H = \psi_1(z)x_2(z) - a\psi_2(z)x_1(z) - u\gamma(z), \\ \gamma(z) = (\lambda_0 + b\psi_2(z))x_1(z), \end{gathered} \tag{3.2}$$

where the constant λ_0 is, by virtue of (2.17), nonpositive: $\lambda_0 \leqslant 0$. The adjoint variables ψ_1 and ψ_2 satisfy the equations

$$\begin{aligned} \frac{d\psi_1(z)}{dz} &= a\psi_2(z) + (\psi_0 + b\psi_2(z))u, \\ \frac{d\psi_2(z)}{dz} &= -\psi_1(z), \qquad \psi_2(L) = 0. \end{aligned} \tag{3.3}$$

In accord with physical considerations the neutron flux $N(z) = x_1(z) \geqslant 0$. Therefore, the problem (3.1) is to be considered only on the interval $[0, L]$, where L is the first root of the equation $x_1(z) = 0$. Hence, it follows from (3.1) that $dx_2/dz < 0$ for $z \in [0, L)$. Consequently, $x_2(z) < 0$ for $z = (0, L]$, and on the segment $[0, L]$ the function $x_1(z)$ decreases monotonically from $x_1(0) = 1$ to $x_1(L) = 0$.

Let us show that

$$\psi_0 < 0. \tag{3.4}$$

Indeed, the optimal value of the Hamiltonian (3.2) is equal to zero:

$$H\big(x_0(z), u_0(z), \psi(z)\big) = 0, \qquad z \in [0, L]. \tag{3.5}$$

If $\psi_0 = 0$, then for $z = L$, noting (3.1) and (3.5), we would have $H = \psi_1(L)x_2(L) = 0$. Hence, $\psi_1(L) = 0$. This would lead to $\psi_0 = 0$, $\psi_1(L) = \psi_2(L) = 0$; this result would contradict the assertion of Theorem 2.2 on the existence of a nonzero vector of adjoint variables.

Applying the condition (2.19) for a maximum to the Hamiltonian (2.24) gives the following formula for the optimal control $u_0(z)$:

$$u_0(z) = \begin{cases} u_2, & \text{for } \gamma(z) < 0, \\ \text{not defined}, & \text{for } \gamma(z) = 0, \\ u_1, & \text{for } \gamma(z) > 0. \end{cases} \tag{3.6}$$

We can show that inside the segment $[0, L]$ there are no segments of nonzero length, where $\gamma(z) = 0$. Suppose that $\gamma(z_0) = 0$ for some $z_0 \in (0, L)$. Then

$$\big(\psi_0 + b\psi_2(z_0)\big)x_1(z_0) = 0, \quad \text{i.e. } \psi_0 + b\psi_2(z_0) = 0.$$

Consequently,

$$\frac{d\gamma(z_0)}{dz} = b\frac{d\psi_2(z_0)}{dz}x_1(z_0) = -b\psi_1(z_0)x_1(z_0).$$

Moreover, it follows from the equality (3.5) that

$$\psi_1(z_0)x_2(z_0) - a\psi_2(z_0)x_1(z_0) = 0;$$

hence

$$\psi_1(z_0) = -a\frac{x_1^2(z_0)}{x_2(z_0)} \times \frac{\psi_0}{b}.$$

Therefore, at the points z_0, where $\gamma(z_0) = 0$, we have

$$\frac{d\gamma(z_0)}{dz} = a\frac{x_1^2(z_0)}{x_2(z_0)} \times \frac{\psi_0}{b} \neq 0.$$

This implies that there are no segments of nonzero length, where $\gamma(z)$ vanishes. Thus, according to (3.6), the optimal control $u_0(z)$ is defined by

$$u_0(z) = \begin{cases} u_2, & \text{for } \gamma(z) < 0, \\ u_1, & \text{for } \gamma(z) > 0. \end{cases} \tag{3.7}$$

In other words, the optimal configuration of a reactor consists of zones in which $u(t)$ takes on the values u_1 and u_2.

Let us determine the number of such zones and their arrangement. Since the derivative $d\gamma/dz$ is distinct from zero at the points z, where $\gamma(z_0) = 0$, it follows that the monotone function $\gamma(z)$, as z varies from 0 to L, can change sign just once, and, consequently, the number of zones is at most two. Then $\gamma(L) = 0$, and the derivative

$$\frac{d\gamma(L)}{dL} = (\psi_0 + b\psi_2(L))\frac{dx_1(L)}{dL} = (\psi_0 + b\psi_2(L))\,x_2(L) = \psi_0 x_2(L) > 0.$$

Hence, in the zone adjoining the point $z = L$ we must have $u_0(z) = u_2$. The second zone adjoins the point $z = 0$. According to (3.5) we have $0 = a\psi_2(0) + \psi_0 u(0) + b\psi_2 u(0)$. Therefore, $\psi_2(0) = -\psi_0 u(0)[a + bu(0)]^{-1}$. Further, we obtain

$$\gamma(0) = \psi_0 + b\psi_2(0) = \psi_0 - b\psi_0 u_0[a + bu(0)]^{-1} = a\psi_0[a + bu(0)]^{-1}.$$

Further, by hypothesis, $a + bu(0) > 0$. Therefore,

$$\operatorname{sgn}\gamma(0) = \operatorname{sgn} a\psi_0 = -\operatorname{sgn} a.$$

Thus, if $a > 0$, then for all $z \in [0, L]$ the control $u_0(z) = u_2$, i.e., the configuration consists of one zone only. But if $a < 0$, then in the zone adjoining the point $z = 0$, we obtain $u_0(z) = u_1$, which indicates a two-zone optimal configuration.

In a similar way, the problem of an optimal configuration of a reactor of given size L is investigated. It turns out that depending on the values of the parameters of the problem (3.1) the optimal number of zones is equal to two or three. This and other problems concerning the configuration of a nuclear reactor are treated in detail in the books by A. P. Rudik cited in § 3 (subsection 2) of Chapter III.

2. Control of motion with regulated friction. Consider a material point of mass m in rectilinear motion along the x-axis subjected to a tractive force u_2 and a frictional force $(k_1 + c_1 u_1)\dot{x}$. Denote by $x_1 = x$ the coordinate of the material point and by $x_2 = \dot{x}$ the velocity of the point. Then the equations of motion are bilinear and have the form

$$\dot{x}_1 = x_2, \qquad m\dot{x}_2 = u_2 - k_1 x_2 - u_1 c_1 x_2, \qquad k_1 > 0,\ c_1 > 0. \tag{3.8}$$

Equations (3.8) are sometimes used as a model of motion for an automobile or a railroad car, where the control u_1 describes the force applied to the brake. We divide both sides of the second of the equations (3.8) by m and set $b = m^{-1}$, $k = k_1 m^{-1}$, $c = c_1 m^{-1}$. Then equations (3.8) take the form

$$\dot{x}_1 = x_2, \qquad \dot{x}_2 = bu_2 - kx_2 - u_1 c x_2. \tag{3.9}$$

Natural control constraints are as follows:

$$0 \leqslant u_1 \leqslant 1, \qquad |u_2| \leqslant 1. \tag{3.10}$$

Note that the problem with control constraints of the form $0 \leqslant u_1 \leqslant v_1$, $|u_2| \leqslant v_2$ reduces to (3.9), (3.10) by the change of variables $u_1 \to u_1 v_1$, $u_2 \to u_2 v_2$, $b \to b v_2$, $c \to c v_1$.

It is required to deduce controls u_1, u_2 in order to transfer the system (3.9) from a prescribed initial state $\big(x_1(0), x_2(0)\big)$ to the origin $(0,0)$ in minimum time T, i.e.,

$$x_1(T) = 0, \qquad x_2(T) = 0, \qquad T \to \min . \tag{3.11}$$

It is readily seen that the moment T coincides with the moment at which the system (3.9) first reaches the origin.

It should be noted that in the optimal control problem (3.9)–(3.11) there exists an admissible control (u_1, u_2) transferring the system (3.9) to the origin in a finite time. Indeed, for an admissible control u_1 we can take, for example, $u_1 \equiv 0$, in which case the system (3.9) becomes linear and takes the form

$$\dot{x}_1 = x_2, \qquad \dot{x}_2 = b u_2 - k_1 x_2, \qquad b > 0,\ |u_2| \leqslant 1.$$

For this linear system there exists a time-optimal control $\bar{u}_2$ that transfers it from an arbitrary initial state to the origin [15]. Therefore, in the original problem (3.9)–(3.11) the controls $u_1 \equiv 0$, $\bar{u}_2$ are admissible.

To solve the problem (3.9)–(3.11), we use Pontryagin's maximum principle. The Hamiltonian for the problem (3.9)–(3.11) is of the form

$$H = \lambda_0 + \psi_1 x_2 + \psi_2 (b u_2 - k x_2 - u_1 c x_2), \qquad \lambda_0 = \text{constant} \leqslant 0. \tag{3.12}$$

The adjoint variables ψ_1, ψ_2 satisfy the equations

$$\dot{\psi}_1 = 0, \qquad \dot{\psi}_2 = -\psi_1 + \psi_2 (k + c u_1). \tag{3.13}$$

Hence, ψ_1 is a constant. It follows from the condition (2.19) for a maximum of the Hamiltonian H that

$$u_1(t) = \tfrac{1}{2}[1 + \operatorname{sgn} \varphi_1(t)] \quad \text{for } \varphi_1(t) \neq 0, \qquad \varphi_1(t) = -x_2(t)\psi_2(t), \tag{3.14}$$

$$u_2(t) = \operatorname{sgn} \psi_2(t) \quad \text{for } \psi_2(t) \neq 0. \tag{3.15}$$

or

$$u_1(t) = \begin{cases} 0, & \text{for } x_2(t)\psi_2(t) < 0, \\ 1, & \text{for } x_2(t)\psi_2(t) > 0. \end{cases}$$

We investigate the behavior of the functions $\varphi_1(t)$ and $\psi_2(t)$. We can show that the function $\varphi_1(t)$ cannot vanish on a whole interval. Assume the contrary, i.e., assume that $\varphi_1(t) \equiv 0$ on some interval. Then either $x_2(t) \equiv 0$ $(t \in S)$ or $\psi_2(t) \equiv 0$ $(t \in S_1)$, one of the intervals S or S_1 having a nonzero length. First, consider the case in which $\psi_2(t) \equiv 0$, $t \in S_1$. Then on the strength of (3.13) we have

$$\dot{\psi}_2(t) \equiv 0 = -\psi_1 + \psi_2 (k + c u_1) = -\psi_1, \qquad t \in S_1.$$

Therefore, $\psi_1 = 0$. Furthermore, by virtue of (2.22) the maximum value of the Hamiltonian is equal to zero:

$$H \equiv 0, \qquad 0 \leqslant t \leqslant T. \tag{3.16}$$

Hence, $\lambda_0 = 0$. Thus, the adjoint variables are all equal to zero:

$$\psi_2(t) = 0, \qquad t \in [0, T], \qquad \psi_1 = 0, \qquad \lambda_0 = 0. \tag{3.17}$$

However, on the basis of the maximum principle relation (3.17) is not possible.

Now suppose that $x_2(t) = 0$ ($t \in S$). Then $\dot{x}_2(t) = 0$ ($t \in S$), i.e., according to (3.9) the control $u_2(t) = 0$ ($t \in S$). It follows from this and from (3.16) that $\lambda_0 = 0$. Therefore, in view of (3.15) the equality $u_2(t) = 0$ ($t \in S$) holds only when $\psi_2(t) \equiv 0$ ($t \in S$). However, as shown above, the last identity does not hold.

Let us show that the function $\psi_2(t)$ cannot have more than one zero on the segment $[0, T]$. Suppose that the converse is true, i.e., $\psi_2(t_1) = \psi_2(t_2) = 0$, $t_1 < t_2$, $t_1, t_2 \in [0, T)$, and t_1 and t_2 are two successive simple zeros of the function $\psi_2(t)$. In that case if at the points t_1 and t_2 the function $\dot{\psi}_2(t) \neq 0$, then by virtue of (3.13) the constant $\psi_1 = -\dot{\psi}_2(t_1)$, $\psi_1 = -\dot{\psi}_2(t_2)$, which cannot be true. Further, suppose that the graph of the function $\psi_2(t)$ touches the t-axis at one of the points t_i, say, at the point t_1. Then $\dot{\psi}_2(t_1) = \psi_2(t_1) = 0$, i.e., on the strength of (3.13) the constant $\psi_1 = 0$. Hence and from (3.16) it follows that $\lambda_0 = 0$. As a consequence, the identity (3.17) must hold, and this cannot be true.. Therefore, the control $u_2(t)$ can have only one switch point.

In order to investigate the control $u_1(t)$, we shall show that $x_2(t)$ cannot have more than one zero on $[0, T)$. Suppose that this is not so. Then there exist points $t_1, t_2 \in [0, T)$ such that $x_2(t_1) = x_2(t_2) = 0$. In addition, suppose that

$$\dot{x}_2(t_1) \neq 0, \qquad \dot{x}_2(t_2) \neq 0. \tag{3.18}$$

In that case if $\lambda_0 = 0$, then on the strength of (3.12), (3.16) and the equalities $u_2(t_1) = 0$, $u_2(t_2) = 0$ that follow from (3.9) we obtain

$$\psi_2(t_1) = \psi_2(t_2) = 0. \tag{3.19}$$

Equalities (3.19) cannot hold, since it has been established earlier that $\psi_2(t)$ can have at most one zero. So, $\lambda_0 < 0$, in which case it follows from (3.9) that

$$\operatorname{sgn} u_2(t_i) = \operatorname{sgn} \dot{x}_2(t_i), \qquad i = 1, 2.$$

Thus, on each of the segments $[t_1, t_2]$ and $[t_2, T]$ at least one switch of the control $u_2(t)$ occurs, and this cannot be true.

Now, let t_1 be a muptiple zero of the function $x_2(t)$, i.e., $x_2(t_1) = \dot{x}_2(t_1) = 0$. Then according to (3.9) $u_2(t_1) = 0$. It follows then from (3.16) that $\lambda_0 = 0$, and from (3.15) that $\psi_2(t_1) = 0$. In a similar way, at the point t_2 we obtain either

$\dot{x}_2(t_2) \neq 0$, $\lambda_0 = 0$ or $\dot{x}_2(t_2) = x_2(t_2) = 0$. In both cases $\psi_2(t_2)$ must vanish. Consequently, the equalities (3.19) must hold, which cannot be true.

Thus, we have established that the control $u_1(t)$ has at most two switch points, and the control $u_2(t)$ at most one. Moreover, the switch point for the control $u_2(t)$ serves as a switch point for the control $u_1(t)$.

The results obtained have a simple physical meaning. Suppose that at the initial moment the point is moving in the direction of the origin, i.e., the condition $x_1(0)x_2(0) < 0$ is satisfied. Then it is readily seen that time optimality is associated with the following controls. On the time segment $[0, t_1]$ the tractive force is maximum in value and points to the origin, and the brake force control $u_1 = 0$. On the segment $[t_1, T]$ the tractive force is also maximum in value but points away from the origin, and the brake force control u_1 is a maximum. This means that on the segment $[0, t_1]$ the material point is accelerated, and on the segment $[t_1, T]$ retarded (see Figure 3.1, a). In this case the controls u_1 and u_2 have one switch point each.

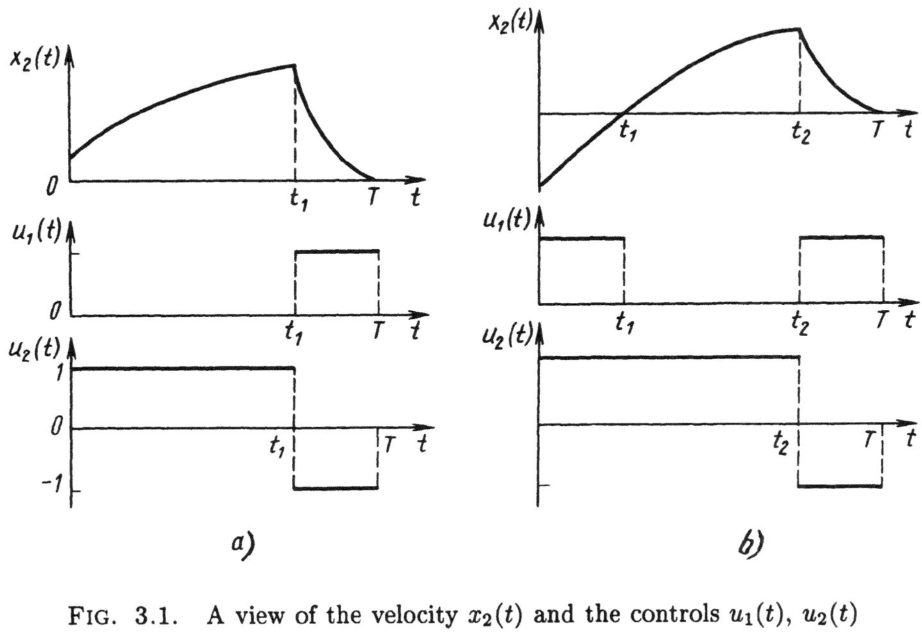

FIG. 3.1. A view of the velocity $x_2(t)$ and the controls $u_1(t)$, $u_2(t)$ for:
$a)$ $x_1(0) < 0$, $x_2(0) \geqslant 0$; $b)$ $x_1(0) \leqslant 0$, $x_2(0) < 0$

Now, suppose that at the initial moment the material point is moving away from the origin, i.e., $x_1(0)x_2(0) > 0$. Then on the segment $[0, t_1]$, i.e., until the moment at which the velocity $x_2(t)$ vanishes, the tractive force u_2 is a maximum and points to the origin, and the frictional force is maximum in value. On the segment $[t_1, t_2]$ the tractive force is the same, but the frictional force vanishes. On the segment $[t_2, T]$ the tractive force u_2 is a maximum and points away from the origin, and the frictional force is maximum in value. In this case the control $u_2(t)$ has one switch point, and the control $u_1(t)$ two switch points (see Figure 3.1, *b*).

In the case $x_2(0) = 0$ and $x_1(0) < 0$ the controls are of the same form as shown in Figure 3.1, *a*. But if $x_2(0) < 0$ and $x_1(0) = 0$ the optimal controls have the form shown in Figure 3.1, *b*. In the case $x_1(0) > 0$ the optimal control of the tractive force $u_2(t)$ is of opposite sign as compared to the case $x_1(0) < 0$, but the frictional force is of the same form.

We turn now to analytical expressions for the switch times. They all can be derived in the same way. Therefore, we restrict ourselves to the case depicted in Figure 3.1, *b*. The expression for t_1 can be derived from the condition $x_2(t_1) = 0$. On the segment $[0, t_1]$ equation (3.9) is of the form

$$\dot{x}_2 = b - (k + c)x_2, \qquad x_2(0) < 0.$$

Integrating this equation gives

$$t_1 = \frac{1}{k+c} \ln \frac{b - (k+c)x_2(0)}{b}, \qquad x_2(0) < 0.$$

To determine the switch time t_2, we integrate the equation

$$\dot{x}_2 = b - kx_2, \qquad x_2(t_1) = 0,$$

over the segment $[t_1, t_2]$, and the equation

$$\dot{x}_2 = -b - (k + c)x_2, \qquad x_2(T) = 0$$

over the segment $[t_2, T]$.

Equating the values of $x_2(t_2)$ obtained from the two foregoing equations leads to one of the relations that connect the unknown moments T and t_2:

$$\frac{2k+c}{k} = e^{(k+c)(T-t_2)} + \frac{k+c}{k} e^{-k(t_2-t_1)}.$$

Another relation connecting T and t_2 follows from the expression

$$\int_0^T \dot{x}_1(t)\, dt = x_1(T) - x_1(0) = -x_1(0) = \int_0^T x_2(t)\, dt.$$

Substituting the above expressions for $x_2(t)$ on the segments $[0, t_1]$, $[t_1, t_2]$, $[t_2, T]$ in the foregoing expression yields

$$x_1(0) = \frac{b}{(k+c)^2} \left[e^{(k+c)t_1} - e^{(k+c)(T-t_2)} \right] + \frac{b}{k^2} \left[-e^{k(t_1-t_2)} + 1 \right] + \frac{b(T - t_1 - t_2)}{k+c} - \frac{b(t_2 - t_1)}{k}.$$

TABLE 3.1

Moments	$x_1(0) = -2$	$x_1(0) = -3$	$x_1(0) = -10$
t_1	1.00000	1.00000	1.00000
t_2	1.21432	1.28512	3.38075
T	1.88461	2.16037	4.09073

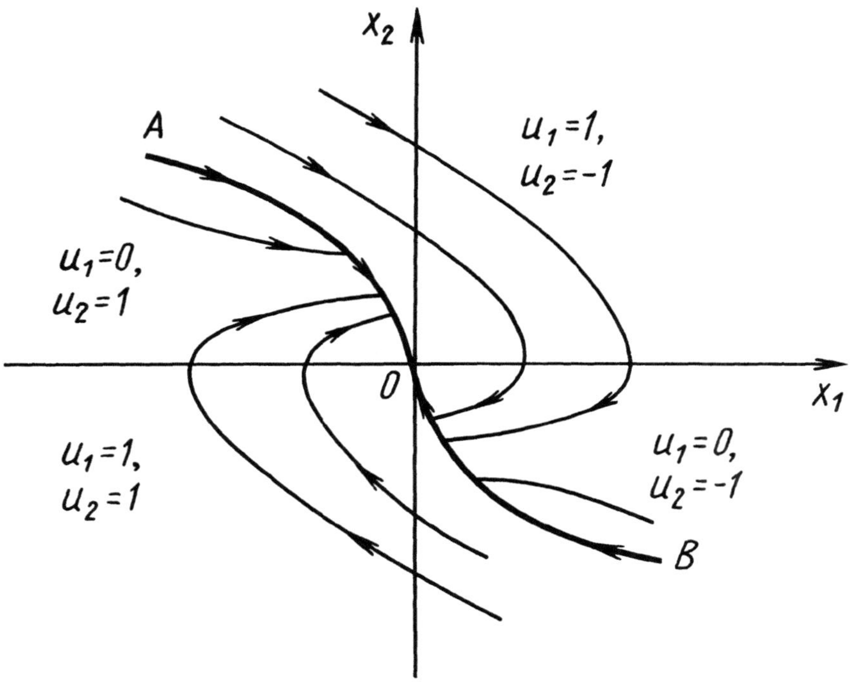

FIG. 3.2. The switching curve for optimal control in the problem (3.9)–(3.11):
$u_1 = 1$, $u_2 = -1$ on the curve AO; $u_1 = 1$, $u_2 = 1$ on the curve OB

From the foregoing equations we can define the moments t_2 and T as functions of the initial state $x(0) = x_1(0)$ and the initial velocity $\dot{x}(0) = x_2(0)$.

As an example, the switch times t_1 and t_2 and also the terminal moment T for the parameters $b = 4$, $k = 1$, $c = 1$, $x_2(0) = 2(1 - e^2) < 0$ were calculated. Using Newton's method, the results given in Table 3.1 were obtained.

Consider now the switching curves for the problem (3.9)–(3.11). Utilizing the obtained results concerning the structure of the optimal controls u_1 and u_2 and comparing the minimal times, we conclude that the switching curve is described by the curve AOB and the x-axis (see Figure 3.2). Moreover,

$$AO: \qquad x_1 = -\frac{x_2}{b} + \frac{a}{b^2}\ln\left(\frac{a+bx_2}{a}\right),$$

$$BO: \qquad x_1 = -\frac{x_2}{b} - \frac{a}{b^2}\ln\left(\frac{a-bx_2}{a}\right).$$

3. Problem of the soft landing on the moon. In S 3 (subsection 3) of Chapter III the problem of a soft landing on the moon was stated as the following optimal problem with an unspecified terminal moment T:

$$\begin{gathered} \dot{y} = v, \qquad \dot{v} = \frac{\beta u(t)}{m} - g_{\mathrm{m}}, \qquad \dot{m} = -u(t), \\ y(0) = y_0 > 0,\ v(0) = v_0 < 0,\ m(0) = m_0 > 0,\ y(T) = 0,\ v(T) = 0, \\ J(u) = m_0 - m(T) \to \inf, \\ 0 \leqslant u(t) \leqslant \gamma,\ m_0 - m(T) \leqslant M,\ \beta = \text{constant} > 0,\ g_{\mathrm{m}} = 1.62\,\mathrm{m}\cdot\mathrm{sec}^{-2} \end{gathered} \tag{3.20}$$

Here the same notation is used as in Chapter III.

Denote $x_1 = y$, $x_2 = v$, $x_3 = m$ and the adjoint variables by ψ_1, ψ_2, ψ_3. Then the Hamiltonian for the problem in hand is of the form

$$\begin{aligned} H(x,u,\psi) &= \psi_1 x_2 + \psi_2\left(\frac{\beta u}{x_3} - g_{\mathrm{m}}\right) - \psi_3 u \\ &= \psi_1 x_2 - \psi_2 g_{\mathrm{m}} + u\rho, \qquad \rho = \frac{\beta\psi_2}{x_3} - \psi_3. \end{aligned} \tag{3.21}$$

It follows from the condition for a maximum of the Hamiltonian (3.21):

$$\max_{0\leqslant u\leqslant\gamma} H(x,u,\psi) = H\big(x_0(t), u_0(t), \psi(t)\big) = 0$$

that the optimal control $u_0(t)$ becomes

$$u_0(t) = \begin{cases} 0, & \text{for } \rho(t) < 0, \\ \gamma, & \text{for } \rho(t) > 0. \end{cases}$$

Let us investigate the behavior of the function $\rho(t)$. The equation for the adjoint variables ψ_1, ψ_2, ψ_3 can be written thus:

$$\begin{gathered} \dot{\psi}_1 = 0, \qquad \dot{\psi}_2 = -\psi_1, \qquad \dot{\psi}_3 = \frac{\beta\psi_2 u}{x_3^2}, \\ \psi_1(0) = \psi_{10},\ \psi_2(0) = \psi_{20},\ \psi_3(T) = -1. \end{gathered} \tag{3.22}$$

It follows from (3.22) that

$$\psi_1(t) \equiv \psi_{10}, \qquad \psi_2(t) = \psi_{20} - \psi_{10}t.$$

Furthermore,

$$\dot{\rho}(t) = \frac{\beta\dot{\psi}_2}{x_3} - \frac{\beta\psi_2\dot{x}_3}{x_3^2} - \dot{\psi}_3 = -\frac{\beta\psi_1}{x_3} = -\frac{\beta\psi_{10}}{x_3}.$$

This equality implies that for $\psi_{10} \neq 0$ the function $\rho(t)$ is strictly decreasing or increasing depending on the sign of ψ_{10} and may vanish at most at one point.

Now suppose that $\psi_{10} = 0$. Then $\rho(t) = \rho_0 =$ constant. Let us demonstrate that $\rho_0 \neq 0$. Assume the contrary: $\rho_0 = 0$. Then it follows from the condition

$$0 = H\big(x(0), u_0(0), \psi(0)\big) = -g_{\mathrm{m}}\psi_{20}$$

that $\psi_{20} = 0$ and $\psi_2(t) \equiv 0$. This leads to a contradiction, since, on the one hand, $\rho(T) = -\psi_3(T) = 1$ and, on the other hand, $\rho(T) = \rho_0 = 0$.

Thus, we have shown that either $\rho(t)$ is a strictly monotonic function, or $\rho(t) = \rho_0 \neq 0$. In the first instance the control $u_0(t)$ has one switch point, and in the second instance there is no switch. In the case of one switch the optimal control becomes

$$u_0(t) = \begin{cases} 0, & \text{for } t < \tau, \\ \gamma, & \text{for } t > \tau. \end{cases}$$

The switch time τ can be determined from the condition $\rho(\tau) = 0$. Then the optimal procedure for a soft landing can be described as follows. Until the moment τ the brake engine is off, and the space vehicle falls freely on the moon. At the moment τ the brake engine starts operating at full power and maintains full power until the touch-down.

Because of the limited fuel resources, it is readily seen that the braking procedure described above cannot ensure a soft landing for any initial heights y_0 and for any initial velocities v_0. We determine now the initial conditions for a space vehicle, whereby a soft landing is possible. On the segment $[0, \tau]$, equations (3.20) for $u(t) \equiv 0$ imply that

$$\begin{aligned} x_1(\tau) &= y_0 + v_0\tau - \tfrac{1}{2}g_{\mathrm{m}}\tau^2 = y(\tau), \\ x_2(\tau) &= v_0 - g_{\mathrm{m}}\tau = v(\tau), \\ x_3(\tau) &= m_0. \end{aligned} \tag{3.23}$$

For $t \geqslant \tau$, equations (3.20) for $u(t) = \gamma$ can be solved to give

$$\begin{aligned} x_3(\tau + s) &= m_0 - \gamma s, \\ x_2(\tau + s) &= v(\tau) - g_{\mathrm{m}}s - c\ln\left(1 - \frac{\gamma s}{m_0}\right), \\ x_1(\tau + s) &= y(\tau) + [v(\tau) + \beta]s - \frac{1}{2}g_{\mathrm{m}}s^2 + \frac{\beta}{\gamma}(m_0 - \gamma s)\ln\left(1 - \frac{\gamma s}{m_0}\right). \end{aligned} \tag{3.24}$$

At the precise moment $(\tau + s)$ of the touch-down the following conditions must be satisfied:

$$x_1(\tau + s) = 0, \qquad x_2(\tau + s) = 0.$$

The parameter s appearing in these conditions is constrained by

$$0 \leqslant \gamma s \leqslant M. \tag{3.25}$$

The parameter s has the meaning of braking time. The meaning of the inequality (3.25) is also clear: The braking time cannot exceed the time $t_{\max}$ of operation of the brake engine at the maximum fuel consumption rate γ, i.e., $t_{\max} = M/\gamma$.

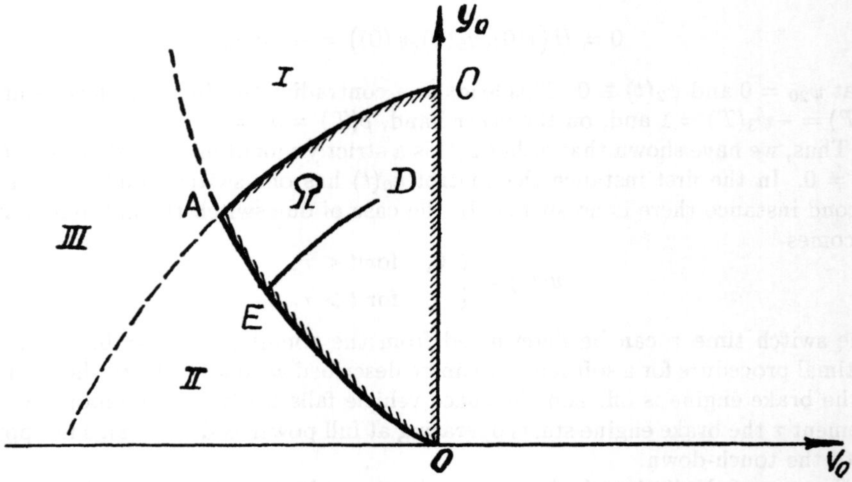

FIG. 3.3. The switching curve in the problem of a soft landing on the moon

Finally, in the plane (y_0, v_0) we obtain the following pattern (see Figure 3.3). The curve OEA has the parametric equation (3.23) with the parameter s varying from 0 to $t_{\max} = M/\gamma$. The curve AC is a free-fall parabola of the form (3.23) passing through the point C. The region Ω is the region of initial data (y_0, v_0) for which a soft landing on the moon is feasible, given the fuel mass M and the maximum rate of fuel consumption γ.

For the initial conditions corresponding to the point D of the region Ω the space vehicle is first falling freely on the moon along the parabola DE. At the moment when the parabola intersects the switching curve OA the brake engine is switched on, and it operates until the fuel is completely burned out. The burn-out moment must coincide exactly with the touch-down moment. If the initial conditions are on the curve OA, then the brake engine is immediately switched on and it operates till the touch-down moment. But if the initial conditions are outside the region Ω, then the soft landing is impossible due to several reasons:

the amount of fuel M is not sufficient (Region I)
the fuel consumption rate γ is not sufficient (Region II)
both M and γ is small (Region III).

4. Problem of the planar transfer of a space vehicle from one circular orbit to another[2]**.** Suppose that a space vehicle is in motion along a prescribed circular orbit. It is required to transfer it to a prescribed circular orbit lying in the same plane provided that the fuel consumption is minimum and the transfer time is unspecified.

We use the following notation: R is the radius of the original orbit, g_R is the acceleration of gravity at the distance R from the earth's center, M is the initial mass of the rocket, $(Rg_R)^{1/2}$ and $R^{1/2}g_R^{-1/2}$ are the velocity and period of revolution of the space vehicle in the original circular orbit. Using these variables, we can write the dimensionless equations of plane motion of the space vehicle in the earth's spherical gravitational field in the form

$$\begin{gathered} \dot{r} = v_r, \qquad \dot{\theta} = v_\theta r^{-1}, \qquad \dot{\mu} = -c^{-1}u_1, \\ \dot{v}_r = v_\theta^2 r^{-1} - r^{-2} + u_1\mu^{-1}\sin u_2, \\ \dot{v}_\theta = -v_r v_\theta r^{-1} + u_1\mu^{-1}\cos u_2. \end{gathered} \tag{3.26}$$

Here r is the dimensionless polar radius, θ is the polar angle, v_r and v_θ are dimensionless radial and tangential velocities of the space vehicle, μ is the dimensionless instantaneous mass, c is the dimensionless gas exhaust velocity, u_2 is the thrust angle relative to the local horizon, u_1 is the thrust magnitude, and $0 \leqslant u_1 \leqslant a$. The thrust magnitude u_1 and the thrust angle u_2 are regarded as control functions in this problem.

The cost functional can be written as

$$J(u_1, u_2) = \mu(0) - \mu(T) \to \min. \tag{3.27}$$

The initial conditions are of the form

$$r(0) = 1,\ v_r(0) = 0,\ v_\theta(0) = 1,\ \theta(0) = \theta_0,\ \mu(0) = \mu_0, \tag{3.28}$$

and the end conditions have the form

$$r(T) = r_1,\ v_r(T) = 0,\ v_\theta(T) = v_1,\ \theta(T) \text{ is arbitrary.} \tag{3.29}$$

The terminal moment is unspecified.

The problem under consideration is a problem in which the initial manifold (the circle of radius $r(0) = 1$) and the terminal manifold (the circle of radius $r(T) = r_1$) are given. Therefore, the necessary conditions (2.17)–(2.29) for optimal control can be utilized.

The Hamiltonian for the problem (3.26)–(3.29) can be expressed by the following relations:

$$\begin{gathered} H = H_0 + u_1 H_1, \\ H_0 = \psi_r v_r + \psi_\theta v_\theta r^{-1} + \psi_{v_r}(r^{-1}v_\theta^2 - r^{-2}) - \psi_{v_\theta} v_r v_\theta r^{-1}, \\ H_1 = \mu^{-1}(\psi_{v_r}\sin u_2 + \psi_{v_\theta}\cos u_2 - \psi_\mu \mu c^{-1}). \end{gathered}$$

[2] Henrion, R., *La Théorie de la Variation Seconde et ses Applications en Commande Optimale*, Académie Royale de Belgique, Bruxelles, (1976); Ivashkin, V.A., *Optimization of Manoeuvres in Space*, Nauka, Moscow, (1975).

The equations for the adjoint variables ψ_r, ψ_θ, ψ_{v_r}, ψ_{v_θ}, ψ_μ are of the form

$$\dot{\psi}_r = -\frac{\partial H}{\partial r} = \psi_\theta v_\theta r^{-2} + \psi_{v_r}(v_\theta^2 r^{-2} - 2r^{-3}) - \psi_{v_\theta} v_r v_\theta r^{-2},$$

$$\dot{\psi}_\theta = -\frac{\partial H}{\partial \theta} = 0, \qquad \dot{\psi}_{v_r} = -\frac{\partial H}{\partial (v_r)} = -\psi_r + \psi_{v_\theta} v_\theta r^{-1},$$

$$\dot{\psi}_{v_\theta} = -\frac{\partial H}{\partial (v_\theta)} = -\psi_\theta r^{-1} - 2\psi_{v_r} v_\theta r^{-1} + \psi_{v_\theta} v_r r^{-1},$$

$$\dot{\psi}_\mu = -\frac{\partial H}{\partial \mu} = u_1 \mu^{-2}(\psi_{v_r} \sin u_2 + \psi_{v_\theta} \cos u_2).$$

It follows from Pontryagin's maximum principle that

$$\sin u_2 = \psi_{v_r}\lambda^{-1}, \qquad \cos u_2 = \psi_{v_\theta}\lambda^{-1},$$

$$\lambda = \left(\lambda_{v_r}^2 + \lambda_{v_\theta}^2\right)^{1/2}, \qquad u_1 = \begin{cases} 0, & \text{for } H_1 < 0, \\ a, & \text{for } H_1 > 0, \end{cases}$$

$$H \equiv 0.$$

The transversality condition is of the form

$$\psi_\theta = 0. \tag{3.30}$$

The condition (3.24) guarantees the orthogonality of the vector $\psi(T) = (\psi_r(T), \psi_\theta(T), \psi_{v_r}(T), \psi_{v_\theta}(T), \psi_\mu(T))$ to the tangent of the circle of radius r_1 at any point of the circle at the moment T.

The problem (3.26)–(3.29) was investigated by many authors. It was shown that the transfer trajectory consists of both active ($u_1 = a$) and ballistic ($u_1 = 0$) parts. Ballistic parts are Kepler elliptic orbits, whose points include the switch-on and switch-off points that are either symmetric or asymmetric with respect to the major axis of the ellipse. The symmetric parts will be denoted by S, and the asymmetric ones by A. A view of the trajectory SA is shown in Figure 3.4.

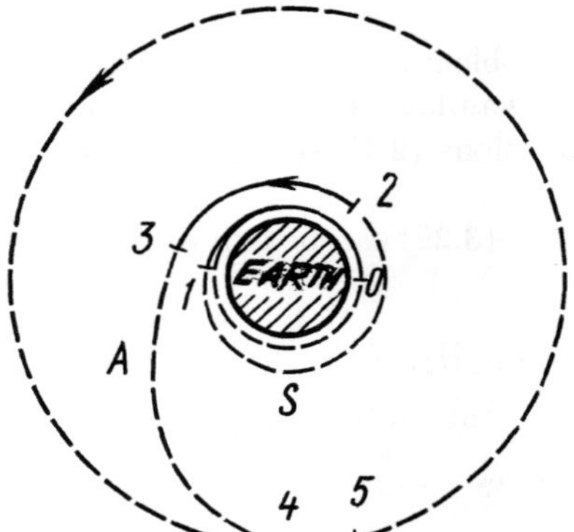

FIG. 3.4. A view of the trajectory SA:
—, active part; - - -, ballistic part

The most interesting result obtained in the solution of this problem is the existence of numerous trajectories that satisfy the necessary conditions for optimal control, Pontryagin's maximum principle. On each of these trajectories a local minimum of the fuel-consumption functional (3.21) is attained. For a given number of active parts, $n = 1, 2, 3$, optimal and nonoptimal trajectories have been obtained; they are given in Table 3.1. None of these trajectories furnishes the global minimum of the functional (3.21). Various locally optimal trajectories can be obtained by numerical calculations; the results are given in Table 3.2.

TABLE 3.1

Number of active parts	Optimal trajectory	Nonoptimal trajectory
1	A	—
2	AS, SA	A_2
3	AS_2, SAS, S_2A	A_3, SA_2, A_2S, ASA

TABLE 3.2

Type of trajectory	A	AS	AS_2	SA	SAS	S_2A
Fuel consumption (dimensionless)	0.630843	0.630828	0.630825	0.625339	0.625198	0.623631

In Table 3.2 the fuel consumption in dimensionless units accompanying the transfer from one circular orbit to another characterized by a radius ratio of 2.2 is given; the acceleration coefficient being a=0.03, and the dimensionless exhaust velocity $c = 0.3228$. For $n = 4$ and $n = 5$ we have corresponding optimal trajectories that require a smaller amount of fuel, but have, in fact, insignificant advantage over the other trajectories. Therefore, from an engineering point of view we need consider only trajectories with 3 to 5 active parts and dispense with more complicated alternatives that are much harder to implement.

§ 4. Control of ecological systems

One of the current important applications of control theory concerns ecology. It is devoted to the investigation of the effect of Man's actions on the processes of interaction between plants and animals, and between them and the world at large.

A peculiar feature of ecological systems is that their behavior is connected with a great number of interdependent factors, which can hardly be taken fully into account. Therefore, when treating particular problems, simplifications often have to be introduced. To describe simplified models, various types of equations are used: ordinary differential equations, equations with time lag, partial differential equations, stochastic equations, and so on. Some of them are given below. It should be noted that all of them can be derived in the same way, utilizing the law of conservation of population size (for a particular species).

1. Equations describing the evolution of a single population. There exist a great number of various models describing the behavior of an isolated population[3] In a number of such models the rate of change of population size can be written in the form of three summands: the first summand involves birth-rate, the second mortality, and the third migration. One of the first such models put forward by T. R. Malthus in 1798 is expressed by the equation

$$\dot{N}(t) = \lambda N(t), \qquad N(0) = N_0 > 0, \tag{4.1}$$

where $N(t)$ is the population size at time t, and $\lambda > 0$ is a constant.

In the Malthus model, migration is not taken into account, the number of births and deaths are proportional to the population size, birth-rate being greater than death-rate. According to the Malthus law, population size must grow exponentially, a fact which is not universally confirmed.

G. Quetelett made an assumption that there must be a kind of saturation effect as far as population size is concerned. To describe the population dynamics on the basis of this assumption, his disciple Verhulst has put forward the following equation:

$$\dot{N}(t) = \lambda N(t)\left(1 - \frac{N(t)}{K}\right), \qquad t \geqslant 0, \tag{4.2}$$
$$N(0) = N_0 > 0.$$

Here the constant K determines the time-invariant population size. The constant K is called the *environment capacity* for a given population. Using, for example, the method of seperation of variables, we can obtain a solution of the problem (4.2) in the form

$$N(t) = \lambda N_0 e^{\lambda t}\left[\lambda + \frac{\lambda}{K} N_0 (e^{\lambda t} - 1)\right]^{-1}. \tag{4.3}$$

It follows from this expression that

$$\lim_{t\to\infty} N(t) = K.$$

Therefore, for the model (4.2) the population size tends to a terminal value K. In addition, according to (4.2) the derivative $\dot{N}(t) > 0$ whenever $N_0 < K$ and $\dot{N}(t) < 0$ whenever $N_0 > K$. Hence, noting (4.3), we see that the steady-state solution $N(t) \equiv K$ of equation (4.2) is asymptotically stable, whereas the zero equilibrium solution $N(t) \equiv 0$ is not stable.

[3]Murray, J. D., *Mathematical Biology*, Springer, New York, (1989).

The hypothesis of an instantaneous reaction of the population to a change in its size represents a significant deficiency of models of the type (4.2). In reality, however, a change in population size does not immediately affect the rate $\dot{N}(t)$. This leads necessarily to *equations with delay* or *retarded equations*. So, if the overall aftereffect can be characterized by a delay time $h > 0$, we can use the corresponding model proposed by G. E. Hutchinson which is described by the equation

$$\dot{N}(t) = \lambda N(t)\left(1 - \frac{N(t-h)}{K}\right). \tag{4.4}$$

A more general model introduced by C. M. Cushing in 1976 takes into account both discrete and distributed delays; it is described by the equation

$$\dot{N}(t) = \lambda N(t)\left[1 - \int_0^\infty N(t-s)\,dK(s)\right], \qquad t \geqslant 0. \tag{4.5}$$

Here $K(s)$ is a nondecreasing bounded function, and the integral on the right side of the equation is of the Stiltjes type. Under a corresponding choice of the function $K(s)$, equation (4.5) can be reduced to equations (4.1), (4.2), (4.4). However, the model (4.5) does not take account of age nonhomogeneity in the population.

Population models that allow for a distribution of life-spans and migration—*McKendrick models*—are derived in the following way. Let $N(t,\tau)$ denote the number of individuals whose age at time t is less than or equal to τ. We denote by $n(t,\tau)$ the age density being the derivative of the function $N(t,\tau)$ with respect to τ: $n(t,\tau) = \partial N(t,\tau)/\partial\tau$. Then the following equation is valid for the function $n(t,\tau)$:

$$\frac{\partial n(t,\tau)}{\partial t} + \frac{\partial n(t,\tau)}{\partial \tau} = -\mu(t,\tau)n(t,\tau) + g(t,\tau). \tag{4.6}$$

Here $\mu(t,\tau)$ denotes the mortality rate, and $g(t,\tau)$ the migration rate. In other words, after a time Δt (considered to be small) the number of deaths at an age of τ at time t is $\mu(t,\tau)n(t,\tau)\Delta t$ and the number of migrants is $g(t,\tau)$. The sign of the function g depends on the type of migration: the function g is positive for immigration and negative for emigration. The boundary condition for $n(t,0)$ is determined by the number of births at time t. It can be represented thus:

$$n(t,0) = u(t)\int_{\tau_1}^{\tau_2} K(t,\tau)n(t,\tau)\,d\tau. \tag{4.7}$$

Here τ_1 and $\tau_2 > \tau_1$ determine the age limits outside of which there is no reproduction. The function $u(t)$ represents birth-rate, and, finally, the function $K(t,\tau)$ describes the age-dependence of reproduction. In addition to the boundary condition (4.7), an initial condition is needed for the solution of equation (4.7). It is taken in the form

$$n(0,\tau) = n_0(\tau), \qquad \tau \geqslant 0, \tag{4.8}$$

where $n_0(\tau)$ is a prescribed function specifying the age distribution at the initial moment $t = 0$. Suppose that the function $\mu(t,\tau)$ does not depend on t, i.e., $\mu(t,\tau) \equiv \mu(\tau)$, and the boundary condition is of the form

$$n(t,0) = u(t)N(t) = \varphi(t), \qquad N(t) = \int_0^\infty n(t,\tau)\,d\tau. \tag{4.9}$$

Then a solution of the problem (4.6), (4.8), (4.9) can be represented by

$$n(t,\tau) = \begin{cases} n_0(\tau - t)\exp\Big(-\int_{\tau-t}^{\tau} \mu(s)\,ds\Big), & \text{for } 0 \leqslant t \leqslant \tau, \\ \varphi(t-\tau)\exp\Big(-\int_0^{\tau} \mu(s)\,ds\Big), & \text{for } \tau \leqslant t. \end{cases}$$

These expressions can be used to analyze the population dynamics as described by equation (4.6).

Another type of nonhomogeneity arises when the population is nonuniformly distributed over the habitat. This results in population diffusion, i.e., the movement of individuals from one region of the habitat to another region. To describe evolution in this case, we introduce the population density $D(t,x,y)$ at the point with coordinates x, y at time t. The function $D(t,x,y)$ is a solution of the Hutchinson equation with diffusion

$$\begin{aligned} \frac{\partial D(t,x,y)}{\partial t} &= \alpha\left(\frac{\partial^2 D(t,x,y)}{\partial x^2} + \frac{\partial^2 D(t,x,y)}{\partial y^2}\right) \\ &+\lambda D(t,x,y)\left(1 - \frac{1}{K}D(t-h,x,y)\right). \end{aligned}$$

Here α is the diffusion coefficient.

An efficient way to describe various types of nonhomogeneity as well as incompleteness of information is provided by stochastic modeling based on the use of stochastic (random) variables and processes. For example, suppose that the growth coefficient in equation (4.2) is a random quantity. Then, to describe the population dynamics, instead of (4.2) we can use the following equation(the stochastic Verhulst equation):

$$dN(t) = \big(\lambda dt + \sigma d\xi(t)\big)N(t)\left(1 - \frac{N(t)}{K}\right),$$

where $\xi(t)$ is a standard Wiener process.

2. Communities of at least two species. There exist a large number of various types of equations modeling the behavior of communities containing at least two species (populations).

One of the first such models (the Lotka–Volterra model, or the "predator–prey" model) serves to describe the behavior of a community composed of two species interpreted as "predator" and "prey". Let $x(t)$ and $y(t)$ denote the respective populations of prey and predators at time t. The Lotka–Volterra model is described by the equations

$$\begin{aligned} \dot{x}(t) &= a_1 x - a_2 x y, \qquad t \geqslant 0,\ x(0) = x_0 > 0, \\ \dot{y}(t) &= a_2 a_3 x y - a_4 y, \qquad y(0) = y_0 > 0, \end{aligned} \tag{4.10}$$

where $a_i > 0$ are some constants.

The system (4.10) has only one nonzero equilibrium state with coordinates $x_1 = a_4(a_2a_3)^{-1}$ and $y_1 = a_1 a_2^{-1}$. Note that the solution $x(t)$, $y(t)$ of the problem (4.10) is positive for any t. We set $x_2(t) = x(t)x_1^{-1}$, $y(t) = y(t)y_1^{-1}$. Then the first integral of the system (4.10) can be written as

$$\left(\frac{e^{x_2}}{x_2}\right)^{a_4} \times \left(\frac{e^{y_2}}{y_2}\right)^{a_1} = C. \tag{4.11}$$

The quantities x_1 and y_1 are the average (over the period) numbers of prey and predators, respectively. The value of the constant C for a specific trajectory is determined by the left side of (4.11) at time $t = 0$.

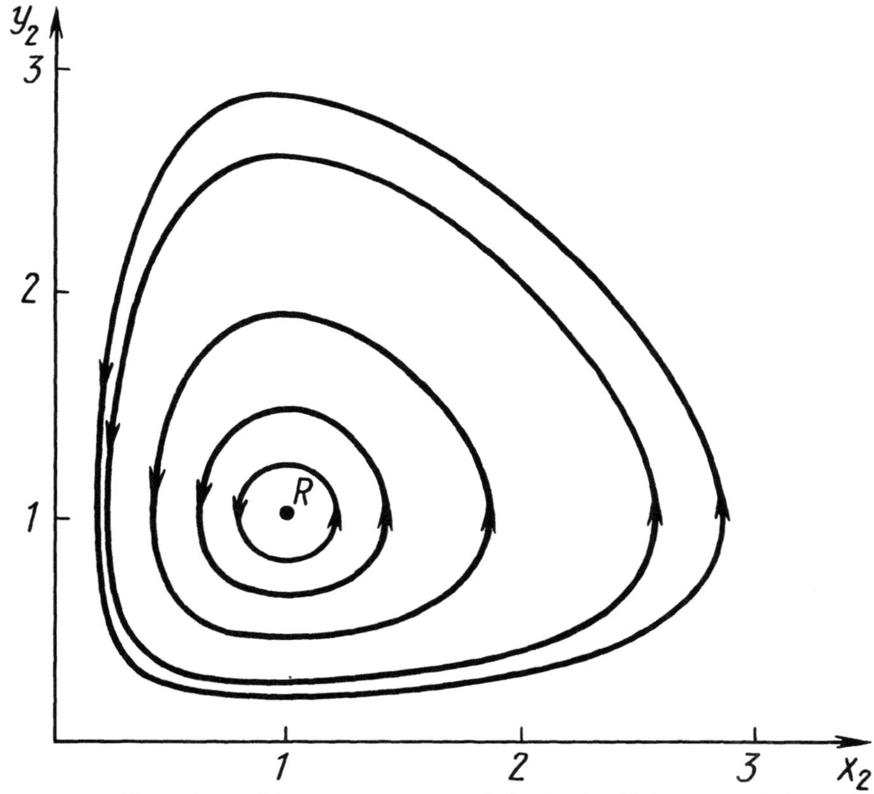

FIG. 4.1. Phase trajectories of the Lotka–Volterra model

It follows from (4.11) that the phase trajectories of the system (4.10) are closed curves that enclose one another. Moreover, the larger the value of C, the larger the maxima of the functions $x(t)$ and $y(t)$ (see Figure 4.1, where the arrows indicate the directions of the evolution of the system as time increases). If the initial state of the system (4.10) coincides with the equilibrium state R, then the constant C in (4.11) takes on the minimum value that is equal to $C_0 = \exp(a_1 + a_4)$. For $C > C_0$ the phase trajectories enclose the point R. Thus, the point R is a stable equilibrium state of the system (4.10) with respect to disturbances of the initial state.

A more detailed analysis shows that the point R is a center. Indeed, by linearizing equation (4.10) in the neighborhood of the point R, we find that in the linear approximation the functions $x_3(t) = x(t) - x_1$, $y_3(t) = y(t) - y_1$ satisfy the relations

$$\dot{x}_3(t) = -a_4 a_3^{-1} y_3(t), \qquad \dot{y}_3(t) = a_3 a_1 x_3(t). \tag{4.12}$$

Let us set up the characteristic equation that corresponds to these relations. We have

$$\det \begin{vmatrix} -\lambda & -a_4 a_3^{-1} \\ a_1 a_3 & -\lambda \end{vmatrix} = \lambda^2 + a_1 a_4 = 0.$$

The roots λ_1 and λ_2 of this equation are pure imaginary: $\lambda_{1,2} = \pm i\sqrt{a_1 a_4}$. Consequently, the point R is a center, and the period of oscillations relative to R in the "predator–prey" system is, in the linear approximation, equal to $2\pi\left(\sqrt{a_1 a_4}\right)$.

The Lotka–Volterra model enables explanation of some of the phenomena of the outside world. However, certain phenomena are unaccounted for in this model. Therefore, the model (4.10) has been refined in several ways with a view to take into account intraspecific competition, delays, random disturbances, age and habitat nonhomogeneities, and so on. For example, the equations of interaction of predators and prey that take account of intraspecific competition are of the form

$$\begin{aligned} \dot{x}(t) &= a_1 x - a_2 xy - \gamma_1 x^2, \\ \dot{y}(t) &= a_2 a_3 xy - a_4 y - \gamma_2 y^2, \end{aligned} \tag{4.13}$$

where $\gamma_1 > 0$, $\gamma_2 > 0$ are given constants. As compared to equations (4.10), equations (4.13) have the additional summands $\gamma_1 x^2$ and $\gamma_2 y^2$ that take account of the competition between among prey and predators, respectively. The problems of the stability of time-invariant states for the system (4.13) are of great interest.[4]

Different models were proposed to describe the evolution of various populations. Let us turn our attention to one of them. Suppose there are n competing species of prey whose numbers are $N_1(t), \dots, N_n(t)$ and m species of predators $P_1(t), \dots, P_m(t)$. The equations of their interaction can be represented in the

[4]Svirizhev, Yu. P., and Logofet, D. O., *Stability of Biological Populations*, Nauka, Moscow, (1978).

form

$$\dot{N}_i(t) = N_i(t)\left[\lambda_i - \sum_{j=1}^{n} \alpha_{ij} N_j(t) - \sum_{k=1}^{m} \beta_{ik} P_j(t)\right], \qquad i = 1, \ldots, n,$$

$$\dot{P}_l(t) = P_l(t)\left[-\gamma_l + \sum_{j=1}^{n} \delta_{lj} N_j(t)\right], \qquad l = 1, \ldots, m.$$

Here λ_j, α_{ij}, β_{ik}, γ_l, δ_{lj} are positive constants.

The foregoing models were discussed in terms of the "predator–prey" system. In should be noted, however, that similar equations are used when modeling the interaction of the animal and the plant world, and also for solving the problems of biotechnology.

As an example, we consider the Monod model, which serves to model the biosynthesis process that takes place in a tank containing microorganisms. It is assumed that nutrition products of concentration $u(t)$ are fed continuously into the tank. The interaction of microorganisms and nutrition products results in an increase in the useful biomass whose concentration is denoted by $x(t)$. The liquid flowing out of the tank contains the useful biomass. The mass of the microorganism population in the tank is assumed to be constant. Further, we denote by $y(t)$ the concentration of nutrition products inside the tank. The biosynthesis process is described by the functions $x(t)$ and $y(t)$ that satisfy the equations

$$\begin{aligned} \dot{x}(t) &= [a(y) - \lambda]x(t), \\ \dot{y}(t) &= -\beta a(y)x - \lambda y + \lambda u(t), \end{aligned} \tag{4.14}$$

where the constant $\lambda > 0$, $\beta > 0$ and the specific reproduction rate $a(y)$ are determined by the particular type of a biosynthesis process.

It should be noted that a number of actual processes are not taken into account in the Monod model (4.14), such as the finiteness of the mixing rate for nutrition products that are fed into the tank, dissimilar age-dependence of the reproduction rate of bacteria and of their reaction to nutrition products, etc. These effects can be taken care of by introducing lumped and distributed delays into the model (4.14).

3. Statements of typical control problems for ecological systems. As is customary in control theory, the statement of a particular control problem for an ecological system incorporates the equations of evolution of the system, the cost functional, and control and state constraints.

In ecology control can be implemented in various forms, such as the use of fertilizers, pesticides and insecticides, irrigation, mowing, fishing, hunting, the laws on migration, etc. These factors can be taken into account in the equations of evolution of the system, the boundary conditions, and the cost functional.

Several typical control problems are stated below.

Control of the "predator–prey" system. Consider a "predator–prey" system modeled by the Lotka–Volterra equations (4.10). The control $u(t)$ for the system may consist in destroying both the prey and the predators with respective probabilities $r_1u(t)x(t)$ and $r_2u(t)y(t)$, where r_1 and r_2 are prescribed nonnegative numbers. In view of (4.10), the equations of the control "predator–prey" system can be taken in the form

$$\begin{aligned} \dot{x}(t) &= x(t)[a_1 - a_2y(t) - r_1u(t)], \qquad t \geqslant 0, \\ \dot{y}(t) &= y(t)[a_2a_3x(t) - a_4 - r_2u(t)]. \end{aligned} \tag{4.15}$$

The initial conditions for the system (4.15) are of the form

$$x(0) = x_0 > 0, \qquad y(0) = y_0 > 0. \tag{4.16}$$

The control $u(t)$ is constrained in a natural way by the following inequalities:

$$0 \leqslant u(t) \leqslant \gamma. \tag{4.17}$$

The goal of the control may be either to maintain the system in the neighborhood of the equilibrium state (x_1, y_1) of the system (4.15) for $u = 0$ or to transfer the system to the equilibrium state in minimum time.

Consider the time-optimal problem. Let $T = T(u)$ denote a first moment such that a solution of the problem (4.15), (4.16) generated by the control u is of the form

$$x(T) = x_1 = a_4(a_2a_3)^{-1}, \qquad y(T) = y_1 = a_1a_2^{-1}. \tag{4.18}$$

Then the goal of the time-optimal control problem is to determine a control $u_0(t)$ such that the relations (4.15)–(4.18) are satisfied and

$$\inf_u T(u) = T(u_0). \tag{4.19}$$

The investigation of the problem (4.15)–(4.19) is carried out below.

Let now $T > 0$ be a fixed moment. To make the system (4.15) evolve in the neighborhood of the state (x_1, y_1) on the interval $[0, T]$, we can use the condition for a minimum of the integral quadratic cost functional to deduce a control of the form

$$\int_0^T [\alpha_1 (x(t) - x_1)^2 + \alpha_2 (y(t) - y_1)^2]\, dt \to \inf_u,$$

where $\alpha_i \geqslant 0$ are weight coefficients.

It should be noted that similar statements of optimal control problems are of interest when considering other control models, such as the Monod model (4.14), in which the control is taken as the concentration $u(t)$ of nutrition products fed into the tank with the biomass.

Optimal catching of a species. In the problem of catching fish (or animals) the objective of control is to maximize the overall catch. Suppose that the species population size is determined by the Verhulst equation (4.2) and the catch rate (i.e., the catch per unit time) is $u(t)$. Then the equation of the dynamics of the species population size is of the form

$$\dot{N}(t) = \lambda N(t)\left(1 - \frac{N(t)}{K}\right) - u(t)N(t), \qquad 0 \leqslant t \leqslant T, \tag{4.20}$$

where T is a specified moment. The control $u(t)$ constrained by (4.17) must satisfy the condition

$$\int_0^T u(t)N(t)\,dt \to \sup_u,$$

i.e., the condition for a maximum catch of the fish.

The use of such a cost functional is not fully justified, since its use may lead to irreversible changes in the species population size. In those cases when we are interested not only in obtaining maximum catches but also in maintaining the species population size at a certain desirible level N_c, the following cost functional is preferable:

$$\int_0^T [\alpha_1 u(t)N(t) - \alpha_2 \big(N(t) - N_0\big)^2]\,dt \to \sup_u,$$

where $\alpha_i \geqslant 0$ are certain weight coefficients.

Control of population size. In many countries the problem of the control of the population size is very important. The main controllable factor is the number of children per family. If the relations (4.6)–(4.8) are to be used to model this problem, then the function $u(t)$ in the boundary condition (4.7) represents the number of children per family. The cost functional in the problem (4.6)–(4.8) is determined by the desirable density $n_0(\tau)$ of the age distribution of the population. For example, the cost functional can be given in the form

$$\int_0^\infty d\tau \int_0^T [n(t,\tau) - n_0(\tau)]^2 h(t,\tau)\,dt \to \inf_u .$$

Here the moment T and the function $h(t,\tau)$ are prescribed. T determines the period of time over which the population size is planned, and the function $h(t,\tau)$ specifies the individual contribution of various groups to the cost functional. The control constraint is of the form $0 \leqslant u(t) \leqslant u_0$, where u_0 is prescribed. Note that the control in this problem appears only in the boundary condition (4.7).

Another way to control population size is connected with migration laws, i.e., the choice of the function $g(t,\tau)$ in equation (4.6).

The foregoing examples of statements of optimal control problems in ecology show that they belong to the framework of general control theory. However, there are some peculiar features, such as the typical unidirectional property of control actions, i.e., as a rule, the constraints are of the form (4.17). Moreover, the phase variables, by their nature, are nonnegative. Therefore, equations modeling the dynamics of ecological phenomena and admissible controls must satisfy these requirements.

4. Time-optimal control of the "predator–prey" system. Consider the control problem for the system (4.15)–(4.18) for $r_2 = 0$, i.e., we assume that the control can act only on the prey. Transforming to new variables according to the formulas

$$t \to a_1^{-1}t, \qquad x \to a_4(a_2a_3)^{-1}x, \qquad y \to a_1a_2^{-1}y,$$
$$b \to a_4a_1^{-1}, \qquad \gamma \to \gamma r_1 a_1^{-1}, \qquad u \to a_1 r_1^{-1} u,$$

we obtain the equations of the system in the form

$$\dot{x}(t) = (1 - y - u)x, \qquad \dot{y}(t) = b(x-1)y, \qquad t > 0, \tag{4.21}$$
$$x(0) = x_0,\ y(0) = y_0,\ \ x_0 > 0,\ y_0 > 0,$$
$$0 \leqslant u \leqslant \gamma,\ \ \gamma > 0.$$

For $u = 0$, this system has only one nonzero equilibrium state R in the plane (x, y) with coordinates (1,1), and the goal is to transfer the system (4.21) from state (x_0, y_0) to state $R(1,1)$ in minimum time.

The existence of an optimal control $u_0(t)$ in the problem under consideration has been established in a recent paper.[5]

We shall use the maximum principle to both seek an optimal solution and investigate it. We introduce the function $H(x, y, \lambda_0, \psi_1, \psi_2, u)$ that is equal to

$$H = \psi_1(1 - y - u)x + \psi_2 b(x-1)y + \lambda_0. \tag{4.22}$$

Here the constant $\lambda_0 \leqslant 0$, and the functions $\psi_1(t)$ and $\psi_2(t)$ satisfy the equations

$$\begin{aligned} \dot{\psi}_1(t) &= \psi_1(t)\big(y(t) - 1 + u(t)\big) - b\psi_2(t)y(t), \\ \dot{\psi}_2(t) &= \psi_1(t)x(t) + b\psi_2(t)\big(1 - x(t)\big). \end{aligned} \tag{4.23}$$

On the strength of the maximum principle there exists a nonzero solution ψ_1, ψ_2 of equations (4.23) such that

$$\max_{0\leqslant u\leqslant\gamma} H(x, y, \lambda_0, \psi_1, \psi_2, u) = H(x, y, \lambda_0, \psi_1, \psi_2, u_0) \equiv 0. \tag{4.24}$$

It follows from equations (4.23) that the functions $\psi_i(t)$ cannot vanish on whole intervals. Indeed, if, for example, $\psi_1(t) \equiv 0$ for $t_1 \leqslant t \leqslant t_2$, $t_1 < t_2$, then we should also have $\dot{\psi}_1(t) \equiv 0$ for $t_1 \leqslant t \leqslant t_2$. Therefore, by the first of the equations (4.23) the function $\psi_2(t) \equiv 0$, $t_1 \leqslant t \leqslant t_2$, i.e., both adjoint variables would vanish simultaneously, but this cannot be true for a time-optimal problem (see Theorem 2.1). By an analogous argument we can prove that the functions ψ_1 and ψ_2 can have only simple zeros. It follows from this and (4.22), (4.24) that the optimal control $u_0(t)$ is piecewise continuous and equal to either zero or γ.

For a further investigation of the optimal control we shall transform the variables $\psi_i(t)$ to the variables $\varphi_i(t)$ according to the formulas

$$\varphi_1(t) = x(t)\psi_1(t), \qquad \varphi_2(t) = y(t)\psi_2(t).$$

Because of the functions $x(t)$ and $y(t)$ being positive, the signs of the functions φ_1 and φ_2 coincide with respective signs of ψ_1 and ψ_2. Equations (4.21), (4.23) lead to the following equations for φ_i:

$$\begin{aligned} \dot{\varphi}_1(t) &= -bx(t)\varphi_2(t), \\ \dot{\varphi}_2(t) &= y(t)\varphi_1(t). \end{aligned} \tag{4.25}$$

[5]Kolmanovskii, V. B., and Spivak, A. K., Time-Optimal Control of the "Predator-Prey" System, *Prikl. Mat. Mekh.*, (1990), vol. 54, no. 3. (Translated into English as *J. Appl. Math. Mech.*)

Let us now split the optimal trajectory into parts that are contained entirely in either the region $x > 1$ or the region $x < 1$.

First, we show that on any portion of the optimal trajectory contained entirely in the region $x > 1$ there can be at most one optimal switch point. Moreover, a switch from $u = 0$ to $u = \gamma$ is only possible.

Indeed, suppose that $x(t) > 1$ for $t_1 \leqslant t \leqslant t_2$, and there is a control jump at the moment τ, where $\tau \in [t_1, t_2]$. Since τ is an optimal switch point, by virtue of (4.24) we have $\varphi_1(\tau) = 0$ at this point. Further, it follows from (4.21), (4.24) that

$$\frac{\varphi_1(t)\dot{x}(t)}{x(t)} + \frac{\varphi_2(t)\dot{y}(t)}{y(t)} + \lambda_0 = 0. \tag{4.26}$$

But by virtue of (4.21) the derivative $\dot{y}(t) > 0$ for $t_1 \leqslant t \leqslant t_2$. Therefore, on the strength of (4.26) we have

$$\varphi_2(t) = \frac{y(t)}{\dot{y}(t)}\left[-\lambda_0 - \varphi_1(t)\frac{\dot{x}(t)}{x(t)}\right], \qquad t_1 \leqslant t \leqslant t_2.$$

Substituting this expression for $\varphi_2(t)$ in the first of the equations (4.25) gives

$$\dot{\varphi}_1(t) = \frac{b}{\dot{y}(t)}[y(t)\dot{x}(t)\varphi_1(t) + \lambda_0 x(t)y(t)], \qquad t_1 \leqslant t \leqslant t_2, \tag{4.27}$$
$$\varphi_1(\tau) = 0.$$

We regard equation (4.27) as a Cauchy problem for the function $\varphi_1(t)$. Then according to Cauchy's formula (1.6) of Chapter IV the sign of the function $\varphi_1(t)$ for $t_1 \leqslant t \leqslant t_2$ is determined by the sign of λ_0, since the values b, $\dot{y}(t)$, $x(t)$ and $y(t)$ are positive for $t_1 \leqslant t \leqslant t_2$. But if $\psi_1(0) = 0$, then in view of (4.27) the function $\varphi_1(t) \equiv 0$ for $t_1 \leqslant t \leqslant t_2$, and this cannot be true as has been established before. But $\psi_0 \leqslant 0$. Therefore, the constant $\psi_0 < 0$; hence, $\varphi_1(t) < 0$ for $\tau \leqslant t \leqslant t_2$ and $\varphi_1(t) > 0$ for $t_1 \leqslant t \leqslant \tau$. Consequently, on the strength of the maximum principle we conclude that as long as the optimal trajectory remains in the region $x > 1$ the optimal control is $u_0 = \gamma$ for $t > \tau$ and $u_0 = 0$ for $t < \tau$.

In a similar way we can prove that on any portion of the optimal trajectory contained entirely in the region $x < 1$ there can be at most one optimal switch point; moreover, a switch from $u_0 = 0$ to $u_0 = \gamma$ is only possible. The shape of the switching curve for optimal control depends on the value of γ.

First, suppose that $\gamma = 1$. The switching curve $APRSB$ for optimal control is given in Figure 4.2, the optimal control being $u_0 = 1$ to the right of this curve, and $u_0 = 0$ to the left of the curve. The part APR of this curve represents the trajectory of the system (4.21) for $u_0 = \gamma = 1$ passing through the point $R(1, 1)$.

The part RSB of the switching curve, where $u_0 = 0$, was obtained numerically. The algorithm for numerical solution for any γ is as follows. We write equations (4.21), (4.25) in inverse time $t \to -t$

$$\dot{x}(t) = [u + y(t) - 1]x(t), \qquad \dot{y}(t) = b[1 - x(t)]y(t), \tag{4.28}$$

$$\begin{aligned} \dot{\varphi}_1(t) &= b\varphi_2(t)x(t), \\ \dot{\varphi}_2(t) &= -\varphi_1(t)y(t), \qquad t > 0. \end{aligned} \tag{4.29}$$

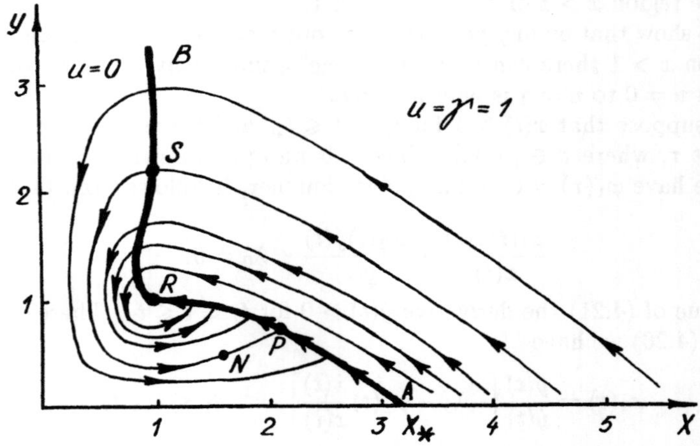

FIG. 4.2. Switching curve for the optimal control in the time-optimal problem for the Lotka–Volterra model in the case $\gamma = 1$

Since the transfer of the system (4.21) to the target point $R(1,1)$ on the optimal trajectory can occur only for the control $u = \gamma$, it follows that the initial condition $\varphi_1(0)$ for the system (4.29) must be of the form $\varphi_1(0) < 0$. Consider now the system (4.28), (4.29) for $u = \gamma = 1$ and the initial conditions

$$x(0) = 1, \qquad y(0) = 1, \qquad \varphi_1(0) < 0, \qquad \varphi_2(0) \text{ arbitrary}, \tag{4.30}$$

where $\varphi_1(0)$, $\varphi_2(0)$ are certain prescribed numbers. We solve the problem (4.28)–(4.30) for $u = 1$ on the segment $[0, \tau_1]$, where τ_1 is the first zero of the function $\varphi_1(t)$. At the moment τ_1 a switch of the control to $u = 0$ occurs. Further, we look for a solution of the problem (4.28), (4.29) for $u = 0$ with the initial conditions at the point τ_1 that have been determined at the preceding step of the algorithm. The solution is sought on the interval $[\tau_1, \tau_2]$, where τ_2 is the next (after τ_1) zero of the function φ_1. The point with coordinates $x(\tau_2)$, $y(\tau_2)$ lies on the switching curve RSB. Assigning other values to the variables $\varphi_1(0)$ and $\varphi_2(0)$, we obtain other points of the curve RSB.

The part APR of the switching curve is the trajectory of the system (4.28) for $u = 1$ with the initial conditions $x(0) = 1$, $y(0) = 1$. In other words, the curve APR is described by the equations

$$\dot{x} = xy, \qquad \dot{y} = b(1 - x)y, \qquad x(0) = 1,\ y(0) = 1,\ t \geqslant 0. \tag{4.31}$$

The first integral of the system (4.31) is of the form

$$\left(xe^{-x}\right)^b e^{-y} = C.$$

Thus, noting the initial conditions $x(0) = 1$, $y(0) = 1$, we can write the equation of the curve APR as

$$y = b \ln x - bx + b + 1. \tag{4.32}$$

By virtue of (4.31) the derivative $\dot{x}(t) > 0$ and $\dot{y}(t) < 0$ for $t > 0$. Moreover, the functions $x(t)$ and $y(t)$ are positive. Hence, the function $x(t)$ increases monotonically as $t \to \infty$ and must tend to a finite limit, and the function $y(t)$ decreases monotonically as $t \to \infty$ and tends to zero. Indeed, if the limit of the function $x(t)$ as $t \to \infty$ were equal to $+\infty$, then by virtue of (4.32) the function $y(t)$ would be negative for sufficiently large t, and this cannot be true. Furthermore, if the limit of the function $y(t)$ as $t \to \infty$ were equal to some positive number, then by virtue of the first of the equations (4.31) the limit of the function $x(t)$ as $t \to \infty$ would be equal to $+\infty$, and this cannot be true, as shown above. Thus, to solve the problem (4.31) and determine the switching curve APR, we have

$$\lim_{t\to\infty} x(t) = x_*, \qquad \lim_{t\to\infty} y(t) = 0.$$

The graph of x_* as a function of the parameter b is given in Figure 4.3. Note that the curves with arrows in Figure 4.2 show optimal trajectories of the system (4.21).

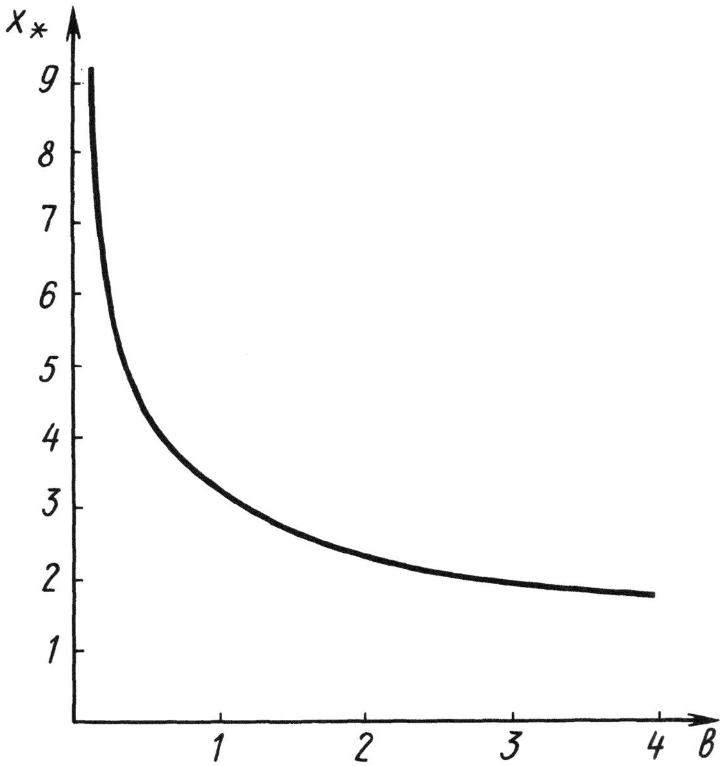

FIG. 4.3. Graph of x_* as a function of the parameter b in the time-optimal problem for the Lotka–Volterra model

Consider now the case $\gamma > 1$. The switching curve $APRSB$ for $\gamma > 1$ is given in Figure 4.4. The part APR of this curve is determined by the equations

$$\dot{x}(t) = (\gamma - 1 + y)x, \qquad \dot{y}(t) = b(1 - x)y, \qquad t \geqslant 0 \tag{4.33}$$
$$x(0) = 1, \; y(0) = 1.$$

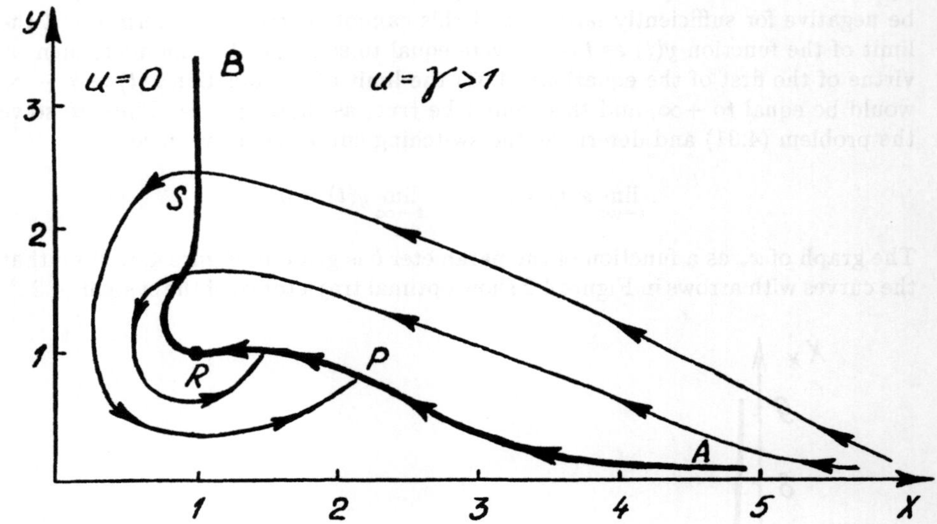

FIG. 4.4. Switching curve for the optimal control in the time-optimal problem for the Lotka–Volterra model in the case $\gamma > 1$

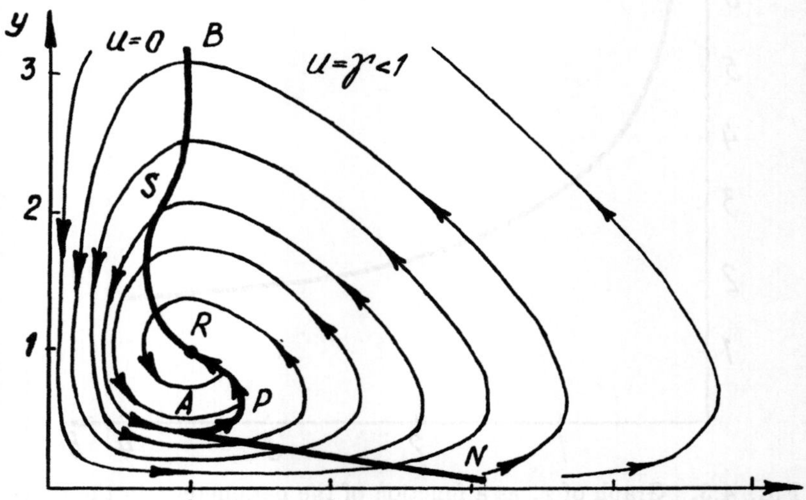

FIG. 4.5. Switching curve for the optimal control in the time-optimal problem for the Lotka–Volterra model in the case $\gamma < 1$

The first integral of the system (4.33) is of the form

$$\left(xe^{-x}\right)^b \left(y^{1-\gamma}e^{-y}\right) = C.$$

Hence, the equation of the curve APR has the form

$$bx - b\ln x + (\gamma - 1)\ln y + y = b + 1.$$

Further, note that the solution of the problem (4.33) satisfies the relations

$$\lim_{t\to\infty} x(t) = \infty, \qquad \lim_{t\to\infty} y(t) = 0.$$

Therefore, the ordinate of the switching curve APR tends to zero as $x \to \infty$. A portion of the switching curve RSB has been obtained numerically with the help of the same algorithm as in the case $\gamma = 1$.

Finally, in the case $\gamma < 1$ the switching curve $NAPRSB$ is shown in Figure 4.5. As before, the part APR of the switching curve represents the trajectory of the problem (4.33), and the parts NA and RSB have been obtained numerically.

A comparison of Figures 4.2, 4.4 and 4.5 shows that in the case $\gamma \geqslant 1$ the number of optimal control switch points is at most two, whereas in the case $\gamma < 1$ it can be made arbitrarily large for sufficiently large population size of both the prey (x_0) and the predators (y_0).

Main Results and Formulas of Chapter V

Pontryagin's maximum principle for a Mayer problem. Let $u_0(t)$ and $x_0(t)$ be the optimal control and optimal trajectory in the problem

$$\dot{x}(t) = f\big(t, x(t), u(t)\big), \qquad t \leqslant t_0 \leqslant T,\ x(t_0) = x_0$$
$$J(u) = \varphi\big(x(T)\big) \to \inf_u, \qquad u(t) \in U.$$

Then $u_0(t)$ satisfies the following condition for a maximum

$$\max_{u\in U} H\big(t, x_0(t), u, \psi(t)\big) = H\big(t, x_0(t), u_0(t), \psi(u)\big),$$

where

$$H = \psi'(t) f\big(t, x(t), u(t)\big),$$
$$\dot{\psi}(t) = -H_x\big(t, x_0(t), u_0(t), \psi(t)\big), \qquad \psi(T) = -\varphi_x\big(x_0(T)\big).$$

The maximum principle for a Boltz problem:

$$\max_{u\in U} H\big(t, x_0(t), u, \psi(t)\big) = H\big(t, x_0(t), u_0(t), \psi(t)\big),$$

where

$$H\big(t, x(t), u, \psi(t)\big) = \psi'(t) f\big(t, x(t), u\big) - F_0\big(t, x(t), u(t)\big).$$

The maximum principle for a problem with a variable terminal end:

$$\max_{u\in U} H\big(t, x_0(t), u, \psi(t)\big) = H\big(t, x_0(t), u_0(t), \psi(t)\big),$$

where

$$H\big(t, x(t), u, \psi(t)\big) = \sum_{i=1}^{n} \psi_i(t) f_i(t, x, u) + \psi_{n+1} F_0(t, x, u),$$
$$\psi_{n+1} = \text{constant} \leqslant 0, \qquad \dot{\psi}(t) = -H_x\big(t, x_0(t), u_0(t), \psi(t)\big).$$

For time-invariant problems

$$H\big(t, x_0(t), u_0(t), \psi(t)\big) \equiv 0.$$

The maximum principle for a discrete system. Suppose that the set

$$Q(x) = \bigcup_{u \in U} \left(F_0(x,u), f(x,u)\right)$$

is convex for any $x \in R^n$ and consider the system

$$x(j+1) = f\big(x(j), u(j)\big), \qquad j = 0, 1, \ldots, N-1, \; x \in R^n,$$
$$x(0) = x_0, \; u(j) \in U \subset R^m, \; x(N) = x_N,$$
$$J(u) = \sum_{j=0}^{N-1} F_0\big(x(j), u(j)\big) \to \inf_u .$$

Then the necessary conditions for an optimal control $u_0(j)$ and a corresponding optimal trajectory $x_0(j)$ are that there exist a nonzero vector $\big(\psi_{n+1}, \psi(j)\big)$, where the constant $\psi_{n+1} \leqslant 0$ and the vector $\psi(j)$ satisfies the equation

$$\Psi(j-1) = \frac{\partial H\big(x(j), u(j), \psi(j)\big)}{\partial x}, \qquad j = (N-1), \ldots, 0,$$

such that for any $j = 0, 1, \ldots, (N-1)$ the following relations are valid:

$$\max_{u \in U} H\big(x(j), u, \psi(j)\big) = H\big(x_0(j), u_0(j), \psi(j)\big),$$
$$H(x, u, \psi) = \psi_{n+1} F_0(x, u) + \psi' f(x, u).$$

Ecological Models

Malthus model:

$$\dot{N}(t) = \lambda N(t), \qquad N(0) = N_0 > 0.$$

Verhulst model:

$$\dot{N}(t) = \lambda N(t)\left(1 - \frac{N(t)}{K}\right), \qquad t \geqslant 0,$$
$$N(0) = N_0 > 0.$$

Hutchinson equation:

$$\dot{N}(t) = \lambda N(t)\left(1 - \frac{N(t-h)}{K}\right).$$

Cushing model:

$$\dot{N}(t) = \lambda N(t)\left[1 - \int_0^\infty N(t-s)\, dK(s)\right], \qquad t \geqslant 0.$$

Model allowing for a distribution of life-spans and migration (the McKendrick model):

$$\frac{\partial n(t,\tau)}{\partial t} + \frac{\partial n(t,\tau)}{\partial \tau} = -\mu(t,\tau)n(t,\tau) + g(t,\tau),$$
$$n(t,0) = u(t)\int_{\tau_1}^{\tau_2} K(t,\tau)n(t,\tau)\,d\tau.$$

Hutchinson equation with diffusion:

$$\frac{\partial D(t,x,y)}{\partial t} = \alpha\left(\frac{\partial^2 D(t,x,y)}{\partial x^2} + \frac{\partial^2 D(t,x,y)}{\partial y^2}\right)$$
$$+\lambda D(t,x,y)\left(1 - \frac{1}{K}D(t-h,x,y)\right).$$

Stochastic Verhulst equation:

$$dN(t) = \big(\lambda dt + \sigma d\xi(t)\big)N(t)\left(1 - \frac{N(t)}{K}\right).$$

Lotka–Volterra or "predator–prey" model:

$$\dot{x}(t) = a_1 x - a_2 xy, \qquad t \geqslant 0,\ x(0) = x_0 > 0,$$
$$\dot{y}(t) = a_2 a_3 xy - a_4 y, \qquad y(0) = y_0 > 0.$$

"Predator–prey" model with intraspecific competition:

$$\dot{x}(t) = a_1 x - a_2 xy - \gamma_1 x^2,$$
$$\dot{y}(t) = a_2 a_3 xy - a_4 y - \gamma_2 y^2.$$

Multispecies "predator–prey" model:

$$\dot{N}_i(t) = N_i(t)\left[\lambda_i - \sum_{j=1}^{n}\alpha_{ij}N_j(t) - \sum_{k=1}^{m}\beta_{ik}P_j(t)\right], \qquad i = 1,\dots,n,$$
$$\dot{P}_l(t) = P_l(t)\left[-\gamma_l + \sum_{j=1}^{n}\delta_{lj}N_j(t)\right], \qquad l = 1,\dots,m.$$

Monod model:

$$\dot{x}(t) = [a(y) - \lambda]x(t),$$
$$\dot{y}(t) = -\beta a(y)x - \lambda y + \lambda u(t).$$

Cost functional for optimal catching of a species:

$$\int_0^T [\alpha_1 u(t) N(t) - \alpha_2 \big(N(t) - N_0\big)^2] \, dt \to \sup_u .$$

Cost functional for control of population size:

$$\int_0^\infty d\tau \int_0^T [n(t,\tau) - n_0(\tau)]^2 h(t,\tau) \, dt \to \inf_u .$$

CHAPTER VI

LINEAR CONTROL SYSTEMS

For linear systems the problems of time optimality, controllability, observability, and the problem of constructing observers of different type are treated.

§ 1. A Time-Optimal Problem

1. Evaluation of the number of switch points. Consider the problem of transferring a linear system from an initial state $x(t_0) = x_0$ to a terminal state $x(T) = x_1$ in minimum time.

Let the equations of the system be of the form

$$\dot{x}(t) = Ax(t) + Bu(t), \qquad x \in R^n,\ u \in R^m, \qquad (1.1)$$
$$x(t_0) = x_0,\ x(T) = x_1,$$

where A and B are constant matrices.

Here $u = (u_1, u_2, \ldots, u_m)$ is a control vector subject to the constraints

$$|u_i(t)| \leqslant 1, \qquad i = 1, \ldots, m. \qquad (1.2)$$

The control $u(t)$ is chosen from the condition for a minimum transfer time $T - t_0$, which is equal to

$$\int_{t_0}^{T} dt \to \inf . \qquad (1.3)$$

Necessary conditions for the optimality of the control $u_0(t)$ for the problem (1.1)–(1.3) can be derived from the maximum principle stated in § 2 of Chapter V.

The Hamiltonian H for the problem (1.1)–(1.3) is of the form

$$H(x, u, \psi) = \psi'(Ax + Bu), \qquad \psi \in R^n. \qquad (1.4)$$

The maximum principle requires the existence of a nonzero solution of the adjoint system

$$\dot{\psi}(t) = -\frac{\partial H}{\partial x} = -A'\psi(t), \qquad (1.5)$$

such that the maximality condition

$$\max_{|u_i| \leqslant 1} \psi'(t)Bu = \psi'(t)Bu_0(t). \qquad (1.6)$$

is satisfied.

Let b_{ij} $(i = 1, \ldots, n;\ j = 1, \ldots, m)$ denote the elements of the matrix B, and b_j $(j = 1, \ldots, m)$ denote its columns.

Note that

$$\psi'(t)Bu = \sum_{k=1}^{m} u_k \sum_{j=1}^{n} \psi_j b_{jk}. \tag{1.7}$$

It is readily seen from equalities (1.6) and (1.7) that the component $u_{0,k}(t)$ of the optimal control $u_0(t)$ must be of the form

$$u_{0,k}(t) = \operatorname{sgn} \sum_{i=1}^{n} \psi_i(t) b_{ik} = \operatorname{sgn} \psi'(t) b_k, \qquad \psi'(t) b_k \neq 0. \tag{1.8}$$

Thus, the components of the optimal control $u_{0,k}$ assume only the values 1 and -1. The points at which the value of $u_{0,k}$ changes from -1 to 1 and vice versa are called the *control switch points.*

Let us evaluate the number of switch points for the optimal control $u_0(t)$. Henceforth, we assume that the following condition called the *condition of generic position* is satisfied: all matrices of the form

$$G_j = (b_j, Ab_j, \ldots, A^{n-1}b_j) \qquad (j = 1, \ldots, m)$$

are nondegenerate, i.e.,

$$\det G_j \neq 0 \qquad (j = 1, \ldots, m). \tag{1.9}$$

Theorem 1.1. *Let the condition of generic position* (1.9) *be satisfied. Then each component* $u_{0,k}(t)$ *of the optimal control* $u_0(t)$ *has a finite number of switch points and can assume only the values* 1 *and* -1.

Proof. By equality (1.8) the number of switch points for $u_{0,k}(t)$ is determined by the number of zeros of the function $\psi'(t)b_j$, where $\psi(t) \not\equiv 0$.

Suppose that the function $\psi'(t)b_j$ has an infinite number of zeros on the segment $[t_0, T]$. By Cauchy's formula the solution of equation (1.5) can be written as

$$\psi(t) = e^{-A'(t-t_0)}\psi(t_0), \qquad \psi'(t) = \psi'(t_0)e^{-A(t-t_0)}.$$

The function

$$\psi'(t)b_j = \psi'(t_0)e^{-A(t-t_0)}b_j \qquad (j = 1, \ldots, m)$$

is analytic for each j; by hypothesis it has infinitely many zeros on $[t_0, T]$. Hence, this function is identical to zero, i.e.,

$$\psi'(t)b_j \equiv 0 \qquad (j = 1. \ldots, m). \tag{1.10}$$

Differentiating the identity (1.10) $n - 1$ times with respect to t gives

$$\psi'(t_0)e^{-A(t-t_0)}G_j \equiv 0, \qquad (j = 1, \ldots, m).$$

By the condition of generic position (1.9) it follows that $\psi(t_0) = 0$, which gives

$$\psi(t) \equiv 0. \tag{1.11}$$

The condition (1.11) contradicts the maximum principle and thus the theorem is proved. □

Theorem 1.2. *Let the condition of generic position* (1.9) *be satisfied and, moreover, all eigenvalues of the matrix A be real-valued. Then the number of switch points of each component $u_{0,k}(t)$ of the optimal control $u_0(t)$ is at most $n-1$.*

Proof. Denote by $\lambda_1, \ldots, \lambda_k$ distinct real eigenvalues of the matrix A and by $q_1, \ldots, q_k$ their multiplicities. Then any component $\psi(t)$ of the solution of the system (1.5) can be represented in the form

$$\psi(t) = \gamma_1(t)e^{\lambda_1 t} + \cdots + \gamma_k(t)e^{\lambda_k t}, \tag{1.12}$$

where $\gamma_i(t)$ are polynomials whose degree is at most $q_i - 1$.

Let us evaluate the number of real zeros of the function (1.12). It can be shown by induction that the function

$$\gamma_1(t)e^{\lambda_1 t} + \cdots + \gamma_j e^{\lambda_j t}, \qquad j = 1, \ldots, k,$$

possesses at most $q_1 + \cdots + q_j - 1$ real zeros. For $j = 1$ the function $\gamma_1(t)\exp(\lambda_1 t)$ has at most $q_1 - 1$ zeros, since the degree of the polynomial $\gamma_1(t)$ is at most $q_1 - 1$. Suppose that this assertion holds for $j = l - 1$ and does not apply to $j = l$. Then the function $\varphi(t)$, which is equal to

$$\varphi(t) = \gamma_1(t)e^{\lambda_1 t} + \cdots + \gamma_l(t)e^{\lambda_l t},$$

has at least $q_1 + \cdots + q_l$ real zeros. Consequently, the function $\varphi_1(t)$ equal to

$$\varphi_1(t) = \varphi(t)e^{-\lambda_l t} = \gamma_1(t)e^{(\lambda_1 - \lambda_l)t} + \cdots + \gamma_{l-1}(t)e^{(\lambda_{l-1} - \lambda_l)t} + \gamma_l(t)$$

also has at least $q_1 + \cdots + q_l$ real zeros. But between any two zeros of the function $\varphi_1(t)$ there exists at least one zero of the derivative $\dot{\varphi}_1(t)$. Therefore, on differentiating the function $\varphi_1(t)$ q_l times, we conclude that the number of zeros of the function $d^{q_l}\varphi_1(t)/dt^{q_l}$ is at least $q_1 + \cdots + q_{l-1}$. However, $d^{q_l}\varphi_1(t)/dt^{q_l}$ is of the form

$$\bar{\gamma}_1(t)e^{\lambda_1 t} + \cdots + \bar{\gamma}_{l-1}(t)e^{\lambda_{l-1} t},$$

where $\bar{\gamma}_i(t)$ are polynomials of degree at most $q_i - 1$.

By the induction hypothesis the number of zeros of $d^{q_l}\varphi_1(t)/dt^{q_l}$ is at most $q_1 + \cdots + q_{l-1} - 1$. The contradiction obtained indicates that the function (1.12) has at most $n - 1$ real zeros. The functions $\psi'(t)b_j$ are also of the form (1.12); hence, the number of real zeros of each function $\psi'(t)b_j$ $(j = 1, \ldots, m)$ is at most $n - 1$. □

Note the following assertions (with no proof given) [15]. The existence of an admissible control implies the existence and uniqueness of optimal control in the problem (1.1)–(1.3). Here the term "admissible" denotes a control such that it transfers the system from state x_0 to state x_1 in finite time. Furthermore, if the matrix A is stable (i.e., all eigenvalues of the matrix A have negative real parts) and the condition of generic position is satisfied, then there exists a time-optimal control transferring the system from any initial state x_0 to the origin.

2. The damping of the material point. Consider the time-optimal problem of transferring a material point of unit mass that is in rectilinear motion with initial velocity $\dot{x}(0)$ from an arbitrary initial state $x(0)$ to the origin $x(T) = 0$ with the zero terminal velocity $\dot{x}(T) = 0$.

Denote by $x_1(t)$ the coordinate of the point, and by $x_2(t)$ its velocity. Then the equations of motion are of the form

$$\dot{x}(t) = x_2(t), \qquad \dot{x}_2(t) = u(t), \qquad t \geqslant 0, \tag{1.13}$$
$$x_1(0) = x(0), \; x_2(0) = \dot{x}(0),$$

where $u(t)$ is a control (acting force) constrained by

$$|u(t)| \leqslant 1. \tag{1.14}$$

The control $u(t)$ is to be chosen so that the system (1.13) is transferred to the origin in minimum time T, i.e., it is required that

$$x_1(T) = x_2(T) = 0, \qquad T \to \min. \tag{1.15}$$

By virtue of the equality (1.4) the Hamiltonian H is of the form

$$H(x, u, \psi) = \psi_1 x_2 + \psi_2 u.$$

Equations (1.5) now become

$$\dot{\psi}_1 = 0, \qquad \dot{\psi}_2 = -\psi_1.$$

Hence,

$$\psi_1 = C_1, \qquad \psi_2 = -C_1 t + C_2, \qquad C_1, C_2 = \text{constant}.$$

According to (1.8), the optimal control $u_0(t)$ in the problem (1.13)–(1.15) can be written in the form

$$u_0(t) = \operatorname{sgn} \psi_2(t) = \operatorname{sgn}(-C_1 t + C_2).$$

Since the function $-C_1 t + C_2$ is linear, it follows that the optimal control $u_0(t)$ has only one switch point and two constancy intervals, on which u_0 takes on the values -1 or 1. This also follows from Theorem 1.2, since the matrix

$$A = \begin{pmatrix} 0 & 1 \\ 0 & 0 \end{pmatrix}$$

has two real eigenvalues $\lambda_1 = 0$ and $\lambda_2 = 0$.

If on some interval $u_0(t) = 1$, then in this time interval the system (1.13) is in motion along the parabolas

$$x_1(t) = C_3 + \frac{x_2^2(t)}{2}.$$

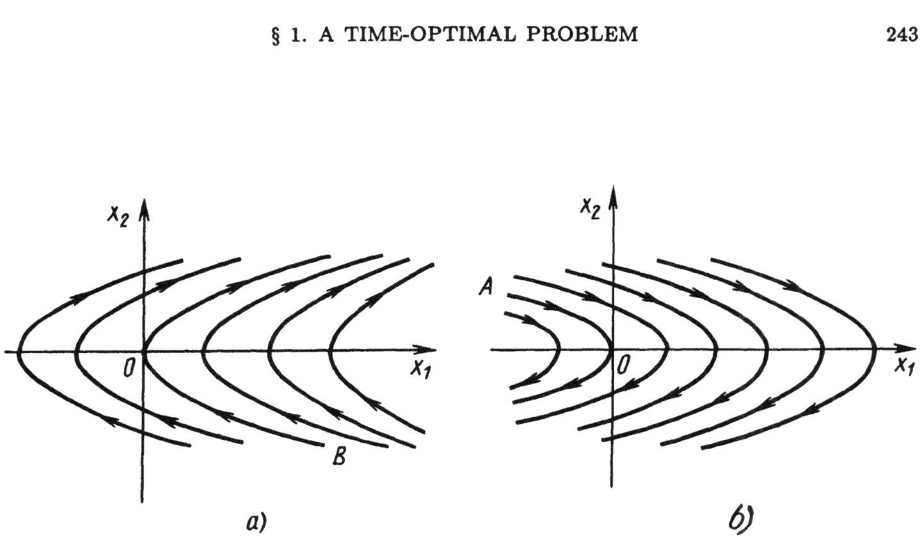

FIG. 1.1. Phase trajectories of the material point:
a, for $u = 1$; *b*, for $u = -1$

The motion is upwards, since $\dot{x}_2(t) = 1 > 0$ (see Figure 1.1, *a*).

But if $u_0(t) = -1$, the motion is in the downward direction along the parabolas

$$x_1(t) = C_4 - \frac{x_2^2(t)}{2}$$

(see Figure 1.1, *b*). The constants C_3, C_4 are determined by the initial conditions for the system (1.13).

Let us now determine the optimal trajectories. We can reach the origin by moving only along the curves AO and OB (see Figure 1.2). By Theorem 1.2 the optimal control may only have one switch point. If the initial state $(x_1(0), x_2(0))$ of the system (1.13) is above the curve AOB, then $u_0(t) = -1$ until the moment τ when the phase point first enters the arc BO. At the moment τ the control $u_0(t)$ switches to $u_0(t) = 1$, $t \geqslant \tau$, and maintains this value until the moment the phase point arrives at the point $(0,0)$. Here the trajectory coincides with the arc BO. But if the initial point $(x_1(0), x_2(0))$ is below the curve AOB, then the control $u_0(t) = 1$ until the moment τ_1 when the phase point first enters the arc AO. For $t \geqslant \tau_1$ the control becomes $u_0(t) = -1$ and this value is maintained along the arc AO until the moment the phase point arrives at the point $(0,0)$. If there is at most one switch point, there is no other way of reaching the origin. Therefore, the curve AOB is a switching curve. The optimal trajectories are shown in Figure 1.2.

The family of trajectories shown in Figure 1.2 are obtained by means of the necessary conditions for optimal control; only these trajectories can be optimal. To prove the optimality of the trajectories, it is still necessary to prove the existence of solutions of the original optimal control problem. For the proof of the optimality of these trajectories, see, for example, the book [15], where the formulas for the minimum transfer time are also given.

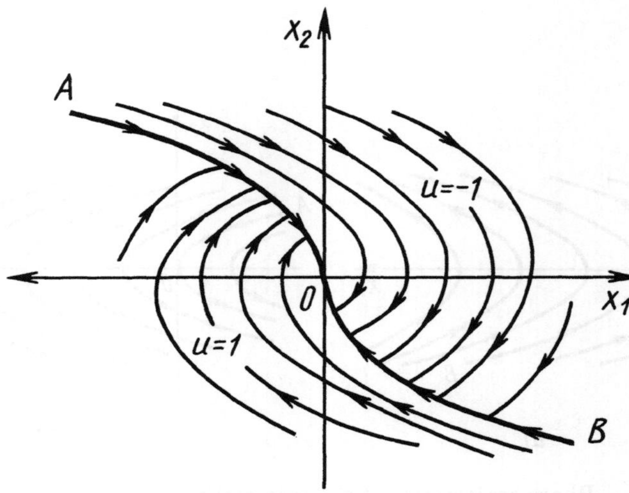

FIG. 1.2. The switching curve AOB and optimal trajectories in the problem of the optimal damping of a material point

Denote by $V\big(x_1(0), x_2(0)\big)$ the minimal time of the transfer of the system from the state $\big(x_1(0), x_2(0)\big)$ at time $t = 0$ to the origin $(0,0)$. Then if the point $\big(x_1(0), x_2(0)\big)$ is below the curve AOB, then

$$V\big(x_1(0), x_2(0)\big) = x_2(0) + [x_1(0) + x_2^2(0)/2]^{1/2};$$

but if the point $\big(x_1(0), x_2(0)\big)$ is above the curve AOB, then we have

$$V\big(x_1(0), x_2(0)\big) = -x_2(0) + 2[-x_1(0) + x_2^2(0)/2]^{1/2}.$$

3. The damping of a pendulum. The motion of a pendulum in reduced variables in the neighborhood of the lower stable equilibrium state is described by the linearized equation

$$\ddot{x}(t) + \pi^2 x = \pi^2 u, \qquad |u| \leqslant 1,$$

which is equivalent to the system

$$\dot{x}_1 = \pi x_2, \qquad \dot{x}_2 = -\pi x_1 + \pi u, \qquad t \geqslant 0,\ x_1(0) = x_0,\ x_2(0) = \dot{x}_0. \tag{1.16}$$

It is required to deduce a control $u(t)$ that transfers the system from an initial state $(x_0, \dot{x}_0)$ to a terminal state $x_1(T) = x_2(T) = 0$ in minimum time.

The matrix of the system (1.16) can be written as

$$A = \begin{pmatrix} 0 & \pi \\ -\pi & 0 \end{pmatrix}.$$

It has two imaginary eigenvalues $\lambda_1 = -\pi i$, $\lambda_2 = \pi i$. We can apply Theorem 1.1 to the system (1.16) but not Theorem 1.2.

Let us determine an optimal control $u_0(t)$ for the system (1.16). The Hamiltonian H and the equations for the adjoint variables are of the form

$$H = \psi_1 \pi x_2 - \psi_2 \pi x_1 + \psi_2 \pi u, \qquad \dot{\psi}_1 = \pi \psi_2, \qquad \dot{\psi}_2 = -\pi \psi_1.$$

Hence, $\psi_2(t) = C_1 \sin \pi(t-\alpha)$, and, moreover,

$$u_0(t) = \operatorname{sgn} \psi_2(t) = \operatorname{sgn}[C_1 \sin \pi(t-\alpha)]. \tag{1.17}$$

Therefore, $u_0(t)$ takes on the values 1 or -1, and switching occurs every second.

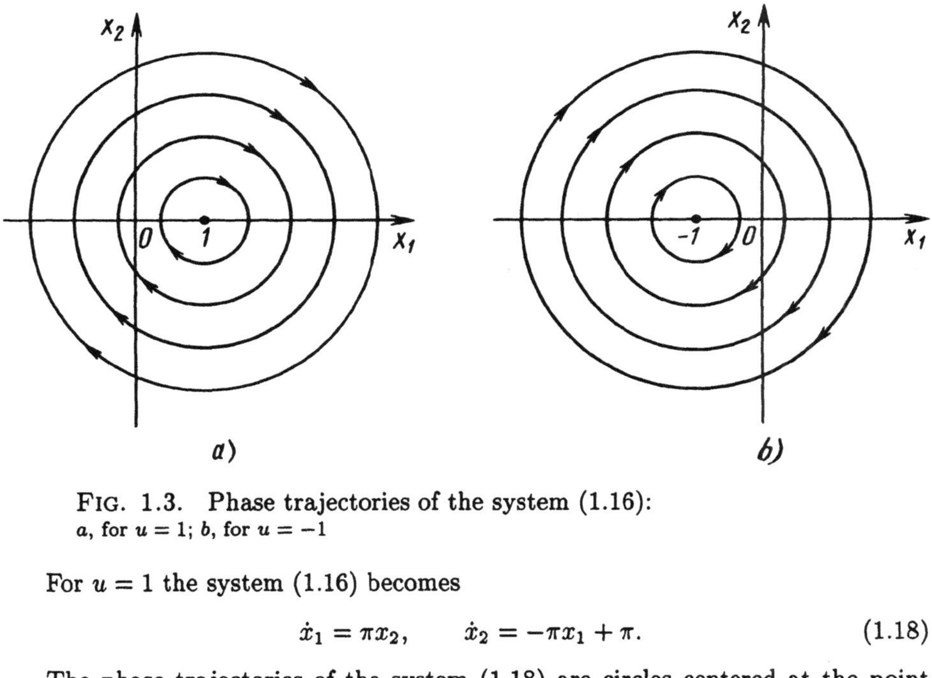

FIG. 1.3. Phase trajectories of the system (1.16): a, for $u = 1$; b, for $u = -1$

For $u = 1$ the system (1.16) becomes

$$\dot{x}_1 = \pi x_2, \qquad \dot{x}_2 = -\pi x_1 + \pi. \tag{1.18}$$

The phase trajectories of the system (1.18) are circles centered at the point $(1, 0)$ having the equations

$$(x_1 - 1)^2 + x_2^2 = C_2^2. \tag{1.19}$$

The phase point is moving clockwise along any of the circles (1.19) traversing the whole circle in 2 sec (see Figure 1.3, a). In a similar way, for $u = -1$ the phase trajectories are circles centered at the point $(-1, 0)$ having the equations

$$(x_1 - 1)^2 + x_2^2 = C_3^2;$$

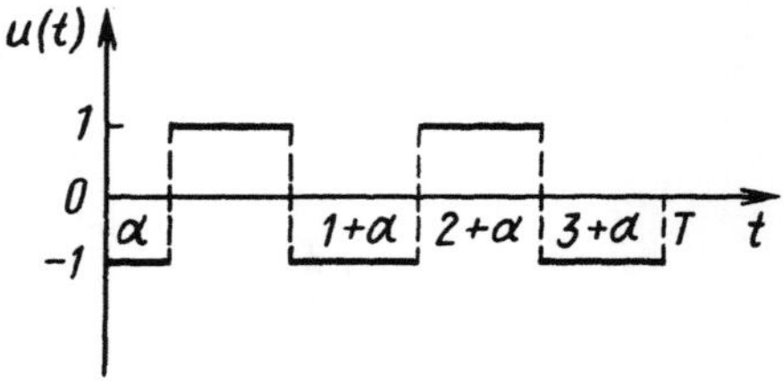

FIG. 1.4. Optimal control for the system (1.16) for $\alpha > 0$

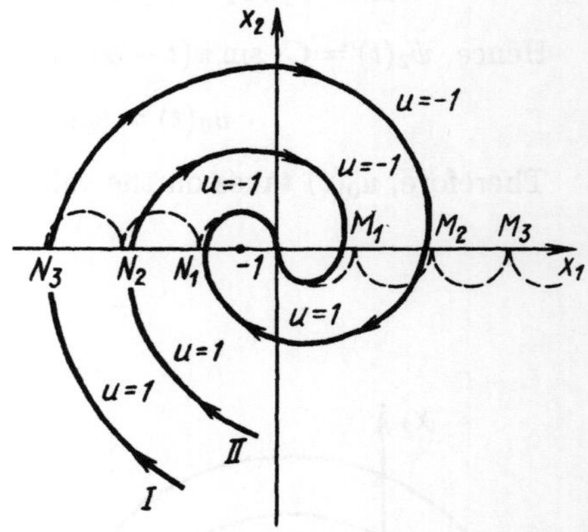

FIG. 1.5. Optimal trajectories and the switching curve $N_3N_2N_1OM_1M_2M_3$ for the system (1.16)

any of these circles are traversed clockwise by the phase point in 2 sec (see Figure 1.3, *b*).

According to formula (1.17), an optimal control is constructed as shown in Figure 1.4. A corresponding optimal trajectory can be obtained in the following way. On a finite interval $[3+\alpha, T]$ whose length is less than 1 the point moves along the portion N_1O of the circle, since it is the only way for the phase point to arrive at the origin for the control $u = -1$.

On the segment $[2+\alpha, 3+\alpha]$ the control is $u = 1$, and the phase point executes a half revolution until it intersects the circle M_1M_2. Further, it makes another half revolution until it intersects the circle N_3N_2, etc. Finally, moving from $t = 0$ to $t = T$, the point reaches the origin (the trajectory I in Figure 1.5). Clearly, each optimal trajectory has a specific finite number of switch points; this number depends on the initial data $(x_0, \dot{x}_0)$. The larger the value $x_0^2 + \dot{x}_0^2$, the larger the number of switches before the phase point has reached the origin. In Figure 1.5 the trajectory I terminates after three switches, and the trajectory II after two.

4. Control of the rotation of an axially symmetric space vehicle [3]. The motion of a space vehicle with three pairs of steering jet engines is described by Euler's equations

$$A\dot{x}_1 = (B - C)x_2x_3 + M_1(t), \qquad B\dot{x}_2 = (C - A)x_1x_3 + M_2(t),$$
$$C\dot{x}_3 = (A - B)x_1x_2 + M_3(t), \tag{1.20}$$

where A, B, C are the principal moments of inertia of the space vehicle, $M_i(t)$ are the torques produced by the jet engines, $x_i(t)$ are angular rotational velocities.

There exist many space vehicles having one symmetry axis, such as "Soyuz", "Mercury", "Gemini", "Apollo".[1] For a space vehicle with one symmetry axis two moments of inertia, say, A and B, coincide: $A = B$. Then equations (1.20) take the form

$$\begin{aligned} \dot{x}_1 &= \frac{A-C}{A} x_2 x_3 + \frac{M_1(t)}{A}, \\ \dot{x}_2 &= \frac{C-A}{A} x_1 x_3 + \frac{M_2(t)}{A}, \\ \dot{x}_3 &= \frac{M_3(t)}{C}. \end{aligned} \tag{1.21}$$

These equations imply that the angular velocity $x_3(t)$ can be independently controlled, but due to the gyroscopic effect its changes affect the velocities $x_1(t)$ and $x_2(t)$.

Consider the following problem. For the space vehicle to be stable and to create an artificial gravity, we must maintain a prescribed value $x_3(t) = \xi_3$ of the angular velocity of rotation about its symmetry axis, while two other rotations must be decreased to zero in minimum time, i.e., $x_1(T) = 0$, $x_2(T) = 0$. Setting

$$\omega = \frac{A-C}{A}\xi_3, \qquad u_1(t) = \frac{M_1(t)}{A}, \qquad u_2(t) = \frac{M_2(t)}{A},$$

we write the first two equations (1.21) in the form

$$\begin{aligned} \dot{x}_1(t) &= \omega x_2(t) + u_1(t), \qquad & |u_1(t)| \leqslant 1, \\ \dot{x}_2(t) &= -\omega x_1(t) + u_2(t), \qquad & |u_2(t)| \leqslant 1. \end{aligned} \tag{1.22}$$

The time-optimal problem for the system (1.22) is as follows: determine controls $u_1(t)$ and $u_2(t)$ such that the system (1.22) is transferred from an arbitrary initial state $x_1(0) = \xi_1$, $x_2(0) = \xi_2$ to the terminal state $x_1(T) = 0$, $x_2(T) = 0$. The solution of this problem is in many ways similar to that of the problem of damping the pendulum, but here we have two controls instead of one as in the previous case.

The Hamiltonian for the time-optimal problem for the system (1.22) is of the form

$$H = \psi_1(\omega x_2 + u_1) + \psi_2(-\omega x_1 + u_2).$$

The adjoint variables ψ_1 and ψ_2 satisfy the equations $\dot{\psi}_1 = \omega\psi_2$, $\dot{\psi}_2 = -\omega\psi_1$ and are of the form

$$\psi_1(t) = C_1 \cos\omega t + C_2 \sin\omega t, \qquad \psi_2(t) = -C_1 \sin\omega t + C_2 \cos\omega t.$$

---

[1] Rauschenbach, B. V., and Tokar', E. N., *Orientation control for space vehicles*, Nauka, Moscow, (1974).

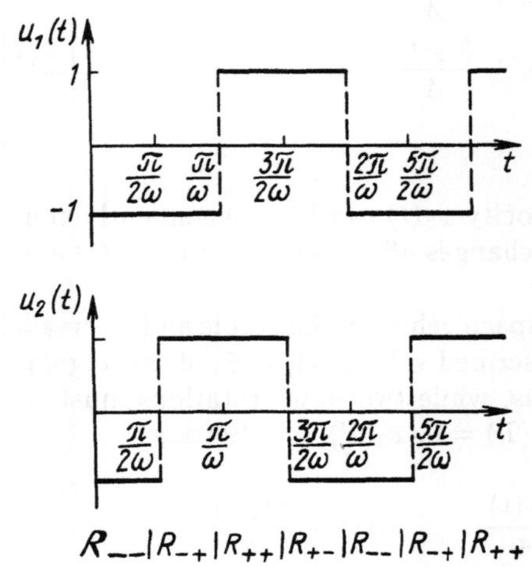

FIG. 1.6. Optimal controls u_1 and u_2 in the problem of the control of the rotation of the space vehicle

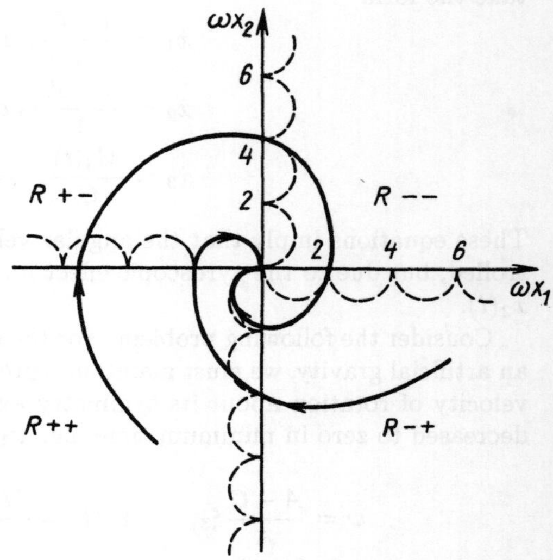

FIG. 1.7. Optimal trajectories (solid lines) and switching curves (dotted lines) in the problem of the control of the rotation of the space vehicle

It follows from the condition for a maximum that the optimal controls $u_{1,0}(t)$ and $u_{2,0}(t)$ can be written as

$$u_{1,0}(t) = \operatorname{sgn} \psi_1 = \operatorname{sgn}[C_1 \cos \omega t + C_2 \sin \omega t],$$
$$u_{2,0}(t) = \operatorname{sgn} \psi_2 = \operatorname{sgn}[-C_1 \sin \omega t + C_2 \cos \omega t].$$

The optimal controls $u_{1,0}(t)$ and $u_{2,0}(t)$ are illustrated in Figure 1.6. As shown in the figure, there exist four regions in the plane x_1, x_2, where the controls are constant. We denote these regions as follows:

$$R_{--} = \{u_{1,0} = -1, \quad u_{2,0} = -1\},$$
$$R_{+-} = \{u_{1,0} = 1, \quad u_{2,0} = -1\},$$
$$R_{++} = \{u_{1,0} = 1, \quad u_{2,0} = 1\},$$
$$R_{-+} = \{u_{1,0} = -1, \quad u_{2,0} = 1\}.$$

Then, the same argument as in the preceding subsection indicates that the dotted curves shown in Figure 1.7 are the switching curves between the regions R_{--}, R_{+-}, etc. The first quadrant corresponds to the region R_{--}, the second one to the region R_{+-}, the third one to the region R_{++}, and the fourth one to the region R_{-+}.

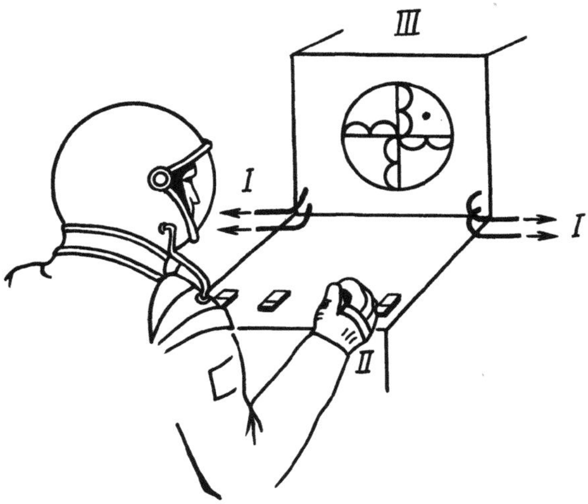

FIG. 1.8. Manual control of the rotation of the space vehicle: I, inputs for x_1- and x_2-deviations; II, manual controls; III, display unit

Furthermore, in Figure 1.7 the optimal trajectories composed of arcs of circles are depicted, and the transfer times for the phase point are indicated.

Figure 1.8 shows a manual control system governing the rotation of a space vehicle. It was shown experimentally in laboratory conditions that after some training the astronaut can control the damping of the space vehicle optimally, provided the natural frequency of the space vehicle satisfies the condition $\omega < 3\,\mathrm{rad\,sec}^{-1}$ [3].

5. The controllability set. The time-optimal problem (1.1)–(1.3) can be solved, in general, not for all vectors x_0 and x_1. Suppose that for some x_1 the transfer of the system (1.1) to the terminal state x_1 in finite time is not possible for all initial states x_0. Then, we are up against the problem of describing the *controllability set* $M(x_1)$ composed of initial states such that this transfer is feasible. It is considered that $x_0 \in M(x_1)$ if there exists a control $u(t)$ subject to the constraint (1.2) and a terminal moment T such that $x(T) = x_1$, where $x(t)$ is the solution of equation (1.1) generated by a control function $u(t)$ for a given initial condition $x(t_0) = x_0$. Moreover, an optimal control in the problem (1.1)–(1.3) may exist only for $x_0 \in M(x_1)$.

For an illustration of the above statement we now turn to the problem of damping the pendulum in a neighborhood of the upper equilibrium state. The equations of motion in dimensionless variables, linearized in the neighborhood of this state, are of the form

$$\dot{x}_1(t) = x_2, \qquad \dot{x}_2(t) = x_1 + u, \qquad |u| \leqslant 1,\ t \geqslant 0. \tag{1.23}$$

Note that equation (1.23) differs from equation (1.15) by the sign of x_1. It is required to deduce a control u such that it steers the system (1.23) from a prescribed initial state to the origin in minimum time. In other words, an optimal control is deduced from the following conditions:

$$x_1(T) = 0, \qquad x_2(T) = 0, \qquad T \to \min .$$

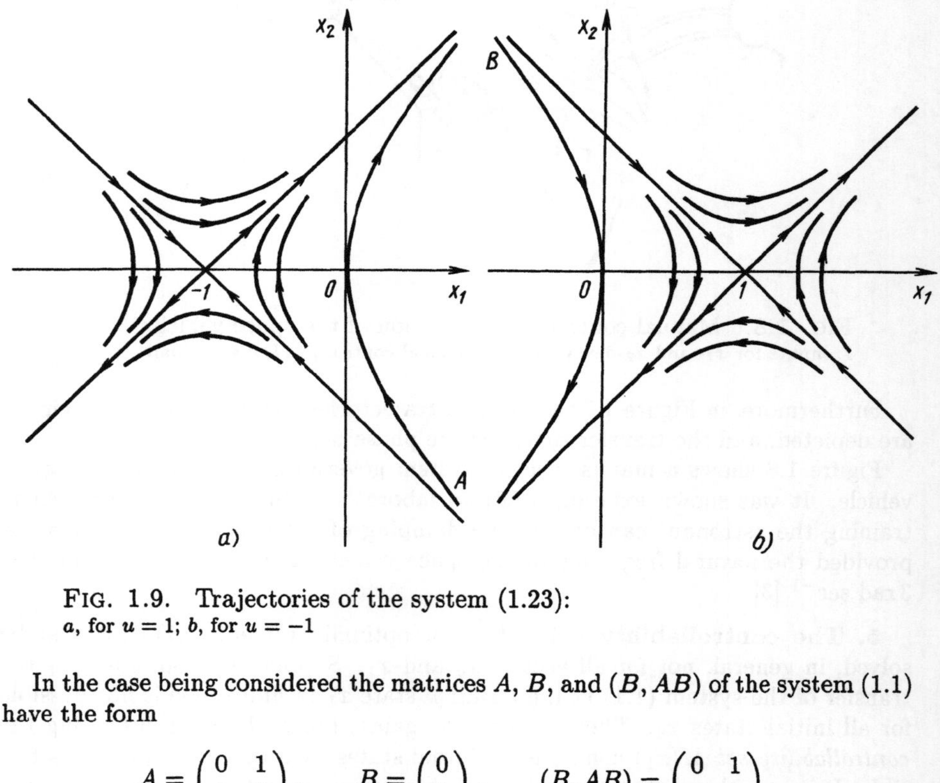

FIG. 1.9. Trajectories of the system (1.23):
a, for $u = 1$; *b*, for $u = -1$

In the case being considered the matrices A, B, and (B, AB) of the system (1.1) have the form

$$A = \begin{pmatrix} 0 & 1 \\ 1 & 0 \end{pmatrix}, \qquad B = \begin{pmatrix} 0 \\ 1 \end{pmatrix}, \qquad (B, AB) = \begin{pmatrix} 0 & 1 \\ 1 & 0 \end{pmatrix}.$$

The eigenvalues of the matrix A are real: $\lambda_1 = 1$, $\lambda_2 = -1$. In addition, the condition of generic position is satisfied, since the rank of the matrix (B, AB) equals two. Therefore, by Theorem 1.2 the optimal control is piecewise constant; it has at most two constancy intervals and on each one is equal to either 1 or -1. Besides, the properties of an optimal control can also be obtained directly from the maximum principle, whereby an optimal control is determined from the condition for a maximum of the function $H = \psi_1 x_2 + \psi_2(x_1 + u)$ with respect to u, i.e., $u = \operatorname{sgn}\psi_2$. Here the adjoint variables ψ_1 and ψ_2 satisfying the equations $\dot\psi_1 = -\psi_2$, $\dot\psi_2 = -\psi_1$ are of the form

$$\psi_1 = -C_1 e^t + C_2 e^{-t}, \qquad \psi_2 = C_1 e^t + C_2 e^{-t}, \qquad |C_1| + |C_2| \neq 0,$$

where C_1, C_2 are constants. Hence, the function $\psi_2(t)$ can have at most one zero, and, consequetly, the optimal control can have at most one switch point.

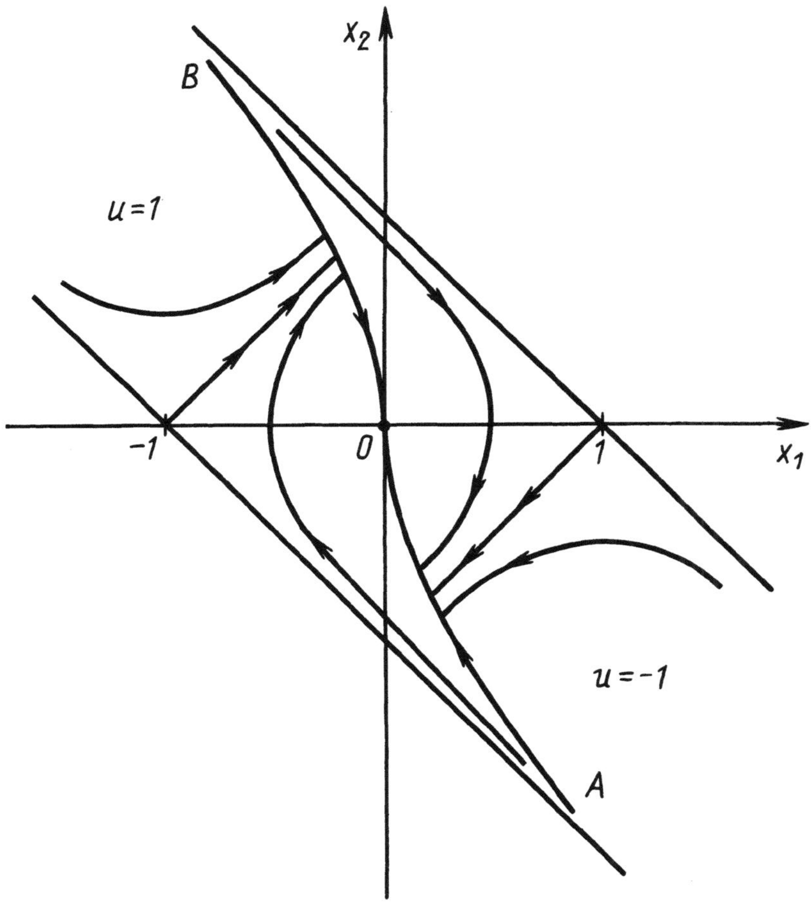

FIG. 1.10. The set of reachability and the switching curve OAB for the system (1.23)

With the aim to determine the switching curve for optimal control, we turn to constructing the field of phase trajectories for the system (1.23) for $u = 1$ and for $u = -1$. If $u = 1$, then the family of phase trajectories of equation (1.23) is of the form $(x_1 + 1)^2 - x_2^2 = C$, where C is an arbitrary constant. The family in question consists of hyperbolas (for $C \neq 0$) and two straight lines $x_2 = x_1 + 1$ and $x_2 = -x_1 - 1$ (for $C = 0$), which are, in fact, asymptotes of these hyperbolas (see Figure 1.9, *a*). In the figure the arrows indicate the direction of motion of the phase point, which represents the solution of the system (1.23) as the time t increases. The critical point $(-1, 0)$ of the system (1.23) for $u = 1$ is a saddle. Analogously, for $u = -1$ the family of phase trajectories is described by the equation$(x_1 - 1)^2 - x_2^2 = C$ and is shown in Figure 1.9, *b*. The critical point $(1, 0)$ in this case is also a saddle. Since the optimal control has at most one switch point and the last portion of the optimal trajectory must terminate at the origin, it follows from the comparison of the optimal trajectories in Figures 1.9, *a* and *b*, that the switching curve is composed of arcs of the hyperbolas BO (for $u = -1$) and of OA (for $u = 1$), which pass through the origin.

Finally, we obtain that the controllability set is the strip between the straight lines $x_2 = x_1 - 1$ and $x_2 = x_1 + 1$ that does not contain these lines. A view of the optimal trajectories and optimal control is given in Figure 1.10. The optimal control is equal to -1 above the curve AOB and on the arc BO and is equal to 1 under the curve AOB and on the arc OA.

§ 2. Controllability of Linear Systems

1. Controllability of linear time-invariant systems. Consider a linear control system

$$\dot{x}(t) = Ax(t) + Bu(t), \qquad t \geqslant t_0,\ x \in R^n,\ u \in R^m, \tag{2.1}$$

where A and B are $n \times n$ and $n \times m$ constant matrices, respectively. There are no control constraints. The system (2.1) is said to be *controllable* on the segment $[t_0, t_1]$, $t_1 > t_0$, if for any vectors $x_0, x_1 \in R^n$ there exists a solution of equation (2.1) satisfying the conditions $x(t_0) = x_0$, $x(t_1) = x_1$. In other words, a controllable system can be transferred from an arbitrary state x_0 at the moment t_0 to the state x_1 at the moment t_1, provided a suitable control is chosen. The conditions for the controllability of the system (2.1) depend on the rank of the *controllability matrix*, which is of the form

$$K = (B, AB, \ldots, A^{n-1}B). \tag{2.2}$$

The pair of matrices (A, B) is said to be *controllable* if the rank of the matrix (2.2) is equal to n.

Theorem 2.1. *For the system* (2.1) *to be controllable it is necessary and sufficient that the rank of the controllability matrix K be equal to n (the dimension of the vector x).*

Proof. The necessity. Suppose that the system (2.1) is controllable, but the rank of the matrix K is less than n. Then there exists a nonzero vector $y \in R^n$ such that it is orthogonal to all columns of the matrix K, i.e., $y'K = 0$. Hence,

$$y'B = 0, \qquad y'AB = 0, \ldots, y'A^{n-1}B = 0. \tag{2.3}$$

Now we write the characteristic equation of the matrix A:

$$\det(A - \lambda I) = (-1)^n(\lambda^n + \alpha_1\lambda^{n-1} + \cdots + \alpha_n) = 0,$$

where α_i are some numbers. By the Hamilton-Cayley theorem the matrix A satisfies its own characteristic equation. Therefore,

$$A^n = -\alpha_1 A^{n-1} - \cdots - \alpha_n I. \tag{2.4}$$

Multiplying both sides of the equality (2.4) by B on the right and by y' on the left and noting formulas (2.3) gives

$$y'A^nB = -\alpha_1 y'A^{n-1}B - \cdots - \alpha_n y'B = 0.$$

For $j \geqslant n+1$ we can show in a similar way that $y'A^jB = 0$. Thus, we finally obtain for all $j \geqslant 0$

$$y'A^jB = 0 \qquad (j = 0, 1, \ldots). \tag{2.5}$$

According to Cauchy's formula the solution of equation (2.1) with zero initial condition $x(t_0) = 0$ may be represented by

$$x(t) = \int_{t_0}^{t} e^{A(t-s)}Bu(s)\,ds. \tag{2.6}$$

Expanding $\exp\big(A(t-s)\big)$ in series and recalling the equality (2.5), we obtain for all controls $u(t)$ the relation

$$y'x(t) = 0, \qquad t_0 \leqslant t \leqslant t_1. \tag{2.7}$$

Therefore, for all controls the trajectories $x(t)$ of the system (2.1) lie in the subspace of R^n that is orthogonal to the nonzero vector y. It means that not all points can be attained by the system (2.1), and this contradicts the controllability hypothesis for the system (2.1).

The sufficiency. Let the rank of the controllability matrix K be equal to n. Then it can be shown that the system (2.1) is controllable. First, we show that the set of reachability Q_1 for the moment $t_1 > t_0$ with the initial condition $x(t_0) = 0$ and with the control constraint $|u_i(t)| \leqslant 1$ $(i = 1, \ldots, m)$ has dimension n. As was shown in § 2 of Chapter V, the set Q_1 is compact and convex. Assume now that the dimension of Q_1 is less than n. Then there exists a nonzero vector $y \in R^n$ such that $y'x(t_1) = 0$ for all $x(t_1) \in Q_1$. In other words,

$$y' \int_{t_0}^{t_1} e^{A(t_1-s)}Bu(s)\,ds = 0. \tag{2.8}$$

Thus, noting that the control $u(t)$ is arbitrary, we can write

$$y'e^{A(t_1-s)}B = 0, \qquad t_0 \leqslant s \leqslant t_1. \tag{2.9}$$

Differentiating successively this expression with respect to s and setting $s = t_1$, we obtain

$$y'B = 0, \qquad y'AB = 0, \ldots, y'A^{n-1}B = 0. \tag{2.10}$$

However, the relations (2.10) contradict the assumption about the rank of the matrix K being equal to n. Further, the set Q_1 is symmetric with respect to the origin, since the controls $u(t)$ and $-u(t)$ are both admissible. Hence, Q_1 contains a ball centered at the origin.

Denote by Q_r the set of reachability for the system (2.1) for the moment $t_1 > t_0$ with the initial condition $x(t_0) = 0$ and with the control constraint $|u_i(t)| \leqslant r$. It is readily seen that $Q_r = rQ_1$. As $r \to \infty$ the set Q_r contains any ball from R^n. Hence, the set of reachability for the system (2.1) coincides with the whole space R^n. □

Example 2.1. Consider a controllable rectilinear motion of a material point subjected to a scalar control force $u(t)$. The equations of motion are of the form $\ddot{x}(t) = u(t)$. Let $x_1(t)$ denote the coordinate of the material point and $x_2(t)$ its velocity. Let us write the equations of motion in the form

$$\dot{x}_1(t) = x_2(t), \qquad \dot{x}_2(t) = u(t). \tag{2.11}$$

The equations (2.11) can be represented in the form (2.1), provided that

$$A = \begin{pmatrix} 0 & 1 \\ 0 & 0 \end{pmatrix}, \qquad B = \begin{pmatrix} 0 \\ 1 \end{pmatrix}.$$

By virtue of the equality (2.2) the matrix K has the form

$$K = \begin{pmatrix} 0 & 1 \\ 1 & 0 \end{pmatrix}.$$

Hence, the rank of the matrix K is two, i.e., the motion of the material point is controllable.

Remarks. 1. Let us assume additionally that the rank of the matrix B is equal to k. Then for the system (2.1) to be controllable it is necessary and sufficient that the rank of the matrix $(B, AB, \ldots, A^{n-k}B)$ be equal to n.

2. The controllability condition in Theorem 2.1 is, in general, weaker than the condition (1.9) of generic position. For a scalar control these two conditions are equivalent.

3. The criterion for controllability embodied in Theorem 2.1 was established under the assumption that there is no control constraint. Analogs of the criterion for controllability are also of interest in cases where there are control constraints. Below we consider one such case in which the control $u(t)$ is subject to the constraint $|u(t)| \leqslant C$, where $C > 0$ is a given constant. It should be noted that in this case the conditions of Theorem 2.1 are not sufficient for the system (2.1) to be controllable. Indeed, if all eigenvalues of the matrix A lie in the left half-plane, then for any initial state x_0 and any control the system (2.1) remains inside

some bounded region. Conversely, if all eigenvalues of the matrix A lie in the right half-plane, then for any control the origin x_0 cannot be reached, provided that the initial state x_0 is sufficiently removed from it. In both these cases the system (2.1) is uncontrollable.

Consider a criterion for controllability when there is a control constraint.[2]

Proposition. *For the system* (2.1) *with control constraint* $|u(t)| \leqslant C$, $C > 0$, *to be controllable, it is necessary and sufficient that the rank of the matrix* (2.2) *be equal to* n, *and, moreover, that all eigenvalues of the matrix* A *lie on the imaginary axis.*

2. Controllability of linear time-varying systems. We are concerned with establishing the sufficient conditions for the controllability of linear time-varying systems of the form

$$\dot{x}(t) = A(t)x(t) + B(t)u(t), \qquad x \in R^n,\ u \in R^m \tag{2.12}$$

The matrices $A(t)$ and $B(t)$ are assumed to be continuously differentiable up to order $n - 1$ inclusive in the neighborhood of some point $\tau \in [t_0, t_1]$. Let the matrices $K_i(t)$ be defined by the following relations:

$$K_1(t) = B(t), \qquad K_i(t) = A(t)K_{i-1}(t) - \frac{dK_{i-1}(t)}{dt}, \qquad i = 2, \ldots, n.$$

Theorem 2.2 [11]. *If there exists a point* $\tau \in [t_0, t_1]$ *at which the rank of the matrix*

$$K = (K_1, \ldots, K_n) \tag{2.13}$$

is equal to n, *then the system* (2.12) *is controllable on the interval* $[t_0, t_1]$.

Proof. The proof of this theorem is identical with the proof of the sufficiency in Theorem 2.1 except for formulas (2.8)–(2.10) that must be modified as follows. According to Cauchy's formula, instead of (2.8) we can write the following equality

$$\int_{t_0}^{t_1} y' Z(t, s)B(s)u(s)\, ds = 0,$$

where $Z(t, s)$ is the fundamental matrix of solutions for the homogeneous equation $\dot{x}(t) = A(t)x(t)$. In this case the relation (2.9) takes the form

$$y' Z(t_1, s)B(s) \equiv 0, \qquad t_0 \leqslant s \leqslant t_1.$$

Differentiating successively this identity with respect to s and setting $s = \tau$, we obtain the sequence $y_1' K_1(\tau) = 0, \ldots, y_1' K_n(\tau) = 0, y_1 = Z'(t_1, s)y$ that contradicts the assumption about the matrix (2.13) being of full rank. Using the

[2]Ovseyevich, A. I., On the total controllability of linear systems, *Prikl. Mat. Mekh.*, (1989), vol. 53, no. 5, 845–848. (Translated into English as *J. Appl. Math. Mech.*)

similar argument as in the proof of Theorem 2.1, we complete the proof of Theorem 2.2. □

We now need to focus upon the necessary and sufficient conditions for the controllability of the system (2.12) on the segment $[t_0, t_1]$ assuming that the matrices $A(t)$ and $B(t)$ are only continuous on this segment. These conditions are stated in terms of the matrix $W(t_0, t)$ that is equal to

$$W(t_0, t) = \int_{t_0}^{t} Z(t_0, s)B(s)B'(s)Z'(t_0, s)\, ds,$$

where the Cauchy matrix $Z(t, s)$ is defined by the relation (1.5) of Chapter IV. It follows from the definition of the matrix $W(t_0, t)$ that it is always positive semidefinite for $t_0 \leqslant t$.

Theorem 2.3. *For the system* (2.12) *to be controllable on the segment* $[t_0, t_1]$, *it is necessary and sufficient that the matrix* $W(t_0, t_1)$ *be positive definite.*

Proof. The sufficiency. Let x_0 and x_1 be two arbitrary points in R^n. We try to determine the control $u(t)$ that effects the transfer of the system (2.12) from the state $x_0 = x(t_0)$ to the state $x_1 = x(t_1)$. By Cauchy's formula (1.7) of Chapter IV we have

$$x(t) = Z(t, t_0)x_0 + \int_{t_0}^{t} Z(t, s)B(s)u(s)\, ds.$$

Since by hypothesis the matrix $W^{-1}(t_0, t_1)$ exists, we set

$$u(t) = -B'(t)z'(t_0, t)W^{-1}(t_0, t_1)[x_0 - Z(t_0, t_1)x_1].$$

We can show that this control induces the transfer of the system (2.12) from the state $x_0 = x(t_0)$ to the state $x_1 = x(t_1)$. Taking account of formulas (1.8) of Chapter IV, we obtain

$$\begin{aligned} x(t_1) = Z(t_1, t_0)x_0 - \int_{t_0}^{t_1} & Z(t_1, s)B(s)B'(s)Z'(t_0, s)\, ds \\ & \times W^{-1}(t_0, t_1)[x_0 - Z(t_0, t_1)x_1] = Z(t_1, t_0)x_0 - Z(t_1, t_0) \\ & \times \int_{t_0}^{t_1} Z(t_0, s)B(s)B'(s)Z'(t_0, s)\, ds\, W^{-1}(t_0, t_1) \\ & \times [x_0 - Z(t_0, t_1)x_1] = x_1. \end{aligned}$$

The necessity. By the definition of the matrix $W(t_0, t_1)$ it is positive semidefinite for $t_1 > t_0$. Therefore, to establish that it is, in fact, positive definite, we need only to prove that the equality $x_0'W(t_0, t_1)x_0 = 0$ is possible only for $x_0 = 0$. Let us assume the contrary, i.e., we assume that the system (2.12) is controllable on the segment $[t_0, t_1]$, but the matrix $W(t_0, t_1)$ is degenerate. Then there exists a nonzero vector $x_0 \in R^n$ such that $x_0'W(t_0, t_1)x_0 = 0$. Since the system (2.12) is controllable on the segment $[t_0, t_1]$, there exists a control $u(t)$ that transfers the system from the state $x(t_0) = x_0$ to the state $x(t_1) = 0$. Therefore,

$$x(t_1) = Z(t_1, t_0)x_0 + \int_{t_0}^{t_1} Z(t_1, s)B(s)u(s)\,ds = 0.$$

Hence, recalling the relations (1.8) of Chapter IV, we can write

$$\begin{aligned} x_0 &= -Z^{-1}(t_1, t_0) \int_{t_0}^{t_1} Z(t_1, s)B(s)u(s)\,ds \\ &= -\int_{t_0}^{t_1} Z(t_0, s)B(s)u(s)\,ds. \end{aligned} \tag{2.13}$$

Now note that

$$\begin{aligned} 0 = x_0' W(t_0, t_1)x_0 &= x_0' \int_{t_0}^{t_1} Z(t_0, s)B(s)B'(s)Z'(t_0, s)\,ds\, x_0 \\ &= \int_{t_0}^{t_1} \|B'(s)Z'(t_0, s)x_0\|^2\,ds. \end{aligned}$$

Therefore, the vector $B'(s)Z'(t_0, s)x_0 \equiv 0$, $t_0 \leqslant s \leqslant t_1$. It follows that $x_0' Z(t_0, s)B(s)u(s) \equiv 0$. Thus,

$$0 = x_0' \int_0^T Z(t_0, s)B(s)u(s)\,ds.$$

It is easily shown from this and from the relation (2.13) that

$$x_0' x_0 = -x_0' \int_{t_0}^{t_1} Z(t_0, s)B(s)u(s)\,ds = 0.$$

However, this equality cannot be true, since it contradicts the assumption that x_0 is a nonzero vector. □

It follows from Theorem 2.3 and from the definition of the matrix $W(t_0, t_1)$ that if the system (2.12) is controllable on some segment $[t_0, t_1]$, then it is also controllable on any other segment $[t_0, t_2]$ for $t_2 > t_1$.

Differentiating both sides of the formula defining the matrix $W(\tau, t)$ with respect to τ and taking account of the relations (1.8) of Chapter IV yields the following equations:

$$\frac{\partial W(\tau, t)}{\partial \tau} = A(\tau)W(\tau, t) + W(\tau, t)A'(\tau) - B(\tau)B'(\tau), \qquad W(t, t) = 0.$$

3. Canonical form of linear time-invariant control systems. Consider again the linear system (2.1) with constant matrices A and B. Assume that the rank of the matrix (2.2) is equal to j, where $j \leqslant n$. We show that in this case there exists a change of variable $x = Ty$ (where T is a nondegenerate $n \times n$ matrix) such that the last $n - j$ coordinates $y_{j+1}, \dots, y_n$ of the vector y do not depend on the control and the preceding coordinates $y_1, \dots, y_j$. Denote by $k_1, \dots, k_j$ linearly independent columns of the matrix K. The transformation matrix T has the vectors $k_1, \dots, k_j$ for its first j columns. The other $n - j$ columns of this matrix are chosen in an arbitrary way, provided that the resulting matrix is nondegenerate. It follows from equation (2.1) that

$$\dot{y}(t) = T^{-1}ATy + T^{-1}Bu. \tag{2.14}$$

Theorem 2.4. *For the matrices $T^{-1}AT$ and $T^{-1}B$ in equation (2.14) the following expressions are valid:*

$$T^{-1}AT = \begin{pmatrix} A_1 & A_2 \\ 0_1 & A_3 \end{pmatrix}, \qquad T^{-1}B = \begin{pmatrix} B_1 \\ 0_2 \end{pmatrix}. \tag{2.15}$$

Here A_1 is a $j \times j$ matrix, A_2 is a $j \times (n-j)$ matrix, A_3 is an $(n-j) \times (n-j)$ matrix, and B_1 is a $j \times m$ matrix. The matrices 0_1 and 0_2 in (2.15) denote $(n - j) \times j$ and $(n - j) \times m$ zero matrices. Moreover, the rank of the matrix

$$(B_1, A_1B_1, \dots, A_1^{j-1}B_1) \tag{2.16}$$

is equal to j.

Proof. Let N_j denote the subspace of the space R^n that is a linear hull of the vectors $k_1, \dots, k_j$. The subspace N_j is invariant with respect to the system (2.1) (see [11]). The latter indicates that if $x(t_0) \in N_j$, then the solution $x(t)$ of equation (2.1) with initial condition $x(t_0)$ belongs to N_j for all t and for an arbitrary control $u(t)$. The equations of the space N_j for the new variables $y(t)$ are of the form

$$y_{j+1} = 0, \dots, y_n = 0.$$

Indeed, if the vector $y = (y_1, \dots, y_j, 0, \dots, 0)$, then by the definition of the transformation matrix T the corresponding vector x is equal to $x = y_1k_1 + \dots + y_jk_j$, i.e., $x \in N_j$. Thus, noting the invariancy of the set N_j, we see that if

$$y_i(t_0) = 0, \qquad i = j+1, \dots, n, \tag{2.17}$$

then for all t and for any control there exists a solution $y(t)$ of the system (2.14), provided that

$$y_i(t) = 0, \qquad i = j+1, \dots, n. \tag{2.18}$$

Now suppose that the expressions (2.15) are not valid. Then either the (i, q)th element of the matrix $T^{-1}AT$ is distinct from zero for $i \geqslant j+1$, $q < j+1$ or the (i, l)th element of the matrix $T^{-1}B$ is distinct from zero for $i \geqslant j+1$, $l = 1, \dots, m$.

In either case there exists a control $u(t)$ such that for some moment $t \neq t_0$ the corresponding solution of equation (2.14) with initial condition (2.17) satisfies the inequality $y_j(t) \neq 0$ for some $i > j$. However, this statement contradicts the equality (2.18). Thus, we have proved that the expressions (2.15) are valid.

Further, we demonstrate that the rank of the matrix (2.16) is j. Consider the matrix $Q_1 = T^{-1}K$. On account of the relations (2.3), (2.15) we have

$$Q_1 = \begin{pmatrix} B_1 & A_1B_1 & \dots & A_1^{j-1}B_1 & \dots & A_1^{n-1}B_1 \\ 0 & 0 & \dots & 0 & \dots & 0 \end{pmatrix},$$

where 0 are zero matrices.

Since the matrix T is nondegenerate, it follows that the rank of the matrix Q_1 is equal to that of the matrix K, i.e., is equal to j. Moreover, j linearly independent columns of the matrix K are contained in the matrix $(B, AB, \dots, A^{j-1}B)$ [11].

Therefore, by the definition of the matrix Q_1 its j linearly independent columns must be contained in the matrix

$$\begin{pmatrix} B_1 & \dots & A_1^{j-1}B_1 \\ 0 & \dots & 0 \end{pmatrix}.$$

Since the rank of the matrix Q_1 is equal to j, it follows that the rank of the matrix K is also equal j. □

Theorem 2.3 implies that if the vector y is expressed as $y = (z_1, z_2)$, where $z_1 \in R_j$, $z_2 \in R_{n-j}$, then on the strength of (2.14), (2.15) the following equations for the components z_1 and z_2 are valid:

$$\dot{z}_1 = A_1z_1 + A_2z_2 + B_1u, \qquad z_2 = A_3z_2,$$

the initial conditions being

$$\begin{pmatrix} z_1(t_0) \\ z_2(t_0) \end{pmatrix} = T^{-1}x(t_0).$$

Thus, the components z_2 of the vector y do not depend on either the control u or z_1 and can be calculated in advance as certain functions of time. Substituting these functions in the equation for z_1 yields a nonhomogeneous equation.

Suppose that the system (2.1) is controllable, $u(t)$ is a scalar control, and B is a column vector. We express the characteristic polynomial of the matrix A as

$$\det(\lambda I - A) = \lambda^n + \alpha_1\lambda^{n-1} + \dots + \alpha_n.$$

Further, consider a sequence of vectors $\gamma_1, \dots, \gamma_n$, where

$$\gamma_1 = (A^{n-1} + \alpha_1A^{n-2} + \dots + \alpha_n)B, \dots, \gamma_{n-1} = (A + \alpha_1)B, \qquad \gamma_n = B.$$

The assumption that the system (2.1) is controllable implies that the vectors $\gamma_1, \dots, \gamma_n$ form a basis in the space R^n.

Let L be a nondegenerate transfer matrix from the original basis to the basis $\gamma_1, \ldots, \gamma_n$. Put $x(t) = Ly(t)$. Then

$$\dot{y}(t) = L^{-1}ALy + L^{-1}Bu, \tag{2.19}$$

where

$$L^{-1}AL = \begin{pmatrix} 0 & 1 & 0 & 0 & \cdots & 0 \\ 0 & 0 & 1 & 0 & \cdots & 0 \\ \cdots & \cdots & \cdots & \cdots & \cdots & \cdots \\ \cdots & \cdots & \cdots & \cdots & \cdots & \cdots \\ 0 & 0 & 0 & 0 & \cdots & 1 \\ -\alpha_n & -\alpha_{n-1} & -\alpha_{n-2} & -\alpha_{n-3} & \cdots & -\alpha_1 \end{pmatrix}, \qquad L^{-1}B = \begin{pmatrix} 0 \\ 0 \\ \vdots \\ 0 \\ 1 \end{pmatrix}.$$

In other words, the system (2.1) is equivalent to the following equation of the nth order for $y_1(t)$:

$$y_1^{(n)}(t) + \alpha_1 y_1^{(n-1)}(t) + \cdots + \alpha_n y_1(t) = u(t), \qquad y_1^{(i)}(t) = \frac{d^i y_1(t)}{dt^i}. \tag{2.20}$$

Thus, the system (2.1) with scalar control, satisfying the condition for controllability, can be reduced to one nth-order equation of the form (2.20).

4. Canonical form of linear time-varying control systems. Let us describe the canonical form of the system (2.12) with variable coefficients assuming that $u(t)$ is a scalar control, B is a column vector, and the rank of the matrix K defined by Theorem 2.2 is equal to n for all $t \in [t_0, t_1]$. Let us define successively the following operators and matrices:

$$\Delta = A(t) - \frac{d}{dt}, \qquad \alpha = K^{-1}\Delta^n B,$$

$$e_1 = (1, 0, \ldots, 0)', \qquad e_{n-1} = (0, \ldots, 0, 1, 0)', \qquad e_i \in R^n \qquad i = 1, \ldots, n-1,$$

$$A_1 = (\alpha, e_1, \ldots, e_{n-1}), \qquad \Delta_1 = A'_1 + \frac{d}{dt},$$

$$D = (e_1, \Delta_1 e_1, \ldots, \Delta_1^{n-1} e_1), \qquad F(t) = DK^{-1}.$$

Further, we set $z(t) = F(t)x(t)$. Then the following equation for the function $z(t)$ holds:

$$\dot{z}(t) = A_0(t)z(t) + B_0(t)u, \qquad z \in R^n,\ u \in R^1,$$

where

$$A_0(t) = \begin{pmatrix} 0 & 1 & 0 & \cdots & 0 \\ 0 & 0 & 1 & \cdots & 0 \\ \vdots & \vdots & \vdots & \ddots & \vdots \\ 0 & 0 & 0 & \cdots & 1 \\ a_1(t) & a_2(t) & a_3(t) & \cdots & a_n(t) \end{pmatrix}, \qquad B_0(t) = (0, 0, \ldots, 0, 1)',$$

or

$$A_0(t) = F(t)A(t)F^{-1}(t) + \dot{F}(t)F^{-1}(t).$$

As in subsection 3, this system can be reduced to one nth-order equation for $z_1(t)$ of the form

$$z_1^{(n)}(t) + a_1(t)z_1^{(n-1)}(t) + \cdots + a_n(t)z_1(t) = u(t).$$

5. The Hautus criterion for controllability[3]. In a number of problems the Hautus criterion for the controllability of systems can be applied.

Theorem 2.5[4]. *For the system* (2.1) *to be controllable, it is necessary and sufficient that for any complex number* λ *the rank of the matrix* $(A - \lambda I, B)$ *be equal to* n.

Remark. This criterion for controllability can be stated in the following equivalent form: *for the system (2.1) to be controllable, it is necessary and sufficient that for any complex number* λ *the equation*

$$x'(A - \lambda I, B) = 0, \qquad x \in R^n,$$

or the equivalent system of two equations

$$x'A = \lambda x', \qquad x'B = 0$$

have a unique solution $x = 0$.

Proof. The necessity. Suppose that the system (2.1) is controllable, but our theorem does not hold true. Then there exists a number λ and a nonzero vector $x \in R^n$ such that

$$x'A = \lambda x', \qquad x'B = 0, \qquad x' \neq 0.$$

Hence, we have the equalities

$$\begin{aligned} &x'AB = \lambda x'B = 0, \\ &x'A^2B = x'AAB = \lambda x'AB = 0, \\ &\dots\dots\dots\dots\dots\dots\dots\dots \\ &\dots\dots\dots\dots\dots\dots\dots\dots \\ &x'A^{n-1}B = x'AA^{n-2}B = \lambda x'A^{n-2}B = 0. \end{aligned}$$

Thus, noting formula (2.2), we can write

$$x'\big(B, AB, \dots, A^{n-1}B\big) = 0 = x'K.$$

However, the last equation $x'K = 0$ must only have the zero solution $x = 0$, since by Theorem 2.1 the rank of the controllability matrix K is equal to n. This contradiction proves the necessity.

---

[3]This subsection has been written in collaboration with S. M. Lavrenov.

[4]Hautus, M. L. J., Controllability and Observability Conditions of Linear Autonomous Systems, *Proc. Kon. Ned. Akad. Wetensch, Ser. A*, (1969), vol. 72, 443–448.

The sufficiency. Suppose that the conditions of the theorem are satisfied, but the system (2.1) is uncontrollable. Then the rank of the matrix (2.2) is equal to j, where $j < n$. By the nondegenerate transformation T we can reduce the system (2.1) to the form (2.14), (2.15). Consider the $(n-j) \times (n-j)$ matrix A_3 in the expression (2.15). Let $x_0 \in R^{n-j}$ denote a nonzero eigenvector of the matrix A_3' and λ the corresponding eigenvalue, i.e., $A_3'x_0 = \lambda x_0$, $x_0 \neq 0$. Further, let $(0, x_0')$ be a row vector whose first j components are zero. Then, recalling the expressions (2.15), we can write

$$(0, x_0') \begin{pmatrix} A_1 & A_2 \\ 0_1 & A_3 \end{pmatrix} = (0, x_0'A_3) = (0, \lambda x_0'); \qquad (0, x_0') \begin{pmatrix} B_1 \\ 0_2 \end{pmatrix} = 0.$$

We rewrite these relations in the form

$$(0, x_0')T^{-1}AT = \lambda(0, x_0'), \qquad (0, x_0') = T^{-1}B = 0.$$

Hence, there exist a nonzero vector $x' = (0, x_0')T^{-1}$ and a number λ satisfying the equalities $x'A = \lambda x'$, $x'B = 0$. However, these equalities contradict the hypotheses of the theorem. The proof of the theorem is complete. □

Remark. Theorem 2.5 can be restated thus: *for the system (2.1) to be controllable, it is necessary and sufficient that for all eigenvalues λ_i $(i_1, \ldots, i_n)$ of the matrix A the rank of the matrix $(A - \lambda I, B)$ be equal to n.*

This assertion holds true because the equation $x'A = \lambda x'$ can have a nonzero solution if and only if λ coincides with one of the eigenvalues λ_i $(i_1, \ldots, i_n)$.

Consider examples in which Theorem 2.5 is used.

Example 2.2. Suppose that A is a diagonal matrix having elements λ_i in the principal diagonal. It can be shown that the system $\dot{x} = Ax + Bu$ is controllable if and only if all the rows of the matrix B are distinct from zero, and the rows of the matrix B corresponding to identical diagonal elements of the matrix A are linearly independent.

We introduce the matrix

$$S(\lambda) = (A - \lambda I, B) = \begin{pmatrix} \lambda_1 - \lambda & 0 & \ldots & 0 & b_{11} & \ldots & b_{1m} \\ 0 & \lambda_2 - \lambda & \ldots & 0 & b_{21} & \ldots & b_{2m} \\ \vdots & \vdots & \ddots & \vdots & \vdots & \ddots & \vdots \\ 0 & 0 & \ldots & \lambda_n - \lambda & b_{n1} & \ldots & b_{nm} \end{pmatrix}.$$

By Theorem 2.5 a necessary and sufficient condition for controllability is that the rank of the matrix $S(\lambda)$ for any complex λ be equal to n. Let $\lambda = \lambda_i$. If the corresponding row of the matrix B is zero, then the whole row $S(\lambda_i)$ becomes zero, and the rank of $S(\lambda_i)$ is necessarily less than n. But if $\lambda_i = \lambda_j$ $(i \neq j)$, then for $\lambda = \lambda_i = \lambda_j$ the rows become linearly independent whenever the ith and jth rows of the matrix B are linearly independent.

Example 2.3. We can show that the controllability of the pair (A, B) implies the controllability of the pair $(A + \eta I, B)$ for any number η.

According to Theorem 2.5, it is sufficient to show that for any complex eigenvalue λ_1 the rank of the matrix $(A+\eta I-\lambda_1 I, B)$ is equal to n. The controllability of the pair (A, B) implies that for any λ the rank of the matrix $(A-\lambda I, B)$ is also equal to n. But for $\lambda = \lambda_1$ we have $(A + \eta I - \lambda_1 I, B) = (A - \lambda I, B)$. Hence, the rank of the matrix $(A+\eta I-\lambda_1 I, B)$ for any λ_1 is equal n, i.e., the pair $(A-\lambda I, B)$ is controllable.

Example 2.4. We shall show that the controllability of the system

$$\dot{x} = Ax + bu, \qquad x \in R^n,\ u \in R^1,$$

implies the controllability of the following system:

$$\dot{x} = Ax + bv, \qquad \dot{v} = C'x + \alpha v + u, \qquad v \in R^1,$$

where A is a constant matrix, b and C are constant vectors from R^n, and α is a prescribed number.

Let S_n denote the $n \times (n+1)$ matrix $S_n(\lambda) = (A - \lambda I, b)$. It can be shown that the rank of the matrix

$$S_{n+1}(\lambda) = \begin{pmatrix} S_n(\lambda) & & \vdots & 0 \\ \cdots & \cdots & \cdots & \cdots \\ C & \alpha - \lambda & \vdots & 1 \end{pmatrix} = \begin{pmatrix} A - \lambda I & b & \vdots & 0 \\ \cdots & \cdots & \cdots & \cdots \\ C & \alpha - \lambda & \vdots & 1 \end{pmatrix}$$

is equal to $n+1$. Here the $(n+1) \times (n+2)$ matrix $S_{n+1}(\lambda)$ is obtained from S_n by adding a single row composed of the components $(c_1, \dots, c_n)$ of the vector C, of the numbers $\alpha - \lambda$ and 1, and also of one column whose elements are all zero except for the last one equal to unity.

Since the pair (A, b) is controllable, the rank of the matrix $S_n(\lambda)$ is equal to n. Hence, the rows of the matrix $S_n(\lambda)$ are linearly independent and, consequently, the first n rows of the matrix $S_{n+1}(\lambda)$ are also linearly independent. The last row of the matrix $S_{n+1}(\lambda)$ is linearly independent from the first n rows, since its last elements are zeros and the last element of the $(n+1)$th row is unity. Therefore, the rank of the matrix $S_{n+1}(\lambda)$ is equal to $n+1$.

6. Controllability of a two-link manipulator. Consider a two-link manipulator (see Figure 2.1) consisting of two rigid bodies: Q_1 of mass m_1 and Q_2 of mass m_2, which are fixed together by means of the hinge O_2 and made fast to the immovable base via the hinge O_1. The axes of the hinges are parallel. The motion of the manipulator is in the plane perpendicular to the axes of the hinges. The manipulator is controlled by two torques u_1 and u_2 applied to the axes of the hinges O_1 and O_2. Suppose that the link Q_2 of the manipulator is balanced statically, i.e., its mass center is located on the axis O_2. Then the equations of motion of the manipulator are of the form

$$(I_1 + m_2 L^2)\ddot{\varphi}_1 = u_1 - u_2, \qquad I_2 \ddot{\varphi}_2 = u_2. \tag{2.21}$$

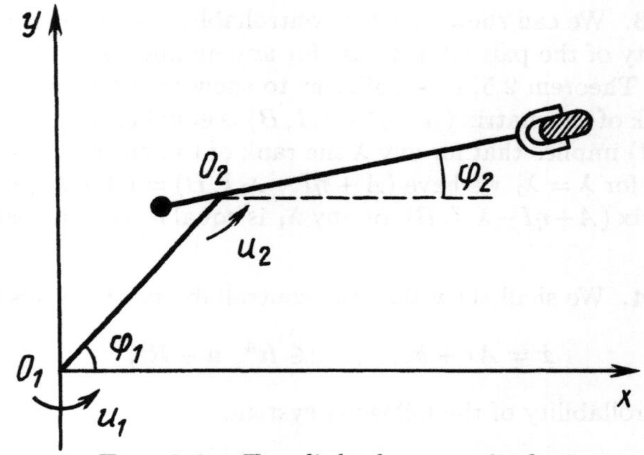

FIG. 2.1. Two-link plane manipulator

Here φ_1 is the angle between the link Q_1 and the axis O_1x of the fixed coordinate system O_1xy, φ_2 is the angle between Q_2 and Q_1x, L is the distance between the axes of the hinges O_1 and O_2, I_1 and I_2 are the moments of inertia of the links Q_1 and the Q_2 about the axes O_1 and O_2, respectively.

We introduce the variables

$$x_1 = (I_1 + m_2L^2)^{1/2}\varphi_1, \qquad x_3 = I_2^{1/2}\varphi_2,$$
$$x_2 = (I_1 + m_2L^2)^{1/2}\dot{\varphi}_1, \qquad x_4 = I_2^{1/2}\dot{\varphi}_2,$$

and write the system (2.21) in the form

$$\dot{x}_1 = x_2, \qquad \dot{x}_2 = u_1 - u_2, \qquad \dot{x}_3 = x_4, \qquad \dot{x}_4 = u_2. \tag{2.22}$$

The system (2.22) has the form (2.1), where the matrices A and B are the following:

$$A = \begin{pmatrix} 0 & 1 & 0 & 0 \\ 0 & 0 & 0 & 0 \\ 0 & 0 & 0 & 1 \\ 0 & 0 & 0 & 0 \end{pmatrix}, \qquad B = \begin{pmatrix} 0 & 0 \\ 1 & -1 \\ 0 & 0 \\ 0 & 1 \end{pmatrix}.$$

The controllability matrix K of the system (2.22) is of dimensions 4×8:

$$K = (B, AB, A^2B, A^3B) = \begin{pmatrix} 0 & 0 & 1 & -1 & 0 & 0 & 0 & 0 \\ 1 & -1 & 0 & 0 & 0 & 0 & 0 & 0 \\ 0 & 0 & 0 & 1 & 0 & 0 & 0 & 0 \\ 0 & 1 & 0 & 0 & 0 & 0 & 0 & 0 \end{pmatrix}.$$

The first four columns of this matrix are linearly independent because

$$\det \begin{pmatrix} 0 & 0 & 1 & -1 \\ 1 & -1 & 0 & 0 \\ 0 & 0 & 0 & 1 \\ 0 & 1 & 0 & 0 \end{pmatrix} = -1.$$

Since the rank of the matrix B is equal to 2, we would need to confine ourselves, according to Remark 1 of subsection 1, to finding the rank of the matrix (B, AB), which is equal to 4. Hence, the system (2.22) is controllable.

The same conclusion can be drawn by using the criterion for controllability embodied in Theorem 2.5. Indeed, all eigenvalues of the matrix A are zero. Further, the matrix $(A - \lambda I, B)$ for $\lambda = 0$ becomes

$$(A, B) = \begin{pmatrix} 0 & 1 & 0 & 0 & 0 & 0 \\ 0 & 0 & 0 & 0 & 1 & -1 \\ 0 & 0 & 0 & 1 & 0 & 0 \\ 0 & 0 & 0 & 0 & 0 & 1 \end{pmatrix}.$$

Because of the linear independent of the 2nd, 4th, 5th, and 6th columns the rank of this matrix is equal to 4. Therefore, by the remark to Theorem 2.5 the system (2.22) is controllable.

We can show that the system (2.22) is uncontrollable if only the torque u_2 is used to control the system. In this case the matrices B and $K = (B, AB, A^2B, A^3B)$ are of the form

$$B = \begin{pmatrix} 0 \\ -1 \\ 0 \\ 1 \end{pmatrix}, \qquad K = \begin{pmatrix} 0 & -1 & 1 & -1 \\ -1 & 0 & 0 & 0 \\ 0 & 1 & -1 & 1 \\ 1 & 0 & 0 & 0 \end{pmatrix}.$$

Since the 2nd and 4th columns of the matrix K are identical, its rank is less than 4. Hence, the system (2.22) with one control u_2 is uncontrollable.

Here we can also use Theorem 2.5 to test for the controllability of the system (2.22). Since all eigenvalues of the matrix A are zero, it is sufficient to determine the rank of the matrix (A, B), which is of the form

$$(A, B) = \begin{pmatrix} 0 & 1 & 0 & 0 & 0 \\ 0 & 0 & 0 & 0 & -1 \\ 0 & 0 & 0 & 1 & 0 \\ 0 & 0 & 0 & 0 & 1 \end{pmatrix}.$$

The rank of this matrix is equal to 3. Hence, by Theorem 2.5 the system (2.22) is uncontrollable whenever only one torque u_2 is applied.

§ 3. Observation in Linear Systems. Observers

1. Statement of an observation problem. Duality of control and observation problems. When investigating the evolution of a dynamical system or constructing an F-control, we must know how to determine the state vector (phase vector) of the system. However, in practical situations a direct measurement of the phase vector is difficult due to various reasons, such as the technical impossibility or the excessive cost of carrying out observations, etc. One must also take into account the occurrence of errors in the observation channel hindering an accurate determination of the phase vector. Problems of estimating the state vector of a system in the presence of noise are dealt with in Part Three of the book. Here we treat the case in which there are no observation errors.

Let the motion of the system be described by a linear differential equation

$$\dot{x}(T) = A(t)x(t), \qquad x(t) \in R^n, \tag{3.1}$$

where the matrix $A(t)$ is known a priori, and the vector $x(t)$ is not directly observable. The observed vector $y(t)$ is connected with the vector $x(t)$ by

$$y(t) = Q(t)x(t), \qquad y \in R^l. \tag{3.2}$$

Here the prescribed matrix $Q(t)$ is called the *measurement matrix*. It determines the observable linear combinations of the components of the state vector $x(t)$. The observation problems to be discussed in this section deal with the determination of certain functions of the components of x, given the measurement data (3.2).

To be definite, let us consider the following problem. Suppose that the measurement of the vector $y(t)$ is carried out on the segment $[t_0, t_1]$. It is required to construct a linear functional of $y(t)$, where $t_0 \leqslant t \leqslant t_1$, that serves as an estimate for a scalar quantity $\alpha' x(t_1)$, where α is a prescribed vector from R^n. In other words, it is required to find a function $\gamma(t) \in R^l$ such that

$$\int_{t_0}^{t_1} \gamma'(t)y(t)\, dt = \alpha' x(t_1). \tag{3.3}$$

For each prescribed value of the vector α the equality (3.3) yields a corresponding linear combination of the components of the state vector $x(t_1)$. In particular, if the ith component of the vector α is $\alpha_i = 1$ and all other $\alpha_j = 0$ $(j \neq i)$, then the relation (3.3) yields an explicit value for the ith component $x_i(t)$ of the phase vector $x(t)$.

Denote by $Z(t,s)$ the Cauchy matrix for the equation $\dot{x}(t) = A(t)x(t)$. Then, according to Cauchy's formula, we have

$$x(t) = Z(t, t_1)x(t_1).$$

Therefore,

$$y(t) = Q(t)x(t) = Q(t)Z(t, t_1)x(t_1).$$

Substituting this expression in the relation (3.3) gives the following relation that is valid for any $x(t_1)$:

$$\int_{t_0}^{t_1} \gamma'(t)y(t)\, dt = \int_{t_0}^{t_1} \gamma'(t)Q(t)Z(t, t_1)\, dt\, x(t_1) = \alpha' x(t_1).$$

Since $x(t_1)$ is an arbitrary vector, the function $\gamma(t)$ must satisfy the equation

$$\int_{t_0}^{t_1} Z'(t, t_1)Q'(t)\gamma(t)\, dt = \alpha. \tag{3.4}$$

Consider now an auxiliary linear control system

$$\dot{\beta}(t) = A_1(t)\beta(t) + B(t)u(t), \qquad t_0 \leqslant t \leqslant t_1, \tag{3.5}$$
$$\beta(t_0) = \beta_0,\ \beta(t_1) = \beta_1, \quad \beta \in R^n,\ u \in R^l.$$

Here $A_1(t)$ and $B(t)$ are certain matrices that will be determined in the sequel. It is required to transfer the system (3.5) from the given initial state $\beta(t_0) = \beta_0$ to the given terminal state $\beta(t_1) = \beta_1$.

The problem under discussion can be reduced to the determination of a control $u(t)$ from the equation

$$\int_{t_0}^{t_1} Z_1(t_1, t)B(t)u(t)\,dt = \beta_1 - Z(t_1, t_0)\beta_0, \tag{3.6}$$

where $Z_1(t, s)$ is the Cauchy matrix for the solutions of the equation $\dot{\beta}(t) = A_1(t)\beta(t)$. Thus, the observation problem has been reduced to one of solving the integral equation (3.4) for the function $\gamma(t)$ and the auxiliary control problem to one of solving the integral equation (3.6) for the control function $u(t)$. Choose now the matrices $A_1(t)$ and $B(t)$ in (3.5) so that equations (3.4) and (3.6) become equivalent. Comparison of equations (3.4) and (3.6) immediately yields the relations $B(t) = Q'(t)$ and $Z'(t, t_1) = Z_1(t_1, t)$, provided that $\alpha = \beta_1 - Z(t_1, t_0)\beta_0$. The latter relation gives the expression for $A_1(t)$. Invoking the properties of the Cauchy matrix as embodied in formula (1.8) of Chapter IV, we can write

$$Z'(t, t_1) = Z_1(t_1, t) = Z_1^{-1}(t, t_1).$$

Differentiating both sides of this equality and invoking the properties indicated above gives

$$\frac{\partial Z'(t, t_1)}{\partial t} = Z'(t, t_1)A'(t) = \frac{\partial Z_1^{-1}(t, t_1)}{\partial t} = -Z_1^{-1}(t, t_1)A_1(t).$$

Consequently,

$$Z'(t, t_1)A'(t) = -Z_1(t_1, t)A_1(t) = -Z'(t, t_1)A_1(t).$$

Thus, we obtain finally

$$A_1(t) = -A'(t), \qquad B(t) = Q'(t). \tag{3.7}$$

It should be noted that the right sides of equations (3.4) and (3.6) can be made equal by an appropriate choice of the vectors $\beta(t_0)$ and $\beta(t_1)$. Then the solutions of the integral equations (3.4) and (3.6) are equivalent, and hence the solutions of the observation problem (3.1)–(3.3) and of the control problem (3.5) are equivalent. Therefore, these problems are called *dual.*

Definition 3.1. The system (3.1) is said to be *observable on the interval* $[t_0, t_1]$, given the measurement data (3.2), if for any vector $\alpha \in R^n$ there exists a function $\gamma(t)$ satisfying the equality (3.4).

In other words, for the system (3.1), which is observable on the interval $[t_0, t_1]$, it is possible to accurately determine the desired quantity $\alpha' x(t_1)$ through the use of a linear transformation of the measurement data $y(t)$ given on the interval $[t_0, t_1]$.

The established duality of the observation problem (3.1)–(3.3) and of the control problem (3.5) allows the definite conclusion that the definition (3.1) of observability conforms to that of controllability of the system (3.5). Hence, the conditions for the observability of the system (3.1) coincide with those for the controllability of the system (3.5), (3.7). The conditions for controllability as embodied in Theorems 2.1 and 2.2 (and also the fact that the rank of a matrix does not change under the multiplication of any column by any nonzero quantity) lead to the following conditions for observability.

Theorem 3.1. *If the matrices A and Q are constant, then for the system* (3.1) *to be observable, given the measurement data* (3.2), *it is necessary and sufficient that the rank of the observability matrix*

$$(Q', A'Q', \dots, (A')^{n-1}Q') \tag{3.8}$$

be equal to n.

Theorem 3.2. *Suppose there exists a point $\tau \in [t_0, t_1]$ at which the derivatives of the matrices $A(t)$ and $Q(t)$ are continuous up to order $n-1$ inclusive. Moreover, let the rank of the matrix $(K_1, \dots, K_n)$, where*

$$K_1(\tau) = Q'(\tau), \qquad K_i(\tau) = A'(\tau)K_{i-1}(\tau) + \frac{dK_{i-1}(\tau)}{d\tau}, \qquad i \geqslant 2.$$

be equal to n at the point τ. Then the system (3.1) *is observable, given the measurement data* (3.2).

The conditions for observability are expressed by the following theorem, which is analogous to the Hautus criterion given by Theorem 2.5.

Theorem 3.3. *If the matrices A and Q are constant, then for the system* (3.1) *to be observable, given the measurement data* (3.2), *it is necessary and sufficient that for any complex number λ the rank of the matrix $(A' - \lambda I, Q')$ or the rank of the matrix $\binom{A-\lambda I}{Q}$ be equal to n.*

The pair of matrices (A, Q) is said to be *observable* if the rank of the matrix (3.8) is equal to n.

Let us turn to examples of the use of Theorem 3.3.

Example 3.1. Suppose that the pair of matrices (A, Q) is observable and the matrix Q can be expressed by $Q = (O_1, I_1)$, where O_1 is an $l \times (n-l)$ matrix having zeros as its elements and I_l is an $l \times l$ unit matrix. Let us write the matrix A as

$$A = \begin{pmatrix} A_{11} & A_{12} \\ A_{21} & A_{22} \end{pmatrix}.$$

Here A_{11} and A_{22} are $(n-l) \times (n-l)$ and $l \times l$ matrices, respectively, A_{12} is an $(n-l) \times l$ matrix, and A_{21} is an $l \times (n-l)$ matrix. We seek to prove that the pair (A_{11}, A_{21}) is observable.

To obtain the proof, we use Theorem 3.3 according to which it is sufficient to establish that for any λ the rank of the matrix

$$P_1(\lambda) = \begin{pmatrix} A_{11} - \lambda I_{n-l} \\ A_{21} \end{pmatrix}.$$

is equal to $n - l$. Let P_2 denote an $n \times l$ matrix of the form

$$P_2(\lambda) = \begin{pmatrix} A_{12} \\ A_{22} - \lambda I_l \end{pmatrix}.$$

By Theorem 3.3 for any λ the rank of the matrix

$$\begin{pmatrix} A - \lambda I_n \\ Q \end{pmatrix} = \begin{pmatrix} P_1(\lambda) & P_2(\lambda) \\ O_1 & I_l \end{pmatrix}.$$

is equal to n. Since the rank of the matrix $\begin{pmatrix} P_2(\lambda) \\ I_l \end{pmatrix}$ is equal to l (because of the submatrix I_l), the rank of the matrix $\begin{pmatrix} P_1(\lambda) \\ O_l \end{pmatrix}$ must be equal to $n - l$. Therefore, the rank of the matrix P_1 is equal to $n - l$.

Example 3.2. Consider the system

$$\dot{x}_1 = Ax_1 + Fx_2, \qquad x_1 \in R^n$$
$$\dot{x}_2 = Dx_2, \qquad x_2 \in R^m.$$

Only the first component of the vector x_1 is directly observable, i.e., the observation equation is of the form

$$y = Qx_1, \qquad Q = (1, 0, \dots, 0),\ y \in R^1.$$

Assume that the matrices A, D, and F of respective dimensions $n \times n$, $m \times m$, and $n \times m$ are expressed by

$$A = \begin{pmatrix} -a_1 & 1 & 0 & \dots & 0 \\ -a_2 & 0 & 1 & \dots & 0 \\ \dots & \dots & \dots & \dots & \dots \\ \dots & \dots & \dots & \dots & \dots \\ -a_{n-1} & 0 & 0 & \dots & 1 \\ -a_n & 0 & 0 & \dots & 0 \end{pmatrix}, \qquad D = \begin{pmatrix} -\alpha & 1 & 0 & \dots & 0 \\ 0 & -\alpha & 1 & \dots & 0 \\ \dots & \dots & \dots & \dots & \dots \\ \dots & \dots & \dots & \dots & \dots \\ 0 & 0 & 0 & \dots & 1 \\ 0 & 0 & 0 & \dots & -\alpha \end{pmatrix},$$

$$F = \begin{pmatrix} F_1 & 0 & \dots & 0 \\ F_2 & 0 & \dots & 0 \\ \dots & \dots & \dots & \dots \\ \dots & \dots & \dots & \dots \\ F_n & 0 & \dots & 0 \end{pmatrix},$$

where α and a_i $(i = 1, \dots, n)$ are prescribed quantities.

We seek to determine the conditions for the elements F_1 of the matrix F that assure the observability of the system. It should be noted that this problem arises when synthesizing adaptive observers.

Let A_0 denote an $N \times N$ matrix $(N = n + m)$ and $S(\lambda)$ an $(N+1) \times N$ matrix, where

$$A_0 = \begin{pmatrix} A & F \\ O_1 & D \end{pmatrix}, \qquad S(\lambda) = \begin{pmatrix} A_0 & -\lambda I_N \\ Q & O_2 \end{pmatrix}.$$

Here O_1 and O_2 are $m \times n$ and $1 \times m$ zero matrices. By Theorem 3.3 for the pair (A_0, Q) to be observable, it is necessary and sufficient that for any λ the rank of the matrix $S(\lambda)$ be equal to N. We set

$$P_1(\lambda) = \begin{pmatrix} A - \lambda I_n \\ O_1 \\ Q \end{pmatrix}, \qquad P_2(\lambda) = \begin{pmatrix} F \\ D - \lambda I_m \\ O_2 \end{pmatrix}.$$

Then $S(\lambda) = (P_1, P_2)$. It should be noted that for any λ the rank of the matrix P_1 is equal to n, since the first $(n-1)$ rows of the matrix P_1 and its last row are linearly independent.

Consider separately the cases in which $\lambda \neq -\alpha$ and $\lambda = -\alpha$. If $\lambda \neq -\alpha$, then the rang of the matrix P_2 is equal to the rang of the matrix $P_2 - \lambda I_m$ and is equal to m for any matrix F, since the matrix $D - \lambda I_m$ is nondegenerate. Hence, for $\lambda \neq -\alpha$ the rang of the matrix $S(\lambda)$ is equal to N.

Now let $\lambda = -\alpha$. Then the matrix $S(-\alpha)$ is of the form

$$S(-a) = \begin{pmatrix} -\alpha_1 + \alpha & 1 & 0 & \dots & 0 & F_1 & 0 & 0 & \dots & 0 \\ -a_2 & \alpha & 1 & \dots & 0 & F_2 & 0 & 0 & \dots & 0 \\ \dots & \dots & \dots & \dots & \dots & \dots & \dots & \dots & \dots & \dots \\ \dots & \dots & \dots & \dots & \dots & \dots & \dots & \dots & \dots & \dots \\ -a_n & 0 & 0 & \dots & \alpha & F_n & 0 & 0 & \dots & 0 \\ 0 & \dots & \dots & \dots & 0 & 0 & 1 & 0 & \dots & 0 \\ 0 & \dots & \dots & \dots & 0 & 0 & 0 & 1 & \dots & 0 \\ 0 & \dots & \dots & \dots & 0 & 0 & 0 & 0 & \dots & 1 \\ 1 & 0 & 0 & \dots & 0 & 0 & 0 & 0 & \dots & 0 \end{pmatrix}.$$

The last m rows of the matrix $S(-\alpha)$ are linearly independent. It is readily seen that the rank of the matrix $S(-\alpha)$ is equal to $N = m + n$, provided that its columns, starting from the 2nd and ending with the $(n+1)$th (containing the components F_i), are linearly independent. The last condition is met if the determinant

$$\Delta = \begin{vmatrix} 1 & 0 & \dots & 0 & F_1 \\ \alpha & 1 & \dots & 0 & F_2 \\ \dots & \dots & \dots & \dots & \dots \\ \dots & \dots & \dots & \dots & \dots \\ 0 & 0 & \dots & \alpha & F_n \end{vmatrix}$$

is distinct from zero.

Further, we compute the determinant Δ. Let Δ_{n-i} denote the determinant obtained from the original one Δ by striking out its first i rows and i columns. Then $\Delta = \Delta_n$, and developing the determinant Δ with respect to the elements in the first row, we obtain

$$\Delta = \Delta_n = \Delta_{n-1} + (-1)^{n+1}\alpha^{n-1}F_1 = \Delta_{n-1} + (-\alpha)^{n-1}F_1.$$

It is readily seen that the following recurrence relation holds:

$$\Delta_{i+1} = \Delta_i + (-\alpha)^i F_{n-i}, \qquad \Delta_1 = F_n.$$

Therefore, we obtain the following expression for the determinant Δ:

$$\Delta = \sum_{i=1}^{n-1}(-\alpha)^i F_{n-i}.$$

Thus, the condition for observability $\Delta \neq 0$ becomes

$$\sum_{i=1}^{n-1}(-\alpha)^i F_{n-i} \neq 0.$$

2. On a method of determining the state vector. Consider the observation problem (3.1), (3.2) with constant matrices A and Q assuming that the rank of the observability matrix (3.8) is maximum, i.e., is equal to n. Then we can show that the phase vector $x(t_0)$ (and hence also, by Cauchy's formula, the vector $x(t)$ for all t) can be reconstructed on an arbitrarily small interval $[t_1, t_2]$ using the measurement data (3.2). To this end, we can use the following relation which is a consequence of formulas (3.1), (3.2):

$$y(t) = Qx(t) = Q[\exp A(t - t_0]x(t_0). \tag{3.9}$$

Differentiating successively the equality (3.9) with respect to t and putting $t = t_0$, we obtain the system of equations in $x(t_0)$:

$$y(t_0) = Qx(t_0), \qquad \dot{y}(t_0) = QAx(t_0), \dots, y^{(n-1)}(t_0) = QA^{n-1}x(t_0). \tag{3.10}$$

The system (3.10) consists of nl equations with n unknown components of the vector $x(t_0)$. By their construction equations (3.10) are compatible, and they have a unique solution for $x(t_0)$ if and only if the rank of the matrix (3.8) is equal n. Thus, whenever the conditions for observability of Theorem 3.1 are satisfied, by solving equations (3.10) for $x(t_0)$, we can reconstruct the phase vector $x(t)$ with the help of the formula

$$x(t) = [\exp A(t - t_0)]x(t_0).$$

It should be noted that all the subsequent derivatives $y^{(i)}(t_0)$ $(i \geqslant n)$ are linear combinations of the preceding derivatives $y^{(j)}(t_0)$ $(j = 0, \dots, n-1)$. Indeed, multiplying the equality (2.4) by Q on the left and by $x(t)$ on the right, we obtain by virtue of (3.9)

$$\begin{aligned} (QA^n)x(t) &= (-\alpha_1 QA^{n-1} - \dots - \alpha_n Q)x(t), \\ y^{(n)}(t_0) &= -\alpha_1 y^{(n-1)}(t_0) - \dots - \alpha_n y(t_0). \end{aligned} \tag{3.11}$$

In a similar way, multiplying the equality (2.4) by Q on the left and by $Ax(t)$ on the right, we obtain

$$\begin{aligned} (QA^{n+1})x(t) &= (-\alpha_1 QA^{n} - \dots - \alpha_n QA)x(t), \\ y^{(n+1)}(t_0) &= -\alpha_1 y^{(n)}(t_0) - \dots - \alpha_n \dot{y}(t_0). \end{aligned} \tag{3.12}$$

Substituting the expression for $y^{(n)}(t_0)$ (3.11) in the right side of (3.12), we establish that $y^{(n+1)}(t_0)$ is a linear combination of the functions $y^{(i)}(t_0)$ $(i = 0, \dots, n-1)$. In a similar way, multiplying the equality (2.4) by Q on the left and by $A^{n+1}x(t)$ $(i = 2, 3, \dots)$ on the right, we find that all $y^{(i)}(t_0)$ are linear combinations of the functions $y^{(j)}(t_0)$ $(j = 0, \dots, n-1)$. Therefore, to express $x(t_0)$ in terms of the derivatives $y^{(i)}(t_0)$, we need only to take account of the values $i = 0, \dots, n-1$, i.e., we can confine ourselves to the system of equations (3.10).

3. Observer of maximum order. The procedure for determining a state vector that was discussed in subsection 2 is based on the calculation of derivatives of the function $y(t)$ with all the inherent difficulties encountered in such calculations. Hence, the implementation of this procedure is troublesome. We proceed now to an outline of an approach to the problem of estimating a state vector $x(t)$ through the use of an equation called an *observer* such that its solution $m(t)$ converges to $x(t)$ for any initial conditions $x(0)$ and $m(0)$ in the following sense:

$$\lim_{t \to \infty} |x(t) - m(t)| = 0. \tag{3.13}$$

Consider the observation problem (3.1), (3.2) with constant matrices A and Q. Let the rank of the matrix (3.8) be maximum. We define the observer by the equation

$$\dot{m}(t) = Am(t) + V(t)\big(y(t) - Qm(t)\big), \qquad t \geqslant 0. \tag{3.14}$$

Here the $(n \times l)$ matrix $V(t)$ is to be determined, subject to the condition (3.13). Let $\varepsilon(t)$ denote the estimation error:

$$\varepsilon(t) = x(t) - m(t). \tag{3.15}$$

It follows from (3.1), (3.14) that the error $\varepsilon(t)$ satisfies the equation

$$\dot{\varepsilon}(t) = [A - V(t)Q]\varepsilon(t). \tag{3.16}$$

As is known [12], if the rank of the matrix (3.8) is maximum, i.e., is equal to n, then we can choose the matrix $V(t)$ to be a constant such that the spectrum of the matrix $(A - VQ)$ takes on prescribed values. In other words, in the case under scrutiny we can choose a constant matrix V so that the relation (3.13) is satisfied.

Remark. Suppose that the right side of the equation (3.1) contains a control term of the form $B(t)u(t)$, where $B(t)$ is a given matrix and $u(t) \in R^m$ is a known function. So, the equation for the system becomes

$$\dot{x}(t) = A(t)x(t) + B(t)u(t). \tag{3.17}$$

In that case the conditions for observability preserve the form (see Theorems 3.1 and 3.2), and the observer of maximum order is described by the equations

$$\dot{m}(t) = A(t)m(t) + B(t)u(t) + V(t)\big(y(t) - Qm(t)\big). \tag{3.18}$$

In this case the estimation error (3.15) satisfies equation (3.16) as before.

4. Observer of reduced order (the Luenberger observer). The observer (3.18) is of the same order as the phase vector $x(t)$. Moreover, on account of the measurement data (3.2) for equation (3.17) we can reduce the order of the equations for the observer. Suppose that the matrices A and Q in (3.17), (3.2) are constant, the rank of the matrix (3.8) is maximum and equals n, and the rank of the matrix Q is equal to $l > 0$. Choose any matrix W of dimensions $(n - l) \times n$ so that the square matrix L of order n equal to

$$L = \begin{pmatrix} Q \\ W \end{pmatrix} \tag{3.19}$$

be nondegenerate. Let E and D denote blocks of the inverse matrix L^{-1}, i.e., $L^{-1} = (ED)$, where E is an $n \times l$ matrix and D is an $n \times (n - l)$ matrix. Clearly, $EQ + DW = I$, where I is a unit matrix of order n.

Consider a vector $\alpha(t) \in R^{n-l}$ connected with $x(t)$ by the relation

$$x(t) = L^{-1}\begin{pmatrix} y(t) \\ \alpha(t) \end{pmatrix} = Ey(t) + D\alpha(t). \tag{3.20}$$

Note that $\big(y(t), \alpha(t)\big)' = Lx(t)$. On the right side of the relation (3.20) only the vector $\alpha(t)$ needs to be estimated. Therefore, in view of (3.17), (3.19) the function $\alpha(t) = Wx(t)$ satisfies the equation

$$\dot{\alpha}(t) = W\big(AD\alpha(t) + AEy(t) + Bu(t)\big) \tag{3.21}$$
$$\alpha(0) = \alpha_0$$

The initial condition α_0 for equation (3.21) may be arbitrary. If $\alpha_0 = Wx_0$, then the equality (3.20) holds. But if $\alpha_0 \neq Wx_0$, then the equality (3.20) does not hold any longer and the error $\varepsilon(t) = x(t) - Ey(t) - D\alpha(t) \not\equiv 0$ occurs.

It would seem in thwe latter case that we can find an estimate for the vector $x(t)$ by solving equation (3.21) and substituting the obtained solution in the right side of (3.20). In this case the estimation error $\varepsilon(t) = x(t) - \big(Ey(t) + D\alpha(t)\big)$ satisfies the equation

$$\begin{aligned}\dot\varepsilon(t) &= \dot x(t) - E\dot y(t) - D\dot\alpha(t)\\ &= Ax + Bu - EQ(Ax + Bu) - DW(AD\alpha + AEy + Bu)\\ &= Ax - DWAD\alpha - DWAEy - EQAx\\ &= (EQ + DW)Ax - DWAD\alpha - DWAEy - EQAx\\ &= DWAx - DWA(Ey + D\alpha) = DWA\varepsilon.\end{aligned}$$

The ranks of the matrices D and W are at most $n - l$; hence, by Sylvester's inequality the rank of the matrix DWA is also at most $n - l$.

Since the determinant of the matrix is equal to the product of its eigenvalues and the matrix DWA is degenerate, it cannot be a Hurwitz matrix. Hence, in general, $\varepsilon(t)$ does not tend to zero as $t \to \infty$. This means that the estimation procedure (3.20), (3.21) does not meet the basic condition (3.13) that all observers must satisfy.

We turn now to a discussion of an alternative estimation procedure. Let us define an estimate $\alpha_1(t) \in R^{n-l}$ of the vector $\alpha(t)$ by the following equation derived from (3.21) and the expression for $\dot y(t)$:

$$\begin{aligned}\dot\alpha_1(t) = WA\big(D\alpha_1(t) + Ey(t)\big) + V\big[\dot y(t) - QA\big(Ey(t) + D\alpha_1(t)\big)\big]\\ + (W - VQ)Bu(t), \qquad \alpha_1(0) = \alpha_1^0.\end{aligned} \tag{3.22}$$

Here the matrix V must be chosen so that $|\alpha(t) - \alpha_1(t)| \to 0$ as $t \to \infty$. The initial value of the estimate α_1^0 is prescribed in an arbitrary way. In view of (3.21), (3.22) the estimation error $\delta(t) = \alpha(t) - \alpha_1(t)$ satisfies the equation

$$\dot\delta(t) = (W - VQ)AD\delta(t). \tag{3.23}$$

Indeed, it follows from (3.22), (3.2), (3.17), (3.20) that

$$\begin{aligned}\dot\alpha_1(t) &= WA(D\alpha_1 + Ey) + VQ[\dot x - A(Ey + D\alpha_1) - Bu] + WBu\\ &= WA(D\alpha_1 + Ey) + VQA(x - Ey - D\alpha_1) + WBu\\ &= WA(D\alpha_1 + Ey) + VQAD(\alpha - \alpha_1) + WBu.\end{aligned} \tag{3.24}$$

Subtracting equation (3.24) from (3.21) yields equation (3.23). Under the assumptions made the constant matrix V of dimensions $(n - l) \times l$ can be chosen so that the trivial solution of equation (3.23) is exponentially stable with respect to the initial disturbances [12]. It follows that

$$\lim_{t\to\infty} \delta(t) = 0. \tag{3.25}$$

Now that the estimate $\alpha_1(t)$ of the vector $\alpha(t)$ has been constructed, it is natural, in view of (3.20), to define the estimate $m(t)$ of the vector $x(t)$ by the equality

$$m(t) = Ey(t) + D\alpha_1(t). \tag{3.26}$$

Moreover, we can show that the condition (3.13) is satisfied as well. In fact,

$$\begin{aligned} x(t) - m(t) &= Ey(t) + D\alpha(t) - \big(Ey(t) + D\alpha_1(t)\big) \\ &= D\big(\alpha(t) - \alpha_1(t)\big) \to 0, \qquad t \to \infty. \end{aligned}$$

In certain cases it turns out to be difficult to construct the estimate $\alpha_1(t)$ through the use of equation (3.22) whose right side depends on the derivative of the observation process $y(t)$. However, we can circumvent this difficulty by effecting the change of variable

$$\alpha_2(t) = \alpha_1(t) - Vy(t).$$

Then on the strength of (3.26) we obtain

$$m(t) = D\alpha_2(t) + (E + DV)y(t). \tag{3.27}$$

With regard to (3.22), (3.2), (3.17) the equation for $\alpha_2(t)$ can be written as

$$\begin{gathered} \dot{\alpha}_2(t) = (W - VQ)[AD\alpha_2 + Bu + A(DV + E)y], \\ \alpha_2(0) = \alpha_2^0. \end{gathered} \tag{3.28}$$

The initial vector α_2^0 can be chosen in an arbitrary way. It follows from the relations (3.20), (3.26), (3.25) that the estimation error $\delta_1(t) = x(t) - m(t)$ satisfies the equality

$$\lim_{t\to\infty} \delta_1(t) = \lim_{t\to\infty} [x(t) - m(t)] = \lim_{t\to\infty} D\delta(t) = 0.$$

Thus, formula (3.27), in fact, determines an observer for the system (3.17). Since in order to implement the observer (3.27) we need to determine the vector α_2 of dimension $n - l$, the resulting observer is called the *observer of reduced order* (the *Luenberger observer*).

Let us derive the equation for the estimation error $\delta_1(t)$. Differentiating both sides of the equality $\delta_1(t) = D\delta(t)$ with respect to t we obtain in view of (3.23)

$$\dot{\delta}_1 = D\dot{\delta} = D(W - VQ)AD\delta = D(W - VQ)A\delta_1.$$

In conclusion, we summarize the main steps of the procedure to construct the Luenberger observer:

1°. The matrix W is chosen so that the matrix L defined by (3.19) is nondegenerate.

2°. The matrix L^{-1} is expressed as $L^{-1} = (E, D)$, where E is an $n \times l$ matrix and D is an $n \times (n-l)$ matrix.

3°. The matrix V is defined in a way such that the matrix $(W - VQ)AD$ is a Hurwitz matrix, i.e., all of its eigenvalues lie in the left half-plane of the complex plane.

4°. The solution $\alpha_2(t)$ of the problem (3.28) is obtained.

5°. The estimate $m(t)$ of the vector $x(t)$ is determined by means of formula (3.27).

In those cases in which we can calculate the derivative $\dot{y}(t)$ the steps 4° and 5° can be replaced by the following steps:

4°) The problem (3.22) is solved for $\alpha_1(t)$.

5°) The estimate $m(t)$ of the vector $x(t)$ is determined by means of formula (3.26).

Remark. If the measurement data (3.2) also contain noise, then the observer of maximum order (3.14) with the matrix $V(t)$ obtained from the condition for minimum mean-square error is called the *Kalman filter.* A detailed treatment of the Kalman filter is given in Chapter XI. Experimental studies have been carried out to compare the performance of the Kalman filter and the Luenberger observer in the problem of the estimation of the state of a double reversible pendulum for inaccurately specified initial conditions and in the presence of noise.[5] The Luenberger observer has turned out to be more sensitive to the presence of noise as compared to the Kalman filter.

5. Observer of reduced order in the stabilization system of an aircraft[6]. The longitudinal motion of an aircraft can be described by the equations

$$\begin{gathered} \ddot{\vartheta} = d_1\alpha - d_2\delta, \qquad \dot{\theta} = d_3\alpha, \\ \alpha = \vartheta - \theta, \\ T\dot{\delta} + \delta = K_s u, \end{gathered} \tag{3.29}$$

where ϑ is the angle of pitch, θ is the slope of the flight path, α is the angle of attack, δ is the magnitude of elevator deflection, d_i are aerodynamic coefficients, T and K_s are the characteristics of the electro-hydraulic controller, and u is the control input (see Figure 3.1). Using the variables

$$x_1 = \dot{\vartheta}, \qquad x_2 = \vartheta, \qquad x_3 = \delta, \qquad x_4 = \theta,$$

we can write equations (3.29) in the form

$$\dot{x} = Ax + Bu,$$

[5]Andreyev, Yu. N., State Space Algebraic Methods in the Theory of Control of Linear Objects, *Avtomat. i Telemekh.*, (1977), no. 3, 5–50. (Translated into English as *Automat. Remote Control*)

[6]Kuzovkov, N. T., *Modal Control and Observers*, Mashinostroenie, Moscow, (1976).

where

$$A = \begin{pmatrix} 0 & -d_1 & -d_2 & d_1 \\ 1 & 0 & 0 & 0 \\ 0 & 0 & -T^{-1} & 0 \\ 0 & d_3 & 0 & -d_3 \end{pmatrix}, \qquad B = \begin{pmatrix} 0 \\ 0 \\ K_s T^{-1} \\ 0 \end{pmatrix}, \qquad x = \begin{pmatrix} x_1 \\ x_2 \\ x_3 \\ x_4 \end{pmatrix}.$$

Here the parameters d_1, d_2, d_3, K_s, T are all positive quantities. Although any of the state variables $x_1, \ldots, x_4$ can be mesured, this is not always expedient. For example, only two of the state variables can be measured, and the other two determined through the use of an observer.

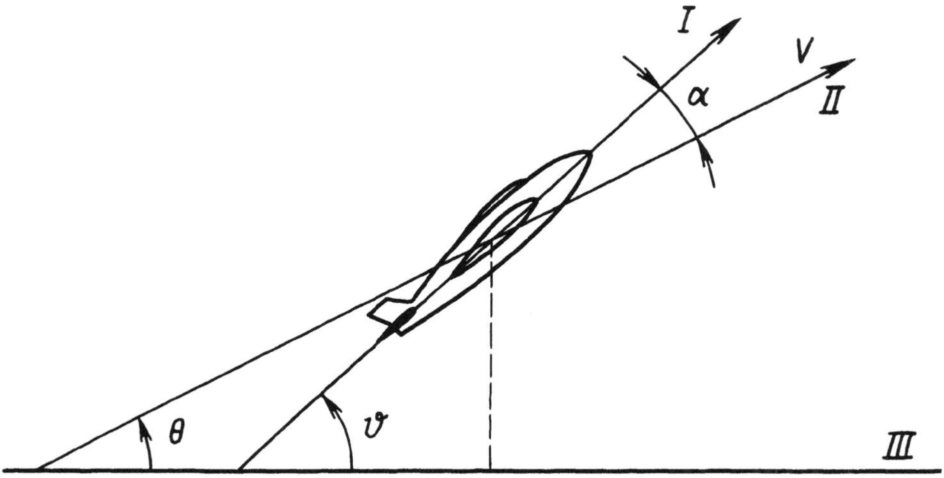

FIG. 3.1. The coordinates of the longitudinal motion of an aircraft: I, the axis of the aircraft; II, the direction of the instantaneous velocity V; III, the horizontal; ϑ, the angle of pitch; θ, the slope of the flight path; $\alpha = \vartheta - \theta$, the angle of attack

In the case when the variables $x_3 = \delta$ and $x_4 = \theta$ are measured, the measurement matrix Q has the form

$$Q = \begin{pmatrix} 0 & 0 & 1 & 0 \\ 0 & 0 & 0 & 1 \end{pmatrix},$$

and the observability matrix $\left(Q', A'Q', (A')^2 Q', (A')^3 Q'\right)$ is of the form

$$\begin{pmatrix} 0 & 0 & 0 & 0 & 0 & d_3 & 0 & -d_3^2 \\ 0 & 0 & 0 & d_3 & 0 & -d_3^2 & 0 & -d_1 d_3 + d_3^3 \\ 1 & 0 & -T^{-1} & 0 & T^{-2} & 0 & -T^{-3} & -d_2 d_3 \\ 0 & 1 & 0 & -d_3 & 0 & d_3^2 & 0 & d_1 d_3 - d_3^3 \end{pmatrix}.$$

The rank of the latter matrix is equal to 4, since the 2nd, 4th, 6th, and 8th columns of this matrix are linearly independent. Hence, when the measured parameters are thus chosen, the system is observable.

In an analogous way, it can be proved that if the measured parameters are chosen to be $x_2 = \vartheta$ and $x_3 = \delta$, the system is also observable.

However, if the measured parameters are chosen to be $x_1 = \dot{\vartheta}$ and $x_3 = \delta$, the system is unobservable. In this case the measurement matrix Q is the following:

$$Q = \begin{pmatrix} 1 & 0 & 0 & 0 \\ 0 & 0 & 1 & 0 \end{pmatrix}.$$

Writing out the observability matrix (3.8), we can establish that its rank is equal to 3. We can also prove through the use of Theorem 3.3 that the system is unobservable. In the case under consideration the matrix $(A' - \lambda I, Q')$ has the form

$$\begin{pmatrix} -\lambda & 1 & 0 & 0 & 1 & 0 \\ -d_1 & -\lambda & 0 & d_3 & 0 & 0 \\ -d_2 & 0 & -\frac{1}{T} - \lambda & 0 & 0 & 1 \\ d_1 & 0 & 0 & -d_3 - \lambda & 0 & 0 \end{pmatrix}.$$

For $\lambda = 0$ the 2nd and 4th rows of this matrix are linearly dependent. Therefore, the system (3.29) is unobservable if the measured variables are x_1 and x_3.

When any three state variables are measured, the system (3.29) is observable. But if only one state variable is known, the system (3.29) is unobservable.

We turn now to constructing an observer of reduced order subject to the condition that the measured variables are $x_1 = \dot{\vartheta}$, $x_2 = \vartheta$, $x_3 = \delta$ and the variable $x_4 = \theta$ is to be reconstructed by the observer. The measurement matrix in this case is of the form

$$Q = \begin{pmatrix} 1 & 0 & 0 & 0 \\ 0 & 1 & 0 & 0 \\ 0 & 0 & 1 & 0 \end{pmatrix}.$$

The matrix W is taken as $W = (0\ 0\ 0\ 1)$. It is readily seen that the matrix L^{-1} now becomes

$$L^{-1} = \begin{pmatrix} 1 & 0 & 0 & 0 \\ 0 & 1 & 0 & 0 \\ 0 & 0 & 1 & 0 \\ 0 & 0 & 0 & 1 \end{pmatrix}.$$

Hence, we have

$$E = \begin{pmatrix} 1 & 0 & 0 \\ 0 & 1 & 0 \\ 0 & 0 & 1 \\ 0 & 0 & 0 \end{pmatrix}, \qquad D = \begin{pmatrix} 0 \\ 0 \\ 0 \\ 1 \end{pmatrix}.$$

For V we take the matrix $V = (0\ 0\ -1)$. For the Luenberger observer of the first order, given the explicit expressions for the matrices Q, W, E, D, and V, equation (3.28) becomes

$$\dot{\alpha}_2 = -d_3 \alpha_2 + \frac{K_s}{T} u + (d_3 - \frac{1}{T}) x_3.$$

For the estimate of the state of the system (3.29) we now take the vector $(x_1, x_2, x_3, \alpha_2)$ where x_1, x_2, x_3 are the measured variables.

Let us now assume that the variables x_3 and x_4 are measured. Then the measurement matrix Q is

$$Q = \begin{pmatrix} 0 & 0 & 1 & 0 \\ 0 & 0 & 0 & 1 \end{pmatrix}.$$

The matrix W may be taken as

$$W = \begin{pmatrix} 1 & 0 & 0 & 0 \\ 0 & 1 & 0 & 0 \end{pmatrix}.$$

The matrices L^{-1}, E, D, WAD are of the form

$$L^{-1} = \begin{pmatrix} 0 & 0 & 1 & 0 \\ 0 & 0 & 0 & 1 \\ 1 & 0 & 0 & 0 \\ 0 & 1 & 0 & 0 \end{pmatrix}, \qquad E = \begin{pmatrix} 0 & 0 \\ 0 & 0 \\ 1 & 0 \\ 0 & 1 \end{pmatrix},$$

$$D = \begin{pmatrix} 1 & 0 \\ 0 & 1 \\ 0 & 0 \\ 0 & 0 \end{pmatrix}, \qquad WAD = \begin{pmatrix} 0 & d_1 \\ 1 & 0 \end{pmatrix}.$$

We choose a 2×2 matrix V as

$$V = \begin{pmatrix} 1 & 0 \\ 0 & 1 \end{pmatrix}.$$

Then, after a series of computations, we obtain

$$W - VQ = \begin{pmatrix} 1 & 0 & -1 & 0 \\ 0 & 1 & 0 & -1 \end{pmatrix}, \qquad AD = \begin{pmatrix} 0 & -d_1 \\ 1 & 0 \\ 0 & 0 \\ 0 & d_3 \end{pmatrix},$$

$$(W - VQ)AD = \begin{pmatrix} 0 & -d_1 \\ 1 & -d_3 \end{pmatrix}, \qquad (W - VQ)B = \begin{pmatrix} -K_s T^{-1} \\ 0 \end{pmatrix},$$

$$(W - VQ)A(DV + E) = \begin{pmatrix} -d_2 + T^{-1} & 0 \\ 1 & 0 \end{pmatrix}.$$

Thus, we obtain an equation of the form (3.28) for the observer of the second order:

$$\begin{pmatrix} \dot\alpha_2^1 \\ \dot\alpha_2^2 \end{pmatrix} = \begin{pmatrix} 0 & -d_1 \\ 1 & -d_3 \end{pmatrix} \begin{pmatrix} \alpha_2^1 \\ \alpha_2^2 \end{pmatrix} - \begin{pmatrix} K_s T^{-1} \\ 0 \end{pmatrix} u + \begin{pmatrix} -d_2 + T^{-1} & 0 \\ 1 & 0 \end{pmatrix} \begin{pmatrix} x_3 \\ x_4 \end{pmatrix}. \tag{3.30}$$

Here α_2^1 and α_2^2 denote the coordinates of the vector α_2. It should be noted that equation (3.30) is asymptotically stable, since its characteristic equation has the form $\lambda^2 + d_3\lambda + d_1 = 0$, where $d_1 > 0$, $d_3 > 0$ and both its roots are located in the left half-plane.

For an estimate $m(t)$ of the vector $x(t)$, in view of equation (3.27), we take the vector

$$m(t) = D\alpha_2(t) + (E + DV)y(t) = \begin{pmatrix} \alpha_2^1(t) \\ \alpha_2^2(t) \\ x_3(t) \\ x_4(t) \end{pmatrix}.$$

Thus, an observer of reduced order in the stabilization system of an aircraft has been constructed.

§ 4. Linear Feedback Control Systems

1. Various ways of describing feedback control systems. The realization problem. The feedback control systems are of major importance in engineering. For such systems there exists special terminology and specific methods of investigation have been evolved. Most feedback control systems can be described schematically by means of the block diagram composed of two units: the generalized plant, G, to be controlled and the controller, K (see Figure 4.1). The goal of the design of a feedback control system is to realize the "least" error z at the output end of the plant by optimizing the structure and parameters of the controller K.

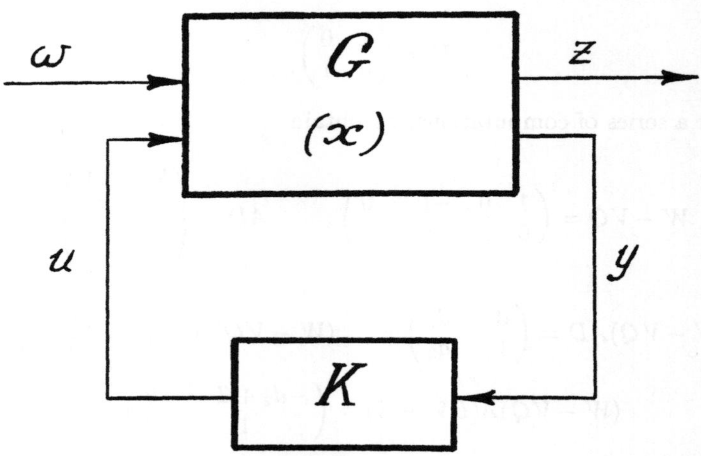

FIG. 4.1. A generalized block diagram of a feedback control system: G, generalized plant; K, controller; ω, all external inputs (including disturbances, sensor noise, and commands); z, output or error signal; y, measured variables; u, control input; x, state variables

The following terms are used. A system with one (scalar) input $\omega(t)$ and one (scalar) output $z(t)$ is called a *single input–single output system*, or *SISO-system*, for short. If there are many inputs and outputs ($\omega(t)$ and $u(t)$ being vectors of appropriate, in general, different, dimensions), then the system is called a *multiple input–multiple output system* (*multi-input–multi-output system*), or *MIMO-system*, for short.

A time-invariant linear SISO-system is described in one of the following two ways. One way is to use the *state-space description*. It is generally assumed for a SISO-system that $x(t) = z(t) = y(t)$ and $\omega(t) = 0$. Then a direct relation of the output $z(t)$ of the closed-loop system and its input $u(t)$ can be given. This relation is described by a linear differential equation with constant coefficients:

$$\sum_{i=0}^{n} a_i z^{(i)}(t) = \sum_{i=0}^{m} b_i u^{(i)}(t), \qquad z^{(i)}(t) = \frac{d^i z}{dt^i}, \tag{4.1}$$
$$a_i = \text{constant},\ a_n \neq 0,\ b_i = \text{constant}.$$

It is usually assumed that the following *condition for realizability* is valid:

$$n > m. \tag{4.2}$$

Under the condition (4.2) an arbitrary system (4.1) can be realized as a combination of ideal amplifiers and ideal integrators.

We assume that $u(t) = z(t) = 0$ for $t \leqslant 0$. Taking the Laplace transform of both sides of equation (4.1) gives

$$\sum_{i=0}^{n} a_i s^i \bar{z}(s) = \sum_{i=0}^{m} b_i s^i \bar{u}(s), \tag{4.3}$$

where $\bar{z}(s)$ and $\bar{u}(s)$ are the Laplace transforms of $z(t)$ and $u(t)$, respectively. Formula (4.3) yields

$$\bar{z}(s) = G(s)\bar{u}(s). \tag{4.4}$$

In the relation (4.4) the function $G(s)$ is a rational function of the complex variable s of the form

$$G(s) = \frac{\sum\limits_{i=0}^{m} b_i s^i}{\sum\limits_{i=0}^{n} a_i s^i} \tag{4.5}$$

A function $G(s)$ of the form (4.5) is called a *transfer function from u to z*. If a SISO-system is described by the relation (4.4), then we have a so-called *"input-output" description*, or a *frequency-domain description*. For a SISO-system both descriptions are equivalent. The transfer function of the system can be uniquely determined from equation (4.1). Conversely, if the denominator and numerator of the function $G(s)$ are coprime, then only one differential equation of the form (4.3) corresponds to the transfer function $G(s)$.

MIMO-systems can also be described either in the state space or in the frequency domain (by means of the "input-output" operator). In the *state space* approach the time-invariant linear MIMO-system shown in Figure 4.1 is described by the equations

$$\begin{aligned} \dot{x}(t) &= Ax(t) + B_1\omega(t) + B_2u(t), \\ z(t) &= C_1x(t) + D_{11}\omega(t) + D_{12}u(t), \\ y(t) &= C_2x(t) + D_{12}\omega(t) + D_{22}u(t), \\ x \in R^n,\ \omega &\in R^q,\ u \in R^m,\ z \in R^l,\ y \in R^p. \end{aligned} \tag{4.6}$$

The *transfer matrix of the system* (4.6) *from the input* $\omega(t)$ *to the output* $z(t)$ can be expressed as a lower linear-fractional transform (LLFT):

$$\begin{aligned} \bar{z}(s) &= G(s)\bar{\omega}(s), \\ G(s) = G_L(P, K) &= P_{11} + P_{12}K(I - P_{22}K)^{-1}P_{21}. \end{aligned} \tag{4.7}$$

Here $P_{ij} = D_{ij} + C_i(sI - A)^{-1}B_j$, $i, j = 1, 2$, $K = K(s)$ is the transfer matrix of the controller, and $\bar{u}(s) = K(s)\bar{y}(s)$. Moreover, the following condition is to be satisfied:

$$\det\bigl(I - P_{22}(s)K(s)\bigr)^{-1} \neq 0.$$

Given the state-space description (4.6), the transfer matrix $G(s)$ is determined uniquely from (4.7).

The inverse transition from the frequency-domain description to the one in the state space constitutes the so-called "realizability problem". Suppose that some matrix $\tilde{G}(x)$ is given. We must introduce the state variables x and define matrices A, B, C, and D such that the transfer matrix $G(s)$ from u to y of the system

$$\begin{aligned} \dot{x} &= Ax + Bu, \\ y &= Cx + Du, \end{aligned} \tag{4.8}$$

which is equal to $G(s) = D + C(sI - A)^{-1}B$, would coincide with $\tilde{G}(s)$: $G(s) = \tilde{G}(s)$.

The realizability problem does not have a unique solution (see Exercises of Part Two). Various realizations may differ not only by the form of the matrices A, B, C, and D but also by the dimension of the state vector x. A realization of the MIMO-system is called *minimal* if the dimension of the vector x in it is not greater than that in any other system. The following *criterion of the minimality of a realization* is known:[7]

A realization (A, B, C, D) *is minimal if and only if the pair* (A, B) *is controllable, and the pair* (A, C) *is observable.*

Even the minimal realization is not determined uniquely. It is easily seen that if (A, B, C, D) is a minimal realization of some system, then for any nondegenerate matrix T the realization $(TAT^{-1}, TB, CT^{-1}, D)$ is also minimal.

[7] Kailath, T., *Linear Systems*, Prentice-Hall, Englewood Cliffs, N. J., (1980).

2. Criteria of performance for SISO-systems. In the construction of a feedback control system the controller must be chosen so that the resulting closed-loop system has a "good" performance. We can make this restriction precise in a number of ways, including the following one. The main problems for SISO-systems are understood to be the following two:

the *regulation problem* which consists in maintaining the output variable $y(t)$ in the proximity of a prescribed constant level $y_d =$ constant;

the *tracking problem* which consists in maintaining the output variable $y(t)$ in the proximity of a prescribed function $y_d(t)$.

The difference $\varepsilon(t) = y_d(t) - y(t)$ is called a *dynamic error*, and the quantity $\varepsilon(t) = y_d - y(t)$ is called a *control error*.

A feedback control system in which the error $\varepsilon(t) = 0$ for all $t \geqslant 0$ is said to be an ideal performance system. Such systems are called *invariant*.[8] The first conditions for the invariance of control systems were obtained by G. V. Shchipanov and N. N. Luzin in the late thirties. However, in actual practice it proved impossible to implement invariant systems owing to the fact that the invariance property is violated under small perturbations of the parameters of the system.

In engineering practice it is often assumed that a *system has good performance* if it satisfies the following constraints:

—the closed-loop system is stable, i.e., to each bounded input there corresponds a bounded output;

—linear ($y_d(t) = a_0 + a_1 t$, $|a_1| \leqslant \bar{a}_1$) and harmonic low-frequency signals ($y_d(t) = b\cos\omega t$, $|b| \leqslant \bar{b}$, $0 \leqslant \omega \leqslant \omega_1$) are reproduced with a small limiting error ε_0, i.e.,

$$\varepsilon_\infty = \lim_{t\to\infty} |\varepsilon(t)| \leqslant \varepsilon_0;$$

—high-frequency disturbances acting on the system ($N(t) = a\cos wt$, $|a| \leqslant \bar{a}$, $w \geqslant w_2 \geqslant w_1$) are suppressed.

For linear SISO-systems these constraints suffice to formulate constraints on the transfer function. The requirement of stability is equivalent to the restriction that the denominator of the transfer function have no zeros in the right half-plane. This condition, together with the condition for realizability, suggests that the transfer function $G(s)$ is bounded on the imaginary axis:

$$\sup_\omega |G(i\omega)| \leqslant \infty.$$

It can be shown that the other constraints lead to the conclusion that the ideal transfer function $G_{\mathrm{id}}(s)$ must coincide with that of an *ideal low-pass filter* (see Figure 4.2). We can now state the *main criterion for choosing a controller for a SISO-system*:

The transfer function $K(s)$ of the controller is chosen so that the transfer function $G(s)$ of the closed-loop system approximates as close possible the ideal transfer function $G_{\mathrm{id}}(s)$.

---

[8] Kukhtenko, A. I., *Problem of Invariance in Automatic Control*, Gostekhizdat Ukr. SSR, Kiev, (1963).

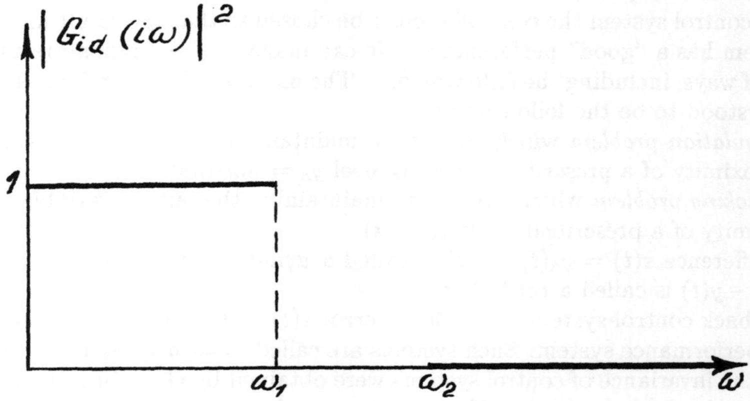

FIG. 4.2. Gain-frequency characteristic of an ideal low-pass filter: $[0, \omega_1]$, pass band; ω_2, cut-off frequency

For more details on the techniques of synthesizing SISO systems, the reader is referred to several excellent books on the subject.[9]

3. Criteria of performance for MIMO-systems. The Hardy spaces H_2 and H_∞. For MIMO-systems the choice of criteria of performance presents some difficulties. We can, as in § 2 of Chapter IV, introduce a quadratic cost functional and try to determine a controller in the state space by solving an appropriate LQ-problem. Such an approach is widely used. However, the choice of a quadratic functional is more or less arbitrary, and hence the performance of the resulting system can turn out to be "middling".

Other criteria of performance are stated in more conventional (from an engineering point of view) terms of the transfer matrix $G(s)$. For a realizable and stable system its transfer matrix $G(s)$ is analytical in the right half-plane $\operatorname{Re} z > 0$ and bounded on the imaginary axis. For such matrices this enables us to determine two norms: $\|G(s)\|_2$ and $\|G(s)\|_\infty$.

In the frequency domain the norm $\|G(s)\|_2$ is determined by the equality

$$\|G(s)\|_2 = \left(\frac{1}{2\pi}\int_{-\infty}^{\infty} \operatorname{tr}\left[G^*(i\omega)G(i\omega)\right] d\omega\right)^{1/2}, \tag{4.9}$$

where $i^2 = -1$ and $G^*(i\omega)$ is a transpose and an adjoint matrix to $G(i\omega)$. In the time domain the norm $\|G(s)\|_2$ for the system (4.8) is given by

$$\begin{aligned}\|G(s)\|_2 &= \operatorname{tr}\left[CL_cC'\right] = \operatorname{tr}\left[B'L_0B\right] \\ &= \operatorname{tr}\left[\sqrt{BB'}L_0\sqrt{BB'}\right],\end{aligned} \tag{4.10}$$

[9]Zadeh, L. A., and Desoer, C. A, *Linear System Theory: the State-Space Approach*, McGraw-Hill, New York, (1963); Pervozvanskii, A. A., *Automatic Control*, Nauka, Moscow, (1986).

where L_c denotes the controllability Gramian of (A, B) and L_0 the observability Gramian of (A, C), i.e.,

$$AL_c + L_cA' + BB' = 0, \qquad A'L_0 + L_0A + C'C = 0. \tag{4.11}$$

The norm $\|G(s)\|_\infty$ is determined by

$$\|G(s)\|_\infty = \sup_\omega \sigma_{\max}\big(G(i\omega)\big). \tag{4.12}$$

Here $\sigma_{\max}\big(G(i\omega)\big)$ is the maximum singular number of the matrix $G(i\omega)$. It will be recalled that

$$\sigma_{\max}\big(G(i\omega)\big) = \Big(\lambda_{\max}\big(G^*(i\omega)G(i\omega)\big)\Big)^{1/2}, \tag{4.13}$$

where $\lambda_{\max}$ is the maximum eigenvalue of the matrix $G^*(i\omega)G(i\omega)$. Since the matrix $G^*(i\omega)G(i\omega) \geqslant 0$ (i.e., positive semidefinite), it follows that all its eigenvalues are real and nonnegative. Therefore, $\sigma_{\max}\big(G(i\omega)\big)$ is a real nonnegative number.

For a SISO-system $\|G(s)\|_\infty = \sup_\omega |G(i\omega)|$ is the maximum of the gain-frequency characteristic of the system.

The proof of the fact that the expressions (4.9) and (4.13) are norms, together with basic information about the spaces H_2 and H_∞ is given in a book by W. Rudin.[10]

The Hardy spaces H_2 and H_∞ appear in a natural way when we consider MIMO-systems. Let us show this. Consider first a signal $u(t) \in L_2$ of finite energy:

$$\|u(t)\|_2^2 = \int_0^\infty u'(t)u(t)\,dt \leqslant \infty. \tag{4.14}$$

As is known, the Laplace transform $\bar{u}(s)$ of such a signal is a complex-valued function analytical in the right half-plane $\operatorname{Re} s > 0$ and such that

$$\sup_{\gamma\to+0}\left(\int_{-\infty}^{\infty} \bar{u}'(\gamma - i\omega)\bar{u}(\gamma + i\omega)\,d\omega\right) \leqslant \infty.$$

In other words, $\bar{u}(s)$ belongs to the Hardy space H_2, in which the norm $\|\bar{u}(s)\|_2$ is defined by the relation

$$\|\bar{u}(s)\|_2 = \left(\frac{1}{2\pi}\int_{-\infty}^{\infty} \bar{u}'(-i\omega)\bar{u}(i\omega)\,d\omega\right)^{1/2}. \tag{4.15}$$

In this case, Parseval's equality holds true: $\|u(t)\|_2 = \|\bar{u}(s)\|_2$.

We now assume that the input signal to a realizable and stable MIMO-system is a signal $u(t) \in L_2$. Then the output signal $y(t)$ of this system satisfies the inequality

$$\|y(t)\|_2 = \|y(s)\|_2 \leqslant \|G(s)\|_\infty\|\bar{u}(s)\|_2 = \|G(s)\|_\infty\|u(t)\|_2. \tag{4.16}$$

[10]Rudin, W., *Real and Complex Analysis*, McGraw-Hill, New York, (1970).

The norms appearing in the inequality (4.16) are defined by formulas (4.14), (4.15), and (4.12). Thus, the output energy $\|y(t)\|_2$ decreases with $\|G(s)\|_\infty$.

We can now state the following *criterion of performance for a MIMO-system*:

The performance of the controller K is the better, the less is the H_∞-norm of the transfer matrix $G(s)$ of the closed-loop system.

This criterion of performance is due to G. Zames[11] and is now one of the most widely applied criteria of performance for a MIMO-system. If the input signal of the system is taken as a disturbance, and the output signal as an error, then by minimizing the H_∞-norm of the transfer matrix, we minimize the energy of the error for the worst input disturbance.

§ 5. Fundamentals of H_∞-Theory

1. Statement of the problem of constructing an H_∞-optimal controller. The main problem in H_∞-theory is the *problem of determining an H_∞-optimal controller*:

In the class of admissible controllers K (i.e., ensuring the stability of a closed-loop system) choose a controller K_{opt} so that the following equality holds:

$$\|G_L(P, K_{\text{opt}})\|_\infty = \inf_K \|G_L(P, K)\|_\infty = \gamma_{\min}. \tag{5.1}$$

Another major problem in H_∞-theory is the *problem of determining an H_∞-suboptimal controller*:

In the class of admissible controllers K, for a given admissible level $\gamma > \gamma_{\min}$, choose a controller K_γ so that the following inequality is valid:

$$\|G_L(P, K_\gamma)\|_\infty \leqslant \gamma \tag{5.2}$$

An H_∞-optimal controller can also be constructed in the frequency domain by means of certain procedures of the theory of analytical functions and of functional analysis (using the theory of Hankel operators, the solution of the Nehari problem, and various methods of factorization of the transfer matrix). However, such an approach leads to complicated calculations.

At present, the iteration approach in the state space is the one that is most often used. In this approach, first, for some $\gamma > \gamma_{\min}$ an H_∞-suboptimal controller is constructed directly in the state space. Then the values of γ are successively decreased. Ultimately, we obtain an H_∞-optimal controller. For reasons to be explained later such an approach is called the *2-Riccati method*.

Thus, in this approach the most important problem is that of constructing an H_∞-suboptimal controller.

[11] Zames, J., Feedback and Optimal Sensitivity: Model Reference Transformations, Multiplicative Seminorms, and Approximate Inverses, *IEEE Trans. Automat. Control*, (1981), vol. AC-26, 301–320.

2. Estimates of the H_∞- and H_2-norms of the transfer matrix of an auxiliary system. Consider an auxiliary system

$$\begin{aligned} \dot{x}(t) &= Ax(t) + Bu(t), \\ y(t) &= Cx(t), \\ x \in R^n, \; & u \in R^m, \; y \in R^p. \end{aligned} \tag{5.3}$$

Suppose that the matrix A is stable, i.e., all eigenvalues $\lambda_i(A)$ of the matrix A lie in the left half-plane, $\operatorname{Re}\lambda_i(A) < 0$. The transfer matrix from $u(t)$ to $y(t)$ of the system (5.3) has the form

$$G(s) = C(sI - A)^{-1}B. \tag{5.4}$$

Lemma 5.1 (the estimate of the H_∞-norm). *Let there exist a constant number $\gamma > 0$ and a matrix $Q \geqslant 0$ satisfying an algebraic Riccati equation of the form*

$$AQ + QA' + \gamma^{-2}QC'CQ + BB' = 0. \tag{5.5}$$

Then

$$\|G(s)\|_\infty \leqslant \gamma. \tag{5.6}$$

Proof. Adding and subtracting $i\omega IQ$ on the left-hand side of the algebraic Riccati equation (5.5), we obtain the equation

$$BB' = (i\omega I - A)Q + Q(-i\omega I - A)' - \gamma^{-2}QCC'Q.$$

Multiplying both sides of this equation by $C(i\omega I - A)^{-1}$ on the left and $\left(-i\omega I - A'\right)^{-1}C'$ on the right gives

$$\begin{aligned} C(i\omega I - A)^{-1}BB'\left(-i\omega I - A'\right)^{-1}C' & \\ = C(i\omega I - A)^{-1}QC' + CQ\left(-i\omega I - A'\right)^{-1}C' & \\ - \gamma^{-2}C(i\omega I - A)^{-1}QC'CQ\left(-i\omega I - A'\right)^{-1}C'. & \end{aligned} \tag{5.7}$$

Note that the left side of (5.7) is equal to $G(i\omega)G^*(i\omega)$. Introducing the matrix $S = C(i\omega I - A)^{-1}QC'$, we can write the relation (5.7) in the form

$$\begin{aligned} G(i\omega)G^*(i\omega) &= S + S^* - \gamma^{-2}SS^* \\ &= \gamma^2 I - (\gamma I - \gamma^{-1}S)(\gamma I - \gamma^{-1}S)^* \end{aligned} \tag{5.8}$$

Since for any matrix S

$$(\gamma I - \gamma^{-1}S)(\gamma I - \gamma^{-1}S)^* \geqslant 0,$$

it follows from (5.8) that

$$G(i\omega)G^*(i\omega) \leqslant \gamma^2 I, \qquad \lambda_{\max}(GG^*) \leqslant \gamma^2.$$

Therefore, $\|G(s)\|_\infty \leqslant \gamma$. □

Remark 5.1. It can be shown that for $\|G(s)\|_\infty < \gamma$ there exists a unique matrix $Q \geqslant 0$ satisfying equation (5.5) and such that $\operatorname{Re}\lambda(A + \gamma^{-2}QCC') \leqslant 0$.

We now estimate the norm $\|G(s)\|_2$ of the matrix (5.4). It will be recalled that

$$\|G(s)\|_2^2 = \operatorname{tr}(CL_cC') = \operatorname{tr}(B'L_0B) = \operatorname{tr}\left[\sqrt{BB'}L_0\sqrt{BB'}\right].$$

Lemma 5.2 (the estimate of the H_2-norm). *Let there exist a number $\gamma > 0$ and a matrix $P \geqslant 0$ satisfying an algebraic Riccati equation of the form*

$$A'P + PA + \gamma^{-2}PBB'P + C'C = 0. \tag{5.9}$$

Then

$$\|G(s)\|_2^2 \leqslant \operatorname{tr}(PBB').$$

Proof. We define a functional J by the formula

$$J = \operatorname{tr}(PBB') = \operatorname{tr}\left[\sqrt{BB'}P\sqrt{BB'}\right],$$

where P is a solution of the algebraic Riccati equation (5.9). Further, subtracting equation (4.10) from equation (5.9) gives

$$A'(P - L_0) + (P - L_0)A + \gamma^{-2}PBB'P = 0.$$

Since the matrix A is stable and the matrix $\gamma^{-2}PBB'P \geqslant 0$ for all B and P, it follows from Theorem 1.7.2 of Chapter I (on the solvability of the Lyapunov equation) that $P - L_0 \geqslant 0$. Therefore,

$$\sqrt{BB'}P\sqrt{BB'} \geqslant \sqrt{BB'}L_0\sqrt{BB'},$$

and hence

$$J = \operatorname{tr}\left[\sqrt{BB'}P\sqrt{BB'}\right] \geqslant \operatorname{tr}\left[\sqrt{BB'}L_0\sqrt{BB'}\right] = \|G(s)\|_2^2.$$

This completes the proof. □

3. The H_∞-problem for a static controller. Consider the system

$$\begin{gathered}\dot{x} = A_1x + B_1\omega(t) + B_2u(t),\\ z(t) = C_1x(t) + D_{12}u(t), \qquad D_{12}'C_1 = 0,\\ x \in R^n,\ w \in R^q,\ u \in R^m,\ z \in R^l.\end{gathered} \tag{5.10}$$

We set the problem of constructing a static controller

$$u = Kx, \qquad K \in \mathcal{L}(R^n, R^m), \tag{5.11}$$

where K is a numerical matrix of appropriate dimensions such that the output $z(t)$ is small for any arbitrary disturbance $\omega(t)$.

The system (5.10) closed by the controller (5.11) will be described by equations (5.3) in which

$$A = A_1 + B_2K, \qquad B = B_1, \qquad C = C_1 + D_{12}K.$$

The problem stated above can now be made more precise in the following way: *determine a matrix* $K \in \mathcal{L}(R^n, r^m)$ *such that*

(1) $\operatorname{Re}\lambda_i(A) = \operatorname{Re}\lambda_i(A_1 + B_2K) < 0$, $i = 1, \dots, n$;

(2) $\|G\|_\infty < \gamma$, *where* γ *is prescribed, and* $G(s) = C(sI - A)^{-1}B$;

(3) *the functional* $J = \operatorname{tr}(PBB')$, *where* P *is a solution of the algebraic Riccati equation (5.9) attains its minimum value.*

Theorem 5.1. *Suppose that for some* $\gamma > 0$ *there exists a matrix* K *satisfying the conditions* (1)–(3). *Then the matrix* K *is of the form*

$$K = -\left(D_{12}'D_{12}\right)^{-1}B_2'P, \tag{5.12}$$

where the matrix $P \geqslant 0$ *satisfies an algebraic Riccati equation of the form*

$$A_1'P + PA_1 + C_1'C_1 - PB_2\left(D_{12}'D_{12}\right)^{-1}B_2'P + \gamma^{-2}PB_1B_1'P = 0. \tag{5.13}$$

Proof. It should be noted first that the set of matrices K such that $\operatorname{Re}\lambda_i(A) = \operatorname{Re}\lambda_i(A_1 + B_2K) < 0$ is an open set. Therefore, to determine the matrix K that renders a minimum of the functional J, we can use the method of indeterminate multipliers of Lagrange. Let λ_0 and Q denote the set of Lagrange multipliers, where λ_0 is some number, and Q is an $n \times n$ matrix. We introduce the Lagrange function by the formula

$$\begin{aligned} L = \lambda_0 \operatorname{tr}(PB_1B_1') + \operatorname{tr} Q\big[&(A_1 + B_2K)'P + P(A_1 + B_2K) \\ &+ \left(C_1'C_1 + K'D_{12}'D_{12}K\right) + \gamma^{-2}PB_1B_1'P\big]. \end{aligned}$$

Since $\operatorname{Re}\lambda_i(A_1 + B_2K) < 0$, the equality $\lambda_0 = 0$ cannot hold. Hence, we can set $\lambda_0 = 1$. Taking the partial derivatives of the function L with respect to the matrices Q, K, we have

$$\begin{aligned} \frac{\partial L}{\partial Q} &= (A_1 + B_2K)'P + P(A_1 + B_2K) \\ &\qquad + \left(C_1'C_1 + K'D_{12}'D_{12}K\right) + \gamma^{-2}PB_1B_1'P = 0, \\ \frac{\partial L}{\partial K} &= 2B_2'PQ + 2D_{12}'D_{12}KQ = 0. \end{aligned} \tag{5.14}$$

This yields

$$K = -\left(D_{12}'D_{12}\right)^{-1}B_2'P.$$

Substituting this expression for the matrix K in equation (5.14), we can verify that the matrix P satisfies equation (5.13). □

Remark 5.1. The converse statement is also true. A controller of the form (5.12) with matrix $P \geqslant 0$ satisfying equation (5.13) ensures that the conditions (1)–(3) hold in the system (5.10) closed by this controller.

Remark 5.2. Subsequently, in § 3 (subsection 3) of Chapter VII the following time-invariant linear-quadratic problem will be treated:

$$\begin{gathered}\dot{x}(t) = Ax(t) + Bu(t), \quad x(0) = 0,\\ J(u) = \int_0^\infty [x'(t)N_1x(t) + u'N_0u]\,dt,\\ N_1 > 0,\ N_0 > 0.\end{gathered} \tag{5.15}$$

The optimal control in the problem (5.15) is of the form

$$u_0(t,x) = -N_0^{-1}B'Px, \tag{5.16}$$

where the matrix P is a solution of the algebraic Riccati equation

$$A'P + PA + N_1 - PBN_0^{-1}B'P = 0. \tag{5.17}$$

Note that in this LQ-problem the optimal controller (5,16), (5.17) coincides exactly with the controller described in Theorem 5.1 if we set

$$N_0 = D_{12}'D_{12}, \qquad N_1 = C_1'C_1, \qquad B_1 = 0.$$

4. The general case of a dynamic controller. Consider the general MIMO-system

$$\begin{gathered}\dot{x}(t) = A_1x(t) + B_1\omega(t) + B_2u(t),\\ z(t) = C_1x(t) + D_{12}u(t),\\ y(t) = C_2x(t) + D_{21}\omega(t),\\ x \in R^n,\ \omega \in R^q,\ u \in R^m,\ z \in R^l,\ y \in R^p.\end{gathered} \tag{5.18}$$

We shall try to determine a dynamic controller of the form

$$\begin{gathered}\dot{x}_c(t) = A_cx_c(t) + B_cy(t),\\ u(t) = C_cx_c(t).\end{gathered} \tag{5.19}$$

Suppose that the following conditions are met:

(a) the matrix triples (A_1, B_1, C_1) and (A_1, B_2, C_2) are stabilizable and detectable,

(b) $D_{12}'(C_1 \quad D_{12}) = (0 \quad 1)$,

(c) $\begin{pmatrix} B_1 \\ D_{21}\end{pmatrix} D_{21}' = \begin{pmatrix} 0 \\ 1\end{pmatrix}$.

The condition (a) is usually met in optimal control problems. The conditions (b) and (c) can always be satisfied due to previous linear transformations of the system (5.18).

Theorem 5.2[12] (the existence of an H_∞-suboptimal controller). *Suppose that*

(1) *there exists a unique real solution* $P \geqslant 0$ *of the algebraic Riccati equation*

$$A_1'P + PA_1 + C_1'C_1 + \gamma^{-2}PB_1B_1'P - PB_2\left(D_{12}'D_{12}\right)^{-1}B_2'P = 0 \qquad (5.20)$$

such that

$$\operatorname{Re}\lambda_i\Big(A_1 - B_2\left(D_{12}'D_{12}\right)^{-1}B_2'P\Big) \leqslant 0;$$

(2) *there exists a unique real solution* $Q \geqslant 0$ *of the algebraic Riccati equation*

$$A_1Q + QA_1' + B_1B_1' + \gamma^{-2}QC_1'C_1Q - QC_2'\left(D_{21}D_{21}'\right)^{-1}C_2Q = 0 \qquad (5.21)$$

such that

$$\operatorname{Re}\lambda_i\Big(A_1' - C_2'\left(D_{21}D_{21}'\right)^{-1}C_2Q\Big) \leqslant 0;$$

(3) *the spectral radius of the product* PQ *is less than* γ^2*:*

$$\rho(PQ) < \gamma^2.$$

Then there exists an H_∞*-suboptimal controller* K *of the form* (5.19) *such that*

$$\|G_L(P, K)\|_\infty \leqslant \gamma.$$

In general, there exist infinitely many H_∞-suboptimal controllers. One of these is described by equations (5.19), in which

$$\begin{aligned}
B_c &= QC_2'\left(D_{21}D_{21}'\right)^{-1};\\
C_c &= -\left(D_{12}'D_{12}\right)^{-1}B_2'P(I - \gamma^{-2}QP)^{-1};\\
A_c &= A - B_cC_2 + B_2C_c + \gamma^{-2}QC_1'C_1 - B_cD_{12}C_c.
\end{aligned}$$

In view of the fact that two algebraic Riccati equations (5.20) and (5.21) are to be solved in order to determine an H_∞-suboptimal controller based on Theorem 5.2, this method is called the *2-Riccati method.*

5. Robust stability. Concepts connected with the H_∞-norm can be conveniently used in robust stability problems. Such problems arise in cases in which the parameters of the plant are not known precisely but with error.

A more precise statement of one of such problems is the following. Let G_{nom} be the transfer function of a nominal model of the plant and G_Δ be the transfer function of a real (perturbed) plant. We assume that the disturbances are of additive type, i.e.,

$$G_\Delta = G_{\text{nom}} + \Delta. \qquad (5.22)$$

---

[12]Doyle, J. C., Glover, K., Khargonekar, P. P., and Francis, B. A., State-Space Solutions to Standard H^2 and H^∞ Control Problems, *IEEE Trans. Automat. Control*, (1989), vol. AC-34, no. 8, 831–847.

Here $\Delta = \Delta(s)$ is an arbitrary complex uncertainty, i.e., the coefficients of $\Delta(s)$ may be arbitrary complex numbers. We further assume that the H_∞-norm of the disturbances Δ is small:

$$\|\Delta\|_\infty < \varepsilon. \tag{5.23}$$

The class of perturbed plants of the form (5.22), (5.23) will be called the *class* D_ε.

Let us set the following problem.

Determine a controller K which stabilizes any of the plants of class D_ε.

Note that other descriptions of disturbances can be used, such as the multiplicative representation of disturbances $G_\Delta = G_{\text{nom}}(1+\Delta)$, and so on. For each representation of disturbances a corresponding criterion of robust stability is obtained. One of the problems of robust stability is treaeted in § 5 of Chapter I, and its solution is given as the Kharitonov theorem. We shall cite without proof the following theorem on robust stability.

Theorem 5.3[13] (robust stability for H_∞-bounded uncertainty). *A controller K stabilizes any of the plants of class D_ε if and only if*

(*a*) *the controller K stabilizes the nominal plant with transfer matrix G_{nom};*

(*b*) *the H_∞-norm of the nominal plant closed by the controller K is less than ε^{-1}, i.e.,*

$$\|G_{\text{nom}}(P,K)\|_\infty \leqslant \varepsilon^{-1}.$$

§ 6. Zeros of a Linear Time-Invariant System and Their Use

In this section we discuss several results of the theory of MIMO-systems which are connected with the concept of zeros of the system in question.[14] Suppose that a MIMO-system is described by the equations

$$\begin{aligned} \dot{x}(t) &= Ax(t) + Bu(t), \qquad x(0) = x_0,\ x \in R^n,\ u \in R^m, \\ y(t) &= Cx(t), \qquad y \in R^l. \end{aligned} \tag{6.1}$$

An $(n+l)\times(n+m)$ block-matrix $P(\lambda)$ of the form

$$P(\lambda) = \begin{pmatrix} \lambda I_n - A & -B \\ C & 0 \end{pmatrix} \tag{6.2}$$

is called the *matrix of the system* (6.1). In the expression (6.2) λ denotes an arbitrary complex number and I_n an $n \times n$ unit matrix. For $\lambda = 0$ the matrix $P(0)$ is assumed to be of maximum rank, i.e.,

$$\operatorname{rank} P(0) = \begin{pmatrix} -A & -B \\ C & 0 \end{pmatrix} = \min(n+l, n+m).$$

---

[13] Glover, K., and MacFarlane, D., Robust Stabilization of Normalized Coprime Factor Plant Description with H_∞-Bounded Uncertainty, *IEEE Trans. Automat. Control*, (1989), vol. AC-34, no. 8, 821–830.

[14] This section has been written in collaboration with E. M. Smagina and uses some of the results published in her book: Smagina, E. M., *Analysis of Linear Multidimensional Plants Using the Concept of a Zero of the System*, Tomsk University Publishing House, Tomsk, (1990).

By an *invariant zero of the system* (6.1) we shall mean a complex number $\lambda = \alpha$ for which the rank of the matrix decreases, i.e.,

$$\operatorname{rank} P(\lambda)\big|_{\lambda=\alpha} < \min(n+l, n+m). \tag{6.3}$$

Theorem 6.1. *Let α be an invariant zero of the system* (6.1), *not coinciding with any of the eigenvalues λ_i ($i = 1, \dots, n$) of the matrix A. Then there exists an initial condition x_0 and an input*

$$u(t) = u_0 e^{\alpha t} 1(t), \qquad u_0 = \textit{constant}, \ u_0 \in R^m, \tag{6.4}$$
$$1(t) = \begin{cases} 1, & \textit{for } t > 0, \\ 0, & \textit{for } t \leqslant 0, \end{cases}$$

such that the output of the system (6.1) *is identically zero.*

Proof. Taking the Laplace transform of both sides of equation (6.1) gives

$$\begin{gathered} \lambda \bar{x}(\lambda) - x_0 = A\bar{x}(\lambda) + B\bar{u}(\lambda), \\ \bar{y}(\lambda) = C\bar{x}(\lambda), \qquad \bar{x}(\lambda) = \int_0^\infty e^{-\lambda t} x(t)\, dt. \end{gathered} \tag{6.5}$$

Here $\bar{x}(\lambda)$, $\bar{u}(\lambda)$, $\bar{y}(\lambda)$ are vectors whose components are the Laplace transforms of the components of the vectors $x(t)$, $u(t)$, $y(t)$, respectively.

If $\lambda \neq \lambda_i$, then

$$\bar{x}(\lambda) = (\lambda I_n - A)^{-1}\big(x_0 + B\bar{u}(\lambda)\big). \tag{6.6}$$

The Laplace transform $\bar{u}(\lambda)$ of the input (6.4) is computed in explicit form:

$$\bar{u}(\lambda) = \frac{u_0}{\lambda - \alpha}. \tag{6.7}$$

It follows from (6.5)–(6.7) that

$$\begin{aligned} \bar{y}(t) &= C(\lambda I_n - A)^{-1}\left(x_0 + B\frac{u_0}{\lambda - \alpha}\right) \\ &= \frac{1}{\lambda - \alpha} C(\lambda I_n - A)^{-1}\big((\lambda - \alpha) I_n x_0 + B u_0\big) \\ &= \frac{1}{\lambda - \alpha}\big[C x_0 - C(\lambda I_n - A)^{-1}\big((\alpha I_n - A)x_0 - B u_0\big)\big]. \end{aligned} \tag{6.8}$$

In the derivation of this formula the following relation has been used:

$$(\lambda - \alpha) I_n = (\lambda I_n - A) - (\alpha I_n - A).$$

Let us impose the condition $y(t) \equiv 0$, $t \geqslant 0$. We now have

$$y(t_0) = C x_0 = 0, \qquad \bar{y}(\lambda) \equiv 0. \tag{6.9}$$

In view of (6.8) we obtain from the relations (6.9)

$$C(\lambda I_n - A)^{-1}\big((\alpha I_n - A)x_0 - Bu_0\big) \equiv 0. \tag{6.10}$$

The identity (6.10) holds true for all λ if and only if

$$(\alpha I_n - A)x_0 - Bu_0 = 0. \tag{6.11}$$

The first of the equalities (6.9) and the equality (6.11) can be combined to give

$$\begin{pmatrix} \alpha I_n - A & -B \\ C & 0 \end{pmatrix} \begin{pmatrix} x_0 \\ u_0 \end{pmatrix} = 0. \tag{6.12}$$

Let us consider equation (6.12) as one in x_0, u_0. A nontrivial solution of this equation exists if the matrix

$$P(\alpha) = \begin{pmatrix} \alpha I_n - A & -B \\ C & 0 \end{pmatrix}$$

does not have maximum rank. For $l \geqslant m$ this constitutes a necessary and sufficient condition, and for $l < m$ a sufficient condition.

Thus, we have established that if α is an invariant zero of the system, then there exist an input of the form (6.4) and an initial condition x_0 such that there is no transfer through the system, i.e., $y(t) \equiv 0$. □

Example 6.1. Consider the system of the second order with one input and one output:

$$\dot{x} = \begin{pmatrix} 2 & 0 \\ 1 & 1 \end{pmatrix} x + \begin{pmatrix} 1 \\ 0 \end{pmatrix} u, \qquad y = (1 \quad 1)x, \tag{6.13}$$

where

$$A = \begin{pmatrix} 2 & 0 \\ 1 & 1 \end{pmatrix}, \qquad B = \begin{pmatrix} 1 \\ 0 \end{pmatrix}, \qquad C = (1 \quad 1).$$

The matrix A has two distinct eigenvalues $\lambda_1 = 1$, $\lambda_2 = 2$. Let us determine an invariant zero of the system (6.13). To this end, we construct the matrix of the system

$$P(\lambda) = \begin{pmatrix} \lambda I_2 - A & -B \\ C & 0 \end{pmatrix} = \begin{pmatrix} \lambda - 2 & 0 & -1 \\ -1 & \lambda - 1 & 0 \\ 1 & 1 & 0 \end{pmatrix}.$$

For $\lambda = 0$ the rank decreases from 3 to 2. Hence, $\alpha = 0$ is an invariant zero of the system (6.13). Since the rank of the matrix $P(\lambda)\big|_{\lambda=0}$ decreases, we can use the following equation of the form (6.12):

$$\begin{pmatrix} -2 & 0 & -1 \\ -1 & -1 & 0 \\ 1 & 1 & 0 \end{pmatrix} \begin{pmatrix} x_{10} \\ x_{20} \\ u_0 \end{pmatrix} = 0 \tag{6.14}$$

to obtain the vector $(x_0, u_0) = (x_{10}, x_{20}, u_0) = (1, -1, -2)$ for which the equality (6.14) is satisfied. It follows that if the constant signal

$$u(t) = -2e^0 1(t) = -2 \cdot 1(t) \tag{6.15}$$

is applied to the system, then, given the initial condition $x(0) = (1, -1)$, the output of the system is zero.

This can be directly verified by determining the solution of the system (6.13) for the obtained initial conditions and for the input (6.15). To this end, the system (6.13) with the input (6.15) can be written as

$$\begin{aligned} \dot{x}_1 &= 2x_1 - 2, & x_1(0) &= 1, \\ \dot{x}_2 &= x_1 + x_2, & x_2(0) &= -1. \end{aligned} \tag{6.16}$$

It can be readily verified that the solution of the system (6.16) is of the form $x_1(t) = 1$, $x_2(t) = -1$. Hence, the output of the system (6.16) is

$$y(t) = (1 \quad 1)\begin{pmatrix} 1 \\ -1 \end{pmatrix} = 0.$$

The $l \times m$ matrix

$$G(\lambda) = C(\lambda I_n - A)^{-1}B \tag{6.17}$$

is called the *matrix transfer function of the system* (6.1).

Let us assume that $\operatorname{rank} G(0) = \min(l, m)$. A complex number $\lambda = \alpha$ is called a *transfer zero of the system* if

$$\operatorname{rank} G(\lambda)\big|_{\lambda=\alpha} < \min(l, m).$$

It can be shown that the set of transfer zeros of the system (6.1) is contained in the set of invariant zeros of the same system (the former set being, in general, a subset of the latter set).[15] At the same time the following assertion holds.

Theorem 6.2. *If the system* (6.1) *is completely controllable and observable, then the sets of its transfer and invariant zeros coincide.*

Proof. Let $l \geqslant m$. Then the rank of the matrix $P(\lambda)$ of the system (6.1) can change only in the case when for some $\lambda = \alpha$ some columns of the matrix $P(\alpha)$ become linearly dependent. The rank of the $(n+l) \times (n+m)$ matrix $P(\lambda)$ does not change if it is multiplied on the right by a nondegenerate $(n+m) \times (n+m)$ matrix $L_1(\lambda)$ of the form

$$L_1(\lambda) = \begin{pmatrix} I_n & (\lambda I_n - A)^{-1}B \\ 0 & I_m \end{pmatrix}, \qquad \lambda \neq \lambda_i.$$

---

[15]See Exercise 36 of Part Two.

Hence we obtain the following sequence of equalities:

$$\operatorname{rank} P(\lambda) = \operatorname{rank}\big(P(\lambda)L_1(\lambda)\big)$$
$$= \operatorname{rank}\begin{pmatrix} \lambda I_n - A & 0 \\ C & C(\lambda I_n - A)^{-1}B \end{pmatrix} = \operatorname{rank}\begin{pmatrix} \lambda I_n - A & 0 \\ C & G(\lambda) \end{pmatrix}.$$

Owing to the Hautus criterion for observability (see Theorem 3.3), we have

$$\operatorname{rank}\begin{pmatrix} \lambda I_n - A \\ C \end{pmatrix} = n.$$

for all complex numbers λ. Therefore, in this case the rank of the matrix $P(\lambda)$ can decrease for $\lambda = \alpha$ if and only if the rank of the matrix $G(\lambda)$ decreases for $\lambda = \alpha$. Thus, in the case under consideration the invariant and transfer zeros coincide.

Let now $l < m$. If the matrix $P(\lambda)$ is multiplied on the left by a nondegenerate $(n+l) \times (n+l)$ matrix $L_2(\lambda)$ of the form

$$L_2(\lambda) = \begin{pmatrix} I_n & 0 \\ -C(\lambda I_n - A)^{-1} & I_l \end{pmatrix}, \qquad \lambda \neq \lambda_i,$$

then

$$\operatorname{rank} P(\lambda) = \operatorname{rank}\big(L_2(\lambda)P(\lambda)\big) = \operatorname{rank}\begin{pmatrix} \lambda I_n - A & -B \\ 0 & G(\lambda) \end{pmatrix}.$$

Owing to the Hautus criterion for controllability (see Theorem 2.5), we have

$$\operatorname{rank}(\lambda I_n - A \quad -B) = n.$$

for all complex values λ. Therefore, for $\lambda = \alpha$ the rows of the matrix $P(\lambda)$ can become linearly dependent if the rank of the matrix $G(\lambda)$ decreases for this $\lambda = \alpha$. This completes the proof of the theorem. □

Example 6.2. For the system (6.13) the transfer function $G(\lambda)$ is of the form

$$G(\lambda) = C(\lambda I_2 - A)^{-1}B = (1 \quad 1)\begin{pmatrix} \lambda - 2 & 0 \\ -1 & \lambda - 1 \end{pmatrix}^{-1}\begin{pmatrix} 1 \\ 0 \end{pmatrix}$$
$$= \frac{\lambda}{(\lambda - 1)(\lambda - 2)}. \tag{6.18}$$

The only transfer zero of the system (4.13) is $\lambda = 0$.

Furthermore, the system (6.13) is controllable. Indeed, in the matrix

$$\begin{pmatrix} \lambda I_2 - A \\ Q \end{pmatrix} = \begin{pmatrix} \lambda - 2 & 0 \\ -1 & \lambda - 1 \\ 1 & 1 \end{pmatrix}$$

for $\lambda = 0$ the first and third rows are linearly independent, whereas for $\lambda \neq 0$ the second and third rows are linearly independent.

Thus, the system (6.13) has only one invariant zero $\lambda = 0$, which coincides with the transfer zero.

For some other properties of the invariant and transfer zeros the reader is referred to Exercises of Part Two and also to the book by E. M. Smagina cited at the beginning of this section.

Main Results and Formulas of Chapter VI

The condition of generic position. For the system (2.1):

$$\dot{x}(t) = Ax(t) + Bu(t), \qquad x \in R^n,\ u \in R^m,$$

all matrices $G_j = (b_j, Ab_j, \dots, A^{n-1}b_j)$ are nondegenerate; here, $j = 1, \dots, m$, and the b_j are the columns of the matrix B.

Theorem on the finite number of switch points. Let the condition of generic position (1.9) be satisfied. Then each component $u_{0,k}(t)$ of the optimal control $u_0(t)$ has a finite number of switch points. Moreover, if all eigenvalues of the matrix A are real, then the number of such switch points is at most $n-1$.

Controllability of linear systems. For linear time-invariant systems to be controllable, it is necessary and sufficient that the rank of the matrix $K = (B, AB, \dots, A^{n-1}B)$ be equal to n (the dimension of the vector x).

For a linear system (2.1), subject to the control constraint $|u(t)| \leqslant C$, $C > 0$, to be controllable, it is necessary and sufficient that the rank of the matrix $K = (B, AB, \dots, A^{n-1}B)$ be equal to n, and, moreover, that all eigenvalues of the matrix A lie on the imaginary axis.

The Hautus criterion. For the system (2.1) to be controllable, it is necessary and sufficient that for any complex number λ the rank of the matrix $(A - \lambda I, B)$ be equal to n (the dimension of the vector x).

Controllability of linear time-varying systems. For a linear time-varying system of the form (2.12): $\dot{x}(t) = A(t)x(t) + B(t)u(t)$ to be controllable, it is sufficient that there exists a point $\tau \in [t_0, t_1]$ at which the rank of the matrix $K = (K_1, \dots, K_n)$ is equal to n; here

$$K_1(t) = B(t), \qquad K_i(t) = A(t)K_{i-1}(t) - \frac{dK_{i-1}(t)}{dt}.$$

For the system (2.12) to be controllable on the interval $[t_0, t_1]$, it is necessary and sufficient that the matrix $W(t_0, t_1)$ be positive definite.

Canonical form of linear time-invariant control systems. There exists a matrix T such that $Ty(t) = x(t)$, $\dot{y}(t) = T^{-1}ATy + T^{-1}Bu$. Moreover,

$$T^{-1}AT = \begin{pmatrix} A_1 & A_2 \\ 0_1 & A_3 \end{pmatrix}, \qquad T^{-1}B = \begin{pmatrix} B_1 \\ 0_2 \end{pmatrix},$$

where 0_1 and 0_2 are zero matrices of respective dimensions $(n-j) \times j$ and $(n-j) \times m$, and the rank of the matrix $(B_1, A_1B_1, \dots, A_1^{j-1}B_1)$ is equal to j.

The linear time-invariant system (2.1) with scalar control can be reduced to one equation of the nth order of the form (2.20).

The linear time-varying system (2.12) with scalar control and nondegenerate matrix $K(t)$ defined in Theorem 2.2 can be reduced to one equation of the nth order.

Observability of linear systems. For the system $\dot{x}(T) = A(t)x(t)$ to be observable, given the measurement data $y(t) = Q(t)x(t)$, it is necessary and sufficient that the rank of the observability matrix $\left(Q', A'Q', \ldots, (A')^{n-1}Q'\right)$ be equal to n.

For the same system to be observable, it is necessary and sufficient that for any complex number λ the rank of the matrix $(A' - \lambda I, Q')$ be equal to n.

Observer of maximum order:

$$\dot{m}(t) = Am(t) + V[y(t) - Qm(t)].$$

Observer of reduced order (the Luenberger observer):

$$\begin{aligned} m(t) &= D\alpha_2(t) + (E + DV)y(t), \\ \dot{\alpha}_2(t) &= (W - VQ)[AD\alpha_2(t) + Bu(t) + A(DV + E)y(t)], \end{aligned}$$

where

$$L = \begin{pmatrix} Q \\ W \end{pmatrix}, \qquad L^{-1} = (E \quad D), \qquad (W - VQ)AD$$

are stable matrices.

Otherwise,

$$\begin{aligned} m(t) &= Ey(t) + D\alpha_1(t), \\ \dot{\alpha}_1(t) &= WA\big(D\alpha_1(t) + Ey(t)\big) \\ &\qquad + V\left[\dot{y}(t) - QA\big(Ey(t) + D\alpha_1(t)\big)\right] + (W - VQ)Bu(t). \end{aligned}$$

Linear time-invariant feedback systems.

SISO-system:

state-space description:

$$\sum_{i=0}^{n} a_i z^i(t) = \sum_{i=0}^{m} b_i u^i(t);$$

transfer function from $\bar{u}(s)$ to $\bar{z}(s)$:

$$G(s) = \frac{\sum_{i=0}^{m} b_i s^i}{\sum_{i=0}^{n} a_i s^i};$$

condition for realizability: $n > m$.

MIMO-system:

state-space description:

$$\begin{aligned}\dot{x}(t) &= Ax(t) + B_1\omega(t) + B_2u(t),\\ z(t) &= C_1x(t) + D_{11}\omega(t) + D_{12}u(t),\\ y(t) &= C_2x(t) + D_{12}\omega(t) + D_{22}u(t),\\ x \in R^n,\ \omega &\in R^q,\ u \in R^m,\ z \in R^l,\ y \in R^p;\end{aligned}$$

transfer matrix from input $\bar{\omega}(s)$ to output $\bar{z}(s)$ for MIMO-system P closed by a controller K:

$$\begin{gathered}\bar{z}(s) = G(s)\bar{w}(s),\\ G(s) = G_L(P,K) = P_{11} + P_{12}K(I - P_{22}K)^{-1}P_{21},\\ P_{ij} = D_{ij} + C_i(sI - A)^{-1}B_j, \qquad i,j = 1,2,\\ \det(I - P_{22}(s)K(s))^{-1} \neq 0.\end{gathered}$$

Norms in Hardy spaces H_2 and H_∞.

transfer matrix $G(s)$:

$$\begin{aligned}\|G(s)\|_2 &= \Big(\frac{1}{2\pi}\int_{-\infty}^{\infty} \operatorname{tr}\big[G^*(i\omega)G(i\omega)\big]\,d\omega\Big)^{1/2},\\ \|G(s)\|_\infty &= \sup_\omega\Big(\lambda_{\max}\big(G^*(i\omega)G(i\omega)\big)\Big)^{1/2};\end{aligned}$$

input-output estimate:

$$\|y(t)\|_2 \leqslant \|G\|_\infty\|u(t)\|_2.$$

H_∞-theory.

H_∞-criterion for the choice of controller K:

$$\|G_l(P,K_{\text{opt}})\|_\infty = \inf_K \|G_L(P,K)\|_\infty;$$

existence of an H_∞-suboptimal controller: if $P \geqslant 0$ and $Q \geqslant 0$ are solutions of the algebraic Riccati equations

$$\begin{gathered}A_1'P + PA_1 + C_1'C_1 - PB_2\big(D_{12}'D_{12}\big)^{-1}B_2'P + \gamma^{-2}PB_1B_1'P = 0,\\ A_1Q + QA_1' + B_1B_1' + \gamma^{-2}QC_1'C_1Q - QC_2'\big(D_{21}D_{21}'\big)^{-1}C_2Q = 0,\\ \rho(PQ) < \gamma^2,\end{gathered}$$

then there exists a dynamic γ-suboptimal H_∞-controller;

robust stabilization: a controller K stabilizes the family D_ε of plants if K stabilizes the nominal plant and

$$\|G_{\text{nom}}(P,K)\|_\infty \leqslant \varepsilon^{-1}.$$

Zeros of linear time-invariant feedback control systems. A complex number $\lambda = \alpha$ for which the rank of the matrix

$$P(\lambda) = \begin{pmatrix} \lambda I_n - A & -B \\ C & 0 \end{pmatrix}$$

decreases is called an *invariant zero of the system* (6.1):

$$\begin{aligned} \dot{x}(t) &= Ax(t) + Bu(t), \qquad x(0) = x_0, \; x \in R^n, \; u \in R^m, \\ y(t) &= Cx(t), \qquad y \in R^l. \end{aligned}$$

If α is an invariant zero of the system (6.1), not coinciding with any of the eigenvalues of the matrix A, then there exists an initial condition x_0 and an input $u(t)$ such that the output of the system (6.1) is identically zero, i.e., $y(t) \equiv 0$.

A complex number $\lambda = \alpha$ for which the rank of a matrix transfer function of the form

$$G(\lambda) = C(\lambda I_n - A)^{-1}B$$

decreases is called a *transfer zero of the system* (6.1).

If the system (6.1) is completely controllable and observable, then the sets of its transfer and invariant zeros coincide.

CHAPTER VII

DYNAMIC PROGRAMMING APPROACH. SUFFICIENT CONDITIONS FOR OPTIMAL CONTROL

Various evolutionary processes can be treated by investigating the properties of individual trajectories or of the whole set of trajectories. The necessary conditions for optimal control discussed in Chapters IV and V were derived from the study of individual trajectories. In this chapter we are concerned with the theory of dynamic programming, which is based on the study of the whole set of optimal trajectories.

§ 1. The Bellman Equation and its Properties

1. The principle of dynamic programming. Heuristic derivation of the Bellman equation. The dynamic programming approach is one of the most widely used approaches in synthesizing optimal control for a variety of objects of diverse nature. It is based on the *principle of dynamic programming*, which is stated as follows [4]:

PRINCIPLE OF OPTIMALITY. An optimal control has the property that whatever the initial state and the initial control are, the remaining control must constitute an optimal control with regard to the state resulting from the initial control.

We shall give a detailed explanation of the dynamic programming approach with reference to the problem of the optimal control of the following system on a finite time interval:

$$\dot{x}(t) = f\big(t, x(t), u\big), \qquad t_0 \leqslant t \leqslant T,\ x(t_0) = x_0,\ u \in U, \tag{1.1}$$

$$J(u) = \varphi\big(x(T)\big) + \int_{t_0}^{T} F\big(t, x(t), u\big)\, dt \to \inf_{u \in U}. \tag{1.2}$$

We introduce a scalar function $V(s, x)$ called the *Bellman function*, which is defined as follows. Suppose that the motion of the system (1.1) is on the interval $t_0 \leqslant s \leqslant t \leqslant T$ and the initial condition is of the form $x(s) = x$, $x \in R^n$, i.e.,

$$\dot{x}(t) = f\big(t, x(t), u\big), \qquad s \leqslant t \leqslant T,\ x(s) = x. \tag{1.3}$$

For the system (1.3) we consider the problem of minimizing the functional

$$\varphi\big(x(T)\big) + \int_{s}^{T} F\big(t, x(t), u\big)\, dt \to \inf_{u \in U}. \tag{1.4}$$

By an *admissible control* we mean any measurable function $u(t,x) \in U$ such that for any $s \in [t_0, T]$ and $x \in R^n$ there exists a solution of the problem (1.3), and the functional (1.2) is finite.

It should be noted that for $s = t_0$ and $x = x_0$ the problem (1.3), (1.4) is identical with the problem (1.1), (1.2). Clearly, the minimum value of the functional (1.4) on the trajectories of the system (1.3) depends on the initial moment s and the initial state x. By the *Bellman function* we designate the function that is equal to the infimum of the functional (1.4) on the trajectories of the system (1.3) with respect to all admissible controls, i.e.,

$$V(s,x) = \inf_{\substack{u\in U,\\ s\leqslant t\leqslant T}} \Big[\varphi(x(T)) + \int_s^T F(t,x(t),u)\,dt\Big]. \tag{1.5}$$

Consider now arbitrary moments s and τ such that $t_0 \leqslant s \leqslant \tau \leqslant T$. Then on the basis of the above-stated principle of dynamic programming we have

$$\begin{aligned}
V(s,x) &= \inf_{\substack{u\in U,\\ s\leqslant t\leqslant T}} \left[\varphi(x(T)) + \int_s^T F(t,x(t),u)\,dt\right]\\
&= \inf_{\substack{u\in U,\\ s\leqslant t\leqslant \tau}} \inf_{\substack{u\in U,\\ \tau\leqslant t\leqslant T}} \left[\varphi(x(T)) + \int_s^\tau F(t,x(t),u)\,dt + \int_\tau^T F(t,x(t),u)\,dt\right]\\
&= \inf_{\substack{u\in U,\\ s\leqslant t\leqslant \tau}} \left[\int_s^\tau F(t,x(t),u)\,dt + \inf_{\substack{u\in U,\\ s\leqslant t\leqslant \tau}} \left[\int_\tau^T F(t,x(t),u)\,dt + \varphi(x(T))\right]\right]\\
&= \inf_{\substack{u\in U,\\ s\leqslant t\leqslant \tau}} \left[\int_s^\tau F(t,x(t),u)\,dt + V(\tau,x(\tau))\right].
\end{aligned} \tag{1.6}$$

Note that $x(t)$ in the equality (1.6) denotes the solution of the problem (1.3). The determined control $u(t)$, $s \leqslant t \leqslant \tau$, affects both the value of the integral on the right side of (1.6) and the value of the argument $x(\tau)$ of the function $V(\tau, x(\tau))$. Equation (1.6) can be rewritten as follows:

$$\inf_{\substack{u\in U,\\ s\leqslant t\leqslant \tau}} \left[\int_s^\tau F(t,x(t),u)\,dt + V(\tau,x(\tau)) - V(s,x)\right] = 0. \tag{1.7}$$

Suppose that the Bellman function $V(t,x)$ is continuously differentiable. Then for small $\tau - s$ we obtain up to terms of higher order in $\tau - s$

$$V(\tau,x(\tau)) - V(s,x) = \dot V(s,x)(\tau - s). \tag{1.8}$$

Here $\dot V(s,x)$ is the total derivative of the function V along the trajectories of the system (1.1) equal to

$$\dot V(s,x) = V_s(s,x) + V_x'(s,x)(f(t,x,u)), \tag{1.9}$$

$$V_s(s,x) = \frac{\partial V(s,x)}{\partial s}, \qquad V_x(s,x) = \frac{\partial V(s,x)}{\partial x}.$$

Now, divide both parts of equality (1.7) by $\tau - s$ and pass to the limit as $\tau \to s+0$. Then, noting equality (1.8), we obtain an equation called the *Bellman equation*:

$$\inf_{u\in U}[V_t(t,x) + V_x'(t,x)f(t,x,u) + F(t,x,u)] = 0, \tag{1.10}$$

$$t_0 \leqslant t \leqslant T, \ x \in R^n.$$

According to the definition (1.5) of the Bellman function we have

$$V(T,x) = \varphi(x), \qquad x \in R^n. \tag{1.11}$$

Thus, if the solution of the original optimal control problem (1.1), (1.2) exists and the Bellman function $V(t,x)$ is continuously differentiable, then the relations (1.10), (1.11) hold true, Note that the infimum (the greatest lower bound) in equation (1.10) is calculated with respect to the vector parameter $u \in U \subset R^m$.

2. Determining an F-control with the help of the dynamic programming approach. If the control u yielding the infimum of expression (1.10) exists, then it is the function of the time t and the phase control x, i.e., $u = u(x,t)$. Thus, the control can be determined as an F-control through the use of the dynamic programming approach.

However, it is necessary to keep the following in mind:

1°. A given class of admissible controls does not sometimes contain the control that furnishes the infimum in (1.15) (recall that a control $u(t,x)$ is called *admissible* if this control generates a solution of equation (1.1), and the functional (1.2) is finite).

2°. The Bellman function $V(t,x)$ does not always possess the kind of smoothness that was used when deducing the Cauchy problem (1.10), (1.11). In other words, the Bellman function does not always satisfy the appropriate Bellman equation or else satisfies it in a certain generalized sense. Therefore, the solution of the Bellman equation does not necessarily coincide with the corresponding Bellman function.

3°. If the Bellman function $V(t,x)$ satisfies the Bellman equation, then this does not imply that the control furnishing an infimum in (1.10) is optimal. In particular, given this control, the solutions of the equations of motion (1.1) may, in general, not exist.

4°. The solution of the Cauchy problem (1.10), (1.11) may turn out not to be unique. In this case an additional investigation is required to establish which solution is a Bellman function of the original optimal control problem.

At the same time the dynamic programming approach leads to a solution of the problem (1.10), (1.11). We present the appropriate theorem.

Theorem 1.1. *Suppose there exists a unique continuously differentiable solution V_0 of the problem* (1.10), (1.11) *and an admissible solution $u_0(t,x)$ such that*

$$\inf_{u\in U}\left[f'(t,x,u)\frac{\partial V_0(t,x)}{\partial x} + F(t,x,u)\right]$$

$$= f'(t,x,u_0)\frac{\partial V_0(t,x)}{\partial x} + F(t,x,u_0). \tag{1.12}$$

Then the F-control $u_0(t,x)$ is optimal, and the corresponding Bellman function is $V_0(t,x)$.

Proof. We compute the total derivative along the trajectories of the system (1.1) generated by the control u_0. By virtue of equalities (1.9) we have

$$\dot{V}_0(t,x) = \frac{\partial V_0(t,x)}{\partial t} + f'(t,x,u_0)\frac{\partial V_0(t,x)}{\partial x}.$$

Further, using the relations (1.12) and (1.10), we obtain

$$\dot{V}_0(t,x) = -F(t,x,u_0), \qquad t_0 \leqslant t \leqslant T,\ x \in R^n. \tag{1.13}$$

The equality (1.13) holds true for any $x \in R^n$. In particular, it is valid if we substitute for x the trajectory $x_0(t)$ that corresponds to the F-control u_0, i.e.,

$$\dot{V}_0\big(t,x_0(t)\big) = -F\Big(t,x_0(t),u_0\big(t,x_0(t)\big)\Big). \tag{1.14}$$

Integration of both sides of (1.14) with respect to t from t_0 to T gives

$$V_0(t_0,x_0) = V\big(T,x(T)\big) + \int_{t_0}^{T} F\Big(t,x_0(t),u_0\big(t,x_0(t)\big)\Big)\,dt. \tag{1.15}$$

But $V\big(T,x(T)\big) = \varphi\big(x(T)\big)$ due to the condition (1.11). Thus, noting the relations (1.2) and (1.15) we can write

$$J(u_0) = V_0(t_0,x_0). \tag{1.16}$$

Now let $u(t,x)$ be an admissible F-control, and $x(t)$ be the corresponding solution of equation (1.1). Then, using the equality (1.10) as in the derivation of relation (1.14), we establish that

$$\dot{V}\big(t,x(t)\big) \geqslant -F\Big(t,x(t),u\big(t,x(t)\big)\Big).$$

Integration of this inequality from t_0 to T, by virtue of (1.11), gives

$$J(u) \geqslant V_0(t_0,x_0). \tag{1.17}$$

Comparison of (1.16) and (1.17) shows that $J(u_0) = V_0(t_0,x_0) \leqslant J(u)$. Thus, it has been established that the control u_0 is optimal. □

Let us enumerate the main steps in deducing an F-control on the basis of the dynamic programming approach:

1°. Determine the control $u = u(t,x,V)$ that furnishes the minimum of the left side of equation (1.10).

2°. By substituting this control in equation (1.10), find a nonlinear partial differential equation for the function $V(t,x)$. The boundary condition has the form (1.11).

3°. The resulting problem is solved and $V(t,x)$ is determined.

4°. Substitute the function $V(t,x)$ in the expression for $u(t,x,V)$ derived in step 1° yields the optimal F-control $u(t,x) = u\big(t,x,V(t,x)\big)$.

Moreover, according to Theorem 1.1 we may state that if the solution $V(t,x)$ of the Bellman equation is unique, and the derived control $u(t,x)$ is admissible, then $u(t,x)$ is an optimal F-control and $V(t,x)$ is a Bellman function.

3. Connection between the dynamic programming approach and the maximum principle. Pontryagin's maximum principle can be derived from the dynamic programming approach if certain additional conditions are imposed on the smoothness of the Bellman function.

We shall illustrate this by using the optimal control problem (1.1), (1.2) as an example. The dynamic programming approach enables reduction of this problem to the solution of the Cauchy problem (1.10), (1.11). Suppose there exists a unique continuous solution $V(t,x)$ of this problem with continuous derivatives V_t, V_x, V_{tx}, V_{xx} and an admissible control $u_0(t,x)$ satisfying the equality (1.12):

$$\inf_{u\in U}[f'(t,x,u)V_x + F(t,x,u)] = f'(t,x,u_0)V_x + F(t,x,u_0). \tag{1.18}$$

Then, as was shown in subsection 2, the control u_0 is optimal. Let $x_0(t)$ denote the optimal trajectory generated by the F-control u_0.

We introduce the functions $\psi(t) \in R^n$ and $H \in R^1$ by the equations

$$\begin{aligned} \psi(t) &= -V_x(t,x(t)), \\ H(t,x,u,\psi(t)) &= \psi'(t)f(t,x,u) - F(t,x,u). \end{aligned} \tag{1.19}$$

Then the equality (1.18) implies that

$$\begin{aligned} &f'(t,x_0(t),u_0)V_x(t,x_0(t)) + F(t,x_0(t),u_0) \\ &\quad= \min_{u\in U}\big[f'(t,x_0(t),u)V_x(t,x_0(t)) + F(t,x_0(t),u)\big]. \end{aligned}$$

Hence, noting equations (1.19), we obtain

$$\begin{aligned} &\max_{u\in U}\big[-f'(t,x_0(t),u)V_x(t,x_0(t)) - F(t,x_0(t),u)\big] \\ &\quad= \max_{u\in U}\big[\psi'(t)f(t,x_0(t),u) - F(t,x_0(t),u)\big] \\ &\quad= \max_{u\in U} H(t,x_0(t),u,\psi(t)) = H(t,x_0(t),u_0,\psi(t)). \end{aligned} \tag{1.20}$$

Further, by virtue of (1.10) and (1.18) we have

$$V_t(t,x) = -V_x'(t,x)f(t,x,u_0) - F(t,x,u_0) = H(t,x,u,\psi(t)).$$

We differentiate both sides of this expression with respect to x and set $x = x_0(t)$. This can be done, since V_{tx} exists and is continuous. Taking account of equations (1.19), we can write

$$V_{tx}(t,x_0(t)) = -\dot{\psi}(t) = \frac{\partial H}{\partial x}(t,x_0(t),u_0(t),\psi(t)). \tag{1.21}$$

In a similar way, differentiating both sides of (1.11) with respect to x and putting $x = x_0(T)$, we obtain, in view of (1.19),

$$V_x(T,x_0(T)) = -\psi(T) = \varphi_x(x_0(T)).$$

Thus, noting the relation (1.21), we find that the function $\psi(t)$ is a solution of the Cauchy problem

$$\dot{\psi}(t) = -H_x\big(t, x_0(t), u_0(t), \psi(t)\big), \qquad \psi(T) = -\varphi_x\big(x_0(T)\big). \tag{1.22}$$

The conditions (1.20), (1.22) coincide with the relations (1.6), (1.7), (1.8) of the maximum principle from Chapter V.

It should be noted that the dynamic programming approach enables reduction of the optimal control problem to the investigation of the Cauchy problem (1.10), (1.11) for a nonlinear partial differential equation. At the same time, with the help of the maximum principle the solution of control problems can be reduced to the investigation of boundary-value problems for systems of ordinary differential equations, the latter problem being easier to solve.

4. Determining F-control in the problem of damping the motion of a rigid body. Consider the problem of the quickest damping of a rigid body which is in rotation relative to its mass center O. Let x_1, x_2, x_3 denote the projections of the vector of angular momentum on the respective axes Ox_1, Ox_2, Ox_3, where the coordinate axes Ox_i are the principal central axes of inertia for the body. We denote by u_i the projections of the control torque on the same axes. It will be recalled that if $\omega = (\omega_1, \omega_2, \omega_3)$ is the angular velocity of a rigid body, then $x_1 = A\omega_1$, $x_2 = B\omega_2$, $x_3 = C\omega_3$, where A, B, C are the principal central moments of inertia of the rigid body. Using this notation, we can write Euler's equations in the form

$$\begin{aligned} &\dot{x}_1(t) + \frac{C-B}{CB} x_2 x_3 = u_1, \\ &\dot{x}_2(t) + \frac{A-C}{AC} x_1 x_2 = u_2, \qquad |u| = (u_1^2 + u_2^2 + u_3^2)^{\frac{1}{2}} \leqslant b, \\ &\dot{x}_3(t) + \frac{B-A}{AB} x_1 x_2 = u_3, \qquad t \geqslant 0. \end{aligned} \tag{1.23}$$

At the initial moment $t_0 = 0$ the values of the components of the angular momentum $x_i(0)$ are given, with $|x(0)| \geqslant R$. We denote by $T \geqslant 0$ the first moment at which $|x(T)| = R$, where $R \geqslant 0$ is a given number. The control torque u must be determined in a way such that it minimizes the damping time T.

To solve this problem, we use the dynamic programming approach. Let V be the Bellman function for the problem; it denotes the minimum damping time. Since the equations of motion (1.23) and the control constraint do not depend explicitly on time, the damping time depends solely on the initial vector x but does not depend on the initial moment. Hence, the Bellman function also depends solely on x, i.e., $V = V(x_1, x_2, x_3)$. Since in time-optimal problems $F \equiv 1$, the Bellman equation becomes

$$\begin{aligned} \inf_{u, |u| \leqslant b} \Big[&V_{x_1}\Big(u_1 - x_2 x_3 \frac{C-B}{CB}\Big) \\ &+ V_{x_2}\Big(u_2 - x_1 x_3 \frac{A-C}{AB}\Big) + V_{x_3}\Big(u_3 - x_1 x_2 \frac{B-A}{AB}\Big)\Big] = -1, \end{aligned} \tag{1.24}$$

$$V(x) = 0 \quad \text{for } |x| = R. \tag{1.25}$$

We try to determine a control that furnishes the infimum of the expression

$$u_1 V_{x_1} + u_2 V_{x_2} + u_3 V_{x_3}, \qquad |u| \leqslant b. \tag{1.26}$$

It should be noted that the expression (1.26) is a scalar product of the vectors u and V_x. Here the vector u can be rotated through an arbitrary angle relative to the vector V_x because of the constraint $|u| \leqslant b$. Hence, noting the definition of the scalar product of two vectors (being the product of their lengths and the cosine of the angle they enclose), we find that the value of the control in question must be a maximum, and its direction must be opposite to that of the vector V_x. A unit vector directed along the vector V_x is equal to $V_x|V_x|^{-1}$. It follows that the control u_i that yields the infimum of the expression (1.26) is of the form

$$u_i = -b\frac{\partial V}{\partial x_i}\left(\sum_{i=1}^{3}\left(\frac{\partial V}{\partial x_i}\right)^2\right)^{-1/2}. \tag{1.27}$$

Substituting the control (1.27) in equation (1.24) gives

$$\begin{aligned} x_2 x_3 \frac{C-B}{CB} V_{x_1} &+ x_1 x_3 \frac{A-C}{AC} V_{x_2} \\ &+ x_1 x_2 \frac{B-A}{BA} V_{x_3} - b\left(\sum_{i=1}^{3}\left(\frac{\partial V}{\partial x_i}\right)^2\right)^{-1/2} = -1. \end{aligned} \tag{1.28}$$

Because of the spherical symmetry of the problem (1.24), (1.25) we try to find its solution in the form $V(x) = b^{-1}\omega(|x|)$, where ω is a scalar function, and $|x| = \sqrt{x_1^2 + x_2^2 + x_3^2}$. The relations (1.24), (1.25) yield the following equation for the function ω:

$$\frac{d\omega(r)}{dr} = -1, \qquad \omega(R) = 0.$$

Hence, $\omega(|x|) = |x| - R$. Thus, we find the optimal control u_i and the damping time $V(x)$:

$$u_i = -bx_i|x|^{-1}, \quad |u| = b, \qquad V(x) = \frac{1}{b}(|x| - R). \tag{1.29}$$

The expressions (1.29) indicate that the magnitude of the optimal control vector is maximum, and its direction is opposite to that of the vector of angular momentum.

5. Optimal procedure for reducing the power of a nuclear reactor. We shall be concerned with the determination of an optimal control for reducing the power of a nuclear reactor in the one-group approximation (see § 3 of Chapter III). Let $n(t)$ be the number of neutrons in the reactor at time t, and the control $u(t)$ be proportional to the change in the neutron multiplication coefficient. Then

$$\dot{n}(t) = u(t)n(t), \qquad 0 \leqslant t \leqslant T, \; n(0) = n_0 > 0, \tag{1.30}$$

where T is a prescribed power-reduction time. Let us assume that the cost functional to be minimized is of the form

$$J(u) = \alpha \ln^2 n(T) + \beta \int_0^T u^2(t)\, dt \to \inf_u, \tag{1.31}$$

where $\alpha \geqslant 0$ and $\beta > 0$ are prescribed constants, and the integral is proportional to the square of the induced absorption. The control is subject to the natural constraint $u(t) \geqslant 0$. The Bellman equation for the problem (1.30), (1.31) can be written in the form

$$V_t + \inf_u (unV_n + \beta u^2) = 0, \qquad V(T,n) = \alpha \ln^2 n. \tag{1.32}$$

The control that minimizes the left side of (1.32) is

$$u = -\frac{n}{2\beta} V_n. \tag{1.33}$$

Substituting this control in equation (1.32) to determine the function $V(t,n)$, we obtain

$$V_t - \frac{n^2}{4\beta}(V_n)^2 = 0, \qquad V(T,n) = \alpha \ln^2 n.$$

We can verify by differentiation that the solution of this equation is the function

$$V(t,n) = \frac{\alpha\beta \ln^2 n}{\beta + \alpha(T-t)}. \tag{1.34}$$

It follows from (1.33) and (1.34) that the optimal F-control is of the form

$$u(t,n) = -[\beta + \alpha(T-t)]^{-1} \alpha \ln n.$$

6. The linear-quadratic problem. Using the dynamic programming approach, we will try to find the solution of the linear-quadratic problem (called the LQ-problem, for short) and to compare it with the solution that has been obtained before, in § 2 of Chapter IV. Suppose that the equations of motion of the system are of the form

$$\dot{x}(t) = Ax + Bu, \qquad t \geqslant t_0,\ x \in R^n,\ u \in R^m,\ x(t_0) = x_0. \tag{1.35}$$

We consider a quadratic cost functional of the form

$$J(u) = x'(T)N_1x(T) + \int_{t_0}^T (x'(s)N_2(s)x(s) + u'(s)N_3(s)u(s))\, ds, \tag{1.36}$$
$$N_1 \geqslant 0,\ N_2(s) \geqslant 0,\ N_3(s) > 0.$$

For the problem (1.35), (1.36) the Bellman equation (1.10) becomes

$$\inf_u [V_t + V_x'(Ax + Bu) + x'N_2x + u'N_3u] = 0. \tag{1.37}$$

The boundary condition (1.11) for equation (1.37) takes the form

$$V(T, x) = x'N_1x. \tag{1.38}$$

We try to determine the control u yielding the infimum of the expression (1.37). Differentiating the expression in brackets in equation (1.37) and equating the resulting derivative to zero gives in turn

$$B'V_x + 2N_3(t)u = 0, \qquad u = -\tfrac{1}{2}N_3^{-1}(t)B'V_x. \tag{1.39}$$

Next, substituting the expression (1.39) for u in (1.37) yields the following equation for V:

$$V_t + x'N_2x + V_x'Ax - \tfrac{1}{4}V_x'BN_3^{-1}B'V_x = 0. \tag{1.40}$$

Thus, the Bellman function satisfies the nonlinear partial differential equation (1.40) with boundary condition (1.38). We seek a solution to this equation in the form

$$V(t, x) = x'P(t)x, \tag{1.41}$$

where the matrix $P(t)$ is to be determined. Substituting the expression (1.41) in equation (1.40), we find that the matrix $P(t)$ is a solution of the problem

$$\dot{P}(t) + N_2 + A'P + PA - \tfrac{1}{4}PBN_3^{-1}B'P = 0, \qquad t_0 \leqslant t \leqslant T, \tag{1.42}$$
$$P(T) = N_1.$$

Comparing equation (1.42) with equation (2.8) of Chapter IV, we see that they coincide fully. Further investigation of the problem (1.42) can be carried out along similar lines as in § 2 of Chapter IV and the same results obtained. Moreover, by virtue of equations (1.39) and (1.41) the F-control can be expressed by the formula

$$u(t, x) = -N_3^{-1}(t)B'P(t)x,$$

and the minimum value of the cost functional (1.36) is given by

$$J(u_0) = x_0'P(t_0)x_0 = V(t_0, x_0).$$

§ 2. Control on an Unbounded Time Interval. Stabilization of Dynamical Systems

1. Problem statements. Consider the control system described by the following equations:

$$\dot{x}(t) = f\big(t, x(t), u\big), \qquad t \geqslant 0, \ f(t, 0, 0) = 0 \tag{2.1}$$
$$x(0) = x_0, \ x(t) \in R^n, \ u \in R^m.$$

The control u of the system (2.1) is sought in the class of functions depending on time t and the moving coordinates $x(t)$, i.e., is sought as an F-control of the form $u = u(t, x(t))$ subject to the constraint $u = u(t, x(t)) \in U$, where U is a given set in R^m. Moreover, it is assumed that $u(t, 0) = 0$.

We can point out the following control problem statements for an unbounded time interval.

1°. *Find an F-control* $u(t, x)$ *such that the system* (2.1) *becomes asymptotically stable in some sense.*

Depending on the definition of stability we distinguish between *asymptotic stabilization*, *exponential stabilization*, *global stabilization*, and so on, according to the type of stability of the trivial solution of the system (2.1), which is generated by the control $u(t, x)$. Such control is said to be *stabilizing* (or, to be more precise, *asymptotically stabilizing*, *exponentially stabilizing*, or *asymptotically globally stabilizing*, and so on).

2°. *Find an F-control* $u(t, x)$ *that minimizes the cost functional*

$$J(u) = \int_0^\infty F_0\big(t, x(t), u(t, x)\big)\, dt \to \inf_{u \in U}. \tag{2.2}$$

Here F_0, $F_0 \colon R^1 \times R^n \times R^m \to R^1$, is a prescribed function.

It should be noted that these two statements of the stabilization problem are not always equivalent to each other. So, let the function F_0 in the functional (2.2) be such that $F_0(t, x, u) \gg 1$ for $|x| \leqslant Nt$ and $F_0(t, x, u) \equiv 0$ for $|x| > N(t + 1)$, where $N > 0$ is a given number. Then the optimal control in the problem (2.1), (2.2) may turn out to be such that it generates a solution of equation (2.1) whose time of being outside the set $|x| \leqslant Nt$ is a maximum. Under such control the system (2.1) is, in general, unstable.

However, under certain additional assumptions given below the optimal control in the problem (2.1), (2.2) converts the trivial solution of the system (2.1) into an asymptotically stable one.

In this connection a third statement of the stabilization problem given below is of interest.

3°. *Find an F-control* $u(t, x)$ *that minimizes the functional* (2.2) *and simultaneously converts the system* (2.1) *into a stable system* (*in some sense*).

In this case the stabilization problem is called the *optimal stabilization problem*, and the corresponding control is called *optimal stabilizing*. In what follows, the main attention is focused on the optimal stabilization problem.

2. Lyapunov's direct method for the optimal stabilization problem. We shall state the conditions for the solvability of the optimal stabilization problem in terms of scalar Lyapunov functions $V(t, x)$ possessing certain specific properties (see in Chapter I about Lyapunov functions and their properties).

As in Chapter I we denote by $\omega_i(r)$ ($r \geqslant 0$) some wedges, i.e., scalar continuous nondecreasing functions such that $\omega_i(r) > 0$ for $r > 0$ and $\omega_i(0) = 0$. Further, let L_u denote the operator

$$L_u V(t, x) = \frac{\partial V(t, x)}{\partial t} + f'(t, x, u) \frac{\partial V(t, x)}{\partial x}.$$

Note that $L_u V(t,x)$ is the total derivative of the function $V(t,x)$ along the trajectories of the system (2.1) generated by the control u.

Now we present the conditions under which an F-control $u(t,x)$ is stabilizing.

Theorem 2.1. *The control $u(t,x)$ is stabilizing for the system (2.1) if there exists a continuously differential Lyapunov function $V(t,x)$ such that*

$$\omega_1(|x|) \leqslant V(t,x) \leqslant \omega_2(|x|), \tag{2.3}$$

$$L_u V(t,x) \leqslant -\omega_3(|x|). \tag{2.4}$$

The trivial solution of the system (2.1) generated by this control is uniformly asymptotically stable.

The proof of the theorem follows from the theorem of Lyapunov on asymptotic stability (see Theorem 2.3 of Chapter I).

Consider the optimal stabilization problem. The control $u \in U$ is said to be *admissible* in the optimal stabilization problem if it is globally stabilizing and the integral (2.2) exists.

Theorem 2.2 [10]. *Suppose there exist a Lyapunov function $V(t,x)$ and an admissible control $u_0(t,x)$ such that the inequalities (2.3) are satisfied and the following relation is valid:*

$$\inf_{u\in U}[L_u V(t,x) + F_0(t,x,u)] = L_{u_0} V(t,x) + F_0(t,x,u_0) = 0, \tag{2.5}$$

where the infimum is calculated with respect to the vector parameter $u \in U \subset R^m$. Then the control $u_0(t,x)$ is an optimal stabilizing one, with

$$\inf \int_0^\infty F_0(t,x(t),u(t,x))\,dt$$
$$= \int_0^\infty F_0\Big(t,x_0(t),u_0\big(t,x_0(t)\big)\Big)\,dt = V(0,x_0), \tag{2.6}$$

where the infimum is calculated with respect to all admissible controls $u(t,x)$.

Proof. The relation (2.5) implies the equality

$$L_{u_0} V(t,x) = -F_0\big(t,x,u_0(t,x)\big).$$

Substituting the trajectory $x_0(t)$ for x in this relation and integrating the resulting relation with respect to t from 0 to $T > 0$ gives

$$\int_0^T L_{u_0} V(t,x_0(t))\,dt = V(T,x_0(T)) - V(0,x_0)$$
$$= -\int_0^T F_0\Big(t,x_0(t),u_0\big(t,x_0(t)\big)\Big)\,dt \tag{2.7}$$

Since the control u_0 is admissible, we have

$$\lim_{t\to\infty} x_0(t) = 0. \tag{2.8}$$

It follows from the inequalities (2.3) that

$$0 \leqslant V\big(T, x_0(T)\big) \leqslant \omega_2(|x_0(t)|). \tag{2.9}$$

Passing to the limit in the expression (2.7) as $T \to \infty$ and taking account of (2.8) and (2.9), we obtain

$$\int_0^\infty F_0\Big(t, x_0(t), u_0\big(t, x_0(t)\big)\Big)\, dt = V(0, x_0). \tag{2.10}$$

Let now $u(t, x)$ be an arbitrary admissible control. Then by virtue of (2.5) the following inequality holds:

$$L_u V(t, x) \geqslant -F_0\big(t, x, u(t, x)\big).$$

Substituting the trajectory $x(t, u)$ (generated by the control u) for x in this relation, integrating the resulting relation with respect to t from 0 to $T > 0$, and passing to the limit as $T \to \infty$ gives

$$V(0, x_0) \leqslant \int_0^\infty F_0\Big(t, x(t, u), u\big(t, x(t, u)\big)\Big)\, dt.$$

By comparing this inequality with equality (2.10), we verify the relations (2.6). The proof is complete. □

3. Exponential stabilization. Both the conditions and the assertions of Theorem 2.2 can be modified for other constraints on V, u_0 and the right sides of the system (2.1). We present one such modification.

In this subsection by an *admissible* control we mean an F-control $u(t, x) \in U$ such that there exist a solution of equation (2.1) and the integral in (2.2). It should be noted that, generally speaking, the convergence of the integral (2.2) does not imply the convergence of $x(t)$ to zero as $t \to \infty$. The subsequent conditions ensure the asymptotic stability of the trivial solution of the system (2.1). Let us state first the following lemma.

Lemma 2.1. *Suppose that a nonnegative scalar differentiable function $f(t)$ satisfies the conditions*

$$\begin{aligned} &\left|\frac{d}{dt} f(t)\right| \leqslant C_1 f(t), \qquad t \geqslant 0,\ C_1 \geqslant 0, \\ &\int_0^\infty f(t)\, dt \leqslant C_2 < \infty, C_2 > 0. \end{aligned} \tag{2.11}$$

Then

$$\lim_{t\to\infty} f(t) = 0.$$

Proof. Integrating the inequality (2.11) gives

$$f(t) \leqslant f(0) + C_1 \int_0^t f(s)\,ds \leqslant f(0) + C_1C_2,$$
$$f(t) \geqslant f(0) - C_1 \int_0^t f(s)\,ds \geqslant f(0) - C_1C_2.$$

These inequalities imply that $f(t)$ is a uniform bounded function for $t \geqslant 0$. It follows from the inequalities (2.11) that $|f(t)|$ is also bounded on the half-axis $[0,\infty)$, i.e., $|f(t)| \leqslant C_3$ and $|df(t)/dt| \leqslant C_1C_3$.

Now suppose that the relation $\lim_{t\to\infty} f(t) = 0$ is not valid. Then there exists a sequence of points t_i and a number $\varepsilon > 0$ such that

$$f(t_i) \geqslant \varepsilon, \qquad t_i \to \infty,\ i \to \infty,\ |t_i - t_{i-1}| > \frac{2\varepsilon}{C_1C_3}.$$

Moreover,

$$f(s) \geqslant \frac{\varepsilon}{2}, \qquad s \in L_i = \left[t_i - \frac{\varepsilon}{2C_1C_3},\ t_i + \frac{\varepsilon}{2C_1C_3}\right].$$

The segments L_i and L_j do not intersect for any i and j $(i \neq j)$. Therefore, the integral (2.11) must diverge, since

$$\int_0^\infty f(t)\,dt \geqslant \int_0^T f(t)\,dt \geqslant \sum_{t_j \leqslant T} \int_{L_i} f(t)\,dt \geqslant \frac{\varepsilon}{2}\frac{\varepsilon}{C_1C_3}N(T). \tag{2.12}$$

Here $N(T)$ is the number of points t_i not exceeding T. Since $N(T) \to \infty$ as $T \to \infty$, it follows that the left integral in (2.12) is divergent, and this contradicts the conditions of the lemma. This contradiction proves the lemma. □

Theorem 2.3. *Suppose that for any admissible control $u(t,x)$ the following inequalities hold true:*

$$F_0\big(t,x,u(t,x)\big) \geqslant C_1|x|^\alpha, \qquad t \geqslant 0,\ C_1, \alpha > 0, \tag{2.13}$$
$$|f(t,x,u(t,x))| \leqslant C_2|x|, \qquad t \geqslant 0,\ C_2 > 0. \tag{2.14}$$

Further, suppose that some admissible control $u_0(t,x)$ and a continuously differentiable function $V(t,x)$ satisfy equation (2.5), with

$$0 \leqslant V(t,x) \leqslant \omega_2(|x|). \tag{2.15}$$

Then $u_0(t,x)$ is an optimal stabilizing control, and $V(t,x)$ is a Bellman function in the problem (2.1), (2.2). Hence, the trivial solution of the system (2.1) generated by the control $u_0(t,x)$ is globally exponentially stable and the equality (2.6) holds.

Proof. For any admissible control $u(t,x)$ we have by virtue of the inequality (2.13)

$$C_1 \int_0^t |x(s)|^\alpha \, ds \leqslant \int_0^t F_0(s,x,u(s,x))\, ds.$$

Passing to the limit as $t \to \infty$ in this inequality, we obtain

$$\int_0^\infty |x(s)|^\alpha \, ds \leqslant C_1^{-1} J(u) < \infty.$$

Further, we find through the use of the inequality (2.14)

$$\frac{d}{dt}|x(t)|^\alpha = \alpha |x(t)|^{\alpha-2} x'(t) f(t,x,u),$$
$$\left| \frac{d}{dt}|x(t)|^\alpha \right| \leqslant \alpha C_2 |x(t)|^\alpha.$$

Applying Lemma 2.1 to the function $|x(t)|^\alpha$, we can write

$$\lim_{t\to\infty} |x(t)|^\alpha = 0, \qquad \lim_{t\to\infty} x(t) = 0. \tag{2.16}$$

Using the relation (2.16) and repeating the proof of Theorem 2.2, we verify that the control $u_0(t)$ satisfying the equality (2.5) is optimal, i.e.,

$$V(0,x_0) = J(u_0) \leqslant J(u). \tag{2.17}$$

To complete the proof of Theorem 2.3, it remains to establish that the trivial solution of the system (2.1) generated by the control $u_0(t,x)$ is globally exponentially stable. It can be shown that the function $V(t,x)$ satisfies the estimate

$$C_3 |x|^\alpha \leqslant V(t,x), \qquad t \geqslant 0,\ x \in R^n. \tag{2.18}$$

Let $y(s,t,x)$ denote the solution of the problem

$$\dot{y}(s) = f\big(s,y,u_0(s,y)\big), \qquad s \geqslant t,\ y(t) = x. \tag{2.19}$$

As before, we can show that the solution of the problem (2.19) has the following properties:

$$V(t,x) = \int_t^\infty F_0\Big(s, y(s), u_0\big(s,y(s)\big)\Big)\, ds,$$
$$-\alpha C_2 |y(s)|^\alpha \leqslant \frac{d}{ds}|y(s)|^\alpha \leqslant \alpha C_2 |y(s)|^\alpha,$$
$$y(s) \to 0,\ s \to \infty.$$

Hence, there exists a T such that

$$|y(T)|^\alpha \leqslant \tfrac{1}{2}|x|^\alpha.$$

It follows that

$$\begin{aligned} V(t,x) &= \int_t^\infty F_0\big(s, y(s), u_0(s, y(s))\big)\, ds \geqslant C_1 \int_t^\infty |y(s)|^\alpha\, ds \\ &\geqslant C_1 \int_t^T |y(s)|^\alpha\, ds \geqslant -\frac{C_1}{2C_2} \int_t^T \frac{d|y(s)|^\alpha}{ds}\, ds \\ &= \frac{C_1}{\alpha C_2}\left[|x|^\alpha - |y(T)|^\alpha\right] \geqslant \frac{C_1}{2\alpha C_2}|x|^\alpha. \end{aligned}$$

Thus, the estimate (2.18) is established. Further, it follows from equation (2.5) that

$$L_{u_0} V(t,x) = \dot V\big(t, x_0(t)\big) = -F_0\big(t, x_0(t), u_0(t,x)\big). \tag{2.20}$$

In view of the inequalities (2.13) and (2.18), it follows from the relation (2.20) that

$$\dot V\big(t, x_0(t)\big) \leqslant -C_1|x_0(t)|^\alpha \leqslant -C_3 V\big(t, x_0(t)\big).$$

Integrating this inequality gives

$$V\big(t, x_0(t)\big) \leqslant -C_4 \exp(-C_3 t).$$

Utilizing the inequality (2.18) again gives

$$|x_0(t)| \leqslant C_4 \exp(-C_3 \alpha^{-1} t). \tag{2.21}$$

It follows that any solution $x_0(t)$ of the system (2.1) generated by the control$u_0(t,x)$ satisfies the inequality (2.21), i.e., the trivial solution of equation (2.1) is globally exponentially stable. □

4. Stabilization of the motion of a manipulator robot[1]. One of the important problems in the theory of manipulator robots is the problem of the synthesis of the control loop. If the characteristics of the environment and of the robot are known precisely, then this problem is, in fact, one of determining the program control of motion. Moreover, if some parameters of the law of motion are not known precisely, then a number of synthesis problems arise, in which the control takes the form depending on the actual values of the measured coordinates.

Consider an n-link manipulator (see Figure 2.1). The manipulator links are absolutely rigid bodies that are connected in series by means of hinges. The first link is connected by a hinge to an immovable base, and at the end of the last link there is a gripper with a moveable weight m. We denote by $\varphi = (\varphi_1, \dots, \varphi_n)$ the vector of hinge angles, by $H(\varphi)$ the matrix of inertia, by T the kinetic energy of

[1] Ananievskii, I. M., and Kolmanovskii, V. B., Stabilization of Some Nonlinear Hereditary Mechanical Systems, *Nonlinear Anal.*, (1990), vol. 15, no. 2, 101–114.

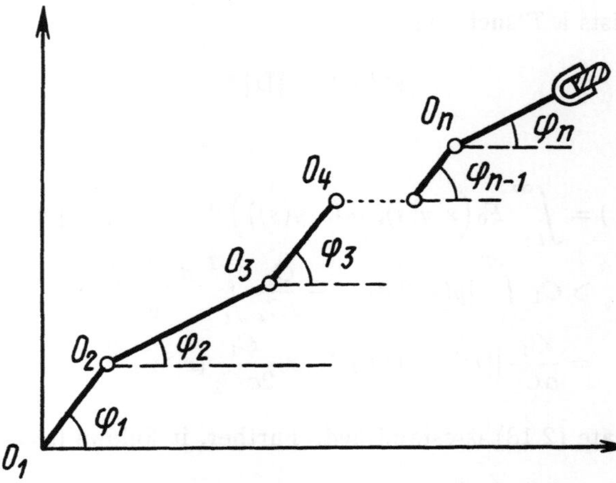

FIG. 2.1. An n-link manipulator robot

the manipulator equal to $T = (1/2)\dot{\varphi}' H(\varphi)\dot{\varphi}$. In the latter expression the prime indicates transposition and $\dot{\varphi} = d\varphi/dt$ is the vector of generalized velocities. Let $\psi = H(\varphi)\dot{\varphi}$ be the generalized momentum vector. Then the equations of motion (the Lagrange equations) are of the form

$$\begin{aligned} \dot{\psi}(t) &= T_\varphi - \gamma H^{-1}(\varphi)\psi + u, \qquad T_\varphi = \frac{\partial T}{\partial \varphi}, \\ \dot{\varphi}(t) &= H^{-1}(\varphi)\psi, \qquad t \geqslant 0, \end{aligned} \tag{2.22}$$

where H^{-1} is a matrix that is inverse to H (recall that $H(\varphi)$ is a continuously differentiable positive definite uniformly bounded matrix), and γ is a diagonal positive semidefinite matrix of coefficients $\gamma_i \geqslant 0$ of viscous friction in the manipulator hinges. The actual values of m and γ_i are unknown; it is known only that $0 \leqslant m \leqslant m_0$ and $0 \leqslant \gamma_i \leqslant \gamma_i^0$, where m_0 and γ_i^0 are prescribed. The control torques act in the manipulator hinges and for a proportional-differential controller (PD-controller) are of the form

$$u = -A\dot{\varphi} - B\varphi. \tag{2.23}$$

We consider the problem of determining diagonal matrices $A > 0$ and $B > 0$ such that the corresponding control (2.23) is a stabilizing one for the system (2.22).

Let the scalar Lyapunov function be of the form

$$V = T + \tfrac{1}{2}\varphi' B\varphi + \varepsilon\psi'\varphi, \tag{2.24}$$

where the parameter ε is determined in the sequel. We calculate the total derivative $\dot{V}$ of the function (2.24) along the trajectories of the system (2.22), (2.23). Since $T = (1/2)\psi' H^{-1}\psi$, we have

$$\dot{V} = \dot{\psi}' H^{-1}\psi + \tfrac{1}{2}\psi'\left(\frac{dH^{-1}}{dt}\right)\psi + \dot{\varphi}' B\varphi + \varepsilon\dot{\psi}'\varphi + \varepsilon\psi'\dot{\varphi}. \tag{2.25}$$

We shall transform several terms in the expression (2.25). It should be noted that

$$\frac{dH^{-1}}{dt} = -\sum_{i=1}^{n} \dot{\varphi}_i H^{-1}\left(\frac{\partial H}{\partial \varphi_i}\right)H^{-1}, \qquad \dot{\psi}' = \psi' H^{-1}.$$

Hence,

$$\begin{aligned}
\frac{1}{2}\psi'\left[\frac{dH^{-1}(\varphi)}{dt}\right]\psi &= \frac{1}{2}\psi'\sum_{i=1}^{n}\dot{\varphi}_i\frac{\partial H^{-1}(\varphi)}{\partial \varphi_i}\psi \\
&= -\frac{1}{2}\psi'\sum_{i=1}^{n}\dot{\varphi}_i H^{-1}\left(\frac{\partial H}{\partial \varphi_i}\right)H^{-1}\psi \\
&= -\frac{1}{2}\sum_{i=1}^{n}\varphi_i\dot{\varphi}'\left(\frac{\partial H}{\partial \varphi_i}\right)\dot{\varphi} = -\frac{1}{2}\sum_{i=1}^{n}(H^{-1}\psi)_i\frac{\partial}{\partial \varphi_i}(\dot{\varphi}'H\dot{\varphi}) \\
&= -\sum_{i=1}^{n}(H^{-1}\psi)_i\frac{\partial T}{\partial \varphi_i} = -\psi'H^{-1}T_{\varphi}, \\
\dot{\psi}'H^{-1}\psi &= [T_{\varphi} - (A-\gamma)H^{-1}\psi - B\varphi]'H^{-1}\psi.
\end{aligned} \tag{2.26}$$

Moreover,

$$\begin{gathered}
\dot{\varphi}'B\varphi = \psi'H^{-1}B\varphi, \qquad \psi'\dot{\varphi} = \psi'H^{-1}\psi, \\
\dot{\psi}'\varphi = [T_{\varphi} - (A+\gamma)H^{-1}\psi - B\varphi]'\varphi.
\end{gathered} \tag{2.27}$$

Substituting formulas (2.26) and (2.27) in equation (2.25) and collecting like terms, we can write

$$\begin{aligned}
\dot{V} = \varepsilon\varphi'T_{\varphi} &- \psi'[H^{-1}(A+\gamma)H^{-1} - \varepsilon H^{-1}]\psi \\
&- \varepsilon\varphi'B\varphi - \varepsilon\varphi'(A+\gamma)H^{-1}\psi.
\end{aligned} \tag{2.28}$$

Further, we can provide the following estimates for the summands in equation (2.28):

$$\begin{gathered}
\varepsilon\varphi'(A+\gamma)H^{-1}\psi \leqslant \frac{1}{2}\varepsilon^2\varphi'(A+\gamma)\varphi + \frac{1}{2}\psi'H^{-1}(A+\gamma)H^{-1}\psi, \\
\varepsilon\varphi'T_{\varphi} = -\frac{\varepsilon}{2}\psi'H^{-1}\sum_{i}\varphi_i\left(\frac{\partial H}{\partial \varphi_i}\right)H^{-1}\psi.
\end{gathered} \tag{2.29}$$

Equations (2.28) and (2.29) imply the inequality

$$\dot{V} \leqslant W,$$

where

$$\begin{aligned}
W = -\varepsilon\varphi'\left[B - \frac{\varepsilon}{2}(A+I\gamma)\right]\varphi - \psi'H^{-1} \\
\times\left[\frac{A+\gamma}{2} - \varepsilon H + \frac{\varepsilon}{2}\sum_{i}\varphi_i\frac{\partial H}{\partial \varphi_i}\right]H^{-1}\psi.
\end{aligned} \tag{2.30}$$

It follows from formulas (2.29) that $\dot{V}$ is negative definite for all sufficiently small $\varepsilon > 0$ in any bounded region $(\varphi, \dot{\varphi})$. Thus, the trivial solution of the system (2.22) generated by the control (2.23) is globally asymptotically stable.

Remark. The problem treated above is closely connected with the problem of the transfer (with damping) of the manipulator from a given initial state $(\varphi(0), \dot{\varphi}(0))$ to a prescribed terminal state $(\bar{\varphi}, 0)$. However, since the values of the parameters m and γ_i are unknown, it follows that, in general, it is impossible to ensure the transfer of the manipulator to a prescribed terminal state. Consequently, we can further formalize the control problem statement by requiring that the relations

$$\lim_{t\to\infty} \varphi(t) = \bar{\varphi}, \qquad \lim_{t\to\infty} \dot{\varphi}(t) = 0$$

be satisfied. This can always be implemented by a control of the form $u = -A\dot{\varphi} - B(\varphi - \bar{\varphi})$, where $A > 0$, $B > 0$ are arbitrary diagonal matrices.

By way of example, consider a plane two-link manipulator connected to an immovable base (see Chapter VI, § 2, subsection 6). The expression for the kinetic energy T of the two-link manipulator is of the form

$$T = \tfrac{1}{2}\big(H_1\dot{\varphi}_1^2 + H_2\dot{\varphi}_2^2 + 2H_3\dot{\varphi}_1\dot{\varphi}_2\cos(\varphi_1 - \varphi_2)\big), \tag{2.31}$$

where φ_1 and φ_2 are hinge angles of manipulator links in the absolute coordinate system. The mass-inertia characteristics of the two-link manipulator are the following: $H_1 = 42.7$, $H_2 = 9.4$, $H_3 = 5$ (kg·m^2).

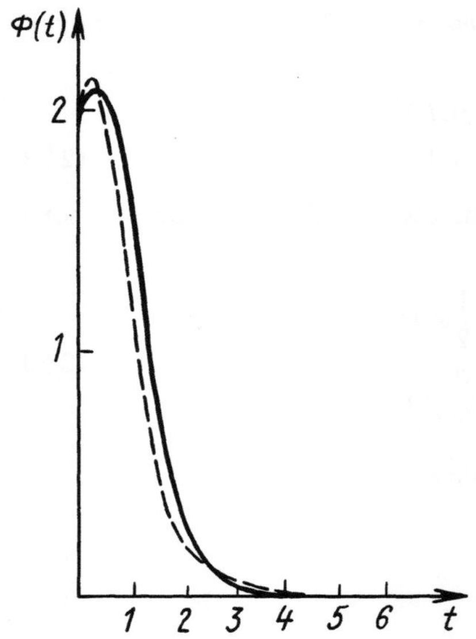

FIG. 2.2. Transient processes for an aggregate of angular variables and angular velocities

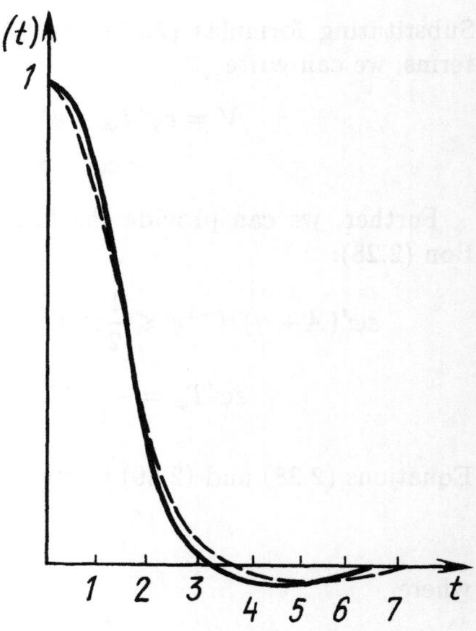

FIG. 2.3. Transient processes for the hinge angle φ_1 of the first link of the two-link manipulator

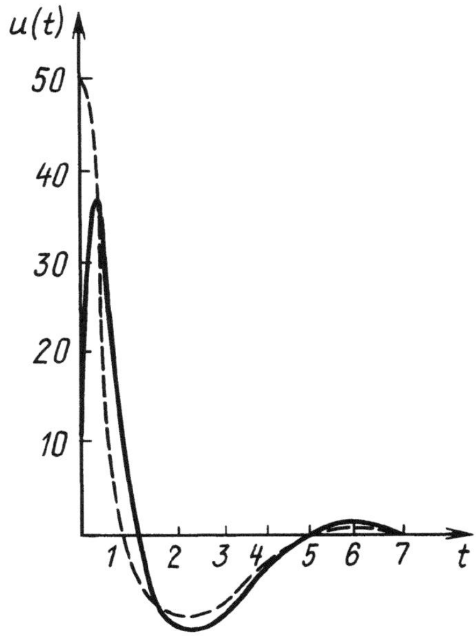

FIG. 2.4. The time-dependence of the control torque in the first link of the two-link manipulator

It is required to transfer the two-link manipulator from the state $\varphi_i = 1\,\mathrm{rad}$, $\dot\varphi_i = 0$ to the state $\varphi_i = 0$, $\dot\varphi_i = 0$ $(i = 1, 2)$ using the following proportional-differential controller (PD-controller):

$$u = -75\dot\varphi - 50\varphi. \tag{2.32}$$

For comparison, the same problem was solved using a proportional-integral-differential controller (PID-controller) of the form

$$u = -75\dot\varphi - 500\int_0^t e^{-10(t-s)}\varphi(s)\,ds. \tag{2.33}$$

The equations of motion (2.22), (2.31), (2.32) and (2.22), (2.31), (2.33) were integrated numerically. In Figures 2.2–2.4 the solid lines correspond to the PID-controller (2.33), and the dashed lines correspond to the PD-controller (2.32). The graphs of the function $\Phi = \varphi_1^2 + \varphi_2^2 + \dot\varphi_1^2 + \dot\varphi_2^2$ that illustrate how the trajectories of motion tend to zero are shown in Figure 2.2. The graphs of the hinge angle φ_1 (expressed in radians) for the first link versus time are given in Figure 2.3. The time-dependence of the control torque (expressed in newton·m) in the first link is shown in Figure 2.4. As shown by the calculation, for roughly the same characteristics of the transient processes the PID-controller (2.33) has the advantage over the PD-controller (2.32) in that it requires a substantially smaller value of the maximum control torque.

§ 3. Stabilization of Linear Systems

1. Time-varying linear-quadratic problems. One of the thoroughly-studied stabilization problems is the linear-quadratic problem (LQ-problem), i.e., the control problem for linear systems with a quadratic cost functional. We present here some pertinent results for linear systems

$$\dot{x}(t) = A(t)x(t) + B(t)u, \qquad t \geqslant 0,\ x \in R^n,\ u \in R^m,\ x(0) = x_0 \tag{3.1}$$

with the quadratic cost functional

$$J(u) = \int_0^\infty [x'(t)N_1(t)x(t) + u'N_0(t)u]\,dt \to \inf_u . \tag{3.2}$$

The prescribed matrices A, B, N_1, N_0 consist of bounded continuous elements. Moreover, the matrices N_1 and N_0 are uniformly positive definite, i.e., $N_1(t) \geqslant CI$, $N_0(t) \geqslant CI$, where $C > 0$ is a constant, and I is a unit matrix. It follows from the last inequalities that for any $x \in R^n$, $u \in R^m$, and $t \geqslant 0$ the following estimates hold: $x'N_1(t)x \geqslant Cx'x$, $u'N_0(t)u \geqslant Cu'u$. The control u in the problem (3.1), (3.2) is sought in the form $u = u(t, x(t))$. No other constraints on the control are imposed. To solve the problem (3.1), (3.2), according to Theorem 2.1, we must construct the function $V(t,x)$, which is, clearly, to be sought in the form

$$V(t,x) = x'P(t)x,$$

where the symmetric matrix $P(t) > 0$ is to be determined.

It should be noted at this stage that the Bellman equations for the LQ-problem on a finite time interval described by relations (2.1), (2.2) of Chapter IV and for the problem (3.1), (3.2) on an infinite time interval are, in fact, identical. Therefore, using formulas (2.8), (2.11) of Chapter IV, we find that the optimal control u_0 and the matrix $P(t)$ for the problem (3.1), (3.2) satisfy the relations

$$u_0(t,x) = -N_0^{-1}(t)B'(t)P(t)x, \qquad t \geqslant 0, \tag{3.3}$$

$$\dot{P}(t) + A'(t)P(t) + P(t)A(t) - P(t)B(t)N_0^{-1}(t)B'(t)P(t) + N_1(t) = 0. \tag{3.4}$$

According to Theorem 2.3, if there exists a solution $P(t)$ of equation (3.4) such that

$$P(t) \geqslant \alpha I, \qquad \|P(t)\| < C, \qquad \alpha > 0,\ C > 0, \tag{3.5}$$

then the system (3.1) with control (3.3) is exponentially stable. Moreover, the control (3.3) minimizes the cost functional (3.2) that is equal to $V(0, x_0)$.

We are now concerned with the sufficient conditions for the existence and uniqueness of the solution $P(t)$, $t \geqslant 0$, satisfying the conditions (3.5). Suppose that the elements of the matrices $A(t)$ and $B(t)$ for $t \geqslant 0$ have continuous bounded derivatives up to order $n-1$ inclusive. We now introduce n matrices

$Q_1(t), \dots, Q_n(t)$ of the same dimensions $n \times m$ and the $n \times nm$ matrix $Q(t)$ with the help of the relations

$$Q_1(t) = B(t), \quad Q_2(t) = \frac{dQ_1(t)}{dt} - A(t)Q_1(t), \dots,$$
$$Q_i(t) = \frac{dQ_{i-1}(t)}{dt} - A(t)Q_{i-1}(t); \qquad i = 2, \dots, n,$$
$$Q(t) = \big(Q_1(t), Q_2(t), \dots, Q_n(t)\big).$$

Theorem 3.1 [10]. *Suppose that the parameter constraints for the problem* (3.1), (3.2) *stated above are satisfied. Moreover, suppose there exists a number* $\Delta > 0$ *such that on any segment* $[t, t+\Delta]$, $t \geqslant 0$, *there exists a point* $\tau(t)$ *at which the matrix* $Q(t)$ *has rank* n, *i.e.,*

$$\operatorname{rank} Q\big(\tau(t)\big) = n, \qquad t \geqslant 0. \tag{3.6}$$

Then for $t \geqslant 0$ *there exists a unique solution* $P(t)$ *satisfying the conditions* (3.5). *The optimal control* $u_0(t)$ *has the form* (3.3), *the system* (3.1) *with this control is exponentially stable, and the cost functional* $J(u_0) = V(0, x_0)$.

2. Use of the method of successive approximations for determining optimal control. Theorem 3.1 furnishes certain sufficient conditions for the existence of a positive definite solution of equation (3.4). For an efficient construction of this solution one can use the following *method of successive approximations*, which is close to the original Newton method in the Bellman interpretation. With reference to the problem (3.1), (3.2) this method can be stated as the following theorem.

Theorem 3.2. *Suppose there exists a control of the form* $u = R(t)x$ *such that equation* (3.1) *is exponentially stable and the following integral is convergent:*

$$P_1(t) = \int_t^\infty Z_1(t,s)\big(N_1(s) + R'(s)N_0(s)R(s)\big)Z_1'(t,s)\,ds, \qquad t \geqslant 0,$$

where $Z_1(t,s)$ *is the Cauchy matrix for the linear equation*

$$\dot{x}(t) = -[A(t) + B(t)R(t)]'x(t).$$

Further, let the sequence of matrices $P_i(t) \geqslant 0$ $(i \geqslant 2)$ *be defined by the relations*

$$P_i(t) = \int_t^\infty Z_i(t,s)\big[N_1(s) + P_{i-1}(s)B(s)N_0^{-1}(s)B'(s)P_{i-1}(s)\big]$$
$$\times Z_i'(t,s)\,ds, \qquad i \geqslant 2,$$

where $Z_i(t,s)$ *are the fundamental matrices for the linear equations*

$$\dot{x}(t) = -\big[A'(t) - P_{i-1}(t)B(t)N_0^{-1}(t)B'(t)\big]x(t), \qquad i \geqslant 2.$$

Then the sequence of matrices $P_i(t)$ is monotone nonincreasing (i.e., $P_{i+1}(t) - P_i(t) \leqslant 0$, $t \geqslant t_0$) and converges to the unique solution $P(t)$ of equation (3.4) satisfying the estimates (3.6). Here the matrices $P_i(t)$ ($i \geqslant 2$) are the solution of a linear equation of the form

$$\dot{P}_i(t) + [A' - P_{i-1}BN_0^{-1}B']P_i + P_i[A - BN_0^{-1}B'P_{i-1}] + N_1 + P_{i-1}BN_0^{-1}B'P_{i-1} = 0, \qquad i \geqslant 0.$$

The successive approximations $u_i(t)$ to the optimal control are expressed by the formula

$$u_i(t,x) = -N_0^{-1}B'(t)P_{i-1}(t)x.$$

3. Time-invariant linear-quadratic equation. We shall discuss a number of results concerning the optimal stabilization problem (3.1), (3.2) for constant matrices A, B, N_1, N_0, i.e., we shall be concerned with the problem

$$\dot{x}(t) = Ax(t) + Bu, \qquad x(0) = x_0, \tag{3.7}$$

$$J(u) = \int_0^\infty [x'(t)N_1x(t) + u'N_0u]\,dt, \qquad N_1 > 0,\ N_0 > 0. \tag{3.8}$$

The value of the cost functional (3.8) does not change when the initial moment $t = 0$ is replaced by a arbitrary moment $t = t_0$. Therefore, the Bellman function V depends only on x. Moreover, let us assume that

$$V(x) = x'Px. \tag{3.9}$$

Then the corresponding optimal control is

$$u_0(t,x) = u_0(x) = -N_0^{-1}B'Px. \tag{3.10}$$

and the constant symmetric matrix $P > 0$ in formulas (3.9), (3.10) is the solution of the equation

$$A'P + PA - PBN_0^{-1}B'P + N_1 = 0 \tag{3.11}$$

which is called the *algebraic Riccati equation.*

The theorem that follows furnishes the conditions under which equation (3.11) has a unique solution $P > 0$, and, consequently, the relations (3.9), (3.10) will be valid.

Theorem 3.3 [10]. *Suppose that in the stationary LQ-problem* (3.7), (3.8) *the following conditions are satisfied*:

$$N_0 > 0, \qquad N_1 > 0, \qquad \operatorname{rank}(B, AB, ..., A^{n-1}B) = n. \tag{3.12}$$

Then equation (3.11) *has a unique solution $P > 0$, and the relations* (3.9), (3.10) *hold. The system* (3.10) *with the control* (3.10) *is exponentially stable and*

$$\min_u J(u) = J(u_0) = x_0'Px_0.$$

Note that the last condition in formula (3.12) coincides with the controllability condition discussed in § 2 of Chapter VI.

Thus, for constant matrices A, B, N_1, N_0 solution of the optimal stabilization problem (3.1), (3.2) has been reduced to determining a positive definite solution of equation (3.11).

4. The algebraic Riccati equation. Consider some methods of solving Riccati's equation (3.11).

One of these, sometimes called the *stabilization method*, is based on approximation to the LQ-problem (3.7), (3.8) on an infinite time interval by the LQ-problem on a finite time interval $0 \leqslant t \leqslant T$ for the system (3.7) with the cost functional

$$J_1(u) = \int_0^T [x'(t)N_1x(t) + u'N_0u]\,dt. \tag{3.13}$$

If u_1 denotes the optimal control in the problem (3.7), (3.13), then it is readily seen that $J_1(u_1) \leqslant J_1(u_0)$, where u_0 is the optimal control for the problem (3.7), (3.8). Moreover, intuitively, we can assume that $J_1(u_1) \to J_1(u_0)$ as $T \to \infty$. Thus, recalling the relations (2.8) and (2.11) of Chapter IV that furnish the solution of the problem (3.7), (3.13), we see that if there exists a positive definite solution P of equation (3.11), then it represents the limit as $T \to \infty$ of solutions $\alpha(t)$ of the Cauchy problem

$$\dot{\alpha}(t) + \alpha(t)A + A'\alpha(t) - \alpha(t)BN_0^{-1}B'\alpha(t) + N_1 = 0, \qquad 0 \leqslant t \leqslant T,$$
$$\alpha(T) = 0.$$

It is convenient to set $\beta(t) = \alpha(T - t)$. Then the function $\beta(t)$ is a solution of the equation

$$\dot{\beta} = \beta A + A'\beta - \beta BN_0^{-1}B'\beta + N_1, \qquad \beta(0) = 0,$$

whence we can derive the relation $\lim_{t\to\infty} \beta(t) = P$. This result is justified in [10].

For some optimal stabilization problems of the form (3.7), (3.8) the optimal control coefficients can be directly calculated without first determining the optimal function $V_0(t, x)$. We describe certain methods of computing these coefficients.

Suppose that in the system (3.7) u is a scalar control, and the matrix A and the column vector B are constant. In the cost functional (3.8) let $N_0 = I$, where I is a unit matrix. The optimal control in this problem is of the form $u_0 = Q'x$, where the elements of the constant column vector $Q \in R^n$ are to be defined. To define the vector Q, consider the determinant $D(\lambda)$ of order $2n$:

$$D(\lambda) = \begin{vmatrix} A - \lambda I & BB' \\ N_1 & -A' - \lambda I \end{vmatrix}.$$

This determinant being a polynomial of degree $2n$ can be expressed as the product

$$D(\lambda) = (-1)^n D_1(\lambda) D_1(-\lambda).$$

All roots of the polynomial $D_1(\lambda)$ of order n lie in the left half-plane. It can be shown that the determinant of the matrix $A + BQ' - \lambda I$ is equal to $(-1)^n D_1(\lambda)$, i.e.

$$\det |A + BQ' - \lambda I| - (-1)^n D_1(\lambda) = 0. \tag{3.14}$$

Thus, the proposed algorithm of determining the optimal control coefficients Q consists of the following steps:

1°. The roots $\lambda_1, \ldots, \lambda_n$ of the equation $D(\lambda) = 0$, that have a negative real part, are found and the polynomial $D_1(\lambda)$ is formed:

$$D_1(\lambda) = (\lambda - \lambda_1) \cdot \ldots \cdot (\lambda - \lambda_n).$$

2°. The coefficients of like powers of the argument λ in equation (3.14) are equated.

3°. The resulting system of linear equations is solved for the components of the vector Q, and the optimal F-control $u_0 = Q'x$ is determined.

To determine the vector Q, another method arising from the maximum principle has been proposed.[2] Suppose that $u_0 = Q'x$ be a scalar F-control in the problem (3.7), (3.8) for $N_0 = I$. Let N_{ij} denote the elements of the matrix N_1. In equation (3.7) B is a column vector.

We denote by $d(\lambda)$ the characteristic determinant of the system (3.7) with the zero control $u = 0$, by $d_0(\lambda, Q)$ the characteristic determinant of the system (3.7) with the control $u_0 = Q'x$, and by $d_k(\lambda)$ the determinant that is obtained from the determinant $d(\lambda)$ by substituting the vector B for its kth column. Then the equation

$$d(\lambda)d(-\lambda) + \sum_{i,j=1}^{n} N_{ij} d_i(\lambda) d_j(-\lambda) = 0$$

has precisely n roots $\lambda_1, \ldots, \lambda_n$ with a positive real part. Moreover, for all λ the following equality is valid:

$$(\lambda + \lambda_1) \cdot \ldots \cdot (\lambda + \lambda_n) = (-1)^n d_0(\lambda, Q),$$

which is a linear equation in the components of the vector Q and can serve to determine the optimal control $u_0 = Q'x$.

Remark. Suppose that for the system (3.7) there exists an admissible continuous control $u(t, x)$ with a continuous bounded derivative with respect to x such that the system (3.7) is exponentially stable. Then there exists a linear control that furnishes the solution of the optimal stabilization problem (see [7] of Part One).

Let us state the necessary and sufficient conditions for the existence of the above-mentioned admissible control for the problem (3.7) in the case when the controllability condition (3.12) is no longer valid.

We reduce equation (3.7) to the canonical form (2.14), (2.15) of Chapter VI by the change of variable $x = Ty$. Suppose that $y = (y_1, y_2)$, $y_1 \in R^k$, $y_2 \in R^{n-k}$, where k is the rank of the controllability matrix (3.12). Then

$$\dot{y}_1 = A_1 y_1 + A_2 y_2 + B_1 u, \qquad \dot{y}_2 = A_3 y_2.$$

---

[2]Lurie, A. I., The Minimum Quadratic Cost Functional for a Control System, *Izv. Akad. Nauk SSSR Tekhn. Kibernet.*, (1963), no. 4. (Translated into English as *Soviet J. Comput. Systems Sci.*)

Further, note the following equality:

$$\det(A - \lambda I) = \det(A_1 - \lambda I_1)\det(A_3 - \lambda I_2).$$

Moreover, by virtue of the condition $\operatorname{rank}(B_1, A_1B_1, \dots, A_1^{k-1}B_1) = k$ the optimal stabilization problem for the system $\dot y_1 = A_1y_1 + B_1u$ can be solved. Thus, the necessary and sufficient condition for the existence of an admissible control such that the system (3.7) is exponentially stable is that all eigenvalues of the matrix A_3 lie in the left half-plane.

§ 4. Stabilization of Quasilinear Systems

Real systems are usually nonlinear ones. If nonlinearities are small, these systems can be approximated by linear ones. Otherwise, in determining control we have to take account of nonlinear summands in the equations of motion.

1. Quasioptimal stabilization and estimation of its error. In what follows, we propose a procedure for quasioptimal stabilization and give error estimates for arbitrary initial disturbances. The control system under consideration is of the form

$$\dot x(t) = \varepsilon f\big(t, x(t)\big) + B(t)u, \qquad t \geqslant 0,\ x \in R^n,\ u \in R^m,\ x(0) = x_0. \tag{4.1}$$

and is called a *quasilinear system.*

Here the function $f(t,x)$ is continuous in all arguments and satisfies the condition $|f(t,x)| \leqslant Cx$ $(C > 0)$, the matrix $B(t)$ is continuous and bounded on the interval $[0,\infty)$, the parameter $\varepsilon \geqslant 0$ and the vector x_0 are specified. It is required to determine an F-control $u = u\big(t, x(t)\big)$ such that the cost functional is minimized, i.e.,

$$J(u) = \int_0^\infty \big(x'(t)N_1(t)x(t) + u'N_0(t)u\big)\,dt \to \inf. \tag{4.2}$$

It should be noted that if the equation of motion contains a linear summand $A(t)x(t)$, then it can be reduced to the form (4.1) by means of a nondegenerate transformation of coordinates. Moreover, under the assumptions made, if for some control u the functional $J(u) < \infty$, then the solution of the system (4.1) generated by this control tends to zero as $t \to \infty$.

We assume that for the values of ε under consideration there exists a solution of the problem (4.1), (4.2). Let $v(t,x)$ denote the optimal control and $V(t,x)$ the corresponding Bellman function. Occasionally, to stress the dependence of the solution of equation (4.1) on the control v, we will denote it by $x(t,v)$.

First, we shall give a heuristic description of the *quasioptimal stabilization algorithm.* We write the Bellman equation for the problem (4.1), (4.2):

$$\begin{aligned}\frac{\partial V(t,x)}{\partial t} &+ \varepsilon f'(t,x)\frac{\partial V(t,x)}{\partial x} + x'N_1x \\ &= \frac{1}{4}\Big(\frac{\partial V(t,x)}{\partial x}\Big)' B_1\Big(\frac{\partial V(t,x)}{\partial x}\Big), \qquad B_1 = BN_0^{-1}B'.\end{aligned} \tag{4.3}$$

The optimal control v is of the form

$$v(t,x) = -\frac{1}{2}N_0^{-1}B'\frac{\partial V(t,x)}{\partial x}. \tag{4.4}$$

We express V as a series in powers of ε:

$$V(t,x) = V_0(t,x) + \varepsilon V_1(t,x) + \dots. \tag{4.5}$$

To determine the functions $V_i(t,x)$, we substitute the expression (4.5) in equation (4.3) and equate to zero the coefficients of like powers of ε. Then we obtain that V_0 satisfies equation (4.3) for $\varepsilon = 0$, and for V_i $(i \geqslant 1)$ we have a sequence of equations that are linear in V_i:

$$\frac{\partial V_i(t,x)}{\partial t} + f'(t,x)\frac{\partial V_{i-1}(t,x)}{\partial x} = \frac{1}{4}\sum_{j=0}^{i}\frac{\partial V_j'(t,x)}{\partial x}B_1\frac{\partial V_{i-j}(t,x)}{\partial x}. \tag{4.6}$$

We look for a solution of equation (4.6) in the class of continuously differentiable functions bounded below and satisfying the estimate $V_i(t,x) \leqslant C|x|^2$. The ith approximation $u_i(t,x)$ to the optimal control v is of the form

$$u_i(t,x) = -\frac{1}{2}N_0^{-1}B'\frac{\partial W}{\partial x}, \qquad W_i(t,x) = \sum_{j=0}^{i}\varepsilon^j V_j(t,x). \tag{4.7}$$

Thus, the essential part of the algorithm is connected with the necessity of solving the problem (4.6). It will be recalled that if the problem (4.6) has a solution for $\varepsilon = 0$, then $V_0 = x'P(t)x$, where the matrix $P(t)$ is a unique bounded positive definite solution of the Riccati equation (3.4) for $A(t) \equiv 0$, i.e., of the following equation

$$\dot{P}(t) - P(t)B_1P(t) + N_1(t) = 0, \qquad t \geqslant 0.$$

The other approximations under the appropriate assumption of the smoothness of $f(t,x)$ are expressed by the formula

$$V_i(t,x) = \int_t^{\infty}\Bigg[f'(s,x(s))\frac{\partial V_{i-1}(s,x(s))}{\partial x} - \frac{1}{4}\sum_{j=1}^{i-1}\frac{\partial V_j'(s,x(s))}{\partial x}B_1(s)\frac{\partial V_{i-j}(s,x(s))}{\partial x}\Bigg]ds, \qquad i \geqslant 1. \tag{4.8}$$

Here the sum for $i = 1$ is zero, and $x(s)$ denotes the solution of equation (4.1) for $\varepsilon = 0$, $s \geqslant t$, for the control $u_0 = -N_0^{-1}(s)B'(s)P(s)x(s)$, and for the initial condition $x(t) = x$.

Let us find an estimate for the zero approximation. The zero approximation control $u_0(t,x)$ is defined by the expression (4.7) for $i=0$.

Let us assume that the original control problem (4.1), (4.2) has a solution for a given value of ε and also for $\varepsilon=0$; moreover, the following inequality is valid:

$$x'N_1x - 2\varepsilon f'(t,x)P(t)x \geqslant C|x|^2. \tag{4.9}$$

Then for some constant $C>0$ defined by the parameters of the problem (4.1), (4.2) the following inequality holds:

$$0 \leqslant J(u_0) - J(v) \leqslant C\varepsilon. \tag{4.10}$$

Let us prove this inequality, which indicates that controlling the system (4.1) by means of u_0 leads to an error in the cost functional (4.2) of the order of ε. Consider the functional

$$\begin{aligned} J_0(u) &= J(u) - \varepsilon\int_0^\infty f'(t,x(t))\frac{\partial V_0(t,x(t))}{\partial x}\,dt \\ &= \int_0^\infty \left[x'(t)N_1(t)x(t) + u'N_0(t)u - \varepsilon f'(t,x)\frac{\partial V_0(t,x)}{\partial x}\right]dt. \end{aligned} \tag{4.11}$$

In view of the condition (4.9) the integrand of the functional $J_0(u)$ is positive definite with respect to the phase coordinates, i.e., a condition of the form

$$x'N_1x + u'N_0u - \varepsilon f'(t,x)\frac{\partial V_0(t,x)}{\partial x} \geqslant C|x|^2. \tag{4.12}$$

is satisfied.

Thus, by virtue of Theorem 2.3 the control $u_0(t,x)$ is the solution of the problem of minimizing the functional (4.11) along the trajectories of the system (4.1), and the corresponding Bellman function is $V_0(t,x)$. Hence, noting the inequality (4.12), we have

$$\int_0^\infty |x(t,u_0)|^2\,dt \leqslant CV_0(0,x_0). \tag{4.13}$$

It follows from this inequality and from the expression (4.7) that $J(u_0)<\infty$, i.e., the control u_0 is admissible also for the original problem (4.1), (4.2). The difference $J(u_0)-J(v)$ in the inequality (4.10) can be expressed as

$$J(u_0) - J(v) = [J_0(u_0) - J(v)] + [J(u_0) - J_0(u_0)]. \tag{4.14}$$

Let us estimate the differences on the right side of (4.14). Because of the optimality of the control v and of the admissibility of the control u_0 we obtain

$$J(v) \leqslant J(u_0) = J_0(u_0) + J(u_0) - J_0(u_0). \tag{4.15}$$

Further, using the equality (4.11), the boundedness of P, and the assumptions about f, we can write

$$\begin{aligned} |J(u_0) - J_0(u_0)| &\leqslant 2\varepsilon \int_0^\infty \left| f'(t, x(t, u_0)) P(t) x(t, u_0) \right| dt \\ &\leqslant 2\varepsilon C \int_0^\infty |x(t, u_0)|^2 \, dt. \end{aligned} \tag{4.16}$$

The inequalities (4.13) and (4.16) imply the estimate

$$|J(u_0) - J_0(u_0)| \leqslant \varepsilon C_1. \tag{4.17}$$

Thus, in view of (4.15) we obtain

$$J(v) \leqslant J_0(u_0) + \varepsilon C_1. \tag{4.18}$$

In a similar way, because of the optimality of the control u_0 in the problem (4.1), (4.11) we obtain

$$J_0(u_0) \leqslant J_0(v) = J(v) + [J_0(v) - J(v)].$$

Now we estimate the difference $J_0(v) - J(v)$. On the strength of relation (4.11) we obtain in a similar way as for (4.16)

$$[J_0(v) - J(v)] \leqslant 2\varepsilon C \int_0^\infty |x(t, v)|^2 \, dt. \tag{4.19}$$

But in view of the esimate (4.18) and the fact that the matrix N_1 is uniformly positive definite, the following inequality holds

$$\int_0^\infty |x(t, v)|^2 \, dt \leqslant C J(v) \leqslant C V_0(0, x_0) + \varepsilon C C_1.$$

Thus, noting the inequality (4.19), we can write

$$|J_0(v) - J(v)| \leqslant \varepsilon C_2.$$

Therefore,

$$J_0(u_0) \leqslant J(v) + |J_0(v) - J(v)| \leqslant J(v) + \varepsilon C_2. \tag{4.20}$$

Comparing the relations (4.18) and (4.20), we find that $|J_0(u_0) - J(v)| \leqslant \varepsilon C$. This inequality and the inequality (4.17), in view of (4.14), imply that

$$|J_0(u_0) - J(v)| \leqslant \varepsilon C.$$

Thus, the proof of the estimate (4.10) for the zero approximation is complete.

Consider the estimates for higher approximations. When considering the ith approximation u_i to the optimal control, we assume that there exist continuously differentiable functions $V_j(t,x)$, $j \leqslant i$, that are bounded below and satisfy equation (4.6) and also the estimates

$$|W_j(t,x)| \leqslant C|x|^2, \qquad |\frac{\partial W_j(t,x)}{\partial x}| \leqslant C|x|, \qquad j \leqslant i. \tag{4.21}$$

It follows from the definition (4.7) of controls $u_j(t,x)$ and from the esimates (4.21) that $u_j(t,0) \equiv 0$.

In justifying the estimate for the zero approximation, we have used the trick of interpreting the zero approximation $u_0(t,x)$ to the control as the optimal control for the original system (4.1) with the cost functional that differs from the original functional (4.2) by a quantity of the order of ε. We shall use the same trick in the investigation of the controls $u_i(t,x)$, $i \geqslant 1$, defined by the equality (4.7). To set up the auxiliary stabilization problem in which $u_i(t,x)$ plays the role of the optimal control, we write the equation for the function $W(t,x)$ that occurs in the relation (4.7). Multiplying equation (4.6) by ε^i and summing the resulting equations with respect to the i, we obtain

$$\frac{\partial W_i}{\partial t} + \varepsilon f'(t,x)\frac{\partial W_i}{\partial x} + x'N_1x + \delta_i\varepsilon^{i+1} = \frac{1}{4}\left(\frac{\partial W_i}{\partial x}\right)' B_1 \left(\frac{\partial W_i}{\partial x}\right), \tag{4.22}$$

$$\delta_i = -f'\frac{\partial V_i}{\partial x} + \frac{1}{4}\sum_{l=1}^{i}\sum_{j=0}^{l-1} \varepsilon^j \left(\frac{\partial V_l}{\partial x}\right) B_1 \left(\frac{\partial V_{i+1+j-l}}{\partial x}\right).$$

Consider the functional

$$J_i(u) = J(u) + \varepsilon^{i+1}\int_0^T \delta_i\big(t, x(t,u)\big)\,dt. \tag{4.23}$$

As in the case of (4.9), we assume that the problem (4.1), (4.2) has a solution for a given value of ε and also for $\varepsilon = 0$; moreover, the following inequality is valid:

$$x'N_1x - \varepsilon^{i+1}\delta_i \geqslant C|x|^2. \tag{4.24}$$

In view of the condition (4.24) the functional $J_i(u)$ is positive definite with respect to the phase coordinates. Therefore, applying Theorem 2.3 and noting equations (4.22) and the estimates (4.21), we establish that in the problem (4.1), (4.23) the control $u_i(t,x)$ is optimal, and the Bellman function is equal to $W_i(t,x)$.

Thus, noting the inequality (4.24), we find that the integral

$$\int_0^\infty |x(t,u_i)|^2\,dt \leqslant CW_i(0,x_0) \tag{4.25}$$

is finite. In addition,

$$2u_i(t,x) = -N_0^{-1}(t)B'(t)\frac{\partial W_i}{\partial x}.$$

Hence, on the strength of (4.21) and (4.7) we have $|u_i(t,x)| \leqslant C|x|$. In view of (4.25) it follows that $J(u_i) < \infty$, i.e., u_i is an admissible control in the stabilization problem (4.1), (4.2). Further justification of the estimates for the ith approximation is similar to the proof of the inequality (4.10), namely we must show that

$$0 \leqslant J(u_i) - J(v) \leqslant C\varepsilon^{i+1}. \tag{4.26}$$

Obviously,

$$J(u_i) - J(v) = [J_i(u_i) - J(v)] + [J(u_i) - J_i(u_i)].$$

Let us estimate the differences on the right side of this relation. By virtue of (4.24) and (4.25) we have

$$\begin{aligned} J(v) &\leqslant J(u_i) \leqslant J_i(u_i) + |J(u_i) - J_i(u_i)| \\ &\leqslant J_i(u_i) + \varepsilon^{i+1}\left|\int_0^\infty \delta_i(t, x(t,u_i))\,dt\right| \\ &\leqslant J_i(u_i) + \varepsilon^{i+1}C\int_0^\infty |x(t,u_i)|^2\,dt \\ &\leqslant J_i(u_i) + \varepsilon^{i+1}C\gamma_i, \qquad \gamma_i = \int_0^\infty |x(t,u_i)|^2\,dt. \end{aligned} \tag{4.27}$$

To establish the inequality that is opposite to (4.27), we use the optimality of the control u_i in the stabilization problem (4.1), (4.23). Then

$$\begin{aligned} J_i(u_i) &\leqslant J_i(v) \leqslant J(v) + |J_i(v) - J(v)| \\ &\leqslant J(v) + \varepsilon^{i+1}\left|\int_0^\infty \delta_i(t, x(t,v))\,dt\right|^2 \\ &\leqslant J(v) + \varepsilon^{i+1}C\int_0^\infty |x(t,v)|^2\,dt \\ &\leqslant J(v) + \varepsilon^{i+1}CW(0,x_0). \end{aligned} \tag{4.28}$$

It follows from the inequalities (4.27) and (4.28) that

$$|J_i(u_i) - J(v)| \leqslant \varepsilon^{i+1}C\gamma_i. \tag{4.29}$$

Moreover, in deducing the estimate (4.27), we established the relation

$$|J(u_i) - J_i(u_i)| \leqslant \varepsilon^{i+1}CC_i. \tag{4.30}$$

The inequalities (4.29) and (4.30) imply that

$$0 \leqslant J(u_i) - J(v) \leqslant |J_i(u_i) - J(v)| + |J(u_i) - J_i(u_i)| \leqslant \varepsilon^{i+1}CC_i.$$

Thus, the estimate for the ith approximation (4.26) has been established.

Remark. Let us discuss the sufficient conditions for the estimates (4.21). Suppose that the problem (4.1), (4.2) has a solution for $\varepsilon = 0$, the function $f(t,x)$ and its derivatives $\partial^j f(t,x)/\partial x^j$ are continuous, with $|\partial^j f/\partial x^j| \leqslant C$, $j = 1, \dots, i$. Then there exists a unique solution of equation (4.6) that satisfies the estimates (4.21). This solution may be represented in the form (4.8).

2. Adaptive stabilization. Throughout this subsection we assume that the value of the parameter ε in equation (4.1) is unknown. The *adaptive stabilization problem* is to determine a control such that the trivial solution of the system (4.1) is asymptotically stable. Such control called an *adaptive control* is sought in the form

$$B\omega(t,x) - k(t)f(t,x).$$

The function $\omega(t,x)$ and the scalar coefficient $k(t)$ of an additional control loop are chosen in a way so as to ensure the stability of the trivial solution of the system

$$\dot{x}(t) = \big(\varepsilon - k(t)\big)f\big(t,x(t)\big) + B(t)\omega\big(t,x(t)\big). \tag{4.31}$$

Lemma 4.1. *Suppose that for some fixed value of ε_0 there exists a control $\omega_0(t,x)$, $\omega_0(t,0) = 0$, such that the trivial solution of the equation*

$$\dot{y}(t) = \varepsilon_0 f\big(t,y(t)\big) + B\omega_0\big(t,y(t)\big), \qquad t \geqslant 0,$$

is asymptotically stable uniformly with respect to the initial data. Then the adaptive stabilization problem has a solution for any ε.

Proof. The existence of a Lyapunov function $V(t,x)$ follows from the hypotheses of the lemma and also from the appropriate inversion theorem (see Remark 2 of § 2, Chapter I). The function V is positive definite, has an infinitesimal upper limit, and the sum

$$V_t(t,x) + V_x'(t,x)\big(\varepsilon_0 f(t,x) + B\omega_0(t,x)\big) = R(t,x)$$

is negative definite. We define the control coefficient $k(t)$ in equation (4.31) by means of the relations

$$\dot{k}(t) = -\tfrac{1}{2}f'\big(t,x(t)\big)V_x\big(t,x(t)\big), \qquad t \geqslant 0, \tag{4.32}$$
$$V_x = \frac{\partial V}{\partial x}, \qquad k(0) = 0.$$

Consider the function

$$R^1 = V(t,x) + \big(\varepsilon - \varepsilon_0 - k(1)\big)^2.$$

Calculating the total derivative $\dot{R}_1$ of the function $R_1(t,x)$ along the trajectories of the system (4.31), (4.32) leads to $\dot{R}_1 = R$.

Therefore, by virtue of Theorem 4.5 of Chapter I on partial stability the trivial solution of the system (4.31) is asymptotically stable. Thus, the control $\omega_0(t,x)$

and the controller (4.32) ensure stabilization of the system (4.1) for any *a priori* unknown value of the parameter ε. □

Lemma 4.1 enables reduction of the adaptive stabilization problem for the system (4.1) to the task of determining the Lyapunov function V and the control ω_0. Moreover, the constructions of subsection 1 can be used. Specifically, if the system (4.1) is stabilized for $\varepsilon = 0$, then we may set $V = V_0$, $\omega_0 = u_0$, where $V_0 = x'P(t)x$ and $u_0 = -N_0^{-1}B'P(t)x$.

But if the conditions of subsection 1 for $\varepsilon = \varepsilon_0$ are satisfied, then we can set $V = W(t,x)$, $\omega_0 = u_i(t,x)$, where the functions W and u_i for $\varepsilon = \varepsilon_0$ have the form (4.7). In both cases the coefficient $k(t)$ is defined by equation (4.32). At the same time, when considering particular systems, we can take account of their peculiar features and, by means of Lemma 4.1, determine a stabilizing control depending on part of the phase coordinates. This feature is important, since it enables minimization of the number of coordinate transducers. It should be noted that the solution of the adaptive stabilization problem for a prescribed programmed motion of the system (4.1) can be stated in an analogous way.

§ 5. Sufficient Conditions for Optimal Control Using Auxiliary Functions

In this section we give the sufficient conditions for optimal control that have been obtained in [13]. These conditions are stated in terms of the concepts that are close to Lyapunov's direct method and the dynamic programming approach.

1. Conditions for optimal control. Consider the control system

$$\dot{x}(t) = f\big(t, x(t), u(t)\big), \qquad 0 \leqslant t \leqslant T,\ x \in R^n,\ u \in R^m. \tag{5.1}$$

The initial state x_0 of the system (5.1) is given for $t = 0$

$$x(0) = x_0. \tag{5.2}$$

By the solution $x(t)$ of the problem (5.1), (5.2) generated by the control $u(t)$ we mean an absolutely continuous function satisfying equation (5.1) for all $t \in [0, T]$ and the control $u(t)$.

At the prescribed terminal moment T the following terminal condition must be met:

$$x(T) = x_1 \in X_1 \subset R^n. \tag{5.3}$$

In formula (5.3) X_1 is a given set of admissible terminal values of the process $x(t)$.

The control $u(t)\colon R^1 \to R^m$ is a measurable function of time which assumes values in the prescribed set U for almost all t:

$$u(t) \in U \subset R^m. \tag{5.4}$$

The cost functional is given by

$$J(u) = \varphi_0\big(x(T)\big) + \int_0^T F_0\big(t, x(t), u(t)\big)\, dt. \tag{5.5}$$

In the derivation of the conditions for optimality it is assumed that the functions $F_0(t,x,u)$ and $f(t,x,u)$ are measurable with respect to all variables (t,x,u), are continuous in u, and are continuously differentiable with respect to x, and the function $\varphi_0(x)$ is continuously differentiable with respect to x.

The control $u(t)$ is said to be *admissible* if this control generates a unique solution of the problem (5.1), the functional (5.5) is finite, and the conditions (5.2)–(5.4) are satisfied. The set of all admissible controls on the segment $[0,T]$ is denoted by W.

The optimal control problem for the system (5.1) is to determine an admissible control u_0 such that it furnishes an absolute minimum of the cost functional (5.5) on the set W. But if this minimum is not attained, then it is worth considering the problem of constructing a minimizing sequence of admissible controls $u_i \in W$ such that

$$\lim_{i\to\infty} J(u_i) = \inf_{u\in W} J(u).$$

Let us state sufficient conditions for optimal control in terms of an auxiliary function $V(t,x)$ that is, in a sense, analogous to the Lyapunov function introduced in Chapter I. The function $V(t,x)\colon [0,T]\times R^n \to R^1$ is continuously differentiable with respect to the independent variables $t\in[0,T]$ and $x\in R^n$.

Consider an arbitrary admissible control $u(t)$ and a corresponding trajectory $x(t)$ of the system (5.1). Let $\dot V_u(t,x)$ denote the total derivative of the function V along the trajectories of the system (5.1) generated by the control $u(t)$. Under these assumptions we have

$$\begin{gathered}\dot V_u(t,x) = V_t + V_x' f(t,x,u),\\ V_t = \frac{\partial V(t,x)}{\partial t}, \qquad V_x = \frac{\partial V(t,x)}{\partial x},\\ V\big(t_1,x(t_1)\big) - V\big(t,x(t)\big) = \int_t^{t_1} \dot V_u\big(s,x(s)\big)\,ds. \end{gathered}\tag{5.6}$$

The relations (5.5), (5.6) imply the following expression for the functional $J(u)$ that is valid for any admissible control u:

$$\begin{aligned} J(u) = \int_0^T &\big[F_0\big(t,x(t),u(t)\big) - \dot V_u\big(t,x(t)\big)\big]\,dt\\ &+ \varphi_0\big(x(T)\big) + V\big(T,x(T)\big) - V(0,x_0). \end{aligned}\tag{5.7}$$

Theorem 5.1. *Suppose there exists a scalar continuously differentiable function $V(t,x)$ and an admissible control $u_0(t)\in W$ satisfying the following conditions for all $u(t)\in W$:*

$$\begin{aligned} &\int_0^T \big[F_0\big(t,x(t),u(t)\big) - \dot V_u\big(t,x(t)\big)\big]\,dt\\ &\qquad \geqslant \int_0^T \big[F_0\big(t,x_0(t),u_0(t)\big) - \dot V_{u_0}\big(t,x_0(t)\big)\big]\,dt = 0, \end{aligned}\tag{5.8}$$

$$\begin{aligned}\varphi_0(x(T)) + V(T, x(T)) &\geqslant \varphi_0(x_0(T)) + V(T, x_0(T)) \\ &= \inf_{x \in X_1} [\varphi_0(x) + V(T, x)] = q, \end{aligned} \tag{5.9}$$

where $x(t)$ and $x_0(t)$ are the solutions of the problem (5.1) generated by the controls $u(t)$ and $u_0(t)$. Then the control $u_0(t)$ is optimal, and the minimum value $J(u_0)$ of the cost functional (5.7) is

$$J(u_0) = q - V(0, x_0). \tag{5.10}$$

Proof. For a proof it is sufficient to note that

$$\begin{aligned} J(u) - J(u_0) = &\left\{ \int_0^T [F_0(t, x(t), u(t)) - \dot{V}_u(t, x(t))]\, dt \right. \\ &\left. + \varphi_0(x(T)) + V(T, x(T)) - V(0, x_0) \right\} \\ &- \left\{ \int_0^T [F_0(t, x_0(t), u_0(t)) - \dot{V}_{u_0}(t, x_0(t))]\, dt \right. \\ &\left. + \varphi_0(x_0(T)) + V(T, x_0(T)) - V(0, x_0) \right\} \\ = &\left\{ \int_0^T [F_0(t, x(t), u(t)) - \dot{V}_u(t, x(t))]\, dt \right. \\ &\left. - \int_0^T [F_0(t, x_0(t), u_0(t)) - \dot{V}_{u_0}(t, x_0(t))]\, dt \right\} \\ &+ \left\{ \varphi_0(x(T)) + V(T, x(T)) - \varphi_0(x_0) - V(T, x_0(T)) \right\} \geqslant 0. \end{aligned}$$

Thus, noting the conditions (5.8), (5.9), we have $J(u) - J(u_0) \geqslant 0$ for any $u \in W$. Therefore, the optimality of the control u_0 is proved. Formula (5.10) follows immediately from the relations (5.7), (5.9). □

Remark. Occasionally, the condition (5.8) of Theorem 5.1 is expressed in terms of the function R defined by

$$\begin{aligned} R(t, x, u) &= -F_0(t, x, u) + V_t(t, x) + V_x'(t, x) f(t, x, u) \\ &= -F_0(t, x, u) + \dot{V}_u(t, x). \end{aligned}$$

The condition (5.8) expressed in terms of the function R is of the form

$$\int_0^T R(t, x(t), u(t))\, dt \leqslant \int_0^T R(t, x_0(t), u_0(t))\, dt = 0.$$

Theorem 5.1 does not provide a procedure for determining the optimal control u_0 and the function $V(t, x)$. One of such procedures that we discuss below is based on a somewhat different expression for the functional $J(u)$.

Consider the function

$$H(t,x,u,V_x) = -F_0(t,x,u) + V_x'(t,x)f(t,x,u).$$

The upper limit of the function $H(t,x,u,V_x)$ with respect to the independent variable $u \in U$ will be denoted by $\bar{H}$, i.e.,

$$\bar{H}(t,x,V_x) = \sup_{u\in U} H(t,x,u,V_x).$$

It follows from the relations (5.6) that the functional (5.5) can be expressed in the form

$$J(u) = \varphi_0\big(x(T)\big) + V\big(T,x(T)\big) - V(0,x_0) + \int_0^T \big[(\bar{H}-H) - (V_t+\bar{H})\big]\,dt. \tag{5.11}$$

Here the independent variables are $\big(t,x(t),u(t)\big)$.

Theorem 5.2. *Suppose there exist a continuously differentiable function $V(t,x)$ and an admissible control $u_0(t)$ satisfying the condition (5.9) and the following ones:*

$$\bar{H}\Big(t,x_0(t),V_x\big(t,x_0(t)\big)\Big) = H\Big(t,x_0(t),u_0(t),V_x\big(t,x_0(t)\big)\Big), \tag{5.12}$$

$$V_t(t,x) + \bar{H}\big(t,x,V_x(t,x)\big) \leqslant V_t\big(t,x_0(t)\big) + \bar{H}\Big(t,x_0(t),V_x\big(t,x_0(t)\big)\Big) = 0, \tag{5.13}$$

$$0 \leqslant t \leqslant T,\ x \in R^n.$$

Then $u_0(t)$ is an optimal control, and $J(u_0)$ is expressed by formula (5.10).

Example 5.1. Consider the scalar control system

$$\dot{x}(t) = u(t), \qquad 0 \leqslant t \leqslant 1, \tag{5.14}$$
$$x(0) = x(1) = 0,\ |u| \leqslant 1.$$

The cost functional to be minimized is of the form

$$J(u) = \int_0^1 \big(x^2(t) - u(t)\big)\,dt \to \inf_u. \tag{5.15}$$

We shall show that in this problem the optimal control $u_0 \equiv 0$ and the optimal trajectory $x_0(t) \equiv 0$, $0 \leqslant t \leqslant 1$.

It can be verified that these functions satisfy the hypotheses of Theorem 5.1 for $V(t,x) = -x$. For any admissible control $u(t)$, noting (5.14), (5.15), we have

$$F_1\big(t,x(t),u(t)\big) - \dot{V}_u\big(t,x(t)\big) = x^2(t) \geqslant 0. \tag{5.16}$$

Moreover, for the control $u_0(t) \equiv 0$ and the trajectory $x_0(t) \equiv 0$ the following equality holds:

$$F_0\big(t, x_0(t), u_0(t)\big) - \dot{V}_{u_0}\big(t, x_0(t)\big) = 0. \tag{5.17}$$

Comparing (5.16) and (5.17), we find that the condition (5.8) of Theorem 5.1 is satisfied. We turn now to the verification of the condition (5.9). In our case the set X_1 consists of one point $x = 0$, and the function $\varphi_0 = 0$. It follows that

$$V(T, x_0(T)) = V(1, x_0(1)) = 0,$$
$$\inf_{x \in X_1} V(T, x) = \inf_{x \in X} (-x) = 0.$$

Hence, the condition (5.9) also holds. Thus, by virtue of Theorem 5.1 the optimal control $u_0(t) \equiv 0$.

Example 5.2. Consider the scalar linear-quadratic problem

$$\dot{x}(t) = u, \qquad 0 \leqslant t \leqslant T,\ x_0 = x(0),$$
$$J(u) = \int_0^T u^2(t)\,dt + \alpha x^2(T) \to \inf_u . \tag{5.18}$$

Here x_0, T, α are prescribed constants. By means of Theorem 5.1 we prove that the optimal control u_0 in the problem (5.18) is of the form

$$u_0(t) = \alpha x(t)[\alpha(t - T) - 1]^{-1}. \tag{5.19}$$

For $V(t, x)$ we take the function

$$V(t, x) = \alpha x^2[\alpha(t - T) - 1]^{-1}. \tag{5.20}$$

Now we verify that the condition (5.7) holds. Since in our case the function F_0 is $F_0 = u^2$, we have

$$\begin{aligned} F_0(t, x, u) - \dot{V}_u(t, x) &= u^2 - 2\alpha x u[\alpha(t - T) - 1]^{-1} + \alpha^2 x^2 [\alpha(t - T) - 1]^{-2} \\ &= [\alpha(t - T) - 1]^{-2} \big[u\big(\alpha(t - T) - 1\big) - \alpha x\big]^2 \geqslant 0. \end{aligned} \tag{5.21}$$

At the same time, by virtue of (5.19) we have

$$F_0(t, x, u_0) - \dot{V}_{u_0}(t, x) = [\alpha(t - T) - 1]^{-2} \big[u_0\big(\alpha(t - T) - 1\big) - \alpha x\big]^2 = 0. \tag{5.22}$$

It follows from the expressions (5.21), (5.22) that the condition (5.8) of Theorem 5.1 holds true. Moreover, it follows from the expressions (5.18), (5.20) for all x that $\varphi_0(x) + V_0(T, x) = \alpha x^2 - \alpha x^2 = 0$. Hence, the condition (5.9) is also satisfied. Therefore, by Theorem 5.1 the control (5.19) is optimal.

2. Sufficient conditions for the existence of a minimizing sequence. As for Theorem 5.1, we can state sufficient conditions that some sequence of admissible controls u_i be minimizing. In the conditions (5.8), (5.9) it is sufficient, formally, to replace the control u_0 by u_i, the trajectory x_0 by x_i, and pass to the limit as $i \to \infty$. In this way, we can prove the following assertion.

Theorem 5.3. *Suppose there exists a continuously differentiable function $V(t,x)$ and a sequence of admissible controls $u_i(t)$ such that*

$$\lim_{i\to\infty} \left[\varphi_0\big(x_i(T)\big) + V\big(T, x_i(T)\big)\right] = \inf_{x\in X_1} \left[\varphi_0(x) + V(T,x)\right] = q, \tag{5.23}$$

where $x_i(t)$ is a solution of the problem (5.1), (5.2) *generated by the control $u_i(t)$. Further, suppose that for all $u \in W$ the following relation holds true:*

$$\int_0^T \left[F_0\big(t, x(t), u(t)\big) - \dot V_u\big(t, x(t)\big)\right] dt$$
$$\geqslant \lim_{i\to\infty} \int_0^T \left[F_0\big(t, x_i(t), u_i(t)\big) - \dot V_{u_i}\big(t, x_i(t)\big)\right] dt = 0. \tag{5.24}$$

Then the sequence of controls u_i is minimizing for the problem (5.1)–(5.5), *and the following relation is valid:*

$$\lim_{i\to\infty} J(u_i) = q - V(0, x_0).$$

3. Problems with unspecified time. The foregoing procedure for obtaining sufficient conditions for optimality in terms of auxiliary functions V can also be used in some other problems (for example, in problems with phase constraints in the form of various equalities and inequalities, in problems, where the set U depends on coordinates, and so on). We shall discuss the resulting modifications of the conditions for optimality for the problem (5.1)–(5.5), in which it is additionally assumed that the terminal moment $T \geqslant 0$ is unspecified. In other words, the functional (5.5) is minimized by choosing an appropriate control $u(t)$ and moment $T \geqslant 0$. Let θ be the set of all admissible moments T. The aggregate of the moment $T \in \theta$ and the control $u(t)$, where $0 \leqslant t \leqslant T$ and θ is a prescribed set, is called *admissible* if these quantities satisfy the relations (5.1)–(5.4), and the functional (5.5) is finite.

Theorem 5.4. *Suppose there exist a continuously differentiable function $V(t,x)$, an admissible control $u_0(t)$ and terminal moment $T_0 \in \theta$ such that*

1°) *for any admissible $\big(T, u(t)\big)$ the following inequality holds true:*

$$\int_T^{T_0} \left[F_0\big(t, x(t), u(t)\big) - \dot V_u\big(t, x(t)\big)\right] dt \leqslant 0; \tag{5.25}$$

2°) *the condition* (5.8) *is satisfied for $T = T_0$;*

3°) *the following inequality holds true*:

$$\varphi_0(x(t)) + V(T, x(T)) \geqslant \varphi_0(x_0(T_0)) + V(T_0, x_0(T_0)) = \inf_{\substack{x \in X_1, \\ T \in \Theta}} [\varphi_0(x) + V(T, x)]. \tag{5.26}$$

Then T_0, $u_0(t)$, where $0 \leqslant t \leqslant T_0$, are optimal.

Proof. To prove that the control u_0 is optimal, it is sufficient to verify that $J(u,T) - J(u_0,T_0) \geqslant 0$. In view of the expression (5.7) we have

$$\begin{aligned} J(u,T) - J(u_0,T_0) &= \int_0^T [F_0(t, x(t), u(t)) - \dot{V}_u(t, x(t))]\, dt \\ &+ \varphi_0(x(T)) + V(T, x(T)) - V(0, x_0) \\ &- \int_0^{T_0} [F_0(t, x_0(t), u_0(t)) - \dot{V}_{u_0}(t, x_0(t))]\, dt \\ &- V(T_0, x_0(T)) + V(0, x_0) - \varphi_0(x_0(T)). \end{aligned} \tag{5.27}$$

Consider certain summands on the right side of the expression (5.27). In view of the condition (5.26) we obtain

$$V(T, x(T)) + \varphi_0(x(T)) - V(T_0, x_0(T)) - \varphi_0(x_0(T)) \geqslant 0. \tag{5.28}$$

Further, it should be noted that

$$\begin{aligned} &\int_0^T [F_0(t, x(t), u(t)) - \dot{V}_u(t, x(t))]\, dt \\ &\quad = \int_0^{T_0} [F_0(t, x(t), u(t)) - \dot{V}_u(t, x(t))]\, dt \\ &\quad\quad - \int_T^{T_0} [F_0(t, x(t), u(t)) - \dot{V}_u(t, x(t))]\, dt. \end{aligned} \tag{5.29}$$

It follows from the expressions (5.27)–(5.29) and the conditions (5.25), (5.26) that

$$\begin{aligned} J(u,T) - J(u_0,T_0) \geqslant &\int_0^{T_0} [F_1(t, x(t), u(t)) - \dot{V}_u(t, x(t))]\, dt \\ &- \int_0^{T_0} [F_0(t, x_0(t), u_0(t)) - \dot{V}_{u_0}(t, x_0(t))]\, dt \\ &- \int_T^{T_0} [F_0(t, x(t), u(t)) - \dot{V}_u(t, x(t))]\, dt \geqslant 0. \end{aligned}$$

The proof is complete. □

Main Results and Formulas of Chapter VII

For the Boltz problem

$$\dot{x}(t) = f(t, x, u), \qquad x(t_0) = x_0,$$
$$\int_{t_0}^{T} F(t, x(t), u)\, dt + \varphi(x(T)) \to \inf_{u \in U}$$

Bellman function

$$V(s, x) = \inf_{u \in U} \left[\varphi(x(T)) + \int_s^T F(t, x(t), u)\, dt \right],$$

Bellman equation

$$\inf_{u \in U} [V_t(s, x) + V_x'(t, x) f(t, x, u) + F(t, x, u)] = 0, \qquad t_0 \leqslant t \leqslant T,$$
$$V(T, x) = \varphi(x),$$

optimal control $u_0(t, x)$

$$\inf_{u \in U} \left[f'(t, x, u) \frac{\partial V_0(t, x)}{\partial x} + F(t, x, u) \right]$$
$$= f'(t, x, u_0) \frac{\partial V_0(t, x)}{\partial x} + F(t, x, u_0),$$

where V_0 is a unique solution of the Bellman equation.

For the linear-quadratic problem the Bellman function is $V(t, x) = x' P(t) x$, where $P(t)$ is a solution of Riccati's matrix equation, and the minimum value of the cost functional $J(u_0) = V(t_0, x_0)$.

Stabilization. The control $u(t, x)$ is stabilizing if

$$\omega_1(|x|) \leqslant V(t, x) \leqslant \omega_2(|x|), \qquad L_u V(t, x) \leqslant -\omega_3(|x|).$$

The control $u_0(t, x)$ is optimal stabilizing if

$$\omega_1(|x|) \leqslant V(t, x) \leqslant \omega_2(|x|),$$
$$\inf_{u \in U} [L_u V(t, x) + F_0(t, x, u)] = L_{u_0} V(t, x) + F_0(t, x, u_0) = 0.$$

In the time-invariant linear-quadratic problem

$$\dot{x} = Ax + Bu, \qquad x(0) = x_0,$$

$$J(u) = \int_0^\infty [x'(t)N_1x(t) + u'(t)N_0u(t)]dt \to \inf_u, \qquad N_0 > 0,\ N_1 > 0,$$

the optimal control $u_0(t,x) = -N_0^{-1}B'Px$, where P satisfies the *algebraic Riccati equation:*

$$A'P + PA - PBN_0^{-1}B'P + N_1 = 0.$$

For a controllable system there exists a unique solution $P > 0$ of the algebraic Riccati equation, the system is exponentially stable, and $J(u_0) = x_0'Px_0$.

In the quasilinear stabilization problem

$$\dot{x} = \varepsilon f(t,x) + Bu,$$

$$J(u) = \int_0^\infty [x'(t)N_1(t)x(t) + u'(t)N_0(t)u(t)]\, dt \to \inf_u$$

the ith approximation $u_i(t,x)$ to the optimal control v is of the form

$$u_i(t,x) = -\frac{1}{2}N_0^{-1}B'\frac{\partial W_i}{\partial x}, \qquad W_i(t,x) = \sum_{j=0}^{i} \varepsilon^j V_j(t,x),$$

where

$$V_0(t,x) = x'P(t)x, \qquad \dot{P}(t) - P(t)BN_0^{-1}(t)B'P(t) + N_1(t) = 0,$$

$$V_i(t,x) = \int_0^\infty \Bigg[f'(s,x(s))\frac{\partial V_{i-1}(s,x(s))}{\partial x} - \frac{1}{4}\sum_{j=1}^{i-1} \frac{\partial V_j'(s,x(s))}{\partial x} BN_0^{-1}(t)B'\frac{\partial V_{i-j}(s,x(s))}{\partial x}\Bigg] ds, \qquad i \geqslant 1.$$

The following relations hold:

$$0 \leqslant J(u_0) - J(v) \leqslant C\varepsilon, \qquad 0 \leqslant J(u_i) - J(v) \leqslant C\varepsilon^{i+1}.$$

Sufficient conditions for optimal control using auxiliary functions. If for the problem (5.1)–(5.5) there exist a scalar continuously differentiable function $V(t,x)$ and an admissible control $u_0(t) \in W$ satisfying the following conditions for all $u(t) \in W$:

$$\int_0^T [F_0(t,x(t),u(t)) - \dot{V}_u(t,x(t))]\, dt \geqslant \int_0^T [F_0(t,x_0(t),u_0(t)) - \dot{V}_{u_0}(t,x_0(t))]\, dt = 0,$$

$$\varphi_0(x(T)) + V(T, x(T)) \geqslant \varphi_0(x_0(T)) + V(T, x_0(T)) = \inf_{x \in X_1} [\varphi_0(x) + V(T, x)] = q,$$

then the control $u_0(t)$ is optimal, and $J(u_0) = q - V(0, x_0)$.

The control $u_0(t)$ is optimal if in place of the above conditions the following conditions are satisfied:

$$\bar{H}\Big(t, x_0(t), V_x\big(t, x_0(t)\big)\Big) = H\Big(t, x_0(t), u_0(t), V_x\big(t, x_0(t)\big)\Big),$$
$$V_t(t, x) + \bar{H}\big(t, x, V_x(t, x)\big) \leqslant V_t\big(t, x_0(t)\big) + \bar{H}\Big(t, x_0(t), V_x\big(t, x_0(t)\big)\Big) = 0,$$
$$0 \leqslant t \leqslant T, \ x \in R^n.$$

If there exists a function $V(t, x)$ and a sequence of admissible controls $u_i(t)$ such that for any $u \in W$ the following conditions are satisfied:

$$\lim_{i \to \infty} \big[\varphi_0\big(x_i(T)\big) + V\big(T_0, x_i(T)\big)\big] = \inf_{x \in X_1} [\varphi_0(x) + V(T, x)] = q,$$
$$\int_0^T \big[F_0\big(t, x(t), u(t)\big) - \dot{V}_u\big(t, x(t)\big)\big]\, dt \geqslant \lim_{i \to \infty} \int_0^T \big[F_0\big(t, x_i(t), u_i(t)\big) - \dot{V}_{u_i}\big(t, x_i(t)\big)\big]\, dt = 0,$$

then the sequence of controls u_i is minimizing for the problem (5.1)–(5.5), and

$$\lim_{i \to \infty} J(u_i) = q - V(0, x_0).$$

CHAPTER VIII

SOME ADDITIONAL TOPICS OF OPTIMAL CONTROL THEORY

In this chapter we are concerned with existence conditions for optimal control. In addition, some cases are discussed, in which the application of the maximum principle is connected with certain peculiar difficulties.

§ 1. Existence of Optimal Control

1. Problem statement and the main assumptions. When investigating control problems, it is natural to pose the question about whether the optimal control exists, i.e., a control that renders the minimum of the cost functional. The existence conditions for optimal control are often justified by constructing a minimizing control sequence and passing to the limit when necessary. In the proof of this, some notions of function theory and functional analysis are utilized (see, for example, [9] of Part Three).

In the sequel the main stages of the corresponding constructions are given for the problem of the optimal control of the system

$$\dot{x}(t) = f_0(t, x(t)) + f_1(t, x(t))u(t), \tag{1.1}$$

$$t_0 \leqslant t \leqslant T,\ x(t) \in R^n,\ u(t) \in U \subset R^m, x(t_0) = x_0,\ x(T) = x_1. \tag{1.2}$$

Here the initial t_0 and terminal T moments, the control set U, and the initial x_0 and terminal x_1 states are given. The cost functional to be minimized is of the form

$$J(u) = \int_{t_0}^{T} F_0(t, x(t))\, dt. \tag{1.3}$$

It is assumed that the functions f_0, F_0, and the matrix f_1 having the form

$$f_0(t,x)\colon R^1 \times R^n \to R^n,$$
$$f_1(t,x)\colon R^1 \times R^n \to Z(R^m, R^n),$$
$$F_0(t,x)\colon R^1 \times R^n \to R^1$$

are continuous in all arguments. Moreover, the functions f_0 and f_1 satisfy the Lipschitz conditions

$$|f_i(t,x) - f_i(t,y)| \leqslant C_i|x - y|, \qquad i = 0, 1,\ t_0 \leqslant t \leqslant T,\ x, y \in R^n, \tag{1.4}$$

where $C_i \geqslant 0$ are some constants, and the symbol $|f_1|$ denotes some norm of the matrix.

We now define the class W of admissible controls which contains the optimal control $u(t) \in W$ if at each time instant t ($t_0 \leqslant t \leqslant T$) it belongs to the set U and is Lebesgue measurable as a function of time. Moreover, it is assumed that this control generates a solution of equation (1.1) satisfying the prescribed boundary conditions (1.2). It will be recalled that by a solution of equation (1.1) associated with the measurable function $u(t)$ we mean an absolutely continuous function $x(t)$ satisfying the appropriate integral identity

$$x(t) = x(t_0) + \int_{t_0}^{t} [f_0(s, x(s)) + f_1(s, x(s))u(s)]\, ds, \qquad t_0 \leqslant t \leqslant T \tag{1.5}$$

The infimum of the functional (1.3) with respect to the set of all admissible controls is denoted by J_0. Thus,

$$J_0 = \inf_{u(t) \in W} J(u).$$

2. Main theorem. We shall consider the following important theorem that plays a major role in optimal control theory.

Theorem 1.1[1]**.** *Let the assumptions of subsection* 1 *be satisfied, and let there exist at least one admissible control in the problem* (1.1)–(1.3). *Moreover, let the set* $U \in R^n$ *be compact and convex. Then in the problem* (1.1)–(1.3) *there exists an optimal control* $u_0(t)$.

Proof. The proof of this theorem is very complicated; so it will be carried out in several stages.

1°. *Properties of the minimizing control sequence.* It follows from the existence of an admissible control and from the definition of an infimum that there exists a control sequence $u_i(t) \in W$ ($i = 1, 2, \ldots$) called *minimizing* such that

$$\inf_{u(t) \in W} J(u) = \lim_{i \to \infty} J(u_i) = J_0.$$

Let $x_i(t)$ denote the sequence of solutions of the problem (1.1)–(1.3) generated by the corresponding control sequence $u_i(t)$. We verify that the sequence of solutions $x_i(t)$ is uniformly bounded and equicontinuous.

Since the set U is compact in a finite-dimensional space R^m, it is closed and bounded. Hence, the aggregate of all admissible controls W is uniformly bounded on the segment $t_0 \leqslant t \leqslant T$ by some constant C, i.e., in particular,

$$|u_i(t)| \leqslant C, \qquad t_0 \leqslant t \leqslant T,\ i = 1, 2, \ldots . \tag{1.6}$$

---

[1]Filippov, A. F., On Some Problems of Optimal Control Theory, *Vestnik Moskov. Univ. Ser. I Mat. Mekh.*, (1959), no. 2, 25–32. (Translated into English as *Moscow Univ. Math. Bull.*)

Moreover, by the Lipschitz condition (1.4) the following estimates hold:

$$\begin{aligned} |f_j(t,x)| = |f_j(t,x) - f_j(t,0) + f_j(t.0)| \leqslant |f_j(t,0)| \\ + |f_j(t,x) - f_j(t,0)| \leqslant |f_j(t,0)| + C_j|x|, \qquad j = 0,1. \end{aligned} \tag{1.7}$$

The identity (1.5), together with (1.6), (1.7), implies that for all $i \geqslant 1$ the following inequalities hold true:

$$\begin{aligned} |x_i(t)| \leqslant |x_0| + \int_{t_0}^{t} [|f_0(s,0)| + C|f_1(s,0)|]\, ds \\ + (C_0 + CC_1) \int_{t_0}^{t} |x_i(s)|\, ds. \end{aligned}$$

Thus, noting the Gronwall–Bellman lemma (see Lemma 6.1 of Chapter I), we can find that

$$\begin{aligned} |x_i(t)| \leqslant \left[|x_0| + \int_{t_0}^{t} (|f_0(s,0)| + C|f_1(s,0)|)\, ds \right] \\ \times \exp\big((C_0 + CC_1)(t - t_0)\big). \end{aligned} \tag{1.8}$$

But the functions $f_j(t,0)$ are continuous and, consequently, bounded for $0 \leqslant t \leqslant T$. Therefore, by virtue of (1.8) the sequence $x_j(t)$ is uniformly bounded, i.e., there exists a constant C_2 such that for all $i \geqslant 1$ and $t_0 \leqslant t \leqslant T$ we can assert that

$$|x_i(t)| \leqslant C_2. \tag{1.9}$$

For the proof of the equicontinuity of the set of functions $x_i(t)$ $(i = 1, 2, \dots)$ we take two arbitrary points t_1 and t_2 of the segment $[t_0, T]$. Taking account of (1.5), we have

$$x_i(t_2) - x_i(t_1) = \int_{t_1}^{t_2} [f_0(s, x_i(s)) + f_1(s, x_i(s)) u_i(s)]\, ds. \tag{1.10}$$

Further, note that by virtue of (1.7)

$$|f_j(s, x_i(s))| \leqslant |f_j(s,0)| + C_j|x_i(s)|, \qquad j = 0,1. \tag{1.11}$$

It follows from the relations (1.11), (1.9) and (1.6) that the integrand in (1.10) satisfies the estimate

$$|f_0(s, x_i(s))| + |f_1(s, x_i(s)) u_i(s)| \leqslant C_3, \qquad i \geqslant 1,\ t_0 \leqslant s \leqslant T,$$

where C_3 is some constant. Hence, in view of (1.10), the functions $x_i(t)$ satisfy the Lipschitz condition:

$$|x_i(t_2) - x_i(t_1)| \leqslant C_3|t_2 - t_1|, \qquad i \geqslant 1,\ t_0 \leqslant t_1,\ t_2 \leqslant T. \tag{1.12}$$

The last inequality, which is valid for all $i \geqslant 1$ and any points t_1 and t_2 on the segment $[t_0, T]$, means that the sequence $x_i(t)$ is equicontinuous.

It will be recalled that by virtue of the Arczela lemma the uniformly bounded and equicontinuous sequence of continuous functions $x_i(t)$ is compact in the space of continuous functions on the segment $[t_0, T]$. Hence, it contains a subsequence that is uniformly convergent on $[t_0, T]$. Without loss of generality, we may assume that the sequence $x_i(t)$ itself is uniformly convergent. The corresponding limit is denoted by $x^0(t)$. Thus, we have established that

$$\lim_{i\to\infty} \max_{t_0 \leqslant t \leqslant T} |x^0(t) - x_i(t)| = 0. \tag{1.13}$$

2°. *Properties of the function $x^0(t)$.* It follows from the equalities (1.3), (1.13) and the definition of the number J_0 that

$$J_0 = \lim_{i\to\infty} J(u_i) = \lim_{i\to\infty} \int_{t_0}^{T} F_0\big(t, x_i(t)\big)\, dt = \int_{t_0}^{T} F_0\big(t, x^0(t)\big)\, dt.$$

Hence, the function $x^0(t)$ renders a minimum of the cost functional (1.3). Therefore, to prove that $x^0(t)$ is an optimal trajectory of the system (1.1), (1.2), it is sufficient to verify that there exists a control $u^0(t) \in W$ such that the pair $\big(x^0(t), u^0(t)\big)$ is a solution to the problem (1.1), (1.2).

We give this proof. First of all, observe that the function $x^0(t)$ satisfies the Lipschitz condition (1.12) for $i = 0$, since all elements of the sequence $x_i(t)$ satisfy the inequality (1.12) and the relation (1.13) is valid.

Therefore, by Lebesgue's theorem the function $x^0(t)$ is absolutely continuous, i.e., it is differentiable almost everywhere, and its derivative $\dot{x}^0(t)$ is integrable on $[t_0, T]$.

Consider the set $D(t) \subset R^n$, $t_0 \leqslant t \leqslant T$. For a given t the set $D(t)$ consists of all points $f_0\big(t, x^0(t)\big) + f_1\big(t, x^0(t)\big)u$ of the space R^n that result from u ranging over the set U. Since U is convex, it follows that $D(t)$ is also convex. We shall show that for values of t such that $\dot{x}^0(t)$ exists, the inclusion

$$\dot{x}^0(t) \in D(t), \qquad t_0 \leqslant t \leqslant T, \tag{1.14}$$

is valid.

Observe that if formula (1.14) holds true for some t, then it immediately follows from the definition of the set $D(t)$ that there exists a vector $u_0(t) \in U$ such that for this t the following equation is valid:

$$\dot{x}^0(t) = f_0\big(t, x^0(t)\big) + f_1\big(t, x^0(t)\big)u_0(t). \tag{1.15}$$

Consider a number $\Delta > 0$. In order to justify the inclusion (1.14), we approximate the derivative $\dot{x}^0(t)$ by the expression $[x_i(t+\Delta) - x_i(t)]\Delta^{-1}$, which converges to $\dot{x}^0(t)$ as $i \to \infty$ and $\Delta \to 0$. The functions $x_i(t)$ are absolutely continuous.

Since they may be represented as integrals of their derivatives, we have according to equation (1.1)

$$\frac{1}{\Delta}[x_i(t+\Delta) - x_i(t)] = \frac{1}{\Delta}\int_t^{t+\Delta} \left[f_0\big(s, x_i(s)\big) + f_1\big(s, x_i(s)\big)u_i(s)\right] ds. \quad (1.16)$$

Take an arbitrary number $\varepsilon > 0$. We define the convex compact set $D_\varepsilon(t) \subset R^n$ as the set of all points in R^n whose Euclidean distance from the set $D(t)$ does not exceed ε.

Let us verify the fact that the right side of the expression (1.16) for sufficiently small Δ and large i belongs to the set $D_\varepsilon(t)$. To this end, because of the convexity of $D_\varepsilon(t)$ it is sufficient to establish that the integrand function in (1.16) has the same property. The function $f_0(s,x) + f_1(s,x)u$ is continuous and, moreover, is uniformly continuous on the compact set $[t_0, T] \times \{x\colon |x| \leqslant C_2\} \times U$. Therefore, there is a $\delta(\varepsilon) > 0$ such that for any $u \in U$ the vector $f_0(s,x) + f_1(s,x)u \in D_\varepsilon$ whenever

$$|s-t| + |x - x_0(t)| < \delta(\varepsilon).$$

In particular, assuming here that $x = x_i(s)$, $u = u_i(t)$, we obtain that the vector $f_0\big(s, x_i(s)\big) + f_1\big(s, x_i(s)\big)u_i(s) \in D_\varepsilon$ whenever

$$|s-t| + |x_i(s) - x_0(t)| < \delta(\varepsilon). \quad (1.17)$$

Further, we have

$$|x_i(s) - x_0(t)| \leqslant |x_i(s) - x_0(s)| + |x_0(s) - x_0(t)|. \quad (1.18)$$

Taking account of the fact that the function $x^0(t)$ is uniformly continuous, we choose a small Δ_0 so that for all $s \in [t, t+\Delta]$, $\Delta \leqslant \Delta_0$, the following inequality is satisfied:

$$|s-t| + |x_0(s) - x_0(t)| < \delta/2. \quad (1.19)$$

Further, noting the relation (1.13), we take an i_0 such that for all $i \geqslant i_0$ the following inequality holds:

$$|x_i(s) - x_0(s)| < \delta/2. \quad (1.20)$$

Formulas (1.18)–(1.20) show that, given i_0 and Δ_0, for all $i \geqslant i_0$, $\Delta \leqslant \Delta_0$, and $s \in [t, t+\Delta]$ the inequality (1.17) holds true. Thus, for the indicated values of i and Δ it has been proved that the right side of the expression (1.16) belongs to the set $D_\varepsilon(t)$.

Therefore, by virtue of the relations (1.2), (1.16) the functions

$$\Delta^{-1}[x_i(t+\Delta) - x_i(t)]. \quad (1.21)$$

belong to the set $D_\varepsilon(t)$. Since the set $D_\varepsilon(t)$ is closed, the limit of the functions (1.21) as $i \to \infty$ also belongs to $D_\varepsilon(t)$. In view of (1.13), this limit is equal to

$$\Delta^{-1}[x_0(t+\Delta) - x_0(t)].$$

Passing to the limit as $\Delta \to 0$ in the last expression and taking account of the closure of $D_\varepsilon(t)$, we find that $\dot{x}^0(t) \in D_\varepsilon(t)$. Thus, noting that the number ε is arbitrary, it follows that the inclusion (1.14) and also the equality (1.15) are valid.

3°. *Measurability of the function $u^0(t)$.* To complete the proof, it remains to be seen that the function $u^0(t)$ can be chosen to be a measurable one. To this end, we shall use the following lemma.

Lemma 1.1. *Suppose that the set $Q \subset [t_0, T] \times U$ can be expressed as*

$$Q = \bigcup_{i=1}^{\infty} \alpha_i,$$

where α_i are compact subsets in $[t_0, T] \times U$. Then there exists a measurable function $u(t)$ such that $(t, u(t)) \in Q$ for almost all $t \in N \subset [t_0, T]$. Here N is the set of all those points of the segment $[t_0, T]$ for which the set $Q_i = \{u : (t, u) \in Q\}$ is nonempty.

In addition to this lemma we shall use the following theorem.

Theorem (N. N. Luzin). *For a function $u^0(t) \in R^m$, $t_0 \leqslant t \leqslant T$, to be measurable* [9] *(of Part Three), it is necessary and sufficient that for any $\varepsilon > 0$ there exists a closed set $\tau \subset [t_0, T]$, whose measure is greater than $T - t_0 - \varepsilon$, such that the function $u^0(t)$ is continuous on τ.*

By virtue of this theorem and the measurability of the function $\dot{x}^0(t)$, $t_0 \leqslant t \leqslant T$, there exist closed sets $\tau_i \subset [t_0, T]$ on which the function $\dot{x}^0(t)$ is continuous, with $\tau_1 \subset \tau_2 \subset \tau_3 \subset \ldots$, and so the measure of the set $[t_0, T] \setminus \bigcup_{i=1}^{\infty} \tau_i$ is equal to zero.

For an arbitrary fixed i we define the set α_i as a set of points $t \in \tau_i$ and $u \in U$ for which the following equality is valid:

$$\dot{x}_0(t) = f_0\bigl(t, x^0(t)\bigr) + f_1\bigl(t, x^0(t)\bigr)u. \tag{1.22}$$

In other words,

$$\alpha_i = \bigl\{(t, u) \, : \, t \in \tau_i, \; u \in U, \\ \dot{x}_0(t) = f_0\bigl(t, x^0(t)\bigr) + f_1\bigl(t, x^0(t)\bigr)u\bigr\}. \tag{1.23}$$

As shown above, for any $t \in \tau_i$ the set of points $u \in U$ satisfying equation (1.22) is nonempty. It is also clear that this set is compact. Therefore, the sets α_i defined by formula (1.23) are compact and nonempty. Hence, with regard to Lemma 1.1 it follows that there exists a measurable function $u^0(t)$ that satisfies equation (1.15) for almost all t.

The proof of the main theorem is complete. □

Remark. In a similar way, under the appropriate additional restrictions sufficient conditions for optimal control can also be established for other problems, such as problems with unspecifed terminal conditions, problems with unspecified time, with other cost functionals and control constraints, and so on.

3. Analysis of the conditions of the main theorem. We turn to a discussion of the conditions of the main theorem. Obviously, if in the problem (1.1)–(1.3) there is no admissible control, it follows that there is no optimal control. The requirements that the set U be compact and convex are also important. We shall cite a few examples.

Example 1.1. Consider the problem

$$\begin{gathered} \dot{x}(t) = u, \qquad 0 \leqslant t \leqslant 1, \\ x(0) = x(1) = 0, \\ J(u) = \int_0^1 x(t)\, dt \to \sup_{u(t)\in W}. \end{gathered} \tag{1.24}$$

The set U is taken as the interval

$$U = (-1, 1). \tag{1.25}$$

The set U defined by (1.25) is convex but not compact.

If in the problem (1.24) instead of the set U we take its closure, i.e., $U = [-1, 1]$, then in this problem the unique optimal control is of the form

$$\begin{gathered} u^0(t) = \begin{cases} 1, & \text{for } 0 \leqslant t < \frac{1}{2}, \\ -1, & \text{for } \frac{1}{2} \leqslant t \leqslant 1; \end{cases} \\ x^0(t) = \begin{cases} t, & \text{for } 0 \leqslant t < \frac{1}{2}, \\ \frac{1}{2} - t, & \text{for } \frac{1}{2} \leqslant t \leqslant 1. \end{cases} \end{gathered} \tag{1.26}$$

However, the function $u^0(t)$ defined by (1.26) does not belong to the set (1.25). On the other hand, in the problem (1.24), (1.25) there exists a sequence

$$u_i(t) = \begin{cases} 1 - \frac{1}{i}, & \text{for } 0 \leqslant t < \frac{1}{2}, \\ -1 + \frac{1}{i}, & \text{for } \frac{1}{2} \leqslant t \leqslant 1 \end{cases}$$

converging to the function $u^0(t)$ so that

$$J(u_i) < J(u_0) = \sup_{u(t)\in W} J(u), \qquad J(u_i) \to J(u_0), \quad i \to \infty.$$

Thus, there is no optimal control in the problem (1.24), (1.25).

Example 1.2. Consider the problem

$$\begin{gathered} \dot{x} = u, \qquad 0 \leqslant t \leqslant 1, \\ x(0) = x(1) = 0, \\ J(u) = \int_0^1 x^2(t)\, dt \to \inf_{u(t)\in W} \end{gathered} \tag{1.27}$$

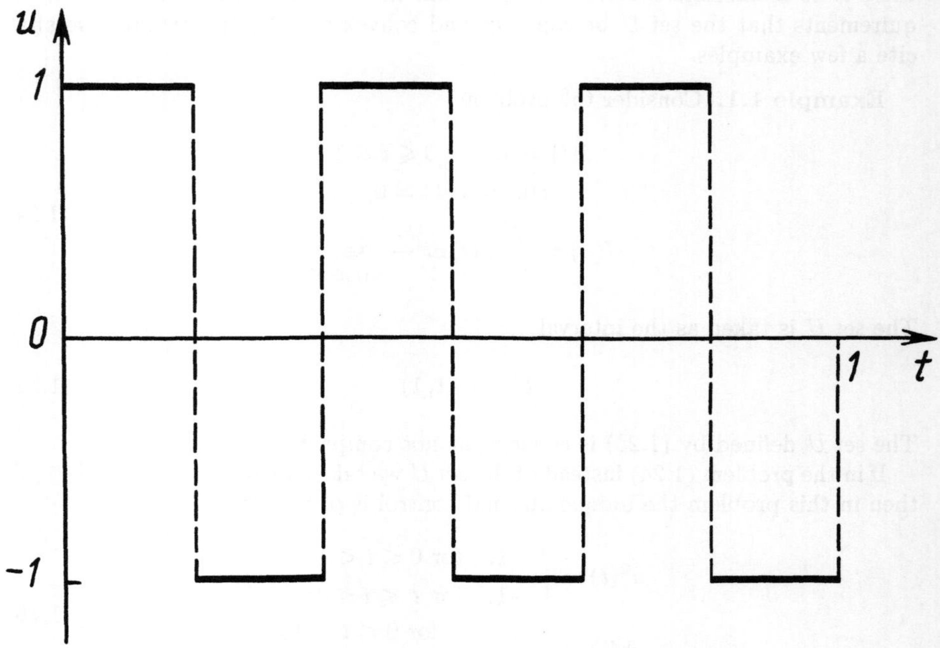

FIG. 1.1. A view of the control function $u_3(t)$

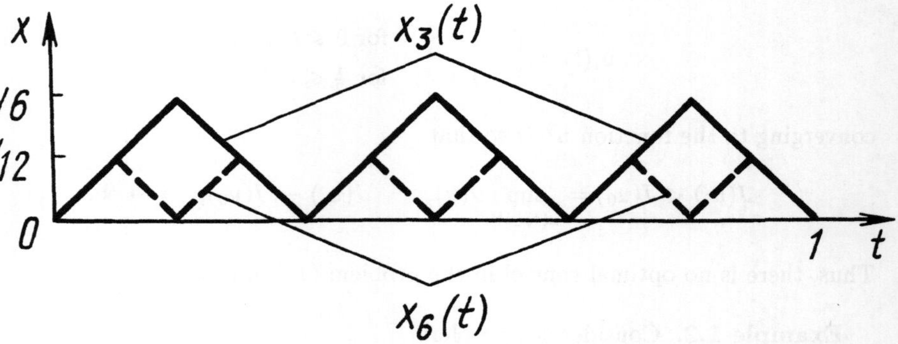

FIG. 1.2. A view of the trajectories $x_3(t)$ and $x_6(t)$

Suppose that the set U consists only of points -1 and 1, i.e.,

$$U = \{-1\} \cup \{1\}. \tag{1.28}$$

Clearly, the set U is compact but not convex. Let us show that in the problem (1.27), (1.28) there is no optimal control. Indeed, for any admissible control u we have $J(u) \geqslant 0$, and there exists a minimizing sequence such that $J(u_i) \to 0$. For u_i we can take a sequence of the form

$$u_i(t) = \begin{cases} 1, & \text{for } \frac{2k}{2i} \leqslant t < \frac{2k+1}{2i}, \\ -1, & \text{for } \frac{2k+1}{2i} \leqslant t < \frac{2(k+1)}{2i}, \end{cases} \quad k = 0, 1, \ldots, i-1. \tag{1.29}$$

The sequence u_i generates the sequence of solutions $x_i(t)$ defined by

$$x_i(t) = \begin{cases} t - \frac{k}{i}, & \text{for } \frac{2k}{2i} \leqslant t < \frac{2k+1}{2i}, \\ -t + \frac{k+1}{i}, & \text{for } \frac{2k+1}{2i} \leqslant t < \frac{2(k+1)}{2i}, \end{cases} \quad k = 0, 1, \ldots, i-1. \tag{1.30}$$

The graphs of the control function $u_3(t)$ and of the trajectories $x_3(t)$ and $x_6(t)$ are shown in Figures 1.1 and 1.2.

By calculation we obtain that

$$J(u_i) = \frac{1}{4i} \to 0, \qquad i \to \infty.$$

It follows from formula (1.29) (see also Figure 1.2) that the sequence $x_i(t)$ for $0 \leqslant t \leqslant 1$ uniformly converges to zero: $x_i(t) \to x^0(t) \equiv 0$, $0 \leqslant t \leqslant 1$. At the same time, the control sequence $u_i(t)$ does not converge pointwise for $t \in (0, 1)$ (Figure 1.1).

Further, suppose that for some t_1 the equality (1.15) holds true; in our case this equality can be written as

$$0 = \dot{x}^0(t_1) = u^0(t_1). \tag{1.31}$$

Then it follows from the equality (1.31) that at the moment t_1 the optimal control $u^0(t)$ must be zero. However, this cannot be true, since zero does not belong to the set U. Hence, there is no optimal control in this problem.

Note that if we take the convex closed hull of the set U for the set of admissible controls U_1, i.e., $U_1 = [-1, 1]$, then there is an optimal control in the problem (1.27). In fact, in this case $u^0(t) \equiv 0$ is the optimal control.

Some other examples in which there is no optimal control are given in the sequel.

§ 2. Singular Optimal Control

1. The definition and determination of singular controls. When the conditions of existence theorems for optimal control are satisfied, the question arises as to whether it can be determined through the use of various conditions for an extremum, such as the maximum principle. When the maximum principle is used to determine an optimal control, it is desirable that it be uniquely determined by the condition for a maximum of the function H with respect to u (see, for example, formula (2.28) of Chapter V). However, for some problems this condition may not be satisfied, i.e., a maximum of the function H may be attained on the time interval at more than one point of the set U. Let us cite an example.

Example 2.1. Consider the problem of minimizing the functional

$$J(u) = \int_0^1 x^2(t)\,dt$$

with respect to the solutions of the equation

$$\dot{x}(t) = u(t), \qquad 0 \leqslant t \leqslant 1,$$
$$x(0) = 0,\ x(1) = 0,$$

subject to the constraint $|u| \leqslant 1$. By Theorem 1.1 there exists an optimal control in this problem.

Clearly, in this problem the optimal control is $u_0(t) \equiv 0$, and the optimal trajectory is $x_0(t) \equiv 0$. On the other hand, the Hamiltonian H for the problem in hand is of the form

$$H = \psi(t)u - x^2.$$

It follows from the maximum principle that the optimal control is $\operatorname{sgn}\psi(t)$ for $\psi(t) \neq 0$. Further, since the optimal control $u_0(t) \equiv 0$, the maximum of the Hamiltonian H can be attained only for $\psi(t) \equiv 0$. However, in this case the maximum of the Hamiltonian H can also be attained for any other control u. Therefore, the condition for a maximum does not allow to distinguish between optimal and nonoptimal control.

In similar cases we need a singular optimal control, which cannot be determined solely from the maximum principle but other conditions for optimality have to be invoked. Some of these conditions are given below. A detailed account of the theory of singular optimal control is given in the book,[2] on which the subject matter of subsections 1 and 2 is based.

It should be noted that the actual determination of a singular control and the form of the conditions for the optimality of such a control essentially depend on the character of control constraints, the dimension of the control vector, and the type of variations used when deducing the conditions for optimality.

The main constructions used for determining a singular control are to be given for a system with a scalar control u with values in an open region. Consider the system

$$\dot{x}(t) = f_0\big(x(t)\big) + f_1\big(x(t)\big)u, \qquad 0 \leqslant t \leqslant T, \tag{2.1}$$
$$x \in R^n,\ u \in U \subset R^1,\ x(0) = x_0.$$

[2]Gabasov, R., and Kirillova, F. M., *Singular Optimal Control*, Nauka, Moscow, (1973).

Here the moment T, the set U, and the vector x_0 are given, the functions $f_0(x)$ and $f_1(x)$ are well-behaved. The cost functional $J(u)$ is defined by

$$J(u) = \varphi\big(x(T)\big) \to \inf_{u \in U}. \tag{2.2}$$

By virtue of Theorem 1.2 of Chapter V the equations of the maximum principle for the problem (2.1), (2.2) are of the form

$$\begin{aligned} H(x,\psi,u) &= H_0(x,\psi) + H_1(x,\psi)u, \\ H_0 &= f_0'(x)\psi, \qquad H_1 = f_1'(x)\psi, \\ \dot{x}(t) &= \frac{\partial H}{\partial \psi}, \qquad x(0) = x_0, \\ \dot{\psi}(t) = -\frac{\partial H}{\partial x} &= -\left[\frac{\partial f_0(x)}{\partial x} + u\frac{\partial f_1(x)}{\partial x}\right]'\psi, \qquad 0 \leqslant t \leqslant T, \\ \psi(T) &= -\frac{\partial \varphi\big(x(T)\big)}{\partial x}, \qquad \psi \in R^n. \end{aligned} \tag{2.3}$$

Further, if $u_0(t)$ is an optimal control, and $\big(x_0(t), \psi_0(t)\big)$ is the corresponding solution of the problem (2.3), then for any moment $t \in [0,T]$ we have the equality

$$H\big(x_0(t), \psi_0(t), u_0(t)\big) = \max_{u \in U} H\big(x_0(t), \psi_0(t), u\big) \tag{2.4}$$

We turn now to the task of defining singular control. Consider a group of functions $\big(u(t), x(t), \psi(t)\big)$ satisfying the relations (2.3), (2.4) and also some subset U_1 of the set U containing more than one point. Suppose that on some time segment $[\tau_0, \tau_1]$, where $0 \leqslant \tau_0 \leqslant \tau_1 \leqslant T$, the following identity holds for any $u \in U_1$:

$$\begin{aligned} H\big(x(t), \psi(t), u(t)\big) &= \max_{u \in U} H\big(x(t), \psi(t), u\big) \\ &\equiv H\big(x(t), \psi(t), u\big), \qquad \tau_0 \leqslant t \leqslant \tau_1,\ u \in U_1. \end{aligned} \tag{2.5}$$

Then the time segment $[\tau_0, \tau_1]$ is called a *singular mode interval*, and the control $u(t)$, $\tau_0 \leqslant t \leqslant \tau_1$, is called a *singular control*, the trajectory $x(t)$ a *singular mode trajectory*, and, finally, the aggregate $\big(u(t), x(t)\big)$ is called a *singular mode* on the time segment $[\tau_0, \tau_1]$. Note that in the case under consideration, in view of equations (2.3), the identity (2.5) holds true only for

$$H_1\big(x(t), \psi(t)\big) \equiv 0, \qquad \tau_0 \leqslant t \leqslant \tau_1. \tag{2.6}$$

Consider the problem of determining a singular control. For this, taking into account the system (2.3), we compute successively the total derivatives with respect to time of the function $H_1(x,\psi)$. In view of the identity (2.6) they are zero for $\tau_0 \leqslant t \leqslant \tau_1$. This differentiation is carried out until a control with a nonzero multiplier appears in one of the derivatives. To facilitate the computation work, we use the Poisson brackets $\{H_0, H_1\}$, which are a scalar function of the form

$$\{H_0(x,\psi), H_1(x,\psi)\} = \left[\frac{\partial H_0(x,\psi)}{\partial x}\right]' \frac{\partial H_1(x,\psi)}{\partial \psi} - \left[\frac{\partial H_0(x,\psi)}{\partial \psi}\right]' \frac{\partial H_1(x,\psi)}{\partial x}.$$

With regard to (2.3) we have

$$\dot{H}_1 = -\{H_0, H_1\} = 0, \qquad \tau_0 \leqslant t \leqslant \tau_1. \tag{2.7}$$

This expression for $\dot{H}_1$ does not contain a control. Therefore, we continue with computation and find the second derivative $\ddot{H}_1$ of the function H_1 according to the system (2.3). In view of the expression (2.7) for $\tau_0 \leqslant t \leqslant \tau_1$ we obtain

$$\ddot{H}_1 = \{H_0, \{H_0, H_1\}\} + u(t)\{H_1, \{H_0, H_1\}\} = 0. \tag{2.8}$$

Observe that the equalities (2.7), (2.8) are, in general, valid along the relevant trajectories of the system (2.3). Hence, at the points $t \in [\tau_0, \tau_1]$, where the multiplier $\{H_1, \{H_0, H_1\}\}$ for $u(t)$ is not zero, the singular control is

$$u(t) = -\frac{\{H_0, \{H_0, H_1\}\}}{\{H_1, \{H_0, H_1\}\}}. \tag{2.9}$$

But if in formula (2.8) the summand containing $u(t)$ is zero for $\tau_0 \leqslant t \leqslant \tau_1$, then we compute the subsequent total derivatives of the function H_1 according to the system (2.3). It turns out that the control $u(t)$ may occur only in the derivatives of even order $2k$, which have the form

$$\begin{aligned} \frac{d^{2k}}{dt^{2k}} H_1 = 0 = &\Big\{H_0, \{H_0, \ldots, \{H_0, H_1\}, \ldots\}\Big\} \\ &+ u(t)\Big\{H_1, \{H_0, \ldots, \{H_0, H_1\} \ldots\}\Big\}, \qquad \tau_0 \leqslant t \leqslant \tau_1. \end{aligned} \tag{2.10}$$

In the right side of the expression (2.10), in the first summand, there are $2k$ symbols H_0 preceding H_1, and in the second summand there are $2k - 1$ symbols H_0 between two H_1's. The relations (2.8), (2.9) follow from (2.10), provided that $k = 1$. The singular control can be determined from the identity (2.10).

Remark. Suppose that in some computation step the coefficient of $u(t)$ in (2.10) is not zero on some portion Δ of the interval $[\tau_0, \tau_1]$. Then further computation of the total derivatives must be carried out only on the remaining portion of the interval $[\tau_0, \tau_1]$ that does not contain Δ.

2. Optimality of singular control. The proof of the maximum principle (see Theorem 1.1 of Chapter V) was based on the use of needle-shaped variations of the control function and on the use of a condition for optimal control of the form $J(u) - J(u_0) \geqslant 0$, where u is any admissible control and u_0 is an optimal control. Moreover, the difference $J(u) - J(u_0)$ was expressed in the form (1.19) (Chapter V), i.e., the conditions for optimal control were obtained from the analysis of the first variation of the functional J.

In the proof of the necessary conditions for the optimality of singular control the next term of the expansion for $J(u) - J(u_0)$ is taken into account as well as its being nonnegative for any admissible variations of the control u_0. In other words, the necessary conditions for the optimality of singular control can be obtained from the requirement that the second variation of the functional be nonnegative. Choosing appropriate variations of the control function, we obtain the corresponding necessary conditions for the optimality of singular control.

We shall cite such conditions for the problem (2.1), (2.2).

Suppose that $u_0(t)$ is an optimal singular control in the problem (2.1), (2.2), the functions f_0 and f_1 are continuously differentiable at least $2l+1$ times, where $l \geqslant 1$, and the following equalities hold along the control $u(t)$:

$$(-1)^k \frac{\partial}{\partial u} \frac{d^{2k}}{dt^{2k}} \frac{\partial H}{\partial u} = 0, \qquad k = 0, 1, \ldots, l-1. \tag{2.11}$$

Then for the control $u_0(t)$ to be optimal, it is necessary that the following inequality be valid:

$$(-1)^l \frac{\partial}{\partial u} \frac{d^{2l}}{dt^{2l}} \frac{\partial H}{\partial u} \leqslant 0, \tag{2.12}$$

given the control $u_0(t)$. In formulas (2.11), (2.12) the function H is defined by the relation (2.3), the symbol d/dt denotes, as above, the total derivative according to the system (2.3), given the control $u_0(t)$ (the phrase "given the control $u(t)$" means that the derivative with respect to u is calculated at the point $u(t)$). In particular, it follows that the requirements (2.11), (2.12) for $l > 1$ imply the appropriate smoothness with respect to t of the singular control $u(t)$.

For $l = 1$ the conditions (2.11), (2.12) go over into the inequality

$$\frac{\partial}{\partial u} \frac{d^2}{dt^2} \frac{\partial H}{\partial u} \geqslant 0, \tag{2.13}$$

which is called the *Kelley necessary condition for the optimality of singular control.*

For $l = 2$ the conditions (2.11), (2.12) yield the *Kopp–Moyer necessary conditions for the optimality of singular control*:

$$\frac{\partial}{\partial u} \frac{d^2}{dt^2} \frac{\partial H}{\partial u} = 0, \qquad \frac{\partial}{\partial u} \frac{d^4}{dt^4} \frac{\partial H}{\partial u} \leqslant 0. \tag{2.14}$$

In specific problems situations may arise when the optimal control consists of singular and nonsingular portions. The foregoing conditions for optimality are also valid for singular portions. At the same time the problem may arise as to the conditions for sewing together the singular and nonsingular portions of the optimal control, meaning the conditions at the points of separation of singular and nonsingular portions of the optimal control. The sewing together of the singular and nonsingular portions of the optimal control may result in the appearance of a so-called *chattering mode*.

Example 2.2. Consider the optimal control problem

$$\begin{gathered} \dot{x}_1 = x_2, \qquad \dot{x}_2 = u, \qquad \dot{x}_3 = -x_1^2, \qquad 0 \leqslant t \leqslant 1, \\ x_1(0) = x_2(0) = x_3(0) = 0, \ |u(t)| < 1, \\ J(u) = x_3(1) \to \inf_{|u|<1} . \end{gathered} \tag{2.15}$$

By virtue of (2.3), (2.15) the functions H, H_0, H_1 become

$$\begin{aligned} H &= \psi_1 x_2 + \psi_2 u - \psi_3 x_1^2, \\ H_0 &= \psi_1 x_2 - \psi_3 x_1^2, \\ H_1 &= \psi_2, \end{aligned} \tag{2.16}$$

and the equations for the adjoint variables ψ_i take the form

$$\begin{gathered} \dot{\psi}_1 = -\frac{\partial H}{\partial x_1} = 2\psi_3 x_1, \qquad \dot{\psi}_2 = -\frac{\partial H}{\partial x_2} = -\psi_1, \qquad \dot{\psi}_3 = -\frac{\partial H}{\partial x_3} = 0, \\ \psi_1(1) = 0, \ \psi_2(1) = 0, \ \psi_3(1) = -1. \end{gathered} \tag{2.17}$$

We proceed now with determining the singular control. For this, on the strength of the system (2.15), (2.17) we compute in turn the total derivatives with respect to of the function H_1 until the control u appears. With regard to (2.16), (2.17) we have

$$\begin{aligned} \dot{H}_1 &= \dot{\psi}_2 = -\psi_1, \\ \ddot{H}_1 &= \ddot{\psi}_2 = -\dot{\psi}_1 = -2\psi_3 x_1, \\ \dddot{H}_1 &= -2(\dot{\psi}_3 x_1 + \psi_3 \dot{x}_1) = -2\psi_3 x_2, \\ \frac{d^4 H_1}{dt^4} &= -2(\dot{\psi}_3 x_2 + \psi_3 \dot{x}_2) = -2\psi_3 u. \end{aligned} \tag{2.18}$$

Since on the strength of (2.17) the adjoint variable $\psi_3(t) \equiv -1$, it follows from (2.18) and (2.10) that the singular control $u(t) \equiv 0$. Moreover, given this control, the functions

$$x_i(t) \equiv 0, \quad i = 1, 2, 3, \qquad \psi_1(t) = \psi_2(t) = 0, \quad 0 \leqslant t \leqslant 1,$$

i.e., H does not depend on the control u. Further, in view of (2.18), the Kelley condition (2.13) is satisfied, since

$$\frac{\partial}{\partial u} \frac{d^2}{dt^2} \frac{\partial H}{\partial u} = \frac{\partial}{\partial u} \frac{d^2 H_1}{dt^2} = 0.$$

Therefore, the Kelley condition does not imply that the singular control $u(t) \equiv 0$ is not optimal. However, the Kopp–Moyer condition (2.14) is not satisfied. Indeed, by virtue of (2.17), (2.18) we have

$$\frac{\partial}{\partial u}\frac{d^4}{dt^4}\frac{\partial H}{\partial u} = \frac{\partial}{\partial u}\frac{d^4 H_1}{dt^4} = -2\psi_3 = 2.$$

This implies that the singular control $u(t) \equiv 0$ is not optimal.

However, it should be noted that the nonoptimality of the singular control $u(t) \equiv 0$ can be established directly, since for any admissible control $u(t)$ that is nonzero on some interval the value $x_3(1) < 0$ and for $u(t) \equiv 0$ the equality $x_3(t) = 0$ holds.

Example 2.3. Consider an optimal control problem in which the system is described by the same equations as in (2.15), and the cost functional is of the form

$$J_1(u) = x_3(1) \to \sup_{|u|<1} .$$

In order to use the necessary conditions for optimal control, we must replace the functional $J_1(u)$ by the functional $-J_1(u)$, which is to be minimized. Thus, in the problem under consideration the system is described by equations (2.15), and the cost functional is of the form

$$J_2(u) = -J_1(u) = -x_3(1) \to \inf_{|u|<1} . \tag{2.19}$$

It follows immediately from equations (2.15) that for any admissible control u the value of the cost functional $J_2(u) \geqslant 0$, and for $u(t) \equiv 0$ we have $J_2(0) \equiv 0$. Hence, $u(t) \equiv 0$ is the optimal control in the problem (2.13), (2.19).

Repeating the same computation as in Example 2.2, we obtain that in the problem (2.15) with functional $J_2(u)$ there exists the singular control $u(t) \equiv 0$, which is associated with the functions

$$x_1(t) = x_2(t) = x_3(t) = \psi_1(t) = \psi_2(t) = 0,$$
$$\psi_3(t) \equiv 1, \qquad 0 \leqslant t \leqslant 1.$$

Here as in the preceding example, the Kelley condition is satisfied:

$$\frac{\partial}{\partial u}\frac{d^2}{dt^2}\frac{\partial H}{\partial u} = \frac{\partial}{\partial u}\frac{d^2 H_1}{dt^2} = 0.$$

The Kopp–Moyer condition is also satisfied:

$$\frac{\partial}{\partial u}\frac{d^4}{dt^4}\frac{\partial H}{\partial u} = \frac{\partial}{\partial u}\frac{d^4 H_1}{dt^4} = -2\psi_3(t) = -2 \leqslant 0.$$

3. Generalization of the Kelley and Kopp–Moyer conditions[3]**.** For the control system (2.1) with the boundary conditions

$$x(0) = x_0, \qquad x(T) = x_1 \tag{2.20}$$

consider the time-optimal problem

$$J(u) = T \to \min_{u_0 \leqslant u \leqslant u_1} . \tag{2.21}$$

Suppose there exists a solution of the problem (2.1), (2.20), (2.21), and this solution has a singular mode portion corresponding to a singular mode segment $[\tau_0, \tau_1]$, $0 \leqslant \tau_0 \leqslant \tau_1 \leqslant T$. In view of the condition (2.6) the multiplier $H_1(x(t), \psi(t))$ associated with the control u in the Hamiltonian H of the system (2.1) is identically zero on the interval $[\tau_0, \tau_1]$, i.e.,

$$H_1(x(t), \psi(t)) = f_1'(x(t)\psi(t))\psi(t) \equiv 0, \qquad \tau_0 \leqslant t \leqslant \tau_1.$$

By virtue of (2.9) the singular control on the segment $[\tau_0, \tau_1]$ is expressed by the formula

$$u^0(t) = -\frac{\{H_0, \{H_0, H_1\}\}}{\{H_1, \{H_0, H_1\}\}}, \qquad \{H_1, \{H_0, H_1\}\} \neq 0, \ H_0 = f_0\psi. \tag{2.22}$$

For the existence of a control singular mode portion, it is necessary that the control (2.22) be constrained by $u_0 \leqslant u^0(t) \leqslant u_1$. The analysis of this constraint enables us to write necessary conditions for optimal control in a more compact form.

We introduce the following functions of the variables x and ψ:

$$H^0 = f_0'(x)\psi + u_0 f_1'(x)\psi, \qquad H^1 = f_0'(x)\psi + u_1 f_1'(x)\psi.$$

The quantities H^0, H^1 are Hamiltonians corresponding to the regular trajectories of the maximum principle with control bounds u_0 and u_1. We introduce the functions

$$A_0 = \{H^1, \{H^1, H^0\}\}, \qquad A_1 = \{H^0, \{H^0, H^1\}\}.$$

A simple calculation immediately gives

$$A_0 + A_1 = -(u_1 - u_0)^2\{H_1, \{H_0, H_1\}\},$$
$$u_0 A_0 + u_1 A_1 = (u_1 - u_0)^2\{H_0, \{H_0, H_1\}\}.$$

Then the expression (2.22) for the singular control can be written as

$$u^0(t) = \frac{u_0 A_0 + u_1 A_1}{A_0 + A_1}, \qquad A_0 + A_1 \neq 0. \tag{2.23}$$

[3]Melikyan, A., A., Necessary Optimality Conditions for a Singular Surface in the Form of Synthesis, *J. Optim. Theory Appl.*, (1994), vol. 82, no. 2, 208–217.

If $A_0 + A_1 > 0$, then it follows from the inequality $u_0 \leqslant u^0(t) \leqslant u_1$ that $A_0 \geqslant 0$ and $A_1 \geqslant 0$. But if $A_0 + A_1 < 0$, then we must have $A_0 \leqslant 0$ and $A_1 \leqslant 1$. Thus, the quantities A_0, A_1 are of the same sign as their sum $A_0 + A_1$, i.e., the point (A_0, A_1) can only be in quadrant I or III. Further, it follows from the Kelley condition (2.13) that quadrant III must be excluded. Indeed, by the Kelley condition we have

$$\frac{\partial}{\partial u}\frac{d^2}{dt^2}\frac{\partial H}{\partial u} = \{H_1, \{H_0, H_1\}\} = \frac{A_0 + A_1}{(u_1 - u_0)^2} \geqslant 0. \tag{2.24}$$

Hence,

$$A_0 \geqslant 0, \qquad A_1 \geqslant 0. \tag{2.25}$$

Further, since $\{H^0, H^1\} = (u_0 - u_1)\{H_0, H_1\}$, we obtain, with regard to (2.25), the following *system of necessary conditions for the singular mode portion of the optimal control*:

$$\{H^0, H^1\} = 0, \qquad \{H^0, \{H^0, H^1\}\} \geqslant 0, \qquad \{H^1, \{H^0, H^1\}\} \geqslant 0. \tag{2.26}$$

Note some specific features of the system of necessary conditions for optimal control (2.26). First, they are expressed only in terms of the regular controls u_0, u_1. Secondly, the inequalities (2.26) lead to the Kelley condition (2.24) and the singular control constraint $u_0 \leqslant u^0(t) \leqslant u_1$.

In a more general situation, when A_0, A_1, viewed as functions of time, can become identically zero, we would have an augmented system of necessary conditions for optimal control. Specifically, for some odd number $m = 2l + 1 \geqslant 3$ we may consider the following equalities and inequalities, which hold on some special hypersurface in the phase space:

$$\begin{gathered}\{H^0, H^1\} = 0, \qquad \Big\{H^{i_k}, \{\dots, H^{i_3}, \{H^0, H^1\}, \dots\}\Big\} = 0, \qquad k = 1, \dots, 2l, \\ (-1)^p\Big\{H^{i_m}, \{\dots, H^{i_3}\{H^0, H^1\}, \dots\}\Big\} \geqslant 0, \qquad m = 2l + 1 \geqslant 3.\end{gathered} \tag{2.27}$$

For $m = 3$ the conditions (2.27) coincide with (2.26). Here the indices i_k ($k = 3, \dots$) assume one of the values 0 or 1; p is the number of zero components of the integer-valued vector $(0, 1, i_3, \dots, i_m)$. Thus, there are as many equalities and inequalities in the relations (2.27) as there exist similar integer-valued vectors. The conditions (2.27) are called *generalized Kopp–Moyer conditions*.

Under certain assumptions similar results hold true also for nonautonomous systems with functionals of more general form. For more details, see the book cited at the beginning of this section.

§ 3. Chattering Mode

In the foregoing examples the optimal control could be taken as a smooth function or as having a finite number of switch points. At the same time Theorem 1.1 guarantees the existence of measurable control only. The question arises as to whether there exists an optimal control having a more complicated structure, such as having an infinite number of switch points on a finite time interval. Such situations occur in practice, and the corresponding optimal control is usually referred to as a *chattering mode*.

Chattering modes are often conected with singular control. If there exist singular portions of the optimal control, this may lead to chattering modes, i.e., modes in which the optimal control has infinitely many switch points on a finite time interval.

We consider this question in greater detail in the following control problem:[4]

$$\dot{x}_1 = x_2, \qquad \dot{x}_2(t) = u, \qquad 0 \leqslant t \leqslant T = 8, \tag{3.1}$$

$$x_1(0) = x_1(T) = 2,\ x_2(0) = -2,\ x_2(T) = 2,\ |u| \leqslant 1, \tag{3.2}$$

$$J(u) = \int_0^T x_1^2(t)\,dt \to \inf_{|u|\leqslant 1}. \tag{3.3}$$

The existence of an optimal solution (x_1, x_2, u_0) of the problem (3.1)–(3.3) is a consequence of Theorem 1.1.

Let us prove the uniqueness of the solution. Suppose there exists another solution y_1, y_2, v of the problem (3.1)–(3.3) which renders the same value $J(u_0)$ of the cost functional (3.3) as the optimal solution (x_1, x_2, u_0). Consider a continuous function $z_1(t) = (x_1 + y_1)/2$ that satisfies the relations (3.1), (3.2), provided that the control is $w = (u_0 + v)/2$. Further, observe that

$$4z_1^2 = (x_1 + y_1)^2 = x_1^2 + y_1^2 + 2x_1y_1 = 2x_1^1 + 2y_1^2 - (x_1 - y_1)^2.$$

Hence,

$$4J(\omega) = 2J(u_0) + 2J(v) - \int_0^T \big(x_1(t) - y_1(t)\big)^2\,dt. \tag{3.4}$$

Moreover, because of the optimality of the control u_0 the inequality $J(w) \geqslant J(u_0)$. Hence, in view of (3.4), it follows that

$$\begin{aligned} 2J(v) &= 4J(\omega) - 2J(u_0) + \int_0^T \big(x_1(t) - y_1(t)\big)^2\,dt \\ &\geqslant 2J(u_0) + \int_0^T \big(x_1(t) - y_1(t)\big)^2\,dt > 2J(u_0). \end{aligned}$$

However, the last inequality cannot be true, since it contradicts the assumption of the optimality of the control v. Thus, the uniqueness of the optimal solution of the problem (3.1)–(3.3) is established.

[4] Marchal, C., Chattering Arcs and Chattering Controls, *J. Optim. Theory Appl.*, (1973), vol. 11, no. 5, 441–468.

We turn now to necessary conditions for optimal control in the form of the maximum principle for the problem (3.1)–(3.3). In view of formulas (2.26), (2.28) of Chapter V, they are of the form

$$\dot{\psi}_1 = -2\lambda_0 x_1, \qquad \dot{\psi}_2 = -\psi_1, \tag{3.5}$$

$$\begin{aligned} &\max_u H(t, x_1, x_2, \psi_1, \psi_2, u) = H(t, x_1, x_2, \psi_1, \psi_2, u_0), \\ &H(t, x_1, x_2, \psi_1, \psi_2, u) = \lambda_0 x_1^2(t) + \psi_1(t) x_2(t) + \psi_2(t) u. \end{aligned} \tag{3.6}$$

Here the constant $\lambda_0 \leqslant 0$. It follows from the relations (3.6) that for values of t satisfying the inequality $\psi_2(t) \neq 0$ the optimal control is

$$u_0(t) = \operatorname{sgn} \psi_2(t). \tag{3.7}$$

At the points t, where $\psi_2(t) = 0$, the optimal control $u_0(t)$ can take on any values in the segment $[-1, 1]$.

We shall show that $\lambda_0 < 0$. Indeed, if $\lambda_0 = 0$, then on the strength of (3.5) the function ψ_1 would be constant, and the function $\psi_2(t)$ linear and, moreover, distinct from zero, since the variables $(\lambda_0, \psi_1, \psi_2)$ do not vanish simultaneously. However, whenever the linear function $\psi_2(t)$ is not zero, the optimal controlcan have, in view of (3.7), at most one switch point, and no control of this type can satisfy all of the boundary conditions (3.2). Therefore, $\lambda_0 < 0$, and hence λ_0 can be taken as any negative constant.

We set $\lambda_0 = -1/2$. Then the relations (3.5) become

$$\dot{\psi}_1 = x_1, \qquad \dot{\psi}_2 = -\psi_1. \tag{3.8}$$

The existence of a solution of the boundary-value problem (3.1), (3.2), (3.7), (3.8) is a consequence of the existence of a solution of the optimal control problem (3.1)–(3.3). Moreover, since the maximum principle in this case is not only a necessary but also a sufficient condition for optimality (see subsection 7, § 2, Chapter V) and the solution of the control problem (3.2), (3.3) is unique, it follows that the solution of the boundary-value problem (3.1), (3.2), (3.7), (3.8) is also unique.

Let us investigate this solution. Differentiating the function $\psi_2(t)$ several times along the trajectories of the system (3.1), (3.8), we obtain

$$\dot{\psi}_2 = -\psi_1, \qquad \ddot{\psi}_2 = -\dot{\psi}_1 = -x_1, \qquad \dddot{\psi}_2 = -\dot{x}_1 = -x_2, \tag{3.9}$$

$$\frac{d^4 \psi_2(t)}{dt^4} = -u_0(t) = -\operatorname{sgn} \psi_2(t). \tag{3.10}$$

With regard to equations (3.1), (3.9) and (3.10) we can write

$$\frac{d^4 \psi_2(t)}{dt^4} = -1, \qquad \ddot{x}_1(t) = 1, \qquad \text{if } \psi_2(t) > 0, \tag{3.11}$$

$$\frac{d^4 \psi_2(t)}{dt^4} = 1, \qquad \ddot{x}_1(t) = -1, \quad \text{if } \psi_2(t) < 0. \tag{3.12}$$

But if on some interval the function $\psi_2(t) \equiv 0$, then all of its derivatives are zero on this interval. Hence, in view of (3.9), we have

$$\frac{d^4\psi_2(t)}{dt^4} \equiv 0, \qquad x_1(t) = x_2(t) = \ddot{x}_1(t) \equiv 0, \quad \text{if } \psi_2(t) \equiv 0. \tag{3.13}$$

By virtue of (3.13) only the points, where $\psi_2(t) = 0$, can be points of discontinuity for the control function $u_0(t)$. The intervals, where $\psi_2(t) \equiv 0$, are, in view of (3.6), singular mode intervals. It is not difficult to verify that the Kelley and Kopp–Moyer conditions (2.13), (2.14) are satisfied on these intervals. We can show that the intervals, where $\psi_2(t) > 0$ or $\psi_2(t) < 0$, cannot be singular mode intervals. Indeed, suppose that the relations (3.11) hold on $[\gamma_0, \gamma_1]$, and the relations (3.13) hold on $[\gamma_1, \gamma_2]$. Let $\gamma_0 < \gamma_1 < \gamma_2$. Then, integrating the relation (3.11), we obtain for $t \in [\gamma_0, \gamma_1]$

$$\begin{aligned}\psi_2(t) = -\frac{(t-\gamma_1)^4}{24} &+ \alpha_1(t-\gamma_1)^3 + \alpha_2(t-\gamma_1)^2 \\ &+ \alpha_3(t-\gamma_1) + \alpha_4, \qquad \gamma_0 < t \leqslant \gamma_1,\end{aligned} \tag{3.14}$$

where α_i are some constants. But on the interval $[\gamma_1, \gamma_2)$ the following relation holds:

$$\psi_2(t) \equiv \dot{\psi}_2(t) \equiv \ddot{\psi}_2(t) \equiv \dddot{\psi}_3(t) \equiv 0.$$

Since, in view of equations (3.9), the function $\psi_2(t)$ and its three derivatives are continuous at the point $t = \gamma_1$, it follows that in formula (3.14) all $\alpha_i = 0$. Hence, according to (3.14), the function $\psi_2(t)$ must be strictly negative, i.e., $\psi_2(t) < 0$ for $\gamma_0 < t < \gamma_1$, which contradicts the definition of the interval (γ_0, γ_1).

As shown in the work of C. Marchal referred to at the beginning of this section, in the problem (3.1)–(3.3) under consideration a singular mode interval of the form (3.13) is realized only through the use of a control with infinitely many switch points. In other words, in the problem (3.1)–(3.3) we encounter a chattering mode.

To justify the foregoing assertion, it is sufficient to establish that the function $\psi_2(t)$ has the following properties:

1°. $\psi_2(t)$ is symmetric with respect to the point $T/2$;

2°. $\psi_2(t) \equiv 0$ on some segment $[t_0, T - t_0]$;

3°. Suppose that τ_i $(i \geqslant 0)$ are zeros of the function $\psi_2(t)$ on $(T - t_0, T]$ such that $\tau_0 > \tau_1 > \tau_2 > \dots$, where τ_0 is the zero closest to T. Then for some $q > 1$ the following relations are valid:

$$\tau_{i+1} = (T - t_0)\left(1 - \frac{1}{q}\right) + \frac{\tau_i}{q}, \qquad s_{i+1} = \frac{s_i}{q} = \frac{s_0}{q^{i+1}},$$

where s_i denote the quantities $s_i = \tau_i - T + t_0$.

4°. For $0 \leqslant t < t_0$ the zeros of the function $\psi_2(t)$ are symmetric with respect to point $T/2$ with zeros τ_i.

5°. The points t_0 and $(T - t_0)$ are the accumulation points for the zeros of the function $\psi_2(t)$.

6°. The following relations hold true:

$$\begin{aligned} \psi_2\left(\frac{t-T+t_0}{q}\right) &= -\frac{1}{q^4}\psi_2(t-T+t_0), \qquad t \geqslant T-t_0, \\ \psi_2\left(\frac{s}{q}\right) &= -\frac{1}{q^4}\psi_2(s), \end{aligned} \tag{3.15}$$

where s denotes the quantity $s = (t-T+t_0)$, $0 \leqslant s \leqslant t_0$.

7°. The function $\psi_2(t)$ depends on three parameters t_0, τ_0 and q.

Using the properties 1°–7°, we can derive the equations for the parameters q, t_0 and τ_0 and find their numerical values.

Consider the function $\psi_2(s)$ of the argument $s = t-T+t_0$. If the relations (3.11) hold true on the interval (s_1, s_0), then

$$\psi_2(s) = -\frac{s^4}{24} + \alpha_1 s^3 + \alpha_2 s^2 + \alpha_3 s + \alpha_4, \qquad s_1 \leqslant s \leqslant s_0. \tag{3.16}$$

Observe that for $s_2 \leqslant s \leqslant s_1$ the value of qs is changing on the interval $[s_1, s_0]$. Therefore, it follows from formulas (3.15) and (3.16) that for $s_2 \leqslant s \leqslant s_1$ the function $\psi_2(s)$ is of the form

$$\begin{aligned} \psi_2(s) &= -\frac{1}{q^4}\psi_2(qs) = -\frac{1}{q^4}\left[-\frac{q^4 s^4}{24} + \alpha_1 q^3 s^3 + \alpha_2 q^2 s^2 + \alpha_3 qs + \alpha_4\right] \\ &= -\left[-\frac{s^4}{24} + \alpha_1 \frac{s^3}{q} + \alpha_2 \frac{s^2}{q^2} + \alpha_3 \frac{s}{q^3} + \alpha_4 \frac{1}{q^4}\right]. \end{aligned} \tag{3.17}$$

Similarly, for $s_{i+1} \leqslant s \leqslant s_i$ we obtain

$$\psi_2(s) = (-1)^i \left[-\frac{s^4}{24} + \alpha_1 q^{-i} s^3 + \alpha_2 q^{-2i} s^2 + \alpha_3 q^{-3i} s + \alpha_4 q^{-4i}\right]. \tag{3.18}$$

On the strength of (3.9) the function $\psi_2(t)$ and its first three derivatives are continuous. Hence, equating their values for $s = s_1$ obtained from (3.16) and (3.17), we derive the equations

$$\begin{aligned} \psi_2(s_1) &= -\frac{s_1^4}{24} + \alpha_1 s_1^3 + \alpha_2 s_1^2 + \alpha_3 s_1 + \alpha_4 \\ &= \frac{s^4}{24} - \alpha_1 q^{-1} s_1^3 - \alpha_2 q^{-2} s^2 - \alpha_3 q^{-3} s_1 - \alpha_4 q^{-4}, \\ \dot{\psi}_2(s_1) &= -\frac{s_1^3}{6} + 3\alpha_1 s_1^2 + 2\alpha_2 s_1 + \alpha_3 \\ &= \frac{s_1^3}{6} - 3\alpha_1 q^{-1} s_1^2 - 2\alpha_2 q^{-2} s_1 - \alpha_3 q^{-3}, \\ \ddot{\psi}_2(s_1) &= -\frac{s_1^2}{2} + 6\alpha_1 s_1 + 2\alpha_2 \\ &= \frac{s_1^2}{2} - 6\alpha_1 q^{-1} s_1 - 2\alpha_2 q^{-2}, \\ \dddot{\psi}_2(s_1) &= -s_1 + 6\alpha_1 = s_1 - 6\alpha_1 q^{-1}. \end{aligned}$$

Solving these equations for α_i, we find

$$\alpha_1 = \frac{s_1 q}{3(1+q)}, \qquad \alpha_2 = -\frac{s_1^2 q^2}{2(1+q^2)},$$
$$\alpha_3 = \frac{s_1^3 q^3}{3(1+q^3)}, \qquad \alpha_4 = -\frac{s_1^4 q^4}{12(1+q^4)}. \tag{3.19}$$

Substituting these expressions for α_i in the expression (3.16) and using the equality $\psi_2(s_0) = 0$, we find that the quantity q is a root of the equation

$$\frac{1}{2} - \frac{4}{1+q} + \frac{6}{1+q^2} - \frac{4}{1+q^3} + \frac{1}{1+q^4} = 0. \tag{3.20}$$

It can be shown that equation (3.20) has a unique real solution $q > 1$.

With q known, the values of the parameters t_0 and τ_0 can be determined from the condition for the minimum of the functional (3.3). In the work by C. Marchal cited at the beginning of this section the values of the parameters q, t_0 and τ_0 are given with five decimal places: $q = 4.13016$, $t_0 = 3.43039$ and $\tau_0 = 7.94223$.

We can now write an exact formula for the function $\psi_2(s)$. For $s_{i+1} \leqslant s \leqslant s_i$ the function $\psi_2(s)$ is expressed by

$$\psi_2(s) = \frac{(-1)^i}{12}\Big[-\frac{s^4}{2} + 4\frac{s_{i+1}s^3 q}{1+q} - 6\frac{s_{i+1}^2 s^2 q^2}{1+q^2} + 4\frac{s_{i+1}^3 s q^3}{1+q^3} - \frac{s_{i+1}^4 q^4}{1+q^4}\Big], \tag{3.21}$$

where

$$s_0 = \tau_0 - T + t_0 = 3.48816,$$
$$s_i = \frac{s_0}{q^{i+1}} = \frac{3.48816}{(4.13016)^{i+1}}, \qquad i \geqslant 1.$$

Let us describe the resulting control $u_0(t)$ and trajectory $x_1(t)$. Utilizing the symmetry of the function $\psi_2(t)$, we find that the first switch of the control $u_0(t)$ occurs for $t = T - \tau_0 = 0.05777$, with $u_0(t) = -1$ for $0 \leqslant t \leqslant 0.05777$. Then until the next zero of the function $\psi_2(t)$ the control $u_0(t) = 1$, then again $u_0(t) = -1$, and so on. Thus, on the segment $[0, 3.43039)$ the control has infinitely many switch points, and the control values are 1 or -1. This represents the chattering mode portion of the control function. On the segment $[3.43039, 4.56961]$ the control $u_0(t) \equiv 0$. This is the singular mode portion. On the segment $(4.56961, 8]$ we again encounter a chattering mode.

The optimal trajectory $x_1(t)$ is composed of alternating parabola arcs $\ddot{x} = 1$ and $\ddot{x} = -1$. A view of the optimal trajectory $x_1(t)$ is shown in Figure 3.1. In approaching the point 3.43039 on the left and the point 4.56961 on the right, the trajectory $x_1(t)$ oscillates with vanishing amplitude and with frequency tending to infinity. The missing portion of the graph of $x_1(t)$ in Figure 3.1 is indicated by dots.

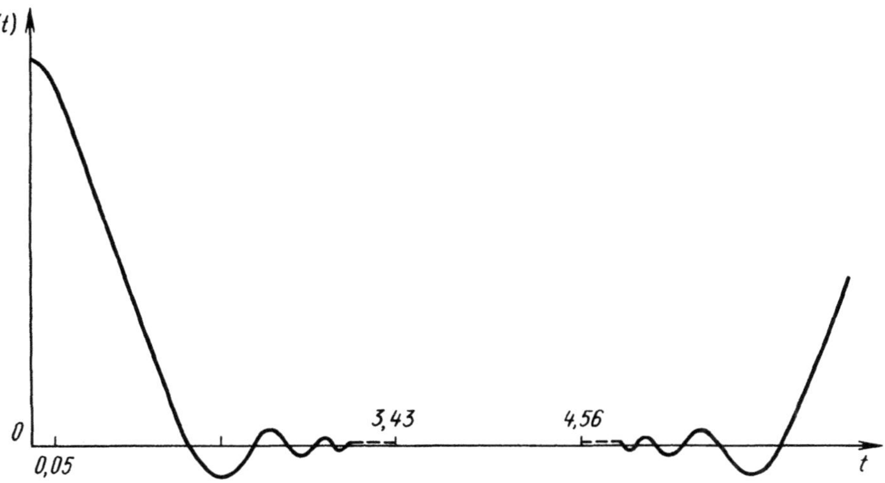

FIG. 3.1. A view of the trajectory of the system (3.1) generated by the optimal control having chattering mode portions

§ 4. Sliding Optimal Mode

In a number of practical situations there is no optimal control. Then one can construct a minimizing sequence (which always exists), i.e., a sequence such that the corresponding values of the cost functional converge to its infimum. However, it is not always possible to indicate a limiting admissible control that renders the infimum. In such cases the more switches the control has, the better (costwise) it is. The investigation of such problems leads to the concept of a sliding optimal mode. There is a definite affinity between a sliding and a chattering mode. However, there exists a fundamental difference between them. In the chattering mode there exists an optimal control, though of a complicated structure, whereas in the sliding mode there is no optimal control.

Let us now cite some relevant examples.

Example 4.1. Let us construct a minimizing sequence for a scalar problem with a variable terminal point of the form

$$\dot{x}(t) = u(t), \qquad 0 \leqslant t \leqslant 1,\ x(0) = 0,\ |u(t)| \leqslant 1, \tag{4.1}$$

$$J(u) = \int_0^1 \left(x^2(t) - u^2(t)\right) dt \to \inf. \tag{4.2}$$

Consider the sequence of admissible controls $u_i(t)$ $(1 = 1, 2, \dots)$ that are defined by

$$u_i(t) = \begin{cases} 1, & \text{for } \frac{k}{i} < t \leqslant \frac{2k+1}{2i}, \\ -1, & \text{for } \frac{2k+1}{2i} < t \leqslant \frac{k+1}{i}. \end{cases} \tag{4.3}$$

Here the index k for each fixed i runs from 0 to $i-1$: $k = 0, \dots, i-1$. The corresponding sequence $x_i(t)$ of the solutions of the problem (4.1), (4.2) is of the form

$$x_i(t) = \begin{cases} t - \frac{k}{i}, & \text{for } \frac{k}{i} < t \leqslant \frac{2k+1}{2i}, \\ -t + \frac{k+1}{i}, & \text{for } \frac{2k+1}{2i} < t \leqslant \frac{k+1}{i}, \\ k = 0, 1, \dots, i-1. \end{cases} \tag{4.4}$$

Note that $u_i(t)$ and $x_i(t)$ coincide with the controls (1.29) and the trajectories (1.30) of Example 1.2. In particular, a view of the control $u_3(t)$ and the trajectories $x_3(t)$ and $x_6(t)$ is given in Figures 1.1 and 1.2.

Further, we evaluate the cost functional (4.2), given the control function $u_i(t)$. By virtue of (4.3) and (4.4) we obtain

$$J(u_i) = \frac{1}{12i^2} - 1.$$

It follows from formula (4.2) that the functional $J(u) \geqslant -1$ for any admissible control u $(|u| \leqslant 1)$. Therefore,

$$J(u_i) \underset{i \to \infty}{\longrightarrow} \inf_{|u| \leqslant 1} J(u) = -1.$$

As is seen from formula (4.4) (see also Figure 1.2), the sequence of trajectories $x_i(t)$ uniformly converges to zero for $0 \leqslant t \leqslant 1$. At the same time the sequence of controls $u_i(t)$ does not converge pointwise.

But if for some t_1 the equality $\dot{x}_0(t) = u_0(t_1)$ holds true, then at this moment $u_0(t_1) = 0$. However, this cannot be true, since the value $u = 0$ does not belong to the set U.

Thus, in the problem (4.1), (4.2) there is no optimal control. Note that this does not contradict the assertion of Theorem 1.1, since in our case the cost functional (4.2) depends explicitly on $u(t)$.

In such cases the limiting trajectory is called a *sliding mode trajectory.*

Example 4.2. Consider the scalar control problem

$$\dot{x}(t) = u(t), \qquad 0 \leqslant t \leqslant 4, \ x(0) = x(4) = 1,$$

$$J(u) = \int_0^4 |x(t)| \, dt \to \inf_u, \qquad u(t) \in U.$$

The set U contains two points 1 and -1, i.e., $|u(t)| = 1$.

We construct a sequence of controls $u_i(t)$ $(i = 1, 2, \dots)$ in the following way. All controls $u_i(t) \equiv -1$ for $0 \leqslant t \leqslant 1$ and $u_i(t) \equiv 1$ for $3 \leqslant t \leqslant 4$. Further, we divide the segment $[1, 3]$ into i equal parts and set

$$u_i(t) = \begin{cases} -1, & \text{for } 1 + \frac{2j}{i} \leqslant t \leqslant 1 + \frac{2j+1}{i}, \\ 1, & \text{for } 1 + \frac{2j+1}{i} \leqslant t < 1 + \frac{2(j+1)}{i}, \end{cases}$$

where $j = 0, \dots, i - 1$.

Clearly, in our example the infimum J_0 of the cost functional $J(u)$ is

$$J_0 = \inf_{u \in U} J(u) = 1.$$

Moreover, after a short calculation we find that $J(u_i) = 1 + 1/i$. Hence, $J(u_i) \to J_0$ for $i \to \infty$. Thus, the sequence u_i is minimizing; no admissible control can render the value J_0 of the cost functional. The limiting trajectory is of the form

$$x(t) = \begin{cases} 1 - t, & \text{for } 0 \leqslant t \leqslant 1, \\ 0, & \text{for } 1 \leqslant t \leqslant 3, \\ t - 3, & \text{for } 3 \leqslant t \leqslant 4. \end{cases}$$

It is a sliding mode trajectory.

Since it is difficult to describe sliding modes in terms of admissible controls, we define them in terms of the trajectories that are generated by these controls.[5]

We shall give the definition of a sliding mode for a control system with a bounded closed set U of the form

$$\begin{gathered} \dot{x}(t) = f\big(x(t), u(t)\big), \qquad 0 \leqslant t \leqslant T,\ x \in R^n,\ u \in U \subset R^m,\ x(0) = x_0, \\ J(u) = \int_0^T F\big(t, x(t), u(t)\big)\, dt + \varphi\big(x(T)\big) \to \inf_{u \in U}. \end{gathered} \tag{4.5}$$

Here T and x_0 are prescribed.

Let $u_i(t)$ be a minimizing sequence of admissible controls, i.e., of measurable functions $u_i(t) \in U$ for each t belonging to the segment $[0, T]$, such that

$$J(u_i) \underset{i \to \infty}{\longrightarrow} J_0 = \inf_{u \in U} J(u),$$

and let $x_i(t)$ be the corresponding sequence of trajectories of the system (4.5); moreover, $x_i(t)$ converges pointwise for $0 \leqslant t \leqslant T$ to the function $x_0(t)$.

---

[5]Gamkrelidze, R. V., On Sliding Optimal Modes, *Dokl. Akad. Nauk SSSR*, (1962), vol. 143, no. 6, 1243–1245. (Translated into English as *Soviet Math. Dokl.*)

Definition 4.1. Suppose that one of the following two conditions is satisfied:

1°) the limiting trajectory $x_0(t)$ is not a solution of equation (4.5) for any admissible control v;

2°) the limiting trajectory $x_0(t)$ is a solution of equation (4.5) for some admissible control v, but $J(v) > \inf_{u \in U} J(u)$.

Then the trajectory $x_0(t)$ is called a *sliding optimal mode trajectory.*

It should be noted that the sliding optimal mode in Example 4.1 corresponds to the case 2° in the Definition 4.1. Indeed, in this example the sliding mode trajectory $x_0(t) \equiv 0$ can be generated by the admissible control $v = 0$. However, at the same time

$$J(v) = 0 > \inf_{u \in U} J(u) = -1.$$

In Example 4.2 the sliding mode trajectory $x_0(t) \equiv 0$, $1 \leqslant t \leqslant 3$, cannot be generated by any admissible control.

As is seen from the foregoing, the case 2° can only occur if the cost functional depends explicitly on the control $u(t)$.

We can describe the set of functions being the limits of the sequences of the trajectories $x_i(t)$ of the system (4.5) that are generated by the sequences of the controls $u_i(t)$ in the following way.

Consider the auxiliary control system

$$\dot{y}(t) = \sum_{i=0}^{n} \alpha_i(t) f\big(y(t), v_i(t)\big), \qquad 0 \leqslant t \leqslant T,\ y \in R^n. \tag{4.6}$$

In equation (4.6) we have as controls: $n+1$ vectors $v_i(t) \in R^m$ and $n+1$ scalar functions $\alpha_i(t)$, subject to the constraints

$$v_i(t) \in U \subset R^m, \qquad \alpha_i(t) \geqslant 0, \qquad \alpha_0(t) + \cdots + \alpha_n(t) \equiv 1. \tag{4.7}$$

In other words, the system (4.6) depends on $(m+1)(n+1)$ control functions $\big(v_0(t), \ldots, v_n(t), \alpha_0(t), \ldots, \alpha_n(t)\big)$.

The systems (4.5), (4.6) are related thus. Take an arbitrary fixed point $x \in R^n$ and form the sets $A(x)$ and $B(x)$. The set $A(x)$ is the totality of all points $f(x, u)$ that are obtained as u ranges over the domain U, and the set $B(x)$ is the totality of the points $\sum_{i=0}^{n} \alpha_i f(x, v_i)$ that are obtained as α_i and v_i range over the domain of definition (4.7). Thus, $A(x)$ is the set of all phase velocities of the system (4.5) that can be realized at the point x, and $B(x)$ is the analogous set of phase velocities for the system (4.6). As is seen from the construction, the set $B(x)$ constitutes a convex closed hull of the set $A(x)$.

Further, suppose that the sequence $\{x_i(t)\}$ of solutions of equation (4.5) converges to the function $y(t)$ at each point $t \in [0, T]$. Then $y(t)$ is a solution of the system (4.6), (4.7). Conversely, any solution of the system (4.6), (4.7) is the limit of a sequence of solutions of equation (4.5) that converges pointwise at each point $t \in [0, T]$.

Optimal sliding modes can be defined in a similar way. Their description can be given in terms of the optimal trajectories of the corresponding control problem for the system (4.6).

For example, consider the time-optimal problem for the system (4.5), i.e., the problem of the transfer of this system from a prescribed initial state x_0 to a given terminal state x_1 in minimum time. The equations describing this problem have the form

$$\dot{x}(t) = f\big(x(t), u(t)\big), \qquad x \in R^n,\ u \in U \subset R^m,\ 0 \leqslant t \leqslant T,$$
$$x(0) = x_0,\ x(T) = x_1,\ J(u) = T \to \inf_{u \in U}.$$

In this case the cost functional does not depend explicitly on u. Therefore, as noted above, a sliding optimal mode may occur if the limiting trajectory $x_0(t)$ is not a solution of the system (4.5) for any admissible control.

Suppose there exists an admissible control that transfers the system (4.5) from x_0 to x_1 in finite time. The following definition of a time-optimal sliding mode seems reasonable.

Definition 4.2. Let $\big(u_i(t), x_i(t)\big)$, $0 \leqslant t \leqslant T_i$, be a sequence of admissible controls and the corresponding trajectories of the system (4.5) such that

$$x_i(0) = x_0, \qquad x_i(T) = x_1, \qquad \lim_{i \to \infty} T_i = T = \inf_{u \in U} J(u). \tag{4.8}$$

Here T denotes the infimum of the transfer time of the system from x_0 to x_1. Further, let the sequence $x_i(t)$, $0 \leqslant t \leqslant T$, converge pointwise as $i \to \infty$ to a function $x_0(t)$ that is not a solution of the system (4.5) for any admissible control. Then the trajectory $x_0(t)$ is called a *time-optimal sliding mode trajectory* for the system (4.5).

We now try to find time-optimal sliding modes for the system (4.5) among the solutions of the time-optimal problem for the system (4.6) with the same initial and terminal states x_0 and x_1. First of all, it should be noted that the transfer time T_1 for the system (4.6) is not greater than the lower bound for the transfer time T from x_0 to x_1 for the system (4.5). However, it may happen that the infimum for the transfer time T from x_0 to x_1 for the system (4.5) is greater than the transfer time T_1 for the system (4.6).[6]

In this connection it is expedient to redefine the sliding mode trajectory so that the transfer time from x_0 to x_1 along this trajectory be equal to T_1.

Definition 4.3. Suppose that $\big(u_i(t), x_i(t)\big)$, $0 \leqslant t \leqslant T_1$, is the sequence of controls and the corresponding trajectories for the system (4.5); moreover, $x_i(t)$ converges pointwise for $0 \leqslant t \leqslant T_1$ to the function $x_0(t)$ satisfying the conditions

[6]Vapnyarski, I. B., The Existence Theorem for Optimal Control in the Boltz Problem, Some of its Applications, and Necessary Conditions for the Optimality of Sliding and Basic Modes, *Zh. Vychisl. Mat. i Mat. Fiz.*, (1979), vol. 7, no. 2, 259–283. (Translated into English as *U.S.S.R. Comput. Math. and Math. Phys.*)

$x_0(0) = x_0$, $x_0(T_1) = x_1$. If the function $x_0(t)$ is not a solution of the problem (4.5) for any admissible control, then it is called a *time-optimal sliding mode trajectory* for the system (4.5).

Note the difference between Definitions 4.2 and 4.3 of an optimal sliding mode. It is required in Definition 4.2 that the sequence of trajectories $x_i(t)$ satisfy the end conditions (4.8), whereas this is not required in Definition 4.3. All optimal sliding mode trajectories of the system (4.5) in the sense of Definition 4.3 are contained among time-optimal trajectories of the system (4.6). The latter can be determined using the maximum principle (see Theorem 2.2 of Chapter V).

We turn to a discussion of necessary conditions for the solvability of the time-optimal problem for the system (4.6), (4.7). Suppose that there exists a solution $(y(t), \alpha_0(t), \dots, \alpha_n(t), v_0(t), \dots, v_n(t))$ of this problem. Then there exists a nonzero solution $\psi(t) \in R^n$ of the equation

$$\dot{\psi}(t) = -\sum_{i=0}^{n} \alpha_i(t) \frac{\partial}{\partial y} H(\psi, y(t), v_i(t)), \qquad 0 \leqslant t \leqslant T_1,$$

$$H(\psi, y, v_i) = \psi' f(y, v_i)$$

such that for all t satisfying the inequality $\alpha_i(t) > 0$ $(i = 0, \dots, n)$ the following equality holds:

$$H(\psi(t), y(t), v_i(t)) = \max_{u \in U} H(\psi(t), y(t), u). \tag{4.9}$$

It follows from this equality that the sliding mode is due to the fact that the maximum of the function $H(\psi, y, u)$ with respect to u is not attained at a single point. But if equation (4.9) has a unique solution for $v_i(t)$, then the function $y(t)$, $0 \leqslant t \leqslant T_1$, is a time-optimal trajectory of the system (4.5).

Example 4.3. Consider the time-optimal problem for the system

$$\dot{x}_1 = u, \qquad \dot{x}_2 = |u| - x_1^2. \tag{4.10}$$

The initial and terminal states are given by the relations

$$x_1(0) = x_2(0) = 0, \tag{4.11}$$

$$x_1(T) = 0, \qquad x_2(T) = 1. \tag{4.12}$$

Finally, the constraints on the control and the cost functional are of the form

$$|u| \leqslant 1, \qquad J(u) = T \to \inf_u. \tag{4.13}$$

We shall construct a sequence $u_i(t)$ of measurable functions satisfying the relations (4.13) and such that the limit of the corresponding sequence of solutions of the problem (4.10), (4.11) converges uniformly to the trajectory that satisfies the conditions (4.12) for $T = 1$ on the segment $[0, 1]$. We set

$$u_i(t) = (-1)^j, \qquad \frac{j}{i} \leqslant t \leqslant \frac{j+1}{i}, \ j = 0, \dots, i-1,$$

and find the limit of solutions of the problem (4.10), (4.11) as $i \to \infty$:

$$x_1(t) \equiv 0, \qquad x_2(t) = t. \tag{4.14}$$

Hence, the time of transfer of the system (4.10) from the initial state (4.11) to the terminal state (4.12) along the trajectory (4.14) is $T = 1$. To prove that the transfer time $T = 1$ is a minimum, we consider the corresponding time-optimal problem for the system (4.6), (4.7):

$$\begin{gathered}
\dot{y}_1(t) = \alpha_1(t)v_1(t) + \alpha_2(t)v_2(t), \\
\dot{y}_2(t) = \alpha_1(t)|v_1(t)| + \alpha_2(t)|v_2(t)| - y_1^2, \\
|v_i(t)| \leqslant 1, \quad \alpha_i \geqslant 0,\ i = 1,2,\ \alpha_1(t) + \alpha_2(t) = 1, \\
y_1(0) = y_2(0) = 0, \quad y_1(T) = 0,\ y_2(T) = 1,\ T \to \inf_{\alpha_i, v_i}.
\end{gathered} \tag{4.15}$$

The optimal control in the problem (4.15) can be taken as

$$\alpha_1(t) = \alpha_2(t) \equiv \tfrac{1}{2}, \qquad v_1(t) \equiv -1,\ v_2(t) \equiv 1. \tag{4.16}$$

The optimality of the control follows from the fact that, given this control, the right side of equation (4.15) in y_2 attains a maximum value 1. Since, with regard to the relations (4.16), the transfer time in the problem (4.15) is equal to 1, this means that we have established that the transfer time $T = 1$ in the original problem (4.10)–(4.13) is a minimum.

Main Results and Formulas of Chapter VIII

Existence of an optimal control. Suppose that for the problem (1.1), (1.2) there exists at least one admissible control, and the set $U \subset R^n$ is compact and convex. Then in the problem (1.1), (1.2) there exists an optimal control.

Singular optimal control. Let U_1 be a subset of the set U containing more than one point, and let $[\tau_0, \tau_1]$, where $0 \leqslant \tau_0 \leqslant \tau_1 \leqslant T$, be a time segment such that

$$\begin{aligned} H\big(x(t), \psi(t), u(t)\big) &= \max_{u\in U} H\big(x(t), \psi(t), u\big) \\ &\equiv H\big(x(t), \psi(t), u\big), \qquad \tau_0 \leqslant t \leqslant \tau_1,\ u \in U_1. \end{aligned}$$

Then the time segment $[\tau_0, \tau_1]$ is called a singular mode interval, and the control $u(t)$, $\tau_0 \leqslant t \leqslant \tau_1$, is called a singular control, the trajectory $x(t)$ is a singular mode trajectory, and, finally, the aggregate $\big(u(t), x(t)\big)$ is called a singular mode on the time segment $[\tau_0, \tau_1]$.

The Kelley necessary condition for the optimality of singular control:

$$\frac{\partial}{\partial u}\frac{d^2}{dt^2}\frac{\partial H}{\partial u} \geqslant 0.$$

The Kopp–Moyer necessary conditions for the optimality of singular control:

$$\frac{\partial}{\partial u}\frac{d^2}{dt^2}\frac{\partial H}{\partial u} = 0, \qquad \frac{\partial}{\partial u}\frac{d^4}{dt^4}\frac{\partial H}{\partial u} \leqslant 0.$$

Chattering mode. The chattering mode is an optimal mode such that the optimal control has an infinite number of switch points on a finite time interval.

In the problem

$$\begin{gathered} \dot{x}_1 = x_2, \qquad \dot{x}_2(t) = u, \qquad 0 \leqslant t \leqslant 8, \\ x_1(0) = x_1(8) = x_2(8) = 2,\ x_2(0) = -2,\ |u| \leqslant 1, \\ J(u) = \int_0^8 x_1^2(t)\,dt \to \inf_{|u|\leqslant 1}. \end{gathered}$$

the chattering mode exists on the segments $[0, 3.43039)$ and $(4.56961, 8]$, whereas the segment $[3.43039, 4.56961]$ is a singular mode segment.

Sliding optimal mode. Suppose that $u_i(t)$ is a minimizing sequence of admissible controls in the problem (4.5), and $x_i(t)$ is the corresponding sequence of trajectories of the system (4.5), with the sequence $x_i(t)$, $0 \leqslant t \leqslant T$, converging pointwise to the function $x_0(t)$. Then the trajectory $x_0(t)$ is called a *sliding optimal mode trajectory* if one of the following two conditions is satisfied:

1°) the limiting trajectory $x_0(t)$ is not a solution of equation (4.5) for any admissible control v;

2°) the limiting trajectory $x_0(t)$ is a solution of equation (4.5) for some admissible control v, but $J(v) > \inf_{u \in U} J(u)$.

Suppose that the sequence $\{x_i(t)\}$ of solutions of equation (4.5) converges to the function $y(t)$ at each point $t \in [0, T]$. Then $y(t)$ is a solution of the system

$$\dot{y}(t) = \sum_{i=0}^{n} \alpha_i(t) f\big(y(t), v_i(t)\big), \qquad 0 \leqslant t \leqslant T,\ y \in R^n.$$

$$v_i(t) \in U \subset R^m,\ \alpha_i(t) \geqslant 0,\ \alpha_0(t) + \cdots + \alpha_n(t) \equiv 1.$$

Conversely, any solution of this system is the limit of a sequence of solutions of equation (4.5) that converges pointwise at each point $t \in [0, T]$.

Let $\big(u_i(t), x_i(t)\big)$, $0 \leqslant t \leqslant T_i$, be a sequence of admissible controls and corresponding trajectories of the system (4.5) be such that $x_i(0) = x_0$, $x_i(T) = x_1$, $\lim_{i \to \infty} T_i = T$, where T denotes the infimum of the transfer time of the system from x_0 to x_1. Further, let the sequence $x_i(t)$, $0 \leqslant t \leqslant T$, converge pointwise as $i \to \infty$ to the function $x_0(t)$ that is not a solution of the system (4.5) for any admissible control. Then the trajectory $x_0(t)$ is called a time-optimal sliding mode trajectory for the system (4.5).

EXERCISES OF PART TWO

1. Suppose that the control of the motion of a point described by the equation

$$\ddot{x} = u(t), \qquad 0 \leqslant t \leqslant 2,\ x(0) = 0,\ \dot{x}(0) = 1,$$

is exercised by means of the PD-controller:

$$u(t) = -C_1 x(t) - C_2 \dot{x}(t).$$

Evaluate the coefficients C_1 and C_2 of the controller that render a minimum of the cost functional

$$J(u) = x^2(2) + \int_0^2 u^2(t)\, dt.$$

Solve this problem under the assumption that either the coordinate can solely be controlled (i.e., $C_2 = 0$) or the velocity (i.e., $C_1 = 0$). Compare the resulting values of the cost functional.

2. Suppose that the controlled small oscillations of the pendulum in the vicinity of the lower stable equilibrium position are described by the equation

$$\ddot{x}(t) = -x(t) + u(t), \qquad x(0) = 1,\ \dot{x}(0) = 0.$$

Evaluate the coefficients C_1 and C_2 of the controller

$$u(t) = -C_1 x(t) - C_2 \dot{x}(t)$$

that render a minimum of the cost functional

$$J(u) = x^2(2) + \dot{x}^2(2) + \int_0^2 u^2(t)\, dt.$$

3. Consider Exercise 2 for the motion of the pendulum in the vicinity of the upper unstable equilibrium position when the equation of motion is the following:

$$\ddot{x} = x(t) + u(t), \qquad x(0) = 1,\ \dot{x}(0) = 0.$$

4. The process of temperature change in a sealed-off room can be described by the equation

$$\dot{x}_1(t) = -\tfrac{1}{Cm}[k(x_1(t) - x_0) + u], \qquad t \geqslant 0,$$

where x_0 is the outside temperature, C is the specific heat of air, m is the mass of the air in the room, $k > 0$ is a prescribed quantity that characterizes the loss by convection, u is the control of the heat released by a room heater, $0 \leqslant u \leqslant u_1$. It is required to heat the room to the temperature x_T in a prescribed time T with minimum control cost.

(a) Find optimal control by minimizing the functional

$$J(u) = \alpha_1 (x_1(T) - x_T)^2 + \int_0^T u^2(t)\, dt,$$

where $\alpha_1 > 0$ is a constant.

(b) Verify that in the absence of the heater (i.e. $u = 0$) the room temperature tends to the outside temperature.

(c) Determine the control that stabilizes the room temperature at the prescribed level x_T by minimizing the functional

$$J(u) = \int_0^\infty [(x_1(t) - x_T)^2 \alpha_1 + u^2]\, dt.$$

(d) Deduce optimal control such that $x_1(T) = x_T$ and the cost functional $\int_0^T |u(t)|\, dt$ is a minimum.

(e) Consider problems (a)–(d) for the case of a bang-bang control $u(t)$ when the heater is either switched on for maximum power or switched off. This corresponds to either $u(t) = u_1$ or $u(t) = 0$ at any instant t.

5. Consider the process of temperature change in a two-room apartment with neighboring rooms and a heater in the first room. Let $x_1(t)$, $x_2(t)$ denote the temperatures and m_1, m_2 the air masses in the first and second rooms, respectively; let k_1 denote the coefficient of heat exchange between the rooms; the constants C and k are the same as in Exercise 4. The equations in x_1 and x_2 are of the form

$$\dot{x}_1 = -\tfrac{1}{Cm_1}[k(x_1(t) - x_0) + u] + k_1(x_2(t) - x_1(t)),$$
$$\dot{x}_2 = -\tfrac{k}{Cm_2}(x_2(t) - x_0) + k_1(x_1(t) - x_2(t)).$$

(a) Show that for $u = 0$ the room temperatures tend to the outside temperature.

(b) Answer the questions of Exercise 4 applied to the scheme under consideration.

6. Suppose that the rocket ascends vertically under the action of gravitational and control forces. The equations of motion are of the form

$$\ddot{x}(t) = -g + u,$$

where g is the gravitational constant and u (the control) is the thrust acceleration. It is required to determine the control u such that the height achieved by the rocket is a maximum, subject to one of the following constraints:

$$\int_0^T |u(t)|\,dt = u_0, \qquad \int_0^T u^2(t)\,dt = u_0,$$

where u_0 is a prescribed constant and T is the time of achieving the maximum height. Work out this exercise under an additional constraint $0 \leqslant u \leqslant u_1$, where u_1 is a prescribed constant.

7. Suppose that a rocket experiences a resistance force depending on the altitude x of the rocket and its velocity $\dot{x}$: $F = f(x, \dot{x})$, where

(a) $f(x, \dot{x}) \to 0$ as $x \to \infty$ ($\dot{x}$ fixed);

(b) $f(x, \dot{x}) \to \infty$ as $\dot{x} \to \infty$ (x fixed).

Discuss the qualitative differences in control this leads to and the physical significance of the results.

8. The control process for a high-temperature furnace can be described by the following equations:

$$\dot{x}(t) = -\alpha_1\big(x(t) - x_0\big) - \alpha_2\big(x^4(t) - x_0^4\big) + \alpha_3(u), \qquad t \geqslant 0.$$

Here, $x(t)$ is the temperature inside the furnace, x_0 is the outside temperature, and u is the control over the amount of heat released to the furnace. The summand $\alpha_1(x - x_0)$ characterizes the loss by convection, and the summand $\alpha_2(x^4 - x_0^4)$ the loss by radiation. The control u is determined in a way such that the furnace has a prescribed temperature x_T at a given moment T, and the functional

$$J(u) = \int_0^T u^2(t)\,dt$$

attains a minimum. Write the necessary conditions for optimal control in the form of the maximum principle.

9. The two-chamber model of "medicine absorbtion and metabolism" can be described by the equations

$$\dot{x}_1 = -\alpha_1 x_1 + u, \qquad \dot{x}_2 = \alpha_1 x_1 - \alpha_2 x_2, \qquad t \geqslant 0,\ 0 \leqslant u \leqslant u_0.$$

Here, x_1 is the mass of medicine in the alimentary canal, x_2 is the mass of medicine in the blood, the control u is the amount (mass) of medicine taken, and $\alpha_1 > 0$, $\alpha_2 > 0$ are given constants. The goal of the control is to attain the desired amount $x_2(T)$ of medicine in the blood in minimum time. Write the necessary conditions for optimal control in the form of the maximum principle.

10. Determine which of the systems $\dot{x} = Ax + Bu$ is controllable if

(a) $A = \begin{pmatrix} 0 & 1 \\ 0 & 0 \end{pmatrix}, \qquad B = \begin{pmatrix} 0 \\ 1 \end{pmatrix};$

(b) $A = \begin{pmatrix} 0 & 1 \\ 1 & 0 \end{pmatrix}, \qquad B = \begin{pmatrix} 0 \\ 1 \end{pmatrix};$

(c) $A = \begin{pmatrix} 0 & 1 \\ -1 & 0 \end{pmatrix}, \qquad B = \begin{pmatrix} 0 \\ 1 \end{pmatrix};$

(d) $A = \begin{pmatrix} 0 & 1 \\ 1 & 1 \end{pmatrix}, \qquad B = \begin{pmatrix} 0 \\ 1 \end{pmatrix};$

(e) $A = \begin{pmatrix} 1 & 0 \\ 0 & 2 \end{pmatrix}, \qquad B = \begin{pmatrix} 1 \\ 1 \end{pmatrix};$

(f) $A = \begin{pmatrix} 1 & 0 \\ 0 & 1 \end{pmatrix}, \qquad B = \begin{pmatrix} 0 \\ 1 \end{pmatrix};$

(g) $A = \begin{pmatrix} 0 & 1 & 0 \\ 0 & 1 & 1 \\ 0 & 0 & 1 \end{pmatrix}, \qquad B = \begin{pmatrix} 0 \\ 0 \\ 1 \end{pmatrix};$

(h) $A = \begin{pmatrix} 0 & 1 & 0 \\ 0 & 1 & 1 \\ 0 & 0 & 1 \end{pmatrix}, \qquad B = \begin{pmatrix} 0 \\ 1 \\ 0 \end{pmatrix}.$

11. Determine which of the systems $\dot{x} = Ax$, $y = B'x$ is controllable, where the matrices A and B are the same as in Exercise 9.

12. Given the scalar system $\dot{x}(t) = ax(t) + bu$, $t \geqslant 0$, with the cost functional

$$J(u) = \lim_{t\to\infty} \frac{1}{t} \int_0^t \left(\alpha x^2(s) + \beta u^2(s)\right) ds \to \inf,$$

deduce optimal control and investigate how it depends on the parameters a, b, $\alpha > 0$, $\beta > 0$.

13. Consider the scalar system $\dot{x} = ax + bu$, $t \geqslant 0$, with the cost functional

$$J(u) = \int_0^\infty \left(\alpha x^2(s) + \beta u^2\right) dt \to \inf,$$

where $a, b \neq 0$, $\alpha > 0$, $\beta > 0$ are given constants. Verify that the optimal control u_0 is of the form

$$u_0 = -\frac{1}{b}[a + \sqrt{a^2 + b^2\alpha\beta - 1}]x,$$

and the Bellman function V has the form

$$V(t, x) = x^2\beta b^{-2}[a + \sqrt{a^2 + b^2\alpha\beta - 1}].$$

14. Consider the scalar system

$$\dot{x}(t) = -ax(t) + bu, \qquad t > 0,$$

(a and b being prescribed coefficients) with the cost functional

$$J(u) = \int_0^\infty [x^2(t) + \alpha x(t)u + u^2]\, dt \to \inf .$$

Determine the range of variation of the parameter α such that the system closed by optimal control is asymptotically stable.

15. For the system $\ddot{x} = ax + bu$, $a, b =$ constant, $t \geqslant 0$, consider exercises similar to Exercises 12–14.

16. Determine the Bellman function and optimal control for the following scalar systems on the segment $0 \leqslant t \leqslant 1$:

$$\begin{aligned}
&\text{(a)} \quad \dot{x} = u, \\
&\qquad\ x(1) \to \min, \qquad |u| \leqslant 1; \\
&\text{(b)} \quad \dot{x} = u, \\
&\qquad\ |x(1)| \to \inf, \qquad |u| \leqslant 1; \\
&\text{(c)} \quad \dot{x} = ax + u, \\
&\qquad\ x^2(1) \to \inf, \qquad |u| \leqslant 1; \\
&\text{(d)} \quad \dot{x} = ax + u, \\
&\qquad\ x(1) \to \inf, \qquad |u| \leqslant 1; \\
&\text{(e)} \quad \dot{x}(t) = u + t, \\
&\qquad\ J(u) = x^2(1) + \int_0^1 u^2(t)\, dt \to \inf .
\end{aligned}$$

17. Verify that the system $\dot{x} = \Lambda x + Bu$, where Λ is a diagonal matrix, is controllable if and only if there is no zero row in the matrix B. Carry out the proof:

(a) directly,

(b) as a consequence of the Hautus controllability criterion (see Theorem 2.5 of Chapter VI).

18. Verify that for the system $\dot{x} = Jx + Bu$, where J is a Jordan matrix, to be controllable, it is necessary and sufficient that at least one element of the matrix B in the row corresponding to the lower row of each Jordan box and at least one element of the matrix B in any other row be distinct from zero.

19. The system

$$\dot{x} = Ax + By, \qquad x \in R^n,$$
$$y = Cx, \qquad y \in R^q,$$

is said to be output-controllable if it can be transferred from any initial state $y(t_0) = Cx(t_0)$ to any terminal state $y(t_1)$ in finite time by the action of piecewise continuous control $u(t)$, $t_0 \leqslant t \leqslant t_1$. Verify that for the system to be output-controllable it is necessary and sufficient that the rank of the matrix

$$M = [CB, CAB, CA^2B, \ldots, CA^{n-1}B]$$

be equal to q.

20. Verify that the system

$$\dot{x} = \Lambda x + Bu,$$
$$y = Cx$$

with diagonal matrix Λ is controllable if and only if there is no zero column in the matrix C. Carry out the proof:

(a) directly,

(b) as a consequence of the Hautus controllability criterion (see Theorem 2.5 of Chapter VI).

21. For the time-varying system

$$\dot{x} = A(t)x, \qquad y = Q(t)x$$

state conditions for observability similar to the conditions for controllability given by Theorem 2.3 of Chapter VI.

22. Determine conditions on the parameters of the scalar system

$$\dot{x} = a(t)x + b(t)u, \qquad x \in R^1, u \in R^1,$$
$$y = c(t)x$$

for the reduction of this system to the system with constant coefficients (with control v and observation $\tilde{y}$) of the form

$$\dot{z} = k_1 z + k_2 v, \tilde{y} = z.$$

23. Reduce the system

$$\dot{x}_1 = x_2 + u,$$
$$\dot{x}_2 = -x_1 + e^t u$$

to a time-invariant system.

Hint. The change of variables $x_1 = y_1 + e^t y_2$, $x_2 = y_3 + e^t y_4$ reduces the system to the form

$$\dot{y}_1 = y_3 + u,$$
$$\dot{y}_2 = -y_2 + y_4,$$
$$\dot{y}_3 = -y_1,$$
$$\dot{y}_4 = -y_2 - y_4 + u.$$

24. The equations of motion of an axially symmetric rigid body stabilized by its own spin, with motors securely fastened to the casing, are of the form

$$\dot{x}_1 = -ax_2 + u\cos\omega t,$$
$$\dot{x}_2 = ax_1 - u\sin\omega t.$$

Here x_1, x_2 are the projections of the angular velocuty of the body on the axes bound with the body, ω is the angular velocity, u is the control torque, and $a > 0$ is a configuration constant.

Reduce this system to a time-invariant control system and investigate it.

Hint. The change of variables

$$x_1 = y_1 \sin\omega t + y_2 \cos\omega t,$$
$$x_2 = y_3 \sin\omega t + y_4 \cos\omega t$$

reduces the original system to the form

$$\dot{y}_1 = \omega y_2 - ay_3,$$
$$\dot{y}_2 = -\omega y_1 - ay_4 - u,$$
$$\dot{y}_3 = \omega y_4 + ay_1 - u,$$
$$\dot{y}_4 = -\omega y_3 + ay_2.$$

(See the book cited in Exercise 54 of Part One.)

25. The motion of a material point controlled by a force pointing in the same direction is described by the equations

$$\ddot{x}(t) = u(t), \qquad t \geqslant 0,\ 0 \leqslant u(t) \leqslant 1,$$
$$x(0) = x_0,\ \dot{x}(0) = \dot{x}_0.$$

Determine the region of controllability. Explain the physical significance of the result obtained.

26. Show that small oscillations of the pendulum described by the equations

$$\begin{gathered} \ddot{x}(t) + x = u(t), \qquad t \geqslant 0,\ 0 \leqslant u(t) \leqslant 1, \\ x(0) = x_0,\ \dot{x}(0) = \dot{x}_0. \end{gathered}$$

can be eliminated for any initial conditions x_0 and $\dot{x}_0$ only by using a force $u(t)$, $0 \leqslant u(t) \leqslant 1$, pointing in the same direction.

27. Show that the system

$$\ddot{x}(t) + x = u(t), \qquad t \geqslant 0,\ 0 \leqslant u(t) \leqslant 1,$$

is controllable.

28. Determine the minimum transfer time in the time-optimal problem

$$\begin{gathered} \ddot{x}(t) + x = u(t), \qquad 0 \leqslant t \leqslant T,\ 0 \leqslant u(t) \leqslant 1, \\ x(0) = x_0,\ \dot{x}(0) = \dot{x}_0, \quad x(T) = \dot{x}(T) = 0, \\ J(u) = T \to \inf_{0 \leqslant u \leqslant 1}. \end{gathered}$$

29. Determine the switching curve in the time-optimal problem

$$\begin{gathered} \ddot{x}(t) = u, \qquad 0 \leqslant t \leqslant T,\ -2 \leqslant u \leqslant 4, \\ x(0) = x_0,\ \dot{x}(0) = \dot{x}_0, \quad x(T) = \dot{x}(T) = 0, \\ J(u) = T \to \inf_{-2 \leqslant u \leqslant 4}. \end{gathered}$$

30. Determine the minimum transfer time in the time-optimal problem

$$\begin{gathered} \dddot{x}(t) = u, \qquad 0 \leqslant t \leqslant T,\ |u| \leqslant 1, \\ x(0) = x_0,\ \dot{x}(0) = \dot{x}_0,\ \ddot{x}(0) = \dot{x}_0, \\ x(T) = \dot{x}(T) = \ddot{x}(T) = 0, \\ J(u) = T \to \inf_{|u| \leqslant 1}. \end{gathered}$$

31. Linearize the bilinear control system

$$\begin{aligned} \dot{x}_1 &= x_2, \\ \dot{x}_2 &= -2x_1 - 3x_3^2 + 2 + u_2, \\ \dot{x}_3 &= -4x_1x_2 - x_3 + 1 + 2x_3u_3 \end{aligned}$$

in the vicinity of the equilibrium state. Verify the controllability of the linearized system.

32. Consider a system "predator-prey" of the form

$$\dot{x}(t) = x(t)[a_1 - a_2 y(t) - r_1 u(t)],$$
$$\dot{y}(t) = y(t)[a_2 a_3 x(t) - a_4 - r_2 u(t)],$$
$$t \geqslant 0,\ 0 \leqslant u(t) \leqslant \gamma.$$

Linearize this system for $u = 0$ in the vicinity of the equilibrium state $x(t) = a_4(a_2 a_3)^{-1}$, $y(t) = a_1 a_2^{-1}$, and express the system as

$$\dot{z} = Az + Bu, \qquad z' = (x, y), \qquad 0 \leqslant u(t) \leqslant \gamma.$$

Construct the switching curves in this linearized system and define the regions from which the system can be transferred to the equilibrium state. Compare the results obtained with those of § 4 of Chapter V.

33. For the SISO-system

$$\sum_{i=0}^{n} a_i y^{(i)}(t) = \sum_{i=0}^{m} b_i u^{(i)}(t),$$

construct any of its realizations in the form

$$\dot{x} = Ax + Bu,$$
$$y = Cx.$$

34. For the SISO-system

$$2y''(t) + 3y'(t) + 2y(t) = u(t) + 5u'(t),$$

verify if there exist two realizations of the same type as in Exercise 33, with matrices (A_1, B_1, C_1) and (A_2, B_2, C_2), where

$$A_1 = \begin{pmatrix} 0 & 1 \\ -1 & -\frac{1}{3} \end{pmatrix}, \qquad B_1 = \begin{pmatrix} 1 \\ 5 \end{pmatrix}, \qquad C_1 = (\,1 \quad 5\,),$$

$$A_2 = \begin{pmatrix} -\frac{3}{2} & 1 \\ -1 & 0 \end{pmatrix}, \qquad B_2 = \begin{pmatrix} 5 \\ 1 \end{pmatrix}, \qquad C_2 = (\,\tfrac{1}{2} \quad 0\,).$$

Indicate other possible realizations of this system.

35. Show that any SISO-system

$$\sum_{i=0}^{n} a_i y^{(i)}(t) = \sum_{i=0}^{m} b_i u^{(i)}(t), \qquad a_n \neq 0,\ b_n \neq 0,$$

is controllable.

36. For a MIMO-system with two inputs and two outputs and with transfer matrix

$$G(s) = \begin{pmatrix} \frac{1}{(s-1)^2} & \frac{1}{(s-1)(s+3)} \\ \frac{-6}{(s-1)(s+3)^2} & \frac{s-2}{(s+3)^2} \end{pmatrix},$$

verify that one of the possible realizations (A, B, C) is of the form

$$A = \begin{pmatrix} -4 & -4 & 2 & 2 & 12 & 12 & -9 & -9 \\ -4 & -4 & 2 & 2 & 12 & 12 & -9 & -9 \\ 0 & 0 & 1 & 1 & 0 & 0 & 0 & 0 \\ 0 & 0 & 1 & 1 & 0 & 0 & 0 & 0 \\ 0 & 0 & 0 & 0 & 1 & 1 & 0 & 0 \\ 0 & 0 & 0 & 0 & 1 & 1 & 0 & 0 \\ 0 & 0 & 0 & 0 & 0 & 0 & 1 & 1 \\ 0 & 0 & 0 & 0 & 0 & 0 & 1 & 1 \end{pmatrix},$$

$$B = \begin{pmatrix} 1 & 1 \\ 1 & 1 \\ 0 & 0 \\ 0 & 0 \\ 0 & 0 \\ 0 & 0 \\ 0 & 0 \\ 0 & 0 \end{pmatrix}, \qquad C = \begin{pmatrix} 0 & 0 & 1 & 1 & 6 & 2 & 9 & -3 \\ 0 & 1 & 0 & -4 & -6 & 5 & 6 & -2 \end{pmatrix}.$$

37. Verify the controllability and observability of the realization from the preceding exercise.

38. For the system in Exercise 34 determine the set of its invariant zeros.

39. Find the H_∞-norm of transfer matrices in Exercises 34 and 36.

40. Find the H_2-norm of transfer matrices in Exercises 34 and 36.

41. For the system with transfer matrix

$$G(s) = \begin{pmatrix} \frac{1}{s(s+1)^2} & \frac{s^2+2s-1}{s(s+1)^2} & \frac{s+2}{s+1} \\ 0 & \frac{s+2}{(s+1)^2} & 0 \\ 0 & 0 & \frac{3(s+2)}{s+1} \\ \frac{s+3}{s(s+1)^2} & \frac{2s^2+2s-3}{s(s+1)^2} & \frac{s+2}{s+1} \end{pmatrix}.$$

compute the H_∞- and the H_2-norm.

42. For the MIMO-system of the preceding exercise construct any of its realizations.

43. Determine the sets of transfer and invariant zeros of the system described in Exercise 36.

44. For the MIMO-system

$$\dot{x} = Ax + Bu, \qquad x \in R^4,\ u \in R^2,$$

where

$$A = \begin{pmatrix} 2 & 1 & 0 & 0 \\ 0 & 1 & 0 & 1 \\ 0 & 2 & 0 & 0 \\ 1 & 1 & 0 & 0 \end{pmatrix}, \qquad B = \begin{pmatrix} 1 & 0 \\ 0 & 0 \\ 0 & 0 \\ 0 & 1 \end{pmatrix},$$

determine the H_∞-norm of the transfer matrix from u to x.

45. For the system described in Exercise 44 determine an arbitrary H_∞-suboptimal static controller of the form $u = Kx$.

46. For the system described in Exercise 41 determine an arbitrary H_∞-suboptimal dynamic controller of the form (5.16) of Chapter VI.

47. For the system described in Exercise 44 determine the sets of transfer and invariant zeros.

48. Show that the transfer zeros of the system $\dot{x} = Ax + Bu$, $y = Hx$ are invariant zeros of this system. The converse is, in general, not true. Give an example.

49. Show that a nondegenerate transformation of the state variables $\hat{x} = Nx$, $\det N \neq 0$ does not effect any change in the invariant and transfer zeros of the system.

50. Show that a nondegenerate transformation of the input variables $\hat{u} = Mu$, $\det M \neq 0$ does not effect any change in the invariant and transfer zeros of the system.

51. Show that a nondegenerate transformation of the ouput variables $\hat{y} = Ty$, $\det T \neq 0$ does not effect any change in the invariant and transfer zeros of the system.

52. Show that a proportional feedback with respect to the state $u = Kx$ (or with respect to the output $u = Ky$) does not effect any change in the invariant and transfer zeros of the system.

53. Show that a proportional-integral feedback with respect to the output

$$u = K_1 x + K_2 \int_0^t y(s)\, ds + K_3 y(0) + v$$

does not effect any change in the invariant and transfer zeros of the system.

54. Consider the problem of transferring the scalar system

$$\dot{x}(t) = ax(t) + bu, \qquad t \geqslant 0,\ x(0) = x_0,$$

where a, $b > 0$ are specified constants and x_0 is a prescribed initial state, to the origin in minimum time, subject to the control constraint $|u| \leqslant 1$. Show that
(a) the Bellman equation for this problem is of the form

$$\inf_{u, |u| \leqslant 1} \left[\frac{\partial V(x)}{\partial x}(ax + bu) \right] = -1, \qquad V(0) = 0;$$

(b) if $a \leqslant 0$, the problem can be solved for any x_0, and the minimum transfer time $V(x)$ is

$$V(x) = -\frac{1}{a} \ln\left(1 - \frac{a}{b}|x|\right), \qquad a \neq 0,$$
$$V(x) = \frac{|x|}{b}, \qquad a = 0,$$

(c) if $a > 0$, the problem can be solved for x satisfying the inequality $|x| < ba^{-1}$, i.e., in the region, where $V(x)$ has the same expression as in the case $a \leqslant 0$.
(d) the optimal control is $u(x) = -\operatorname{sgn} x$.

55. Consider a weight of mass m in rectilinear motion subjected to a force $u(t)$ and a certain resistance. Let $x_1(t)$ denote the coordinate of the mass center of the weight, and $x_2(t)$ denote its velocity. Under certain assumptions the equations of motion may be expressed as

$$\dot{x}_1(t) = x_2(t), \qquad m\dot{x}_2(t) = u(t) - x_2^2(t) \operatorname{sgn} x_2(t), \qquad 0 \leqslant t \leqslant T.$$

Here, the terminal moment T, the initial state $x_1(0)$ and velocity $x_2(0)$ are prescribed.
The cost functional to be minimized is of the form

$$J(u) = \int_0^T F\big(t, x_1(t), x_2(t), u(t)\big)\, dt \to \min_u,$$

where the scalar function $F(t, x_1, x_2, u)$ is continuously differentiable with respect to x_1, x_2, u. The control u is unconstrained. Show that the necessary conditions for optimal control can be derived from the original equations of motion and the following relations:

$$H = -F(t, x_1, x_2, u) + \psi_1 x_2 + \frac{1}{m}\psi_2(u - x_2^2 \operatorname{sgn} x_2), \qquad \frac{\partial H}{\partial u} = 0$$
$$\dot{\psi}_1(t) = -\frac{\partial H}{\partial x_1}, \qquad \dot{\psi}_2(t) = -\frac{\partial H}{\partial x_2}, \qquad \psi_1(T) = 0,\ \psi_2(T) = 0.$$

Noting the expression for the function H, we can write

$$\psi_2(t) = m\frac{\partial F}{\partial u},$$
$$\dot{\psi}_1(t) = \frac{\partial F}{\partial x_1}, \qquad \psi_1(T) = 0,$$
$$\dot{\psi}_2(t) = \frac{\partial F}{\partial x_2} + \frac{2}{m}\psi_2 x_2 \operatorname{sgn} x_2 - \psi_1, \qquad \psi_2(T) = 0.$$

Here, all derivatives of the function F are calculated at the "point" $\big(t, x_1(t), x_2(t), u(t)\big)$

56. The operation of a synchronous motor can be described by

$$\dot{x}_1 = x_2, \qquad 0 \leqslant t \leqslant T,$$
$$\dot{x}_2 = a_1 - a_2 x_2 - a_3 x_3 \sin x_1 - a_4 \sin 2x_1,$$
$$\dot{x}_3 = u - a_5 x_3 + a_6 \cos x_1, \qquad 0 \leqslant u \leqslant u_1.$$

Here, x_1 is the angle of rotation of the rotor, x_2 is the angular velocity of the rotor, the variable x_3 is a characteristic of magnetic-flux linkage. The control is implemented by changing the voltage that creates the rotating field in the motor. The initial states $x_i(0)$, the terminal time T, and the constants a_i are prescribed. The cost functional is the same as in the preceding exercise. Write the necessary conditions for optimal control.

Consider the case of a bang-bang control in which $u(t)$ is equal to either 0 or u_1 at each time instant. Write the necessary conditions for optimal control and the Bellman equation for this problem.

57. Consider the problem of asymptotic stabilization of the motion of a rigid body described by equations (4.2.3). Use the Lyapunov function that is equal to the square of the absolute value of the angular momentum:

$$V = A^2p^2 + B^2q^2 + C^2r^2$$

to show that the problem of the asymptotic stabilization of the motion of a rigid body can be solved in the class of control torques, which are proportional to angular velocities:

$$M_1 = \alpha_1 p, \qquad M_2 = \alpha_2 q, \qquad M_3 = \alpha_3 r.$$

58. Consider the problem of the asymptotic stabilization of the motion of an axially symmetric rigid body for $A = B$ (see subsection 4 of § 1 of Chapter VI). Show that stabilization can only be achieved by using two control torques.

59. On the specified segment $[0,T]$ consider the linear control system

$$\dot{x}(t) = A(t)x(t) + B(t)u, \qquad x \in R^n,\ u \in R^m,$$
$$x(0) = x_0,\ x(T) = x_1.$$

It is required to determine control that generates the transfer from the initial state x_0 to the terminal state x_1, such that the cost functional is minimized:

$$\int_0^T |u(t)|^2\, dt \to \min_u$$

Show that the optimal control u_0 is equal to

$$u_0(t) = B'(t)Z'(T,t)\left[\int_{t_0}^T Z(t,s)B(s)B'(s)Z(T,s)\, ds\right]^{-1} [x_1 - Z(T,t_0)x_0].$$

Here, $Z(t,s)$ is the Cauchy matrix of the system $\dot{x}(t) = A(t)x(t)$.

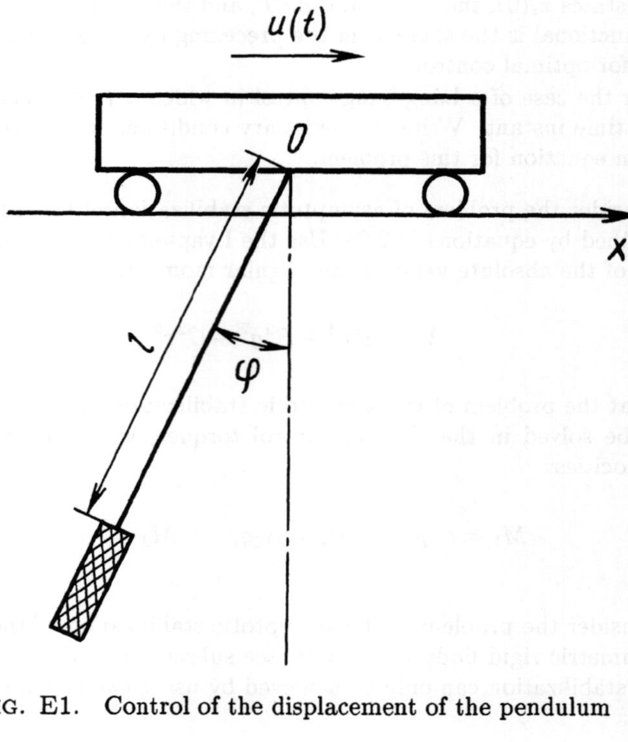

FIG. E1. Control of the displacement of the pendulum

60. A weight suspended by a rope from a trolley must be transferred from one position to another so that it is stationary in the new position (see Figure E.1). This problem can be stated in the following way. The physical pendulum performs small plane oscillations about the axis, which moves with velocity $u(t)$ along the horizontal axis. The equations of the small oscillations of the pendulum are of the form

$$J\ddot{\varphi}(t) + mgl\varphi = mlw, \qquad \dot{x}(t) = u, \qquad \dot{u}(t) = w, \qquad 0 \leqslant t \leqslant T.$$

The control of the pendulum is implemented by changing the velocity $u(t)$ of the suspension point O along the horizontal axis, subject to the constraint $|u(t)| \leqslant u_0$. It is required to transfer the pendulum from a given initial state

$$\varphi(0) = \dot{\varphi}(0) = x(0) = u(0) = 0$$

to a given terminal state

$$\varphi(T) = \dot{\varphi}(T) = u(T) = 0, \qquad x(T) = s > o,$$

where $T = \sqrt{J(mgl)^{-1}}$ is the period of small oscillations of the pendulum.

Show that the optimal control $u_0(t)$ assumes only the values u_0 and $-u_0$.[1] The number n of the intervals of constancy of the control $u_0(t)$ is equal to

$$n = \begin{cases} 1, & \text{whenever } s(2\pi u_0 T)^{-1} \text{ is an integer,} \\ 2\left[s(2\pi u_0 T)^{-1}\right] + 3, & \text{otherwise.} \end{cases}$$

Here, $[b]$ denotes the integral part of the number b. When $n = 1$, the optimal control is equal to

$$u_0(t) \equiv u_0, \qquad 0 < t < T,$$

and the transfer time is su_0^{-1}.

[1]Chernous'ko, F., L., Akulenko, L., D., and Sokolov, B. N., *Control of Oscillations*, Nauka, Moscow, (1980).

PART THREE

OPTIMAL CONTROL OF DYNAMICAL SYSTEMS UNDER RANDOM DISTURBANCES

CHAPTER IX

CONTROL OF STOCHASTIC SYSTEMS. PROBLEM STATEMENTS AND INVESTIGATION TECHNIQUES

This chapter is devoted to basic control problem statements for stochastic systems, to deriving the Bellman equation, to a discussion of linear-quadratic problems under various conditions of *a priori* knowledge, to the investigation of stabilization problems, and to a discussion of various approximation techniques for the determination of optimal control.

§ 1. Statements of Control Problems for Stochastic Systems

As for deterministic systems, specific statements of control problems for stochastic systems depend on the type of equations of motion, on the cost functional to be minimized, on the control and state constraints, on the duration of the control time interval, and so on. We consider each of these problem statements in greater detail.

1. Equations of motion of a system. Equations of motion of a stochastic system differ from those of a deterministic system by the appearence of terms taking account of random disturbances. In what follows, we consider, basically, Ito equations of controlled motion of the form

$$dx(t) = a\big(t, x(t), u\big)\, dt + \sigma\big(t, x(t), u\big)\, d\xi(t), \qquad t \geqslant 0. \tag{1.1}$$

Here $x(t) \in R^n$ is the phase vector, $u \in R^m$ is the control, and $\xi(t) \in R^k$ is a standard Wiener process. The vector function $a\colon [0, \infty) \times R^n \times R^m \to R^n$, and the $n \times k$ matrix σ are prescribed.

It will be recalled that

$$\xi(0) = 0, \qquad E\xi(t) = 0, \qquad E\xi(t)\xi'(t) = It,$$

where E is (mathematical) expectation, I is a unit matrix. Equations (1.1) are taken in the sense of Ito (see, for example, § 3 of Chapter II). The solution of equation (1.1) for $t \geqslant 0$ is determined by the initial condition

$$x(0) = x_0, \tag{1.2}$$

where x_0 is a deterministic or random vector in R^n.

If the time T of motion of the system (1.1) is prescribed, then one speaks of a *control problem on a finite time interval* $[0, T]$. At the same time in a number of cases the control of the system is terminated once the phase trajectory reaches a prescribed target set. Then the terminal moment (moment of termination of motion) is not fixed in advance and is a random variable specified by the actual trajectory. By way of example, we can cite stochastic time-optimal problems, which are sometimes called *problems with random stopping time.* Also considered are *problems on an infinite time interval.* As an example, we can cite the problem of optimal stochastic stabilization.

2. Control constraints. Control problems for stochastic systems are solved subject to control constraints that may depend on available resources and *a priori* knowledge. An example of the control constraint of the first type is

$$u(t) \in U, \tag{1.3}$$

where $U \subset R^m$ is a given set. The constraint (1.3) means that all control realisations belong to the set U.

Along with (1.3), we may encounter integral constraints of the form

$$\int_0^T |u(t)|^\sigma \, dt \leqslant u_0, \tag{1.4}$$

where $\sigma > 0$, $u_0 > 0$ are given constants. The condition (1.4) for $\sigma = 1$ can be interpreted as a restriction on the total control momentum, and for $\sigma = 2$ it can be regarded as an energy restriction. Integral control constraints appear in a perfectly natural way in a number of practical problems (see examples in § 3 of Chapter III). Integral constraints can be of more general form

$$\int_0^T F\big(t, x(t), u(t)\big) \, dt \leqslant u_0.$$

Another important factor is the type of available information. If all phase coordinates of the system cannot be measured, then the optimal control can be determined as a function of time only. Such optimal control is referred to as *program control.* Henceforth, this type of control will be called *P-control.* P-control $u(t)$, based on the *a priori* information about the characteristics of the system, of input and output signals, is usually computed in advance at the stage of designing the control system. However, P-control does not allow to take account of the changes in the characteristics of a controlled object caused by the action of uncontrolled (and unpredicted) internal and external disturbances and also by the errors of final-control elements (actuators).

The control that employs the feedback principle, will be designated as *feedback control* or *F-control.* It is more flexible compared to P-control. If the phase vector $x(s)$ can be measured accurately on the interval of motion $s \in [0, t]$, i.e., from the initial moment $s = 0$ up to the current moment t, then the optimal F-control u

must be constructed as a functional $u = u(t, x_t)$. Here x_t denotes the segment of the trajectory $x(s)$, $0 \leqslant s \leqslant t$. It should be noted that as a result of the substitution of this F-control in the system (1.1), it is described by the *equation with aftereffect*

$$dx(t) = a\big(t, x(t), u(t, x_t)\big)\,dt + \sigma\big(t, x(t), u(t, x_t)\big)\,d\xi(t).$$

In some situations an optimal F-control chosen from among the functionals $u(t, x_t)$ is realized as a function $u\big(t, x(t)\big)$ depending only on current time t and the value of the phase coordinates. Such F-controls are called *Markov controls*.

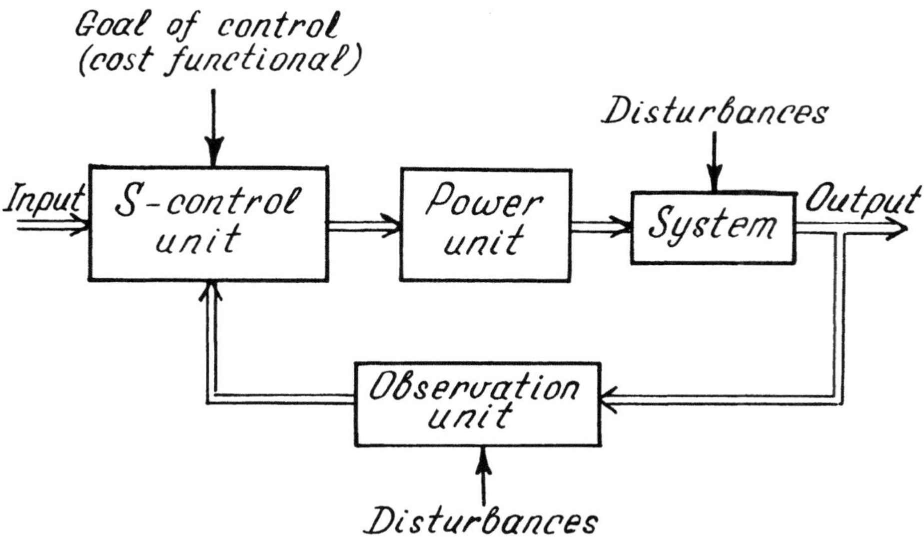

FIG. 1.1. Block-diagram of F-control

In a number of practical situations it is impossible to carry out an accurate measurement of the phase vector $x(t)$, but some other vector $y(t)$ depending on $x(t)$ can be measured accurately. A typical assumption is that the measurement data (observations) are connected with the vector $x(t)$ by a linear relation of the form

$$dy(t) = Q(t)x(t)\,dt + \sigma_0(t)\,d\xi_0(t), \qquad y(0) = 0. \tag{1.5}$$

Here $Q(t)$ is an $l \times n$ matrix defining the structure of measurement data, the matrix $\sigma_0(t)$ determines the measurement accuracy, and the vector $\xi_0(t)$ of arbitrary dimension represents noise (disturbance) in the measurement channel. The observed vector $y(t)$ contains all the information available about the system (1.1); thus, the control is constructed as the functional $u(t, y_t)$, where $y_t = y(s)$, $0 \leqslant s \leqslant t$ (see Figure 1.1).

This problem statement for the optimal control of stochastic systems includes both problems with optimal program control and problems involving the feedback optimal control. Indeed, if the measurement data (1.5) do not possess any information about the state of the system (1.1), then the optimal control is P-control, since it will be a function of time t solely. It should be noted that this is possible in the following two cases:

either the intensity σ_0 of noise in the observation channel is infinitely greater the legitimate signal Qx,

or $Q \equiv 0$ and the noises ξ and ξ_0 are statistically independent.

The problems of the determination of an optimal F-control are obtained if we insert $\sigma_0 \equiv 0$ in equation (1.5) and assume the matrix $Q(t)$ to be nondegenerate for all t. Then the phase vector $x(t)$ can be reconstructed exactly from the observation data $y(t)$. Hence, in this case the optimal control can be sought as a functional $u(t, x_t)$.

Other problem statements for optimal control rely on the possibility of the variation of either the measurement matrix $Q(t)$ or the measurement accuracy defined by the matrix $\sigma_0(t)$. If the purpose of the variation of the matrices $Q(t)$ and $\sigma_0(t)$ is to minimize the estimation error of the phase vector $x(t)$, then such problems are called *optimizaion of observation problems.* Problems in which we can simultaneously vary the control $u(t, y_t)$, the measurement matrix $Q(t)$, and the accuracy matrix $\sigma_0(t)$ are called *problems of optimally combining control and observation.* It should be noted that optimal P- and F-controls in stochastic systems, differ, in general, from one another (in the sense that they furnish different values to the cost functional) as opposed to the deterministic case where optimal P- and F-controls are equivalent (see § 2 of Chapter III).

3. Cost functional. The goal of control is usually formalized by assignment of a cost functional. For stochastic systems the cost functional, as a rule, is the expectation of given functionals depending on the control and the trajectory of the system. In other words, an optimal control displays its optimality with respect to the set of realizations of the control system, rather than to any particular realization (trajectory).

Let us give a few examples. First, we assume as given the time interval $[0, T]$ on which the system functions. In that case the cost functional can be written in the form

$$J(u) = E\left[F(x(T)) + \int_0^T F_1(t, x(t), u)\, dt\right], \tag{1.6}$$

where F and F_1 are prescribed functions, i.e.,

$$F\colon R^n \to R^1, \qquad F_1\colon [0, \infty] \times R^n \times R^m \to R^1.$$

As in the deterministic case, the control problem (1.1), (1.6) for $F = 0$ is called the *Lagrange problem*, for $F_1 = 0$ the *Mayer problem*, and in the general case the *Boltz problem.* The control problem for the system (1.1), (1.6) can always be reduced to a Mayer problem (one of minimizing the terminal functional) if we introduce a new phase variable $x_{n+1}(t)$ by the relations

$$\dot{x}_{n+1}(t) = F_1\big(t, x(t), u\big), \qquad x_{n+1}(t_0) = 0. \tag{1.7}$$

Then $x_{n+1}(t)$ is equal to the second summand in the expression (1.6), and the cost functional (1.6) becomes

$$J_1 = E\big[F\big(x(T)\big) + x_{n+1}(T)\big],$$

i.e., the original control problem (1.1), (1.6) has been reduced to a Mayer control problem for the system (1.1), (1.7) with cost functional J_1.

In a number of problems we are interested in the probability that the system reaches the given target set G. The appropriate cost functional can be obtained from (1.6). We set $F_1 = 0$ and $F(x) = 1$ for $x \in G$ and $F(x) = 0$ for $x \in R^n \setminus G$, where G is a given set in R^n. Then the functional (1.6) is equal to the probability of the system (1.1) reaching the target set G at a finite moment T. If the system functions normally only when its coordinates belong to the set G during the time of motion, then the cost functional can be chosen thus:

$$P\{x(t) \in G,\ 0 \leqslant t \leqslant T\}. \tag{1.8}$$

Here P denotes the probability of the event enclosed in braces.

Consider now some problem statements where the terminal moment is a random variable specified by a particular trajectory. Let $\tau_x(u)$ denote the *first passage time* which it takes the system (1.1) to reach the target set G, given the control u and the initial condition $x(0) = x$. The *stochastic time-optimal problem* is to minimize with respect to u the functional $E\tau_x(u)$ being the mean first passage time to the target set G. The problem of minimizing the functional

$$E\int_0^{\tau_x(u)} F_1\big(t, x(t), u\big)\, dt \tag{1.9}$$

is a generalization of the problem stated above. In the problems where it is required to hold the system within the set G we can also take, along with the relation (1.8), the functional $E\tau_x(u)$, which is to be maximized in such problems.

If the motion of the system (1.1) occurs on an unbounded time interval, then problems can be characterized by the cost functional

$$E\int_0^{\infty} F_1\big(t, x(t), u\big)\, dt. \tag{1.10}$$

These problems are often called *optimal stochastic stabilization problems.* This designation is justified by the fact that under certain conditions optimal control in the problem (1.1), (1.10) makes the system (1.1) asymptotically stable.

§ 2. Dynamic Programming Approach

One of the main approaches to constructing control in systems under random disturbances is *Bellman's dynamic programming approach.* Under this approach the control is synthesized on the basis of a certain partial differential equation called the *Bellman equation.* It should be noted that this approach is particularly efficient for systems without aftereffect, whose optimal control is sought in the class of Markov controls.

1. The Bellman function. For each of the problems stated in § 1 we can construct the corresponding scalar Bellman function $V(t,x)$. This function represents the minimum or maximum value of the cost functional depending on the problem in hand, subject to the condition that the motion of the system (1.1) starts from the state x at time $t \geqslant 0$.

Let us describe in more detail the Bellman function $V(t,x)$ by the example of the problem (1.1), (1.3), (1.6), in which the optimal F-control is sought in the form $u\big(t,x(t)\big)$. An optimal F-control is said to be *admissible* if the function $u(t,x)$, $0 \leqslant t \leqslant T$, $x \in R^n$, satisfies the constraint (1.3), and for this control there exists a solution of the problem (1.1), (1.2) for any $x_0 \in R^n$.

Let $X_{t,x}(s,u)$, $s \geqslant t$, denote the solution generated by an F-control $u = u\big(s,x(s)\big)$ of equation (1.1) with initial condition $X_{t,x}(t,u) = x$.

The Bellman function $V(t,x)$ in the problem (1.1), (1.3), (1.6) is defined by the relation

$$V(t,x) = \inf_u E\Big[F\big(X_{t,x}(T,u)\big) + \int_t^T F_1\Big(s, X_{t,x}(s,u), u\big(s, X_{t,x}(s,u)\big)\Big)\,ds\Big]. \tag{2.1}$$

Here the infimum is calculated over the set U_a of all admissible controls. It follows from the definition (2.1) of the function $V(t,x)$ that at the terminal moment the following boundary condition holds:

$$V(T,x) = F\big(X_{T,x}(T,u)\big) = F(x), \qquad u \in R^n. \tag{2.2}$$

The Bellman function is defined in a similar way also for other problems stated in § 1. Thus, for example, the function $V(t,x)$ for the stochastic time-optimal problem (1.1), (1.3), (1.9) is of the form

$$V(t,x) = \inf_u E\tau_{t,x}(u), \qquad u \in U_a, \tag{2.3}$$

where $\tau_{t,x}(u)$ is the first passage time which it takes the solution $X_{t,x}(s,u)$, $s \geqslant t$, to reach the target set G.

Example 2.1. Let us construct the Bellman function for the scalar noncontrollable system

$$dx(t) = ax(t)\,dt + \sigma\,d\xi(t), \qquad 0 \leqslant t \leqslant T,\ x \in R^1,\ \xi \in R^1, \tag{2.4}$$

where a and σ are some constants.

In the functional (1.6) we set the functions $F(x) = x^2$, $F_1 = 0$. Then according to (2.1) we have $V(t,x) = EX_{t,x}^2(T)$, where $X_{t,x}(s)$ is a solution of equation (2.4) with initial condition $X_{t,x}(t) = x$ for $s \geqslant t$. On the strength of the Ito formula (3.12) of Chapter II we have

$$X_{t,x}(s) = xe^{a(s-t)} + \int_t^s e^{a(\tau-t)}\sigma\, d\xi(\tau).$$

Hence, by virtue of the properties of the Ito stochastic integral, we can write

$$\begin{aligned} V(t,x) = EX_{t,x}^2(T) &= x^2 e^{2a(T-t)} + \int_t^T e^{2a(T-t)}\sigma^2\, d\tau \\ &= x^2 e^{2a(T-t)} + \frac{\sigma^2}{2a}[e^{2a(T-t)} - 1]. \end{aligned}$$

Example 2.2. Consider the scalar controllable system

$$dx(t) = \big(u + ax(t)\big)\, dt + \sigma\, d\xi(t), \qquad 0 \leqslant t \leqslant T, \tag{2.5}$$

where a and σ are constants, the control $u \in R^1$ is constrained by $|u| \leqslant 1$. In the cost functional (1.6) to be minimized the functions $F(x) = x$, $F_1 = 0$. It is readily seen that in this problem the optimal control is $u_0 \equiv -1$, i.e., it is equal to a minimum admissible value. Hence, the Bellman function $V(t,x)$ is of the form

$$V(t,x) = EX_{t,x}(T,-1).$$

It follows from equation (2.5) for $s \geqslant t$, for the control $u \equiv -1$, and for the initial condition $X_{t,x}(t_1,-1) = x$ that

$$X_{t,x}(s,-1) = xe^{a(s-\tau)} - \int_t^s e^{a(s-\tau)}\, d\tau + \int_t^s e^{a(s-\tau)}\sigma\, d\xi(\tau).$$

Hence,

$$V(t,x) = xe^{a(T-t)} - \int_t^T e^{a(T-\tau)}\, d\tau = xe^{a(T-t)} - \frac{1}{a}[e^{a(T-t)} - 1].$$

2. The Bellman equation. We shall deduce the Bellman equation, which constitutes the necessary condition for optimal control. It can serve to determine the Bellman function $V(t,x)$ and the optimal F-control. The deduction of the Bellman equation is based on the *principle of dynamic programming*, which is stated as follows:

PRINCIPLE OF DYNAMIC PROGRAMMING. Let $u_0(t,x)$ be an optimal F-control on the whole interval of motion $[0,T]$ and for all $x \in R^n$. Then the control u_0 has the property that whatever the moment $s \in [0,T]$ and the admissible *control v on the segment $[0,s]$ are, the control u_0 must constitute an optimal control on the segment $[s,T]$ with regard to the state $X_{0,x_0}(s,v)$ resulting from the control v on the segment $[0,s]$.*

Let $P_u(t,x,s,A)$ denote the transition probability function for the Markov process $X_{t,x}(s,u)$, where A is an arbitrary Borel set in the σ-algebra of the subsets of the space R^n. The transition probability function is defined by the expression

$$P_u(t,x,s,A) = P\{X_{t,x}(s,u) \in A\}.$$

Hence, noting that the process X is Markovian, we have

$$\begin{aligned} P_u(t,x,s,A) &= \int_{R^n} P\{(X_{t,x}(\tau,u) \in dz) \cap (X_{t,x}(s,u) \in A)\} \\ &= \int_{R^n} P(X_{t,x}(\tau,u) \in dz) P(X_{\tau,z}(s,u) \in A), \qquad t \leqslant \tau \leqslant s. \end{aligned}$$

The resulting equation for the transition probability function

$$P_u(t,x,s,A) = \int_{R^n} P(X_{t,x}(\tau,u) \in dz) P(X_{\tau,z}(s,u) \in A), \qquad t \leqslant \tau \leqslant s, \tag{2.6}$$

is called the *Kolmogorov-Chapman equation.*

Consider now the control problem (1.1), (1.3), (1.6) for $F_1 \equiv 0$. It follows from the definition (2.1) of the Bellman function that

$$V(t,x) = \inf_v \int_{R^n} F(y) P_v(t,x,T,dy), \tag{2.7}$$

where the infimum is taken over the set of all admissible controls v of the problem (1.1), (1.3), (1.6) on the segment $[t,T]$. Take a number $\Delta \geqslant 0$ such that $t+\Delta \leqslant T$. Let u denote an admissible control on the segment $[t, t+\Delta]$, and v_1 an admissible control on the segment $[t+\Delta, T]$. Then, on the strength of relations (2.6), (2.7) and the principle of dynamic programming we obtain

$$\begin{aligned} V(t,x) &= \inf_u \inf_{v_1} \int_{R^n}\int_{R^n} P_u(t,x,t+\Delta,dz) P_{v_1}(t+\Delta,z,T,dy) F(y) \\ &= \inf_u \int_{R^n} P_u(t,x,t+\Delta,dz) V(t+\Delta,z) \\ &= \inf_u EV(t+\Delta, X_{t,x}(t+\Delta,u)). \end{aligned}$$

Hence,

$$\inf_u E[V(t+\Delta, X_{t,x}(t+\Delta,u)) - V(t,x)] = 0. \tag{2.8}$$

Further, we assume that the function $V(t,x)$ is twice continuously differentiable with respect to x and once with respect to t. Then it follows from the Ito formula for small values of $\Delta \geqslant 0$ that the following relation holds:

$$E[V(t+\Delta, X_{t,x}(t+\Delta,u)) - V(t,x)] = L_u V(t,x) + O(\Delta). \tag{2.9}$$

Here the generator L_u for equation (1.1) is defined by the relation

$$L_u V(t,x) = V_t(t,x) + V_x'(t,x)a(t,x,u) + \tfrac{1}{2}\operatorname{tr}[\sigma(t,x,u)\sigma'(t,x,u)V_{xx}(t,x)],$$

where

$$V_t = \frac{\partial V(t,x)}{\partial t},\ V_x = \left\| \frac{\partial V(t,x)}{\partial x_i} \right\|,\ V_{xx} = \left\| \frac{\partial^2 V}{\partial x_i\,\partial x_j} \right\|,\ i,j = 1,\dots,n,$$

and tr is the trace of a matrix. Substituting (2.9) in (2.8) and passing to the limit as $\Delta \to +0$, we obtain the Bellman equation

$$\inf_{u\in U} L_u V(t,x) = 0, \qquad x \in R^n,\ 0 \leqslant t \leqslant T. \tag{2.10}$$

The boundary condition for this equation is of the form (2.2), i.e., $V(T,x) = F(x)$. It should be stressed that the infimum in (2.10) is taken over u which is considered as a parameter from the set $U \subset R^m$.

In a similar way, we can obtain the Bellman equation for the problem (1.1), (1.3), (1.6), where the function F_1 is arbitrary. The resulting Bellman equation is of the form

$$\inf_{u\in U}[L_u V(t,x) + F_1(t,x,u)] = 0, \qquad x \in R^n,\ 0 \leqslant t \leqslant T, \tag{2.11}$$
$$V(T,x) = F(x).$$

We shall give a few more Bellman equations for some other problems. Thus, for the stochastic time-optimal problem (1.1), (1.3), (1.9) we have the following Bellman equation:

$$\inf_{u\in U} L_u V(t,x) = -1, \qquad t \geqslant 0,\ x \in R^n \setminus G,$$
$$V(t,x) = 0,\ x \in G.$$

For the problem of maximizing the stochastic cost functional (1.8) along the trajectories of the system (1.1) the appropriate Bellman equation is of the form

$$\sup_{u\in U} L_u V(t,x) = 0, \qquad 0 \leqslant t \leqslant T,\ x \in G,$$
$$V(t,x) = 0,\ 0 \leqslant t \leqslant T,\ x \in R^n \setminus G,$$
$$V(T,x) = 1,\ x \in G.$$

For modeling the disturbances acting on the system, some other processes besides the Wiener process are used. So, for example, if there are impulse (shock) disturbances, then these disturbances can be modeled by means of a Poisson process $\eta(t)$. The process $\eta(t) \in R^r$ has components $\eta_1, \dots, \eta_r$, where $\eta_i(t)$ are scalar stochastically independent *Poisson processes* with parameter $\lambda_i(t)$, $i = 1, \dots, r$. Moreover, the processes $\eta(t)$ and $\xi(t)$ are statistically independent. The equations of the system are of the form

$$\begin{aligned} dx(t) = a\big(t, x(t), u\big)\,dt + \sigma\big(t, x(t), u\big)\,d\xi(t) \\ + \sigma_1\big(t, x(t), u\big)\,d\eta(t), \qquad t \geqslant 0. \end{aligned} \tag{2.12}$$

The generator L_u for the process (2.12) that corresponds to the admissible F-control $u\big(t, x(t)\big)$ is the following:

$$\begin{aligned} L_u V = V_t(t,x) + a'(t,x,u)V_x(t,x) \\ + \tfrac{1}{2}\,\mathrm{tr}[\sigma(t,x,u)\sigma'(t,x,u)V_{xx}(t,x)] \\ + \sum_{i=1}^{r} \lambda_i(t)\big[V\big(t, x + \sigma_1(t,x,u)e_i\big) - V(t,x)\big]. \end{aligned} \tag{2.13}$$

Here the vector $e_i \in R^r$ has unity as the ith component, and all the other components of the vector e_i are zero. For the control system (2.12) the same problem statements are valid as for the system (1.1). Moreover, the corresponding Bellman equation does not change form, and its generator L_u is expressed by formula (2.13).

Finally, we shall write the Bellman equation for the problem of optimally combining control and observation. Suppose that an object is described by the system of linear (in phase coordinates) equations of the form (1.1)

$$\begin{aligned} dx(t) = [A(t,u)x(t) + B(t,u)]\,dt + \sigma(t,u)\,d\xi(t), \qquad 0 \leqslant t \leqslant T, \\ x(0) = x_0. \end{aligned} \tag{2.14}$$

The coordinates $x(t)$ of the system (2.14) cannot be measured accurately, but the vector $y(t) \in R^l$ can be directly observed. It is described by the relation that is linear in x:

$$dy(t) = Q(t,u)x(t) + \sigma_0(t,u)\,d\xi_0(t), \qquad t \geqslant 0,\ y(0) = 0. \tag{2.15}$$

In equations (2.14), (2.15) $\xi(t)$ and $\xi_0(t)$ denote standard Wiener processes of arbitrary dimension; the random processes $\xi(t)$, $\xi_0(t)$, and x_0 are statistically independent for all $0 \leqslant t \leqslant T$; the random vector $x_0 \in R^n$ has a nondegenerate Gaussian probability distribution with given mean and covariance matrix D_0.

We are looking for the optimal control as a functional $u(t, y_t)$ such that the constraint (1.13) is satisfied and the cost functional $EF\big(x(T)\big)$ minimized. Suppose that the matrix $\sigma_0(t,u)\sigma_0'(t,u)$ is positive definite uniformly in t. Let $m(t)$ and $D(t)$ denote the mean vector and the covariance matrix, respectively, of the conditional distribution of $x(t)$, given y_t. Note [2, 4, 5, 6, 12] that this conditional distribution is Gaussian, with $m(t)$ and $D(t)$ satisfying the relations

$$\begin{aligned}\dot{m}(t) &= A(t,u)m(t) + B(t,u) \\ &\quad + D(t)Q'(t,u)[\sigma_0(t,u)\sigma_0'(t,u)]^{-1} \\ &\quad \times [y(t) - Q(t,u)m(t)], \qquad m(0) = m_0, \end{aligned} \tag{2.16}$$

$$\begin{aligned}\dot{D}(t) &= A(t,u)D(t) + D(t)A'(t,u) \\ &\quad - D(t)Q'(t,u)[\sigma_0(t,u)\sigma_0'(t,u)]^{-1} \\ &\quad \times Q(t,u)D(t) + \sigma(t,u)\sigma'(t,u), \qquad D(0) = D_0. \end{aligned} \tag{2.17}$$

The difference $y(t) - Q(t,u)m(t)$ is the so-called *innovation process*. It is a Gaussian white noise with covariance matrix $\sigma_0(t,u)\sigma_0'(t,u)\delta(t-s)$, where $\delta(t)$ is a delta-function.

It follows from the foregoing and equations (2.16), (2.17) that the Bellman function for the problem in hand is of the form $V = V(t,m,D)$. It depends on $1+n+n(n+1)/2 = (n+1)(n+2)/2$ arguments (on the scalar argument t, on the vector m of dimension n and on the symmetric $n \times n$ matrix D). The Bellman equation can be written in the form

$$\begin{aligned} V_t(t,m,D) &+ \inf_{u\in U}\Big\{ \big(A(t,u)m + B(t,u)\big)' V_m(t,m,D) \\ &+ \operatorname{tr}\big[V_D(t,m,D)\big(A(t,u)D + DA'(t,u) \\ &- DQ'(t,u)\big(\sigma_0(t,u)\sigma_0'(t,u)\big)^{-1}Q(t,u)D + \sigma(t,u)\sigma'(t,u)\big)\big] \\ &+ \tfrac{1}{2}DQ'(t,u)\big(\sigma_0(t,u)\sigma_0'(t,u)\big)^{-1}Q(t,u)DV_{mm}(t,m,D)\Big\} = 0. \end{aligned} \tag{2.18}$$

The boundary condition for equation (2.18) is expressed by

$$V(T,m,D) = [(2\pi)^n \det D]^{-1/2} \int_{R^n} F(x) \exp\Big[-\tfrac{1}{2}(x-m)'D^{-1}(x-m)\Big]\, dx.$$

In equation (2.18) V_D denotes the matrix with elements $\partial V/\partial d_{ij}$, where d_{ij} are the elements of the matrix D, and V_{mm} is the matrix with elements $\partial^2 V/(\partial m_i \partial m_j)$, $i,j = 1,\dots,n$.

3. Connection between the Bellman function and the Bellman equation. When using the Bellman equation the following questions arise:

1°) Does the Bellman function always satisfy the Bellman equation for the problem under consideration?

2°) Is the control obtained through the use of the Bellman equation an optimal control?

3°) Does the solution of the Bellman equation always coincide with the Bellman function?

In the general case the answer to these three questions is negative for the following reasons:

1°. In general, the Bellman function $V(t, x)$ does not possess the kind of smoothness in t and x that was used when deducing the Bellman equation. In other words, the Bellman function does not always satisfy the Bellman equation for the problem in hand.

2°. Even if the Bellman equation has a smooth solution, the corresponding control may not be an optimal one, since it may not be admissible for various reasons; for example, for this controlthere may be no solution to the stochastic equation of motion. Moreover, there does not always exist a control in the prescribed class of admissible controls such that it renders the infimum of the cost functional.

3°. The solution of the Bellman equation may not be unique. In this case an additional investigation is often required to establish which of the solutions available can play the role of the Bellman function for the original optimal control problem.

These features must be kept in mind when dealing with specific control problems. At the same time, for a number of problems the Bellman equation makes it possible to obtain both the optimal F-control and the associated minimum value of the cost functional.

For example, on a finite time interval $[0, T]$ let us consider the control problem (1.1) with cost functional (1.6), subject to the control constraints (1.3). The Bellman equation that is associated with the problem (1.1)–(1.3), (1.6) has the form (2.11). Here we give sufficient conditions such that the boundary-value problem (2.11) furnishes the solution of the original optimal control problem (1.1)–(1.3), (1.6).

Theorem 2.1. *Let the boundary-value problem* (2.11) *have a unique solution that is continuously differentiable once with respect to t and twice with respect to x, with*

$$\left|\left(\frac{\partial V(t,x)}{\partial x}\right)' \sigma(t,x,u)\right| \leqslant C < \infty$$

uniformly in $t \in [0,T]$, $x \in R^n$, $u \in U$. Further, let the F-control $u_0(t,x)$ that minimizes the expression (2.11) *be admissible. Then the control $u_0(t,x)$ is optimal, and the minimum value of the cost functional* (1.6) *is $J(u_0) = V(0, x_0)$.*

Proof. Let $x_0(t)$ denote the solution of equation (1.1) generated by the control u_0 for the deterministic initial condition (1.2). The solution $x_0(t)$ exists, since $u_0(t,x)$ is an admissible control. Applying the Ito formula to the function $V(t, x_0(t))$ gives

$$dV\big(t, x_0(t)\big) = L_{u_0} V\big(t, x_0(t)\big)\, dt + \left(\frac{\partial V\big(t, x_0(t)\big)}{\partial x}\right)' \sigma\big(t, x_0(t), u_0\big)\, d\xi(t). \tag{2.19}$$

Hence, noting the Bellman equation (2.11), we can write

$$dV\big(t, x_0(t)\big) = -F_1\big(t, x_0(t), u_0\big)\, dt + \left(\frac{\partial V\big(t, x_0(t)\big)}{\partial x}\right)' \sigma\big(t, x_0(t), u_0\big)\, d\xi(t). \tag{2.20}$$

Next, we integrate both sides of (2.20) with respect to t from $t = 0$ to $t = T$ and calculate the expectation of the resulting expression. By virtue of the boundary condition $V\big(T, x_0(T)\big) = F\big(x_0(T)\big)$ we obtain

$$EF\big(x_0(T)\big) - V(0, x_0) = -E\int_0^T F_1\big(t, x_0(t), u_0\big)\, dt. \tag{2.21}$$

Thus, we have established the equality

$$J(u_0) = V(0, x_0). \tag{2.22}$$

Now let u be any admissible control and $x(t)$ be the corresponding solution of the problem (1.1), (1.2). Then, as in the proof of (2.19), we have

$$dV\big(t, x(t)\big) = L_u V\big(t, x(t)\big)\, dt + \left(\frac{\partial V\big(t, x(t)\big)}{\partial x}\right)' \sigma\big(t, x(t), u\big)\, d\xi(t).$$

Hence, in view of equation (2.11), we can write

$$dV\big(t, x(t)\big) \geqslant -F_1\big(t, x(t), u\big)\, dt + \left(\frac{\partial V\big(t, x(t)\big)}{\partial x}\right)' \sigma\big(t, x(t), u\big)\, d\xi(t).$$

Integrating both sides of this inequality with respect to t from $t = 0$ to $t = T$ and calculating the expectation of the resulting expressions in (2.21) gives

$$EF\big(x(T)\big) - V(0, x_0) \geqslant -E\int_0^T F_1\big(t, x(t), u\big)\, dt.$$

Thus, for any admissible control u we have

$$J(u) \geqslant V(0, x_0).$$

Hence, in view of (2.22), the optimality of the control u_0 is established. □

Thus, the dynamic programming approach is sometimes not only a necessary but also a sufficient condition for optimal control.

§ 3. Stochastic Linear-Quadratic Problem on a Finite Time Interval

In this section we consider the application of the dynamic programming approach to the problem of optimal control for a linear stochastic system with a quadratic cost functional to be minimized. This problem is called a *stochastic LQ-problem.*

1. Linear-quadratic problems in the case of an accurate measurement of phase coordinates. Suppose that the control system (1.1) is linear in the coordinates and the control, and has the form

$$dx(t) = [A(t)x(t) + f(t) + B(t)u]\,dt + \sigma_1(t)x(t)\,d\xi_1(t) + \sigma_2(t)u\,d\xi_2(t) + \sigma(t)\,d\xi(t), \qquad 0 \leqslant t \leqslant T,\ x(0) = x_0. \tag{3.1}$$

Here $x \in R^n$ is a vector, $u \in R^m$ is a control, the scalar standard Wiener processes ξ_1, ξ_2 and the standard Wiener process $\xi(t) \in R^l$ are statistically independent, the elements of the matrices A (of dimensions $n \times n$), B $(n \times m)$, σ_1 $(n \times n)$, σ_2 $(n \times m)$, σ $(n \times l)$ are σ-measurable bounded functions, and the function $f(t) \in R^n$ with measurable bounded elements is prescribed. We assume that the phase coordinates $x(t)$ at any time instant can be measured accurately. The F-control of the system (3.1) is sought in the class of Markov controls (i.e., as a function $u(t, x(t))$) through the use of the condition for the maximum of the quadratic cost functional

$$J(u) = E\Big[x'(T)N_1x(T) + 2h_1'x(T) + \int_0^T (x'(t)N_2(t)x(t) + u'(t)N_3(t)u(t) + 2x'(t)N_4(t)u(t) + 2h_2'(t)x(t) + 2h_3'(t)u(t))\,dt\Big]. \tag{3.2}$$

Here $N_1 \geqslant 0$, $N_2(t) \geqslant 0$, $N_3(t) > 0$. The elements of the matrices N_2, N_3, N_4 and of the vectors $h_2 \in R^n$, $h_3 \in R^m$ are given measurable bounded functions.

The Bellman equation (2.11), in view of (3.1), (3.2), becomes

$$\inf_u [L_uV(t,x) + \lambda(t,x,u)] = 0, \qquad 0 \leqslant t \leqslant T,\ x \in R^n, \tag{3.3}$$

$$L_uV = V_t + V_x'(Ax + f + Bu) + \tfrac{1}{2}(x'\sigma_1'V_{xx}\sigma_1x + u'\sigma_2'V_{xx}\sigma_2u + \operatorname{tr}[\sigma\sigma'V_{xx}]),$$

$$\lambda = x'N_2x + u'N_3u + 2x'N_4u + 2h_2'x + 2h_3'u.$$

By virtue of (3.2), the boundary condition for equation (3.3) is expressed by

$$V(T,x) = x'N_1x + 2h_1'x. \tag{3.4}$$

In equation (3.3) the expression

$$(V_x'B + u'N_3 + 2x'N_4 + 2h_3' + \tfrac{1}{2}u'\sigma_2'V_{xx}\sigma_2)u \tag{3.5}$$

depends on the choice of the parameter $u \in R^m$. This expression is a quadratic form in u. If the matrix $2N_3 + \sigma_2' V_{xx} \sigma_2 > 0$, then the quadratic form (3.5) is positive-definite and has a unique minimum with respect to u, which is rendered by the control

$$u_0(t,x) = -\big(2N_3(t) + \sigma_2'(t) V_{xx}(t,x) \sigma_2(t)\big)^{-1} \times \big(2h_3 + 2N_4'(t)x + B'(t)V_x(t,x)\big). \tag{3.6}$$

To obtain formula (3.6), it is necessary to differentiate the expression (3.5) with respect to u and equate the derivative to zero. In the deduction the following relations should be used:

$$\frac{\partial}{\partial u} u' N_3 u = 2N_3 u, \qquad \frac{\partial}{\partial u} h_3' u = h_3, \qquad \frac{\partial}{\partial u} x' N_4 u = N_4' x.$$

Let us clarify the meaning of these relations. On differentiating a scalar function with respect to a vector argument, we obtain a column vector. Hence, for example, the vector $\partial(x' N_4 u)/\partial u$ has as the ith component the expression

$$\frac{\partial}{\partial u_i} \sum_{j,k} x_j N_{jk} u_k = \sum_j x_j N_{ji},$$

where N_{jk} are the elements of the matrix N_4. Thus, $\partial(x' N_4 u)/\partial u = N_4' x$.

The solution of the boundary-value problem (3.3), (3.4) is sought in the form

$$V(t,x) = x' P(t) x + 2G'(t)x + g(t). \tag{3.7}$$

Here the $n \times n$ matrix $P(t) \geqslant 0$, the vector $G(t) \in R^n$, and the scalar function $g(t)$ must be chosen so that equation (3.3) and the boundary conditions (3.4) are satisfied. It should be noted that if $V(t,x)$ has the form (3.7), then the quadratic form (3.5) is positive definite with respect to u and, therefore, in view of (3.6), the optimal control u_0 is determined by the expression

$$u_0(t,x) = -K(t,P)\big[B'(t)P(t)x + N_4'(t)x + B'(t)G(t) + h_3(t)\big], \tag{3.8}$$
$$K(t,P) = \big(N_3(t) + \sigma_2'(t) P(t) \sigma_2(t)\big)^{-1} > 0.$$

Since the control u_0 minimizes the expression (3.5), it can be shown by virtue of equation (3.3) that the following relation holds:

$$L_{u_0} V(t,x) + \lambda(t,x,u_0) = 0. \tag{3.9}$$

Let us substitute in (3.9) the expressions (3.7), (3.8) for V and u_0. As a result, we obtain an identity which holds for all $x \in R^n$, $0 \leqslant t \leqslant T$; the left side of this identity is a quadratic form in x, and the right side is zero. Therefore, this identity holds if and only if all coefficients of the quadratic form are zero. Let us equate the coefficients of the summands quadratic in x of this form to zero, then we obtain the following equations:

$$\begin{gathered}\dot{P}(t)+A'(t)P(t)+P(t)A(t)\\ +\big(P(t)B(t)+N_4(t)\big)K(t,P)\big(P(t)B(t)+N_4(t)\big)'\\ +N_2(t)+\sigma_1'(t)P(t)\sigma_1(t)=0,\qquad 0\leqslant t\leqslant T,\\ P(T)=N_1.\end{gathered}\tag{3.10}$$

In a similar way, equating to zero the coefficients of the summands of the quadratic form that are linear in x, it follows that

$$\begin{gathered}\dot{G}(t)+A'(t)G(t)+P(t)f(t)+h_2(t)+\big(N_4(t)+P(t)B(t)\big)K(t,P)\\ \times\big(h_3(t)+B'(t)G(t)\big)=0,\qquad 0\leqslant t\leqslant T,\\ G(T)=h_1.\end{gathered}\tag{3.11}$$

Finally, equating to zero the constant term of the quadratic form, we determine the scalar function $g(t)$ in expression (3.7) as a quadrature

$$\begin{aligned}g(t)=\int_t^T&\big\{2G'(s)f(s)+\big(h_3'(s)+G'(s)B(s)\big)K(s,P)\\ &\times\big(h_3(s)+B'(s)G(s)\big)+\operatorname{tr}[P(s)\sigma(s)\sigma'(s)]\big\}\,ds.\end{aligned}\tag{3.12}$$

The obtained relations (3.10)–(3.12) serve to determine the coefficients of the Bellman function (3.7) for the optimal control (3.8). Let us consider the nonlinear equation (3.10) for the matrix $P(t)$. Under the assumptions made about the parameters of the system (3.1) and the functional (3.2) there exists a unique matrix $P(t)\geqslant 0$ that satisfies the relations (3.10). To calculate this matrix, we can use the method of successive approximations. This method is, essentially, an extension of Newton's method of tangents modified by R. Bellman for optimal control problems. For the case being considered the method consists in the following. Take an arbitrary control u_0 that is linear in x. Further, evaluate the cost functional (3.2) associated with this control. It coincides with the solution V_1 of the linear equation (3.9) with boundary condition (3.4). The next approximation u_1 to the optimal control is defined by the equations

$$\begin{gathered}L_{u_0}V_1(t,x)+\lambda(t,x,u_0)=0,\\ \inf_u[L_uV_1(t,x)+\lambda(t,x,u)]=L_{u_1}V_1(t,x)+\lambda(t,x,u_1)=0.\end{gathered}$$

Continuing with this procedure, we can construct the sequences of functions $V_i(t,x)$ and controls u_i. It turns out that all controls u_i $(i\geqslant 1)$ are defined by formula (3.6) whose right side contains $V_i(t,x)$ instead of $V(t,x)$. All functions $V_i(t,x)$ $(i\geqslant 1)$ are defined by formula (3.7) which in place of P, G, g involves P_i, G_i, g_i, respectively. The functions $G_i(t)$ and $g_i(t)$ are defined by the relations

(3.11), (3.12), in which $P(t)$ is replaced everywhere by $P_i(t)$. Finally, the matrices $P_i(t)$, $i \geqslant 1$, are the solution of the Cauchy problem for the system of linear equations

$$\begin{gathered}\dot{P}_i(t) + A'P_i + P_iA + N_2 + \sigma_1'P_i\sigma_1 - (N_4 + P_{i-1}B)K(t, P_{i-1})K^{-1}(t, P_i) \\ \times K(t, P_{i-1})(N_4 + P_{i-1}B)' + (P_iB + N_4)K(t, P_{i-1})(N_4 + P_{i-1}B)' \\ + (P_{i-1}B + N_4)K(t, P_{i-1})(N_4 + P_iB)' = 0, \qquad i \geqslant 1, \\ P_i(T) = N_1, \; P_0(t) = N_1, \; 0 \leqslant t \leqslant T.\end{gathered}$$

Under the assumptions made, the sequence (P_i, G_i, g_i) converges to the solution of the problem (3.10)–(3.12) uniformly on $[0, T]$, with[1]

$$\|P - P_i\| + \|Q - G_i\| + \|g - g_i\| = O\left(\frac{1}{i!}\right).$$

Here $\|f(t)\|$ denotes the uniform norm of the function $f(t)$ equal to $\|f(t)\| = \sup_{0 \leqslant t \leqslant T} |f(t)|$, and $O(1/i!)$ denotes the quantities of the order of $1/i!$. The resulting solution P, Q, g of the system of equations (3.10)–(3.12) completely determines the solution of the original problem of optimal control(3.1), (3.2). Indeed, the control (3.8) being linear in x is admissible, since for this control there exists a unique solution of stochastic equations (3.1), and the solution of the Cauchy problem for the Bellman equation (3.3), (3.4) is unique in the class of functions increasing as $|x| \to \infty$ not faster than a power of $|x|$. Thus, in stochastic LQ-problems the dynamic programming approach enables construction of the optimal F-control and determination of the corresponding extremal value of the cost functional. It should be further noted that for $\sigma_1 = 0$, $\sigma_2 = 0$ equation (3.10) coincides with a matrix Riccati equation (see § 2 of Chapter IV), which occurs in the theory of deterministic LQ-problems. Therefore, in the case under study the problem (3.10) can be solved through the use of the methods of the theory of deterministic LQ-problems given in § 2 of Chapter IV.

2. Linear-quadratic problems under incomplete information. Let us consider the LQ-problem in which, as distinct from the foregoing, the phase vector $x(t) \in R^n$ cannot be measured accurately, and the observable quantity $y(t)$ is described by the relations (1.5). The equations of the system are assumed to be of the form (3.1) for $\sigma_1 = 0$, $\sigma_2 = 0$, i.e.,

$$dx(t) = [A(t)x(t) + B(t)u + f(t)]\,dt + \sigma(t)\,d\xi(t), \qquad 0 \leqslant t \leqslant T, \quad (3.13)$$
$$x(0) = x_0.$$

For the observed vector $y(t) \in R^l$ we have in view of (1.5)

$$dy(t) = Q(t)x(t)\,dt + \sigma_0(t)\,d\xi_0(t), \qquad 0 \leqslant t \leqslant T, \; y(0) = 0. \quad (3.14)$$

[1] Kolmanovskii, V. B., On the Approximation of Linear Control Systems with Aftereffect, *Problems Control Inform. Theory/Problemy Upravlen. Teor. Inform.*, (1974), vol. 3, no. 1.

The initial vector x_0 has a nondegenerate Gaussian probability distribution with given mean m_0 and covariance matrix D_0. The measurement matrix $Q(t)$ and the measurement accuracy matrix $\sigma_0(t)$ are prescribed. They have measurable bounded elements, with the matrix $\sigma_0(t)\sigma_0'(t)$ nondegenerate for all $t \in [0, T]$. Finally, $\xi(t)$ and $\xi_0(t)$ are standard independent Wiener processes of arbitrary dimension. In order to minimize the quadratic functional (3.2), the F-control is taken in the form of a functional $u(t, y_t)$, where $y_t = y(s)$, $0 \leqslant s \leqslant t$, is the aggregate of the results of the preceding measurements.

Let $m(t)$ and $D(t)$ denote the mean and the covariance matrix of the conditional distribution of $x(t)$ given y_t. The functions $m(t)$ and $D(t)$ are described by equations (2.16), (2.17). We transform the functional $J(u)$ (3.2) expressing the moment of the coordinate x in terms of m and D. For this we use the equalities

$$\begin{aligned} E\{x'(t)N_2x(t)\} &= E\{[x(t) - m(t) + m(t)]'N_2[x(t) - m(t) + m(t)]\} \\ &= E\{[x(t) - m(t)]'N_2[x(t) - m(t)]\} + E\{m'(t)N_2m(t)\} \\ &\quad + 2E\{[x(t) - m(t)]'N_2m(t)\}, \\ Ex(t) &= E\{E\{\, x(t) \mid y_t \,\}\} = Em(t), \\ E\{\, u(t, y_t) \mid y_t \,\} &= u(t, y_t). \end{aligned} \tag{3.15}$$

Further, note that

$$\begin{aligned} E\{x(t) - m(t)]'N_2m(t)\} &= E\{\, E(x(t) - m(t))'N_2m(t) \mid y_t \,\} \\ &= E\Big\{\{E(x(t) - m(t))' \mid y_t\,\}N_2m(t)\Big\} \\ &= E\big\{(m(t) - m(t))\,N_2m(t)\big\} = 0. \end{aligned}$$

Hence, in view of (3.15), we have

$$\begin{aligned} E\{x'(t)N_2x(t)\} = E\{\operatorname{tr}\big[(x(t) - m(t))\,(x(t) - m(t))'N_2\big]\} \\ + E\{m'(t)N_2x(t)\} = E\operatorname{tr}\{D(t)N_2\} + E\{m'(t)N_2m(t)\}, \end{aligned} \tag{3.16}$$

where tr is the trace of a matrix (that is equal to the sum of its diagonal elements). In a similar way, we have

$$\begin{aligned} E\{x'(T)N_1x(T)\} &= \operatorname{tr}\{D(T)N_1\} + E\{m'(T)N_1m(T)\}, \\ E\{x'(t)N_4(t)u(t, y_t)\} &= E\{E\{\, x'(t)N_4(t)u(t, y_t) \mid y_t \,\}\} \\ = E\{E\{\, x'(t) \mid y_t \,\}&N_4(t)u(t, y_t)\} = E\{m'(t)N_4(t)u(t, y_t)\}, \\ E\{h_2'(t)x(t)\} &= E\{h_2'(t)m(t)\}. \end{aligned} \tag{3.17}$$

It follows from (3.15)–(3.17) that the cost functional (3.2) can be expressed in the form $J = J_1 + J_2$, where

$$J_1 = \operatorname{tr}[D(T)N_1] + \int_0^T \operatorname{tr}[D(t)N_2(t)]\,dt,$$

$$J_2 = E\Big[m'(T)N_1m(T) + 2h_1'm(T) + \int_0^T \big(m'(t)N_2(t)m(t) \\ + 2m'(t)N_4(t)u + u'N_3(t)u + 2h_2'(t)m(t) + 2h_3'(t)u\big)\,dt\Big]. \tag{3.18}$$

However, in the problem (3.13), (3.14), (3.2) under examination the F-control is taken in the form of the functional $u(t, y_t)$. Therefore, by virtue of equation (2.17) for the covariance matrix $D(t)$, the functional J_1 does not depend on the control $u(t, y_t)$, since the coefficients A, Q, σ, and σ_0 do not depend on this control. Hence, the problem (3.13), (3.14), (3.2) has been reduced to the LQ-problem of the optimal control of the vector $m(t)$ satisfying a linear equation of the form (2.16) with the quadratic cost functional (3.18) to be minimized. It will be recalled that in equation (2.16) for $m(t)$ the difference $y(t) - Q(t)m(t)$ is a Gaussian white noise with correlation function $\sigma_0(t)\sigma_0'(t)\delta(t-s)$ (see § 2). Thus, equation (2.16) for $m(t)$ admits the representation

$$dm(t) = [A(t)m(t) + f(t) + B(t)u]\,dt + \sigma_3(t)\,d\xi_3(t), \qquad 0 \leqslant t \leqslant T, \tag{3.19}$$
$$m(0) = m_0.$$

Here $\xi_3(t) \in R^l$ is a standard Wiener process, and the matrix $\sigma_3(t)$ satisfies the equality

$$\sigma_3(t)\sigma_3'(t) = D(t)Q'(t)\big(\sigma_0(t)\sigma_0'(t)\big)^{-1}Q(t)D(t). \tag{3.20}$$

Except for the notation, the LQ-problems (3.18), (3.19) and (3.1), (3.2) for $\sigma_1 = 0$, $\sigma_2 = 0$ are equivalent. Therefore, the solution of the problem (3.18), (3.19) can be obtained from the solution (3.7)–(3.12) of the problem (3.1), (3.2) for $\sigma_1 = 0$, $\sigma_2 = 0$. The appropriate formulas for the solution of the LQ-problem (3.18), (3.19) are given below. The Bellman function $V(t, m)$ for the problem (3.18), (3.19), in view of (3.7), is of the form

$$V(t, m) = m'P(t)m + 2G'(t)m + g(t). \tag{3.21}$$

The optimal control $u_0(t, m)$ of the system (3.19) is expressed by the formula

$$u_0(t, m) = -N_3^{-1}(t)\big[\big(B'(t)P(t) + N_4'(t)\big)m + B'(t)G(t) + h_3(t)\big]. \tag{3.22}$$

In the relations (3.21), (3.22) the functions P, G, g are determined by equations (3.10)–(3.12), in which we must put $\sigma_1 = 0$, $\sigma_2 = 0$, $\sigma\sigma' = \sigma_3\sigma_3'$, where $\sigma_3\sigma_3'$ is expressed by formula (3.20).

We now return to the original control problem (3.13), (3.14), (3.2). The optimal control $u(t, m(t))$ of the system (3.13) is determined by the expression (3.22) in which, by virtue of (2.16), $m(t)$ is described by the equation

$$\begin{gathered} dm(t) = [A(t)m(t) + f(t) + B(t)u(t, m(t))]\,dt \\ + D(t)Q'(t)\big(\sigma_0(t)\sigma_0'(t)\big)^{-1}[dy(t) - Q(t)m(t)\,dt], \\ m(0) = m_0. \end{gathered} \tag{3.23}$$

The covariance matrix $D(t)$, on the strength of (2.17), satisfies the equation

$$\dot{D} = AD + DA' - DQ'(\sigma_0\sigma_0')^{-1}QD + \sigma\sigma', \qquad D(0) = D_0. \tag{3.24}$$

It follows that the matrix $D(t)$ is independent of the observations y_t. Therefore, $D(t)$, together with the functions P, G, g, can be computed before the beginning of the control process. The estimate $m(t)$ satisfies the linear (in m) equation (3.23), in which the control $u_0(t, m)$ expressed by formula (3.22) has been substituted.

The solution $m(t)$ of equation (3.23) is a linear functional of the observation data y_t. Finally, the minimum value of the cost functional (3.2) associated with the control of the system (3.13) by $u_0(t, m(t))$ is expressed by the formula

$$\min_u J = V(0, m_0) + \operatorname{tr}\big(D(T)N_1\big) + \int_0^T \operatorname{tr}\big(D(t)N_2(t)\big)\,dt. \tag{3.25}$$

It should be noted that the minimum value of the cost functional (3.2), as in the case of the precise measurement of the coordinates x, can be computed beforehand according to the formula (3.25).

3. Optimal program control in linear-quadratic problems. Consider the LQ-problem assuming that the coordinates of the system are inaccessible for measurement. In this case optimal control is sought in the class of P-controls $u(t)$ depending solely on time. We assume that the equations of the system have the form (3.13), and the cost functional the form (3.2). Then the solution of the LQ-problem (3.13), (3.2) providing an optimal P-control can be derived from the results of subsection 2. Indeed, as was mentioned in § 1, the assumption about the unavailability of the coordinates $x(t)$ for measurement is equivalent to either the condition $Q \equiv 0$ or the condition $\sigma_0\sigma_0' \to \infty$. When either of these conditions is satisfied, formulas (3.21)–(3.25) lead to the following expression for the optimal P-control $u(t)$ of the system (3.13) with the cost functional (3.2):

$$u(t) = -N_3^{-1}(t)[\big(B'(t)P(t) + N_4'(t)\big)m(t) + B'(t)G(t) + h_3(t)]. \tag{3.26}$$

Let us describe the functions entering into this expression. The vector $m(t) = Ex(t)$ satisfies the ordinary differential equation (3.23) for $Q = 0$:

$$\dot{m}(t) = A(t)m(t) + f(t) + B(t)u(t), \qquad m(0) = Ex_0. \tag{3.27}$$

The matrix $P(t)$ and the vector function $G(t)$ are defined by the relations (3.10), (3.11) for $\sigma_1 = 0$, $\sigma_2 = 0$:

$$\dot{P} + A'P + PA - (PB + N_4)N_3^{-1}(PB + N_4)' + N_2 = 0, \qquad P(T) = N_1, \tag{3.28}$$

$$\dot{G} + A'G + Pf + h_2 - (N_4 + PB)N_3^{-1}(h_3 + B'G) = 0, \qquad G(T) = h_1. \tag{3.29}$$

Thus, we have established the relations (3.26)–(3.29) that determine the optimal P-control $u(t)$ of the system (3.13). According to these relations the algorithm for constructing the optimal P-control consists of the following steps:

1°. Find the solution of the problem (3.28).

2°. Find the solution of the problem (3.29).

3°. Substitute the right side of the relation (3.26) in (3.27), and then the solution $m(t)$ of the problem (3.27) is found.

4°. Substituting the obtained functions $P(t)$, $G(t)$, $m(t)$ on the right side of formula (3.26) yields the optimal P-control $u(t)$ of the system (3.13).

We can derive the expression for the minimum value of the cost functional (3.2) associated with the control (3.26). This expression is provided by formula (3.25) for $Q = 0$:

$$\min_{u(t)} J = m'(0)P(0)m(0) + 2G'(0)m(0) + g(0) + \operatorname{tr}\big(D(T)N_1\big) + \int_0^T \operatorname{tr}\big(D(t)N_2(t)\big)\, dt, \tag{3.30}$$

where the covariance matrix $D(t) = E\{\big(x(t) - m(t)\big)\big(x(t) - m(t)\big)'\}$ satisfies the equation

$$\dot{D} = AD + DA' + \sigma\sigma', \qquad D(0) = D_0. \tag{3.31}$$

The solution of this equation is expressed by

$$D(t) = Z(t,0)D_0Z'(t,0) + \int_0^t Z(t,s)\sigma(s)\sigma'(s)Z'(t,s)\, ds,$$

where $Z(t,s)$ is the Cauchy matrix for the solutions of the ordinary differential equation $\dot{x}(t) = A(t)x(t)$.

Thus, for the system (3.13) with cost functional (3.2) we have deduced two optimal controls in accordance with the type of available information. The first control corresponds to the case in which the observation data is described by formula (3.14). In this case the optimal control is given by formula (3.22), and the value of the cost functional is determined by the expression (3.25). In the second case when there are no observations the optimal program control and the cost functional are provided by the expressions (3.26) and (3.27), respectively. It is intuitively clear that the value of the cost functional (3.25) must not be greater than the value given by (3.30).

The foregoing conclusion can be verified by a direct comparison of the expressions (3.25) and (3.30). In fact, let $R(t)$ denote the matrix that is the difference of the solutions of the problems (3.31) and (3.24). Then the difference of the right sides of the expressions (3.30) and (3.25) is of the form

$$\operatorname{tr}\big(R(T)N_1\big) + \int_0^T \operatorname{tr}\big(R(t)N_2(t)\big)\,dt.$$

In this sum it is sufficient to verify that the summands are not negative. Since the matrix N_1 is positive semidefinite, it can be expressed as $N_1 = MM'$. Furthermore, it follows from (3.31) and (3.24) that $R(t)$ is positive semidefinite. Therefore,

$$\operatorname{tr}\big(R(T)N_1\big) = \operatorname{tr}\big(M'R(t)M\big) \geqslant 0.$$

In a similar way, we can verify that the second summand is not negative. Thus, we have established that optimal control is more effective in the case when additional information about the coordinates is available.

The expressions (3.26)–(3.31) for the optimal P-control can also be obtained directly in the following way. Equations (3.27) for $m(t)$ and (3.31) for $D(t)$ are a consequence of the Ito formula. Further, we can effect a transformation (3.15)–(3.17) (in which $u(t, y_t)$ is replaced by $u(t)$, and the conditional expectation by the unconditional) and express the cost functional (3.2) in the form (3.18). As a result, we obtain the problem of minimizing the functional J_2 (3.18) on the trajectories of the deterministic system (3.27). The solution of this problem can, for example, be obtained by solving the problem (3.1), (3.2) for $\sigma_1 = 0$, $\sigma_2 = 0$, $\sigma = 0$. Using this solution, we can verify the relations (3.26)–(3.31), which determine the optimal P-control of the system (3.13) with cost functional (3.2).

Remark. In an analogous way, we can also consider the LQ-problem of the optimal P-control of the linear system (3.1) for $\sigma_2 = 0$ (involving noise depending on the coordinates) that has the form

$$\begin{aligned} dx(t) &= [A(t)x(t) + f(t) + B(t)u(t)]\,dt \\ &\quad + \sigma_1(t)x(t)\,d\xi_1(t) + \sigma(t)\,d\xi(t), \qquad 0 \leqslant t \leqslant T, \qquad (3.32) \\ x(0) &= x_0. \end{aligned}$$

The optimal P-control in the problem (3.32), (3.2) can also be determined from the relations (3.26)–(3.30), in which the covariance matrix $D(t)$ satisfies the equation

$$\dot{D} = AD + DA' + \sigma\sigma' + \sigma_1' D\sigma_1 = 0, \qquad D(0) = D_0$$

(this being the only difference from the foregoing case of $\sigma_1 = 0$).

4. Linear-quadratic problem for Gaussian and Poisson disturbances. The LQ-problem that has been considered above can be generalized in several directions to include more complicated equations of motion, wider classes of random disturbances acting upon the system, various modifications of the cost functional, and so on. It should be stressed that the problem of determining optimal control can be reduced in a number of cases to integration of deterministic equations (such as, ordinary differential equations in the solution of the problem (3.1), (3.2)). We shall elaborate on this statement by assuming that in addition to Gaussian disturbances of the white noise type also impulse (shock) disturbances modeled by the Poisson process act on the system. Equations (2.12) of such a system can be represented, in the linear case, in the form

$$\begin{aligned} dx(t) = [A(t)x(t) + B(t)u + f(t)]\,dt + \sigma_1(t)x(t)\,d\xi_1(t) + \sigma_2(t)u\,d\xi_2(t) \\ + \sigma(t)\,d\xi(t) + \sigma_3(t)\,d\eta(t), \qquad x(0) = x_0,\ 0 \leqslant t \leqslant T. \end{aligned} \tag{3.33}$$

This equation differs from equation (3.1) only by the last summand $\sigma_3(t)\,d\eta(t)$, where $\sigma_3(t)$ is a given matrix with measurable bounded elements, $\eta(t)$ is a Poisson process (discussed in § 2). The other quantities in (3.33) have the same meaning as the identically designated quantities in equation (3.1). The F-control of the system (3.33) will be taken as a Markov control $u\big(t, x(t)\big)$ in order to minimize the quadratic functional (3.2). It is assumed that the coordinates $x(t)$ of the system can be measured accurately. The Bellman equation for the problem (3.33), (3.2) is of the form (3.3), where the operator L_u must be modified for the generator of the system (2.12) according to the general formula (2.13). Thus, we obtain

$$\begin{aligned} \inf_u \Big[L_u V(t,x) + \lambda(t,x,u) + \sum_{i=1}^{r} \lambda_i(t)\big(V(t,x) + \sigma_1(t,x,u)e_i\big) \\ - V(t,x)\Big] = 0, \qquad T \geqslant t \geqslant 0,\ x \in R^n, \\ V(T,x) = x'N_1x + 2h_1'x. \end{aligned} \tag{3.34}$$

Here $\lambda_i(t)$ is the parameter of the ith component of the Poisson process; the vector $e_i \in R^r$ has the ith component equal to unity and all others zero; the generator L_u and the function $\lambda(t,x,u)$ are defined by formula (3.3). The solution $V(t,x)$ of the problem (3.34) is sought in the form

$$V(t,x) = x'P(t)x + 2G_1'(t)x + g_1(t). \tag{3.35}$$

As in the proof of the relations (3.10)–(3.12), we can establish that the matrix $P(t)$ satisfies equation (3.10) and that the following relations for the vector function $G_1(t) \in R^n$ are valid:

$$\begin{aligned} \dot G_1 + A'G_1 + Pf + h_2 + (N_4 + PB)K(h_3 + B'G_1) + \sum_{i=1}^{r} \lambda_i(t) P\sigma_3 e_i = 0, \\ G_1(T) = h_1,\ K = -(N_3 + \sigma_2' P \sigma_2)^{-1}. \end{aligned}$$

The scalar function $g_1(t)$ is determined by the quadrature:

$$\begin{aligned} g_1(t) = \int_t^T \big[&2G_1'(s)f(s) + \big(h_3(s) + G_1'(s)B(s)\big)K(s) \\ &\times \big(h_0(s) + B'(s)G_1(s)\big) + \operatorname{tr}\big(P(s)\sigma(s)\sigma'(s)\big) \\ &+ \sum_{i=1}^{r} \lambda_i(s)\big(2G_1'(s)\sigma_3(s)e_i + e_i'\sigma_3'(s)P(s)\sigma_3(s)e_i\big)\big]\, ds. \end{aligned}$$

It follows from (3.34), (3.35) that the optimal control $u_0(t,x)$ of the system (3.33) is expressed by formula (3.8) in which G is replaced by G_1, and g by g_1, i.e.,

$$u_0(t,x) = K(t)\big[\big(B'(t)P(t) + N_4'(t)\big)x + h_3(t) + B'(t)G_1(t)\big].$$

5. Control of wire reeling with regard to random disturbances. The deterministic version of this problem was discussed in § 2 (subsection 5) of Chapter IV. However, in reality some of the quantities involved in the problem statement vary in a random manner. We now investigate this problem more fully, taking into account that both the input voltage, $U(t)$, and the coefficient of pivoting friction, ψ, are subjected to random disturbances. We assume that the deviation of the input voltage from its nominal (unperturbed) value can be expressed by $u(t) + \sigma_2\dot{\xi}_2(t)$, and the friction coefficient is of the form $\psi - \sigma_1\dot{\xi}_1(t)$. Here σ_1 and σ_2 are known constants, and $\xi_1(t)$, $\xi_1(t)$ are standard Wiener processes. In that case instead of the problem (2.29), (2.30) of Chapter IV we obtain the following stochastic LQ-control problem:

$$\begin{aligned} dx(t) = \big(-\psi g^{-1}(t)x(t) + \varkappa u\big)\, dt &+ \sigma_1 g^{-1}(t)x(t)\, d\xi_1(t) \\ &+ \varkappa\sigma_2\, d\xi_2(t), \qquad t \geqslant 0. \end{aligned} \tag{3.36}$$

The cost functional to be minimized is of the form

$$J(u) = E\int_0^T \big[\big(r(t)g^{-1}(t)x(t)\big)^2 + \rho u^2\big]\, dt. \tag{3.37}$$

According to formula (3.8) the optimal control in the problem (3.36), (3.37) is expressed by the relations

$$u_0(t) = -K_1(t)x(t), \qquad K_1(t) = \varkappa\rho^{-1}P_1(t),$$

where $P_1(T)$ satisfies the relations

$$\begin{aligned} \dot{P}_1 - 2\psi g^{-1}(t)P_1(t) &- \varkappa^2\rho^{-1}P_1^2(t) + r^2(t)g^{-2}(t) \\ &= -\sigma_1^2 g^{-2}(t)P_1(t), \qquad P_1(T) = 0. \end{aligned} \tag{3.38}$$

Comparing equations (2.31) of Chapter IV and (3.38), we find that taking account of random disturbances leads to the additional summand $\sigma_1^2 g^{-2} P_1$ in the Riccati equation (3.38) caused by the fluctuation of the friction coefficient. On the contrary, taking account of the fluctuations of the input voltage does not affect the value of the optimal control but affects only the value $J(u_0)$ of the minimum cost functional. If, for example, $\sigma_1 = 0$, then $J(u_0)$ is greater than the minimum value of the cost functional (2.30) of Chapter IV for the corresponding deterministic problem (i.e., the problem (3.36), (3.37) for $\sigma_1 = \sigma_2 = 0$) by the value $\varkappa^2 \sigma_2^2 \int_0^T P(t)\,dt$. Here $P(t)$ is the solution of the problem (3.38) for $\sigma_1 = 0$, which coincides in this particular case with the problem (2.31) of Chapter IV.

§ 4. Control on an Unbounded Time Interval. Stabilization of Stochastic Control Systems

1. Problem statement. Stabilization problem statements for stochastic systems differ, above all, from those for deterministic systems discussed in Chapter VII by the fact that equations of motion (1.1) include random summands. This, in turn, leads to the corresponding modification of the cost functional and a diversification of problem statements caused by numerous possible definitions of stability (see Chapter II). We will enlarge on this mostly following [7, Part One].

Suppose that the equations of motion are of the form (1.1), i.e.,

$$dx(t) = a\big(t, x(t), u\big)\,dt + \sigma\big(t, x(t), u\big)\,d\xi(t), \qquad t \geqslant 0,\ x(0) = x_0. \tag{4.1}$$

Here $x(t) \in R^n$ is the phase vector, $u \in R^m$ is the control, and $\xi(t) \in R^k$ is a standard Wiener process; the continuous vector function $a \in R^n$, the $n \times k$ matrix σ, and the deterministic vector $x_0 \in R^n$ are prescribed. The F-control of the system (4.1) is sought in the class of Markov controls, i.e., in the clas of functions of the form $u\big(t, x(t)\big) \in U$, where U is a prescribed set in R^m.

As in the deterministic case, the following stabilization problem statements are possible:

1°. Determine the control $u(t,x)$ for which the system (4.1) becomes stable in some sense.

2°. Determine the control $u(t,x)$ that minimizes the cost functional

$$J(u) = E\int_0^\infty F\Big(t, x(t), u\big(t, x(t)\big)\Big)\,dt \to \inf_{u\in U}, \tag{4.2}$$

where E is mathematical expectation, $F(t,x,u), F\colon [0,\infty) \times R^n \times R^m \to R^1$ is a given continuous function.

3°. Determine the control $u(t,x)$ that minimizes the functional (4.2) and makes the system (4.1) stable. As in Chapter VII, the stabilization problem in this setting is called the *stochastic optimal stabilization problem.*

2. Application of Lyapunov's direct method to optimal stabilization problems. An F-control is said to be *admissible* if $u(t,0) = 0$, the functions $a\big(t,x,u(t,x)\big)$, $\sigma\big(t,x,u(t,x)\big)$ are continuous, and their first derivatives with respect to x exist and are uniformly bounded with respect to t. It is also assumed that $a(t,0,0) = 0$, $\sigma(t,0,0) = 0$. Let C_i, P, q denote certain positive constants. The generator L_u is defined by

$$L_u V(t,x) = V_t + a'(t,x,u)V_x + \tfrac{1}{2}\,\mathrm{tr}[\sigma(t,x,u)\sigma'(t,x,u)V_{xx}].$$

Theorem 4.1. *Let there exist an admissible F-control $u_0(t,x)$ and a scalar function $V_0(t,x)$, twice continuously differentiable with respect to x and once with respect to t everywhere except, perhaps, the set $x = 0$ satisfying, for all $t \geqslant 0$, $x \in R^n$ and some positive $p > 0$ and $q > 0$, the conditions*

$$0 \leqslant V_0(t,x) \leqslant C_1|x|^p,$$
$$\left|\frac{\partial V_0(t,x)}{\partial x}\right| \leqslant C_2(1+|x|^q), \qquad p>0,\ q>0; \tag{4.3}$$
$$\inf_{u\in U}[L_u V_0(t,x) + F(t,x,u)]$$
$$= L_{u_0}V_0(t,x) + F\big(t,x,u_0(t,x)\big) = 0; \tag{4.4}$$
$$F(t,x,u) \geqslant C_3|x|^p. \tag{4.5}$$

Then the trivial solution of the system (4.1) controlled by $u_0(t)$ is exponentially p-stable, and

$$J(u_0) = V_0(0,x_0) = \inf_u E\int_0^\infty F\Big(t,x(t,u),u\big(t,x(t,u)\big)\Big)\,dt, \tag{4.6}$$

where $x(t,u)$ denotes the solution of the problem (4.1) and the infimum is taken over the set of all admissible controls u.

Proof. It follows from the Ito formula that for any admissible control $u(t,x)$ the following relation is valid:

$$dV_0\big(t,x(t,u)\big) = L_u V_0\big(t,x(t,u)\big)\,dt$$
$$+\left(\frac{\partial V_0\big(t,x(t,u)\big)}{\partial x}\right)'\sigma\Big(t,x(t,u),u\big(t,x(t,u)\big)\Big)\,d\xi(t).$$

Integrating this relation with respect to t from $t=0$ to $t=T>0$ and taking the expectation yields

$$EV_0\big(T,x(T,u)\big) - V_0(0,x_0) = E\int_0^t L_u V_0\big(t,x(t,u)\big)\,dt$$
$$+\,E\int_0^t\left(\frac{\partial V_0\big(t,x(t,u)\big)}{\partial x}\right)'\sigma\Big(t,x(t,u),u\big(t,x(t,u)\big)\Big)\,d\xi(t). \tag{4.7}$$

It follows from the definition of an admissible control $u(t,x)$ that, given this control, there exists a solution of the problem (4.1) having bounded moments on any finite time interval $[0,T]$. Hence, in view of the constraints (4.3) on the derivative $\partial V_0/\partial x$ the expectation of the stochastic integral in (4.7) is zero. Thus, for any admissible control $u(t,x)$ we have

$$EV_0\big(T, x(T,u)\big) - V_0(0,x_0) = E\int_0^T L_u V_0\big(t, x(t,u)\big)\,dt. \tag{4.8}$$

By virtue of the condition (4.4), this yields for $u = u_0$

$$\begin{aligned} &- EV_0\big(T, x(T,u_0)\big) + V_0(0,x_0) \\ &\quad = E\int_0^T F\Big(t, x(t,u_0), u_0\big(t, x(t,u_0)\big)\Big)\,dt. \end{aligned} \tag{4.9}$$

However, the function $V_0(t,x)$ is bounded below. Therefore, noting the last equality as $T \to \infty$, we find that $J(u_0) < \infty$. Hence, in view of the condition (4.5), we obtain the estimate

$$\int_0^\infty E|x(t,u_0)|^p\,dt < \infty. \tag{4.10}$$

Let us now verify that the function $E|x(t,u_0)|^p$ satisfies the Lipschitz condition with respect to t. Applying the Ito formula to the function $|x(t,u_0)|^p$ gives

$$\begin{aligned} |x(t,u_0)|^p - |x_0|^p &= \int_0^t L_{u_0}|x(s,u_0)|^p\,ds \\ &+ P\int_0^t |x(s,u_0)|^p x'(s,u_0)\sigma\Big(s, x(s,u_0), u_0\big(s, x(s,u_0)\big)\Big)\,ds. \end{aligned}$$

Taking the expectation of both sides of this equality we find that

$$E|x(t,u_0)|^p - |x_0|^p = E\int_0^t L_{u_0}|x(s,u_0)|^p\,ds. \tag{4.11}$$

Moreover, it follows from the definition of an admissible control that for some constant $C > 0$ the following inequalities hold:

$$|a(t,x,u_0)| \leqslant C|x|, \qquad |\sigma(t,x,u_0)| \leqslant C|x|.$$

Thus, noting the form of the generator L_{u_0}, we can write

$$\big|L_{u_0}|x(t,u_0)|^p\big| \leqslant C_4|x(t,u_0)|^p. \tag{4.12}$$

The relations (4.10)–(4.12) imply that

$$\sup_{t\geqslant 0} E|x(t,u_0)|^p < \infty. \tag{4.13}$$

Further, it follows from (4.11) that for any two moments $t_2 \geqslant t_1 \geqslant 0$ the following equality holds:

$$E|x(t_2,u_0)|^p - E|x(t_1,u_0)|^p = E\int_{t_1}^{t_2} L_{u_0}|x(s,u_0)|^p\,ds,$$

so that, with (4.12), (4.13), we obtain

$$\big|E|x(t_2,u_0)|^p - E|x(t_1,u_0)|^p\big| \leqslant C_5(t_2-t_1).$$

This yields

$$\lim_{t\to\infty} E|x(t,u_0)|^p = 0. \tag{4.14}$$

In the expression (4.9) we pass to the limit as $T\to\infty$. Taking the relation (4.14) into account, we find that

$$J(u_0) = V_0(0,x_0). \tag{4.15}$$

Further, we show that $J(u) \geqslant V_0(0,x_0)$ for any admissible F-control $u(t,x)$ for which $J(u) < \infty$. First, by analogy with (4.14), we establish that for these admissible controls $\lim_{t\to\infty} E|x(t,u)|^p = 0$. Therefore, on the strength of the estimate (4.3) for V_0, we have

$$\lim_{t\to\infty} EV_0\big(t,x(t,u)\big) = 0. \tag{4.16}$$

It should be noted that by virtue of the relations (4.8), (4.4) the following inequality is valid:

$$\begin{aligned} &EV_0\big(T,x(T,u)\big) - V_0(0,x_0)\\ &\qquad \geqslant -E\int_0^T F\Big(t,x(t,u),u\big(t,x(t,u)\big)\Big)\,dt. \end{aligned}$$

Passing to the limit as $T\to\infty$, we find, by virtue of (4.16), that $J(u)\geqslant V_0(0,x_0)$. Comparing this relation with the equality (4.15), we prove that formula (4.6) holds true.

On the strength of the results of Chapter II we can prove that the trivial solution of the system (4.1) generated by the control $u_0(t,x)$ is exponentially p-stable. For this it is sufficient to show that $V_0(t,x)\geqslant C_6|x|^p$. Let $y(s)$ denote the solution of equation (4.1) for $s\geqslant t$ with initial condition $y(t)=x$, given the control u_0. By analogy with the relations (4.14), (4.15), we find that

$$\begin{gathered} V_0(t,x) = \int_t^\infty EF\Big(s,y(s),u_0\big(s,y(s)\big)\Big)\,ds,\\ \lim_{s\to\infty} E|y(s)|^p = 0. \end{gathered} \tag{4.17}$$

It follows that there exists a moment T such that

$$E\big|y(T)\big|^p < \tfrac{1}{2}|x|^p. \tag{4.18}$$

Further, as in the proof of the relation (4.12), it is easily shown that

$$L_{u_0}|y(s)|^p \geqslant C_7|y(s)|^p. \tag{4.19}$$

Formulas (4.17)–(4.19) imply, in view of (4.5), that

$$\begin{aligned} V_0(t,x) &\geqslant C_3 \int_t^\infty E|y(s)|^p\,ds \geqslant C_3 \int_t^T E|y(s)|^p\,ds \\ &\geqslant -C_3C_7^{-1} \int_t^T EL_{u_0}|y(s)|^p\,ds \\ &= C_3C_7^{-1}\left[|x|^p - E|y(T)|^p\right] \geqslant \tfrac{1}{2}C_3C_7^{-1}|x|^p. \end{aligned}$$

Thus, the function V_0 satisfies all the conditions of the theorem on exponential p-stability. Hence, the trivial solution of the system (4.1) is exponentially p-stable, given the control $u_0(t,x)$. □

3. Stabilization of linear stochastic systems. We will dwell briefly on some problems of the optimal stabilization of linear stochastic systems having the form

$$\begin{gathered} dx(t) = [A(t)x(t) + B(t)u(t)]\,dt + \sigma(t)x(t)\,d\xi(t) \qquad t \geqslant 0,, \\ x(0) = x_0 \in R^n,\ u \in R^m. \end{gathered} \tag{4.20}$$

In these problems we have to minimize the quadratic cost functional:

$$J(u) = E\int_0^\infty [x'(t)N_1(t)x(t) + u'N_0(t)u]\,dt \to \inf_u. \tag{4.21}$$

In equation (4.20) $\xi(t)$ denotes a scalar standard Wiener process. The matrices A, B, σ, $N_0 > 0$, $N_1 > 0$ are prescribed and have continuous elements. In the problem (4.20), (4.21) F-control is sought in the form $u = u\big(t, x(t)\big)$. To solve this problem, we use Theorem 4.1. According to this theorem, it is necessary to construct a scalar function $V_0(t,x)$ and a control $u_0(t,x)$ satisfying the conditions (4.3)–(4.5). We look for V_0 as $V_0(t,x) = x'P(t)x$, where the symmetric matrix $P(t) \geqslant 0$ is to be determined. Substituting V_0 in equation (4.4) and equating to zero the coefficients of the same powers of x, as in the proof of the relations (3.8), (3.10), we obtain

$$u_0(t,x) = -N_0^{-1}(t)B'(t)P(t)x, \tag{4.22}$$

$$\begin{gathered} \dot P(t) + A'(t)P(t) + P(t)A(t) - P(t)B(t)N_0^{-1}(t)B'(t)P(t) \\ +N_1(t) + \sigma'(t)P(t)\sigma(t) = 0, \qquad t \geqslant 0. \end{gathered} \tag{4.23}$$

According to Theorem 4.1, if there exists a bounded positive definite solution $P(t)$ of equation (4.23), then, given the F-control (4.22), the system is exponentially stable in mean square and minimizes the functional (4.21) equal to $J(u_0) = V_0(0, x_0)$.

In a number of cases when we are looking for a unique positive definite bounded (for $t \geqslant 0$) solution of equation (4.23), the following theorem can be used.

Theorem 4.2 [7, Part One]. *Suppose there exists a linear admissible control $u = R(t)x$ for which the system (4.20) is exponentially stable in mean square and there exists the limit $P_1(t) = \lim_{T\to\infty} P_1(t,T)$. Here the function $P_1(t,T)$ is defined for each fixed T by the relations*

$$\frac{\partial P_1(t,T)}{\partial t} + (A(t)+B(t)R(t))' P_1(t,T) + P_1(t,T)(A(t)+B(t)R(t))$$
$$+ N_1(t) + R'(t)N_0(t)R(t) + \sigma'(t)P_1(t,T)\sigma(t) = 0, \qquad 0 \leqslant t \leqslant T,$$
$$P_1(T,T) = 0.$$

Further, let the sequence $P_i(t)$ $(i \geqslant 2)$ of positive definite matrices be defined in the following way:

$$P_i(t) = \lim_{T\to\infty} P_i(t,T), \qquad i \geqslant 2,\ 0 \leqslant t \leqslant T,$$

where the matrix $P_i(t,T)$ is a solution of the linear equation

$$\frac{\partial P_i(t,T)}{\partial t} + (A'(t) - P_{i-1}(t)BN_0^{-1}B')P_i(t,T)$$
$$+ P_i(t,T)(A - BN_0^{-1}B'P_{i-1}(t)) + N_1(t)$$
$$+ P_{i-1}(t)BN_0^{-1}B'P_{i-1}(t) = 0, \qquad i \geqslant 2,\ 0 \leqslant t \leqslant T,$$
$$P_i(T,T) = 0.$$

Then the sequence $P_i(t)$ converges to the unique positive definite solution of equation (4.23), with $x'P_{i+1}x \leqslant x'P_i x$ for any $x \in R^n$.

Remark. Similar results hold true for linear systems of the form

$$dx(t) = [A(t)x(t) + B(t)u]\,dt + \sum_{i=1}^{j} \sigma_i(t)x(t)\,d\xi_i(t),$$

with cost functional (4.21). Here $\xi_i(t)$ are scalar independent standard Wiener processes. The optimal control u_0 is expressed by formula (4.22), with the matrix $P(t)$ satisfying the equation

$$\dot P + A'P + PA - PBN_0^{-1}B'P + N_1 + \sum_{i=1}^{j} \sigma_i' P \sigma_i = 0.$$

§ 5. Approximate Methods for Determining Optimal Control

One of the effective methods for the determination of optimal control in stochastic systems is the perturbation method which can be used in the case of equations of motion having small parameters. A small parameter ε may be due to the fact that the nonlinear summands are small, the random disturbances acting upon the system are small, the delay in the system is small, the error of implementation of the control function is small, and so on. Note that for $\varepsilon = 0$ the problem is, as a rule, simplified, and the corresponding optimal control can be used as a zero approximation to the optimal control for the original problem.

In this section we are concerned with establishing a scheme for implementing the perturbation method and estimates for the errors in quasilinear systems and for the value of a quadratic cost functional.

1. Description of the algorithm of successive approximations. Let us consider the stochastic nonlinear control system

$$dx(t) = [\varepsilon f(t,x) + B(t)u)]\,dt + \sigma(t)\,d\xi(t), \qquad x(0) = x_0,\ 0 \leqslant t \leqslant T. \tag{5.1}$$

Here $x(t) \in R^n$ is the phase vector, $u \in R^m$ is the control, the matrices B, σ are prescribed and have measurable bounded elements, the vector $f(t,x) \in R^n$ is measurable with respect to t, x and satisfies the conditions

$$\begin{gathered} |f(t,x_1) - f(t,x_2)| \leqslant C|x_1 - x_2|, \\ |f(t,x)| \leqslant C(1 + |x|). \end{gathered} \tag{5.2}$$

Here and subsequently C denotes various positive constants. In the relations (5.1) the number T, the initial state vector x_0 and the parameter $\varepsilon \geqslant 0$ are prescribed; $\xi(t)$ is a standard Wiener process.

It is required to determine an F-control u as a function of time t and the phase vector $x(t)$, a function that minimizes the quadratic functional $J(u)$ for an arbitrary x_0:

$$J(u) = E\left[x'(T)N_2x(T) + \int_0^T \big(x'(t)N_1(t)x(t) + u'N_0(t)u\big)\,dt\right]. \tag{5.3}$$

Here $x(t)$ denotes the solution of the system (5.1) on the segment $[0,T]$, given the control u; the matrices N_i are prescribed: $N_1 \geqslant 0$, $N_2 \geqslant 0$, $N_0(t) > 0$, $0 \leqslant t \leqslant T$; the elements of the matrices N_0, N_1 are measurable and bounded. It should be noted that the problem of the control of the system

$$\dot{y}(t) = A(t)y + \varepsilon f(t,y) + Bu + \sigma(t)\dot{\xi},$$

with functional (5.3), can be reduced to the problem (5.1), (5.3) by the change of variables $y(t) = z(t)x(t)$, where the matrix z is determined by the conditions $\dot{z} = Az$, $z(0) = I$. We assume that the optimal control $u_0(t,x)$ in the problem (5.1), (5.3) exists. Let $V(t,x)$ denote the Bellman function for the problem (5.1), (5.3).

Let us formally write the equation for V. By virtue of (2.11) we have

$$\begin{gathered} LV + \varepsilon V_x'f - \tfrac{1}{4}V_x'B_1V_x + x'N_1x = 0, \qquad 0 \leqslant t \leqslant T,\ x \in R^n, \\ V(T,x) = x'N_2x, \end{gathered} \tag{5.4}$$

where $LV = V_t + (1/2)\,\mathrm{tr}[\sigma\sigma' V_{xx}]$, $B_1 = BN_0^{-1}B'$. Here V_t is a partial derivative with respect to t; V_x is the vector of first partial derivatives; V_{xx} is the matrix of second partial derivatives with respect to the components of the vector x; tr is the trace of a matrix.

If the solution V of the Cauchy problem (5.4) is found, then the optimal control v is given by the formula

$$v(t,x) = v = -\tfrac{1}{2}N_0^{-1}B'V_x(t,x). \tag{5.5}$$

We expand $V(t,x)$ in a series of powers of ε:

$$V = V^0 + \varepsilon V^1 + \dots. \tag{5.6}$$

We obtain the equations for V^i in the expansion (5.6) by substituting the series (5.6) in equation (5.4) and equating to zero the coefficients of the same powers of ε.

The equation for V^0 is given by the relation (5.4) for $\varepsilon = 0$. The equations for V^i for $i \geqslant 1$ are of the form

$$\begin{gathered} LV^i + f'V_x^{i-1} - \frac{1}{4}\sum_{j=0}^{i}(V_x^j)'B_1(V_x^{i-j}) = 0, \\ V^i(T,x) = 0. \end{gathered} \tag{5.7}$$

Let the approximation u_i to the optimal control be given by

$$\begin{gathered} u_i(t,x) = -\tfrac{1}{2}N_0^{-1}B'(V_x^0 + \dots + \varepsilon^i V_x^i), \\ i \geqslant 1,\ 0 \leqslant t \leqslant T,\ x \in R^n. \end{gathered} \tag{5.8}$$

Henceforth we assume that V^i exists, is unique, and

$$|V_x^i| \leqslant C(1+|x|). \tag{5.9}$$

The method (used subsequently) for the justification of error estimates is based on that the ith control approximation is optimal in the control problem for the system (5.1), with cost functional differing from the original cost functional (5.3) by quantities of the order of ε^{i+1}.

2. Zero approximation estimate. Let us determine a zero approximation estimate for the cost functional, i.e., an estimate of the difference $J(u_0) - J(v)$. For $\varepsilon = 0$ the Cauchy problem (5.4) admits an exact solution given in § 3. According to formulas (3.7)–(3.12), we have

$$\begin{gathered} V_0(t,x) = x'P(t)x + \mathrm{tr}\int_t^T \sigma'(s)P(s)\sigma(s)\,ds, \\ \dot{P} - PB_1P + N_1 = 0, \qquad P(T) = N_2. \end{gathered} \tag{5.10}$$

By virtue of equations (5.18), (5.10) the optimal control u_0 for the problem (5.1), (5.3) for $\varepsilon = 0$ is synthesized in the form

$$u_0(t,x) = -N_0^{-1}B'Px.$$

Consider now as an auxiliary problem the control problem for the system (5.1) with cost functional to be minimized

$$J_0(u) = J(u) - 2\varepsilon E\int_0^T f'\big(t,x(t)\big)P(t)x(t)\,dt. \tag{5.11}$$

The Bellman equation for the problem (5.1), (5.11) is of the form

$$\begin{gathered} LW + \varepsilon f'W_x - \tfrac{1}{4}W_x'B_1W_x + x'N_1x - 2\varepsilon f'(t,x)Px = 0, \\ x \in R^n,\ 0 \leqslant t \leqslant T, \\ W(T,x) = x'N_2x. \end{gathered} \tag{5.12}$$

The solution of the boundary-value problem is unique in the class of functions satisfying the inequality (5.9). It can be immediately verified that the function V_0 defined by the equality (5.10) is a solution of the boundary-value problem (5.12). Therefore, the control $u_0(t,x)$ is optimal in the problem (5.1), and the corresponding Bellman function is $V_0(t,x)$.

We now seek an estimate for the difference $J(u_0) - J(v)$. Let $x(t)$ denote the solution of the problem (5.1) generated by the control v, and $y(t)$ the solution of the same problem, given the control u_0. The optimality of v implies that

$$0 \leqslant J(u_0) - J(v) = [J_0(u_0) - J(v)] + [J(u_0) - J_0(u_0)]. \tag{5.13}$$

Let us estimate the differences in brackets on the right side of the equation. Since v is the optimal control in the original problem (5.1), (5.3), it follows, in view of (5.11), that

$$J(v) \leqslant J(u_0) = J_0(u_0) - 2\varepsilon E\int_0^T f'\big(t,y(t)\big)P(t)y(t)\,dt. \tag{5.14}$$

But the control $u_0(t,x)$ is linear in x. Hence, noting the estimate (5.2) for the function $f(t,x)$, we see that $E|y(t)|^2 \leqslant C(1+|x_0|^2)$. Therefore,

$$\left|E\int_0^T f'\big(t,y(t)\big)P(t)y(t)\,dt\right| \leqslant C(1+|x_0|^2). \tag{5.15}$$

Thus, noting the relation (5.14), we can write

$$J(v) \leqslant J_0(u_0) + \varepsilon C(1+|x_0|^2).$$

We now verify that

$$J_0(u_0) \leqslant J(v) + \varepsilon C(1 + |x_0|^2). \tag{5.16}$$

The optimality of the control u_0 in the problem (5.1), (5.11) yields

$$J_0(u_0) \leqslant J_0(v) = J(v) - 2\varepsilon E \int_0^T f'(t, x(t)) P(t) x(t)\, dt. \tag{5.17}$$

To obtain an estimate for the integral term on the right side of (5.17), we first estimate the quantity $E\int_0^T |v(t, x(t))|^2\, dt$. Let μ be the minimum positive eigenvalue of the matrix N_0. Then, by virtue of the optimality of the control v in the problem (5.1), (5.3) the following relation holds true:

$$\mu E \int_0^T |v(t, x(t))|^2\, dt \leqslant J(0) \leqslant C(1 + |x_0|^2).$$

This yields

$$E \int_0^T |v(t, x(t))|^2\, dt \leqslant C(1 + |x_0|^2).$$

Thus, noting equations (5.1) and the conditions (5.2), we obtain the inequality

$$E|x(t)|^2 \leqslant C(1 + |x_0|^2). \tag{5.18}$$

Hence, by analogy with (5.15), we have

$$E \int_0^T f'(t, x(t)) P(t) x(t)\, dt \leqslant C(1 + |x_0|^2).$$

It follows, in view of (5.17), that formula (5.16) holds true. Comparing the relations (5.14) and (5.16) we see that

$$|J_0(u_0) - J(v)| \leqslant \varepsilon C(1 + |x_0|^2). \tag{5.19}$$

Thus, an estimate for the first difference in (5.13) is obtained. For the second difference in (5.13), noting (5.11), (5.15), we have

$$|J(u_0) - J_0(u_0)| = 2\varepsilon \left| E \int_0^T f'(t, y(t)) P(t) y(t)\, dt \right| \leqslant \varepsilon C(1 + |x_0|^2). \tag{5.20}$$

The estimates (5.13), (5.20), (5.19) imply that

$$0 \leqslant J(u_0) - J(v) \leqslant \varepsilon C(1 + |x_0|^2).$$

This inequality means that if the original system is controlled not in an optimal way but by the zero control approximation u_0, then the error of the cost functional (5.3) is of the order of ε. It will be recalled that when justifying the zero approximation estimate we used the letter C to denote various positive constants depending only on the initial data for the problem and independent of ε.

3. First approximation estimate. According to (5.8) the first approximation u_1 to the optimal control is given by the formula

$$u_1(t,x) = -\tfrac{1}{2}N_0^{-1}(t)B'(t)\big[V_x^0 + \varepsilon V_x^1\big], \tag{5.21}$$

where the function V^1, in view of equation (5.7), is the solution of the boundary-valye problem

$$LV^1 + 2x'Pf - x'PB_1V_x^1 = 0, \qquad V^1(T,x) = 0. \tag{5.22}$$

To use the justification procedure for error estimates given in subsection 2, we construct an auxiliary control system for which the control (5.21) is optimal, and the function Q defined by the equality $Q = V^0 + \varepsilon V^1$ is the Bellman function. Equations (5.10), (5.22) provide the following equation for Q:

$$\begin{gathered} LQ + 2\varepsilon x'Pf - \tfrac{1}{4}(V_x^0)'B_1V_x^0 + x'N_1x - \tfrac{1}{2}\varepsilon(V_x^0)'B_1V_x^1 = 0, \\ Q(T,x) = x'N_2x. \end{gathered} \tag{5.23}$$

Clearly,

$$(V_x^0)'B_1V_x^0 = Q_x'B_1Q_x + \varepsilon^2(V_x^1)'B_1V_x^1 - 2\varepsilon(V_x^1)'B_1(V_x^0 + \varepsilon V_x^1). \tag{5.24}$$

It follows from (5.23), (5.24) that

$$\begin{gathered} LQ + \varepsilon f'Q_x - \varepsilon^2 f'V_x^1 - \tfrac{1}{4}Q_x'B_1Q_x + \tfrac{1}{4}\varepsilon^2(V_x^1)'B_1V_x^1 + x'N_1x = 0, \\ Q(T,x) = x'N_2x. \end{gathered} \tag{5.25}$$

Comparing (5.25) and (5.4), we see that u_1 is the optimal control in the following problem: find the control that renders the minimum of the functional $J_1(u)$ on the trajectories of the system (5.1):

$$\begin{aligned} J_1(u) &= J(u) + \alpha_1(u), \\ \alpha_1(u) &= \varepsilon^2 E\int_0^T \Big[\tfrac{1}{4}\Big(V_x^1\big(t,x(t)\big)\Big)'B_1V_x^1\big(t,x(t)\big) \\ &\qquad - f'\big(t,x(t)\big)V_x^1\big(t,x(t)\big)\Big]\,dt. \end{aligned} \tag{5.26}$$

Thus, the auxiliary problem for optimal control has been constructed. This constitutes a control problem for systems of the form (5.1) with cost functional (5.26). It should be stressed that the equations of motion in the auxiliary system have the same form (5.1) as in the original optimal control problem, but the cost functional (5.26) differs from the original one by the quantity $\alpha_1(u)$ having the order of ε^2. Moreover, by the assumption (5.9), given the control u_1, the second moment of the solution $x(t)$ of equation (5.1) satisfies an estimate of the type (5.18). The Bellman function in the auxiliary control problem (5.1), (5.26) is $Q(t,x)$. Further justification of first approximation estimates is similar to the corresponding argument in subsection 2. For completeness, we will give an outline of the derivation.

First of all, as for (5.13), we have

$$0 \leqslant J(u_1) - J(v) = [J_1(u_1) - J(v)] + [J(u_1) - J_1(u_1)]. \tag{5.27}$$

The difference in the first brackets in (5.27) is estimated in a similar way to (5.18). First, according to (5.26), we obtain

$$J(v) \leqslant J(u_1) = J_1(u_1) - \alpha_1(u_1).$$

Moreover, by virtue of (5.26), (5.2), (5.9) we have

$$|\alpha_1(u_1)| \leqslant \varepsilon^2 C(1 + |x_0|^2).$$

Therefore,

$$J(v) \leqslant J_1(u_1) + \varepsilon^2 C(1 + |x_0|^2). \tag{5.28}$$

We now prove the inequality that is the converse of (5.28). It follows from the optimality of the control u_1 in the problem (5.1), (5.26) that

$$J_1(u_1) \leqslant J_1(v) = J(v) + \alpha_1(v).$$

But, given the control v, the second moment of the solution $x(t)$ to equation (5.1) satisfies the estimate (5.18). Hence, in view of the conditions (5.2), (5.9), it follows that $|\alpha_1(v)| \leqslant \varepsilon^2 C(1 + |x_0|^2)$. Therefore,

$$J_1(u_1) \leqslant J(v) + \varepsilon^2 C(1 + |x_0|^2). \tag{5.29}$$

The relations (5.28), (5.29) yield

$$|J_1(u_1) - J(v)| \leqslant \varepsilon^2 C(1 + |x_0|^2). \tag{5.30}$$

For the expression in the second brackets in (5.27), in view of (5.26), we have

$$|J(u_1) - J_1(u_1)| = |\alpha_1(u_1)| \leqslant \varepsilon^2 C(1 + |x_0|^2).$$

Thus, noting the relation (5.30), we finally obtain

$$0 \leqslant J(u_1) - J(v) \leqslant \varepsilon^2 C(1 + |x_0|^2). \tag{5.31}$$

This estimate indicates that if the system (5.1) is controlled by u_1, the error of the cost functional (5.3) is of the order of ε^2,

4. Higher-order approximation estimates. The auxiliary control problem played a major role in justifying the zero and the first approximation estimates. Once this problem has been constructed, the justification of the error estimates is carried out in a uniform way. Therefore, in the process of constructing higher approximations u_j to the optimal control v we confine ourselves to stating the auxiliary control problem.

As above, the auxiliary control problem must be such that the optimal control in this problem coincides with $u_j(t,x)$, and the Bellman function $Q_j(t,x)$ is equal to a segment of the series (5.6):

$$Q_j(t,x) = V^0 + \varepsilon V^1 + \cdots + \varepsilon^j V^j, \qquad j \geqslant 1. \tag{5.32}$$

Next, we sum equation (5.4) taken for $\varepsilon = 0$ and equation (5.7) multiplied by ε^i $(i = 1, \ldots, j)$. In the resulting equality, on adding and subtracting the expression

$$\frac{1}{4}\left(\frac{\partial Q_j(t,x)}{\partial x}\right)' B_1 \frac{\partial Q_j(t,x)}{\partial x},$$

we obtain

$$\begin{aligned} L_u Q_j &+ x' N_1 x + \varepsilon f' \frac{\partial Q_j}{\partial x} \\ &+ \frac{1}{4}\left[\left(\frac{\partial Q_j}{\partial x}\right)' B_1 \frac{\partial Q_j}{\partial x} - \sum_{i=0}^{j} \varepsilon^i \sum_{k=0}^{i} \left(\frac{\partial V^k}{\partial x}\right)' B_1 \frac{\partial V^{i-k}}{\partial x}\right] \\ &- \varepsilon^{j+1} f' \frac{\partial V^j}{\partial x} = \frac{1}{4}\left(\frac{\partial Q_j}{\partial x}\right)' B_1 \frac{\partial Q_j}{\partial x}. \end{aligned}$$

After substitution of the sum (5.32) for Q_j the expression in brackets becomes

$$\begin{aligned} &\sum_{i=0}^{j}\sum_{k=0}^{i} \varepsilon^{i+k} \left(\frac{\partial V^i}{\partial x}\right)' B_1 \frac{\partial V^k}{\partial x} - \sum_{i=0}^{j} \varepsilon^i \sum_{k=0}^{i} \left(\frac{\partial V^k}{\partial x}\right)' B_1 \frac{\partial V^{i-k}}{\partial x} \\ &\quad = \sum_{k=0}^{j}\sum_{i=k}^{k+j} \varepsilon^i \left(\frac{\partial V^k}{\partial x}\right)' B_1 \frac{\partial V^{i-k}}{\partial x} - \sum_{k=0}^{j}\sum_{i=k}^{j} \varepsilon^i \left(\frac{\partial V^k}{\partial x}\right)' B_1 \frac{\partial V^{i-k}}{\partial x} \\ &\quad = \sum_{k=1}^{j}\sum_{i=j+1}^{k+j} \varepsilon^i \left(\frac{\partial V^k}{\partial x}\right)' B_1 \frac{\partial V^{i-k}}{\partial x} = \varepsilon^{j+1}\delta_j, \end{aligned}$$

$$\delta_j = \sum_{k=1}^{j}\sum_{i=0}^{k-1} \varepsilon^i \left(\frac{\partial V^k}{\partial x}\right)' B_1 \frac{\partial V^{i-k}}{\partial x}.$$

Hence, the function Q_j satisfies the equation

$$\begin{aligned} L Q_j &+ \varepsilon f' \frac{\partial Q_j}{\partial x} + x' N_1 x + \varepsilon^{j+1}\delta_j \\ &= \frac{1}{4}\left(\frac{\partial Q_j}{\partial x}\right)' B_1 \frac{\partial Q_j}{\partial x}, \qquad Q_j(T,x) = x' N_2 x. \end{aligned}$$

It follows that $Q_j(t,x)$ is the Bellman function for the problem of the control of the system (5.1) with cost functional

$$J(u) + \varepsilon^{j+1} E \int_0^T \delta_j(t, x(t))\, dt. \tag{5.33}$$

The optimal control in the problem (5.1), (5.33) is $u_j(t,x)$. Using the same argument as in the proof of the zero and the first approximation estimates, we arrive at the relation

$$0 \leqslant J(u_j) - J(v) \leqslant \varepsilon^{j+1} C(1 + |x_0|^2).$$

Main Results and Formulas of Chapter IX

Problem of the optimal control of a stochastic system:
equations of evolution of the system

$$dx(t) = a(t,x,u)\,dt + \sigma(t,x,u)\,d\xi(t);$$

functional to be minimized (the cost functional)

$$J(u) = E\Big[F\big(x(t)\big) + \int_0^T F_1\big(t,x(t),u\big)\,dt\Big] \to \inf_u;$$

control constraint

$$u(t) \in U(t);$$

the Bellman function

$$V(t,x) = \inf_u E\Big[F\big(X_{t,x}(T,u)\big) + \int_t^T F_1\Big(s, X_{t,x}(s,u), u\big(s, X_{t,x}(s,u)\big)\Big)\,ds\Big];$$

the Bellman equation

$$\begin{aligned}\inf_{u\in U}[L_u V(t,x) + F_1(t,x,u)] &= 0, \qquad 0 \leqslant t \leqslant T,\\ V(T,x) &= F(x),\\ L_u V(t,x) &= V_t(t,x) + V_x'(t,x)a(t,x,u)\\ &\quad + \tfrac{1}{2}\operatorname{tr}[\sigma(t,x,u)\sigma'(t,x,u)V_{xx}(t,x)].\end{aligned}$$

Stochastic linear-quadratic problem:

$$\begin{aligned}dx(t) &= [A(t)x(t) + f(t) + B(t)u]\,dt + \sigma_1(t)x(t)\,d\xi_1(t)\\ &\quad + \sigma_2(t)u\,d\xi_2(t) + \sigma(t)\,d\xi(t), \qquad 0 \leqslant t \leqslant T,\ x(0) = x_0,\\ J(u) &= E\Big[x'(T)N_1x(T) + 2h_1'x(T)\\ &\quad + \int_0^T \{x'(t)N_2(t)x(t) + u'(t)N_3(t)u(t)\\ &\quad + 2x'(t)N_4(t)u(t) + 2h_2'(t)x(t) + 2h_3'(t)u(t)\}\,dt\Big],\\ &\quad N_1 \geqslant 0,\ N_2 \geqslant 0,\ N_3 > 0;\end{aligned}$$

the Bellman function

$$V(t,x) = x'P(t)x + 2G'(t)x + g(t);$$

optimal control

$$u_0(t,x) = -K(t,P)\big[B'(t)P(t)x + N_4'(t)x + B'(t)G_1(t) + h_3(t)\big],$$

where

$$K(t,P) = [N_3(t) + \sigma_2'(t)P(t)\sigma_2(t)]^{-1} > 0,$$

$$\dot P(t) + A'(t)P(t) + P(t)A(t) + \big(P(t)B(t) + N_4(t)\big)K(t,P)\big(P(t)B(t) + N_4(t)\big)'$$
$$+ N_2(t) + \sigma_1'(t)P(t)\sigma_1(t) = 0, \qquad 0 \leqslant t \leqslant T,$$
$$P(T) = N_1;$$
$$\dot G(t) + A'(t)G(t) + P(t)f(t) + h_2(t) + \big(N_4(t) + P(t)B(t)\big)$$
$$\times K(t,P)\big(h_3(t) + B'(t)G(t)\big) = 0, \qquad 0 \leqslant t \leqslant T,$$
$$G(T) = h_1;$$

$$g(t) = \int_t^T \big[2G'(s)f(s) + \big(h_3'(s) + G'(s)B(s)\big)K(s,P)$$
$$\times \big(h_3(s) + B'(s)G(s)\big) + \operatorname{tr} P(s)\sigma(s)\sigma'(s)\big]\,ds;$$

the minimum value of the cost functional

$$J(u_0) = V(t_0, x_0).$$

LQ-problem under incomplete information:

$$dx(t) = [A(t)x(t) + B(t)u + f(t)]\,dt + \sigma(t)\,d\xi(t),$$
$$dy(t) = Q(t)x(t) + \sigma_0(t)\,d\xi_0(t);$$

the *mean* $m(t)$ and the *covariance matrix* $D(t)$ of the conditional probability distribution of $x(t)$, given $y(t)$

$$dm(t) = [A(t)m(t) + f(t) + B(t)u\big(t, m(t)\big)]\,dt$$
$$+ D(t)Q'(t)\big(\sigma_0(t)\sigma_0'(t)\big)^{-1}[dy(t) - Q(t)m(t)\,dt],$$
$$0 \leqslant t \leqslant T,\ m(0) = m_0,$$
$$\dot D = AD + DA' - DQ'\big(\sigma_0\sigma_0'\big)^{-1}QD + \sigma\sigma', \qquad D(0) = 0;$$

optimal control

$$u_0(t,m) = -N_3^{-1}(t)\big[\big(B'(t)P(t)x + N_4'(t)\big)m + B'(t)G(t) + h_3(t)\big];$$

minimum value of the cost functional

$$\min_u J(u) = V(0, m_0) + \operatorname{tr}\big(D(T)N_1\big) + \int_0^T \operatorname{tr}\big(D(t)N_2(t)\big)\,dt.$$

Stabilization of stochastic control systems:

$$dx(t) = a\big(t, x(t), u\big)\,dt + \sigma\big(t, x(t), u\big)\,d\xi(t), \qquad t \geqslant 0,\ x(0) = x_0,$$
$$J(u) = E\int_0^\infty F\Big(t, x(t), u\big(t, x(t)\big)\Big)\,dt \to \inf_{u\in U};$$

the Bellman equation

$$\inf_{u\in U}[L_u V(t,x) + F(t,x,u)] = 0,$$

where

$$L_u V(t,x) = \frac{\partial V(t,x)}{\partial t} + a'(t,x,u)\frac{\partial V}{\partial x} + \tfrac{1}{2}\operatorname{tr}\sigma(t,x,u)\sigma'(t,x,u)\frac{\partial^2 V(t,x)}{\partial x^2};$$

the optimal control $u_0(t,x)$ is a solution of the equation

$$L_{u_0}V_0(t,x) + F\big(t,x,u_0(t,x)\big) = 0.$$

Stabilization of stochastic LQ-problems:

$$dx(t) = [A(t)x(t) + B(t)u(t)]\,dt + \sigma(t)x(t)\,d\xi(t), \qquad x(0) = x_0,$$
$$J(u) = E\int_0^\infty [x'(t)N_1(t)x(t) + u'N_0(t)u]\,dt \to \inf_u, \qquad N_0 > 0,\ N_1 > 0;$$

optimal control

$$u_0(t,x) = -N_0^{-1}(t)B'(t)P(t)x,$$

where $P(t)$ is a unique positive definite solution of the Riccati matrix equation

$$\dot P(t) + A'(t)P(t) + P(t)A(t) - P(t)B(t)N_0^{-1}(t)B'(t)P(t) + N_1(t) + \sigma'(t)P(t)\sigma(t) = 0, \qquad t \geqslant 0.$$

Quasilinear system:

$$dx(t) = [\varepsilon f(t,x) + B(t)u]\,dt + \sigma(t)\,d\xi(t), \qquad 0 \leqslant t \leqslant T,\ x(0) = x_0,$$
$$J(u) = E\left[x'(T)N_2x(T) + \int_0^T \big(x'(t)N_1(t)x(t) + u'N_0(t)u\big)\,dt\right];$$

the Bellman function

$$V(t,x) = V^0(t,x) + \varepsilon V^1(t,x) + \dots,$$

where

$$V_t^0 + \tfrac{1}{2}\operatorname{tr}(\sigma\sigma' V_{xx}^0) - \tfrac{1}{4}(V_x^0)' B_1 V_x^0 + x' N_1 x = 0,$$
$$V^0(T,x) = x' N_2 x, \; B_1 = Q N_0^{-1} B',$$
$$V_t^i + \frac{1}{2}\operatorname{tr}(\sigma\sigma' V_{xx}^i) + f' V_x^{i-1} - \frac{1}{4}\sum_{j=0}^{i} (V_x^j)' B_1 V_x^{i-j} = 0,$$
$$V^i(T,x) = 0;$$

the ith *approximation to the optimal solution*

$$u_i(t,x) = -\tfrac{1}{2} N_0^{-1} B' (V_x^0 + \cdots + \varepsilon^i V_x^i);$$

the estimate for the ith *approximation to the cost functional*

$$0 \leqslant J(u_i) - J(v) \leqslant \varepsilon^{i+1} C (1 + |x_0|^2),$$

where v is the optimal control.

CHAPTER X

OPTIMAL CONTROL ON A TIME INTERVAL OF RANDOM DURATION

In this chapter we deal with methods of solving stochastic time-optimal problems and control problems with a stochastic cost functional. A number of examples are considered, namely the problems of the control of a rigid body, of a gyrostat, of a material point, and of a simple pendulum under random disturbances.

§ 1. Time-Optimal Control

1. Statement of time-optimal problems in dynamical systems under random disturbances. Consider the time-optimal problem for autonomous dynamical systems of the form

$$\dot{x}(t) = f\big(x(t)\big) + Bu\big(x(t)\big) + \sigma\big(x(t)\big)\dot{\xi}(t), \qquad t \geqslant 0, \tag{1.1}$$
$$x(0) = x_0 \in R^n,$$

where B is a given constant matrix, the function f and the matrix σ satisfy the conditions assuring the existence of a solution to equation (1.1) for $u = 0$, $\xi(t)$ is a standard Wiener process, and u is an F-controlsatisfying for all x the constraint condition

$$u(x) \in U \tag{1.2}$$

Here U is a prescribed bounded set.

Let there also be given a closed set Q (bounded or unbounded) in the phase space R^n. In what follows, Γ denotes the boundary of Q, and $\tau_x(u)$ is a random variable that is equal to the *first passage time*, i.e., the time which it takes the system (1.1) to first reach the boundary of Q, given the control u and the initial value $x(0) = x$, where the arbitrary initial vector $x \in R^n \setminus Q$. If some realizations of the system (1.1) do not reach Q in any finite time, the time $\tau_x(u)$ for such trajectories is assumed to be equal to infinity.

It is required to determine a control $u = u(x)$ satisfying the constraint (1.2) and minimizing the mean first passage time

$$E\tau_x(u) \to \inf_{u \in U} \tag{1.3}$$

for any $x \in R^n \setminus Q$.

In optimal control theory for deterministic systems, i.e., for systems of the form (1.1), where $\sigma = 0$, the time-optimal problem holds a special place. This problem has been extensively investigated, and a number of results have been obtained. In particular, the existence and uniqueness theorems for time-optimal control and the form of optimal control have been established, as well as various examples have been treated.

As to the problem (stated above) of minimizing the mean first passage time which it takes the controlled stochastic process to reach the prescribed set, the known results can be reduced, in the main, to the following assertion: under certain conditions for the smoothness of the minimum value $V(x)$ of the functional (1.3) the function $V(x)$ satisfies the following partial differential equation (see § 2 of Chapter VIII):

$$\inf_{u\in U}\big(L_u V(x)+1\big)=0, \qquad x\in Q_1=R^n\setminus Q,\ V(x)=0,\ x\in\Gamma, \tag{1.4}$$
$$L_u V=(f+Bu)'V_x+\tfrac{1}{2}\operatorname{tr}(\sigma\sigma' V_{xx}).$$

Here Q_1 is the complement of Q with respect to R^n, a prime denotes transposition, the vector V_x has coordinates $\partial V/\partial x_i$, the matrix V_{xx} has components $\partial^2 V/(\partial x_i \partial x_j)$, and tr is the trace of a matrix.

However, it should be stressed that the problem (1.4) has, in general, a nonunique solution. Therefore, even if $V(x)$ is a sufficiently smooth function, then it is not clear in advance with which solution it coincides.

Let us illustrate this point by an example.

Example 1.1. Suppose that the system (1.1) is a scalar system of the form

$$dx(t)=u\big(x(t)\big)\,dt+\sqrt{2}\,d\xi(t), \qquad |u|\leqslant 1,\ x(0)=x. \tag{1.5}$$

The set Q is the interval $(-\infty,\alpha]$, where α is a given number. Let $x>\alpha$. Then equation (1.4) becomes

$$V_{xx}-|V_x|+1=0, \qquad x>\alpha,\ V(\alpha)=0.$$

Any function of the form $V(x)=x-\alpha+C(e^{x-\alpha}-1)$ can serve as a solution of this equation. Here and subsequently C denotes various nonnegative constants. Observe that the optimal control is $u(x)=-1$ in this example. Indeed, if $x(t,u)$ is the solution of the problem (1.5) generated by the control u, then $x(t,-1)\leqslant x(t,u)$ with probability one for any control u, $|u|\leqslant 1$. Given the control $u=-1$, the optimal transfer time is equal to $x-\alpha$ for any $x\geqslant\alpha$.

2. Existence of an admissible control. We call a control $u(x)$ *admissible* if there exists a solution of equation (1.1) generated by this control so that $E\tau_x(u)<\infty$ for all finite x. We give without proof the conditions for the existence of an admissible control in terms of Lyapunov's direct method.

Lemma 1.1. *Suppose that in Q_1 there exist a function $u(x)$ and a twice continuously differentiable scalar function $W(x)\geqslant 0$ so that, given the control $u(x)$, there exists a solution of the problem* (1.1) *and*

$$L_u W(x)\leqslant -C, \qquad x\in Q_1. \tag{1.6}$$

Then $u(x)$ is an admissible solution, with $E\tau_x(u)\leqslant C^{-1}W(x)$.

Example 1.2. Consider a material point in rectilinear motion subjected to a control force u ($|u| \leqslant 1$) and to a random disturbing force $\sigma\dot{\xi}$, where σ is a given constant and ξ is a scalar Wiener process. The equations of motion are of the form

$$\dot{x}_1(t) = x_2, \qquad dx_2(t) = u\,dt + \sigma\,d\xi(t), \qquad t \geqslant 0. \tag{1.7}$$

It is required to transfer the coordinate x_1 to zero in minimum time for an arbitrary initial condition (i.e., the set Q is the line $x_1 = 0$). Let us show that for any σ the admissible control is

$$u = u(x_1, x_2) = \begin{cases} 1, & \text{for } x_1 < 0, \\ -1, & \text{for } x_1 > 0. \end{cases} \tag{1.8}$$

To prove that the control (1.8) is admissible, we construct a Lyapunov-type function $W(x)$ and verify the inequality (1.6).

Let us fix the constant $k_1 > \sigma^2/2$. Now choose a k_2 so that for all x_1 and x_2 the function

$$W = k_1 x_2 + \tfrac{1}{2}x_2^2 + k_2 + x_1, \qquad x_1 > 0,$$
$$W = -k_1 x_2 + \tfrac{1}{2}x_2^2 + k_2 - x_1, \qquad x_1 < 0,$$

is positive. On carrying out a calculation, we obtain

$$L_u W = -k_1 + \frac{\sigma^2}{2} < 0$$

everywhere outside the line $x_1 = 0$. Hence, by virtue of Lemma 1.1 the control (1.8) is admissible. It is easily shown that this control is optimal. Indeed, for $x_1(0) > 0$ and for any control $u_1(x) \geqslant -1$ the inequality $\tau_{u_1}(x) < \tau_{-1}(x)$ cannot be true, since $x_1(t, -1) \leqslant x_1(t, u_1)$ with probability one. Here $x_1(t, u)$ is a solution of the system (1.7), given the control u and the arbitrary initial data $x_1(0)$, $x_2(0)$. In a similar way, we can establish the optimality of the control $u = 1$ for $x_1 < 0$.

In the investigation of the existence of admissible control the conditions of Lemma 1.1 can be relaxed in some specific cases. More precisely, suppose in Q_1 there exist a function $u(x)$ and a continuous scalar function $W(x) \geqslant 0$ such that, given the control $u(x)$, there is a solution of the system (1.1), and for any $t_1 \leqslant t_2$ the following inequality is valid:

$$EW\big(x(t_2, u)\big) - EW\big(x(t_1, u)\big) \leqslant -C(t_2 - t_1).$$

Then $u(x)$ is an admissible control. In particular, it is sufficient to require that the function W satisfy the relation

$$\limsup_{\Delta \to +0} \frac{1}{\Delta}\big[EW\big(x(\Delta, u)\big) - W(x)\big] \leqslant -1. \tag{1.9}$$

In this case, if W satisfies the conditions of Lemma 1.1, then the limit in the relation (1.9) is $L_u W(x)$.

3. An algorithm for constructing optimal control. We give a method for constructing time-optimal control. Let S_N denote a sequence of concentric balls of radius N in the space R^n such that $Q \subset S_N$. We denote by s_n the boundary of S_N. Let us determine the sequence of the scalar functions $V_N(x)$ which are the solution of the following boundary-value problems:

$$\inf_{u \in U} (L_u V_N(x) + 1) = 0, \qquad x \in Q_1 \cap S_N, \tag{1.10}$$
$$V_N(x) = 0, \ x \in \Gamma \cup s_N.$$

Here the expression $L_u V_N$ is defined by formula (1.4). When certain conditions are met [4,6], there exists a unique solution V_N of the problem (1.10), with $V_N \geqslant 0$, and for any $N_2 \geqslant N_1$ the following inequality holds:

$$V_{N_2}(x) \geqslant V_{N_1}(x), \qquad x \in Q_1 \cap S_{N_1}.$$

Moreover:

(1) there exists a minimum positive solution of the external boundary-value problem (1.4) denoted by $V_0(x)$;

(2) in any bounded region of x the set $V_N(x)$ converges uniformly to $V_0(x)$ as $N \to \infty$, being monotone nondecreasing;

(3) for no admissible control can we obtain a value of the functional (1.3) which is less than $V_0(x)$;

(4) if there exists a solution of equation (1.1) generated by the control $u_0(x)$ that satisfies the condition

$$\inf_{u \in U} L_u V_0(x) = L_{u_0}(x) V_0(x), \tag{1.11}$$

then $u_0(x)$ is an optimal control.

Let us point out some of the distinctive features of the outlined method for constructing optimal control. As noted above, the Bellman equation (1.4) has, in general, no unique solution. In order to select the Bellman function (equal to the mean transfer time (1.3)) from among these solutions, we introduce a sequence of auxiliary control problems in the following way. Consider the system (1.1) in the region $Q_1 \cap S_N$, and let $\tau_x^N(u)$ be the mean first passage time which it takes the system (1.1) to reach the boundaries of the set $Q_1 \cap S_N$, given the control u and the initial condition $x \in Q_1 \cap S_N$.

The auxiliary control problem is the problem of determining a control u for which the time of exit from the region $Q_1 \cap S_N$ is a minimum, i.e., $E\tau_x^N(u)$ is a minimum over u. The Bellman equation (1.10) corresponding to this auxiliary control problem has a unique solution V_N. Let now $N \to \infty$, i.e., the external boundary of the region $Q_1 \cap S_N$ tends to infinity. Then in the auxiliary problems the optimal control as N increases must change in a way such that as many trajectories as possible tend to the inner boundary, since the time of reaching the external boundary of the region $Q_1 \cap S_N$ for any given initial condition becomes greater than the time of reaching the inner boundary.

Let us outline the main steps of constructing a time-optimal control:

1°. The sequence $V_N(x)$ of solutions of the boundary-value problem (1.10) is constructed.

2°. The limit $V_0(x)$ of the sequence $V_N(x)$ as $N \to \infty$ is determined.

3°. The optimal control u_0 is determined via formula (1.11).

4. Time-optimal control of the motion of a rigid body. Consider the controlled motion of a rigid body relative to its mass center subjected to control torques and random disturbances. Let x_i $(i = 1, 2, 3)$ be the components of the angular momentum of the body relative to the coordinate system whose axes coincide with the principal central axes of inertia (see Figure 3.1 of Chapter I). The Euler equations of motion of a rigid body are of the form

$$\begin{aligned} \dot{x}_1(t) &= x_2x_3(a_2 - a_3)(a_2a_3)^{-1} + u_1 + \sigma\dot{\xi}_1, \qquad t \geqslant 0, \\ \dot{x}_2(t) &= x_1x_3(a_3 - a_1)(a_3a_1)^{-1} + u_2 + \sigma\dot{\xi}_2, \\ \dot{x}_3(t) &= x_1x_2(a_2 - a_1)(a_2a_1)^{-1} + u_3 + \sigma\dot{\xi}_3. \end{aligned} \tag{1.12}$$

Here the vector of initial conditions $x(0) = \big(x_1(0), x_2(0), x_3(0)\big)'$ the a_i are the principal central moments of inertia, σ is a constant, the control $u = (u_1, u_2, u_3)$ is constrained by

$$|u| = (u_1^2 + u_2^2 + u_3^2)^{1/2} \leqslant b, \tag{1.13}$$

where b is a constant.

The system (1.12) constrained by (1.13) admits, for example, the following interpretation. The torques are produced by a pair of controlled gas jets that can be directed at arbitrary angles relative to the rigid body, the effect of each jet being the same regardless of its direction.

We set ourselves the problem of constructing control for the system (1.12) such that the absolute value of the angular momentum takes the prescribed value $r > 0$ in minimum mean time, given an arbitrary initial condition $x(0)$, $|x(0)| \geqslant r$. To solve this problem, it is first necessary to construct the sequence of solutions V_N of the boundary-value problem (1.10), then to determine the limit of the sequence of V_N as $N \to \infty$, and, finally, to construct the optimal control. Let us apply equations (1.1) for the case in hand. For $r < |x| < N$ we have

$$\begin{aligned} &\frac{\sigma^2}{2}\left(\frac{\partial^2 V_N}{\partial x_1^2} + \frac{\partial^2 V_N}{\partial x_2^2} + \frac{\partial^2 V_N}{\partial x_3^2}\right) + \min_{u, |u| \leqslant b}\left(u_1\frac{\partial V_N(x)}{\partial x_1} + u_2\frac{\partial V_N(x)}{\partial x_2}\right. \\ &\left. + u_3\frac{\partial V_N(x)}{\partial x_3}\right) + x_2x_3\frac{a_2 - a_3}{a_2a_3}\frac{\partial V_N}{\partial x_1} + x_1x_3\frac{a_3 - a_1}{a_1a_3}\frac{\partial V_N}{\partial x_2} \\ &+ x_1x_2\frac{a_1 - a_2}{a_1a_2}\frac{\partial V_N}{\partial x_3} = -1; \\ &\qquad V_N(x) = 0, \ |x| = r, \ |x| = N. \end{aligned} \tag{1.14}$$

We define the number $r_1(N)$ as a root of the equation

$$\int_r^{r_1} z(t)\,dt \int_t^{r_1} \frac{du}{z(u)} = \int_{r_1}^{N} z_1(t)\,dt \int_{r_1}^{t} \frac{du}{z_1(u)}, \qquad r \leqslant r_1 \leqslant N. \tag{1.15}$$

Here the following notation is used:

$$\begin{aligned} z(t) &= (rt^{-1})^2 \exp[2b\sigma^{-2}(t-r)], \\ z_1(t) &= (rt^{-1})^2 \exp[2b\sigma^{-2}(r-t)]. \end{aligned} \tag{1.16}$$

The left side of (1.15) for $r_1 = r$ is zero, and, by virtue of (1.16), it increases monotonically. But the right side of (1.15) is zero for $r_1 = N$, and it decreases monotonically for $r \leqslant r_1 \leqslant N$. Hence, for any fixed N equation (1.15) has a unique solution, $r_1(N)$, constrained by

$$r < r_1(N) < N. \tag{1.17}$$

Substitution in (1.14) serves to prove that the solution of equation (1.14) is given by the formulas

$$\begin{aligned} V_N(x) &= \frac{2}{\sigma^2} \int_r^{|x|} z(t)\,dt \int_t^{r_1} \frac{dt_1}{z(t_1)}, && r \leqslant |x| \leqslant r_1(N), \\ V_N(x) &= \frac{2}{\sigma^2} \int_{|x|}^{N} z_1(t)\,dt \int_{r_1}^{t} \frac{dt_1}{z_1(t_1)}, && r_1(N) \leqslant |x| \leqslant N. \end{aligned}$$

Next, we show that

$$\lim_{N\to\infty} r_1(N) = \infty. \tag{1.18}$$

To compute the derivative $dr_1(N)/dN$, we differentiate both sides of the equality (1.15) with respect to N:

$$\begin{aligned} \frac{dr_1(N)}{dN} &= \left[z_1(N) \int_{r_1}^{N} \frac{dt}{z_1(t)} \right] \\ &\times \left[\frac{1}{z_1(r_1)} \int_{r_1}^{N} z_1(t)\,dt + \frac{1}{z(r_1)} \int_r^{r_1} z(t)\,dt \right]^{-1} > 0. \end{aligned} \tag{1.19}$$

Hence, in view of (1.17), we find that the sequence $r_1(N)$ increases monotonically. Assume that the equality (1.18) does not hold. This means that there exists a number r_0 ($r < r_0 < \infty$) such that $r_1(N) \leqslant r_0$ for all N. However, we can show that in this case there exists a constant $C_1 < 0$ such that for all $N > r_0$ the following estimate is valid:

$$\frac{dr_1(N)}{dN} \geqslant C_1; \tag{1.20}$$

obviously, this estimate contradicts the inequality $r_1(N) \leqslant r_0$. By virtue of (1.16) the first factor on the right side of (1.19) is expressed by

$$d_1 = \frac{1}{N^2} \exp\left(-\frac{2b}{\sigma^2}N\right) \int_{r_1}^{t} t^2 \exp\left(\frac{2b}{\sigma^2}t\right) dt.$$

By integrating, we establish that for all $N > r_0$ and $C_2 > 0$ the following inequality is valid:

$$d_1 \geqslant C_2 > 0. \tag{1.21}$$

In a similar way, we can show that for all N and $C_3 > 0$ the following inequalities are valid:

$$\begin{aligned} \frac{1}{z_1(r_1)} \int_{r_1}^{N} z_1(t)\, dt &\leqslant e^{2br_1/\sigma^2} \int_{r_1}^{N} e^{-2bt/\sigma^2}\, dt \leqslant C_3, \\ \frac{1}{z(r_1)} \int_{r}^{r_1} z(t)\, dt &\leqslant r_1^2 e^{2br_1/\sigma^2} \int_{r}^{r_1} \left(\frac{1}{t}\right)^2 e^{-2bt/\sigma^2}\, dt \leqslant C_3. \end{aligned} \tag{1.22}$$

The estimate (1.20) follows from (1.21), (1.22), (1.19). Thus, we have proved the validity of the relation (1.18). On any bounded set of values of x we have by virtue of (1.18), (1.17)

$$\lim_{N\to\infty} V_N(x) = \lim_{r_1\to\infty} \frac{2}{\sigma^2} \int_r^{|x|} z(t)\, dt \int_t^{r_1} \frac{dt_1}{z(t_1)}.$$

Substituting in this relation the expression for the function $z(t)$ from (1.16) gives

$$\lim_{N\to\infty} V_N(x) = V_0(x) = \frac{1}{b}\left[|x| - r + \frac{\sigma^2}{b} \ln \frac{|x|}{r} + \frac{\sigma^4}{2b^2}\left(\frac{1}{r} - \frac{1}{|x|}\right)\right]. \tag{1.23}$$

Using the expression (1.23) for V_0, we determine the control $u_0(x)$ that yields the minimum with respect to u, $|u| \leqslant b$, of the expression

$$u_1 \frac{\partial V_0}{\partial x_1} + u_2 \frac{\partial V_0}{\partial x_2} + u_3 \frac{\partial V_0}{\partial x_3}.$$

We have

$$u_0(x) = -bx|x|^{-1}. \tag{1.24}$$

To prove the optimality of the control (1.24), it is sufficient to verify that the control (1.24) is admissible, i.e., that, given this control, there exists a solution of the system (1.12) and $E\tau_x(u_0) < \infty$. However, these facts immediately follow from the relations (1.24), (1.22) and the equality $L_{u_*} V_0 = -1$. Thus, the time-optimal control of the motion of a rigid body represents a quantity whose absolute value is a maximum and whose direction is contrary to that of the angular momentum. The minimum mean time that it takes the angular momentum to reach the prescribed value is given by the expression (1.23), which is independent of the moments of inertia.

5. Numerical construction of the time-optimal control of the motion of a material point. We shall utilize the algorithm given above for the numerical solution of the time-optimal control problem for the motion of a material point. The point is in rectilinear motion subjected to a scalar control force u, $|u| \leqslant b$, and a random disturbing force $\sigma\dot{\xi}(t)$, where $\xi(t)$ is a standard Wiener process, b and σ are given constants. The equations of motion are of the form

$$\dot{x}(t) = y(t), \qquad dy(t) = u\big(x(t), y(t)\big)\,dt + \sigma\,d\xi(t), \qquad t \geqslant 0, \tag{1.25}$$

where $x(t)$ is the coordinate of the point, and $y(t)$ its velocity. Let Q be a prescribed neighborhood of the origin $x = 0$, $y = 0$. We denote by $\tau_u(x, y)$ the time which it takes the point controlled by u and starting from an arbitrary state $(x, y) \in R^2 \setminus Q$ at $t = 0$ to first reach the set Q. The time-optimal problem in this context is the problem of determining a control u_0 such that the mean passage time to the set Q is a minimum, i.e.,

$$E\tau_{u_0}(x, y) = \inf_u E\tau_u(x, y), \qquad |u| \leqslant b,\ u_0 = u_0(x, y).$$

The Bellman equation associated with the problem in hand for $(x, y) \in R^2 \setminus Q$, according to (1.4), is of the form

$$y\frac{\partial V(x,y)}{\partial x} + \inf_{u,\,|u|\leqslant b}\left(u\frac{\partial V(x,y)}{\partial y}\right) + \frac{\sigma^2}{2}\frac{\partial^2 V(x,y)}{\partial y^2} = -1,$$

where the boundary condition is $V(x, y) = 0$ for (x, y) lying on the regular portion of the boundary of the set Q, i.e., on that portion which is reached by the trajectories of the system (1.25), given an arbitrary admissible control. For the other (irregular) portion of the boundary we need not specify a boundary condition. The procedure for the numerical construction of the minimal positive solution V_0 of the Bellman equation and of the optimal control

$$u_0(x, y) = -b\,\mathrm{sgn}\,\frac{\partial V_0(x,y)}{\partial y}$$

is based on the method of fractional steps (see [17] in the bibliography of Part Four).

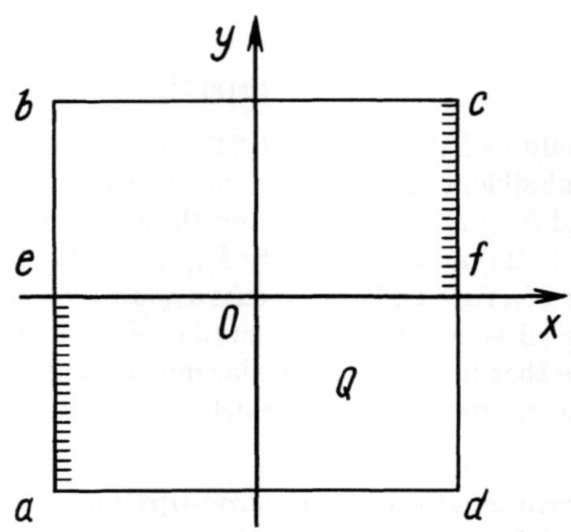

FIG. 1.1. The target set in the time-optimal problem of the control of a material point

Consider the computation results in the case when the region Q is the square $abcd$ (see Figure 1.1). We introduce dimensionless variables according to the formulas $x \to bx$, $y \to by$, $u \to bu$, $b_1 = \sigma^2/2b^2$. The Bellman equation can be expressed in terms of the new variables as

$$y\frac{\partial V(x,y)}{\partial x} - \left|\frac{\partial V(x,y)}{\partial y}\right| + b_1\frac{\partial^2 V(x,y)}{\partial y^2} = -1. \tag{1.26}$$

In Figure 1.1 the hatched portion of the boundary of the set Q indicates the portion that cannot be reached by the point regardless of the value of the control, provided that the motion starts from a position outside the region Q. Therefore, we need not specify a boundary condition for this portion of the boundary. On the remaining part of the boundary the function $V = 0$. For the expanding sequence of regions E_N that enters in the algorithm for constructing an optimal control (see subsection 3) we also take the squares $ABCD$ (see Figure 1.2) of side $2N$.

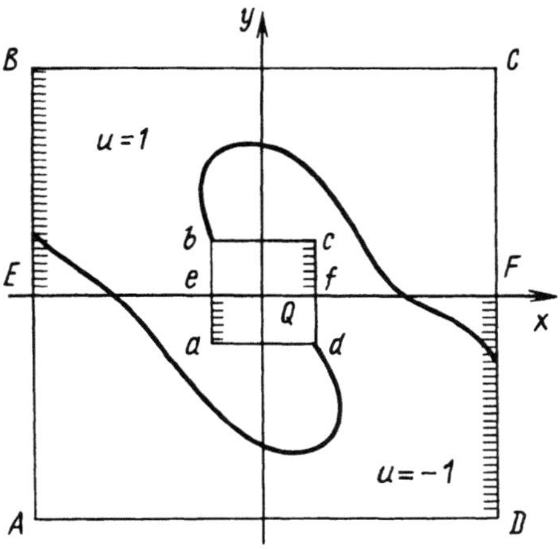

FIG. 1.2. A view of the switching curve in the auxiliary time-optimal problem

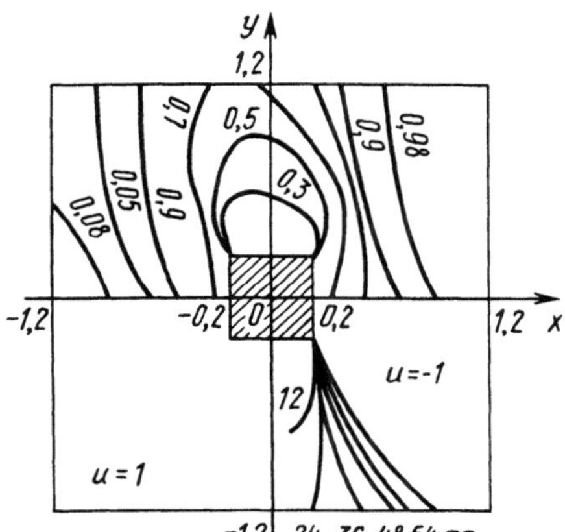

FIG. 1.3. Level curves for the Bellman function and switching curves for optimal control in the time-optimal control problem for the material point

The auxiliary function $V_N(x, y)$ can be determined as a solution of equation (1.25) in the region $E_N \setminus Q$ with zero boundary conditions on the unhatched part of this region. The portions of the boundary EB, FD, ae, and fc of the region $E_N \setminus Q$ are irregular. They cannot be reached by the system evolving from within the region $E_N \setminus Q$ whatever the control may be, and, consequently, boundary conditions for V_N need not be given on them.

A typical switching curve for the control function

$$u_N(x,y) = -\operatorname{sgn} \frac{\partial V_N(x,y)}{\partial y}$$

is shown in Figure 1.2. Some computation results for the original problem are given in Figure 1.3 for the case when *abcd* is a square of side 0.4 (in dimensionless units). In the upper part of the figure the level curves for the Bellman function (equal to the mean passage time) are given. In the lower part of the figure the switching curve for the optimal control ($N = \infty$), together with an approximate view of this curve for different values of the parameter N=12, 24, 36, 48, 54, is shown. For other parts of the figure the switching curve for optimal control and the level curves for the Bellman function can be determined with the help of the relations

$$V(x,y) = V(-x,-y), \qquad u_0(x,y) = -u_0(-x,-y)$$

that follow from the symmetry of the problem. Note that the switching curve for the deterministic time-optimal problem ($\sigma = 0$) designated in Figure 1.3 by the value $N = \infty$ has been determined from the equation $x \operatorname{sgn} y + y^2/2 = -0.18$. The calculations performed show that the switching curve for the stochastic problem ($\sigma \neq 0$) converges as $N \to \infty$ to the switching curve for the deterministic problem. In other words, the time-optimal control is the same for both the deterministic and stochastic problems.

§ 2. Time-Optimality for a Gyrostat

In this section we treat the problem of damping a gyrostat in minimum mean time. This problem belongs, in essence, to one of the central topics of dynamics, the problem of motion of a rigid body about a fixed point, and is connected with the research into ways to control this motion. At present problems of the control of the motion of a rigid body, in particular, problems of the stabilization of equilibrium states and of rotary motion are vigorously investigated, for the most part, in the deterministic setting. Two main control modes are considered: passive control and active control. Passive control is connected with the use of aerodynamic, magnetic, and gravitational forces and is widely used because of the comparative ease of implementing such control. Active control makes use of jet engines and rotating masses (rotors, gyroscopes, and so on). Compared to passive control, it assures a greater rate of decay of the transient process and a smaller value of the cost functional.

1. Problem statement. Consider a mechanical system S consisting of a rigid body with a fixed point O and three homogeneous symmetric rotors (see Figure 2.1). Let $Ox_1x_2x_3$ denote a stationary rectangular coordinate system and $O\eta_1\eta_2\eta_3$ denote a coordinate system whose axes coincide with the principal axes of inertia of the system S relative to the point O. The axes of rotation of the rotors are assumed to coincide with the axes η_1, η_2, η_3. A system of this type is called a *gyrostat.* It is the hallmark of such a system that its mass distribution does not change during the motion. Each rotor of the gyrostat executes a rotation being subjected to a control torque produced by a motor. The gyrostat is acted upon by external random forces, which create random torques.

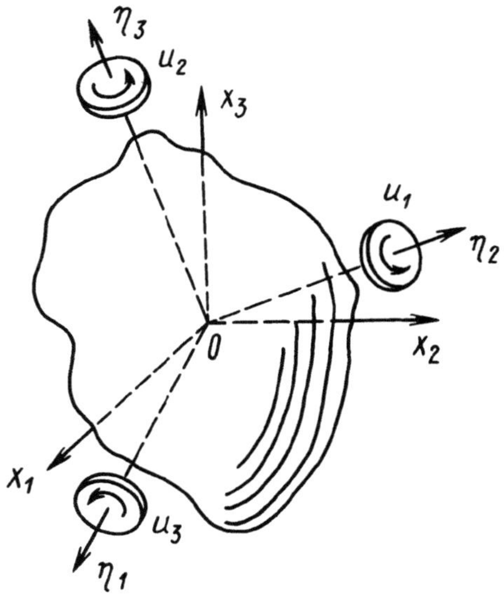

FIG. 2.1. Control of the motion of a gyrostat

Suppose that p is the absolute angular velocity vector of the system S considered as one rigid body, and z is the absolute angular momentum of the gyrostat relative to the point O. The equations of motion referred to the coordinate system $O\eta_1\eta_2\eta_3$ have the form

$$\dot{z} + p \times z = \sigma\dot{\xi}, \qquad t \geqslant 0, \tag{2.1}$$

$$J_i(\dot{w}_i + \dot{p}_i) = -u_i, \qquad i = 1, 2, 3. \tag{2.2}$$

Here the matrix $\sigma = (\sigma_{ij})$, where $i, j = 1, 2, 3$, that characterizes the noise intensity is prescribed, J_i are the axial moments of inertia of the rotors, ω_i are their relative angular velocities, and u_i are the components of the control torque vector u that is due to the motors.

The random disturbing torque $\dot{\xi} = (\dot{\xi}_1, \dot{\xi}_2, \dot{\xi}_3)$ is described by a Gaussian white noise; the symbol $\times$ designates a vector (outer) product. Equations (2.1), (2.2) describe the motion of the mechanical system. Let us eliminate from equations (2.1), (2.2) the angular velocities of the rotors ω_i. Note that $z_i = c_i p_i + J_i\omega_i$, where c_i is the moment of inertia of the gyrostat about the axis η_i. By virtue of

(2.1), (2.2) we have

$$\begin{aligned} A_1\dot{p}_1 &= z_2p_3 - z_3p_2 + u_1 + \sum_{j=1}^{3}\sigma_{1j}\dot{\xi}_j, \\ A_2\dot{p}_2 &= z_3p_1 - z_1p_3 + u_2 + \sum_{j=1}^{3}\sigma_{2j}\dot{\xi}_j, \\ A_3\dot{p}_3 &= z_1p_2 - z_2p_1 + u_3 + \sum_{j=1}^{3}\sigma_{3j}\dot{\xi}_j, \end{aligned} \tag{2.3}$$

where $A_i = c_i - J_i > 0$. Equations (2.1), (2.3) do not contain the rotor angular velocities. It is these equations that are investigated in the sequel.

Take as a target set a ball Q:

$$Q = \{p : |p| = (p_1^2 + p_2^2 + p_3^2)^{1/2} \leqslant \varepsilon,\ \varepsilon > 0\}.$$

We denote by Γ the boundary of this set, and by Q_1 its exterior: $Q_1 = R^3 \setminus Q$ Let $\tau_{pz}(u)$ be the first moment when the angular velocity $p(t) \in \Gamma$, provided that the control u is activated and the motion starts from the arbitrary initial values of angular velocity $p \in Q$ and of angular momentum z.

It is required to determine a control which is constrained by

$$|u| = \left(\sum_{i=1}^{3} u_i^2\right)^{1/2} \leqslant b, \qquad b > 0, \tag{2.4}$$

so as to minimize the expression

$$E\tau_{pz}(u) \to \inf_u, \tag{2.5}$$

i.e., to minimize the mean time which it takes the angular velocity to reach the prescribed neighborhood Q of zero. In this context the damping problem for the gyrostat is to be treated with regard for random disturbances.

2. The existence of an admissible control. It will be recalled that the control u is admissible if, given this control, $E\tau_{pz}(u) < \infty$ for all finite values of p, z, i.e., it is possible to damp the gyrostat in finite mean time.

Note that in the justification of the admissibility of this control u we may consider Γ to be a sphere.

By Lemma 1.1 for a control u to be admissible it is sufficient that there exist a function $W(p,z)$ bounded below in Q_1 and a constant $C > 0$ such that for all $p \in Q_1$ and z the following inequality is valid:

$$L_u W(p,z) \leqslant -C. \tag{2.6}$$

In the inequality (2.6) it is assumed that

$$L_u = (z_2p_3 - z_3p_2)\frac{\partial}{\partial z_1} + (z_3p_1 - z_1p_3)\frac{\partial}{\partial z_2} + (z_1p_2 - z_2p_1)\frac{\partial}{\partial z_3} + A_1^{-1}(z_2p_3 - z_3p_2 + u_1)\frac{\partial}{\partial p_1} + A_2^{-1}(z_3p_1 - z_1p_3 + u_2)\frac{\partial}{\partial p_2} + A_3^{-1}(z_1p_3 - z_3p_1 + u_3)\frac{\partial}{\partial p_3} + \frac{1}{2}\sum_{j,l,k=1}^{3}\frac{1}{A_jA_l}\sigma_{jk}\sigma_{lk} \times \left(\frac{\partial^2}{\partial p_l\,\partial p_j} + A_jA_l\frac{\partial^2}{\partial z_l\,\partial z_j} + 2A_j\frac{\partial^2}{\partial z_j\,\partial p_l}\right). \tag{2.7}$$

The function W for the system (2.1), (2.3) may be taken as

$$W = \varphi(p) - \frac{\alpha}{\varphi^m(p)}, \qquad \varphi(p) = A_1p_1^2 + A_2p_2^2 + A_3p_3^2, \tag{2.8}$$

where α and m are positive constants to be determined later.

Suppose that the matrix σ is nondegenerate and, using the function (2.8), we shall show that the control u_0 being in a reverse direction from that of the angular velocity vector and equal to

$$u_0 = -bp|p|^{-1}, \qquad |p| = \left(\sum_{i=1}^{3}p_i^2\right)^{1/2}. \tag{2.9}$$

is admissible.

Computing L_uW with regard to (2.7)–(2.9) gives

$$L_uW = -2b|p| + \sum_{j,k=1}^{3}\frac{1}{A_j}\sigma_{jk}^2 + \frac{\alpha m}{\varphi^{m+1}(p)} \times \left[-2b|p| + \sum_{j,k=1}^{3}\frac{1}{A_j}\sigma_{jk}^2 - \frac{2(1+m)p'\sigma\sigma'p}{\varphi(p)}\right]. \tag{2.10}$$

Since the matrix σ is nondegenerate for all sufficiently large m, the expression in brackets in (2.10) is negative for all $p \in Q_1$. Let us choose and fix one such m. Further, let us choose a sufficiently large number N so that

$$N > \frac{1}{2b}\sum_{j,k=1}^{3}\frac{1}{A_j}\sigma_{jk}^2 = \frac{1}{2b}\sum_{j=1}^{3}\frac{1}{A_j}\sum_{k=1}^{3}\sigma_{jk}^2.$$

Clearly, for all p, $|p| \geqslant N$ and the chosen m the right side of (2.10) is less than some negative number. Thus, the estimate (2.6) remains to be established only in the region $Q_1 \cap \{|p| \leqslant N\}$. However, for this p and chosen m, N the inequality (2.6) is satisfied, provided we take a sufficiently large number α. Thus, we have established the validity of the relation (2.6).

Since, obviously, the function (2.8) is bounded below in Q_1, the control (2.9) is admissible.

If there are no disturbing torques in equations (2.3), i.e., the matrix $\sigma = 0$, then the control (2.9) is admissible as before; this can be justified as before either with the help of the function (2.8) or through the use of the more simple function $\varphi(p)$.

3. Construction of an optimal control. Consider now certain cases in which the optimal control and the corresponding value of the time (2.5) can be obtained explicitly. Let the geometry of the masses of the gyrostat be such that

$$A_1 = A_2 = A_3 = A. \tag{2.11}$$

We assume that the diffusion matrix σ is diagonal with identical elements in the diagonal which are equal to $\sqrt{\sigma_0}$, where the number $\sigma_0 \geqslant 0$, and the surface Γ is a sphere of given radius r.

Under the assumptions made we further compute the value V_0 of the functional (2.5), given the control (2.9). To this end, we define the sequence of functions W_N from the relations

$$L_{u_0} W_N(p,z) = -1, \qquad r < |p| < N, \ z \in R^3,$$
$$W_N(p,z) = 0, \qquad |p| = r, \ |p| = N, \ z \in R^3.$$

Because of the admissibility of the control u_0 we have

$$V_0 = \lim_{N \to \infty} W_N(p,z).$$

It is readily verified that

$$W_N(p,z) = W_N(p) = \frac{2A^2}{\sigma_0}\left[\int_r^N q(t)\,dt\right]^{-1} \int_r^{|p|} q(t)\,dt \int_{|p|}^N q(t_1)\,dt_1 \int_t^{t_1} \frac{dt_2}{q(t_2)}, \tag{2.12}$$

where

$$q(t) = \left(\frac{r}{t}\right)^2 \exp\left[\frac{2Ab}{\sigma_0}(t-r)\right].$$

Therefore, passing to the limit as $N \to \infty$ gives

$$V_0(p) = \frac{A}{b}\left[|p| - r + \frac{\sigma_0}{Ab}\ln\frac{|p|}{r} + \frac{\sigma_0^2}{2A^2b^2}\left(\frac{1}{r} - \frac{1}{|p|}\right)\right], \qquad |p| \geqslant r. \tag{2.13}$$

Let us show now that the optimal control is given by formula (2.9), and the corresponding minimum value of the time (2.5) by the expression (2.13).

To justify this, in the six-dimensional space of p, z we consider the sequence of functions V_N defined as the solution of the boundary-value problem

$$\inf_{u,|u|\leqslant b} L_u V_N(p,z) = -1, \qquad r < |p| < N,\ z \in R^3, \tag{2.14}$$
$$V_N(p,z) = 0,\ |p| = r,\ |p| = N,\ z \in R^3.$$

By a direct substitution we verify that the solution of equation (2.14) is of the form

$$\begin{aligned} V_N(p,z) = V_N(p) &= \frac{2A^2}{\sigma_0}\int_r^{|p|} q(t)\,dt \int_t^{r_1} \frac{dt_1}{q(t_1)}, \qquad r \leqslant |p| \leqslant r_1 \\ V_N(p,z) = V_N(p) &= \frac{2A^2}{\sigma_0}\int_{|p|}^{N} q_1(t)\,dt_1 \int_{r_1}^{t} \frac{dt_1}{q_1(t_1)}, \qquad r_1 < |p| \leqslant N. \end{aligned} \tag{2.15}$$

Here we have used the notation (2.12) for $q(t)$ and the equality defining the function $q_1(t)$:

$$q_1(t) = \left(\frac{r}{t}\right)^2 \exp\left[\frac{2Ab}{\sigma_0}(r-t)\right]; \tag{2.16}$$

the quantity r_1 denotes the unique root of the equation

$$\int_r^{r_1} q(t)\,dt \int_t^{r_1} \frac{du}{q(u)} = \int_{r_1}^{N} q_1(t)\,dt \int_{r_1}^{t} \frac{du}{q_1(u)}. \tag{2.17}$$

It follows from the expression (2.17) that $\lim_{N\to\infty} r_1(N) = \infty$. Thus, noting the equalities (2.12), (2.15), (2.16), we find that

$$\lim_{N\to\infty} V_N = V_0(p), \tag{2.18}$$

where $V_0(p)$ is defined by the expression (2.13).

Let $\gamma(u) = \tau_{pz}^N(u)$ denote the time which it takes the process (2.1), (2.3) to first reach the boundaries Γ_N of the cylinder $r \leqslant |p| \leqslant N$, given the control u and the initial conditions $r \leqslant |p_0| \leqslant N$, $z_0 \in R^3$. We assume that z_0 is arbitrary, and $p(t,u)$ is a solution of the system (2.3). Using the principle of dynamic programming and the relations (2.14), (2.15), we have

$$V_N = \inf_{u,|u|\leqslant b} E\gamma(u) = \inf_{u,\,|u|\leqslant b} E\tau_{pz}^N(u).$$

Hence, the function V_N defined by formulas (2.15) is the minimum mean first passage time to the boundaries Γ_N.

We now prove the optimality of the control (2.3) and the corresponding time (2.13). Suppose that, given an admissible control v, the following inequality holds for some values of p_0, z_0:

$$E\gamma(v) < V_0(p_0). \tag{2.19}$$

The Ito formula yields

$$
\begin{aligned}
V_N\Big(p(\gamma(v),v),z(\gamma(v),v)\Big) &- V_N(p_0,z_0) \\
&= E\int_0^{\gamma(v)} L_v V_N\left(p(t,v),z(t,v)\right)dt \\
&\geqslant E\int_0^{\gamma(v)} \inf_{u,|u|\leqslant b} L_u V_N\left(p(t,u),z(t,u)\right)dt = -E\gamma(v).
\end{aligned}
$$

Therefore, for all N we have

$$
V_N(p_0,z) \leqslant E\gamma(v). \tag{2.20}
$$

It follows from the relations (2.20) and (2.18) that $V_0(p_0) \leqslant E\gamma(v)$. This contradicts the inequality (2.19). Thus, the proof of the optimality of the control (2.9) is completed.

§ 3. Control Problems with Stochastic Functional

In a number of practical situations the system functions normally only when its coordinates and velocities are within the prescribed limits. This leads to various problems which involve maximizing different cost functionals (such as the probability of reaching the target set, the time of being within a given region of phase coordinates, and so on). Some of these problems are treated in this section.

1. Problem statement and method of solution. For the system (1.1) consider the control problem with stochastic cost functional.

Suppose we have a bounded closed target set $Q \in R^n$ with the boundary Γ. We denote by Q_1 the complement of Q with respect to R^n. Let $\tau_x(u)$ be the time which it takes the system (1.1) to first reach the boundary Γ, given the control u and the initial condition $x(0) = x \in Q_1$. A control (1.2) is said to be *admissible* if, given this control u and an arbitrary $x(0) \in Q_1$, there exists a solution of equation (1.1). Let $P(\cdot)$ denote the probability of the event enclosed in parentheses.

It is required to determine an admissible control such that it yields the maximum of the probability $P\big(\tau_x(u) < \infty\big)$ being the probability of reaching the target set Q in finite time for any $x \in Q_1$.

The algorithm for constructing optimal control in this problem is similar to the procedure for the determination of time-optimal control discussed in § 1. We give it without proof (the proof is given in [4, 6]).

Consider the following boundary-value problem connected with the control problem under consideration:

$$
\sup_{u\in U} L_u V(x) = 0, \quad x \in Q_1; \qquad V(x) = 1, \quad x \in \Gamma. \tag{3.1}
$$

Here the operator L_u is defined by formula (1.4). It can be shown in the usual way that if the function

$$
\sup_{u\in U} P\big(\tau_x(u) < \infty\big), \qquad x \in Q_1 \tag{3.2}
$$

is sufficiently smooth with respect to x, then it satisfies the relations (3.1). However, even in this case these relations, in general, do not define the function (3.2) in view of the fact that the external boundary-value problem (3.1) need not have a unique solution. Further, the regular computational procedure for determining optimal control which involves the solution of the Bellman equation and the subsequent determination of the optimal control is not always feasible in our case, since we may come up against situations in which the function (3.2) is identically equal to unity. As in § 1, we let $S_N \subset R^n$ denote the sequence of concentric spheres of radius N with boundary s_N, with $Q \subset S_N$ and the sequence of functions $V_N(x)$ defined by

$$\sup_{u \in U} L_u V_N(x) = 0, \qquad x \in Q_1 \cap S_N, \tag{3.3}$$
$$V_N(x) = 1, \quad x \in \Gamma, \qquad V_N(x) = 0, \quad x \in s_N.$$

It should be noted that for a number of assumptions there exists a unique solution of the boundary-value problem (3.3) such that

$$1 > V_{N_2}(x) \geqslant V_{N_1}(x) > 0, \qquad x \in Q_1 \cap S_{N_1},\ N_1 \geqslant N_2.$$

The sequence $V_N(x)$ is uniformly convergent to $V_0(x)$ as $N \to \infty$ for any bounded set of values of the argument x, and the limit $V_0(x)$ is a minimum positive solution of the problem (3.1).

Let $u_N(x)$ denote the measurable function that yields the supremum in (3.3). We consider the sequence of controls $v_N(x)$ that are equal to $u_N(x)$ for $x \in Q_1 \cap S_N$ and equal to an arbitrary fixed constant belonging to U for other values of x. Then for any admissible control u we obtain

$$J(x,u) = P\big(\tau_x(u) < \infty\big) \leqslant V_0(x).$$

Further, if the controls $v_N(x)$ are admissible, then $J(x, v_N) \to V_0(x)$ as $N \to \infty$ is uniformly convergent on any bounded set of values of x. Moreover, if the control $v_N(x)$ is used in the system (1.1), then we can approach as much as desired to the optimal value $V_0(x)$ of the functional (3.2) for sufficiently large N.

Consider some properties of the function V_0. As is noted above, if $V_0 \equiv 1$, then the problem (3.1) does not yield an optimal control. The sufficient condition for the identity $V_0 \equiv 1$ is the existence of a unique solution of the boundary-value problem (3.1) in the class of bounded functions, since the bounded function $V_0(x)$ satisfies the relations (3.1), and the function$V \equiv 1$ is also a solution of (3.1).

The other sufficient condition for the identity $V_0 \equiv 1$, given some admissible control u, is the existence of a nonnegative function W such that $L_u W(x) \leqslant -C$ for some constant $C > 0$.

Next, consider the case in which

$$\lim_{|x| \to \infty} V_0(x) = 0. \tag{3.4}$$

We can show that by virtue of the equality (3.4) any admissible control u_0 that yields the maximum of the expression $L_u V_0$ is optimal. To this end, we use the formula

$$EV_0\Big(x\big(\tau_x^N(u_0), u_0\big)\Big) - V_0(x) = 0, \tag{3.5}$$

where $\tau_x^N(u)$ is the time which it takes the process $x(t,u)$ to first leave the region $Q_1 \cap S_N$, given the initial condition $x(0,u) = x \subset Q_1 \cap S_N$. We express the first term in (3.5) as

$$EV_0\Big(x\big(\tau_x^N(u_0), u_0\big)\Big) = P\Big(x\big(\tau_x^N(u_0), u_0\big) \in \Gamma\Big) + E_1 V_0\Big(x\big(\tau_x^N(u_0), u_0\big)\Big), \tag{3.6}$$

where the symbol E_1 indicates that the expectation is taken over all trajectories satisfying the relation $x\big(\tau_x^N(u_0), u_0\big) \in s_N$.

In the equality (3.6) let us pass to the limit as $N \to \infty$. Then, noting the relations (3.4)–(3.6) we obtain $J(x, u_0) = V_0$. Thus, by the foregoing we establish the optimality of the control u_0. We can prove that the sufficient condition for the relation (3.4) is the existence of a positive continuous function W for which the following relations are valid:

$$\sup_{u \in U} L_u W(x) \leqslant 0, \qquad x \in Q_1, \quad \lim_{|x| \to \infty} W(x) = 0.$$

2. Optimal control of the motion of a material point involving a stochastic cost functional. Suppose that the motion of a material point is described by equations (1.25). Consider the problem of determining the optimal control u which renders the maximum of the probability of reaching the boundary $x = 0$ earlier than the boundary $x = q_1$ ($q_1 > 0$ is a given number), provided that at the initial moment the motion of the system (1.25) starts from the arbitrary point (x, y), where $0 < x < q_1$. We denote by τ the time which it takes the system (1.25) to first reach the boundary of the region $0 \leqslant x \leqslant q_1$ and by $V(x,y)$ the Bellman function defined by the relation

$$V(x,y) = \sup_{u, |u| \leqslant b} P\big(x(\tau) = 0\big).$$

By virtue of (3.1) the equation for V is of the form

$$yV_x + \sup_{u, |u| \leqslant b} (uV_y) + \tfrac{1}{2}\sigma^2 V_{yy} = 0, \qquad 0 < x < q_1, \; y \in (-\infty, \infty). \tag{3.7}$$

Observe that there is no control for which the motion of the system (1.25) starting in the strip $0 < x < q_1$ can terminate on the half-lines $x = 0$, $y > 0$ and $x = q_1$, $y < 0$. Therefore, on these half-lines the boundary conditions for equation (3.7) need not be given. On the rays $x = 0$, $y \leqslant 0$ and $x = q_1$, $y \geqslant 0$ we have

$$V(0,y) = 1, \quad y \leqslant 0; \qquad V(q_1, y) = 0, \quad y \geqslant 0. \tag{3.8}$$

It follows from the equations of motion (1.25) that the optimal F-control in the problem under consideration is $u_0(x,y) \equiv u_0 = -b$. Subsituting the control u_0 in (3.7) gives

$$yV_x - bV_y + \tfrac{1}{2}\sigma^2 V_{yy} = 0,$$
$$V(0,y) = 1, \quad y \leqslant 0; \qquad V(q_1,y) = 0, \quad y \geqslant 0.$$

Effecting the change of variables

$$y \to C_1 y, \qquad x \to C_2 x, \qquad C_1 = \frac{\sigma^2}{2} b^{-1},\ C_2 = \left(\frac{\sigma^2}{2}\right)^2 b^{-3}, \tag{3.9}$$

we can write the foregoing equations in the form

$$yV_x - V_y + V_{yy} = 0, \qquad V(0,y) = 1, \quad y \leqslant 0, \tag{3.10}$$
$$V(q,y) = 0, \quad q = q_1 C_2^{-1},\ y \geqslant 0.$$

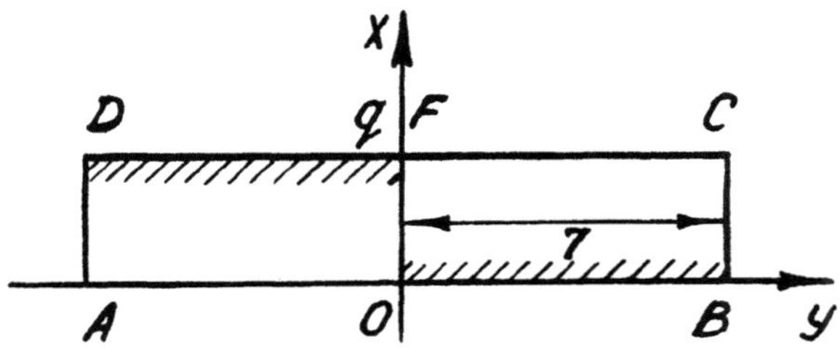

FIG. 3.1. A view of the regular portions of the boundary in the control problem with stochastic functional

For a numerical solution of the problem (3.10) the stabilization method is used. We approximate the function $V(x,y)$ by a function $W(t,x,y)$ satisfying the relations

$$W_t + yW_x - W_y + W_{yy} = 0, \qquad 0 \leqslant x \leqslant q,\ -T \leqslant t \leqslant 0, \tag{3.11}$$
$$W(t,q,y) = 0, \qquad t \leqslant 0,\ y \geqslant 0,$$
$$W(t,0,y) = 1, \qquad t \leqslant 0,\ y \leqslant 0,$$
$$W(0,x,y) = 0, \qquad x > 0,\ -\infty < y < \infty.$$

Note that the function $W(t_0, x_0, y_0)$ is the probability of reaching the straight line $x = 0$ earlier than the line $x = q$ by the system (1.25) being in motion on the segment $[-T, 0]$, given $u = -1$, $\sigma = 1$, if the motion started from the point x_0, y_0 at the moment t_0, where $0 < x_0 < q$. Therefore,

$$W(-T, x, y) \to V(x, y) \quad \text{for } T \to \infty.$$

We can determine $W(t, x, y)$ for $-T \leqslant t \leqslant 0$ and x, y from the rectangle $ABCD$ shown in Figure 3.1, where $BC = OF = AD = q$ and $OB = OA = L$. The region $ABCD$ transforms into the strip $0 \leqslant x \leqslant q$ as $L \to \infty$.

Let us specify the conditions on the boundaries of the region $ABCD$.

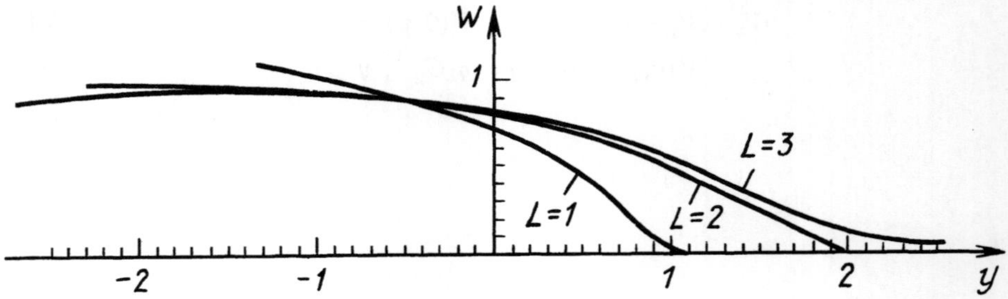

FIG. 3.2. A view of the Bellman function

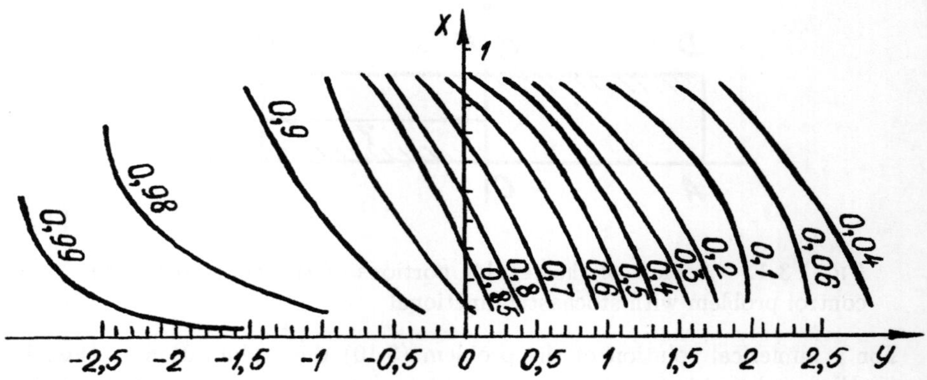

FIG. 3.3. Level curves for the Bellman function

In view of the condition (3.11) and the probabilistic meaning of the function W we have

$$\begin{aligned} W(t, x, y) &= 1, \qquad t < 0, (x, y) \in OAD, \\ W(t, x, y) &= 0, \qquad t \leqslant 0, (x, y) \in BCF. \end{aligned} \tag{3.12}$$

Since the system (1.25) does not reach the rays $x = 0$, $y > 0$ and $x = q$, $y < 0$, no boundary conditions are to be given on these rays. The unattainable portion of the boundary in Figure 3.1 is hatched. For $t > 0$ we have from (3.11)

$$W(0,x,y) = 0, \quad x > 0, \qquad W(0,0,y) = 1, \quad y \leqslant 0. \tag{3.13}$$

The scheme for the numerical solution of the problem (3.10)–(3.11) is the same as given in subsection 5 of § 1. Here we cite the numerical results. Figure 3.2 depicts the convergence of the function W with respect to L after stabilization with respect to T has already taken place for $q = 1$, $x = 0.5$. Figure 3.2 demonstrates that stabilization with respect to L occurs at least for $L = 3$.

In Figure 3.3 level curves of the function $W(-T,x,y)$ are shown for $T = 10$, $L = 3$, $q = 1$ when the solution of the problem (3.11)–(3.13) has stabilized with respect to both T and L. Therefore, the level curves in Figure 3.3 practically coincide with the level curves for the desired Bellman function $V(x,y)$. The numerical results given in Figure 3.3 show that the maximum probability for the material point to reach the straight line $x = 0$ earlier than the line $x = q$ decreases for the increasing velocity, given a fixed coordinate, as well as for the increasing coordinate, given a fixed velocity.

3. Maximization of the probability that a point stays in a given region. Consider a controlled motion of the material point governed by equations (1.25) starting from an arbitrary state $(x,y) \in Q$ at time $t = 0$. It is required to determine the F-control $u = u(x,y)$ that yields the maximum of the probability that the point stays in the region Q over a given time interval $0 \leqslant t \leqslant T_1$. The Bellman equation for this problem is of the form

$$V_t + yV_x + \sup_{u,|u|\leqslant b} (uV_y) + \tfrac{1}{2}\sigma^2 V_{yy} = 0.$$

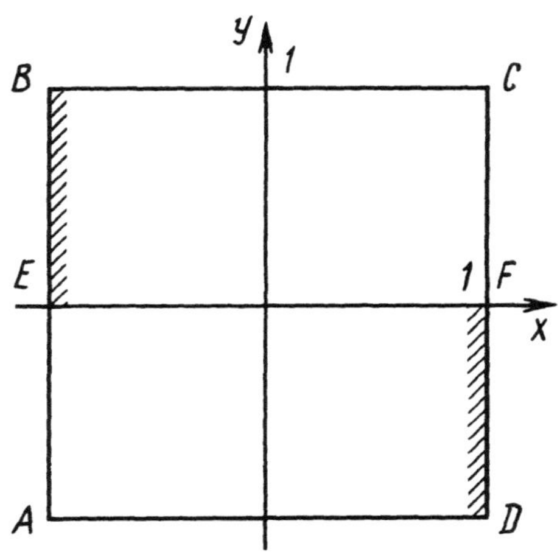

FIG. 3.4. A view of the unattainable portions of the boundary in the problem of maximizing the probability of the point staying inside the given region

We effect a change of phase coordinates according to formulas (3.9) and a change of time according to the formula $t \to 2^{-1}\sigma^2 b^{-2} t$. Suppose that the region Q in the new variables has the form of the square $|x| \leqslant 1$, $|y| \leqslant 1$ (see Figure 3.4). The Bellman equation in the new variables becomes

$$V_t + yV_x + |V_y| + V_{yy} = 0, \qquad 0 \leqslant t \leqslant T = 2b^2\sigma^{-2}T_1. \tag{3.14}$$

For $t = T$ inside the region Q the following initial condition holds:

$$V(T, x, y) = 1, \qquad |x| < 1,\ |y| < 1. \tag{3.15}$$

When the motion starts strictly from the inside of the square Q, the portions BE and FD of the boundary cannot be reached. The unattainable portion of the boundary of the square Q is shown in Figure 3.4 as hatched. The boundary conditions for equation (3.14) need not be given on the hatched portion of the boundary. For points $(x, y) \in BCF \cup EAD$ of the boundary the following relation holds:

$$V(t, x, y) = 0, \qquad 0 \leqslant t \leqslant T. \tag{3.16}$$

On calculation of the solution V of the problem (3.14)–(3.16) the optimal control u_0 is expressed by

$$u_0(t, x, y) = \operatorname{sgn} V_y(t, x, y). \tag{3.17}$$

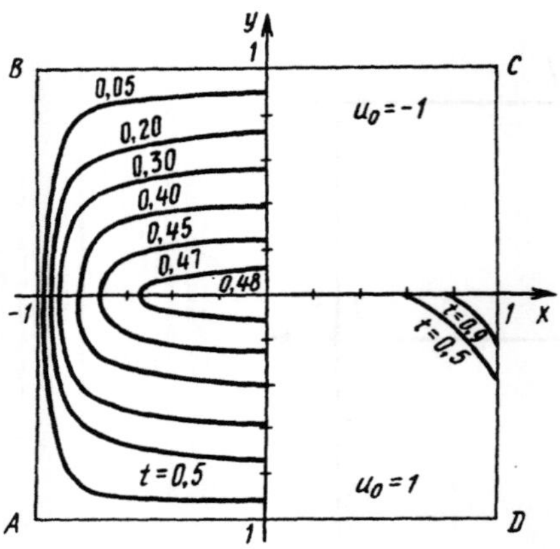

FIG. 3.5. The Bellman level curve (on the left) and the switching curve for optimal control (on the right)

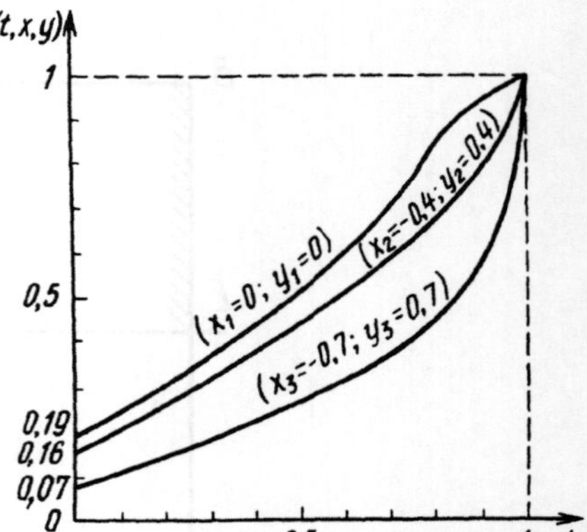

FIG. 3.6. Graphs of the Bellman function

The problem (3.14)–(3.16) has been solved numerically using the method of fractional steps for the following values of the parameters: $T = 1$, the coordinate step 0.1, the time step 0.01. The computation results are shown in Figure 3.5. On the left side of the figure the level curves of the function $V(t,x,y)$ are given for $t = 0.5$; on the right side the switching curves for the optimal control u_0 are given for $t = 0.5$ and $t = 0.9$. By the central symmetry of the problem we can determine V and u_0 in the other part of the square by using the relations

$$V(t,x,y) = V(t,-x,-y);$$
$$u_0(t,x,y) = -u_0(t,-x,-y).$$

In Figure 3.6 the time dependence of the Bellman function V is shown for three fixed points of the square with coordinates: $x_1 = 0$, $y_1 = 0$; $x_2 = -0.4$, $y_2 = 0.4$; $x_3 = -0.7$, $y_2 = -0.7$.

4. Stochastic control of the motion of a simple pendulum. The motion of a simple pendulum subjected to random disturbances is described by the equations

$$\begin{aligned} &\dot{x}(t) = y(t), \\ &dy(t) = [-a^2x(t) + u(x,y)]\,dt + \sigma\,d\xi(t) + \gamma\,d\omega(t), \qquad t \geqslant 0. \end{aligned} \tag{3.18}$$

Here x is the coordinate, y is the velocity, $\xi(t)$ and $\omega(t)$ are independent standard Wiener and Poisson processes, $E\omega(t) = \lambda t$, the control $u(x,y)$ is subject to the condition $|u| \leqslant b$, the positive constants a, b, σ, γ, λ are prescribed. The motion of the system (3.18) starts at the moment $t = 0$ from the state (x,y) belonging to a given bounded region Q. It is required to deduce a control such that the probability of the system (3.18) staying in the region Q on a given time interval $[0,T]$ is a maximum.

Denote by $V(t,x,y)$ the Bellman function for this problem which is equal to the maximum with respect to u of the probability of the system (3.18) staying in Q on the time interval $[t,T]$, provided that $\big(x(t),y(t)\big) = (x,y) \in Q$. We carry out the change of variables according to the formulas

$$\begin{aligned} &t \to \frac{\sigma^2}{2b^2}t, \qquad x \to \frac{\sigma^4}{4b^3}x, \qquad y \to \frac{\sigma^2}{2b}y, \qquad u \to bu, \\ &a^2 \to \frac{4b^4}{\sigma^4}a^2, \qquad \lambda \to \frac{2b^2}{\sigma^4}\gamma, \qquad \gamma \to \frac{\sigma^2}{2b}\gamma, \qquad T \to \frac{\sigma^2}{2b^2}T. \end{aligned}$$

In the new variables the equation for the function V becomes

$$\begin{aligned} &\frac{\partial V}{\partial t} + y\frac{\partial V}{\partial x} - a^2x\frac{\partial V}{\partial y} + \left|\frac{\partial V}{\partial y}\right| + \frac{\partial^2 V}{\partial y^2} \\ &\quad + \lambda\big(V(t,x,y+\gamma) - V(t,x,y)\big) = 0, \qquad t \in [0,T]. \end{aligned} \tag{3.19}$$

Equation (3.19) is solved in the square $Q = ABCD$ (see Figure 3.4). The portions BE and FD of the boundary are unattainable. Therefore, the boundary values for them need not given. For other points of the boundary of the region Q the boundary conditions are of the form

$$V(t,x,y) = 0, \qquad 0 \leqslant t \leqslant T,\ (x,y) \in EAD \cup BCF. \tag{3.20}$$

The initial condition is given by

$$V(T,x,y) = 1, \qquad |x| < 1,\ |y| < 1. \tag{3.21}$$

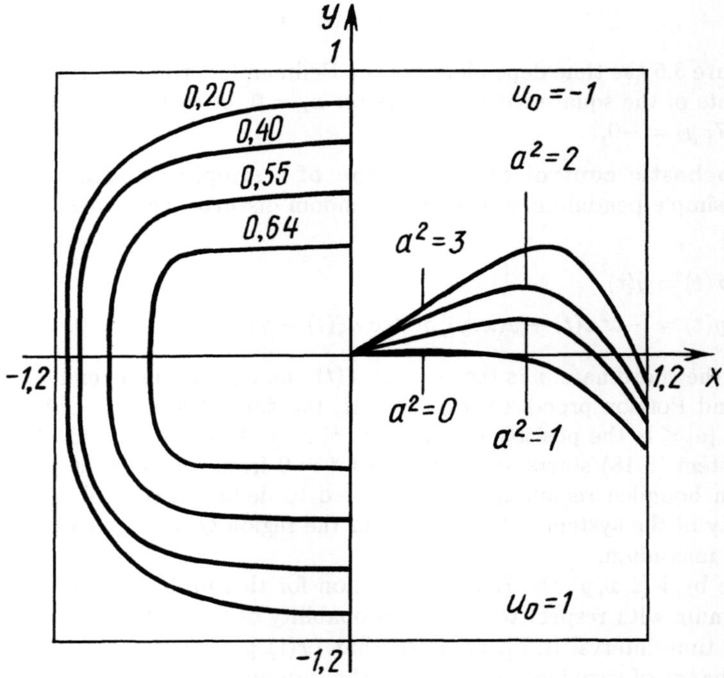

FIG. 3.7. Switching curves and level lines of the Bellman function in the problem of the control of a simple pendulum

The optimal control is defined by the relation $u_0(t,x,y) = \operatorname{sgn} V_y(t,x,y)$ of the form (3.17). For numerical solution of the problem (3.19)–(3.21), as above, the method of fractional steps was used. The computation results for $T = 1$ and the time step 0.01 are given in Figure 3.7, where the switching curves for optimal control (on the right) for $t = 0.6$ and different values of a^2 (a^2=0, 1, 2, 3) and also the level curves of the function $V(t,x,y)$ (on the left) for $t = 0.5$ and $a^2 = 1$ are given.

To construct the switching curves for optimal control and the level lines of the function $V(t,x,y)$ in the other half of the square it is sufficient to note that

$$V(t,x,y) = V(t,-x,-y), \qquad u_0(t,x,y) = -u_0(t,-x,-y).$$

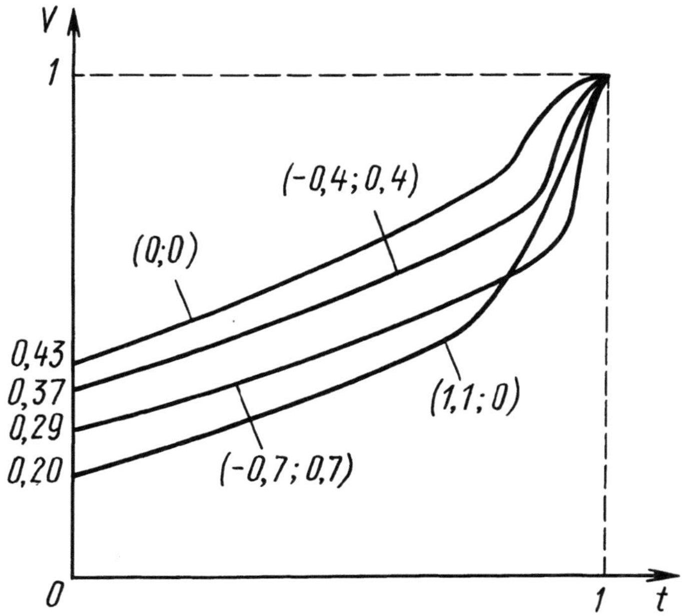

FIG. 3.8. Graphs of the maximum value of probability

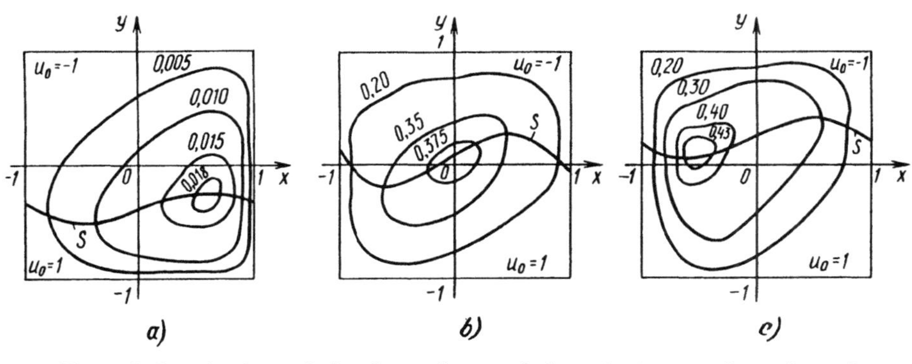

FIG. 3.9. A view of the dependence of the solution on the value of impulse disturbance for $a^2 = 3$, $t = 0.5$, and:
a, $\lambda = 10$, $\gamma = 10$; b, $\lambda = 10$, $\gamma = 2$; c, $\lambda = 1$, $\gamma = 10$

In Figure 3.8 the time-dependence of the function $V(t, x, y)$ is shown for four fixed points with coordinates (0, 0), (-0.4, 0,4), (-0.7, 0.7), (1.1, 0) and for $a^2 = 1$. Note that the level curves of the function $V(t, x, y)$ and its dependence on t for fixed (x, y) and different a^2 remain practically the same.

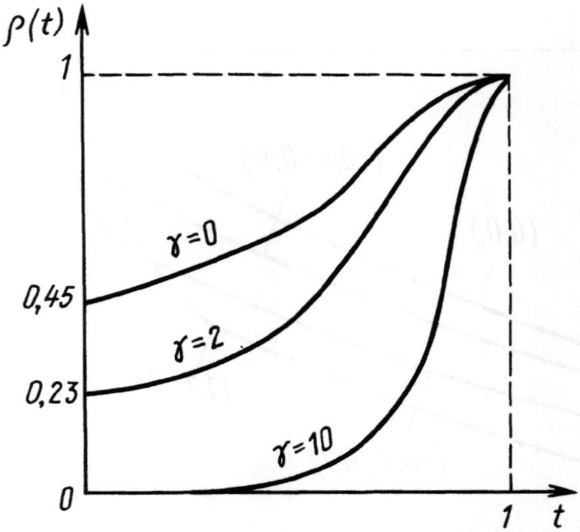

FIG. 3.10. The maximum of the Bellman function as a function of the value of an impulse disturbance

Let now $\gamma \neq 0$. The central symmetry of the level curves and the switching curve is now disturbed (see Figure 3.9, *a*). In Figure 3.9 the closed curves are the level curves of the function V, and the switching curve is designated by S. However, treating the function V as a function of γ yields $V(t,x,y,\gamma) = V(t,-x,-y,-\gamma)$. Thus, when the sign of γ is reversed, the pattern is reflected symmetrically with respect to the origin. For $\gamma > 0$ the switching curve moves downward (relative to the axis Ox), and for $\gamma < 0$ the switching curve moves upward (see Figure 3.9, *b*, and Figure 3.9, *c*). As γ increases (in absolute value) the function $\rho(t) = \sup_{(x,y)\in S_0} V(t,x,y)$ decays at a faster rate as $t \to 0$ (see Figure 3.10, where λ=10). The same effect occurs as λ increases, given a constant γ.

5. Control of a rigid body, given the stochastic cost functional. Consider the controlled motion of a rigid body acted upon by random diturbances. Let x_i be the components of angular momentum x relative to the coordinate system, rigidly bound with the body, whose axes coincide with the principal central axes of inertia of the body. The Euler equations of motion are of the form (1.12):

$$\begin{aligned} \dot{x}_1 &= x_2x_3(a_2-a_3)(a_2a_3)^{-1} + u_1 + \sigma\dot{\xi}_1, \qquad a_i > 0, \\ \dot{x}_2 &= x_1x_3(a_3-a_1)(a_3a_1)^{-1} + u_2 + \sigma\dot{\xi}_2, \\ \dot{x}_3 &= x_1x_2(a_2-a_1)(a_2a_1)^{-1} + u_3 + \sigma\dot{\xi}_3. \end{aligned} \tag{3.22}$$

The control u_i is subject to the constraint $|u_i| \leqslant b_i$, ξ_i denote standard Wiener processes, σ_i and b_i are given constants, $i = 1,2,3$. The motion of the system (3.22) starts from the point x belonging to a bounded region $Q \subset R^3$ at time

$t = 0$. It is required to determine the controls u_i that yield the maximum of the probability of the system (3.22) staying in the region Q over the interval $[0, T]$, where $T > 0$ is prescribed.

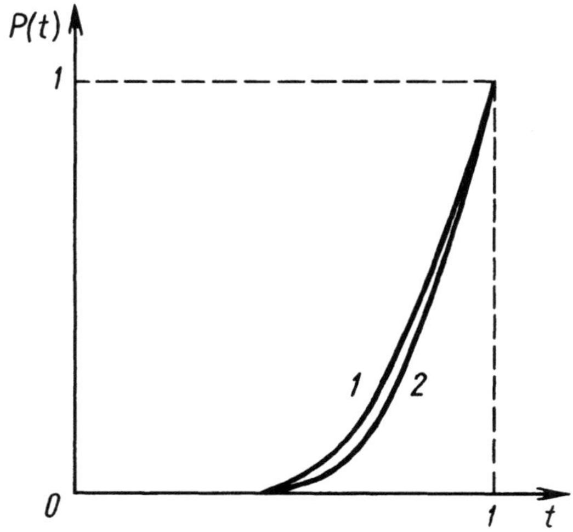

FIG. 3.11. Graph of the maximum of cost functional against time: 1, case 1; 2, case 2

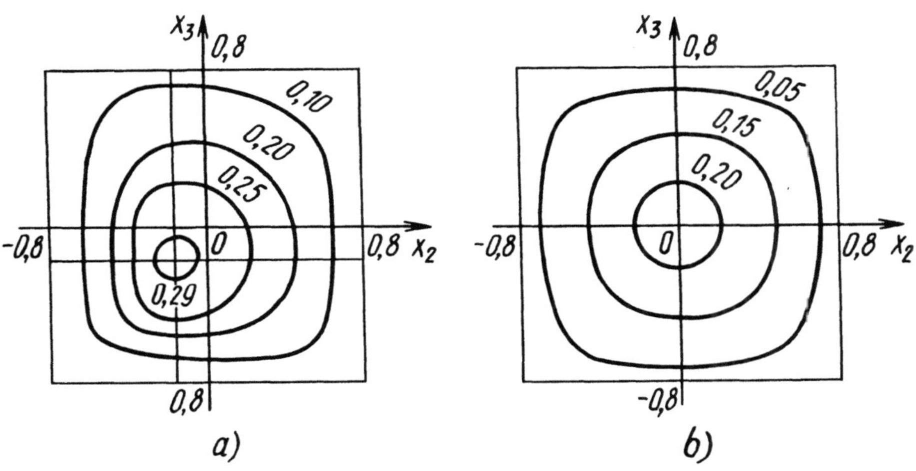

FIG. 3.12. Level curves of the Bellman function: a, case 1; b, case 2

The Bellman equation for the problem under consideration is of the form

$$\frac{\partial V}{\partial t}+(a_2-a_3)(a_2a_3)^{-1}x_2x_3\frac{\partial V}{\partial x_1}+(a_3-a_1)(a_1a_3)^{-1}x_3x_1\frac{\partial V}{\partial x_2}$$

$$+(a_1-a_2)(a_2a_1)^{-1}x_1x_2\frac{\partial V}{\partial x_3}+\sum_{i=1}^{3}\left[\varepsilon_i\frac{\partial^2 V}{\partial x_i^2}+b_i\left|\frac{\partial V}{\partial x_i}\right|\right]=0, \tag{3.23}$$

$$\varepsilon_i=\frac{\sigma_i^2}{2},$$

$$V(t,x)=0,\quad 0\leqslant t\leqslant T,\ x\in\Gamma,\qquad V(T,x)=1,\quad x\in Q\setminus\Gamma.$$

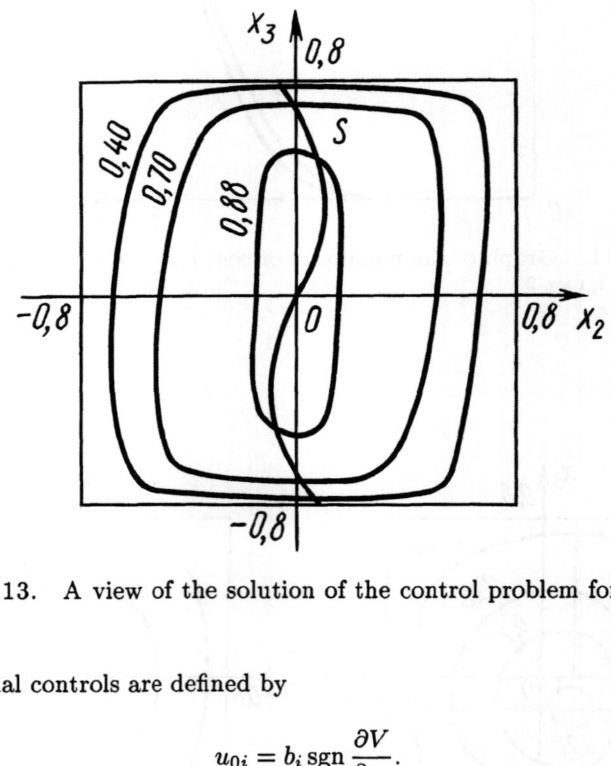

FIG. 3.13. A view of the solution of the control problem for a rigid body

The optimal controls are defined by

$$u_{0i}=b_i\operatorname{sgn}\frac{\partial V}{\partial x_i}.$$

Equation (3.23) has been solved in the cube Q of side 1.6 centered at the origin. The computational algorithm is the same as in subsection 4. The coordinate step was taken as 0.1, and the time step as 0.01.

We bring the calculation results for several special cases.

Case 1. $a_2=2$, $a_3=3$, $a_1=b_i=\varepsilon_i=1$ $(i=1,2,3)$.

Case 2. $a_2 = 2$, $a_1 = a_3 = b_1 = \varepsilon_i = 1$ $(i = 1, 2, 3)$, $b_2 = b_3 = 0$.

In case 1 for the switching surfaces for the controls u_i we have the planes $x_i = -0.2$, where $i = 1, 2, 3$. In case 2 the switch of the control u_1 occurs in the plane $x_1 = -0.2$. Figure 3.11 shows the time-dependence of the function

$$p(t) = \sup_{x_2, x_3} V(t, 0, x_2, x_3).$$

For level surfaces of the function $V(t, x_1, x_2, x_3)$ we have embedded surfaces. Figure 3.12, *a*, shows the sections of these surfaces by the plane $x_1 = 0$ for $t = 0.8$ (case 1), and Figure 3.12, *b*, shows the same for case 2.

Case 3. $a_1 = a_3 = b_2 = \sigma_2 = 1$, $a_2 = 2$, $b_1 = b_3 = \varepsilon_1 = \varepsilon_3 = 0$.

For case 3 Figure 3.13 shows the sections of the level surfaces of the function $V(t, x_1, x_2, x_3)$ for $t = 0.91$, $x_1 = -0.7$ and also the switching curve for optimal control in the plane $x_1 = -0.7$ designated by the letter S.

In case 3 for fixed values of b_i and σ_i calculations have been carried out for different a_i. In all cases the pattern of the level curves and the switching curves is basically the same.

6. Maximization of the mean time for the system staying within a given region. Determination of optimal control for the motion of a rigid body. In the cases where the system (1.1) functions normally only inside some bounded region Q it is of practical interest to consider, in addition to problems with the stochastic cost functional discussed above, also problems in which it is required to maximize the first passage time, i.e., the time which it takes the system to first reach the boundary Γ of the region Q.

A control $u(x)$, $u(x) \in U$, that assures the existence of a solution of equation (1.1) is said to be *admissible*.

Let $\tau_x(u)$ denote the first passage time to the boundary Γ, given the control u and the initial condition $x(0) = x$. For any $x \in Q$ it is required to determine an admissible control so as to maximize the functional $E\tau_x(u)$ being the mean passage time, or the mean time which it takes the system (1.1) to leave the region Q.

The Bellman equation for the problem in question is of the form

$$\sup_{u \in U} L_u V(x) = -1, \qquad x \in Q,\ V(x) = 0,\ x \in \Gamma, \tag{3.24}$$

where the generator L_u is defined by formula (1.4).

Next, we apply the method of successive approximations for solving the boundary-value problem (3.24). Let us take an arbitrary admissible control u_1 (a first approximation to the optimal control) and define V_1 as a solution of the linear problem

$$L_{u_1} V_1 = -1, \quad x \in Q, \qquad V_1(x) = 0, \quad x \in \Gamma.$$

Further, we construct the sequence u_i, V_i by means of the recurrence relations

$$\begin{aligned} L_{u_i} V_i &= -1, \qquad x \in Q, \\ V_i(x) &= 0, \qquad x \in \Gamma, \\ \sup_{u \in U} u' B' \frac{\partial V_i(x)}{\partial x} &= u'_{i+1}(x) B' \frac{\partial V_i(x)}{\partial x}. \end{aligned}$$

It can be shown that the sequence V_i converges to the solution of the problem (3.24), with $V_{i+1}(x) \geqslant V_i(x)$.

Consider the controlled motion of a rigid body, relative to its mass center, acted upon by control torques and by random disturbing forces of the Gaussian white noise type. Equations of motion projected on the axes of inertia have the form (1.12), (1.13). At the initial moment the absolute value of the angular momentum is subject to the condition $|x(0)| < r$. The desired control must be such that it yields the maximum of the mean time which it takes the angular momentum to first reach the boundary $|x| = r$, given an arbitrary $x(0)$, $|x(0)| < r$.

Equations (3.24) defining the maximum mean first passage time and the optimal F-control now become

$$\begin{aligned} A(V) = \Bigg[\frac{\sigma^2}{2} \sum_{i=1}^{3} \frac{\partial^2 V}{\partial x_i^2} &+ \frac{\partial V}{\partial x_1} x_2 x_3 (a_2 - a_3)(a_2 a_3)^{-1} \\ &+ \frac{\partial V}{\partial x_2} x_1 x_3 (a_3 - a_1)(a_1 a_3)^{-1} + \frac{\partial V}{\partial x_3} x_1 x_2 (a_1 - a_2)(a_2 a_1)^{-1} \Bigg], \end{aligned}$$

$$A(V) + \sup_{u,\, |u| \leqslant b} u' \frac{\partial V}{\partial x} = -1, \tag{3.25}$$

$$V(x) = 0 \quad \text{for } |x| = r.$$

The solution of this boundary-value problem is the function

$$V_0(x) = \frac{2}{\sigma^2} \int_{|x|}^{r} z(t)\, dt \int_0^t \frac{ds}{z(s)}, \tag{3.26}$$

where

$$z(t) = \exp\left(\int_t^r \frac{2}{\sigma^2} \left(-b + \frac{\sigma^2}{s} \right) ds \right). \tag{3.27}$$

Substituting the expression (3.26) in (3.25), we determine an $u_0(x)$ such that it yields the maximum of the right side of (3.25). For $x \neq 0$ we have

$$u_0(x) = -bx|x|^{-1}. \tag{3.28}$$

For $x = 0$ the function $V_0(x)$ has a vanishing derivative and, consequently, it does not define a control at the point $x = 0$. However, we can show the following assertions hold.

1°. If $x(0) \neq 0$, then, given the control (3.28), the probability of the system (1.12) reaching zero earlier than the boundary $|x| = r$ is zero.

2°. If $x(0) = 0$, then, given any admissible control, the system (1.12) will leave the point O in a mean time however small, and, therefore, by assertion 1° the system reaches, with probability one, the surface $|x| = r$ earlier than returning to the initial point.

It follows from assertions 1° and 2° that the value of the control at the point O is not relevant, and, thus, the control (3.28) is optimal.

To justify assertion 1°, we take a number $\varepsilon > 0$ and compute the probability $\omega(x)$ that the system (1.12) reaches the surface $|x| = \varepsilon$ earlier than $|x| = r$, given the control (3.28) and the initial condition $x = x(0)$, $\varepsilon \leqslant |x(0)| \leqslant r$. The function $\omega(x)$ is a solution of the Dirichlet problem

$$A(\omega) - b\frac{x'}{|x|}\frac{\partial \omega}{\partial x} = 0, \qquad \omega(x) = \begin{cases} 1, & \text{for } |x| = \varepsilon, \\ 0, & \text{for } |x| = r, \end{cases}$$

where $A(\omega)$ is defined by the operator in brackets in (3.25). On calculation we obtain for $x\colon \varepsilon \leqslant |x(0)| \leqslant r$

$$\omega(x) = \int_{|x|}^{r} z(s)\, ds \left[\int_{\varepsilon}^{r} z(s_1)\, ds_1\right]^{-1}, \tag{3.29}$$

where the function $z(t)$ is defined by the equality (3.27). It follows from (3.27), (3.29) that $\omega(x) \to 0$ for $\varepsilon \to 0$.

We now prove assertion 2°. Let $u(x)$ be any admissible control, and $\tau_u(\varepsilon) = \beta$ be the time which it takes the system (1.12) to leave the sphere $|x| \leqslant \varepsilon$, given the control u and the zero initial condition. Since $E\beta < \infty$, it follows from the Ito formula that

$$\begin{aligned} V_0(\beta) - V_0(0) &= E\int_0^{\beta} L_u V_0\big(x(t,u)\big)\, dt \leqslant E\int_0^{\beta} \sup_{|u|\leqslant b} L_u V_0\big(x(t,u)\big)\, dt \\ &= -E\beta. \end{aligned}$$

Therefore, by virtue of the relation (3.26) and the definition of β we have

$$E\beta \leqslant \frac{2}{\sigma^2}\int_0^{\varepsilon} z(t)\, dt \int_0^{t} \frac{ds}{z(s)}.$$

Hence, noting the equality (3.27) and the fact that the number ε is arbitrary, it follows that assertion 2° is valid. Thus, we have proved that the control of the motion of a rigid body is defined by formula (3.28).

Formula (3.28) shows that the optimal control is maximum in value and directed contrary to the direction of the angular momentum. The mean time which it takes the absolute value of the angular momentum to reach the upper bound r is defined by formula (3.26).

Main Results and Formulas of Chapter X

Time-optimal problem:

$$dx(t) = [f(x) + Bu(x)]\,dt + \sigma(x)\,d\xi, \qquad x(0) = x_0,$$
$$E\tau_x(u) \to \inf_{u\in U}, \; x \in R^n \setminus Q;$$

the Bellman equation

$$\inf_{u\in U}[L_u V(x) + 1] = 0, \qquad x \in R^n \setminus Q,$$
$$V(x) = 0, \qquad x \in \Gamma,$$
$$L_u V = (f + Bu)' V_x + \tfrac{1}{2}\operatorname{tr} \sigma\sigma' V_{xx};$$

optimal control $u_0(t,x)$ is a solution of the equation

$$\inf_{u\in U} L_u V_0(x) = L_{u_0} V_0(x),$$

where $V_0(x)$ is a minimal positive solution of the external boundary-value problem for the Bellman equation.

Problem with stochastic functional:

$$P(\tau_x(u) < \infty) \to \sup_{u\in U};$$

the Bellman equation

$$\sup_{u\in U} L_u V(x) = 0, \qquad x \in R^n \setminus Q$$
$$V(x) = 1, \qquad x \in \Gamma.$$

CHAPTER XI

OPTIMAL ESTIMATION OF THE STATE OF THE SYSTEM

In this chapter we consider problems in optimal mean-square estimation under various assumptions concerning the observed and unobserved processes. The Kalman filter is discussed and its properies investigated. The problems of optimal interpolation and extrapolation are treated.

§ 1. Estimation Problems Involving Random Disturbances

1. Statement of an optimal estimation problem. Optimal estimation problems can be described as follows. We have a random process $x(t) \in R^n$ which models the motion of a system. The process $x(t)$ cannot be directly measured. However, it is possible to measure another process $y(t)$ which contains some information about the state $x(t)$ of the system. Given the actually observed values of the process $y(t)$ over the time interval $[t_0, T]$, where $T > t_0$, it is required to determine an optimal (in the mean-square sense) estimate $m(\tau)$ of the vector $x(\tau)$.

Depending on the relation between the moments τ and T, we distinguish between the following problems: the *filtering problem* if $\tau = T$, the *extrapolation problem* if $\tau > T$, and the *interpolation problem* if $\tau < T$. An estimate $m(\tau)$ is said to be *optimal ("best") in the mean-square sense* if it minimizes the mean-square error:

$$E\|x(\tau) - m(\tau)\|^2 = \min_F E\|x(\tau) - F(y_T)\|^2. \tag{1.1}$$

Here the minimum is taken over all mappings F from the space of observations y_T into the state space of the system, $\|x\|^2 = x'x$, and it is assumed that the random variables $x(\tau)$ and $F(y_T)$ have finite second moments. The symbol y_T designates the aggregate of the actually observed values of $y(t)$ over the interval $0 \leqslant t \leqslant T$.

It follows from the definition (1.1) of an optimal (in the mean-square sense) estimate $m(\tau)$ that

$$m(\tau) = E\{\, x(\tau) \mid y_T \,\}. \tag{1.2}$$

Indeed, set $\gamma(\tau) = E\{\, x(\tau) \mid y_T \,\}$. Then

$$\begin{aligned} E\|x(\tau) - F(y_T)\|^2 &= E\|x(\tau) - F(y_T) + \gamma(\tau) - \gamma(\tau)\|^2 \\ &= E\|x(\tau) - \gamma(\tau)\|^2 + E\|\gamma(\tau) - F(y_T)\|^2 \\ &\quad + 2E\big(x(\tau) - \gamma(\tau)\big)'\big(\gamma(\tau) - F(y_T)\big). \end{aligned} \tag{1.3}$$

Moreover, by the properties of the conditional mean we have

$$\begin{aligned} E\big(x(\tau)-\gamma(\tau)\big)'\big(\gamma(\tau)-F(y_T)\big) &= E\{E\big(x(\tau)-\gamma(\tau)\big)'\big(\gamma(\tau)-F(y_T)\big) \mid y_T\} \\ &= E\Big\{E\{\big(x(\tau)-\gamma(\tau)\big)' \mid y_T\}\big(\gamma(\tau)-F(y_T)\big)\Big\} \\ &= E\big(\gamma(\tau)-\gamma(\tau)\big)'\big(\gamma(\tau)-F(y_T)\big) = 0. \end{aligned}$$

Thus, noting the relation (1.3), we obtain the equality

$$E\|x(\tau)-F(y_T)\|^2 = E\|x(\tau)-\gamma(\tau)\|^2 + E\|\gamma(\tau)-F(y_T)\|^2. \tag{1.4}$$

It should be noted that, in view of (1.1), the functional F is determined from the minimum condition with respect to F for the left side of the equality (1.4), but on the right side of this equality only the second summand depends on F. This summand is nonnegative and vanishes for $F = \gamma(\tau)$. Hence, recalling the definition of $\gamma(\tau)$, we obtain the relation (1.2).

Thus, in the general case the solution of the estimation problem is given by formula (1.2). However, in practical situations it is not easy to utilize this formula, since the formula involves the conditional mean which is difficult to compute. If additional constraints are incorporated in the problem statement, sometimes the solution of the estimation problem can be obtained in closed form. In what follows, we give this solution for the following three cases: linear estimation, estimation of Gaussian variables, Kalman filtering.

2. Linear estimation. A *linear estimation problem* is the problem of determining the optimal (best) linear mapping F which yields the minimum of the mean-square deviation of the vector $x(\tau)$ from its linear estimate provided by the linear estimator $F(y_T)$.

We shall treat this problem under the assumption that $x(\tau)$ and $y(t)$ have finite second moments. Let H_0 denote a Hilbert space of random vectors $\alpha \in R^n$ with finite second moments. The scalar product of two elements α and β of H_0 is equal to $E\alpha'\beta$. It will be recalled that a *Hilbert space* is a complete linear space equipped with a scalar product. To construct the best linear estimator of the vector $x(\tau)$, given the observed values of the process $y(t)$ over the interval $[0,T]$, we consider the Hilbert subspace $H \subset H_0$ which is the closed linear hull of the values of the $y(t)$, $0 \leqslant t \leqslant T$, and of the unit vector. The best linear estimate x_0 of the vector x is the element of the Hilbert space H which is at a minimum distance from x. Thus, the vector x_0 can be determined from the equation

$$\min_{\alpha\in H} E\|x-\alpha\|^2 = E\|x-x_0\|^2. \tag{1.5}$$

According to [3], equation (1.5) has a unique solution x_0, which is the projection of x on H. The element $x_0 \in H$ is determined in a unique way from the equation

$$E(x-x_0)'\alpha = 0, \qquad \forall \alpha \in H. \tag{1.6}$$

The corresponding estimation error, which is equal to the square of the length of the perpendicular dropped from the terminal point of the vector x to H, is of the form $\delta = E|x|^2 - E|x_0|^2$. Equating the element α in (1.6) with the unit vector gives $Ex = Ex_0$. Therefore, the optimal linear estimate is an unbiased one.

In the general case an optimal linear estimate may turn out to be unsatisfactory, as the following example demonstrates. Suppose that at discrete moments $k \neq l$ a random variable $\zeta(k) = \exp(i\varphi k)$ is observed, where $i = \sqrt{-1}$ and the random variable φ is uniformly distributed over the segment $[-\pi, \pi]$. Given the observation data $\zeta(k)$, $k \neq l$, it is required to construct an estimate of $\zeta(l)$. Since $E\zeta(k)\zeta(l) = 0$ for $k \neq l$, the best linear estimate, in view of (1.6), is zero, and, consequently, does not depend on the observation data. At the same time, it is sufficient to make only one measurement $\zeta(k)$ at any moment k, and the whole sequence $\zeta(l)$ is reconstructed according to the formula $\zeta(l) = [\zeta(k)]^{l/k}$.

3. Optimal estimation of Gaussian variables. We assume that in addition to the conditions imposed earlier the random variables x and $y(t)$, where $0 \leqslant t \leqslant T$, have a joint Gaussian probability distribution and zero means. Then equation (1.6) defining the optimal linear estimate x_0 implies that the random variables $x - x_0$ and $y(t)$ are independent. Therefore,

$$\begin{aligned} E\{\, x \mid y_T \,\} &= E\{\, (x - x_0 + x_0) \mid y_T \,\} = E\{\, (x - x_0) \mid y_T \,\} + E\{\, x_0 \mid y_T \,\} \\ &= E(x - x_0) + x_0 = x_0. \end{aligned}$$

Thus, in the Gaussian case, given the observation data y_T, the best (in the mean-square sense) estimate of x (equal to $E\{\, x \mid y_T \,\}$) coincides with the best linear estimate x_0.

4. Linear estimation of stationary processes. The Wiener–Hopf equation. Consider the problem of the optimal linear estimation of the process $x(t + T)$, $T \geqslant 0$, given the actual observed values $y(s)$, $-\infty < s \leqslant t$, assuming that the scalar processes $x(t)$ and $y(t)$ are weakly stationary processes and stationarily related. This means that the expected values of the processes $x(t)$ and $y(t)$ are constant (without loss of generality, they can be assumed to be zero), the second moments of the processes x, y are finite, and the covariance functions depend only on the difference of the arguments, i.e.,

$$\begin{aligned} R_{xx}(t - s) &= Ex(t)x(s), \\ R_{yy}(t - s) &= Ey(t)y(s), \\ R_{xy}(t - s) &= Ex(t)y(s). \end{aligned} \tag{1.7}$$

The functions R_{xx}, R_{yy}, R_{xy} are considered to be given. The best linear estimator $x_0(t + T)$ of the process $x(t + T)$, given the observation data y_t, is sought as

$$x_0(t + T) = \int_{-\infty}^{t} k(t, s)y(s)\, ds. \tag{1.8}$$

The kernel $k(t,s)$ of the estimator (1.8) is to be determined. To this end, we use equation (1.6). According to this equation we have

$$Ex_0(t+T)y(s) = Ex(t+T)y(s), \qquad s \leqslant t.$$

Substituting the relation (1.8) for the estimator x_0 and taking account of the equalities (1.7) gives

$$R_{xy}(t+T-s) = \int_{-\infty}^{t} k(t,\tau)R_{yy}(\tau - s)\,d\tau.$$

After the change of variables $t - s = s_1$, $t - \tau = \tau_1$ we obtain

$$R_{xy}(T+s_1) = \int_0^\infty k(t, t-\tau_1)R_{yy}(-\tau_1 + s_1)\,d\tau_1, \qquad s_1 \geqslant 0. \tag{1.9}$$

However, the left side of equation (1.9) does not depend on t. Therefore, the right side of equation (1.9) also must not depend on t. Hence, the kernel k of the estimator (1.8) is such that $k(t,t-\tau) = \gamma(\tau)$, where the function $\gamma(\tau)$, in view of (1.9), is determined from the *Wiener–Hopf equation*

$$R_{xy}(T+s_1) = \int_0^\infty \gamma(\tau)R_{yy}(s_1 - \tau)\,d\tau, \qquad s_1 \geqslant 0. \tag{1.10}$$

Next, the optimal estimator (1.8) can be expressed as

$$\begin{aligned} x_0(t+T) &= \int_{-\infty}^{t} k(t,s)y(s)\,ds = \int_0^\infty k(t,t-s_1)y(t-s_1)\,ds_1 \\ &= \int_0^\infty \gamma(s)y(t-s)\,ds. \end{aligned} \tag{1.11}$$

It follows from the expression (1.11) and the stationarity of the process $y(t)$ that the estimator x_0 is stationary. Using the equality (1.11) for the optimal linear estimator, we find that the estimation error

$$\begin{aligned} \delta &= E|x(t+T)|^2 - E|x_0(t+T)|^2 \\ &= R_{xx}(0) - \int_0^\infty \int_0^\infty \gamma(s)R_{yy}(s-\tau)\gamma(\tau)\,ds\,d\tau. \end{aligned} \tag{1.12}$$

Thus, to calculate the estimator (1.11) and also the estimation error (1.12), we have to solve equation (1.10). This equation is an integral equation of the first kind with symmetric kernel with respect to the function γ; it is possible that the solution of this equation may not exist. Several conditions for the existence of a solution and a method for its construction due to Wiener is given below.

Suppose that the covariance function $R_{yy}(t)$ can be expressed in the form

$$R_{yy}(t) = \int_{-\infty}^{\infty} |h(i\omega)|^2 e^{-i\omega t}\,d\omega, \qquad i = \sqrt{-1}, \tag{1.13}$$

and there exists a function $a(t)$ satisfying the condition $a(t) = 0$ for $t < 0$ such that

$$h(z) = \int_0^\infty a(t)e^{-zt}\,dt, \qquad \operatorname{Re} z \geqslant 0, \tag{1.14}$$

where z is a complex variable. The expressions (1.13), (1.14) hold for stationary processes with absolutely continuous spectrum and spectral density that allows factorization. Formulas (1.13), (1.14) and Parseval's inequality yield

$$R_{yy}(t) = \int_0^\infty a(t+s)\bar{a}(s)\,ds, \tag{1.15}$$

where $\bar{a}(t)$ is the complex conjugate of $a(t)$.

As in the deduction of formula (1.15), we assume that there exists a function $b(t)$ for which the cross-covariance function $R_{xy}(t)$ may be expressed as

$$R_{xy}(t) = \int_0^\infty b(t+s)\bar{a}(s)\,ds. \tag{1.16}$$

Substituting the relations (1.15), (1.16) in equation (1.10) gives

$$\int_0^\infty \left[b(T+\tau+s) - \int_0^\infty \gamma(t)a(\tau-t+s)\,dt\right]\bar{a}(s)\,ds = 0.$$

A sufficient condition for this equation to be valid is that the function γ satisfy the relation

$$b(T+t) - \int_0^\infty \gamma(\tau)a(t-\tau)\,d\tau = 0, \qquad t > 0. \tag{1.17}$$

Since the function $a(t) = 0$ for $t < 0$, it follows that the equality (1.17) can be written as the convolution

$$b(T+t) = \int_0^t \gamma(\tau)a(t-\tau)\,d\tau, \qquad t > 0. \tag{1.18}$$

To determine the solution γ of equation (1.18), we use the Laplace transform. Set

$$b_0(T,z) = \int_0^\infty b(T+s)e^{-zs}\,ds, \qquad \gamma_0(z) = \int_0^\infty \gamma(s)e^{-zs}\,ds, \tag{1.19}$$

Further, take the Laplace transform of both sides of equation (1.18). Taking account of the equality (1.14) and the designations (1.19), we obtain

$$b_0(T,z) = \gamma_0(z)h(z).$$

Hence, by the Mellin conversion formula we have

$$\gamma(t) = \frac{1}{2\pi i}\int_{-\infty}^\infty e^{i\omega t}\frac{b_0(T,i\omega)}{h(i\omega)}\,d\omega. \tag{1.20}$$

Thus, the solution $\gamma(t)$ of equation (1.10) is obtained. The conditions for the validity of the transformations (1.13)–(1.19) which imply formula (1.20) are stated in [2].

§ 2. The Kalman filter

1. Problem statement. We study the filtering problem under the assumption that the motion of the system $x(t)$ and the observation process $y(t)$ is described by an Ito system of linear stochastic differential equations of the form

$$dx(t) = A(t)x(t)\,dt + \sigma(t)\,d\xi(t), \qquad 0 \leqslant t \leqslant T,\ x(0) = x_0, \tag{2.1}$$

$$dy(t) = Q(t)x(t)\,dt + \sigma_0(t)\,d\xi_0(t), \qquad y(0) = 0. \tag{2.2}$$

Here the vectors $x(t) \in R^n$, $y(t) \in R^m$, the matrices A, σ, Q, σ_0 are prescribed and have measurable bounded elements, ξ and ξ_0 denote standard Wiener processes of respective dimensions n and m, and ξ_0 is a Gaussian vector whose distribution parameters are $Ex_0 = 0$, $Ex_0x_0' = D_0$. The random variables $\xi(t)$, $\xi_0(t)$, x_0 are assumed to be independent, and the covariance matrix D_0 is positive definite. The matrix $\sigma_0(t)\sigma_0'(t)$ is nondegenerate for all $t \in [0,T]$, where T is an arbitrary fixed moment. The filtering problem is the problem of constructing the best (in the mean-square sense) estimator of the vector $x(T)$, given the observed values y_T of the process (2.2) over the interval $[0,T]$.

Let $m(T)$ and $D(T)$ denote the conditional mean and covariance matrix of the vector $x(T)$, given y_T:

$$m(T) = E\{\, x(T) \mid y_T \,\}, \qquad D(T) = E\{ \big(x(T) - m(T)\big)\big(x(T) - m(T)\big)' \mid y_T \}.$$

By virtue of the results of § 1 the vector $m(T)$ is the best (in the mean-square sense) estimate of the vector $x(T)$, and $D(T)$ is the covariance matrix of the difference $x(T) - m(T)$. It follows from the assumptions concerning the parameters of the system (2.1), (2.2) that the joint probability distribution of the process $\big(x(t), y(t)\big)$ is Gaussian. Therefore, the optimal estimate $m(T)$ is a linear functional of the observation data, i.e.,

$$m(T) = \int_0^T u'(t,T)\,dy(t), \tag{2.3}$$

where the matrix $u'(t,T)$ of dimensions $n \times m$ (the kernel of the estimator) can be determined from the condition for the minimum of the expression $E\|x(T) - m(T)\|^2 = J$ with respect to the matrix u. Since the argument T is assumed to be fixed, though arbitrary, it can be dropped, and for brevity we will write $u(t)$ instead of $u(t,T)$. It should be stressed, however, that the dependence of the kernel $u(t,T)$ of the estimator on T may turn out to be crucial when T is varied.

2. The dual problem of optimal control. We shall show that the kernel $u(t)$ of the best estimator (2.3) is, in fact, the optimal control of the linear deterministic system (2.1), (2.2) with quadratic cost functional.

Let $\alpha(t)$ denote a square matrix of order n described by

$$\dot{\alpha}(t) = -A'(t)\alpha(t) + Q'(t)u(t), \qquad 0 \leqslant t \leqslant T,\ \alpha(T) = I, \tag{2.4}$$

where I is a unit matrix of order n, the control $u(t)$ coincides with the kernel of the estimator (2.3), and the matrices $A(t)$ and $Q(t)$ are the same as in equations (2.1), (2.2). Now we choose a cost functional for equation (2.4) so that the optimal control in the optimization problem thus set up is equal to the kernel of the best estimator (2.3). To this end, we multiply equation (2.1) on the left by $\alpha'(t)$, multiply the equation for $\alpha'(t)$ on the right by $x(t)$, and add the obtained equations. Thus, we have

$$\alpha'(t)\,dx(t) + \big(d\alpha'(t)\big)x(t) = \alpha'(t)\sigma(t)\,d\xi(t) + u'(t)Q(t)x(t)\,dt.$$

Substituting the product $Q(t)x(t)$ on the right side of the equation from the observation equation (2.2) yields

$$d\big(\alpha'(t)x(t)\big) = \alpha'(t)\sigma(t)\,d\xi(t) + u'(t)\big(dy(t) - \sigma_0(t)\,d\xi_0(t)\big).$$

Integrating both sides of this relation from $t = 0$ to $t = T$ and noting the condition $\alpha(T) = I$, we obtain

$$\begin{aligned} \alpha'(0)x(0) + \int_0^T \alpha'(t)\sigma(t)\,d\xi(t) - \int_0^T u'(t)\sigma_0(t)\,d\xi_0(t) \\ = x(T) - \int_0^T u'(t)\,dy(t). \end{aligned} \tag{2.5}$$

It should be noted that because of $x(0)$, $\xi(t)$, and $\xi_0(t)$ being independent, in view of the equality $\operatorname{tr} xx' = x'x$, we have

$$\begin{aligned} E\Big[\alpha'(0)x(0) + \int_0^T \alpha'(t)\sigma(t)\,d\xi(t) - \int_0^T u'(t)\sigma_0(t)\,d\xi_0(t)\Big]' \\ \times \Big[\alpha'(0)x(0) + \int_0^T \alpha'(t)\sigma(t)\,d\xi(t) - \int_0^T u'(t)\sigma_0(t)\,d\xi_0(t)\Big] \\ = \operatorname{tr}\Big[\alpha'(0)D_0\alpha(0) + \int_0^T \big(\alpha'(t)N_1(t)\alpha(t) + u'(t)N_0(t)u(t)\big)\,dt\Big], \end{aligned} \tag{2.6}$$

where $N_0 = \sigma_0\sigma_0'$, $N_1 = \sigma\sigma'$.

We now multiply both sides of the equality (2.5) scalarly by itself and take the expectation. In view of the relation (2.6) we obtain

$$\begin{aligned} J = E\|x(T) - m(T)\|^2 = E\Big\|x(T) - \int_0^T u'(t)\,dy(t)\Big\|^2 \\ = \operatorname{tr}\Big[\alpha'(0)D_0\alpha(0) + \int_0^T \big(\alpha'(t)N_1(t)\alpha(t) + u'(t)N_0(t)u(t)\big)\,dt\Big]. \end{aligned} \tag{2.7}$$

It follows from the equality (2.7) that the kernel $u(t)$ of the optimal estimator (2.3) must simultaneously minimize the quadratic (in α) functional (2.7) on the trajectories of the deterministic system (2.4). The optimal control problem (2.4), (2.7) is dual to the original filtering problem (2.1), (2.2). Using the solution of the dual linear-quadratic optimal control problem given in § 3 of Chapter VIII, we can obtain the expressions for the optimal filter (2.3) and the estimation error J.

3. Equation for the estimation error. The problem (2.4), (2.7) differs somewhat from the problem treated in § 3 of Chapter VIII by that it was set up for the matrix-valued process $\alpha(t)$ with an initial condition prescribed for $t = T$.

For an immediate use of the results of § 3 of Chapter VIII, consider the control system (2.4) separately for each column $\alpha_i(t)$, $i = 1, \ldots, n$, of the matrix $\alpha(t)$. It follows from (2.4) that

$$\dot{\alpha}_i(t) = -A'(t)\alpha_i(t) + Q'(t)u_i(t), \qquad \alpha_i(T) = e_i, \tag{2.8}$$

where $u_i(t)$ denotes the ith column of the control matrix $u(t)$, and $e_i \in R^n$ the vector whose ith component is equal to unity and the other are zero. By virtue of the relation (2.7), the control $u_i(t)$ is determined from the condition for the minimum of the functional

$$J_i = \alpha_i'(0)D_0\alpha_i(0) + \int_0^T \big(\alpha_i'(t)N_1(t)\alpha_i(t) + u_i'(t)N_0(t)u_i(t)\big)\,dt. \tag{2.9}$$

Repeating the deduction of formula (2.7) verbatim, we find that

$$\min_{u_i} J_i = d_{ij}(T), \qquad m_i(T) = \int_0^T u_i'(t)\,dy(t), \quad i = 1, \ldots, n. \tag{2.10}$$

Here $u_i(t)$ is the optimal control in the problem (2.8), (2.9); we also set $m(T) = (m_1(T), \ldots, m_n(T))$, and $d_{ij}(T)$ are the elements of the matrix $D(T)$. Finally, to ensure that the motion of the system (2.8) occur in the direction of increasing time (as in § 3 of Chapter IX), we carry out the change of variables $s = T - t$, $\beta_i(s) = \alpha_i(T - s)$, $0 \leqslant s \leqslant T$.

By virtue of (2.8) the equation for $\beta_i(s)$ has the form

$$\dot{\beta}_i(s) = A'(T-s)\beta_i(s) - Q'(T-s)u_i(T-s), \qquad \beta_i(0) = e_i. \tag{2.11}$$

The cost functional (2.9) is expressed in terms of the new variables as follows:

$$J_i = \beta_i'(T)D_0\beta_i(T) + \int_0^T \big(\beta_i'(s)N_1(T-s)\beta_i(s) + u_i'(T-s)N_0(T-s)u_i(T-s)\big)\,ds. \tag{2.12}$$

The solution of the problem (2.10), (2.11) yields formulas (3.7) and (3.8) of Chapter IX, according to which we have

$$\min_{u_i} J_i = e_i'P(0)e_i, \qquad u_i(T-s) = N_0^{-1}(T-s)Q(T-s)P(s)\beta_i(s), \tag{2.13}$$

where the square matrix P of order n satisfies the relations

$$\frac{dP(s)}{ds} + A(T-s)P(s) + P(s)A'(T-s) - N_1(T-s) \\ - P(s)Q'(T-s)N_0^{-1}(T-s)Q(T-s)P(s) = 0, \qquad P(T) = D_0. \tag{2.14}$$

But in view of (2.8), (2.10), (2.13) we have

$$d_{ii}(T) = \min_{u_i} J_i = \beta_i'(0)P(0)\beta_i(0) = e_i'P(0)e_i = P_{ii}(0),$$

where the P_{ij} are the elements of the matrix P. In other words, the diagonal elements of the matrices $D(T)$ and $P(0)$ coincide. Let us show that the other elements of these matrices are equal. To this end, we consider the sum $x_i(T) + x_j(T)$ whose optimal estimate is given by

$$\int_0^T \big(u_i'(t) + u_j'(t)\big)\, dy(t). \tag{2.15}$$

Hence, in view of the definition of the covariance matrix $D(T)$, we conclude that the estimation error of the sum $x_i(T) + x_j(T)$ is

$$J_{ij} = E\left[x_i(T) + x_j(T) - \int_0^T \big(u_i'(t) + u_j'(t)\big)\, dy(t)\right]^2 \\ + d_{ii}(T) + d_{jj}(T) + 2\, d_{ij}(T). \tag{2.16}$$

We now establish another expression for the quantity J_{ij}. Let $\alpha_{ij}(t)$ denote the solution of equation (2.8) in which u_i is replaced by $u_i + u_j$ with initial condition $\alpha_{ij}(T) = e_i + e_j$. Repeating the foregoing argument from formula (2.3) up to formula (2.7), we verify that the kernel of the optimal estimator (2.15) is simultaneously the optimal control of the process $\alpha_{ij}(t)$ with quadratic cost functional (2.9) in which α_i is replaced by α_{ij} and u_i by $u_i + u_j$. The solution of this optimal control problem is also given by (2.13), (2.14) in which α_i is replaced by α_{ij} and e_i by $e_i + e_j$. Thus, in particular, noting the relations (2.13), (2.16), we obtain

$$J_{ij} = 2\, d_{ij}(T) + d_{ii}(T) + d_{jj}(T) = (e_i + e_j)'P(0)(e_i + e_j).$$

Hence, in view of the established fact that the diagonal elements of matrices $D(T)$ and $P(0)$ are equal to one another, it follows that $d_{ij}(T) = P_{ij}(0)$ for any values of the indices $i, j = 1, \dots, n$. Therefore, $D(T) = P(0)$. Further, we find the equation that is satisfied by the matrix $D(t)$ as a function of t, $t \in [0, T]$. We denote by $P(T, s)$, $0 \leqslant s \leqslant T$, the solution of the problem (2.14) so as to stress its dependence on T. Then, the following relation for the covariance matrix $D(T)$

holds true: $D(T) = P(T, 0)$. Thus, noting that T is an arbitrary moment, we can write for any $t \in [0, T]$

$$D(t) = P(t, 0). \tag{2.17}$$

Here the matrix $P(t, \tau)$, $\tau \in [0, t]$, evolves in τ, for any t, according to relations of the type (2.14) in which T is replaced by t, and s by τ. Namely, the matrix $P(t, \tau)$, $\tau \in [0, t]$, evolves according to the relations

$$\frac{\partial P(t,\tau)}{\partial \tau} + A(t-\tau)P(t,\tau) + P(t,\tau)A'(t-\tau) - N_1(t-\tau)$$
$$- P(t,\tau)Q'(t-\tau)N_0^{-1}(t-\tau)Q(t-\tau)P(t,\tau) = 0, \qquad P(t,t) = D_0.$$

Hence, in view of (2.14), we have $P(t, 0) = P(T, T-t)$. Therefore, recalling (2.17), we can write $D(t) = P(T, T-t)$. It follows from this equality and (2.14) that the covariance matrix $D(t)$ satisfies the following equation

$$\dot{D}(t) = A(t)D(t) + D(t)A'(t) - D(t)Q'(t)(\sigma_0\sigma_0')^{-1}(t)Q(t)D(t)$$
$$+ (\sigma\sigma'), \qquad t \geqslant 0,\ D(0) = D_0. \tag{2.18}$$

Equation (2.18) does not depend on the observation data y_t. Thus, the estimation error at time t equal to $\operatorname{tr} D(t)$ can be calculated beforehand before measurement. Equation (2.18) is a matrix Riccati differential equation similar to equation (2.8) of Chapter IV that belongs to the theory of linear-quadratic control problems. Consequently, to solve equation (2.8), we can use methods discussed in § 2 of Chapter IV.

It should be noted that under the assumptions imposed on the parameters of the problem (2.18) its solution exists for all finite $t \geqslant 0$, is unique and positive definite.[1]

4. Equation for the optimal estimate. The optimal estimate $m(T)$ is of the form (2.3), where the kernel $u(t, T)$, by virtue of (2.13), is expressed by

$$u(t,T) = N_0^{-1}(t)Q(t)P(T-t)\alpha(T,t),$$

where the matrix P is given by the relations (2.14). Therefore,

$$u(t,T) = N_0^{-1}(t)Q(t)D(t)\alpha(T,t), \tag{2.19}$$

where $\alpha(t)$ satisfies equation (2.4), given the control (2.19). Note that on the right side of (2.19) only the matrix α depends on the argument T. In the foregoing this argument of the matrix α was dropped, since the moment T was assumed to be fixed, though arbitrary. But now it is expedient to take note of this dependence, since we are going to derive an equation for the estimate $m(T)$ as a function of

[1] Kalman, R., New Methods in Wiener Filtering Theory, *Proc. of the First Symp. on Engineering Appl. of Random Function Theory and Prob.*, J. Wiley, New York-London, (1963).

T. So, let $\alpha(T,t)$ denote the solution of the Cauchy problem (2.4), (2.19). Then, according to (2.3), (2.19), the following expression holds true:

$$m(T) = \int_0^T \alpha'(T,t)D(t)Q'(t)N_0^{-1}(t)\,dy(t). \tag{2.20}$$

Here in view of (2.4), (2.19), the matrix $\alpha'(T,t)$ satisfies the equations

$$\frac{\partial\alpha'(T,t)}{\partial t} = -\alpha'(T,t)[A(t) - D(t)Q'(t)N_0^{-1}(t)Q(t)], \qquad \alpha(T,T) = I, \tag{2.21}$$

where t is variable and T fixed. Comparing the relations (2.21), (2.20) with Cauchy's formula implies that the expression (2.20) is a representation for $t = T$ of the solution $m(t)$ of the equation

$$dm(t) = A(t)m(t)\,dt + D(t)Q'(t)(\sigma_0\sigma_0')^{-1}(t)[dy(t) - Q(t)m(t)\,dt], \tag{2.22}$$
$$m(0) = 0.$$

Equations (2.18), (2.22) for the covariance matrix $D(t)$ and the optimal estimate vector $m(t)$ constitute the *equations of the Kalman filter.*

In the deduction of equation (2.22) it was assumed that $m(0) = 0$. We are going to show that for an arbitrary value of $Ex(0) = m_0$ equations of the optimal filter have the same form (2.18), (2.22), the only difference being that the initial condition for equation (2.22) becomes $m(0) = m_0$. In equation (2.1) we carry out the change of variables $x(t) = Z(t,0)x_1(t)$, where the matrix $Z(t,s)$ is determined by the relations

$$\frac{\partial Z(t,s)}{\partial t} = A(t)Z(t,s) \qquad \frac{\partial Z(t,s)}{\partial s} = -Z(t,s)A(s),\ Z(t,t) = I. \tag{2.23}$$

Thus, we obtain

$$dx_1(t) = Z^{-1}(t,0)\sigma(t)\,d\xi(t), \qquad x_1(0) = x_0,$$
$$dy(t) = Q(t)Z(t,0)x_1(t)\,dt + \sigma_0(t)\,d\xi_0(t), \qquad y(0) = 0.$$

Further, assuming that $x_2(t) = x_1(t) - m_0$, we find

$$dx_2(t) = Z^{-1}(t,0)\sigma(t)\,d\xi(t), \qquad x_2(0) = x_0 - m_0,$$
$$d\left[y(t) - \int_0^t Q(s)Z(s,0)m_0\,ds\right] = Q(t)Z(t,0)x_2(t)\,dt + \sigma_0(t)\,d\xi_0(t). \tag{2.24}$$

Further, designating the observed value by $y_2(t)$, we have

$$y_2(t) = y(t) - \int_0^t Q(s)Z(s,0)m_0\,ds,$$
$$dy_2(t) = Q(t)Z(t,0)x_2(t)\,dt + \sigma_0(t)\,d\xi_0(t). \tag{2.25}$$

The filtering problem (2.25), (2.24) now belongs to the same type of problem as the problem (2.1), (2.2) treated above. The covariance matrix $D_2(t)$ of the conditional distribution of $x_2(t)$, given that the vector $y_2(s)$, $0 \leqslant s \leqslant t$, is measured over the interval $[0, t]$ is described by an equation of the form (2.18). Noting (2.25), (2.24), we can write

$$\dot{D}_2 = -D_2 Z'(t,0)Q'(t)N_0^{-1}(t)Q(t)Z(t,0)D_2, \qquad D_2(0) = D_0. \tag{2.26}$$

Since $D(t) = Z(t,0)D_2(t)Z'(t,0)$, we obtain by differentiating both sides of this equality with respect to t and noting the equalities(2.26) and (2.23) that the matrix $D(t)$ satisfies the relations (2.18). Indeed,

$$\begin{aligned} \dot{D} &= AD + DA' + Z(t,0)\dot{D}_2 Z'(t,0) \\ &= AD + DA' - DQ'N_0^{-1}QD, \qquad D(0) = D_0. \end{aligned}$$

By virtue of equation (2.22) the equation for the optimal estimate $m_2(t)$ of the vector $x_2(t)$, given the measured values $y_2(s)$ $(0 \leqslant s \leqslant t)$, is of the form

$$dm_2(t) = D_2 Z'(t,0)Q'(t)N_0^{-1}[dy_2(t) - QZ(t,0)m_2(t)\,dt], \qquad m_2(0) = 0.$$

By the definition of the function $x_2(t)$, the following expressions are valid:

$$m_2(t) = Z(t,0)^{-1}m(t) - m_0, \qquad m(t) = Z(t,0)[m_2(t) + m_0].$$

Hence,

$$\begin{aligned} dm(t) &= Z(t,0)\,dm_2(t) + AZ(t,0)(m_2(t) + m_0)\,dt \\ &= Am(t)\,dt + Z(t,0)D_2Z'(t,0)Q'N_0^{-1}[dy_1 - QZ(t,0)m_2(t)\,dt]. \end{aligned}$$

This finally gives

$$\begin{aligned} dm(t) &= A(t)m(t)\,dt + DQ'(\sigma_0\sigma_0')^{-1}[dy - Q(t)m(t)]\,dt, \\ m(0) &= m_2(0) + m_0 = m_0. \end{aligned} \tag{2.27}$$

Thus, we have proved the validity of the equations (2.18), (2.27) of the Kalman filter.

5. Stability of the filter. Under certain additional assumptions we can prove that the optimal filter (2.27) is stable with respect to disturbances of the initial conditions m_0. We shall discuss some of these assumptions. Let $Z(t, s)$ denote, as above, the matrix satisfying the relations (2.23). Suppose there exist positive constants τ, C_1, C_2 such that for all $t \geqslant \tau$ the following inequalities are valid:

$$\begin{aligned} C_1 I &\leqslant \omega_1(t) = \int_{t-\tau}^{t} Z'(s,t)Q'(s)\big(\sigma_0(s)\sigma_0'(s)\big)^{-1}Q(s)Z(s,t)\,ds \\ &\leqslant C_2 I, \\ C_1 I &\leqslant \omega_2(t) = \int_{t-\tau}^{t} Z(t,s)\sigma(s)\sigma'(s)Z'(t,s)\,ds \leqslant C_2 I. \end{aligned} \tag{2.28}$$

Then the trivial solution of the homogeneous equation (2.27), i.e. of the equation

$$\dot{m}(t) = [A(t) - D(t)Q'(t)N_0^{-1}(t)Q(t)]m(t), \qquad t \geqslant 0,\ m(0) = m_0, \tag{2.29}$$

is stable with respect to disturbances of the initial conditions. For proof, we note that the estimates

$$[\omega_2^{-1}(t) + \omega_1(t)]^{-1} \leqslant D(t) \leqslant \omega_1^{-1}(t) + \omega_2(t), \qquad t \geqslant \tau. \tag{2.30}$$

for the solution $D(t)$ of the system (2.18) hold true, provided that the constraints (2.28) are satisfied.

Consider now the Lyapunov function

$$V(t, m) = m'D^{-1}(t)m.$$

By virtue of the relations (2.28), (2.30) the function V is positive definite and admits an infinitely small upper limit. Let us compute the derivative of the function V along the trajectories of the system (2.29). We obtain

$$\begin{aligned}\dot{V}\big(t, m(t)\big) &= \dot{m}'(t)D^{-1}(t)m(t) + m'(t)D^{-1}(t)\dot{m}(t)\\ &\quad - m'(t)D^{-1}(t)\dot{D}(t)D^{-1}(t)m(t).\end{aligned}$$

On substitution of the right side of (2.29) for the derivative $\dot{m}$, and of the right side of (2.18) for the derivative $\dot{D}(t)$ we have

$$\begin{aligned}\dot{V} &= m'(A - DQ'N_0^{-1}Q)'D^{-1}m + m'D^{-1}(A - DQ'N_0^{-1}Q)m\\ &\quad - m'D^{-1}(AD + DA' - DQ'N_0^{-1}QD + N_1)D^{-1}m\\ &= -m'Q'N_0^{-1}Qm - m'D^{-1}N_1D^{-1}m. \end{aligned} \tag{2.31}$$

In any case the right side of (2.31) is nonpositive. Therefore, by Lyapunov's stability theorem (Theorem 2.1 of Chapter I) the linear system (2.29) is stable. We can impose further constraints on the matrices $Q'N_0^{-1}Q$ or N_1 of the original filtering problem (2.1), (2.2); for example, one of them may be taken to be positive definite. Then the derivative $\dot{V}$, along with the solution of (2.31), is negative definite, and by Theorem 2.3 of Chapter I the system (2.29) is uniformly asymptotically stable.

Remark. In the case of constant matrices A, σ, Q, σ_0 in equations (2.1), (2.2) it is necessary and sufficient for the validity of the inequalities (2.28) that the matrices

$$(\sigma, A\sigma, \dots, A^{n-1}\sigma), \qquad (Q', A'Q', \dots, (A')^{n-1}Q'). \tag{2.32}$$

are of full rank, i.e., of rank n.

6. Filtering problem with constant parameters. One of the main difficulties of the filtering problem is the necessity of solving the matrix Riccati equation (2.18) for the covariance matrix $D(t)$. In a number of cases an approximate solution of this equation can be obtained. We shall cite one such method for solving the system (2.1), (2.2) with constant matrices A, σ, Q, σ_0 assuming that the matrices (2.32) are of full rank n. Then the solution $D(t)$ of the problem (2.18) converges as $t \to \infty$ to a unique positive definite solution $\bar{D}$ of the algebraic Riccati equation

$$A\bar{D} + \bar{D}A' - \bar{D}Q'(\sigma_0\sigma_0')^{-1}Q\bar{D} + \sigma\sigma' = 0. \tag{2.33}$$

In other words, for t sufficiently large we can replace $D(t)$ approximately by $\bar{D}$.

The algorithm for determining $\bar{D}$ consists of the following steps (see Chapter VII, § 3):

1°. The scalar polynomial $b(\lambda)$ of degree n with zeros having negative real parts is determined from the equation

$$\det(A_1 - \lambda I) = (-1)^n b(\lambda) b(-\lambda),$$

where I is a unit matrix of order $2n$, and the matrix A_1 is of the form

$$A_1 = \begin{pmatrix} -A' & Q'(\sigma_0\sigma_0')^{-1}Q \\ \sigma\sigma' & A \end{pmatrix}.$$

It is often convenient first to find the roots $\lambda_1, \ldots, \lambda_n$ of the equation $\det(\lambda I - A_1) = 0$ having negative real parts, and then express the polynomial $b(\lambda)$ as $b(\lambda) = (\lambda - \lambda_1)\ldots(\lambda - \lambda_n)$.

2°. The matrix $b(A_1)$ is expressed as the block matrix

$$b(A_1) = \begin{pmatrix} b_1 & b_2 \\ b_3 & b_4 \end{pmatrix},$$

where the b_i are square matrices of order n.

3°. The desired solution of equation (2.33) is of the form

$$\bar{D} = b_4 b_2^{-1} = b_3 b_1^{-1}.$$

Further, note that all the eigenvalues of the matrix $A - \bar{D}Q'(\sigma_0\sigma_0')^{-1}Q$ lie in the left half-plane. Consequently, the filter (2.27) obtained on replacing the matrix D in (2.27) by $\bar{D}$ is uniformly asymptotically stable in the case of constant matrices A, σ, Q, σ_0 and of the matrices (2.32) having full rank.

We shall illustrate the foregoing algorithm for defining $\bar{D}$ by the following example.

Example 2.1. Consider the motion of a material point subject to random disturbances

$$\dot{x}_1 = x_2, \qquad dx_2 = \sqrt{2}\, d\xi_1(t).$$

The observation process $y(t)$ satisfies the equation

$$dy(t) = x_1(t)\,dt + \sqrt{\frac{1}{2}}\,d\xi_2(t).$$

Here ξ_1, ξ_2 denote independent standard Wiener processes. We now have

$$A = \begin{pmatrix} 0 & 1 \\ 0 & 0 \end{pmatrix}, \qquad \sigma = \sqrt{2}\begin{pmatrix} 0 & 0 \\ 0 & 1 \end{pmatrix}, \qquad Q = (1 \quad 0), \qquad \sigma_0 = \sqrt{\frac{1}{2}}.$$

Hence,

$$A_1 = \begin{pmatrix} 0 & 0 & 2 & 0 \\ -1 & 0 & 0 & 0 \\ 0 & 0 & 0 & 1 \\ 0 & 2 & 0 & 0 \end{pmatrix}.$$

Further,

$$\det(\lambda I - A_1) - \lambda^4 + 4.$$

The roots λ_1, λ_2 of the equation $\det(\lambda I - A_1) = 0$ having a negative real part are the following: $\lambda_1 = -1 + i$, $\lambda_2 = -1 - i$. It follows that

$$b(\lambda) = (\lambda - \lambda_1)(\lambda - \lambda_2) = \lambda^2 + 2\lambda + 2.$$

Therefore, the matrix $b(A_1)$ is of the form

$$b(A_1) = \begin{pmatrix} 2 & 0 & 4 & 2 \\ -2 & 2 & -2 & 0 \\ 0 & 2 & 2 & 2 \\ -2 & 4 & 0 & 2 \end{pmatrix} = A_1^2 + 2A_1 + 2I,$$

where I is a unit matrix.

Hence, the blocks b_i of the matrix $b(A_1)$ have the form

$$b_1 = \begin{pmatrix} 2 & 0 \\ -2 & 2 \end{pmatrix}, \quad b_2 = \begin{pmatrix} 4 & 2 \\ -2 & 0 \end{pmatrix}, \quad b_3 = \begin{pmatrix} 0 & 2 \\ -2 & 4 \end{pmatrix}, \quad b_4 = \begin{pmatrix} 2 & 2 \\ 0 & 2 \end{pmatrix}$$

Thus, in our case the unique positive definite solution $\bar{D}$ of an equation of the form (2.33) is

$$\bar{D} = b_4 b_2^{-1} = b_3 b_1^{-1} = \begin{pmatrix} 2 & 2 \\ 0 & 2 \end{pmatrix}.$$

7. Filtering problem with degenerate noise in the observation channel. In the deduction of equations (2.18), (2.27) of the optimal filter we have essentially used the assumption of the nondegeneracy of noise in the observation channel, i.e., of the nondegeneracy of the matrix $\sigma_0\sigma_0'$. At the same time there are situations in which the observed signal (the coordinates of the system or their combination) is not contaminated by noise. In such cases certain functions of the state of the system can be measured precisely. There exist different approaches to the investigation of filtering problems with degenerate noise. In the first of these, one part of the coordinates is estimated precisely, and for the estimation of the other coordinates the Kalman filter of reduced order is used [11]. In another approach the original problem is replaced by an approximate one in which the observation equation is supplemented with nondegenerate noise of low intensity.

We shall discuss in detail the first approach using as an example the problem of optimally estimating the state $x(t) \in R^n$ of a system of the form (2.1)

$$\dot{x}(t) = Ax(t), \qquad 0 \leqslant t \leqslant T,\ x(0) = x_0,$$

with constant matrix A. The Gaussian vector x_0 has zero mean and covariance matrix D_0. A vector $y \in R^m$ is observed that satisfies equation (2.2) with constant matrices Q and σ_0 such that

$$dy(t) = Qx(t)\,dt + \sigma_0\,d\xi_0(t), \qquad y(0) = 0.$$

The matrix Q has dimensions $m \times n$, and the matrix σ_0 has dimensions $m \times m$. Let $k(\sigma_0)$ be the rank of the matrix σ_0. Suppose that $k(\sigma_0) = r$, where $r \in [0, m]$ is an integer.

By means of elementary transformations we reduce the matrix σ_0 to the following form:

$$\alpha_1\sigma_0 = \begin{pmatrix} \sigma_1 \\ 0 \end{pmatrix} \tag{2.34}$$

Here α_1 is the nondegenerate real square matrix of order m corresponding to these elementary transformations, σ_1 is an $r \times m$ matrix having rank r, and 0 is an $(m - r) \times m$ zero matrix. Further, let

$$\alpha_1 Q = \begin{pmatrix} Q_1 \\ Q_2 \end{pmatrix}, \tag{2.35}$$

where Q_1 and Q_2 are matrices of dimensions $r \times n$ and $(m - r) \times n$ constituting the first r and the last $m - r$ rows of the matrix $\alpha_1 Q$, respectively.

Consider an $n \times (m - r)n$ matrix $K(Q_2)$ of the form

$$K(Q_2) = (Q_2', A'Q_2', \dots, (A')^{n-1}Q_2'). \tag{2.36}$$

Setting

$$\alpha_1 y = (y_1, y_2), \tag{2.37}$$

we have

$$dy_1(t) = Q_1x(t)\,dt + \sigma_1\,d\xi_0(t), \qquad \dot{y}_2(t) = Q_2x(t).$$

Further, depending on the rank of the matrix $K(Q_2)$, two situations are possible. If the rank of the matrix $K(Q_2)$ is n, then all the coordinates of the system $x(t)$ can be reconstructed precisely using the known values of the vector $y_2(t)$, which are determined in terms of the original observations through the use of formula (2.37). Let now the rank of the matrix $K(Q_2)$ be equal to n_1, where $n_1 < n$. We denote by λ a nondegenerate matrix whose first n_1 columns coincide with any linearly independent columns of the matrix $K(Q_2)$, and the other $n - n_1$ columns of the matrix λ are chosen arbitrarily subject to the only condition that the rank $k(\lambda) = n$. Then (see Chapter VI)

$$\lambda' A(\lambda^{-1})' = B = \begin{pmatrix} B_1 & O_1 \\ B_2 & B_3 \end{pmatrix}, \qquad Q_2(\lambda^{-1}) = (Q_3, Q_2). \tag{2.38}$$

Here the matrices B_1, B_2, B_3 have respective dimensions $n_1 \times n_1$, $(n - n_1) \times n_1$, $(n - n_1) \times (n - n_1)$, the matrix Q_3 has dimensions $(m - r) \times n_1$, and the zero matrices 0_1, 0_2 have respective dimensions $n_1 \times (n - n_1)$, $(m - r) \times (n - n_1)$. The matrix

$$L = \left(Q_3', B_1' Q_3', \ldots, (B_1')^{n_1} Q_3'\right)$$

is of full rank, i.e., $k(L) = n_1$.

We set $z = \lambda' x$. Then

$$\begin{aligned} \dot{z}(t) &= \lambda' \dot{x}(t) = Bz(t), \\ dy_1(t) &= Q_1(\lambda^{-1})' z(t)\, dt + \sigma_1\, d\xi_0(t), \\ \dot{y}_2(t) &= Q_2(\lambda^{-1})' z(t). \end{aligned} \tag{2.39}$$

We express the vector $z(t)$ as $z = (z_1, z_2)$, where $z_1 \in R^{n_1}$ and $z_2 \in R^{n-n_1}$. The component z_2 satisfies the relations

$$\begin{aligned} \dot{z}_2(t) &= B_2 z_1(t) + B_3 z_2(t), \\ dy_1(t) &= \left(Q_5 z_1(t) + Q_4 z_2(t)\right) dt + \sigma_1\, d\xi_0(t). \end{aligned} \tag{2.40}$$

Here the matrices Q_5 and Q_4 consist of the first n_1 and the last $n - n_1$ columns of the matrix $Q(\lambda^{-1})'$. In equations (2.39) the vector $z_1(t)$ may be regarded as a known function of time. Indeed, owing to the fullness of the rank of the matrix L, the vector $z_1(0)$ can be reconstructed precisely in terms of the known values of $y_2(s)$ over an arbitrarily small interval $0 \leqslant s \leqslant \varepsilon$ ($\varepsilon > 0$); as for the values $t > 0$, in view of (2.38), (2.39), the function $z_1(t)$ can be found by solving the equation

$$\dot{z}_1(t) = B_1 z_1(t).$$

Moreover, the vector $y_2(t)$ does not contain any information about the process $z_2(t)$. Thus, all information about $z_2(t)$ is contained in the measured values $y_2(s)$, $0 \leqslant s \leqslant t$, that are related to the original measured values $y(s)$ by formula (2.37).

We denote

$$\begin{aligned} &V = Q_4'(\sigma_1 \sigma_1')^{-1} Q_4, \qquad m_1(t) = E\{\, z_2(t) \mid y_{1t} \,\}, \\ &D_1(t) = E\{\, [z_2(t) - m_1(t)][z_2(t) - m_1(t)]' \mid y_{1t} \,\}. \end{aligned}$$

Since the matrix $\sigma_1\sigma_1'$ is nondegenerate (by construction its rank is r), it follows that m_1 and D_1 satisfy Kalman equations of the form (2.18), (2.27). In the case under consideration described by equations (2.40) we obtain

$$\begin{aligned} dm_1(t) &= (B_3 m_1(t) + B_2 z_1(t))\, dt + D_1(t) Q_4'(\sigma_1\sigma_1')^{-1} \\ &\quad \times [dy_1(t) - Q_5 z_1(t)\, dt - Q_4 m_1(t)\, dt], \qquad m_1(0) = 0, \\ \dot{D}_1 &= B_3 D_1 + D_1 B_3' - D_1 V D_1, \qquad D_1(0) = \Lambda. \end{aligned} \tag{2.41}$$

Here Λ denotes an $(n - n_1) \times (n - n_1)$ matrix whose elements are located at the intersection of the last $n - n_1$ rows and columns of the matrix $(\lambda' D_0 \lambda)$.

Let D_2 denote an $n \times n$ matrix whose first n_1 rows and columns are zero, and the lower right corner is occupied by D_1, i.e.,

$$D_2(t) = \begin{pmatrix} 0 & 0 \\ 0 & D_1(t) \end{pmatrix}. \tag{2.42}$$

This definition implies that D_2 is the covariance matrix of the conditional distribution of $z(t)$, given the observation data y_t.

Since under linear nondegenerate transformation of random variables their estimates are transformed according to the same transformation matrix, it follows that, setting $m(t) = E\{\, x(t) \mid y_t \,\}$, $D(t) = E\{\, [x(t) - m(t)][x(t) - m(t)]' \mid y_t \,\}$,

$$m(t) = (\lambda^{-1})' \begin{pmatrix} z_1(t) \\ m_1(t) \end{pmatrix}, \qquad D(t) = (\lambda^{-1})' D_2(t) \lambda^{-1}. \tag{2.43}$$

Thus, the proposed algorithm for constructing the optimal estimate $m(t)$ of the vector $x(t)$ consists of the following steps:

1°. By means of elementary transformations we reduce the matrices σ and Q to the form (2.34), (2.35) and determine the rank of the matrix (2.36).

2°. If the rank $K(Q_2) = n$, then $x(t)$ can be reconstructed precisely.

3°. If the rank $K(Q_2) < n$, then the best estimate is expressed by formulas (2.41)–(2.43), which involve the function $z_1(t)$ that is reconstructed precisely via the known values $y_2(s)$, $0 \leqslant s \leqslant t$, by use of the procedure described in Chapter VI.

8. Optimal extrapolation. The *optimal extrapolation problem* is the problem of constructing the optimal (in the mean-square sense) estimate of a future state $x(\tau)$ of the system (2.1), given the actually observed values of the process (2.2) over the interval $0 \leqslant s \leqslant T$, where $T \leqslant \tau$. As will be shown, the optimal estimate $m(\tau) = E\{\, x(\tau) \mid y_T \,\}$ and the covariance matrix $D(\tau) = E\{\, [x(\tau) - m(\tau)][x(\tau) - m(\tau)]' \mid y_T \,\}$ can be determined by means of an auxiliary filtering problem.

We introduce the function

$$Q_1(t) = \begin{cases} Q(t), & \text{for } 0 \leqslant t \leqslant T, \\ 0, & \text{for } t > T. \end{cases}$$

Let $\gamma(t)$ denote a process such that $\gamma(t) = y(t)$ for $0 \leqslant t \leqslant T$ and $d\gamma(t) = \sigma_0(t)\, d\xi_0(t)$ for $T < t \leqslant \tau$. In other words,

$$d\gamma(t) = Q_1(t)x(t)\, dt + \sigma_0(t)\, d\xi_0(t) \quad \text{for } 0 < t \leqslant \tau.$$

Since $\xi(t)$, $\xi_0(t)$, and x_0 are independent, we have $m(\tau) = m_1(\tau)$, $D(\tau) = D_1(\tau)$, where

$$m_1(\tau) = E\{\, x(\tau) \mid \gamma_\tau \,\}, \qquad D_1(\tau) = E\{\, [x(\tau) - m_1(\tau)][x(\tau) - m_1(\tau)]' \mid \gamma_\tau \,\}.$$

It follows from the definition of $m_1(\tau)$ and $D_1(\tau)$ that they satisfy equations (2.18), (2.27), in which $Q(t)$ has been replaced by $Q_1(t)$. Consequently, in order to calculate the matrix $D(\tau)$, it is necessary to solve the equations

$$\begin{aligned} \dot D(t) &= A(t)D(t) + D(t)A'(t) - D(t)Q'(t)(\sigma_0\sigma_0')^{-1}(t)Q(t)D(t) \\ &\quad + (\sigma\sigma'), \qquad D(0) = D_0,\ 0 \leqslant t \leqslant T, \end{aligned} \tag{2.44}$$

$$\dot D(t) = A(t)D(t) + D(t)A'(t) + (\sigma\sigma'), \qquad T \leqslant t \leqslant \tau. \tag{2.45}$$

The optimal extrapolation estimate $m(\tau)$ is determined from the equations

$$dm(t) = A(t)m(t)\, dt + DQ'(\sigma_0\sigma_0')^{-1}(dy(t) - Q(t)m(t)\, dt), \qquad 0 \leqslant t < T, \tag{2.46}$$

$$\dot m(t) = A(t)m(t), \qquad T \leqslant t \leqslant \tau, \tag{2.47}$$

$$m(0) = m_0.$$

To compute the matrix $D(\tau)$, we must first solve the nonlinear Riccati equation (2.44) on the interval $[0, T]$, and then the linear equation (2.45) on the interval $[T, \tau]$ with initial condition $D(T)$ calculated in the preceding step. Similarly, to compute the estimate $m(\tau)$, we must first solve the stochastic equation (2.46) and determine $m(T)$, and then the deterministic equation (2.47) on the interval $[\tau, T]$ with initial condition $m(T)$.

9. Optimal interpolation. The *optimal interpolation problem* is the problem of constructing the optimal (in the mean-square sense) estimate of a previous state of the vector $x(\tau)$, given the actually observed values of the process (2.2) over the interval $[0, T]$, where $\tau < T$. As will be shown, the interpolation problem can be reduced to a linear-quadratic control problem. The appropriate results will be given here.

Let $m(\tau, T)$ denote the optimal estimate, and $D(\tau, T)$ the corresponding covariance matrix, i.e.,

$$\begin{aligned} m(\tau, T) &= E\{\, x(\tau) \mid y_T \,\}, \\ D(\tau, T) &= E\{\, [x(\tau) - m(\tau, T)][x(\tau) - m(\tau, T)]' \mid y_T \,\}. \end{aligned} \tag{2.48}$$

The optimal interpolation estimate $m(\tau,T)$ is a linear functional of the observations, i.e.,

$$m(\tau,T)=\int_0^T u'(t,\tau,T)\,dy(t). \tag{2.49}$$

The kernel u of the estimator (2.49) is the optimal control of the linear deterministic system involving a δ-function on the right side

$$\dot\alpha(t)=-A'(t)\alpha(t)+Q'(t)u-\delta(t-\tau),\ 0\leqslant t\leqslant T,\ \alpha(T)=0 \tag{2.50}$$

with quadratic cost functional (2.7). The duality of the interpolation problem (2.48) and the control problem (2.50), (2.7) is established by the same reasoning as in subsection 2. Now, using the solution of the dual control problem (2.50), (2.7) given in subsection 2, we obtain the following formulas for the functions (2.48):

$$\begin{aligned} m(\tau,t)&=m(T)-\int_\tau^T \big[A(s)m(s,T)+\sigma_0(s)\sigma_0'(s)D^{-1}(s)\\ &\qquad\times\big(m(s,T)-m(s)\big)\big]\,ds,\\ D(\tau,T)&=D(T)-\int_\tau^T \big[\big(A(s)+\sigma_0(s)\sigma_0'(s)D^{-1}(s)\big)D(s,T)\\ &\qquad+D(s,t)\big(A(s)+\sigma_0(s)\sigma_0'(s)D^{-1}(s)\big)-\sigma(s)\sigma'(s)\big]\,ds. \end{aligned} \tag{2.51}$$

In formulas (2.51) the following notation is used:

$$m(t)=E\{\,x(t)\mid y_t\,\},\qquad D(t)=E\{\,[x(t)-m(t)][x(t)-m(t)]'\mid y_t\,\}.$$

In other words, the functions $m(t)$ and $D(t)$ constitute the solution of the filtering problem discussed above and, consequently, are determined from equations (2.27), (2.18).

10. The discrete Kalman filter. Consider the filtering problem for the case in which the evolution of the system $x(t_k)$ and the observation process are governed by the equations

$$x(t_{k+1})=A_kx(t_k)+\xi(t_k),\qquad k=0,1,2,\ldots,\qquad (t_k<t_{k+1}) \tag{2.52}$$

$$y(t_k)=Q_kx(t_k)+\eta(t_k). \tag{2.53}$$

Here $x(t_k)\in R^n$, $y(t_k)\in R^m$, the random disturbances $\xi(t_k)\in R^n$ and $\eta(t_k)\in R^m$ are Gaussian with zero means and covariance matrices G_k (of dimensions $n\times n$) and Σ_k (of dimensions $m\times m$), the matrices Σ_k being positive definite. The *a priori* distribution of the vector $x(t_0-0)$ at the time immediately preceding the moment at which the observation sequence begins is assumed to be known. It is Gaussian with prescribed mean m_0 and covariance matrix $D_0>0$. The random vectors $x(t-t_0)$, $\xi(t_k)$, $\eta(t_k)$ are independent for all i. The system (2.52), (2.53) can be used both for discrete-time modeling of the observation process and as an approximation of a continuous-time filtering problem.

Under the assumptions made the conditional probability distribution $x(t_k)$, given the observations $y(t_j)$, $0 \leqslant j \leqslant k$, is normal, i.e., it is described by the conditional mean $m(t_k)$ and covariance matrix $D(t_k)$. It can be shown that these characteristics, $m(t_k)$ and $D(t_k)$, vary on the interval between observations due to equations (2.52), and have to be recalculated at the observation points. The appropriate equations are given below.

Let $m(t_k - 0)$, $D(t_k - 0)$ denote the values of the functions $m(t)$ and $D(t)$ immediately preceding the kth observation, $y(t_k)$, and $m(t_k + 0)$, $D(t_k + 0)$ the values of these functions immediately following the kth observation. By virtue of (2.52) we have

$$m(t_{k+1} - 0) = A_k m(t_k + 0), \tag{2.54}$$

$$D(t_{k+1} - 0) = A_k D(t_k + 0) A_k' + G_k. \tag{2.55}$$

By the theorem on normal correlation we obtain

$$m(t_k + 0) = m(t_k - 0) + D(t_k + 0) Q_k' \Sigma_k^{-1} [y(t_k) - Q_k m(t_k - 0)], \tag{2.56}$$

$$m(t_0 - 0) = m_0,$$

$$D(t_k + 0) = D(t_k - 0) \left[I + Q_k' \Sigma_k^{-1} Q_k D(t_k - 0) \right]^{-1}, \tag{2.57}$$

$$D(t_0 - 0) = D_0.$$

Equations (2.54)–(2.57) jointly describe the evolution of the optimal estimate $m(t)$ and of the covariance matrix $D(t)$, which are due to the motion of the system (2.52) and to taking account of the observations (2.53) of the state variable.

In using the filtering equations (2.56), (2.57), we come up against the necessity of computing an inverse matrix (2.57) of dimensions $n \times n$ in each step. This is by far the most time-consuming operation, especially, for large n. For $n > m$ it is expedient to turn to the following equivalent set of filtering equations which involves the inversion of an $m \times m$ matrix in each step:

$$m(t_k + 0) = m(t_k - 0) + R(t_k + 0)[y(t_k) - Q_k m(t_k - 0)], \tag{2.58}$$

$$m(t_0 - 0) = m_0,$$

$$R(t_k + 0) = D(t_k - 0) Q_k' \left[Q_k D(t_k - 0) Q_k' + \Sigma_k \right]^{-1}, \tag{2.59}$$

$$D(t_k + 0) = [I - R(t_k + 0) Q_k] D(t_k - 0), \tag{2.60}$$

$$D(t_0 - 0) = D_0.$$

Here I is a unit matrix.

We shall prove the equivalence of the systems (2.56), (2.57) and (2.58)–(2.60). Let $R(t_k + 0) = D(t_k + 0) Q_k' \Sigma_k^{-1}$. Then equation (2.58) follows immediately from (2.56) and

$$\begin{aligned} R(t_k + 0) &= D(t_k - 0) \left[I + Q_k' \Sigma_k^{-1} Q_k D(t_k - 0) \right]^{-1} Q_k' \Sigma_k^{-1} \\ &= D(t_k - 0) \left[(Q_k' \Sigma_k^{-1})^{-1} + Q_k D(t_k - 0) \right]^{-1}. \end{aligned}$$

This yields (2.59). It follows from equation (2.57) that

$$D(t_k+0)\big[D^{-1}(t_k-0)+Q'_k\Sigma_k^{-1}Q_k\big]=I,$$

or

$$D(t_k+0)D^{-1}(t_k-0)+R(t_k+0)Q_k=I.$$

Hence we have equation (2.60).

Thus, two sets of equations describing the discrete Kalman filter have been obtained: it is advantageous to use the first set (2.56), (2.57) for $n \leqslant m$ and the second set (2.58)–(2.59) for $n > m$.

§ 3. Some Relations of Nonlinear Filtering Theory

1. Statement of the general filtering problem. In the deduction of the Kalman filtering equations (2.18), (2.27) we have used essentially the linearity of the right sides of equations (2.1), (2.2) with respect to coordinates. In the general case such simple filtering equations cannot be obtained. Let us consider in greater detail the filtering problem for the system described by an equation understood in the sense of Ito

$$dx(t)=f\big(t,x(t)\big)\,dt+\sigma\big(t,x(t)\big)\,d\xi(t),\qquad t\geqslant 0,\qquad (3.1)$$
$$x(0)=x_0,\ x\in R^n.$$

The equations for the observed process are of the form

$$dy(t)=Q\big(t,x(t)\big)\,dt+\sigma_0(t)\,d\xi_0(t),\qquad y\in R^m.\qquad (3.2)$$

The random disturbances $\xi(t)$ and $\xi_0(t)$ are assumed to be independent standard Wiener processes, and the matrix $\sigma_0\sigma_0'$ to be nondegenerate. The functions $f(t,x)$ and $Q(t,x)$ are continuous, with $f(t,x)$ continuously differentiable with respect to x, and $Q(t,x)$ is twice continuously differentiable with respect to x. It should be noted that the functions f, σ, Q, and σ_0 may also depend on the "past" observations $y_t=y(s), 0\leqslant s\leqslant t$.

Eqn (3.2) can be understood either in the sense of Ito or in the symmetrized sense. In the latter case we can assume that the derivatives $\dot y(t)$ and $\dot\xi_0(t)$ exist and model the actual wide-band processes.[2] Substituting $\dot\xi_0(t)$ for the wide-band noise process means that the noise cutoff frequency is not being considered. This is permissible only in the case when it is much higher than the relevant frequencies for a particular problem. In almost all practical cases of filtering the bandwidth of a filtering system is certainly much smaller than the frequency range of the noise. Therefore, it may be considered as "white" and equal to the symmetrized

[2]Kulman, N. K., A Note on the Differential Equations of Conditional Probability Density Functions, *J. Math. Anal. Appl.*, (1966), vol. 14, no. 2, 301–308. See also [21] of Part One.

derivative of the Wiener process $\xi_0(t)$. In this case the observed process is taken to be

$$\dot{y}(t) = Q\big(t, x(t)\big) + \sigma_0(t)\dot{\xi}_0(t), \qquad y \in R^m. \tag{3.3}$$

It is required to construct equations for an optimal estimator of $x(t)$. The following estimators are usually considered: the optimal (in the mean-square sense) estimator of the process $x(t)$, which coincides with the conditional mean $m(t)$, and the conditional mode $x^*(t)$. This estimators can be obtained from an equation for the conditional probability distribution of $x(t)$.

2. The Stratonovich differential equations for the conditional probability distribution. Let $P(t,x) = P\big(x(t) \mid y_t\big)$ denote the conditional probability distribution (density function) of the random variable $x(t)$, given the actually observed values $y_t = y(s)$, $0 \leqslant s \leqslant t$, over the interval $[0,t]$. Then the function $P(t,x)$ will satisfy the following equations:[3]

in the Ito form

$$\begin{aligned} P(\tau, x) - P(t,x) = \int_t^\tau \Bigg[&-\sum_{i=1}^n \frac{\partial}{\partial x_i}\big(f_i(s,x)P(s,x)\big) \\ &+ \frac{1}{2}\sum_{i,j=1}^n \frac{\partial^2}{\partial x_i\, \partial x_j}\big(b_{ij}(s,x)P(s,x)\big)\Bigg]\, ds \\ + \int_t^\tau & P(s,x)\big(Q(s,x) - \bar{Q}(s)\big)' \\ & \times N_0^{-1}(s)\big(dy(s) - \bar{Q}(s)\, ds\big); \end{aligned} \tag{3.4}$$

in the symmetrized form

$$\begin{aligned} \frac{\partial P(t,x)}{\partial t} = &-\sum_{i=1}^n \frac{\partial}{\partial x_i}\big(f_i(t,x)P(t,x)\big) \\ &+ \frac{1}{2}\sum_{i,j=1}^n \frac{\partial^2}{\partial x_i\, \partial x_j}\big(b_{ij}(t,x)P(t,x)\big) \\ &+ [F(t,x) - \bar{F}(t)]P(t,x). \end{aligned} \tag{3.5}$$

Here $\tau \geqslant t$, f_i denotes the ith component of the vector f, b_{ij} designates the ijth element of the matrix $b(t,x) = \sigma(t,x)\sigma'(t,x)$, $N_0(t) = \sigma_0(t)\sigma_0'(t)$, and

$$F(t,x) = Q'(t,x)N_0^{-1}(t)\dot{y}(t) - \tfrac{1}{2}Q'(t,x)N_0^{-1}(t)Q(t,x), \tag{3.6}$$

$$\bar{Q}(t) = \int_{R^n} Q(t,x)P(t,x)\, dx, \qquad \bar{F}(t) = \int_{R^n} F(t,x)P(t,x)\, dx.$$

---

[3]Stratonovich, R. L., Conditional Markov Processes, *Teor. Veroyatnost. i Primenen.*, (1960), vol. 5, no. 2, 172–195. (Translated into English as *Theory Probab. Appl.*). See also [21] of Part One. The equation in the Ito form was also derived by H. J. Kushner and published in the paper "On the Dynamical Equations of Conditional Probability Density Functions", *J. Math. Anal. Appl.*, (1964), vol. 8, no. 2, 332–344.

In other words, $\bar{Q}(t)$ and $\bar{F}(t)$ are the conditional means of the vector $Q(t,x(t))$ and of the scalar quantity $F(t,x(t))$, given the actually observed values y_t.

The initial conditional probability distribution $P(0,x)$ coincides with the *a priori* probability distribution for $x(0) = x_0$: $P(0,x) = P_0(x_0)$. The values of $P(t,x)$ and $\partial P(t,x)/\partial x$ are taken to be zero at infinity.

The desired optimal estimators can be obtained, in principle, using the relations

$$m(t) = \int_{R^n} xP(t,x)\,dx, \qquad x^*(t) = \arg\max P(t,x).$$

The resulting estimators turn out to be nonlinear in the general case. It should be noted that both these estimators coincide in the case of a symmetric probability distribution, such as the Gaussian distribution.

Since the construction of a precise solution of equation (2.52) is difficult, various methods for solving the filtering problem have been proposed for specific cases [12–14].

3. Conditionally Gaussian processes. First, we consider the simplest case of nonlinear filtering, namely that of the filtering of *conditionally Gaussian* processes, i.e., of processes described by stochastic Ito equations which are linear in the phase coordinates $x(t)\beta_{ij}(s,x) in R^n$, but nonlinear in the observations $y(t) \in R^m$. The equations in question are of the form

$$dx(t) = A(t,y_t)x(t)\,dt + \sigma(t,y_t)\,d\xi(t), \qquad x(0) = x_0, \tag{3.8}$$

$$dy(t) = Q(t,y_t)x(t)\,dt + \sigma_0(t,y_t)\,d\xi_0(t), \qquad y(0) = y_0, \tag{3.9}$$

where y_t for all t denotes the aggregate of all measured values: $y_t = y(t+\theta)$, $-t \leqslant \theta \leqslant 0$, the functionals A, σ, Q, σ_0 at each moment t depend on the measurement data y_t; the Wiener processes ξ, ξ_0 and the Gaussian vector x_0 are statistically independent. The mean m_0 and the covariance matrix D_0 of the vector x_0 are prescribed.

It should be noted that the conditional probability distribution of $x(t)$, given y_t, is Gaussian. Let $m(t)$ denote the optimal (in the mean-square sense) estimate of the vector $x(t)$, given the measurement data y_t, and $D(t)$ the corresponding covariance matrix of the conditional distribution. Then the equations for $m(t)$ and $D(t)$ are of the form [12]

$$\begin{aligned} dm(t) &= A(t,y_t)m(t) + D(t)Q'(t,y_t) \\ &\quad \times \big(\sigma_0(t,y_t)\sigma_0'(t,y_t)\big)^{-1}[dy(t) - Q(t,y_t)m(t)], \\ &\quad m(0) = m_0; \end{aligned} \tag{3.10}$$

$$\begin{aligned} \dot{D}(t) &= A(t,y_t)D(t) + D(t)A'(t,y_t) + \sigma(t,y_t)\sigma'(t,y_t) \\ &\quad - D(t)Q'(t,y_t)\big(\sigma_0(t,y_t)\sigma_0'(t,y_t)\big)^{-1}Q(t,y_t)D(t), \\ &\quad D(0) = D_0. \end{aligned} \tag{3.11}$$

When the matrices A, σ, Q, σ_0 do not depend on the observations y_t, equations (2.53), (2.54) go over into equations (2.27), (2.18) of the Kalman filter.

4. Quasioptimal and quasilinear filtering (the scalar case). In the cases of a non-Gaussian process $x(t)$ satisfying equation (3.1) and of an observed process $\dot{y}(t)$ satisfying equation (3.3) and containing a nonlinear useful signal $Q(t, x(t))$ in white noise the conditional probability distribution is not Gaussian. However, if it can be approximated by a function containing a finite number of parameters depending solely on t, then the partial differential equation (3.5) yields a system of a finite number of ordinary differential equations for these parameters. The estimators (3.7) are then expressed in terms of these parameters. A *quasioptimal filter* of an appropriate order (depending on the number of the parameters) is thus obtained.

For the sake of simplicity, let us consider the scalar case: $n = m = 1$. Then instead of (3.5) we have the following equation for $P(t, x)$:

$$\frac{\partial P(t,x)}{\partial t} = -\frac{\partial}{\partial x}[f(t,x)P(t,x)] + \frac{1}{2}\frac{\partial^2}{\partial x^2}[b(t,x)P(t,x)] + [F(t,x) - \bar{F}(t)]P(t,x), \tag{3.12}$$

where

$$F(t,x) = \frac{1}{N_0(t)}[\dot{y}(t)Q(t,x) - \tfrac{1}{2}Q^2(t,x)]. \tag{3.13}$$

Here, $b(t,x) = \sigma^2(t,x)$ and $N_0(t) = \sigma_0^2(t)$.

Following a work by Stratonovich,[4] we shall seek a quasioptimal solution to the filtering problem for the simplified case in which

$$f(t,x) = 0, \qquad b(t,x) = b = \sigma^2,$$

where $\sigma(t,x) = \sigma$ and hence b are prescribed constants. As follows from (3.12), the resulting Stratonovich equation has the form

$$\frac{\partial P(t,x)}{\partial t} = \frac{b}{2}\frac{\partial^2 P(t,x)}{\partial x^2} + [F(t,x) - \bar{F}(t)]P(t,x), \tag{3.14}$$

The solution of this equation is sought in the form

$$P(t,x) = \exp\left\{\frac{C(t)}{\mu^2} - \frac{1}{\mu^2}\sum_{j=2}^{\infty}\frac{1}{j!}a_j(t)\big(x - x^*(t)\big)^j\right\}, \tag{3.15}$$

Here $C(t)$ is connected with the normalization condition and plays no further part. The parameter $x^* = x^*(t)$ is the optimal estimate of $x(t)$ equal to the conditional mode, i.e., it is the coordinate of the maximum of $P(t,x)$. The parameter μ is

[4]Stratonovich, R. L., Application of the Theory of Markov Processes to the Optimal Filtering Problem, *Radiotekhn. i Electron.*, (1960), vol. 5, no. 11, 1751–1763. (Translated into English as *Radio Eng. Electron. Phys.* superseded by *Soviet J. Comm. Tech. Electron.*)

introduced solely for the sake of convenience. In another version of the method it could well be made equal to one. Here it is selected as

$$\mu \sim \left[-\frac{\partial^2 \ln P}{\partial x^2}(t, x^*) \right]^{-1/2}, \tag{3.16}$$

in which case

$$a_2 \sim 1. \tag{3.17}$$

Substituting (3.15) into (3.14) we can form differential equations which define the variation of the parameters $x^*, a_2, a_3, \ldots$. If we could produce a system which would simulate these differential equations, then by taking $x^*(t)$ as the output signal, we would have the optimum nonlinear system treating the input signal $y(t)$. But since it is impracticable to simulate an infinite set of differential equations, the problem arises of replacing this set by a finite set of equations by discarding a number of immaterial terms. To solve this problem, it is possible to develop an asymptotic method in which μ plays the role of the small parameter. This corresponds, obviously, to

Substituting $P(t, x) = \exp\{-v(t, x)/mu^2\}$ in equation (3.14) gives

$$\frac{\partial v(t,x)}{\partial t} = \frac{b}{2}\left[\frac{\partial^2 v(t,x)}{\partial x^2} - \frac{1}{\mu^2}\left(\frac{\partial v(t,x)}{\partial x} \right)^2 \right] - \mu^2 F(t,x) + \mu^2 \bar{F}(t). \tag{3.17}$$

Suppose we put in accordance with the expression (3.15)

$$v(t,x) = \sum_{j=2}^{\infty} \frac{1}{j!} a_j(t) \left(x - x^*(t) \right)^j,$$

rewrite $F(t, x)$ in the form

$$F(t,x) = \sum_{l=0}^{\infty} \frac{1}{l!} F_l(t) \left(x - x^*(t) \right)^l, \qquad F_l = \left. \frac{\partial^l F}{\partial x^l} \right|_{x = x^*}$$

and equate terms with like powers of the difference $x - x^*$, we then get the chain of equations

$$\begin{aligned} -a_2 \dot{x}^* &= \frac{b}{2} a_3 - \mu^2 F_1, \\ \dot{a}_2 - a_3 \dot{x}^* &= \frac{b}{2\mu^2}[\mu^2 a_4 - 2a_2^2] - \mu^2 F_2, \\ \dot{a}_3 - a_4 \dot{x}^* &= \frac{b}{2\mu^2}[\mu^2 a_5 - 6a_2 a_3] - \mu^2 F_3, \\ &\ldots\ldots\ldots\ldots\ldots\ldots\ldots\ldots \\ &\ldots\ldots\ldots\ldots\ldots\ldots\ldots\ldots \end{aligned} \tag{3.19}$$

or

$$\dot{a}_j - a_{j+1}\dot{x}^* = \frac{b}{2\mu^2}\Big[\mu^2 a_{j+2} - j a_1 a_j - \binom{j}{2} a_3 a_{j-1} - \cdots - j a_j a_2 \Big] - \mu^2 F_j, \qquad j = 1, 2, \ldots, \ (a_1 = 0), \tag{3.20}$$

which is mathematically equivalent to (3.14). The initial conditions for this system of equations are the following:

$$x^*(0) = x_0^* = \arg\max P_0(x_0), \qquad a_j(0) = -\mu^2 \frac{\partial^j \ln P_0}{\partial x^j}(x_0^*), \qquad j = 2, \dots . \tag{3.21}$$

The chain of equations (3.19), (3.20) is broken if we discard terms with a_{j+1} and a_{j+2} in one of the equations (3.20). In order to check how far this is possible, let us estimate the order of magnitude of the various terms in the stated equations.

Suppose that l_Q is a scale unit characterizing the rapidity of variation of the function $F(t,x)$ defined by

$$F_l \sim \frac{F_2}{l_Q^{l-2}}, \qquad l = 3, 4, \dots . \tag{3.22}$$

It follows from the expression (3.13) that

$$\begin{aligned} F_1 &= \frac{1}{N_0(t)}[\dot{y}(t) - Q(t,x)]\frac{\partial Q(t,x)}{\partial x}, \\ F_2 &= \frac{1}{N_0(t)}[\dot{y}(t) - Q(t,x)]\frac{\partial^2 Q(t,x)}{\partial x^2} - \frac{1}{N_0(t)}\left[\frac{\partial Q(t,x)}{\partial x}\right]^2, \\ F_3 &= \frac{1}{N_0(t)}[\dot{y}(t) - Q(t,x)]\frac{\partial^3 Q(t,x)}{\partial x^3} - \frac{3}{N_0(t)}\frac{\partial Q(t,x)}{\partial x}\frac{\partial^2 Q(t,x)}{\partial x^2}, \\ &\dots\dots\dots\dots\dots\dots\dots\dots \\ &\dots\dots\dots\dots\dots\dots\dots\dots \end{aligned} \tag{3.23}$$

Then, in accordance with this set of relations, we have

$$l_Q \sim \left| \frac{Q'(t,x)}{Q^{j+1}(t,x)} \right|^{1/j}, j = 1, 2, 3, \dots . \tag{3.24}$$

Substituting the estimate (3.22) in equations $l = 3, 4, \dots$ of the system (3.19) and considering that the equations which are obtained in this case are of identical structure differing only in the factor in front of F_2, we draw the conclusion that

$$a_l \sim \frac{a_2}{l_Q^{l-2}}, \qquad l = 3, 4, \dots . \tag{3.25}$$

or else $a_l \sim l_Q^{-l+2}$ by virtue of (3.16). The foundation for breaking the chain of equations is the condition

$$\mu \ll l_Q, \tag{3.26}$$

which we assume to be fulfilled. This corresponds, obviously, to the requirement that the curvature of the condition probability distribution be sufficiently large at the point $x = x^*$.

Taking the relations (3.25) and (3.16) into account, we come to the conclusion that the term $\mu^2 a_{j+2}$ in (3.20) is smaller than terms of the type

$$a_2 a_j, a_3 a_{j-1}, \dots$$

roughly by a factor of l_Q^2/μ^2. In estimating the magnitude of the term $a_{j+1}\dot{x}^*$, it is not its instantaneous value which is taken into account, but ita integral effect during the period $T_1 = \mu^2/b$. In fact, equation (3.20), which can be written in the form

$$\dot{a}_j + \frac{jb}{\mu^2} a_2 a_j = a_{j+1}\dot{x}^* + \dots,$$

has the solution

$$a_j(t) = \int_{-\infty}^{t} \exp\left[\frac{jb}{\mu^2}\int_{t'}^{t} a_2\, dt''\right](a_{j+1}\dot{x}^* + \cdots)\, dt', \tag{3.27}$$

from which it will be seen that terms on the right side have solely an integral effect on a_j, the integration time constant here being $\mu^2/jba_2 \sim \mu^2/b$. In estimating the order of magnitude of an integral of type (3.27) from $a_{j+1}x^*$, we can substitute the expression

$$\int_{t-T_1}^{t} a_{j+1}\dot{x}^*\, dt' \sim a_{j+1}[x^*(t) - x^*(t-T_1)].$$

Therefore, the term $a_{j+1}\dot{x}^*$ has an effect of the same order of magnitude as $a_{j+1}[x^*(t) - x^*(t-T_1)]/T_1$. If the estimation error is not large, the increment $x^*(t) - x^*(t-T_1)$ is of the same order of magnitude as the *a priori* increment over the same interval described by a purely diffusion law. Its variance is equal to bT_1, and, therefore,

$$a_{j+1}\frac{x^*(t) - x^*(t-T_1)}{T_1} \sim a_{j+1}\sqrt{\frac{b}{T_1}}.$$

Since $T_1 = \mu^2/b$, we have

$$a_{j+1}\frac{x^*(t) - x^*(t-T_1)}{T_1} \sim a_{j+1}\frac{b}{\mu} \qquad \left(\mu a_{j+1} \sim \frac{\mu}{l_Q} a_j\right).$$

Consequently, it can be assumed that the term $a_{j+1}x^*$ is l_Q/μ times greater than $(b/2\mu^2)\mu^2 a_{j+2}$ and as many times less than terms of the type $(b/2\mu^2)a_2 a_j, \dots$, i.e., its order of magnitude is μ.

Let us consider the system (3.20) containing J equations, i.e., $j = 1, \dots, J$, where J is an even number to guarantee the normalization condition: $\int_{\infty}^{\infty} P(t,x)\, dx = 1$. Condition (3.26) allows terms containing a_{J+1} and a_{J+2} to be discarded from equation (3.20) taken for $j = J$. In this case a_J will have a relative error of the order of μ/l_Q. In the preceding equation

$$\dot{a}_{J-1} - a_J\dot{x}^* = \frac{b}{2\mu^2}[\mu^2 a_{J+1} - (J-1)a_2 a_{J-1} - \cdots] - \mu^2 F_{J-1},$$

which defines a_{J-1}, we discard a_{J+1} and retain $a_J\dot{x}^*$. Then a_{J-1} acquires an error of the order of $(\mu/l_Q)^2$. In the equation for a_{j-2} we retain all the terms, but a_{J-2} will have a relative error of the order of $(\mu/l_Q)^3$ by virtue of the inaccuracy of the functions a_{J-1} and a_J. Continuing the reasoning further, we find that a_3 has a relative error of the order of $(\mu/l_Q)^{J-2}$, whilst that of a_2 is of the order of $(\mu/l_Q)^{J-1}$. Thus, by considering a finite set of J equations, given the condition (3.26), we find the estimate $x^*(t)$ with a degree of accuracy which is the greater, the larger the number J. The stated set of equations, given the condition (3.26), describes a *quasioptimal filtering algorithm of order* J.

The quasioptimal algorithm of the second order (J=2) corresponds to a Gaussian approximation of the conditional probability distribution, and is, therefore, often termed the *quasioptimal algorithm in the Gaussian approximation* or the *quasilinear algorithm*, since it is an extension, in a way, of the linear algorithm ("the Kalman filter") characterized by the Gaussian conditional distribution. It is described by the system of two equations

$$-h_2\dot{x}^*(t) = -F_1, \tag{3.28}$$

$$\dot{h}_2 = -bh_2^2 - F_2, \tag{3.29}$$

where $h_2 = a_2/\mu^2$.

Introducing the conditional variance $D(t) = 1/h_2(t)$, we can write the filtering equations (3.28), (3.29) in more conventional form:

$$\dot{x}^*(t) = D(t)F_1, \qquad x^*(0) = x_0^*, \tag{3.30}$$

$$\dot{D}(t) = b + F_2D^2(t), \qquad D(0) = D_0 = \left[-\frac{\partial^2 \ln P_0}{\partial x^2}(x_0^*)\right]^{-1}. \tag{3.31}$$

By virtue of the relation (3.16) and (3.26) the filtering equations (3.28), (3.29) or (3.30), (3.31) are justified whenever

$$\mu = \sqrt{\frac{a_2}{h_2}} \ll l_Q, \quad \text{or} \quad \sqrt{h_2} \gg l_Q.$$

Thus, the curvature h_2 of the condition probability distribution must be sufficiently large at the point $x = x^*$.

In view of the relation $D(t) = 1/h_2(t)$, the condition for the validity of the Gaussian approximation becomes the condition of smallness of the condition variance

$$\sqrt{D(t)} \ll l_Q. \tag{3.32}$$

The method of Gaussian approximation also extends to the case of the more general equation (3.12). Here a number of other conditions must be satisfied besides (3.32). By using power-series expansions of $f(t,x)$ and $b(t,x) = \sigma^2(t,x)$, which are similar to that of $F(t,x)$ (formula (3.18)), we can obtain more general filtering equations. This result is due to N. K. Kulman and is stated in the following theorem [14, § 3.2]: *Let the conditions of smallness of the conditional variance $D(t)$ be satisfied*:

$$\begin{aligned} &\text{(a)} \quad \sqrt{D} \ll l_Q, \\ &\text{(b)} \quad \sqrt{D} \ll l_f, \\ &\text{(c)} \quad \sqrt{D} \ll l_b, \end{aligned} \tag{3.33}$$

where

$$\begin{aligned} l_Q &\sim \left| \frac{\partial Q(t,x*)}{\partial x} \bigg/ \frac{\partial^{j+1} Q(t,x*)}{\partial x^{j+1}} \right|^{1/j}, \\ l_f &\sim \left| \frac{\partial f(t,x*)}{\partial x} \bigg/ \frac{\partial^{j+1} f(t,x*)}{\partial x^{j+1}} \right|^{1/j}, \\ l_b &\sim \left| b(t,x*) \bigg/ \frac{\partial^{j} b(t,x*)}{\partial x^{j}} \right|^{1/j}, \\ &(j = 1,2,3,\dots). \end{aligned} \tag{3.34}$$

are scale units characterizing the rapidity of variation of the functions $Q(t,x)$, $f(t,x)$, $b(t,x)$ with respect to x. Then the conditional probability distribution is nearly Gaussian with the conditional mode $x^*(t)$ and the conditional variance $D(t)$ satisfying the following set of equations:

$$\begin{aligned} \dot{x}^*(t) &= f(t,x^*(t)) - \frac{3}{2}\frac{\partial b}{\partial x}(t,x^*(t)) - D(t)\frac{\partial^2 f}{\partial x^2}(t,x^*(t)) \\ &\quad + \frac{D(t)}{N_0(t)}[\dot{y}(t) - Q(t,x^*(t))]\frac{\partial Q}{\partial x}(t,x^*(t)), \qquad x^*(0) = x_0^*; \end{aligned} \tag{3.35}$$

$$\begin{aligned} \dot{D}(t) &= 2\frac{\partial Q}{\partial x}(t,x^*(t))D(t) + b(t,x^*(t)) \\ &\quad - \frac{1}{N_0(t)}\left[\frac{\partial Q}{\partial x}(t,x^*(t))\right]^2 D^2(t), \qquad D(0) = D_0. \end{aligned} \tag{3.36}$$

These equations describe the quasilinear filter in the scalar case. It should be noted that whereas equation (3.35) involves the stated expression for F_1 (see formula (3.23)), equation (3.36) features an expression for F_2 averaged with respect to noise and simplified by virtue of the condition (33, a). The proof is given in [14].

Example 3.1. Suppose that $x(t)$ is a random process satisfying equation (3.1) (n=1) with coefficients

$$f(t,x(t)) = A(t)x_t, \qquad b(t,x(t)) = \sigma^2(t), \tag{3.37}$$

and $y(t)$ is an observed signal satisfying equation (3.2) (m=1). The coefficients A, σ, Q, and σ_0 may also depend on the "past" observations y_t. Then the quasilinear filter is described by the equations that follow from (3.35), (3.36):

$$\dot{x}^*(t) = A(t)x^*(t) + \frac{D(t)}{N_0(t)}\left[\dot{y}(t) - Q\big(t, x^*(t)\big)\right]\frac{\partial Q}{\partial x}\big(t, x^*(t)\big), \qquad x^*(0) = x_0^*; \tag{3.38}$$

$$\dot{D}(t) = -2A(t)D(t) + \sigma^2(t) - \frac{1}{N_0(t)}\left[\frac{\partial Q}{\partial x}\big(t, x^*(t)\big)\right]^2 D^2(t), \qquad D(0) = D_0. \tag{3.39}$$

The condition for the applicability of this quasilinear algorithm is described by the inequality (3.33, a).

In the case of a linear signal

$$Q\big(t, x(t)\big) = Q(t)x(t) \tag{3.40}$$

(Q(t) may also depend on y_t), equations (3.38), (3.39) go over into equations (3.10), (3.11), which are valid for any values of the coefficients A, σ, Q, and σ_0, since the conditions (3.33) are always satisfied.

Example 3.2. *Quasioptimal filtering of a sinusoidal signal with random phase.*[5] Consider an observed signal of the form (3.3), with the useful signal

$$Q\big(t, x(t)\big) = A_0 \cos\big(\omega_0 + x(t)\big), \tag{3.41}$$

where A_0 is the known amplitude, and ω_0 is the prescribed frequency. The phase $x(t)$ is a Brownian motion satisfying the equation

$$dx(t) = \sigma\, d\xi(t). \tag{3.42}$$

This equation is a particular case of equation (3.1) corresponding to the relations

$$f(t,x) = 0, \qquad \sigma(t,x) = \sigma(t),$$

$\sigma(t)$ being independent of x.

In the case under consideration the filtering equations follow from equations (3.38), (3.39) if we put $A = 0$, and take account of (3.41). Thus, we have

$$\dot{x}^*(t) = -\frac{D(t)}{N_0(t)}\left[\dot{y}(t) - A_0\cos\big(\omega_0 + x^*(t)\big)\right] A_0 \sin\big(\omega_0 + x^*(t)\big), \tag{3.43}$$

$$\dot{D}(t) = \sigma^2(t) - \frac{1}{N_0(t)}\left[A_0 \sin\big(\omega_0 + x^*(t)\big)\right]^2, \tag{3.44}$$

[5]Kulman, N. K., Stratonovich, R. L., Phase Automatic Frequency Control and an Optimal Measurement of Narrow-Band Signal Parameters with Non-Constant Frequency in the Presence of Noise, *Radiotekhn. i Electron.*, (1964), vol. 9, no. 1, 67–77. (Translated into English as *Radio Eng. Electron. Phys.* superseded by *Soviet J. Comm. Tech. Electron.*)

where $x^*(0) = x_0^* = 0$ and $D(0) = D_0$. If in these equations we take into account the well-known relations

$$\sin\alpha\cos\alpha = \tfrac{1}{2}\sin 2\alpha, \qquad \sin^2\alpha = \tfrac{1}{2}(1-\cos 2\alpha),$$

where $\alpha = \omega_0 + x^*$, and discard the terms with double frequency $2\omega_0$, then we have the following equations:

$$\dot{x}^*(t) = -\frac{A_0 D(t)}{N_0(t)}\dot{y}(t)\sin\big(\omega_0 + x^*(t)\big), \qquad x^*(0) = x_0^* = 0; \tag{3.45}$$

$$\dot{D}(t) = \sigma^2(t) - \frac{A_0^2}{2N_0(t)}D^2(t), \qquad D(0) = D_0. \tag{3.46}$$

The condition for the applicability of this quasilinear algorithm can be written as follows:

$$D(t) \ll 1, \tag{3.47}$$

since the scale unit in the inequality (3.33, a) $l_Q \sim 1$.

Consider the filtering error $\varepsilon(t) = x^*(t) - x(t)$. In view of equation (3.42) we can rewrite the first of the filtering equations (equation (3.45)) as a standard phase-locked loop equation:

$$\dot{\varepsilon} + K(t)\dot{y}(t)\sin\big(\omega_0 t + x^*(t)\big) = -\dot{\xi}(t),$$

where $\dot{\xi}(t)$ is the instantaneous deviation of the actual signal frequency from ω_0, $K(t) = A_0D(t)/N_0(t)$ is the feedback circuit gain, and the operation of multiplication plays the role of a phase detector. The quasilinear filter consists of a multiplication unit (a phase detector), a controlled oscillator, and a feedback circuit. Note that the estimate of the phase x_1^* coincides with the phase of the oscillator.

The automatic phase control circuit can be simplified in the case of constant parameters σ and σ_0. Then we can insert for $K(t)$ the steady-state gain $K = A_0D/N_0$ ($N_0 = \sigma_0^2$), where D is the steady-state solution of equation (3.46):

$$D = \frac{1}{A_0}\sqrt{2N_0 b} = \frac{\sqrt{2}}{A_0}\sigma\sigma_0. \tag{3.48}$$

Thus, the error variance is proportional to the product $\sigma\sigma_0$ and inversely proportional to the amplitude, A_0. The condition for the applicability of this filter is

$$\sigma\sigma_0 \ll A_0. \tag{3.49}$$

5. Quasilinear filtering (the multidimensional case). We turn now to the multidimensional problem statement as described by equations (3.1), (3.2) (or (3.3)) and cite the appropriate quasilinear equations, which can be obtained by a similar approach as in the scalar case. Of course, the deduction of the results is much more cumbersome and will not be given here.

In the case of the multivariate Gaussian approximation of the conditional probability distribution $P(t,x)$ we have

$$P(t,x) \simeq \exp\left\{C_0(t) - \frac{1}{2}\sum_{i,j=1}^{n} h_{ij}(t)(x_i - x_i^*)(x_j - x_j^*)\right\}. \tag{3.50}$$

Here $\{x_i^*(t)\} = x^*(t)$ is the optimal estimate rendering the maximum of the conditional probability distribution, $\|h_{ij}(t)\| = \|D_{ij}(t)\|^{-1}$, where $D(t) = \|D_{ij}(t)\|$ is the conditional covariance matrix.

To obtain the approximate solution to the equation (3.5) for the conditional probability distribution $P(t,x)$ in the Gaussian approximation, we can use multivariate power-series expansions of $f(t,x)$, $b(t,x) = \sigma(t,x)\sigma'(t,x)$, $F(t,x)$, and of the solution $P(t,x)$, and obtain a chain of matrix equations similar to the scalar set (3.19), (3.20). Breaking the chain of equations and retaining only the top two matrix equations, which constitute the desired quasilinear filtering equations, leads to certain conditions on the elements of the covariance matrix, which will be discussed below. This result is due to N. K. Kulman and is stated in the following theorem [14, § 3.3]: *Let the conditions of smallness of the conditional variances $D_{ii}(t)$ be satisfied:*

$$\begin{aligned} &\text{(a)} \quad \sqrt{D_{ii}} \ll l_{Qi}, \\ &\text{(b)} \quad \sqrt{D_{ii}} \ll l_{fi}, \\ &\text{(c)} \quad \sqrt{D_{ii}} \ll l_{bi}, \end{aligned} \tag{3.51}$$

where

$$\begin{aligned} l_{Qi} &\sim \min_{\mu,\nu}\left|\frac{\partial^2 F(t,x^*)}{\partial x_\mu \partial x_\nu} \Big/ \frac{\partial^{2+N} F(t,x^*)}{\partial x_\mu \partial x_\nu \partial x_{i_1} \dots \partial x_{i_N}}\right|^{1/N}, \\ l_{fi} &\sim \min_{\mu,\nu}\left|\frac{\partial f_\mu(t,x^*)}{\partial x_\nu} \Big/ \frac{\partial^{1+N} f_\mu(t,x^*)}{\partial x_\nu \partial x_{i_1} \dots \partial x_{i_N}}\right|^{1/N}, \\ l_{bi} &\sim \min_{\mu,\nu}\left|b_{\mu\nu} \Big/ \frac{\partial^N b_{\mu\nu}(t,x^*)}{\partial x_{i_1} \dots \partial x_{i_N}}\right|^{1/N} \\ &(\mu,\nu,i_1,\dots,i_N = 1,2,\dots,n;\ N = 1,2,3,\dots) \end{aligned} \tag{3.52}$$

are scale units characterizing the rapidity of variation of the functions $F(t,x)$, $f(t,x)$, $b(t,x)$ with respect to x. Then the conditional probability distribution is nearly Gaussian with the conditional mode $x^*(t)$ and the conditional covariance *matrix $D(t)$ satisfying the following set of equations:*

$$\dot{x}_i^*(t) = f_i(t,x^*(t)) - \sum_{\mu,\gamma=1}^{n} D_{i\mu}(t)\frac{\partial^2 f_\gamma(t,x^*(t))}{\partial x_\mu \partial x_\nu} - \sum_{\mu=1}^{n}\frac{\partial b_{i\mu}(t,x^*(t))}{\partial x_\mu}$$
$$- \frac{1}{2}\sum_{\mu,\gamma,\delta=1}^{n} D_{i\mu}(t)\|D_{ij}(t)\|^{-1}_{\gamma\delta}\frac{\partial b_{\gamma\delta}(t,x^*(t))}{\partial x_\mu} + \sum_{\mu=1}^{n} D_{i\mu}(t)\frac{\partial F(t,x^*(t))}{\partial x_\mu}, \tag{3.53}$$

$$\dot{D}_{ij}(t) = \sum_{\mu=1}^{n}\left[d_{j\mu}(t,x^*(t))D_{i\mu}(t) + d_{i\mu}(t,x^*(t))D_{j\mu}(t)\right]$$
$$+ b_{ij}(t,x^*(t)) + \sum_{\mu,\nu=1}^{n}\frac{\partial^2 F(t,x^*(t))}{\partial x_\mu \partial x_\nu} D_{i\mu}(t)D_{j\nu}(t), \tag{3.54}$$

where $d_{i\mu} = \partial f_i/\partial x_\mu$ and

$$\frac{\partial F(t,x^*(t))}{\partial x_\mu} = \sum_{\alpha,\beta=1}^{m}\left(N_0^{-1}(t)\right)_{\alpha\beta}\left[\dot{y}_\alpha - Q_\alpha(t,x^*(t))\right]\frac{\partial Q_\beta(t,x^*(t))}{\partial x_\mu}, \tag{3.55}$$

$$\frac{\partial^2 F(t,x^*(t))}{\partial x_\mu \partial x_\nu} = -\sum_{\alpha,\beta=1}^{m}\left(N_0^{-1}(t)\right)_{\alpha\beta}\frac{\partial Q_\alpha(t,x^*(t))}{\partial x_\mu}\,\frac{\partial Q_\beta(t,x^*(t))}{\partial x_\nu}. \tag{3.56}$$

These equations describe the quasilinear filter in the multidimensional case. It should be noted that whereas formula (3.55) follows immediately from the expression (3.6), formula (3.56) features the second derivatives of $F(t,x)$ averaged with respect to noise and simplified by virtue of the condition (3.51, a). The proof is given in [14].

The initial values $x_i^*(0) = x_{0i}^*$ and $D_{ij}(0) = D_{ij}^0$ $(i,j = 1,\dots,n)$ for the filtering equations (3.53), (3.54) can be found from the relations

$$\frac{\partial P_0(x_0^*)}{\partial x_{0i}} = 0, \qquad D_{ij}^0 = -\left\|\frac{\partial^2 \ln P_0(x_0^*)}{\partial x_{0\mu}\partial x_{0\nu}}\right\|^{-1}_{ij}. \tag{3.57}$$

Let us formulate the conditions under which the conditional probability distribution $P(t,x)$ will be strictly Gaussian if the initial probability distribution $P_0(x_0)$ is Gaussian:

(1) the elements of the matrix $b(t) = \sigma(t)\sigma'(t)$ are independent of x;
(2) the function $f(t,x) = A(t)x$ depend linearly on x;
(3) the useful signal $Q(t,x) = Q(t)x$ depends linearly on x
(the function $F(t,x)$ depends on x linearly and quadratically)

With these conditions satisfied, the multidimensional quasilinear filtering equations (3.53), (3.54) go over into the Kalman equations (2.27), (2.18).

Remark. In the case of a scalar (one-dimensional) observed signal (3.3) ($m = 1$) formulas (3.55), (3.56) become

$$\frac{\partial F(t, x^*(t))}{\partial x_\mu} = \frac{1}{N_0(t)}[\dot{y}(t) - Q(t, x^*(t))]\frac{\partial Q(t, x^*(t))}{\partial x_\mu}, \tag{3.58}$$

$$\frac{\partial^2 F(t, x^*(t))}{\partial x_\mu\, \partial x_\nu} = -\frac{1}{N_0(t)}\frac{\partial Q(t, x^*(t))}{\partial x_\mu}\frac{\partial Q(t, x^*(t))}{\partial x_\nu}. \tag{3.59}$$

In the case of a scalar observed signal (3.3) ($m = 1$) and of a scalar process $x(t)$ satisfying equation (3.1) ($n = 1$) the quasilinear filtering equations (3.53), (3.54), in view of (3.58), (3.59), go over into the scalar filtering equations (3.35), (3.36).

Example 3.3. *Quasioptimal filtering of a sinusoidal signal with random frequency.* (See the reference to Example 3.2.) Consider an observed signal of the form (3.3), where $m = 1$, with the useful signal similar to (3.41).

$$Q(t, x_1(t)) = A_0 \cos(\omega_0 + x_1(t)), \tag{3.60}$$

where A_0 is the known amplitude, and ω_0 is the prescribed frequency. The phase $x_1(t)$ and the frequency $x_2(t) = \dot{x}_1(t)$ satisfy the system of two equations of the form (3.1):

$$dx_1 = x_2\, dt, \qquad t \geqslant 0, \tag{3.61}$$

$$dx_2 = -a x_2\, dt + \sigma\, d\xi_2(t), \qquad a > 0. \tag{3.62}$$

The initial values are taken to be $x_i^{(}0) = 0$ ($i = 1, 2$).

The two-dimensional quasilinear filtering equations can be obtained from the general quasilinear equations (3.53), (3.54), where $n = 2$ and $m = 1$, in the form

$$\dot{x}_1^* = x_2^* + D_{11}\frac{\partial F(t, x^*)}{\partial x_1}, \tag{3.63}$$

$$\dot{x}_2^* = -a x_2^* + D_{12}\frac{\partial F(t, x^*)}{\partial x_1}, \tag{3.64}$$

$$\dot{D}_{11} = 2D_{12} + D_{11}^2\frac{\partial^2 F(t, x^*)}{\partial x_1^2}, \tag{3.65}$$

$$\dot{D}_{22} = -2aD_{22} + \sigma^2 + D_{12}^2\frac{\partial^2 F(t, x^*)}{\partial x_1^2}, \tag{3.66}$$

$$\dot{D}_{12} = D_{22} - aD_{12} + D_{11}D_{12}\frac{\partial^2 F(t, x^*)}{\partial x_1^2}, \tag{3.67}$$

In view of (3.58), (3.59) and the fact that Q (and hence F) is independent of x_2, we substitute the following expressions for the derivatives of F in foregoing system:

$$\frac{\partial F(t,x^*)}{\partial x_1} = -\frac{1}{N_0(t)}\left[\dot{y}(t) - A_0\cos\left(\omega_0 + x^*(t)\right)\right] A_0 \sin\left(\omega_0 + x^*(t)\right)$$
$$\to -\frac{A_0}{N_0(t)}\dot{y}(t)\sin\left(\omega_0 + x^*(t)\right), \tag{3.68}$$
$$\frac{\partial^2 F(t,x^*)}{\partial x_1^2} = -\frac{1}{N_0(t)}\left[A_0\sin\left(\omega_0 + x^*(t)\right)\right]^2$$
$$\to -\frac{A_0^2}{N_0(t)}. \tag{3.69}$$

(Here we discard the terms with double frequency $2\omega_0$.)

The condition for the applicability of this quasilinear algorithm can be written in the same way as in Example 3.2:

$$D_{11}(t) \ll 1. \tag{3.70}$$

The quasilinear filter constitutes a phase automatic frequency control system and consists of a multiplication unit (a phase detector), a controlled oscillator, and a feedback circuit. It is remarkable that the frequency estimate x_2^* does not coincide with the actual frequency $\omega_0 + \dot{x}_1^*$ of the controlled oscillator, although the phase estimate x_1^* is, in fact, the phase of the oscillator. This illustrates the interesing fact that the optimal (quasioptimal) estimate of the derivative of an unobserved process does not coincide with the derivative of the optimal (quasioptimal) estimate of the unobserved process.

Main Results and Formulas of Chapter XI

Mean-square optimal estimation:
an estimate $m(\tau)$ is the best in the mean-square sense if

$$E\|x(\tau) - m(\tau)\|^2 = \min_F E\|x(\tau) - F(y_T)\|^2;$$

the expression for $m(\tau)$:

$$m(\tau) = E\{\, x(\tau) \mid y_T \,\}.$$

Linear estimation of stationary processes:
optimal estimate:

$$x_0(t+T) = \int_0^\infty \gamma(s) y(t-s)\, ds,$$

where $\gamma(t)$ is a solution of the Wiener–Hopf equation:

$$R_{xy}(T+s_1) = \int_0^\infty \gamma(\tau) R_{yy}(s_1-\tau)\, d\tau, \qquad s_1 \geqslant 0;$$

the solution of the Wiener–Hopf equation

$$\gamma(t) = \frac{1}{2\pi i}\int_{-\infty}^\infty e^{i\omega t}\frac{b_0(T,i\omega)}{h(i\omega)}\, d\omega,$$

where

$$R_{yy}(t) = \int_{-\infty}^\infty |h(i\omega)|^2 e^{-i\omega t}\, d\omega,$$
$$b_0(T,z) = \int_0^\infty b(T+s)e^{-zs}\, ds.$$

The Kalman filter:

$$dx(t) = A(t)x(t)\, dt + \sigma(t)\, d\xi(t), \qquad 0 \leqslant t \leqslant T,\ x(0) = x_0,$$
$$dy(t) = Q(t)x(t)\, dt + \sigma_0(t)\, d\xi_0(t), \qquad y(0) = 0;$$

the equations of the filter:

$$dm(t) = A(t)m(t)\, dt + DQ'(\sigma_0\sigma_0')^{-1}[dy(t) - Q(t)m(t)\, dt], \qquad m(0) = 0,$$
$$\dot{D}(t) = A(t)D(t) + D(t)A'(t) - D(t)Q'(t)(\sigma_0\sigma_0)^{-1}Q(t)D(t)$$
$$+ \sigma(t)\sigma'(t), \qquad t \geqslant 0,\ D(0) = D_0,$$

where

$$m(t) = E\{\, x(t) \mid y_t \,\}, \qquad D(t) = E\{\, \big(x(t) - m(t)\big)\big(x(t) - m(t)\big)' \mid y_t \,\},$$

the Kalman filter is stable if

$$C_1 I \leqslant \int_{t-\tau}^{t} Z'(s,t)Q'(s)\big(\sigma_0(s)\sigma_0'(s)\big)^{-1}Q(s)Z(s,t)\,ds \leqslant C_2 I,$$
$$C_1 I \leqslant \int_{t-\tau}^{t} Z(t,s)\sigma(s)\sigma'(s)Z'(t,s)\,ds \leqslant C_2 I;$$

necessary and sufficient conditions for the stability of a time-invariant Kalman filter:

$$\operatorname{rank}(\sigma, A\sigma, \dots, A^{n-1}\sigma) = n,$$
$$\operatorname{rank}(Q', A'Q', \dots, (A')^{n-1}Q') = n.$$

Discrete Kalman filter

The evolution of the system $x(t_k)$ and the observation process are governed by the equations

$$x(t_{k+1}) = A_k x(t_k) + \xi(t_k), \qquad k = 0,1,2,\dots, \quad (t_k < t_{k+1})$$
$$y(t_k) = Q_k x(t_k) + \eta(t_k).$$

Form I $(n \leqslant m)$

$$m(t_k+0) = m(t_k-0) + D(t_k+0)Q_k'\Sigma_k^{-1}[y(t_k) - Q_k m(t_k-0)],$$
$$m(t_0-0) = m_0,$$
$$D(t_k+0) = D(t_k-0)\big[I + Q_k'\Sigma_k^{-1}Q_k D(t_k-0)\big]^{-1},$$
$$D(t_0-0) = D_0.$$

Form II $(n > m)$

$$m(t_k+0) = m(t_k-0) + R(t_k+0)[y(t_k) - Q_k m(t_k-0)],$$
$$m(t_0-0) = m_0,$$
$$R(t_k+0) = D(t_k-0)Q_k'\big[Q_k D(t_k-0)Q_k' + \Sigma_k\big]^{-1},$$
$$D(t_k+0) = [I - R(t_k+0)Q_k]D(t_k-0),$$
$$D(t_0-0) = D_0.$$

In both forms the following relations hold:

$$m(t_{k+1}-0) = A_k m(t_k+0),$$
$$D(t_{k+1}-0) = A_k D(t_k+0)A_k' + G_k.$$

Nonlinear filtering.

Conditionally Gaussian processes

For stochastic differential equations which are linear in the phase coordinates $x(t)$ and nonlinear in the observations $y(t)$ the conditional probability distribution is Gaussian and the optimal (in the mean-square sense) estimate $m(t)$ of the vector $x(t)$ and the covariance matrix $D(t)$ of the conditional distribution, given the measurement data y_t, satisfy the equations

$$\begin{aligned}
dm(t) &= A(t, y_t)m(t) + D(t)Q'(t, y_t) \\
&\quad \times \big(\sigma_0(t, y_t)\sigma_0'(t, y_t)\big)^{-1}[dy(t) - Q(t, y_t)m(t)], \\
&\qquad m(0) = m_0; \\
\dot{D}(t) &= A(t, y_t)D(t) + D(t)A'(t, y_t) + \sigma(t, y_t)\sigma'(t, y_t) \\
&\quad - D(t)Q'(t, y_t)\big(\sigma_0(t, y_t)\sigma_0'(t, y_t)\big)^{-1}Q(t, y_t)D(t), \\
&\qquad D(0) = D_0.
\end{aligned}$$

Quasilinear filtering (the scalar case).

For stochastic differential equations which are nonlinear in both the phase coordinates $x(t)$ and the observations $y(t)$ in some cases the conditional probability distribution is nearly Gaussian and the optimal estimate $x^*(t)$ (conditional mode) of the vector $x(t)$ and the covariance matrix $D(t)$ of the conditional distribution, given the measurement data y_t, satisfy the equations

$$\begin{aligned}
\dot{x}^*(t) &= f\big(t, x^*(t)\big) - \frac{3}{2}\frac{\partial b}{\partial x}(t, x^*(t)) - D(t)\frac{\partial^2 f}{\partial x^2}(t, x^*(t)) \\
&\quad + \frac{D(t)}{N_0(t)}\big[\dot{y}(t) - Q\big(t, x^*(t)\big)\big]\frac{\partial Q}{\partial x}\big(t, x^*(t)\big), \qquad x^*(0) = x_0^*; \\
\dot{D}(t) &= 2\frac{\partial Q}{\partial x}\big(t, x^*(t)\big)D(t) + b\big(t, x^*(t)\big) \\
&\quad - \frac{1}{N_0(t)}\left[\frac{\partial Q}{\partial x}\big(t, x^*(t)\big)\right]^2 D^2(t), \qquad D(0) = D_0.
\end{aligned}$$

These equations describe the quasilinear filter in the scalar case.

Conditions for the applicability of the quasilinear filter (the scalar case)

The conditions of smallness of the conditional variance $D(t)$ are to be satisfied:

$$\begin{aligned}
&\text{(a)} \quad \sqrt{D} \ll l_Q, \\
&\text{(b)} \quad \sqrt{D} \ll l_f, \\
&\text{(c)} \quad \sqrt{D} \ll l_b,
\end{aligned}$$

where

$$l_Q \sim \left| \frac{\partial Q(t, x*)}{\partial x} \Big/ \frac{\partial^{j+1} Q(t, x*)}{\partial x^{j+1}} \right|^{1/j},$$

$$l_f \sim \left| \frac{\partial f(t, x*)}{\partial x} \Big/ \frac{\partial^{j+1} f(t, x*)}{\partial x^{j+1}} \right|^{1/j},$$

$$l_b \sim \left| b(t, x*) \Big/ \frac{\partial^{j} b(t, x*)}{\partial x^{j}} \right|^{1/j},$$

$$(j = 1, 2, 3, \dots).$$

are scale units characterizing the rapidity of variation of the functions $Q(t, x)$, $f(t, x)$, $b(t, x)$ with respect to x.

Quasilinear filtering (the multidimensional case).

$$\dot{x}_i^*(t) = f_i\big(t, x^*(t)\big) - \sum_{\mu,\gamma=1}^{n} D_{i\mu}(t) \frac{\partial^2 f_\gamma\big(t, x^*(t)\big)}{\partial x_\mu \, \partial x_\nu} - \sum_{\mu=1}^{n} \frac{\partial b_{i\mu}\big(t, x^*(t)\big)}{\partial x_\mu}$$

$$- \frac{1}{2} \sum_{\mu,\gamma,\delta=1}^{n} D_{i\mu}(t) \|D_{ij}(t)\|_{\gamma\delta}^{-1} \frac{\partial b_{\gamma\delta}\big(t, x^*(t)\big)}{\partial x_\mu} + \sum_{\mu=1}^{n} D_{i\mu}(t) \frac{\partial F\big(t, x^*(t)\big)}{\partial x_\mu},$$

$$\dot{D}_{ij}(t) = \sum_{\mu=1}^{n} \big[d_{j\mu}\big(t, x^*(t)\big) D_{i\mu}(t) + d_{i\mu}\big(t, x^*(t)\big) D_{j\mu}(t) \big]$$

$$+ b_{ij}\big(t, x^*(t)\big) + \sum_{\mu,\nu=1}^{n} \frac{\partial^2 F\big(t, x^*(t)\big)}{\partial x_\mu \, \partial x_\nu} D_{i\mu}(t) D_{j\nu}(t),$$

where $d_{i\mu} = \partial f_i / \partial x_\mu$ and

$$\frac{\partial F\big(t, x^*(t)\big)}{\partial x_\mu} = \sum_{\alpha,\beta=1}^{m} \big(N_0^{-1}(t)\big)_{\alpha\beta} \big[\dot{y}_\alpha - Q_\alpha\big(t, x^*(t)\big)\big] \frac{\partial Q_\beta\big(t, x^*(t)\big)}{\partial x_\mu},$$

$$\frac{\partial^2 F\big(t, x^*(t)\big)}{\partial x_\mu \, \partial x_\nu} = - \sum_{\alpha,\beta=1}^{m} \big(N_0^{-1}(t)\big)_{\alpha\beta} \frac{\partial Q_\alpha\big(t, x^*(t)\big)}{\partial x_\mu} \frac{\partial Q_\beta\big(t, x^*(t)\big)}{\partial x_\nu}.$$

These equations describe the quasilinear filter in the multidimensional case.

The initial values $x_i^*(0) = x_{0i}^*$ and $D_{ij}(0) = D_{ij}^0$ $(i, j = 1, \dots, n)$ are

$$\frac{\partial P_0(x_0^*)}{\partial x_{0i}} = 0, \qquad D_{ij}^0 = - \left\| \frac{\partial^2 \ln P_0(x_0^*)}{\partial x_{0\mu} \, \partial x_{0\nu}} \right\|_{ij}^{-1}$$

Conditions for the applicability of the quasilinear filter (the multidimensional case)

The conditions of smallness of the conditional variances $D_{ii}(t)$ are to be satisfied:

$$\text{(a)}\quad \sqrt{D_{ii}} \ll l_{Qi},$$
$$\text{(b)}\quad \sqrt{D_{ii}} \ll l_{fi},$$
$$\text{(c)}\quad \sqrt{D_{ii}} \ll l_{bi},$$

where

$$l_{Qi} \sim \min_{\mu,\nu} \left| \frac{\partial^2 F(t,x^*)}{\partial x_\mu \, \partial x_\nu} \bigg/ \frac{\partial^{2+N} F(t,x^*)}{\partial x_\mu \partial x_\nu \partial x_{i_1} \dots \partial x_{i_N}} \right|^{1/N},$$
$$l_{fi} \sim \min_{\mu,\nu} \left| \frac{\partial f_\mu(t,x^*)}{\partial x_\nu} \bigg/ \frac{\partial^{1+N} f_\mu(t,x^*)}{\partial x_\nu \partial x_{i_1} \dots \partial x_{i_N}} \right|^{1/N},$$
$$l_{bi} \sim \min_{\mu,\nu} \left| b_{\mu\nu} \bigg/ \frac{\partial^N b_{\mu\nu}(t,x^*)}{\partial x_{i_1} \dots \partial x_{i_N}} \right|^{1/N}$$
$$(\mu, \nu, i_1, \dots, i_N = 1, 2, \dots, n; \; N = 1, 2, 3, \dots)$$

are scale units characterizing the rapidity of variation of the functions $F(t,x)$, $f(t,x)$, $b(t,x)$ with respect to x.

CHAPTER XII

OPTIMAL CONTROL OF THE OBSERVATION PROCESS

This chapter is devoted to problems of the control of an observation process under various assumptions concerning the measurement matrix and the intensity of noise. A general procedure for constructing optimal observation laws is given. Specific procedures are obtained and investigated for particular cases.

§ 1. Optimization of the Observation Process

1. Problem statement. Basic relations. In Chapter XI we considered various problems of the optimal estimation of the state $x(t)$ of the system (2.1), given the measured values of the process (2.2), assuming that the measurement matrix $Q(t)$ and the accuracy characteristic $\sigma_0(t)$ are prescribed and fixed. However, in a number of practical situations during the measurement process it may be possible to vary both the combination of the observed coordinates of the system, and the accuracy of their measurement. This leads to various problems of the control of an observation process involving so-called *observation optimization problems.* Such problems have become very important recently in view of the rapid development of sensors and the implementation of computer systems. This has led to a better functioning of control systems, but at the same time they have become more expensive. An increasing sophistication of measuring equipment and processing algorithms is the cause of an appreciable retardation effect appearing in control systems using measurement data. The general block-diagram of the control of the observation process is shown in Figure 1.1.

In problems of the optimization of an observation process it is usually required to optimize the accuracy of the measured phase coordinates subject to constraints imposed on the observation cost or duration. As a particular case, consider the following equations describing the motion of the system and the evolution of the observation process and having the form (2.1), (2.2) of Chapter XI:

$$dx(t) = A(t)x(t)\,dt + \sigma(t)\,d\xi(t), \qquad 0 \leqslant t \leqslant T,\ x(0) = x_0, \tag{1.1}$$

$$dy(t) = Q(t)x(t)\,dt + \sigma_0(t)\,d\xi_0(t), \qquad 0 \leqslant t \leqslant T,\ y(0) = 0. \tag{1.2}$$

Here $x \in R^n$ and $y \in R^m$, ξ and ξ_0 denote standard Wiener processes, x_0 is a Gaussian vector, and $Ex_0 = m_0$, $Dx_0 = D_0$. The random variables x_0, ξ, ξ_0 are independent. The accuracy of determining the coordinates is characterized by the matrix $D(t)$, and the control of the observation process by the matrix

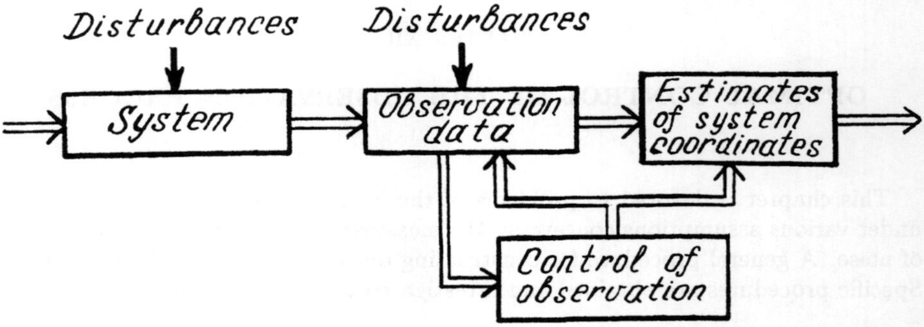

FIG. 1.1. Block-diagram of the control of the observation process

$u = Q'(\sigma_0\sigma_0')^{-1}Q$. These are related by the Kalman filtering equations (2.18) of Chapter XI:

$$\dot{D}(t) = A(t)D(t) + D(t)A'(t) - D(t)u(t)D(t) + \sigma(t)\sigma'(t), \tag{1.3}$$
$$0 \leqslant t \leqslant T,\ D(0) = 0,\ u = Q'(\sigma_0\sigma_0')^{-1}Q.$$

Measurements are often made to optimize the accuracy of determining the value of $q'x(T)$, where q is a given vector in R^n. If, for example, the ith component of the vector q is equal to one, and all others are zero, then $q'x(T) = x_i(T)$, and the problem is reduced to that of optimizing the accuracy of the determination of $x_i(T)$. The accuracy in question can be characterized by the variance of $q'x(T)$. Thus, the cost functional J in observation optimization problems can be given as

$$J = q'D(T)q. \tag{1.4}$$

The control $u(t)$ of an observation process is also subject to a number of constraints. For example, some of the constraints are as follows:

$$u(t) \in U(t), \qquad J_i = \int_0^T F_i\big(t, u(t)\big)\, dt \leqslant C_i, \qquad i = 1, \ldots, r. \tag{1.5}$$

Here $U(t)$ is a given matrix set defined with regard to the feasibility of the control of observations, F_i are prescribed scalar functions, and C_i are prescribed constants. Through the use of functionals of the type J_i account can be taken of various constraints on the cost or duration of measurement. The relations (1.3)–(1.5) show that observation optimization problems are standard problems of the optimal control of deterministic systems (1.3) involving the covariance matrix $D(t)$. This analogy leads to some other problem statements for the control of an observation process.

There are problems in which one of the functionals J_i is minimized, with constraint $J \leqslant c_0$; this corresponds to cases in which it is required to minimize the cost or duration of observation and, at the same time, ensure the accuracy of determining the value of $q'x(T)$ not below the required level. Another class of problems arises when the observations are a sequence of pulses, i.e., the measurements are made at discrete moments t_i. In the latter case we can write formally

$$u(t) = \sum_{i=1}^{r} u_i \delta(t - t_i),$$

where r is a total number of measurements and $\delta(t)$ is the δ-function. The matrices u_i, the moments t_i and their number r are variable parameters to be determined. Furthermore, the *pulse observation law* is of interest because it turns out to be an optimal observation law when the cost functional for the problem under consideration is equivalent to the total number of measurements over the interval $[0, T]$.

Remark. In the case of a pulse observation law it is necessary to make more precise the meaning of equation (1.3) for the covariance matrix $D(t)$. Such refinement can be undertaken because the equation for the inverse matrix D^{-1} remains valid for pulse observation. This equation can be written as

$$\begin{aligned}\frac{dD^{-1}(t)}{dt} &= -D^{-1}(t)A(t) - A'(t)D^{-1}(t) + u(t)\\ &\quad - D^{-1}(t)\sigma(t)\sigma'(t)D^{-1}(t), \qquad D^{-1}(0) = D_0^{-1}.\end{aligned}$$

For pulse observation the random variables $y(t_i)$ are measured at the moments t_i; they are described by the equation

$$y(t_i) = Q(t_i)x(t_i) + \sigma_0(t_i)\zeta_i,$$

where ζ_i is a sequence of independent Gaussian vectors with zero mean and unit covariance matrix.

Given the observed values $y(t_i)$ at the discrete moments t_i $(0 = t_0 < t_1 < t_2 < \dots)$, the optimal estimate $m(t)$ of the phase coordinates of the system (1.1) and the corresponding covariance matrix $D(t)$ at the observation moments are given by the formulas (cf. subsection 11 of Chapter XI)

$$\begin{aligned}m(t_i + 0) &= m(t_i) + D(t_i + 0)Q'(t_i)\big(\sigma_0(t_i)\sigma_0'(t_i)\big)^{-1}[y_i - Q(t_i)m(t_i)],\\ D(t_i + 0) &= \big[D(t_i)^{-1} + Q'(t_i)\big(\sigma_0(t_i)\sigma_0'(t_i)\big)^{-1}Q(t_i)\big]^{-1}.\end{aligned}$$

On the intervals between the moments of actual measurement, i.e., for $t_i < t \leqslant t_{i+1}$, the equations for $m(t)$ and $D(t)$ are of the form

$$\begin{aligned}\dot{m}(t) &= A(t)m(t),\\ \dot{D}(t) &= A(t)D(t) + D(t)A'(t) + \sigma(t)\sigma'(t),\end{aligned}$$

and the initial data are as follows:

$$\lim_{t\to t_i+0} m(t) = m(t_i+0), \qquad m(0) = m_0,$$
$$\lim_{t\to t_i+0} D(t) = D(t_i+0), \qquad D(0) = D_0.$$

Since observation optimization problems are equivalent to those of the optimal control of the deterministic systems, we can use the methods discussed in Chapters V–VII, such as the maximum principle in order to construct optimal observation laws. We shall cite the appropriate relations of the maximum principle for the problem (1.1)–(1.3) for $\sigma = 0$ in which it is required that

$$\int_0^T F_1(t, u(t))\, dt \to \inf_{u\in U}, \qquad J = q'D(T)q = c_1. \tag{1.6}$$

The control $u(t)$ in equation (1.3) involves as a multiplier the quadratic function $D(t)$ of the phase coordinates; this can lead to some difficulties in calculations. Hence, it is expedient to change the variable by setting $\alpha(t) = D^{-1}(t)$. Then in the equation for α the coefficient of the control $u(t)$ is a unit matrix, since by equation (1.3) and the condition $\sigma = 0$ we have

$$\dot\alpha(t) = -A'(t)\alpha(t) - \alpha(t)A(t) + u(t), \qquad \alpha(0) = D_0^{-1}. \tag{1.7}$$

In line with the maximum principle (see Chapter V) we introduce the scalar function

$$H = \sum_{i,k=1}^{n} \psi_{ik}(-A'\alpha - \alpha A + u)_{ik} + \psi_0 f_1(t,u), \tag{1.8}$$

where ψ_0 is a constant, and the ψ_{ik} are the elements of the symmetric matrix of the adjoint variables $\psi(t)$ satisfying the equation

$$\dot\psi(t) = A(t)\psi(t) + \psi(t)A'(t), \qquad 0 \leqslant t \leqslant T. \tag{1.9}$$

At the moment $t = T$ the matrix $\psi(T)$ satisfies the boundary condition (the transversality condition)

$$\psi(T) = \lambda \frac{\partial J}{\partial \alpha},$$

where λ is some constant. The transversality condition written componentwise is of the form

$$\psi_{ik}(T) = \lambda \frac{\partial J}{\partial \alpha_{ik}} = \lambda q' \left[\frac{\partial D}{\partial \alpha_{ik}}\right] q.$$

But

$$\frac{\partial D}{\partial \alpha_{ik}} = -D\left(\frac{\partial \alpha}{\partial \alpha_{ik}}\right)D, \qquad \frac{\partial D_{lj}}{\partial \alpha_{ik}} = -D_{il}D_{kj}.$$

Therefore, the transversality condition can be expressed, finally, as

$$\psi_{ik}(T) = -\sum_{j,l=1}^{n} \lambda D_{ij}(T) D_{kl}(T) q_j q_l = -\lambda \big(D(T)q\big)_i \big(D(T)q\big)_k$$
$$= -\lambda \big(\alpha^{-1}(T)q\big)_i \big(\alpha^{-1}(T)q\big)_k. \tag{1.10}$$

The optimal control $u(t)$ can be determined from the condition for a maximum of the Hamiltonian H with respect to u:

$$\left[\sum_{i,k=1}^{n} \psi_{ik} u_{ik} + \psi_0 F_1(t,u)\right] \to \sup_u, \qquad u \in U,\ 0 \leqslant t \leqslant T. \tag{1.11}$$

Formulas (1.6)–(1.11) form the self-contained system of the relations of the maximum principle. They depend on n^2+2 constants (n^2 elements of the matrix $\psi(T)$ plus the constants λ and ψ_0). But the function H is determined to within a constant multiplier; hence one of the n^2+2 constants may be chosen in an arbitrary way. Usually we can set $\psi_0 = -1$. Then, n^2+1 conditions (1.6), (1.10) remain to determine the other n^2+1 constants.

The following problem is a dual of the above-stated problem:

$$J = q' D(T) q \to \inf_{u \in U}, \qquad \int_0^T F_1\big(t, u(t)\big)\, dt = C_0, \tag{1.12}$$
$$q_i' D(T) q_i = C_i,\ i = 1, \ldots, r,\ r \leqslant n, \tag{1.13}$$

where C_i, $i = 1, \ldots, r$, are given constants, q and q_i are given vectors in R^n. For this problem we write the relations of the maximum principle. The function H preserves the same form (1.8), and the condition (1.11) for the maximum principle, in which ψ_0 is an unknown constant, remains the same.

In a similar way to (1.10), we obtain the transversality condition

$$\psi_{ik}(T) = -\sum_{j=1}^{r} \lambda_i \big[\alpha^{-1}(T) q_j\big]_i \big[\alpha^{-1}(T) q_j\big]_k + \big(D(T)q\big)_i \big(D(T)q\big)_k. \tag{1.14}$$

Next, we can determine n^2+r+1 constants $\psi(T)$, λ_j, and ψ_0 from n^2+r+1 conditions (1.12)–(1.14).

2. Construction of optimal observation laws minimizing terminal variance. Let the equations of motion of the system (1.1) and of the observed process (1.2) be of scalar type and have the form

$$\dot{x}(t) = a x(t), \qquad T \geqslant t \geqslant 0,\ x(0) = x_0,$$
$$dy(t) = Q(t) x(t)\, dt + \sigma_0\, d\xi_0(t), \qquad y(0) = 0.$$

Here a and $\sigma_0 \neq 0$ are prescribed constants, and x_0 is a Gaussian random variable.

The control of the observation process $y(t)$ is implemented by a scalar measurable function $Q(t)$, which for each t can be either zero or one, with the total observation time T_0 ($0 \leqslant T_0 \leqslant T$) prescribed. Owing to the independence of the random variables x_0 and $\xi_0(t)$ the equality $Q(t) = 0$ taken at some instant indicates that at this very instant there is no observation. The restriction on the total duration of the observation process is equivalent to the isoperimetric condition

$$\int_0^T Q(t)\,dt = T_0, \qquad 0 \leqslant T_0 \leqslant T. \tag{1.15}$$

The goal of observation control is to determine the function $Q(t)$ satisfying the above conditions so that it yields the minimum of the variance $D(T)$. Clearly, this problem statement is a special case of the general problem statement for observation control described in subsection 1. Moreover, the set U in (1.5) contains two points 0 and σ_0^{-1}, the function F_1 in (1.12) is such that $F_1(t,0) = 0$ and $F_1(t,\sigma_0^{-1}) = 1$, and, finally, $q = 1$. Equation (1.9) for the adjoint variable takes the form $\dot{\psi}(t) = 2a\psi(t)$. In view of (1.3) the variance $D(t)$ satisfies the equation

$$\dot{D}(t) = 2aD - D^2 u(t), \qquad D(0) = D_0 > 0.$$

The transversality condition (1.14) implies that $\psi(T) = D^2(T)$. Equation (1.7) for the function $\alpha(t) = D_0^{-1}(t)$ can be written as

$$\dot{\alpha}(t) = -2a\alpha(t) + u(t), \qquad \alpha(0) = D_0,$$
$$u(t) = \sigma_0^{-2} Q^2(t) = \sigma_0^{-2} Q(t).$$

The Hamiltonian, H, by virtue of (1.8), is of the form

$$H = \psi\big(-2a\alpha(t) + u\big) + \psi_0 Q(t) = -2a\alpha(t)\psi(t) + u\big(\psi(t) + \sigma_0^2\psi_0\big).$$

It follows from the maximum principle (1.11) that

$$u(t) = \begin{cases} \sigma_0^{-2}, & \text{for } \varphi(t) > 0, \\ 0, & \text{for } \varphi(t) < 0. \end{cases} \tag{1.16}$$

where it has been set that $\varphi(t) = \psi(t) + \sigma_0^2\psi_0$.

It follows from the equation and the boundary condition for $\psi(t)$ that $\psi(t) = D^2(T)\exp[2a(t-T)]$. Since $D(T) > 0$, the adjoint function $\psi(t)$ is monotone increasing for $a > 0$ and monotone decreasing for $a < 0$.

Thus, noting the relations (1.16), we can write:

for $a > 0$

$$u(t) = \begin{cases} 0, & \text{for } 0 \leqslant t \leqslant t_0, \\ \sigma_0^{-2}, & \text{for } t_0 < t \leqslant T; \end{cases}$$

for $a < 0$

$$u(t) = \begin{cases} \sigma_0^{-2}, & \text{for } 0 \leqslant t \leqslant t_0, \\ 0, & \text{for } t_0 < t \leqslant T. \end{cases}$$

Here t_0 denotes the unique switch point for the control $u(t)$, t_0 being equal to the only root of the monotonic function $\psi(t) + \sigma_0^2\psi_0$ for $a \neq 0$. Note that the constant ψ_0 is chosen so that the isoperimetric constraint (1.15) is satisfied. Consequently, $t_0 = T - T_0$ for $a > 0$ and $t_0 = T_0$ for $a < 0$. Finally, if $a = 0$, then the cost functional $D(T)$ does not depend on the specific form of the function $Q(T)$ satisfying the condition (1.15) and is equal to $D(T) = (D_0^{-1} + \sigma_0^{-2}T_0)^{-1}$.

Next, we write the expressions for the minimum value of $D(T)$ $(a \neq 0)$, given the optimal observation law. We have

$$D(T) = \begin{cases} e^{2aT}\Big[D_0^{-1} + \frac{1}{2}\sigma_0^{-2}a^{-1}e^{2aT}(1 - e^{-2aT_0})\Big]^{-1}, & \text{for } a > 0, \\ e^{2aT}\Big[D_0^{-1} + \frac{1}{2}\sigma_0^{-2}a^{-1}(e^{2aT_0} - 1)\Big]^{-1}, & \text{for } a < 0. \end{cases}$$

The obtained solution is not difficult to interpret. For $a > 0$ the energy of the signal x increases with time t and, therefore, it is more advantageous to conduct observations at the end of the interval. But if $a < 0$, the energy of the signal decreases as time t increases and, therefore, it is expedient to conduct observations at the beginning of the interval.

3. An example of observation control with integral cost functional. In the last example the goal of observation control was stated as optimising the determination of the state $x(T)$ of the system at the end of the observation interval. However, if this goal is stated for any moment not just the terminal moment of the observation interval, then we are to deal with an *observation optimization problem with integral cost functional.* Consider one of such problems.

Let the motion of the system and the observation process be described by scalar equations of the form (1.1), (1.2):

$$\begin{aligned} dx(t) &= ax(t)\,dt + \sigma\,d\xi(t), \qquad t \in [0,T],\ x(0) = x_0, \\ dy(t) &= Q(t)x(t)\,dt + \sigma_0\,d\xi_0(t), \qquad y(0) = 0. \end{aligned} \tag{1.17}$$

Here a, σ, σ_0 are constants, the other notation is the same as in subsection 2. In particular, the control $Q(t)$ assumes two values (either zero or one) and satisfies the condition (1.15). By means of an appropriate control, it is required to minimize the cost functional

$$J = \gamma D(T) + \int_0^T D(t)\,dt,$$

where $\gamma \geqslant 0$ is a given constant. We set $\alpha(t) = D^{-1}(t)$. Then, by virtue of (1.3), (1.17), we have

$$\dot\alpha(t) = -2a\alpha(t) - \sigma^2\alpha^2(t) + u(t), \qquad \alpha(0) = D_0^{-1},\ t \geqslant 0, \tag{1.18}$$

where $u(t) = \sigma_0^{-2}Q(t)$. The adjoint variable $\psi(t)$ for $t \in [0,T]$ is determined from the relations

$$\dot\psi(t) = 2\psi(t)\big(a + \sigma^2\alpha(t)\big) - \alpha^{-2}(t), \qquad \alpha(T) = \gamma\alpha^{-2}(T). \tag{1.19}$$

Using the Cauchy formula, we establish that the solution of the problem (1.19) can be expressed as

$$\psi(t) = \gamma\alpha^{-2}(T)\exp\left[2\int_T^t (a+\sigma^2\alpha(s))\,ds\right] + \int_t^T \exp\left[2\int_s^t (a+\sigma^2\alpha(\tau))\,d\tau\right]\alpha^{-2}(s)\,ds.$$

Hence, in view of the estimate $\alpha(t) > 0$, we have the relation $\psi(t) > 0$ for $0 \leqslant t < T$. Differentiation of both sides of the equality (1.19) with respect to t yields

$$\ddot{\psi}(t) = 2\dot{\psi}(t)(a+\sigma^2\alpha(t)) + 2\dot{\alpha}(t)(\alpha^{-3}(t)+\sigma^2\psi(t)). \tag{1.20}$$

The Hamiltonian H is of the same form as in the example of subsection 2, i.e., $H = -2a\alpha(t)\psi(t) + u(\psi(t)+\sigma_0^2\psi_0)$. Hence, in view of the maximum principle the optimal control is again determined by formulas (1.16). Let (t_1, t_2) be an internal subinterval of the segment $[0, T]$, on which the optimal control $u(t)$ is zero. Then, by virtue of (1.16), the following relations hold: $\varphi(t_1) = \varphi(t_2) = 0$ and $\varphi(t) < 0$ for $0 < t_1 < t < t_2 < T$, where the function $\varphi(t) = \psi(t) + \sigma_0^2\psi_0$. Thus, the adjoint variable $\psi(t)$ has a local minimum at some point t_0 $(t_1 < t_0 < t_2)$. It follows that at this point $\dot{\psi}(t_0) = 0$, $\ddot{\psi}(t_0) \geqslant 0$. Moreover, $\alpha(t_0) > 0$, $\psi(t_0) > 0$ hold true at the point t_0. Therefore, by virtue of the equality (1.20) we have $\dot{\alpha}(t_0) \geqslant 0$.

Likewise, it can be established that if $u(t) = \sigma_0^{-2}$ on the subinterval (t_1, t_2), $0 < t_1 < t_2 < T$, then there exists a point $t_0 \in (t_1, t_2)$, at which $\alpha(t_0) \leqslant 0$. By the existence and uniqueness theorem applied to equation (1.18) the function $\alpha(t)$ is monotonic both on the intervals where $u(t) = 0$ and on the intervals where $u(t) = \sigma_0^{-2}$. Therefore, $\dot{\alpha}(t) \leqslant 0$ for all points $t \in (t_1, t_2)$ if $u(t) = \sigma_0^{-2}$ and $0 < t_1 < t < t_2 < T$. At the boundary of two intervals, where the function $\varphi(t)$ changes sign, the control $u(t)$ has a jump. Consequently, the solution $\alpha(t)$ of equation (1.18) is continuous, and its derivative $\dot{\alpha}(t)$ has a jump; moreover, on the strength of (1.18) the derivative $\dot{\alpha}(t)$ on the observation interval (where $u(t) = \sigma_0^{-2}$) is greater by σ_0^{-2}. Therefore, the subinterval where $\dot{\alpha}(t) \leqslant 0$ cannot adjoin the subinterval where $\dot{\alpha}(t) \geqslant 0$.

Thus, if there exists a subinterval (t_1, t_2), $0 < t_1 < t_2 < T$, where the control $u(t)$ is constant, then this subinterval is unique. Moreover, if $\dot{\alpha}(t) \leqslant 0$ for $t \in (t_1, t_2)$, $0 < t_1 < t_2 < T$, then on the boundary subintervals $[0, t_1)$ and $(t_2, T]$ we have if $u(t) = 0$, $\dot{\alpha}(t) < 0$. In a similar way, if $\dot{\alpha}(t) \geqslant 0$ for $t \in (t_1, t_2)$, $0 < t_1 < t_2 < T$, then $\dot{\alpha}(t) > 0$, $u(t) = \sigma_0^{-2}$ for $t \in [0, t_1)$ and $t \in (t_2, T]$. As a result, the following three cases can occur:

1°) there exists only one internal subinterval on which $u(t) = \sigma_0^{-2}$;

2°) there exists only one internal subinterval on which $u(t) = 0$;

3°) there are no such subintervals.

Let us consider each of these cases.

1°. Suppose that $u(t) = \sigma_0^{-2}$ for $0 < t_1 < t < t_1 + T_0 < T$. Then $\dot{\alpha}(t) \leqslant 0$ for $t \in (t_1, t_1 + T_0)$, and outside the segment $[t_1, t_1 + T_0]$ we have $u(t) = 0$, $\dot{\alpha}(t) < 0$. These relations and equation (1.18) for $t = 0$ imply that $2aD_0 + \sigma^2 > 0$.

2°. Suppose that $u(t) = 0$ for $0 < t_1 < t < t_1 + T_0 < T$. Then $\dot{\alpha}(t) \geqslant 0$ for $t \in (t_1, t_1 + T_0)$. Therefore, by virtue of equation (1.18) the following estimate is valid: $2a + \sigma^2\alpha(t_1) \leqslant 0$. Moreover, $\dot{\alpha}(t) > 0$ outside the interval $(t_1, t_1 + T_0)$, i.e., $\alpha(t_1) > \alpha(0) = D_0^{-1}$. Finally, we obtain that $2aD_0 + \sigma^2 < 0$.

3°. The case of no subintervals where $u(t)$ is constant can be interpreted as the limiting case for either case 1° or case 2°. If $2aD_0 + \sigma^2 = 0$, there can be no subintervals where $u(t)$ is constant.

It follows from analysis that the following assertions are valid:

if $a \geqslant 0$ or $a < 0$ but $0 < D_0 \leqslant -(2a)^{-1}\sigma^2$, then there exists a moment t_1, $t_1 \leqslant T - T_0$, such that the optimal observation law is $Q_0(t) = 1$ for $t \in (t_1, t_1 + T_0)$ and $Q_0(t) = 0$ for $t \notin (t_1, t_1 + T_0)$;

if $a < 0$ and $D_0 > -(2a)^{-1}\sigma^2$, then there exists a moment $s_0 \leqslant T_0$, such that $Q_0(t) = 1$ for $0 \leqslant t \leqslant s_0, T - T_0 + s_0 < t \leqslant T$, and $Q_0(t) = 0$ for $t \in (s_0, T - T_0 + s_0)$.

4. Optimal pulse observation laws. Consider the observation optimization problem for a system of the form (1.1) where $\sigma = 0$:

$$\dot{x}(t) = A(t)x(t), \qquad x(0) = x_0,\ 0 \leqslant t \leqslant T. \tag{1.21}$$

Observations of the process (1.21) may be carried out in continuous time $t \in [0, T]$ or in discrete time $t_i \in [0, T]$. Let $y(t)$ denote continuous-time observations and $\tilde{y}(t_i)$ discrete-time observations. We assume that the vectors $y(t)$ and $\tilde{y}(t_i)$ satisfy the equations

$$\begin{aligned} dy(t) &= h(t)Q(t)x(t)\,dt + \sigma_0(t)\,d\xi_0(t), \qquad && y(0) = 0,\ y \in R^m, \\ \tilde{y}(t_i) &= \sqrt{\mu_i}Q(t_i)x(t) + \sigma_0(t_i)\zeta_0(t), && \tilde{y} \in R^m. \end{aligned} \tag{1.22}$$

Here $\xi_0(t)$ is a standard Wiener process, $\zeta(t_i)$ is a sequence of independent identically distributed Gaussian variables with zero mean and unit covariance matrix. The matrices $Q(t)$ and $\sigma_0(t)$ with continuous elements are prescribed, the matrix $\sigma_0(t)\sigma_0'(t)$ is uniformly positive definite, $h(t) \in R^1$ is a bounded scalar measurable function, and the μ_i are positive quantities. The random variables are $\xi_0(t)$, $\zeta(t_i)$, x_0 are independent of one another.

Observation control is realized by a scalar control function

$$\gamma(t) = h^2(t) + \sum_i \mu_i\delta(t - t_i),$$

i.e., by choosing a scalar measurable bounded function $h(t)$, quantities $\mu_i > 0$, and moments $t_i \in [0, T]$ so that

$$\int_0^T \gamma(t)\,dt \leqslant N, \tag{1.23}$$

where N is a given constant. The goal of the control is to minimize the cost functional (1.4). The function $\gamma(t)$ characterizes the observation density. Thus, the inequality (1.23) is a constraint on the total "number" of measurements. This problem will be called *Problem I.* It should be stressed that for any function $\gamma(t)$ having the stated form the following equation for the conditional covariance matrix $D(t)$ holds:

$$\frac{dD^{-1}(t)}{dt} = -D^{-1}(t)A(t) - A'(t)D^{-1}(t) + \gamma(t)Q'(t)\big(sig_0(t)\sigma_0'(t)\big)^{-1}Q(t),$$
$$D^{-1}(0) = D_0^{-1}.$$

Let $Z(t,s)$ denote the Cauchy matrix of the system (1.21) satisfying the relations (1.8) of Chapter IV.

Consider the matrices H and $h(s,t)$:

$$H = \int_0^T h(s,T)\,ds,$$
$$h(s,t) = Z'(s,t)Q'(s)\big(\sigma_0(s)\sigma_0'(s)\big)^{-1}Q(s)Z(s,t).$$

Theorem 1.1. *Let the coefficients of equations (1.21), (1.22) satisfy the foregoing constraints as well as the general conditions of subsection 1, and the matrix $H > 0$ (i.e., positive definite). Then the optimal observation law $\gamma(t)$ furnishing a solution of Problem I is of the form*

$$\gamma(t) = \sum_{i=1}^{m} \mu_i\delta(t - t_i), \qquad 0 \leqslant t_1 < t_2 < \cdots < t_m \leqslant T. \tag{1.24}$$

where μ_i are positive constants, the integer $m \leqslant n(n+1)/2$ and $\mu_1 + \cdots + \mu_m = N$.

Proof. The proof of the theorem will be carried out in four stages.

1°. We denote by V the set of nondecreasing functions $v(t)$ $(0 \leqslant t \leqslant T)$ with bounded variation on $[0,T]$ such that $v(0) = 0$, $v(T) \leqslant N$. The norm $\|u\|$ of an arbitrary function $u(t)$ with bounded variation on $[0,t]$ is defined by the equality $\|u\| = \int_0^T |du(t)|$, where the integral on the right side is understood in the Lebesgue-Stiltjes sense. Note that the norm of a function $v(t) \in V$ is $\|v\| = v(T) \leqslant N$. We define the matrix $r(t,v)$ as follows:

$$r(t,v) = Z'(0,t)D_0^{-1}Z(0,t) + \int_0^t h(s,t)\,dv(s). \tag{1.25}$$

The problem of determining the function $v \in V$ that minimizes the functional $q'r(T,v)^{-1}q$ will be called *Problem II.* Here $q \in R^n$ is the same vector as in (1.4). The goal of the first stage of the proof is to justify the existence of a control $v(t) \in V$ that provides a solution to Problem II. Let $v_i(t) \in V$ $(i = 1, 2, \dots)$ be a minimizing sequence of functions such that

$$\lim_{i\to\infty} q'r(T,v_i)^{-1}q = \inf_{v\in V} q'r^{-1}(T,v)q = \lambda.$$

It follows from the definition of the set of functions V that all $v_i(t)$ $(0 \leqslant t \leqslant T)$ and their norms are uniformly bounded by N. Then by Helly's second theorem [3] there exists a subsequence of the sequence $v_i(t)$ which converges to a function $v_0(t)$ at each point of the segment $[0, T]$. Hence, $v_0(T) \leqslant N$. To keep the notation simple, we assume that the sequence $v_i(t)$ itself converges pointwise to $v_0(t)$. Then, by Helly's first theorem [3] the sequence $v_i(t)$ is weakly convergent to $v_0(t)$. We shall verify that the control $v_0(t)$ furnishes a solution to Problem II.

The monotonicity of $v_i(t)$ implies that for any two points t_1 and t_2 $(0 \leqslant t_1 \leqslant t_2 \leqslant T)$ the following inequality holds:

$$v_0(t_2) - v_0(t_1) \geqslant v_0(t_2) - v_i(t_2) + v_i(t_1) - v_0(t_1).$$

It follows from this inequality and the pointwise convergence of v_i to v_0 that the function $v_0(t)$ is nondecreasing and its norm $\|v_0\| = v_0(T) \leqslant N$, i.e., $v_0 \in V$. To prove the optimality of v_0, let us show that $\lambda = q'r(T, v_0)^{-1}q$. Note that the continuity of the matrix $h(t, T)$ and the weak convergence of v_i to v_0 imply that

$$\lim_{i \to \infty} \int_0^T h(t, T)\, dv_i(t) = \int_0^T h(t, T)\, dv_0(t).$$

Hence, in view of formula (1.25), in which we set successively $v = v_i$ and $v = v_0$, it follows that $r(t, v_i) \to r(t, v_0)$ as $i \to \infty$. Thus, noting the definition of the number λ, we can write

$$\lambda = q'r^{-1}(t, v_0)q.$$

2°. We now prove that $\|v_0\| = N$. Let us assume the contrary, i.e., $\|v_0\| < N$. Let us show that this assumption contradicts the optimality of v_0 in Problem II. To this end, we establish the existence of a control $v_1 \in V$ such that $q'r^{-1}(t, v_1)q < q'r^{-1}(t, v_0)q$. Set

$$v_1(t) = v_0(t) + \varepsilon t, \qquad \varepsilon = T^{-1}(N - \|v_0\|) > 0.$$

Clearly, $v_i \in V$, since $\|v_1\| = \|v_0\| + \varepsilon T = N$. By virtue of the relation (1.25) the matrix $r(t, v_1)$ is positive definite. Therefore,

$$q'r^{-1}(T, v_1)q = \max_{x \in R^n} [2x'q - x'r(T, v_1)x]. \tag{1.26}$$

In the right member of this equationwe now substitute for $r(T, v_1)$ formulas (1.25) for $t = T$ and $v_1 = v_0 + \varepsilon t$. Since the matrix H is positive definite, we have

$$\begin{aligned} q'r^{-1}(T, v_1)q &= \max_{x \in R^n} [2x'q - x'r(T, v_0)x - \varepsilon x'Hx] \\ &< \max_{x \in R^n} [2x'q - x'r(T, v_0)x] = q'r^{-1}(T, v_0)q. \end{aligned}$$

The equality $\|v_0\| = N$ is established.

3°. Consider the auxiliary *Problem III* being the problem of determining a scalar function $\omega(t)$, $\omega(0) = 0$, with bounded variation and minimum norm $\|\omega\|$ for which

$$r(0,\omega) = r(0), \qquad r(T,\omega) = r(T, v_0). \tag{1.27}$$

It should be noted that in Problem III the optimal function $\omega(t)$ is sought among all functions having bounded variation on $[0,T]$, whereas in Problems I and II optimal control is sought only in the class of monotonic functions. In Problem III there exists an admissible control, for example, v_0. Indeed, by the construction of v_0 we have $\|v_0\| = N < \infty$, $r(0, v_0) = D_0^{-1} = r(0)$, and the terminal condition (1.27) is obvious. By constructing a minimizing sequence and passing to the limit (as in stage 1° of the proof), it follows from the existence of the admissible control v_0 that in Problem III there exists an optimal control $\omega_0(t)$ such that $\|\omega_0\| \leqslant N$ (since $\|v_0\|$=N).

First, we show that $\omega_0(t)$ is a nondecreasing function. For the moment, assume the contrary. We introduce the functions ω_{01} and ω_{02}:

$$2\omega_{01}(t) = \int_0^t |d\omega_0(s)| + \omega_0(t), \qquad 2\omega_{02}(t) = \int_0^t |d\omega_0(s)| - \omega_0(t).$$

Clearly, $\omega_{01}(0) = \omega_{02}(0) = 0$, $\omega_0(t) = \omega_{01}(t) - \omega_{02}(t)$, and the functions $\omega_{01}(t)$ and $\omega_{02}(t)$ are nondecreasing. By the assumption the function $\omega_{02}(t)$ has points of increase on $[0,T]$, i.e., $\|\omega_{02}\| > 0$. Hence, $\|\omega_{01}\| < N$. Using the relations (1.25), (1.26) and the fact that $\omega_{02}(t)$ is nondecreasing, we see that

$$\begin{aligned}
\lambda &= q'r^{-1}(T, v_0)q = q'r^{-1}(T,\omega_0)q \\
&= \max_{y\in R^n}\left[2y'q - y'Z'(0,T)r(0)Z(0,T)y - y'\left\{\int_0^T h(s,T)\,d\omega_0(s)\right\}y\right] \\
&\geqslant \max_{y\in R^n}\left[2y'g - y'Z'(0,T)r(0)Z(0,T)y - y'\left\{\int_0^T h(s,T)\,d\omega_{01}(s)\right\}y\right] \\
&= q'r^{-1}(T,\omega_{01})q.
\end{aligned}$$

It follows from this relation that the nondecreasing function $\omega_{01} \in V$ furnishes a solution of Problem II, with $\|\omega_{01}\| < N$. Therefore, repeating the proof in stage 2° verbatim (except for the substitution of ω_{01} for v_0), we establish the existence of a control $v_1 \in V$, $\|v_1\| = N$, such that

$$q'r^{-1}(T,v_1)q < q'r^{-1}(T,\omega_{01})q \leqslant q'r^{-1}(T,v_0)q.$$

The last inequality, however, contradicts the optimality of v_0 in Problem II. Thus, $\omega_0(t)$ is a nondecreasing function. In a similar way, we can prove that in the class V only a function with norm N can furnish a solution of Problem III. Therefore, since the control $v_0(t)$ is admissible in Problem III, it is also optimal in this problem.

4°. On the strength of the results of stages 1° and 2° a nondecreasing function with norm N provides a solution of Problem II. Further, on the strength of the result of stage 3° the solution of Problem II is equivalent to that of Problem III, which can usually be reduced to a moment problem. In the case under consideration this problem is the problem of determining a function $v(t)$ $(0 \leqslant t \leqslant T)$ having minimum norm and satisfying the equation

$$r(T, v_0) - Z'(0,T)D_0^{-1}Z(0,T) = \int_0^T h(s,T)\,dv(s).$$

The solution of the resulting moment problem[1] (consequently, of Problem II) is furnished by the step function $v_0(t)$ whose number of jumps is at most $n(n+1)/2$. It follows from the results of stage 2° that the function $v_0(t)$ is monotone nondecreasing and $\|v_0\| = v_0(T) = N$.

Finally, we try to establish the connection between Problems I and II. The matrix $D^{-1}(T)$, in view of (1.7), can be expressed by

$$D^{-1}(T) = Z'(0,T)D_0^{-1}Z(0,T) + \int_0^T h(s,T)\gamma(s)\,ds. \tag{1.28}$$

Note that any nonnegative function $\gamma(t)$ satisfying the estimate (1.23) is associated with the following function $v(t)$ of bounded variation:

$$v(t) = \int_0^t \gamma(s)\,ds, \qquad 0 \leqslant t \leqslant T,\ v(0) = 0,\ \|v\| = v(T) \leqslant N. \tag{1.29}$$

Furthermore, for functions $v(t)$ expressed as (1.29) equation (1.25) coincides with equation (1.28) for the matrix $D^{-1}(T)$. Let V_1 be the set of all functions $v(t)$ that can be obtained by means of formula (1.25) for various $\gamma(s) \geqslant 0$ satisfying the inequality (1.23). Since $V_1 \subset V$, it follows from comparison of (1.28) with (1.25) that

$$\inf_{v \in V} q'r(T,v)q \leqslant \inf_{v \in V_1} q'D(T)q. \tag{1.30}$$

Strict equality in (1.30) (which means the equality of the minimum values of the cost functionals in Problems I and II) occurs if and only if the optimal control $v_0(t)$ in Problem II can be associated with the function $\gamma(t)$ by means of formula (1.29). This correspondence between $v(t)$ and $\gamma(t)$ takes place, for example, for the step functions $v(t)$. Consequently, the form of the optimal control $v_0(t)$ in Problem II implies that the optimal function $\gamma(t)$ in Problem I is defined by the expression (1.24). □

Remark. The above theorem enables reduction of the problem of determining an optimal observation law to the problem of minimizing a scalar function of a finite number of variables μ_i and t_i, where $i = 1, \dots, m$, $2m \leqslant n(n+1)$, satisfying the conditions of Theorem 1.1. To this end, it is sufficient to determine $D(T)$ from equation (1.28) for $\gamma(t)$ of the form (1.24).

---

[1]Neustadt, W. L., Optimization, a Moment Problem, and Nonlinear Programming, *J. Soc. Industr. and Appl. Math.*, (1964), vol. A2, no. 1.

Let us illustrate the foregoing by an example.

Suppose that (1.21), (1.22) are scalar equations with constant coefficients, i.e.,

$$\dot{x}(t) = ax(t), \qquad x(0) = x_0,$$
$$dy(t) = h(t)x(t)\,dt + \sigma_0\,d\xi_0(t), \qquad y(0) = 0,\ 0 \leqslant t \leqslant T,$$
$$\tilde{y}(t_i) = \sqrt{\mu_i}Q(t_i)x(t) + \sigma_0(t_i)\zeta_0(t), \qquad \tilde{y} \in R^m.$$

It is required to determine a function $h(t)$ constrained by (1.23) such that it minimizes the variance $D(T)$ at the terminal moment of observations. According to Theorem 1.1 whose conditions are assumed to be satisfied, the optimal observation law $\gamma(t) = N\delta(t-t_1)$, i.e., all measurements are made at the unique moment t_1. To determine this moment, we evaluate the cost functional $D(T)$, given the control $\gamma(t) = N\delta(t-t_1)$. The equation for $D^{-1}(t)$ is of the form

$$\frac{dD^{-1}}{dt}(t) = \gamma(t)\sigma_0^{-2} - 2aD^{-1}(t) = N\delta(t-t_1)\sigma_0^{-2} - 2aD^{-1}(t),$$
$$D^{-1}(0) = D_0^{-1}.$$

This yields

$$D(t_1) = D_0e^{2at_1}, \qquad D(t_1+0) = [N\sigma_0^{-2} + D^{-1}(t_1)]^{-1},$$
$$D(T) = D(t_1+0)e^{2a(T-t_1)} = D_0e^{2aT}[1 + N\sigma_0^{-2}D_0e^{2at_1}]^{-1}.$$

Thus, the cost functional $D(T)$ attains a minimum with respect to $t_1 \in [0,T]$ at $t_1 = T$ if $a > 0$, and at $t_1 = 0$ if $a < 0$. Therefore, all observations must be made at the terminal moment of observations if $a > 0$, and at the starting moment of observations if $a < 0$. But if $a = 0$, then the value of the cost functional $D(T)$ is independent of the choice of the observation moment.

5. Pulsed observation of a material point. We shall derive an optimal law of observing a material point, which is in free motion along the line. The coordinate x_1 of the point is amenable to physical measurement. It is required to minimize the variance $d_{22}(T)$ of the velocity $x_2(t)$ at the terminal moment of observations. The equations of motion are

$$\dot{x}_1(t) = x_2(t), \qquad x_2(t) = 0,\ 0 \leqslant t \leqslant T. \tag{1.31}$$

The covariance matrix D_0 of the *a priori* probability distribution of the vector $(x_1(0), x_2(0))$ is of diagonal type with known elements $d_{11}(0)$ and $d_{22}(0)$ in the principal diagonal.

The observation equation is of the form

$$\begin{aligned} dy(t) &= h(t)x_1(t)\,dt + \sigma_0\,d\xi_0(t), \qquad y(0) = 0,\\ \tilde{y}(t_i) &= \sqrt{\mu_i}x_1(t) + \sigma_0(t_i)\zeta_0(t), \qquad \tilde{y} \in R^m. \end{aligned} \tag{1.32}$$

(1.32)

where the constant $\sigma_0 \neq 0$.

The control of observations is realized by a scalar function $h(t)$ constrained by (1.23). Note that equations (1.31), (1.32) can be obtained from the general equations (1.21), (1.22) by setting

$$A = \begin{pmatrix} 0 & 1 \\ 0 & 0 \end{pmatrix}, \qquad Q = (1\ 0).$$

We first verify if the condition $H > 0$ of Theorem 1.1 is satisfied. The matrix $Z(s,t)$ is of the form

$$Z(s,t) = \exp[A(s-t)] = \begin{pmatrix} 1 & s-t \\ 0 & 1 \end{pmatrix}. \tag{1.33}$$

Moreover,

$$Q'(\sigma_0\sigma_0')^{-1}Q = \sigma_0^{-2}\begin{pmatrix} 1 & 0 \\ 0 & 0 \end{pmatrix}. \tag{1.34}$$

Hence,

$$h(s,T) = Z'(s,T)Q'(\sigma_0\sigma_0')^{-1}QZ(s,T) = \sigma_0^{-2}\begin{pmatrix} 1 & s-T \\ s-T & (s-T)^2 \end{pmatrix},$$

and

$$H = \int_0^T h(s,T)\,ds = \sigma_0^{-2}\begin{pmatrix} T & -\frac{1}{2}T^2 \\ -\frac{1}{2}T^2 & \frac{1}{3}T^3 \end{pmatrix}.$$

It is readily seen that the Matrix H is positive definite. According to Theorem 1.1, the optimal observation law $\gamma(t)$ is defined by

$$\gamma(t) = \sum_{i=1}^{3} \mu_i\delta(t-t_i), \qquad 0 \leqslant t_i \leqslant T,\ \mu_1+\mu_2+\mu_3 = N,\ \mu_i \geqslant 0, \tag{1.35}$$

where the moments of measurement t_i and their intensities μ_i are to be determined. To this end, we try to find the expression of the cost functional $d_{22}(T)$ as a function of μ_i and t_i.

An equation for the matrix $D^{-1}(t)$ is of the form

$$\dot{D}^{-1} = -A'D^{-1} - D^{-1}A + \gamma(t)u, \qquad D^{-1}(0) = D_0^{-1},\ u = Q'(\sigma_0\sigma_0')^{-1}Q.$$

This equation has a solution expressed by the formula

$$\begin{aligned} D^{-1}(T) &= e^{-A'T}D_0^{-1}e^{-AT} + \int_0^T e^{-A'(T-t)}\gamma(t)ue^{-A(T-t)}\,dt \\ &= e^{-A'T}\left[D_0^{-1} + \int_0^T e^{A't}\gamma(t)ue^{At}\,dt\right]e^{-AT}. \end{aligned}$$

Using the relations (1.33)–(1.35), we find that

$$D^{-1}(T) = \begin{pmatrix} 1 & 0 \\ -T & 1 \end{pmatrix} \begin{pmatrix} d_{11}^{-1}(0) + N\sigma_0^{-2} & \sigma_0^{-2}\delta_1 \\ \sigma_0^{-2}\delta_1 & d_{22}^{-1}(0) + \gamma_0^{-2}\delta_2 \end{pmatrix}^{-1} \begin{pmatrix} 1 & 0 \\ T & 1 \end{pmatrix},$$

$$\delta_1 = \mu_1 t_1 + \mu_2 t_2 + \mu_3 t_3, \qquad \delta_2 = \mu_1 t_1^2 + \mu_2 t_2^2 + \mu_3 t_3^2.$$

It follows that the cost functional $d_{22}(T)$ is of the form

$$d_{22}(T) = (d_{11}^{-1}(0) + N\sigma_0^{-2})\left[(d_{11}^{-1}(0) + N\sigma_0^{-2})(d_{22}^{-1}(0) + \sigma_0^{-2}\delta_2) - \sigma_0^{-4}\delta_1^2\right]^{-1}.$$

Thus, minimizing the expression for $d_{22}(T)$ with respect to μ_i and t_i is equivalent to maximizing with respect to these variables the following functions

$$(d_{11}^{-1}(0) + N\sigma_0^{-2})\sigma_0^{-2}\delta_2 - \sigma_0^{-4}\delta_1^2 \to \max, \tag{1.36}$$
$$\mu_i \geqslant 0,\ \mu_1 + \mu_2 + \mu_3 = N,\ 0 \leqslant t_i \leqslant T.$$

Note that the function (1.36) satisfies the inequality

$$(d_{11}^{-1}(0) + \sigma_0^{-2}N)\sigma_0^{-2}\delta_2 - \sigma_0^{-4}\delta_1^2 \geqslant d_{11}^{-1}(0)\sigma_0^{-2}\delta_2 \geqslant 0. \tag{1.37}$$

However, for any fixed μ_i the function (1.36) is a quadratic in the variables t_i and is, moreover, positive definite by virtue of (1.37). Therefore, the maximum of the function (1.36) with respect to the variables $t_i \in [0, T]$ can be attained only at one of the vertices of the three-dimensional cube $0 \leqslant t_i \leqslant T$, $i = 1, 2, 3$. Thus, the optimal observation law must have the form $\gamma(t) = \mu_1\delta(t) + \mu_2\delta(t - T)$, where $\mu_2 > 0$, since for $\mu_2 = 0$ all observations are made at the starting point of the process, and this corresponds to the point of the zero minimum of the function (1.36). Substituting the optimal function $\gamma(t)$ given above in (1.36), we see that μ_2 is the maximum point of the function

$$(d_{11}^{-1}(0) + \sigma_0^{-2}N)\mu_2 - \mu_2^2\sigma_0^{-2}, \qquad 0 \leqslant \mu_2 \leqslant N.$$

Consequently, equating to zero the derivative of this function with respect to μ_2, we find that

$$\mu_2 = \min\left[N, \tfrac{1}{2}(d_{11}^{-1}(0)\sigma_0^2 + N)\right].$$

The minimum value of the variance $d_{22}(T)$ of the velocity for the above value of μ_2 is

$$d_{22}(T) = (d_{11}^{-1}(0) + N\sigma_0^{-2})\left[(d_{11}^{-1}(0) + N\sigma_0^{-2}) \times (d_{22}^{-1}(0) + \sigma_0^{-2}\mu_2 T^2) - \sigma_0^{-4}\mu_2^2 T^2\right].$$

6. Optimal noise control in the observation channel. We are concerned with various problem statements that are connected with optimal control of noise contained in observations with the aim to hinder the measurement of the coordinates of the system to the utmost. It is assumed that the motion of the system and the observed process are described by the relations (1.1), (1.2).

The control of noise consists in varying the matrix $\sigma_0(t)$ that characterizes the intensity of noise in the observation channel, so as to maximize some functionals of $D(t)$. The covariance matrix $D(t)$ satisfies equation (1.3) which involves control $u(t)$ depending on the choice of $\sigma_0(t)$.

Clearly, problem statements for optimal noise control differ from those for optimal observation control only in that the cost functional must now be maximized. For example, we may consider problems of maximizing the functional (1.4) on the trajectories of the system (1.3), given the constraint (1.5).

Another problem statement is obtained if it is required to minimize one of functionals J_i of the type (1.5) constrained by $q_i'D(T)q_i \geqslant C_i$, where $C_i > 0$ are prescribed constants, and q_i are prescribed vectors in R^n. The meaning of these constraints is not to allow the "enemy" to determine the linear combinations $q'x(T)$ of the phase coordinates of the system with precision greater than certain critical values C_i.

Just as optimal observation problems, optimal noise control problems are standard optimal control problems for the deterministic system (1.3) in terms of the covariance matrix. Therefore, the methods discussed in Chapters IV–VII can be used for the purpose of determining optimal noise control laws. A few illustrations will make this clear.

1°. *Scalar observation in a scalar system.* Consider scalar equations of motion and a scalar observation of the form (1.1), (1.2):

$$dx(t) = ax(t)\,dt + \sigma\,d\xi(t), \qquad 0 \leqslant t \leqslant T,\ x(0) = x_0,$$
$$dy(t) = x(t) + \sigma_0\,d\xi_0(t), \qquad y(0) = 0.$$

Here a and σ are given constants, x_0 is a Gaussian variable with variance $D_0 > 0$. The control of the noise in observations is utilized by means of the function $\sigma_0(t)$. At each instant of time this function is assumed to take on one of two values $\sqrt{\sigma_1}$ or $\sqrt{\sigma_2}$, where $\sigma_2 > \sigma_1 > 0$, and, in addition,

$$\int_0^T f\big(\sigma_0(t)\big)\,dt = T_0, \qquad 0 \leqslant T_0 \leqslant T, \tag{1.38}$$
$$f\big(\sqrt{\sigma_1}\big) = 0,\ f\big(\sqrt{\sigma_2}\big) = 1.$$

The problem consists in determining a function $\sigma_0(t)$ satisfying the above constraints so that it yields a maximum of the variance $D(T)$ of the state $x(T)$ of the system at the terminal point of the observation. By virtue of (1.3) the variance $D(t)$ satisfies the relations

$$\dot D(t) = 2aD(t) - u(t)D^2(t) + \sigma^2, \qquad D(0) = D_0,\ u(t) = \sigma_0^{-2}(t).$$

As above, in order to express the coefficient of the control $u(t)$ in a form that does not depend on $D(t)$, we make the change of the variable $\alpha(t) = D^{-1}(t)$. Thus, we obtain

$$\dot{\alpha}(t) = -2a\alpha(t) + u(t) - \sigma^2\alpha^2(t), \qquad \alpha(0) = D_0^{-1},\ u(t) = \sigma_0^{-2}(t). \tag{1.39}$$

To determine the optimal control $u_0(t)$ minimizing $\alpha(T)$ subject to the constraints (1.38), we use the relations of the maximum principle. The adjoint variable $\psi(t)$ is a solution of the problem

$$\dot{\psi}(t) = 2\psi(t)\big(a + \sigma_2\alpha(t)\big), \qquad 0 \leqslant t \leqslant T,\ \psi(T) = -1. \tag{1.40}$$

The optimal control $u_0(t)$ is determined from the condition for a maximum of the Hamiltonian H with respect to u:

$$H = \psi(t)\big(-2a\alpha(t) + u - \sigma^2\alpha^2(t)\big) + \psi_0 f(u^{-1/2}) \to \max_u,$$

where ψ_0 is a constant.

Hence, noting the relations (1.38), we find that the optimal control is expressed by

$$u_0(t) = \begin{cases} \sigma_1^{-1}, & \text{for } \psi(t) + C > 0, \\ \sigma_2^{-1}, & \text{for } \psi(t) + C < 0. \end{cases} \tag{1.41}$$

Here the constant C is chosen so that the control satisfies the isoperimetric constraint (1.38). We will analyse the above formula for the optimal control. It follows from the relations (1.40) that $\psi(t) < 0$, $0 \leqslant t \leqslant T$. Therefore, for $a > 0$ and for $a = 0$, $\sigma^2 > 0$ the derivative of the adjoint variable $\dot{\psi}(t) < 0$. Thus, in these cases (i.e., for either $a > 0$ or $a = 0$, $\sigma^2 > 0$) we obtain from (1.41)

$$u_0(t) = \begin{cases} \sigma_1^{-1}, & \text{for } 0 \leqslant t \leqslant T - T_0, \\ \sigma_2^{-1}, & \text{for } T - T_0 < t \leqslant T. \end{cases}$$

In other words, in these cases it is expedient to produce noise of greater intensity on the interval of duration T_0 adjoining the terminal moment. If $a = 0$, $\sigma = 0$, then for any control of the noise (i.e., for any σ_0) constrained by (1.38) the variance $D(T)$ is expressed by

$$D(T) = \left[D_0^{-1} + \sigma_2^{-1}T_0 + \sigma_1^{-1}(T - T_0)\right]^{-1}.$$

If $a = 0$, $\sigma = 0$, then $\dot{\psi}(t) > 0$, in which case

$$u_0(t) = \begin{cases} \sigma_2^{-1}, & \text{for } 0 \leqslant t \leqslant T_0, \\ \sigma_1^{-1}, & \text{for } T_0 < t \leqslant T. \end{cases}$$

It remains to investigate the case $a < 0$, $\sigma^2 > 0$, in which $\ddot{\psi}(t) = 2\psi(t)\big[\big(a + \sigma^2\alpha(t)\big)^2 + a^2 + \sigma^2 u\big] < 0$, $0 \leqslant t \leqslant T$. Hence, $\psi(t)$ is a concave function, i.e., by virtue of (1.40) there exists a moment $\tau \in [0, T_0]$ such that

$$u_0(t) = \begin{cases} \sigma_2^{-1}, & \text{for } 0 \leqslant t \leqslant \tau,\ T + \tau - T_0 < t \leqslant T, \\ \sigma_1^{-1}, & \text{for } \tau < t \leqslant T + \tau - T_0. \end{cases}$$

To determine the moment τ, we try to find a function that characterizes the dependence of $\alpha(T)$ on τ, and then we calculate the minimum of this function with respect to τ. On integrating equation (1.39) successively on each interval of constancy of the control $u_0(t)$, we have

$$\alpha(T) = \sigma^{-2}\big[(\beta - a) - 2\Delta^{-1}\beta\big(\lambda^2\gamma C_1(\lambda_1^2 - 1) + \lambda(\gamma^2 - \lambda_1^2)\big)\big], \tag{1.42}$$

where we set

$$\begin{gathered} \beta = \sqrt{a^2 + \sigma^2\sigma_2^{-1}} > 0, \qquad \lambda = e^{2\beta\tau} \in [1, e^{2\beta T_0}], \\ \gamma = \Big(\beta - \sqrt{a^2 + \sigma^2\sigma_1^{-2}}\Big)\Big(\beta + \sqrt{a^2 + \sigma^2\sigma_1^{-2}}\Big)^{-1}, \qquad \lambda_1 = e^{a_1(T - T_0)}, \\ a_1 = -\sqrt{a^2 + \sigma^2\sigma_1^{-1}} < 0, \\ C_1 = (\beta + a + \sigma^2 D_0^{-1})(\beta - a - \sigma^2 D_0^{-1})^{-1}, \\ \Delta = \lambda^2\gamma C_1(\lambda_1^2 - 1) + \lambda\big[\gamma^2 - \lambda_1^2 + C_1 e^{2\beta T_0}(\gamma^2\lambda_1^2 - 1)\big] - \gamma e^{2\beta T_0}(\lambda_1^2 - 1). \end{gathered} \tag{1.43}$$

The following inequalities hold for the above quantities:

$$\begin{gathered} -1 < \gamma < 0, \; 0 < \lambda_1 < 1, \\ |(\lambda_1 + \gamma)(\lambda_1\gamma + 1)^{-1}| \leqslant 1, \qquad (\lambda_1 - \gamma)(1 - \lambda_1\gamma) \leqslant 1. \end{gathered} \tag{1.44}$$

We proceed with further investigation separately for cases $a_2 = \beta - a - \sigma^2 D_0^{-1} \neq 0$ and $a_2 = 0$. First, let $a_2 \neq 0$. In the expression (1.42) for $\alpha(T)$ only the parameter λ depends on the choice of τ. In considering the dependence of $\alpha(T)$ on λ, it is convenient to denote $\alpha(T) = J(\lambda)$. Let us find the points $\lambda_{\max}$ and $\lambda_{\min}$, where $J(\lambda) = \alpha(T)$ as a function of λ attains its local maximum and minimum, respectively.

For this purpose, we compute the derivative $dJ(\lambda)/d\lambda$. It follows from the relation (1.42) that

$$\begin{aligned} \frac{\partial J(\lambda)}{\partial\lambda} &= -2\beta\sigma^{-2}\Delta^{-2}e^{2\beta T_0}\gamma(\lambda_1^2 - 1) \\ &\quad \times [\lambda^2 C_1(\gamma^2\lambda_1^2 - 1) - 2\lambda C_1\gamma(\lambda_1^2 - 1) - (\gamma^2 - \lambda_1^2)]. \end{aligned}$$

Further, we equate this derivative to zero and find the intervals of sign constancy disregarding temporarily the constraint $\lambda \in [1, \exp(2\beta T_0)]$. As a result, we obtain that for $\lambda < \lambda_{\max}$ and $\lambda > \lambda_{\min}$ the relations $\lambda_{\max} < \lambda_{\min}$ and $\partial J(\lambda)/\partial\lambda > 0$ are always true, and for $\lambda_{\max} < \lambda < \lambda_{\min}$ the derivative $\partial J(\lambda)/\partial\lambda < 0$, with

$$\lambda_{\max} = \begin{cases} (\lambda_1 - \gamma)[C_1(\lambda_1\gamma - 1)]^{-1}, & \text{for } C_1 > 0, \\ (\lambda_1 + \gamma)[C_1(\lambda_1\gamma + 1)]^{-1}, & \text{for } C_1 < 0. \end{cases}$$

$$\lambda_{\min} = \begin{cases} (\lambda_1 + \gamma)[C_1(\lambda_1\gamma + 1)]^{-1}, & \text{for } C_1 > 0, \\ (\lambda_1 - \gamma)[C_1(\lambda_1\gamma - 1)]^{-1}, & \text{for } C_1 < 0. \end{cases}$$

These relations enable us to find the value $\lambda \in [1, \exp(2\beta T_0)]$ which yields the minimum of the function $\alpha(T) = J(\lambda)$. In view of (1.44) we find that for $a < 0$, $\sigma \neq 0$, $\alpha_2 \neq 0$ the following expressions for the duration τ of the first interval of the jamming noise hold true:

$$\tau = \begin{cases} 0, & \text{for } \lambda_{\min} \leqslant 1, \\ 0, & \text{for } 1 < \lambda_{\min} < e^{2\beta T_0}, \quad J(1) < J(\lambda_{\min}), \\ \frac{1}{2\beta} \ln \lambda_{\min}, & \text{for } 1 < \lambda_{\min} \leqslant e^{2\beta T_0}, \quad J(1) > J(\lambda_{\min}), \\ T_0, & \text{for } \lambda_{\min} > e^{2\beta T_0},\ \lambda_{\max} \leqslant 1, \\ 0, & \text{for } \lambda_{\min} > e^{2\beta T_0},\ \lambda_{\max} > 1, \quad J(1) \leqslant J(e^{2\beta T_0}), \\ T_0, & \text{for } \lambda_{\min} > e^{2\beta T_0},\ \lambda_{\max} > 1, \quad J(1) > J(e^{2\beta T_0}). \end{cases}$$

Consider now the case $\alpha_2 = 0$. Then the derivative $dJ(\lambda)/d\lambda$ is of the form

$$\frac{dJ(\lambda)}{d\lambda} = -\frac{2\beta}{\sigma^2} \frac{e^{2\beta T_0}\gamma(\lambda_1^2 - 1)\lambda^2(\gamma^2\lambda_1^2 - 1)}{[\lambda^2\gamma(\lambda_1^2 - 1) + \lambda e^{2\beta T_0}(\gamma^2\lambda_1^2 - 1)]^2}.$$

Thus, in view of the estimates (1.44) for the quantities (1.43) the derivative $dJ(\lambda)/d\lambda > 0$, i.e., $J(\lambda)$ is monotone increasing. Hence, for $\alpha_2 = 0$, $a < 0$, $\sigma \neq 0$ the moment $\tau = 0$. Therefore, the dependence of the optimal control $u_0(t)$ on the problem parameters is fully established.

2°. *Noise control while observing a material point.* Consider a material point in free rectilinear motion. The equations of motion are of the form (1.31). The observed function $y(t)$ is described by an equation of the type (1.2):

$$dy(t) = \big(q_1 x_1(t) + q_2 x_2(t)\big)\, dy + \sigma_0(t)\, d\xi_0(t), \qquad 0 \leqslant t \leqslant T,\ y(0) = 0.$$

Here x_1 is the coordinate of the point, x_2 is its velocity, q_i are given constants. Noise control is realized by choosing the function $\sigma_0(t)$ subject to the same constraints (1.38). It is required to determine a function $\sigma_0(t)$ so that it maximizes the cost functional

$$\beta_1 d_{11}(T) + \beta_2 d_{22}(T) \to \max, \tag{1.45}$$

where $\beta_1 \geqslant 0$, $\beta_2 > 0$ are given constants, and $d_{ii}(T)$ are diagonal elements of the covariance matrix $D(T)$. We write equations for the elements $\alpha_{ij}(t)$ of the matrix $\alpha(t) = D^{-1}(t)$. By virtue of (1.3) we have

$$\begin{aligned} \dot\alpha_{11}(t) &= q_1^2 \sigma_0^{-2}(t), \\ \dot\alpha_{12}(t) &= \dot\alpha_{21}(t) = q_1 q_2 \sigma_0^{-2}(t) - \alpha_{11}(t), \\ \dot\alpha_{22}(t) &= -2\alpha_{12}(t) + q_2^2 \sigma_0^{-2}(t), \\ &0 \leqslant t \leqslant T,\ \alpha_{ij}(0) = \left(D_0^{-1}\right)_{ij}. \end{aligned} \tag{1.46}$$

It follows from these equations that the value of the cost functional for $q_1 = 0$ does not depend on the particular value of the function $\sigma_0(t)$ that is subject to the constraints (1.38). Therefore, we consider only the case $q_1 \neq 0$. It follows from the maximum principle relations (1.7)–(1.9), (1.11) and from the transversality condition (expressed by the equality (1.14) taken with opposite sign) that the optimal noise control $u_0(t)$ is of the form

$$u_0(t) = \begin{cases} \sigma_1, & \text{for } q_1^2\psi_1(t) + q_1q_2\psi_2(t) + C > 0, \\ \sigma_2, & \text{for } q_1^2\psi_1(t) + q_1q_2\psi_2(t) + C < 0. \end{cases}$$

Here the constant C is chosen so that the transversality condition (1.38) is satisfied, and the adjoint variables $\psi_i(t)$ are described by equations of the form (1.9):

$$\dot\psi_1(t) = \psi_2(t), \qquad \dot\psi_2(t) = 2\psi_3(t), \qquad \dot\psi_3(t) = 0, \qquad 0 \leqslant t \leqslant T,$$
$$\psi_3(T) = -\left[\beta_1\alpha_{12}^2(T) + \beta_2\alpha_{11}^2(T)\right]\left[\alpha_{11}(T)\alpha_{22}(T) - \alpha_{11}^2(T)\right]^{-2} < 0.$$

Hence, we see that the adjoint variable $\psi_3(t)$ is constant and negative. Therefore, we have

$$\frac{d^2}{dt^2}\left(q_1^2\psi_1(t) + q_1q_2\psi_2(t)\right) = 2q_1^2\psi_3(t) < 0.$$

It follows that there exists a point $\tau \in [0, T_0]$ such that the optimal noise control $u_0(t) = \sigma_0^{-2}(t)$ is expressed by

$$u_0(t) = \begin{cases} \sigma_2^{-1}, & \text{for } 0 \leqslant t \leqslant \tau,\ T + \tau - T_0 < t \leqslant T, \\ \sigma_1^{-1}, & \text{for } \tau < t \leqslant T + \tau - T_0. \end{cases}$$

Thus, there are at most two intervals of jamming noise which adjoin the initial ($t = 0$) and the terminal ($t = T$) moments. If we integrate equations (1.46) for an optimal noise law and then compute the cost functional (1.45), it proves to be a function of one variable τ. Maximizing this function with respect to τ determines the jamming intervals.

§ 2. Optimal Combination of Control and Observation

Problems of the optimal combination of control and observation are the problems of the simultaneous control of both the motion of the system and the process of observation of its coordinates with the aim to minimize certain cost functionals (see Figure 2.1). The corresponding problem statements as well as the appropriate dynamic programming equations were given in § 2 of Chapter IX. In this section we are concerned with several specific problems.

1. Linear-quadratic problem. Consider a linear control system of the form (3.13 (see Chapter IX):

$$dx(t) = [A(t)x(t) + B(t)u + f(t)]\,dt + \sigma(t)\,d\xi(t), \tag{2.1}$$
$$0 \leqslant t \leqslant T, \qquad x \in R^n,\ x(0) = x_0,$$

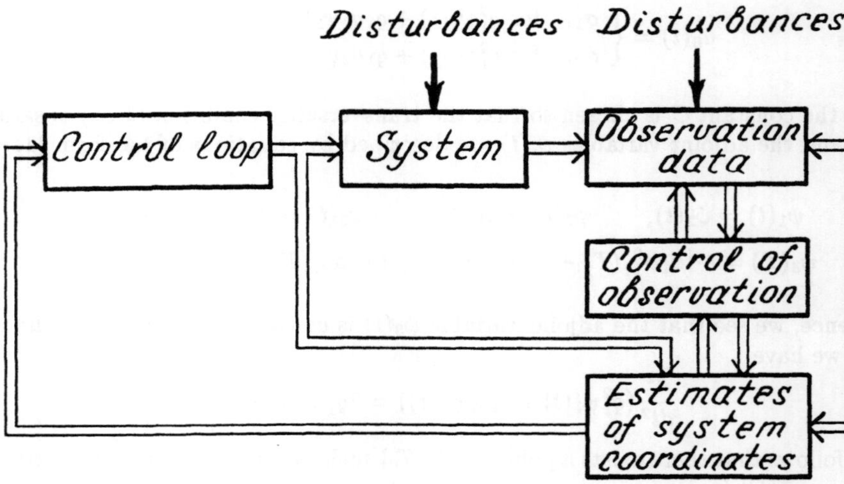

FIG. 2.1. Block-diagram of the optimal combined control of the system and of the observation process

The observed variable $y(t) \in R^l$ is described by the relation (3.14) (see Chapter IX):

$$dy(t) = Q(t)x(t)\,dt + \sigma_0(t)\,d\xi_0(t), \qquad y(0) = 0,\ y \in R^l. \tag{2.2}$$

The cost functional to be minimized is of the form

$$J(u) = Ex'(T)N_1x(T) + \int_0^T \big(x'(t)N_2(t)x(t) + u'N_3(t)u\big)\,dt. \tag{2.3}$$

All designations in formulas (2.1)–(2.3) have the same meaning as in the expressions (3.13), (3.14), (3.2) of Chapter IX. In particular, x_0 is a Gaussian vector with prescribed mean m_0 and covariance matrix D_0; ξ and ξ_0 are standard Wiener processes; x_0, ξ and ξ_0 are independent.

The control of the system is sought in the form of a functional $u(t, y_t)$ depending on time t and the measurement data $y_t = y(s)$, $0 \leqslant s \leqslant t$. The observation control is realized by choosing the measurement matrix $Q(t)$ and the accuracy characteristic $\sigma_0(t)$ subject to the constraint

$$w(t) \in W(t), \qquad w(t) = Q'(t)\big(\sigma_0(t)\sigma_0'(t)\big)^{-1}Q(t), \tag{2.4}$$

where $W(t)$ is a given matrix set.

Thus, the observation control is sought in the class of functions depending only on time t, i.e., in the class of program controls (or P-controls). In general, of interest is also the determination of feedback control (or F-control) of the observation process expressed in terms of the measurement data, i.e., the determination of the matrices Q and σ_0 in the form $Q = Q(t, y_t)$, $\sigma_0 = \sigma_0(t, y_t)$. Generally speaking, widening the class of control functions may lead to reduction in the value of the cost functional (2.3). However, in the case under consideration this does not happen: the minimum value of the cost functional (2.3) is the same both in the class of P-controls of the observation process and in the class of F-controls.[2] Therefore, only observation control of the form (2.4) is considered below. On the strength of the results given in § 3 of Chapter IX, optimal F-control $u_0(t, y_t)$ for any specified observation law can be constructed in the explicit analytical form:

$$u_0(t, y_t) = -N_3^{-1}(t)B'(t)P(t)m(t). \tag{2.5}$$

Here $m(t)$ is an optimal (in the mean-square sense) estimate expressed in terms of the observation data y_t by the relations (2.16), and the positive-definite matrix $P(t)$ satisfies equations of the form (3.10) (see Chapter IX):

$$\dot{P}(t) = -A'(t)P(t) - P(t)A(t) + P(t)B(t)N_3^{-1}(t)B'(t)P(t) - N_2(t), \tag{2.6}$$
$$P(T) = N_1.$$

The minimum value of the cost functional (2.3) associated with the optimal control (2.5) has the form

$$\begin{aligned} J(u_0) &= m_0'P(0)m_0 + \operatorname{tr}[D(T)N_1] \\ &\quad + \int_0^T \operatorname{tr}[D(t)w(t)D(t)P(t) + D(t)N_2(t)]\,dt. \end{aligned} \tag{2.7}$$

Here $D(t)$ is the covariance matrix of the conditional probability distribution of $x(t)$, given $y)t$. The covariance matrix $D(t)$ satisfies the relations (3.24) of Chapter IX:

$$\dot{D}(t) = A(t)D(t) + D(t)A'(t) - D(t)w(t)D(t) + \sigma(t)\sigma'(t), \tag{2.8}$$
$$D(0) = D_0.$$

Thus, the optimal observation law is determined from the condition for a minimum of the functional (2.7) with respect to $w(t) \in W(t)$. On the strength of (2.8) we have

$$\begin{aligned} &\int_0^T \left[A(t)D(t) + D(t)A'(t) - \dot{D}(t) + \sigma(t)\sigma'(t)\right]P(t)\,dt \\ &\qquad = \int_0^T D(t)w(t)D(t)P(t)\,dt. \end{aligned}$$

---

[2] Grigoriev, F. N., Kuznetsov, N. A., and Serebrovsky, A. P., *Control of an Observation Process in Automatic Systems*, Nauka, Moscow, (1986).

Further, integrating by parts gives

$$-\int_0^T \dot{D}(t)P(t)\,dt = -D(T)P(T) + D(0)P(0) + \int_0^T D(t)\dot{P}(t)\,dt$$
$$= -D(T)N_1 + D_0P(0) + \int_0^T D(t)\dot{P}(t)\,dt.$$

The last two equalities yield

$$\int_0^T D(t)w(t)D(t)P(t)\,dt = \int_0^T \big[A(t)D(t) + D(t)A'(t)$$
$$+ \sigma(t)\sigma'(t)\big]P(t)\,dt - D(t)N_2 + D_0P(0) + \int_0^T D(t)\dot{P}(t)\,dt.$$

Hence, on invoking the equality $\mathrm{tr}[A_1A_2] = \mathrm{tr}[A_2A_1]$, which is true for any two square matrices of the same order, we see that the functional (2.7) is expressed by

$$J = m_0'P(0)m_0 + \mathrm{tr}[D_0P(0)] + \int_0^T \mathrm{tr}\big[D(t)\dot{P}(t)$$
$$+ N_2(t)D(t) + \big(A(t)D(t) + D(t)A'(t) + \sigma(t)\sigma'(t)\big)P(t)\big]\,dt.$$

Next, we replace the derivative $\dot{P}(t)$ by the right side of equation (2.6). Then we obtain the following expression for J:

$$J = m_0'P(0)m_0 + \mathrm{tr}[D_0P(0)] + \int_0^T \mathrm{tr}\big[\big(\sigma(t)\sigma'(t)$$
$$+ D(T)P(t)B(t)N_3^{-1}(t)B'(t)\big)P(t)\big]\,dt. \tag{2.9}$$

In this expression the choice of an observation law affects the last summand, which is a linear functional of the covariance matrix $D(t)$. It should be noted that in the deduction of formula (2.9) the particular form of an observation control constraint is not used. It is only essential that observation control is sought as a program control (i.e., as a function of time only not depending on the measurement data). Therefore, various constraints can be imposed on admissible observation laws just as in observation optimization problems considered in § 1.

2. A deterministic scalar system. Suppose that (2.1) and (2.2) are scalar equations of motion and observation:

$$\dot{x}(t) = a(t)x(t) + b(t)u, \qquad 0 \leqslant t \leqslant T,\ x(0) = x_0,$$
$$dy(t) = Q(t)x(t)\,dt + \sigma_0\,d\xi_0(t), \qquad y(0) = 0.$$

Here the constant $\sigma_0 \neq 0$, the continuous functions $a(t)$ and $b(t)$ are prescribed.

The cost functional will be taken as a quadratic functional of the form (2.3):

$$J(u) = E\left[x^2(T) + \int_0^T N_3(t)u^2\,dt\right], \tag{2.10}$$

where $N_3(t)$ is a positive measurable bounded function.

The functional (2.10) is minimized over controls $u = u(t, y_t)$ and measurement laws specified by the scalar measurable function $Q(t)$, which at each instant of time is equal to either zero or one and is subject to the isoperimetric constraint on the duration of the observation process:

$$\int_0^T Q(t)\,dt = T_0, \qquad 0 < T_0 < T. \tag{2.11}$$

We shall determine expressions for optimal control assuming that the function $b(t)$ is distinct from zero almost everywhere; the functions $a(t)$, $b(t)$, and $N_3(t)$ are differentiable and the following inequalities hold true: $\dot{a}(t) \leqslant 0$, $\dot{N}_3(t) \leqslant 0$, $b(t)\dot{b}(t) \geqslant 0$, $N_3(t) > 0$.

By virtue of the relation (3.22) of Chapter IX the optimal control u_0 is of the form

$$u_0 = -N_3^{-1}(t)P(t)b(t)m(t),$$

where the coefficient $P(t)$ satisfies, in view of equation (2.6), the following Bernoulli equation:

$$\dot{P}(t) = -2a(t)P(t) + P^2(t)b^2(t)N_3^{-1}(t), \qquad P(T) = 1.$$

To solve this equation, we introduce a new variable: $\gamma(t) = P^{-1}(t)$. Then the following linear equation is true:

$$\dot{\gamma}(t) = 2a(t)\gamma(t) - b^2(t)N_3^{-1}(t), \qquad \gamma(T) = 1.$$

The solution of this equation is expressed by means of the Cauchy formula (1.7) of Chapter IV in the form

$$P(t) = \left[Z(t,T) + \int_t^T Z(t,s)b^2(s)N_3^{-1}(s)\,ds\right]^{-1},$$

$$Z(t,s) = \exp\left[2\int_s^t a(\tau)\,d\tau\right].$$

The equations for the optimal estimate $m(t)$ and the variance $D(t)$, by virtue of (2.22) of Chapter XI and equation (2.8), are of the form

$$\begin{aligned} \dot{m} &= am + bu + DQ\sigma_0^{-2}(y - Qm), \qquad m(0) = m_0, \\ \dot{D} &= 2aD - D^2Q\sigma_0^{-2}, \qquad D(0) = D_0. \end{aligned} \tag{2.12}$$

In view of the expression (2.9) the optimal observation law $Q_0(t)$ can be determined from the condition for a minimum with respect to $Q(t)$ of the functional

$$\int_0^T P^2(t)b^2(t)N_3^{-1}(t)D(t)\,dt$$

on the trajectories of the system (2.12) subject to the constraint (2.11). Invoking the maximum principle, we can show that the optimal observation law $Q_0(t)$ is determined by a unique parameter t_1 and is of the form

$$Q_0(t) = \begin{cases} 0, & \text{for } t \leqslant t_1, \\ 1, & \text{for } t_1 < t \leqslant t_1 + T_0, \\ 0, & \text{for } t_1 + T_0 < t. \end{cases}$$

In other words, the optimal law $Q_0(t)$ involves only one observation interval. Compared to the case of uncontrolled motion (i.e., $b(t) = 0$), in general, there occurs a displacement of the observation interval.

3. A stochastic scalar system. Suppose that the controlled motion $x(t)$ of the object on the interval $[0, T]$ is described by the scalar equation

$$dx(t) = [ax(t) + bu]\,dt + \sigma\,d\xi, \qquad x(0) = x_0. \tag{2.13}$$

An observed one-dimensional variable $y(t)$ is described by the equation

$$dy(t) = Q(t)x(t)\,dt + \sigma_0\,d\xi_0(t), \qquad t \geqslant 0\ y(0) = 0. \tag{2.14}$$

Here the quantities a, b, σ, $\sigma_0 \neq 0$ are prescribed constants, and $Q(t)$ is a deterministic piecewise continuous function. The random variable x_0 independent of standard Wiener processes $\xi(t)$, $\xi_0(t)$ is normally distributed with parameters $m_0 = Ex_0$, $D_0 = E(x_0 - m_0)^2 > 0$. We also assume that $b \neq 0$, since the problem of optimizing the observation process for the case $b = 0$ corresponding to uncontrolled motion $x(t)$ was discussed above.

Consider the set H of deterministic functions $Q(t)$, equal at each instant of time to either zero or one, so that

$$\int_0^T Q(t)\,dt = T_0, \qquad T_0 < T. \tag{2.15}$$

It is required to determine a control u as a functional depending on t and on actually observed values $y(s)$ over the segment $0 \leqslant s \leqslant t$ and also a function $Q(t) \in H$, so that the following cost functional of the form (1.3) is minimized:

$$E\left(\alpha x^2(T) + \int_0^T u^2\,dt\right) = \min_{u,Q}, \tag{2.16}$$

where the constant $\alpha > 0$.

Let $m(t)$ and $D(t)$ denote the conditional mean and variance, respectively, of the process $x(t)$, given the realization $y(s)$, $0 \leqslant s \leqslant t$. Then it follows from formula (2.8) that the function $D(t)$ is determined by

$$\dot{D}(t) = 2aD(t) - D^2(t)V(t) + \sigma^2, \qquad D(0) = D_0 > 0, \tag{2.17}$$
$$V(t) = Q(t)\sigma_0^{-2}.$$

By virtue of (2.5) the optimal control $u_0(t)$ is of the form

$$u_0(t) = -bg(t)m(t), \tag{2.18}$$

where $g(t)$ is a positive function satisfying the following equations derived from equation (2.6):

$$\dot{g}(t) = -2ag(t) + b^2g^2(t), \qquad t \geqslant 0,\ g(T) = \alpha. \tag{2.19}$$

Finally, the optimal observation law, in view of (2.9), can be determined from the condition for a minimum of the functional

$$\int_0^T g^2(t)z^{-1}(t)\,dt, \tag{2.20}$$

where the function $z(t) = D^{-1}(t)$, in view of equation (2.17), satisfies the following equation

$$\dot{z}(t) = -2az(t) + V(t) - \sigma^2 z^2(t), \qquad t \geqslant 0,\ z(0) = D_0^{-1}. \tag{2.21}$$

The construction of the optimal observation law will be carried out in several steps depending on the initial parameter data of the problem.

Proposition 2.1. *Suppose that the coefficients in equations (2.13), (2.14) satisfy the conditions given above and $a \geqslant 0$. Then there exists a number $t_1 \leqslant T - T_0$ such that the optimal observation law is of the form*

$$Q_0 = \begin{cases} 1, & \text{for } t \in (t_1, t_1 + T_0], \\ 0, & \text{for } t \notin (t_1, t_1 + T_0]. \end{cases} \tag{2.22}$$

Proof. On invoking Pontryagin's maximum principle, we can show that

$$Q_0(t) = \begin{cases} 1, & \text{for } \psi(t) + C > 0, \\ 0, & \text{for } \psi(t) + C < 0, \end{cases} \tag{2.23}$$

where the constant C is chosen so that the condition (2.15) is satisfied, and the adjoint variable $\psi(t)$ is determined by

$$\dot{\psi}(t) = 2\psi(t)\big(a + z(t)\sigma^2\big) - g^2(t)z^{-2}(t), \qquad \psi(T) = 0. \tag{2.24}$$

Using the relations (2.23), (2.24), we can show that there exists only one observation interval. Let us assume the contrary, i.e., we assume that there exist several nonadjacent intervals on the segment $[0, T]$, where the function $Q_0(t)$ given by the relations (2.23) is equal to one. Let $(t_i, s_i]$ denote the ith such observation interval. We shall investigate the behavior of the solution $\psi(t)$ of the problem (1.12) on the segment $[s_i, t_{i+1}]$. Let us show that the function $\psi(t)$ is monotone decreasing on this segment. First, note that since $Q_0(t) = 1$ on the interval $t_i < t \leqslant s_i$ and $Q_0(t) = 0$ for $s_i < t \leqslant t_{i+1}$, we have by virtue of (2.23)

$$r(s_i) = \dot{\psi}(s_i) \leqslant 0. \tag{2.25}$$

Further, find an equation that is satisfied by the derivative $\dot{\psi}(t) = r(t)$ for $t \in (s_i, t_{i+1}]$. Differentiating with respect to t both sides of the equality (2.24) and noting equations (2.19), (2.21) and the equality $Q_0(t) = 0$, $s_i < t \leqslant t_{i+1}$, we obtain

$$\dot{r}(t) - 2r(t)\big(a + z(t)\sigma^2\big) = r_1(t), \tag{2.26}$$

where the expression for the function $r_1(t)$ is of the form

$$r_1(t) = 2\psi(t)\dot{z}(t)\sigma^2 - 2z^{-2}(t)g^2(t)[b^2 g(t) + \sigma^2 z(t)].$$

The last summand on the right side of this equality is negative in view of the fact that the functions $g(t)$ and $z(t)$, as noted above, are positive. Furthermore, by virtue of (2.24) the adjoint variable $\psi(t) \geqslant 0$ for all $0 \leqslant t \leqslant T$, and by virtue of (2.21) and the conditions of the proposition, the derivative $\dot{z}(t) \leqslant 0$ for $s_i < t \leqslant t_{i+1}$. Thus, for all $s_i < t \leqslant t_{i+1}$ the function $r_1(t) < 0$. Hence, in view of (2.25), (2.26), it follows that $r(t)$ is also negative for $s_i < t \leqslant t_{i+1}$. Thus, we have proved that the function $\psi(t)$ is monotone decreasing on the segment $[s_i, t_{i+1}]$, and this, by virtue of (2.23), contradicts the assumption made that $Q_0(t) \neq 0$ for $t_{i+j} < t \leqslant s_{i+j}$ $(j = 0, 1)$ and $Q_0(t) = 0$ for $s_i < t \leqslant t_{i+1}$. This contradiction proves the validity of the relation (2.22). □

Proposition 2.2. *If*

$$a < 0, \qquad 0 < D_0 \leqslant -\frac{3}{4}\frac{\sigma^2}{a},$$

then the relations (2.22) hold true, i.e., there exists a moment $t_1 \leqslant T - T_0$ such that the optimal observation law has the form

$$Q_0 = \begin{cases} 1, & \textit{for } t \in (t_1, t_1 + T_0], \\ 0, & \textit{for } t \notin (t_1, t_1 + T_0]. \end{cases}$$

Proposition 2.3. *If*

$$a < 0, \qquad -\frac{3}{4}\frac{\sigma^2}{a} < D_0,$$

then there exist two nonnegative numbers $s_1 \leqslant T_0$ and $t_2 \leqslant T - T_0$ such that the optimal observation law has the form

$$Q_0 = \begin{cases} 1, & \text{for } 0 \leqslant t \leqslant s_1 \text{ and for } t_2 < t \leqslant t_2 + T_0 - s_1, \\ 0, & \text{for other values of } t. \end{cases}$$

Proof of Proposition 2.2. Consider the solution $z_0(t)$ of equation (2.21) for $V = 0$ and $z_0(0) = D_0^{-1}$. In view of (2.21), it follows that

$$z_0(t) \leqslant z(t), \qquad 0 \leqslant t \leqslant T. \tag{2.27}$$

Next, we show that

$$z(t) \geqslant -\tfrac{4}{3} a\sigma^{-2}. \tag{2.28}$$

By virtue of (2.27) it is sufficient to show that the function $z_0(t)$ satisfies the inequality (2.28). However, it follows from (2.21) that for $z_0(0) \in (0, -2a\sigma^{-2})$ the derivative $\dot{z}_0(0) > 0$ and for $z_0(0) > -2a\sigma^{-2}$ the inequality $\dot{z}_0(0) < 0$ is true. Further, by virtue of (2.21) we obtain

$$\ddot{z}_0(t) = -2\dot{z}_0(t)\big(a + \sigma^2 z_0(t)\big).$$

Thus, for $z_0(0) \in (0, -2a\sigma^{-2})$ the function $z_0(t)$ is monotone increasing, and for $z_0(0) \geqslant -2a\sigma^{-2}$ the solution $z_0(t)$ is monotone nonincreasing; in both cases we have

$$\lim_{t\to\infty} z_0(t) = -2a\sigma^{-2}.$$

Thus, on the strength of the initial condition $z_0(0) \geqslant -4a(3\sigma^2)^{-1}$ the estimate (2.28) is established.

The subsequent proof of Proposition 2.2 is similar to that of Proposition 2.1. Namely, suppose there exist several nonadjacent observation intervals $(t_i, s_i]$, and we study the behaviour of the adjoint variable $\psi(t)$ on the interval $s_i < t \leqslant t_{i+1}$. It follows from equations (2.21), (2.24), (2.26), replacing in (2.26) the function ψ by its expression in terms of $\dot{\psi}$ given by equation (2.24), that the function $r(t) = \dot{\psi}(t)$ satisfies the equality

$$\begin{gathered} g^2(t)z^{-2}(t)\big(a + z(t)\sigma^2\big)^{-1}\big[\gamma(t) + z^{-1}(t)V_0(t)\big(3z(t)\sigma^2 - 2a\big)\big] \\ = \dot{r}(t) - r(t)\big[2a + 2z(t)\sigma^2 - \dot{z}(t)\sigma^2\big(a + z(t)\sigma^2\big)^{-1}\big], \\ V_0 = h_0\sigma_0^{-2}, \end{gathered} \tag{2.29}$$

where

$$\gamma(t) = -z(t)\sigma^2\big(4a + 3z(t)\sigma^2\big) - 2g(t)b^2\big(a + z(t)\sigma^2\big). \tag{2.30}$$

But in view of (2.28) for all $t \geqslant 0$ the function $\gamma(t) < 0$. Hence, noting the relations (2.25), (2.29), and the identity $V_0(t) \equiv 0$ which is true for $s_i < t \leqslant t_{i+1}$, we find that $\psi(t)$ is monotone decreasing on the interval $[s_i, t_{i+1}]$. Further, repeating the same argument as in the proof of Proposition 2.1, we obtain the proof of Proposition 2.2. □

Proof of Proposition 2.3. Let $\tau_0 \in [0,T]$ denote the first point for which $\dot{\psi}(\tau_0) \geqslant 0$. If there is no such point on the segment $[0,T]$, i.e., if $\dot{\psi}(t) < 0$ for all $0 \leqslant t \leqslant T$, then we immediately obtain the proof of Proposition 2.3, since in this case, in view of the fact that $\psi(t)$ is monotone decreasing, the optimal observation law is as follows: $Q_0 = 1$ for $0 \leqslant t \leqslant T_0$ and $Q_0 = 0$ for $T_0 < t \leqslant T$.

So, let such a point $\tau_0 \in [0,T]$ exist. Then, for $0 \leqslant t \leqslant \tau_0$ there can exist, by virtue of (2.23), at most one observation interval. Let us investigate the behavior of $\psi(t)$ for $\tau_0 \leqslant t \leqslant T$. From the fact that the function $\psi(t)$ is nonnegative for $0 \leqslant t \leqslant T$ and from equation (2.24) we find that

$$z(\tau_0) > -a\sigma^2. \tag{2.31}$$

Let $z_0(t,t_0)$, $t_0 \leqslant t \leqslant T$, denote the solution of equation (2.21) for $V = 0$ with initial condition $z_0(t_0,t_0) = z(t_0)$. Similar to (2.27) we have

$$z(t) \geqslant z_0(t,t_0), \qquad t_0 \leqslant t \leqslant T,$$

and, moreover, we derive in a way similar to (2.28) that

$$z_0(t,\tau_0) > -a\sigma^{-2}, \qquad t \geqslant \tau_0.$$

Thus, noting the relation (2.31), we can write

$$z(t) > -a\sigma^{-2}, \qquad \tau_0 \leqslant t \leqslant T. \tag{2.32}$$

Next, consider the function $\gamma(t)$ that enters in formula (2.29). From the point $t = 0$ until the moment at which the solution $z(t)$ of equation (2.21) first reaches the value $z(t) = -a\sigma^{-2}$ we have $\gamma(t) > 0$. From the moment of first reaching the value $z(t) = -a\sigma^{-2}$ until the moment of first reaching the value $z(t) = -4a(3\sigma^2)^{-1}$ the function $\gamma(t)$ is monotone decreasing, since in this time interval

$$\dot{\gamma}(t) = -\dot{z}\sigma^2(4a + 6z\sigma^2) - 2\dot{g}b^2(a + \sigma^2 z) - 2b^2\sigma^2 g\dot{z} < 0.$$

In the deduction of this inequality we have used the fact that the derivative $\dot{g} > 0$ for $a < 0$ by virtue of (2.19) and the derivative $\dot{z} > 0$ for $z(t) < -2a\sigma^{-2}$ by virtue of (2.21).

Finally, for values of t such that $z(t) \geqslant -4a(3\sigma^2)^{-1}$ we have $\gamma(t) < 0$. Further, as in the proof of (2.32) we find that if for some t_0 the function $z(t_0) = -2a\sigma^2$, then $z(t) \geqslant -2a\sigma^{-2}$ for $t \geqslant t_0$. Hence, noting the inequality $\dot{z}(t) > 0$, which holds for points t such that $z(t) < -2a\sigma^{-2}$, and the above-mentioned properties of the function $\gamma(t)$, we see that the function $\gamma(t)$ can have one and only one root τ_1 on the segment $[0,T]$ or else have no root at all on this segment. Moreover,

$$\gamma(t) > 0,\ 0 \leqslant t \leqslant \tau_1, \qquad \gamma(t) < 0,\ \tau_1 < t \leqslant T. \tag{2.33}$$

Let us consider alternate versions for which the relations (2.33) are valid.

First, suppose that for all $t \in [0, T]$ the following inequality is true:

$$\gamma(t) < 0. \tag{2.34}$$

Then, using the argument in the proof of Proposition 2.2, we can prove that there exist at most two observation intervals. Indeed, let (t_0, s_0) be the first observation interval whose initial point is $t_0 < s_0$. If such an observation interval did not exist, then by virtue of the inequality $\dot{\psi}(t) < 0$, $0 \leqslant t < \tau_0$, and of formula (2.23) we would only have one observation interval adjoining zero.

We can show that there are no observations at points $t > s_0$. Indeed, if there existed an observation interval (t_1, s_1), where $t_1 > s_0$, then $\dot{\psi}(s_0) \leqslant 0$. Hence, noting the relations (2.34) and (2.29) for $V = 0$, we see that, as in the proof of Proposition 2.2, the function $\psi(t)$ is monotone decreasing for $s_0 < t < t_1$. However, in view of (2.23) this contradicts the fact that (t_1, s_1) is an observation interval.

Now let

$$\gamma(t) > 0, \qquad 0 \leqslant t \leqslant T. \tag{2.35}$$

We can show that in this case no internal observation intervals can exist. Indeed, if there exists an observation interval (t_0, s_0) such that $t_0 > \tau_0$, then by virtue of (2.23) we have

$$r(t_0) = \dot{\psi}(t_0) > 0.$$

Hence, in view of (2.35) and (2.29) we obtain the inequality $r(t) > 0$, $t_0 < t$. It follows that the function $\psi(t)$ is monotone increasing for $t_0 < t$. Therefore, on the strength of (2.23) the observation interval (t_0, s_0) cannot be an internal one. Moreover, on $[0, \tau_0]$ there can exist at most one observation interval that adjoins zero. Thus, when the condition (2.35) is met, there can be no internal observation intervals, i.e., there can exist only two observation intervals adjoining the points 0 and T.

There remains just one case to be investigated, when $\gamma(t)$ changes sign on $[0, T]$, i.e., the relations (2.33) are satisfied. Suppose now that $\tau_0 \geqslant \tau_1$. The investigation in this case is made along the same lines as in the case (2.34). Repeating the corresponding argument, we find that in this case Proposition 2.3 holds.

Finally, consider the case $\tau_0 < \tau_1$. We will investigate the behavior of the adjoint variable $\psi(t)$ for $\tau_0 \leqslant t \leqslant \tau_1$. Because of formulas (2.24), (2.31) we have

$$\psi(\tau_0) \geqslant \tfrac{1}{2} g^2(\tau_0) z^{-2}(\tau_0) \big(a + z(\tau_0)\sigma^2\big)^{-1}.$$

Hence, in view of (2.24) we obtain that if $t \geqslant \tau_0$, then

$$\begin{aligned} \dot{\psi}(t) \geqslant [2a + 2z(t)\sigma^2] \bigg[\frac{1}{2} g^2(\tau_0) z^{-2}(\tau_0) \big(a + z(\tau_0)\sigma^2\big)^{-1} \mu(\tau_0) \\ - \int_{\tau_0}^{t} g^2(s) z^{-2}(s) \mu(s)\, ds \bigg] - g^2(t) b^2 z^{-2}(t), \end{aligned} \tag{2.36}$$

where

$$\mu(s) = \exp\left[\int_s^t (2a + 2z(s_1)\sigma^2)\, ds_1\right].$$

Let

$$h(t) = -2(a + z(t)\sigma^2) \int_{\tau_0}^t g^2(s) z^{-2}(s) \mu(s)\, ds.$$

Integrating by parts yields

$$\begin{aligned} h(t) &= (a + z(t)\sigma^2) \int_{\tau_0}^t g^2(s) z^{-2}(s) \left(a + z(s)\sigma^2\right)^{-1} d_s \mu(s) \\ &= g^2(t) z^{-2}(t) - (a + z(t)\sigma^2) g^2(\tau_0) z^{-2}(\tau_0) \left(a + z(\tau_0)\sigma^2\right)^{-1} \mu(\tau_0) \\ &\quad - \left(z(t)\sigma^2 + a\right) \int_{\tau_0}^t \mu(s)\, d_s \left[g^2(s) z^{-2}(s) \left(a + z(s)\sigma^2\right)^{-1}\right], \qquad t \geqslant \tau_0. \end{aligned}$$

Thus, noting the relations (2.19), (2.21), (2.31), (2.36) for $t \geqslant \tau_0$, we obtain the estimate

$$\begin{aligned} \dot{\psi}(t) &\geqslant -b^2 \left(a + z(t)\sigma^2\right) \int_{\tau_0}^t \mu(s)\, d_s \left[g^2(s) z^{-2}(s) \left(a + z(s)\sigma^2\right)^{-1}\right] \\ &\geqslant \ b^2 \left(a + z(t)\sigma^2\right) \int_{\tau_0}^t \mu(s) \left(a + z(s)\sigma^2\right)^{-2} g^2(s) z^{-2}(s) \gamma(s)\, ds. \end{aligned} \tag{2.37}$$

By virtue of (2.28), (2.30) and of the definition of τ_1 the inequality (2.23) indicates that the function $\psi(t)$ is monotone increasing on the segment $[\tau_0, \tau_1]$. But for $0 \leqslant t \leqslant \tau_0$ the function $\psi(t)$ is monotone decreasing. Hence, for $0 \leqslant t \leqslant \tau_1$ there can exist at most two observation intervals. Further, by the argument used in the proof of Proposition 2.1 and by the second of the inequalities (2.33), we obtain that if the point τ_1 belongs to the observation interval, then this observation interval is the last one; but if at the moment τ_2 there is no observation, then this means that up to the moment τ_2 there can exist only one observation interval and after this moment there can exist at most one observation interval. Thus, in both cases there exist at most two observation intervals, with the first one adjoining zero. This completes the proof of Proposition 2.3. □

In Propositions 2.1 and 2.2 the problem of finding an optimal observation law reduces to the problem of finding the minimum of a scalar function of one parameter t_1. To accomplish this, we must find the solution of equation (2.21) for $V(t) = Q_0(t)$ and substitute this solution and also the solution of equation (2.19) in the integral (2.22), which on substitution becomes a scalar function of one variable t_1.

In a similar way, optimal control is determined subject to the conditions of Proposition 2.3 by minimizing functions of two variables s_1, t_2. Optimal control for any observation law is determined by the equality (3.6) of Chapter IX.

It should be noted that for $b^2\alpha = 2a$ the solution of the problem (2.19) is constant and is equal to α. Therefore, the problem reduces to that of minimizing the integral $\int_0^T D(s)\,ds$. Thus, in this case the assertion is implied by the results of subsection 3 of § 1.

Main Results and Formulas of Chapter XII

Optimal control of the observation

motion of the system and observation process

$$dx(t) = A(t)x(t)\,dt + \sigma(t)\,d\xi(t), \qquad 0 \leqslant t \leqslant T,\ x(0) = x_0,$$
$$dy(t) = Q(t)x(t)\,dt + \sigma_0(t)\,d\xi_0(t), \qquad 0 \leqslant t \leqslant T,\ y(0) = 0;$$

observation control

$$u = Q'(\sigma_0\sigma_0')^{-1}Q;$$

observation cost functional

$$J(u) = q'x(T) \to \inf_u,$$
$$J(u) = q'D(T)q \to \inf_u;$$

pulse optimal observation law

$$\gamma(t) = h^2(t) + \sum_{i=1}^{r} \mu_i \delta(t - t_i).$$

Optimization of the observation process is equivalent to optimal control in a special deterministic system.

For the scalar observation system

$$\dot{x}(t) = ax(t), \qquad 0 \leqslant t \leqslant T,\ x(0) = x_0,$$

$$dy(t) = Q(t)x(t)\,dt, \qquad \int_0^T Q(t)\,dt = T_0,\ 0 \leqslant T_0 \leqslant T$$

there exists only one observation interval adjoining zero for $a < 0$ or T for $a > 0$, which furnishes the minimum of $D(T)$.

In other cases there exist at most two observation intervals.

EXERCISES OF PART THREE

1. Consider a sequence of scalar random variables x_i with uniformly bounded second moments which converges in mean square as $i \to \infty$ to a random quantity $\bar{x}$. Prove that

$$Ex_i \to Ex_\infty, \qquad Ex_i^2 \to Ex_\infty^2.$$

2. If the equations of motion (2.1)[1] do not contain random disturbances (i.e., $\sigma \equiv 0$), then equation (2.18) for the covariance matrix $D(t)$ is a matrix Bernoulli equation (a particular form of the Riccati equation corresponding to a zero constant term). For $D_0 > 0$ it can be integrated explicitly as in the case of (2.15), (2.17). Noting this, derive an expression for $D(t)$ in the case of scalar x and y satisfying the equations[2]

$$\begin{aligned} \dot{x}(t) &= ax(t), \qquad Ex(0) = 0,\ Ex^2(0) = D_0 > 0, \\ dy(t) &= x(t)\,dt + \sigma_0 d\xi_0(t), \qquad y(0) = 0, \end{aligned}$$

where a and $\sigma_0 \neq 0$ are given constants.

Show that

$$D(t) = \left[D_0^{-1} Z(t,0) + \sigma_0^{-2} \int_0^t Z(t,s)\,ds\right]^{-1},$$
$$Z(t,s) = \exp[2a(s-t)].$$

3. Consider the control system

$$dx(t) = \Big(f\big(x(t)\big) + u\Big)\,dt + \sigma\,d\xi(t).$$

Here $x \in R^n$, $x'f(x) = 0$, σ is a prescribed quantity, $\xi(t) \in R^n$, $u \in R^n$, where the control u is constrained by $u'u \leqslant b^2$. It is required to choose u so that the mean time of reaching the set $x'x \leqslant r$ is minimized. Prove that the optimal control u_0 in this problem is

$$u_0 = \begin{cases} 0, & \text{for } x'x \leqslant r, \\ -bx(x'x)^{-1/2}, & \text{for } x'x > r. \end{cases}$$

Find the corresponding mean time of reaching the set $x'x \leqslant r$ (the Bellman function). Note that for $n = 3$ the rigid body discussed in Chapter X serves as an example of such a system.

---

[1] In this exercise the reader is referred to formulas of Chapter XI.

[2] Here and subsequently the symbols ξ_0, ξ and ξ_i denote independent standard Wiener processes, and their derivatives denote independent standard Gaussian white noise processes.

4. Consider the two-chamber model of "medicine absorbtion and metabolism" described in Exercise 8 of Part Two. The equations of this model are of the form

$$\dot{x}_1(t) = \alpha_1 x_1 + u(t), \qquad \dot{x}_2(t) = \alpha_1 x_1 - \alpha_2 x_2, \qquad t \geqslant 0.$$

At the initial moment $t = 0$ the vector $\big(x_1(0), x_2(0)\big)$ is normally distributed with zero mean and covariance matrix D_0. The observed variable is $y(t) = x_2(t) + \sigma\xi(t)$.

Given σ and P-control $u(t)$, find the covariance matrix $D(t)$ of the estimation error for the vector (x_1, x_2) and determine its limiting values as $t \to \infty$.

5. Find the Bellman function and optimal F-control $u = u(t, x(t))$ in the linear-quadratic problem

$$\ddot{x}(t) + k_1\dot{x}(t) + k_2 x(t) = bu + \sigma\dot{\xi}, \qquad 0 \leqslant t \leqslant T,$$

$$J(u) = E\left[\alpha x^2(T) + \int_0^T u^2\, dt\right] \to \inf,$$

where b, $k_1 > 0$, $k_2 > 0$, $\alpha > 0$ are given constants.

6. Solve Exercise 3 assuming that optimal control is sought as a functional $u = u\big(t, y(t)\big)$, where $y(t) = \dot{x}(t) + \sigma_1\dot{\xi}_1$.

7. Solve Exercise 3 assuming that optimal P-control is sought as a function of time $u = u(t)$. Compare the Bellman functions obtained in Exercises 3–5.

8. Find the Bellman function and optimal control for the following scalar systems on the segment $0 \leqslant t \leqslant T$:

(a)

$$\dot{x}(t) = 1 + u + \dot{\xi},$$

$$E\int_0^T \big(x^2(t) + u^2\big)\, dt \to \inf;$$

(b)

$$\dot{x} = u(1 + \dot{\xi}),$$

$$Ex^2(T) ==\!\!\to \inf.$$

9. Given a scalar system

$$\dot{x}(t) + ax(t) = (b + \sigma\dot{\xi})u$$

with cost functional

$$J(u) = E\int_0^\infty \big(\alpha x^2(t) + u^2\big)\, dt \to \inf.$$

Show that the stabilization condition is of the form $-2\sigma^2 a < b^2$. Find optimal control and a Bellman function under this condition.

10. Find the Bellman function and optimal control in the linear-quadratic problem

$$\dot{x}_1(t) = x_2(t),$$

$$dx_2(t) = -x_2(t)\,dt + bu\big(b\,dt + \sigma\,d\xi(t)\big),$$
$$E\int_{t_0}^{T} (\alpha_1 x_1^2(t) + \alpha_2 x_2^2(t) + u^2)\,dt \to \min,$$

where b, σ, $\alpha_1 > 0$, $\alpha_2 > 0$ are prescribed constants, t_0 and T are prescribed moments.

11. Find the Bellman function and optimal control in the linear-quadratic problem for the following systems:

(a)

$$\dot{x}_1(t) = x_1(t) + x_2(t),$$
$$dx_2(t) = u\big(b\,dt + \sigma\,d\xi(t)\big);$$

(b)

$$\dot{x}_1(t) = u,$$
$$dx_2(t) = -x_1(t)\,dt + \sigma x_2(t)\,d\xi(t);$$

(c)

$$\dot{x}_1(t) = u,$$
$$dx_2(t) = -x_2(t)\big(b\,dt + \sigma\,d\xi(t)\big);$$

(d)

$$dx_1(t) = u\,dt + \sigma\,d\xi(t),$$
$$\dot{x}_2(t) = u.$$

12. Consider a Gaussian parameter θ with zero mean and variance D_0. The observed process $y(t)$ is described by equations of the form (2.2) of Chapter XI:

$$dy(t) = \theta\,dt + d\xi_0(t), \qquad t \geqslant 0,\ y(0) = 0.$$

Evaluate $D(T) = E\{\,(\theta - m(T)^2) \mid y_T\,\}$, where $m(T) = E\{\,\theta \mid y_T\,\}$ and the symbol y_T, as before, denotes the aggregate of observations $y(t)$ for $0 \leqslant t \leqslant T$.

Show that

$$D(T) = D_0[1 + D_0 T]^{-1}.$$

13. Consider a material point in rectilinear motion described by the equation $\ddot{x}(t) = 0$, $t \geqslant 0$. The random initial vector $(x(0), \dot{x}(0))'$ is Gaussian, with prescribed mean $m(0) = E(x(0), \dot{x}(0))'$ and covariance matrix $D(0)$. Evaluate $m(t)$ and $D(t)$ for $t > 0$ if

(1) there are no observations of the state of the system;

(2) there is a noisy observation of the coordinate:

$$dy(t) = x(t)\,dt + \sigma_0\,d\xi_0(t), \qquad t > 0,\ y(0) = 0;$$

(3) there is a noisy observation of the velocity:

$$dy(t) = \dot{x}(t)\,dt + \sigma_0\,d\xi_0(t), \qquad t > 0,\ y(0) = 0;$$

(4) there is a noisy observation of the coordinate and velocity:

$$dy(t) = \big(\alpha_1 x(t) + \alpha_2 \dot{x}(t)\big)\,dt + \sigma_0\,d\xi_0(t), \qquad t > 0,\ y(0) = 0.$$

Here the constants α_1, α_2, and $\sigma_0 \neq 0$ are prescribed.

Compare the covariance matrices $D(t)$ for different types of observation.

14. As in Exercise 13, investigate the same problems (1)–(4) under the condition that the equation of motion is now $\ddot{x} + kx = 0$, $t \geqslant 0$, where k is a given constant.

15. Suppose that $x(t) = x$ is a Gaussian random variable with zero mean and variance D_0, i.e., it satisfies the equation

$$dx(t) = 0, \qquad t \geqslant 0,\ x(0) = x_0 = x.$$

The observed process $y(t)$ is scalar ($m = 1$) and is described by

$$dy(t) = Qx(t)\,dt + \sigma_0\,d\xi_0(t), \qquad y(0) = y_0,$$

where Q and σ_0 are prescribed constants.

Determine the Kalman filter and solve the filtering equations for the optimal estimate $m(t)$ and the error variance $D(t)$.

16. For the problem in Exercise 15 write the stochastic differential equation in the Ito form for the conditional probability distribution $P(t, x)$ and determine the corresponding equation in the symmetrized form.

Hint. Perform the appropriate transformation (see Chapter II) for the non-normalized conditional probability measure $\tilde{P}(t, x)$ satisfying an equation obtained from equation (3.4) of Chapter XI by discarding terms with $\bar{Q}$. Next, write the symmetrized equation for the normalized conditional probability distribution $P(t, x) = C(t)\tilde{P}(t, x)$, where $C(t)$ is the normalization constant.

17. Suppose that $x(t) = x$ is a Gaussian random variable with zero mean and variance D_0, i.e., it satisfies the equation

$$dx(t) = 0, \qquad t \geqslant 0,\ x(0) = x_0 = x.$$

The observed process $y(t)$ is two-dimensional and is described by

$$dy(t) = Qx(t)\,dt + \sigma_0\,d\xi_0(t), \qquad y(0) = y_0,$$

where Q and σ_0 are the prescribed constant matrices:

$$Q = \begin{pmatrix} q_1 \\ q_2 \end{pmatrix}, \qquad \sigma_0 = \begin{pmatrix} \sigma_1 & 0 \\ 0 & \sigma_2 \end{pmatrix},$$

where q_i and σ_i $(i = 1, 2)$ are prescribed constants.

Determine the Kalman filter and solve the filtering equations for the optimal estimate $m(t)$ and the error variance $D(t)$.

18. For the filtering problem in which the unobserved variable is the same as in Exercise 17 and the observed process is described by equation (3.2) of Chapter XI, find the stochastic differential equation in the Ito form for the conditional probability distribution $P(t, x)$ and determine the corresponding equation in the symmetrized form. Write this equation for the particular case of $y(t)$ given in Exercise 17.

See *Hint* in Exercise 16.

19. Suppose that $x(t)$ is a two-dimensional Gaussian process satisfying the equations

$$\begin{aligned} dx_1 &= x_2\,dt, \qquad t \geqslant 0,\ x_1(0) = 0, \\ dx_2 &= -ax_2\,dt + \sigma\,d\xi_2(t), \qquad a > 0,\ x_2(0) = 0. \end{aligned}$$

The observed process $y(t)$ is described by

$$dy(t) = Qx_1(t)\,dt + \sigma_0\,d\xi_0(t), \qquad y(0) = y_0,$$

where Q and σ_0 are prescribed constants.

Determine the Kalman filter and solve the filtering equations for the optimal estimate $m(t)$ and the error covariance matrix $D(t)$.

20. Suppose that $x(t)$ is a two-dimensional Gaussian process satisfying the equations

$$\begin{aligned} dx_1 &= x_2\,dt, \qquad t \geqslant 0,\ x_1(0) = 0, \\ dx_2 &= -ax_2\,dt + \sigma\,d\xi_2(t), \qquad a > 0,\ x_2(0) = 0. \end{aligned}$$

The observed process $y(t)$ is described by

$$dy(t) = Qx_2(t)\,dt + \sigma_0\,d\xi_0(t), \qquad y(0) = y_0,$$

where Q and σ_0 are prescribed constants.

Determine the Kalman filter and solve the filtering equations for the optimal estimate $m(t)$ and the error covariance matrix $D(t)$.

21. Suppose that $x(t)$ is a two-dimensional Gaussian process satisfying the equations

$$dx_1 = -a_1 x_1\,dt + \sigma\,d\xi_1(t), \qquad t \geqslant 0,\ a_1 > 0,\ x_1(0) = 0,$$
$$dx_2 = -a_2 x_2\,dt + dx_1, \qquad a_2 > 0,\ x_2(0) = 0.$$

The observed process $y(t)$ is described by

$$dy(t) = Qx_2(t)\,dt + \sigma_0\,d\xi_0(t), \qquad y(0) = y_0,$$

where Q and σ_0 are prescribed constants.

Determine the Kalman filter and solve the filtering equations for the optimal estimate $m(t)$ and the error covariance matrix $D(t)$.

22. Suppose that $x(t)$ is a two-dimensional Gaussian process satisfying the equations

$$dx_1 = x_2\,dt, \qquad t \geqslant 0,\ x_1(0) = 0,$$
$$dx_2 = -ax_2\,dt + \sigma\,d\xi_2(t), \qquad a > 0,\ x_2(0) = 0.$$

The observed process $y(t)$ is described by

$$dy(t) = (Q_0 + Qx_1)\cos\omega_0 t\,dt + \sigma_0\,d\xi_0(t), \qquad y(0) = y_0,$$

where Q and σ_0 are prescribed constants.

Determine the quasilinear filter and solve the filtering equations for the optimal estimate $x^*(t)$ and the error covariance matrix $D(t)$.

23. Suppose that $x(t)$ is a two-dimensional Gaussian process satisfying the equations

$$dx_1 = x_2\,dt, \qquad t \geqslant 0,\ x_1(0) = 0,$$
$$dx_2 = -ax_2\,dt + \sigma\,d\xi_2(t), \qquad a > 0,\ x_2(0) = 0.$$

The observed process $y(t)$ is described by

$$dy(t) = (Q_0 + Qx_2)\cos\omega_0 t\,dt + \sigma_0\,d\xi_0(t), \qquad y(0) = y_0,$$

where Q and σ_0 are prescribed constants.

Determine the quasilinear filter and solve the filtering equations for the optimal estimate $x^*(t)$ and the error covariance matrix $D(t)$.

24. Suppose that $x(t)$ is a two-dimensional Gaussian process satisfying the equations

$$dx_1 = -a_1 x_1\,dt + \sigma\,d\xi_1(t), \qquad t \geqslant 0,\ a_1 > 0,\ x_1(0) = 0,$$
$$dx_2 = -a_2 x_2\,dt + dx_1, \qquad a_2 > 0,\ x_2(0) = 0.$$

The observed process $y(t)$ is described by

$$dy(t) = (Q_0 + Qx_2)\cos\omega_0 t\,dt + \sigma_0\,d\xi_0(t), \qquad y(0) = y_0,$$

where Q and σ_0 are prescribed constants.

Determine the quasilinear filter and solve the filtering equations for the optimal estimate $x^*(t)$ and the error covariance matrix $D(t)$.

25. Suppose that $x(t)$ is a two-dimensional Gaussian process satisfying the equations

$$dx_1 = x_2\,dt, \qquad t \geqslant 0,\ x_1(0) = 0,$$
$$dx_2 = -ax_2\,dt + \sigma\,d\xi_2(t), \qquad a > 0,\ x_2(0) = 0.$$

The observed process $y(t)$ is described by

$$dy(t) = Q\cos(\omega_0 t + x_2)\,dt + \sigma_0\,d\xi_0(t), \qquad y(0) = y_0,$$

where Q and σ_0 are prescribed constants.

Determine the quasilinear filter and solve the filtering equations for the optimal estimate $x^*(t)$ and the error covariance matrix $D(t)$.

26. Suppose that $x(t)$ is a two-dimensional Gaussian process satisfying the equations

$$dx_1 = -a_1x_1\,dt + \sigma\,d\xi_1(t), \qquad t \geqslant 0,\ a_1 > 0,\ x_1(0) = 0,$$
$$dx_2 = -a_2x_2\,dt + dx_1, \qquad a_2 > 0,\ x_2(0) = 0.$$

The observed process $y(t)$ is described by

$$dy(t) = Q\cos(\omega_0 t + x_2)\,dt + \sigma_0\,d\xi_0(t), \qquad y(0) = y_0,$$

where Q and σ_0 are prescribed constants.

Determine the quasilinear filter and solve the filtering equations for the optimal estimate $x^*(t)$ and the error covariance matrix $D(t)$.

27. Suppose that $x(t)$ is a scalar Gaussian process satisfying the equation

$$dx = -ax\,dt + \sigma\,d\xi(t), \qquad t \geqslant 0,\ a > 0,\ x(0) = 0,$$

The observed process $y(t)$ is two-dimensional and is described by

$$dy(t) = Qx(t)\,dt + \sigma_0\,d\xi_0(t), \qquad y(0) = y_0,$$

where Q and σ_0 are the prescribed constant matrices:

$$Q = \begin{pmatrix} q_1 \\ q_2 \end{pmatrix}, \qquad \sigma_0 = \begin{pmatrix} \sigma_1 & 0 \\ 0 & \sigma_2 \end{pmatrix},$$

where q_i and σ_i $(i = 1, 2)$ are prescribed constants.

Determine the Kalman filter and solve the filtering equations for the optimal estimate $m(t)$ and the error variance $D(t)$.

28. Suppose that $x(t)$ is a two-dimensional Gaussian process satisfying the equation

$$\begin{aligned} dx_1 &= -a_1 x_1\, dt + \sigma_1\, d\xi_1(t), \qquad && t \geqslant 0,\ a_1 > 0,\ x_1(0) = 0, \\ dx_2 &= -a_2 x_1\, dt + \sigma_1\, d\xi_2(t), \qquad && a_2 > 0,\ x_2(0) = 0, \end{aligned}$$

The observed process $y(t)$ is also two-dimensional and is described by

$$dy(t) = Qx(t)\, dt + \sigma_0\, d\xi_0(t), \qquad y(0) = y_0,$$

where Q and σ_0 are the prescribed constant matrices:

$$Q = \begin{pmatrix} q_1 \\ q_2 \end{pmatrix}, \qquad \sigma_0 = \begin{pmatrix} \sigma_1 & 0 \\ 0 & \sigma_2 \end{pmatrix},$$

where q_i and σ_i $(i = 1, 2)$ are prescribed constants.

Determine the Kalman filter and solve the filtering equations for the optimal estimate $m(t)$ and the error covariance matrix $D(t)$.

29. Suppose that $x(t)$ is a two-dimensional Gaussian process satisfying the equation

$$\begin{aligned} dx_1 &= -a_1 x_1\, dt + \sigma_1\, d\xi_1(t), \qquad && t \geqslant 0,\ a_1 > 0,\ x_1(0) = 0, \\ dx_2 &= -a_2 x_1\, dt + \sigma_1\, d\xi_2(t), \qquad && a_2 > 0,\ x_2(0) = 0, \end{aligned}$$

The observed process $y(t)$ is described by

$$dy(t) = (Q_0 + x_1)\cos(\omega_0 t + x_2)\, dt + \sigma_0\, d\xi_0(t), \qquad y(0) = y_0,$$

where Q and σ_0 are prescribed constants.

Determine the quasilinear filter and solve the filtering equations for the optimal estimate $x^*(t)$ and the error covariance matrix $D(t)$.

30. Let the motion of the system be described by the scalar equation

$$\dot{x}(t) = k(t)u, \qquad 0 \leqslant t \leqslant T,\ x(0) = x_0,$$

and the observation process is of the form

$$dy(t) = Q(u)x(t)\, dt + \sigma(t)\, d\xi_0(t).$$

Here $k(t)$ and $\sigma(t)$ are prescribed continuous positive functions, the control u for each moment t is constructed as a function of observations $y(s)$ over the interval $0 \leqslant s \leqslant t$. It is assumed that $u \in U$, where the set U consists of three points: $U = (-1, 0, 1)$, and the function $Q(u)$ is determined by the relations $Q(0) = 1$, $Q(1) = Q(-1) = 0$.

Determine the control u of the motion of the system x and of the observation process y so that $E\{\, x^2(T) \mid y_T \,\}$ is minimized. (For a detailed study of this problem, see [4])

31. Consider the scalar system

$$dx(t) = [-ax(t) + bu]\,dt + a_0 x(t)\,d\xi_0(t) + \sigma u\,d\xi(t), \qquad t > 0,\ x(0) = x_0.$$

where a, b, a_0, σ are prescribed constants. It is required to determine a linear control $u = cx$ so as to minimize the functional

$$E\int_0^\infty \left(\alpha x^2(t) + u^2\right) dt,$$

where $\alpha > 0$ is a constant.

Prove that a necessary and sufficient condition for the solvability of this problem is of the form

$$\left(\frac{b}{\sigma}\right)^2 > -a + \left[a^2 + \left(\frac{b}{\sigma}\right)^2 a_0^2\right]^{1/2}.$$

32. The simplified equations of gyro drift are of the form

$$\dot{x}_1(t) = x_2(t) - x_3(t), \qquad \dot{x}_2 = 0,$$
$$\dot{x}_3(t) = -\tfrac{1}{\tau}x_3(t) + \sigma\sqrt{\tfrac{1}{\tau}}\,\xi, \qquad t > 0,\ x_1(0) = 0,$$

where x_1 is the drift angle, x_2 is the constant drift, and x_3 is the random drift. The *a priori* distribution of the random vector $(x_2(0), x_3(0))$ is Gaussian with known parameters, and let the drift angle $x_1(t)$ be measured accurately.

Solve the equations of the Kalman filter for the estimates of the unknowns $x_2(t)$ and $x_2(t)$.

33. Consider the problem of maximizing the mean passage time to the boundary for the following example of a controlled scalar system

$$\dot{x}(t) = u + \dot{\xi}(t), \qquad x(0) \in [-1, 1],\ |u| \leqslant 1.$$

Let $\tau_x(u)$ denote the moment at which the trajectory of the system first reaches the boundaries of the segment $[-1, 1]$, given the control u and the initial condition $x(0) = x \in [-1, 1]$, and let $V(x)$ denote the corresponding Bellman function

$$V(x) = \sup_{u,\,|u|\leqslant 1} E\tau_x(u).$$

The Bellman equation is of the form

$$\frac{d^2V(x)}{dx^2} + \sup_{u,\,|u|\leqslant 1} u\frac{dV(x)}{dx} = -1, \qquad V(-1) = V(1) = 0.$$

Verify that the symmetric function $V(x)$ equal to

$$V(x) = V(-x) = x - 1 + e - e^x, \qquad 0 \leqslant x \leqslant 1,$$

satisfies the Bellman equation and that the optimal control $u_0(x)$ is of the form $u_0(x) = -\operatorname{sgn} x$.

34. Suppose that a scalar process $x(t)$ and the observation of this process is described by the equations

$$dx(t) = ax(t)\,dt + d\xi(t), \qquad t > 0,\ x(0) = x_0,$$
$$dy(t) = x(t)\,dt + d\xi_0(t),$$

where x_0 is a Gaussian variable with $Ex(0) = m_0$ and $E\big(x(0) - m_0\big)^2 = D_0$.

Show that the estimation error variable described by an equation of the form (2.18) of Chapter XI is equal to

$$D(t) = (\beta + a) + 2\beta\left[\frac{D_0 + \beta - a}{D_0 - \beta - a}e^{2\beta t} - 1\right]^{-1},$$

where $\beta = \sqrt{1 + a^2}$.

35. Consider the scalar linear control system

$$dx(t) = bu\,dt + \sigma\,d\xi(t), \qquad t \geqslant 0,\ x(0) = x_0,$$

with quadratic cost functional

$$J = E\int_0^T \big(gx^2(t) + x^2\big)\,dt \to \min_u .$$

The constants $g > 0$, b, and σ are prescribed.

Show that

(1) the optimal control u_0 is equal to

$$u_0 = -bP(t)x,$$

where

$$P(t) = g\big(1 - e^{-2\beta(T-t)}\big)\left[\beta + \beta e^{-2\beta(T-t)}\right]^{-1},$$
$$\beta = \sqrt{gb^2};$$

(2) the Bellman function $V(t, x)$ is

$$V(t,x) = P(t)x^2 + \sigma^2\int_t^T P(s)\,ds.$$

PART FOUR

NUMERICAL METHODS IN CONTROL SYSTEMS

CHAPTER XIII

LINEAR TIME-INVARIANT CONTROL SYSTEMS

In this chapter we are concerned with certain methods of numerical investigation of linear time-invariant systems. Methods used to determine the stability of closed-loop control systems, to construct Lyapunov functions and stabilizing control are presented as well as methods of numerical investigation of the controlability and observability of time-invariant systems and of solving time-invariant problems of optimal stabilization. The methods discussed are based on computational methods of linear algebra.

§ 1. Stability of Linear Time-Invariant Systems

For linear control systems of the form

$$\dot{x}(t) = A_1 x(t) + B_1 u(t), \qquad x \in R^n,\ u \in R^m, \tag{1.1}$$

the control law is often chosen as a linear law $u = -Kx$. This is implemented by a so-called *proportional controller* or *P-controller*. The system (1.1) closed by means of a P-controller is of the form

$$\dot{x}(t) = (A_1 - B_1 K)x = Ax. \tag{1.2}$$

One of the main requirements on a P-controller is the requirement of the stability of the closed-loop system (1.2).

The investigation of stability of a linear time-invariant system (1.2) reduces to the study of the location of the eigenvalues of the matrix A or the application of the Routh–Hurwitz criterion (see § 5 of Chapter I).

The determination of all eigenvalues (the spectrum) of a general nonsymmetric matrix A constitutes the most complicated problem of computational linear algebra. This problem is called the *complete eigenvalue problem.* To solve this problem, two groups of methods have been developed: precise methods and iterative procedures.

1. Precise methods for solving the complete eigenvalue problem. The determination of eigenvalues of a square matrix A through the use of precise methods is carried out in two steps:

(1) development of the characteristic equation

$$\det|A - \lambda I| = 0, \tag{1.3}$$

i.e., finding the polynomial

$$p(\lambda) = (-1)^n[\lambda^n - p_1\lambda^{n-1} - \cdots - p_n] = 0 \tag{1.4}$$

whose roots are all eigenvalues of the matrix A;

(2) determination of all roots of the algebraic equation (1.4).

The development of the determinant (1.3) can be carried out by methods due to Leverrier, Krylov, Danilevski, Faddeev, and others [6, 9, 23].

The *Krylov method* is based on the use of the Hamilton-Cayley theorem, according to which any matrix A satisfies its own characteristic equation (1.4). In other words,

$$A^n - p_1A^{n-1} - p_2A^{n-2} - \cdots - p_nI = 0. \tag{1.5}$$

Take an arbitrary vector $y^0 \in R^n$ ($y^0 \neq 0$) and multiply by it the equality (1.5):

$$A^ny^0 - p_1A^{n-1}y^0 - \cdots - p_ny^0 = 0. \tag{1.6}$$

Denoting $y^k = A^ky^0 = Ay^{k-1}$, we write the equality (1.6) in the form

$$y^n - p_1y^{n-1} - \cdots - p_ny^0 = 0. \tag{1.7}$$

The relation (1.7) is a system of linear algebraic equations in the unknowns $p_1, \ldots, p_n$. Solving this system by any method (such as the elimination procedure of Gauss [12, 15, 21, 23, 24]) leads to the expression for the characteristic equation (1.4). Next, it is necessary to solve the algebraic equation (1.4) (using the method of parabolas [9, 14], the Bairstow method [6, 18] or other methods [5, 15, 24]).

When using the Krylov method, we may encounter specific cases in which the vectors $y^n, y^{n-1}, \ldots, y^0$ turn out to be linearly dependent, and the system (1.7) has no unique solution. In such cases the Krylov method can also be used, but the algorithm proves to be complicated. It is simpler to take another vector y^0 and repeat the computation. If several such attempts lead to no result, then it is better to use some other method.

Example 1.1. Consider the matrix

$$A = \begin{pmatrix} -1 & 0.01 \\ 0 & -0.99 \end{pmatrix}. \tag{1.8}$$

The eigenvalues of this triangular matrix are, obviously, equal to $\lambda_1 = -1$, $\lambda_2 = -0.99$. The characteristic equation is of the form

$$\lambda^2 + 1.99\lambda + 0.99 = 0, \quad \text{i.e., } p_1 = -1.99;\ p_2 = -0.99.$$

We shall try to find the eigenvalues of the matrix A with the help of the Krylov method. We take $(y^0)' = (1, 1)$ and compute $(y^1)' = (-0.99, -0.99)$, $(y^2)' = (-0.9801, -0.9801)$. Then the system (1.7) becomes

$$-0.99p_1 + p_2 = 0.9801,$$
$$-0.99p_1 + p_2 = 0.9801$$

and is, clearly, degenerate. Further application of the Krylov method is not possible. However, taking $(y^0)' = (0,1)$, we find that $(y^1)' = (0.01, -0.99)$, $(y^2)' = (0.0199, 0.9801)$. The system (1.7) becomes

$$\begin{aligned} 0.01p_1 &= -0.0199, \\ -0.99p_1 + p_2 &= 0.9801. \end{aligned}$$

The solution of this system $p_1 = -1.99$, $p_2 = -0.99$ coincides exactly with the coefficients of the characteristic equation.

The *Leverrier method* is based on Newton's formulas (that generalize Viète's formulas to arbitrary polynomials), according to which the coefficients p_i of the polynomial (1.3) and its roots λ_i are related by

$$p_1 = s_1, \qquad p_2 = \tfrac{1}{2}(s_2 - p_1 s_1), \ \ldots, \ p_n = \tfrac{1}{n}(s_n - p_1 s_{n-1} - p_{n-1}s_1), \tag{1.9}$$
$$s_i = \lambda_1^i + \lambda_2^i + \cdots + \lambda_n^i.$$

Furthermore, it is known that

$$\begin{aligned} s_1 &= \lambda_1 + \cdots + \lambda_n = \operatorname{tr} A, \\ s_2 &= \lambda_1^2 + \cdots + \lambda_n^2 = \operatorname{tr} A^2, \\ &\cdots\cdots\cdots\cdots\cdots\cdots \\ &\cdots\cdots\cdots\cdots\cdots\cdots \\ s_n &= \lambda_1^n + \cdots + \lambda_n^n = \operatorname{tr} A^n. \end{aligned}$$

Thus, the Leverrier method employed for developing a characteristic equation requires the computation of the powers $A^2, \ldots, A^n$ of the matrix A, and of their traces, and the determination of the coefficients according to formulas (1.9).

The Leverrier method is more laborious than the Krylov method, since the computation of the powers $A^2, \ldots, A^n$ requires a number of operations of the order of n^4. At the same time it is insensitive to the disposition of the spectrum (of eigenvalues) of the matrix A.

All precise methods are very economical; when used to determine the eigenvalues of a matrix of order not higher than 10, about 10^3–10^4 operations of multiplication and division are required. For a medium-power computer the evaluation of all eigenvalues of a matrix of order 10 takes 15–30 sec. This is acceptable from an applications point of view.

In [25] the program BASMAT is described which for a given matrix A of order $n \leqslant 10$ enables the determination of its characteristic polynomial, of all eigenvalues λ_i, and of the transition matrix of the system. In this program the Leverrier method is used to develop the characteristic equation and the Bairstow method to find the roots of the corresponding algebraic equation.

Unfortunately, *all precise methods are not stable to a sufficient degree with respect to computational errors (are numerically unstable). Small computational errors may lead to large errors in results.* A few illustrations will make this clear.

Example 1.2. Let us take $(y^0)' = (1, 0.99)$ for the matrix (1.8). Then the system (1.7) becomes

$$\begin{aligned} -0.9901p_1 + p_2 &= 0.980299, \\ -0.9801p_1 + 0.99p_2 &= 0.970299. \end{aligned} \tag{1.10}$$

Suppose that instead of this system as the result of rounding off intermediate and final results we obtain the system

$$\begin{aligned} -0.99p_1 + p_2 &= 0.9800, \\ -0.98p_1 + 0.99p_2 &= 0.9702. \end{aligned} \tag{1.11}$$

The errors in the coefficients of the system (1.11) compared to the system (1.10) do not exceed 0.03%. The system (1.10) has the solution $p_1 = -1.99$, $p_2 = -0.99$,and the system (1.11) has the solution $p_1 = 0$, $p_2 = 0.98$. Thus, the error is about 200%. This very unpleasant result can be explained by the fact that the matrix of the system (1.10) is ill-conditioned. (For more details on matrix conditions, see [5, 12, 15, 21, 23, 24].)

If we continue to determine eigenvalues in this example, we obtain the characteristic equation $\lambda^2 - 0.98 = 0$ and the eigenvalues $\lambda_{1,2} \approx \pm 0.99$. Thus, because of the insignificant errors (of the order of 0.03%) due to the rounding-off of the coefficients, we come to the wrong conclusion that a stable matrix (1.8) (having a good stability margin, at that) is unstable.

We could well imagine that the kind of instability demonstrated above is connected with the need to solve a system of linear differential equations. However, all precise methods involve a second step, namely the solution of an algebraic equation. As the following example shows, the latter problem is also numerically unstable.

Example 1.3. Consider the polynomial

$$p(\lambda) = (\lambda - 1)(\lambda - 2)\dots(\lambda - 19)(\lambda - 20) = \lambda^{20} - 210\lambda^{19} + \dots. \tag{1.12}$$

Suppose that only one of the coefficients of this polynomial is in error, namely instead of the coefficient $p_1 = 210$ of λ^{19} we have obtained $\tilde{p}_1 = 210 + 2^{-23}$. J. Wilkinson [24] has computed the roots of the equation

$$p(\lambda) + 2^{-23}\lambda^{19} = 0$$

to 90 binary digits. Ten roots have become complex with sufficiently large imaginary part, for example, the roots $\lambda_{16} = 16$, $\lambda_{17} = 17$ have gone over into the roots $\tilde{\lambda}_{16,17} \approx 16.731 \pm 2.813i$, and the error in the real root $\lambda_{20} = 20$ has reached almost 5%, with $lam_{20} = 20.847\dots$. It should be stressed that such values $\tilde{\lambda}_i$ have resulted not from computational errors but rather from the actual high sensitivity of the roots of the algebraic polynomial $p(\lambda)$ to the variation of its coefficients.

2. Iterative methods. For dimensions $n > 10$ precise methods, as a rule, do not lead to satisfactory results, and we have to use iterative methods.

The first iterative method used to solve the complete eigenvalue problem for symmetric (or Hermitian) matrices was proposed by C. Jacobi in 1846 before precise methods had been created. However, since the Jacobi method is significantly more labor-consuming (by an order of magnitude compared to a precise method), it had not been applied until computers began to be used for computational purposes. The use of computers led to less stress being laid on the labor content but rather on numerical stability and computational reliability. Wide application of the Jacobi method began since 1960. It has turned out that the Jacobi rotational method possesses good numerical stability and it can be used to determine all eigenvalues of symmetric matrices up to order n=50–100, with about $30n^3$ arithmetic operations.

The idea behind the Jacobi method is to transform the original matrix A to a nearly diagonal form with off-diagonal elements of order $O(\varepsilon)$ by means of orthogonal similarity transformations (rotations). Then eigenvalues practically coincide with diagonal elements, i.e., $\lambda_i = a_{ii} + O(\varepsilon^2)$.

However, for systems of the form (1.2) there is no reason to confine ourselves to only symmetric matrices A; hence, the Jacobi method cannot be considered a general-purpose method for investigating the stability of systems (1.2). For a more detailed discussion of the Jacobi method, see [3, 4, 6, 19].

A numerically stable method for solving the complete eigenvalue problem for nonsymmetric matrices called the *QR-algorithm* was proposed by V. N. Kublanovskaya in 1960 and independently by J. Francis in 1961. The QR-algorithm, just as the Jacobi method, is based on the use of orthogonal transformations. The advantage of orthogonal transformations is in that they do not increase the norms of computational errors.

Let us describe the QR-algorithm in more detail. Let there be given a real matrix A with elements a_{ij}, i.e., $A = (a_{ij})$, $i, j = 1, \dots, n$. The basic idea of the QR-algorithm is to find a similarity orthogonal transformation for the matrix A, $U'AU = S$, such that the matrix S takes the so-called Schur form. A matrix S is said to have the *Schur form* if it is a block-triangular matrix of the form

$$S = \begin{pmatrix} S_{11} & S_{12} & \dots & \dots & S_{1n} \\ 0 & S_{22} & \dots & \dots & S_{2n} \\ \dots & \dots & \dots & \dots & \dots \\ \dots & \dots & \dots & \dots & \dots \\ 0 & \dots & \dots & 0 & S_{nn} \end{pmatrix}.$$

Moreover, the diagonal matrices S_{ii} must be of order 1 or 2. The spectrum of the matrix S (and, consequently, of the similar matrix A) coincides with the union of the spectra of the matrices S_{ii}. If the matrix S_{ii} is of order one, i.e., $S_{ii} = (s_{11}^i)$, then its eigenvalue is $\lambda_i = s_{11}^i$. But if the matrix S_{jj} is of order two, i.e.

$$S_{jj} = \begin{pmatrix} s_{11}^j & s_{12}^j \\ s_{21}^j & s_{22}^j \end{pmatrix},$$

then its eigenvalues can be found from the quadratic equation

$$\lambda^2 - \lambda(s_{11}^j + s_{22}^j) + (s_{11}^j s_{22}^j - s_{21}^j s_{12}^j) = 0.$$

In this case the eigenvalues may turn out to be complex-valued.

Transformation of the matrix A to the Schur form S is carried out in several steps. First, the matrix A is decomposed as

$$A = QR, \tag{1.13}$$

where Q is an orthogonal matrix (i.e., $Q' = Q^{-1}$), and R is an upper triangular matrix. Such a decomposition can be carried out in a number of ways. In one approach the orthogonal matrix Q is expressed as a product of $n-1$ elementary reflections H_i. The matrix H_1 is taken as

$$H_1 = E - 2\beta^2 uu', \qquad 2\beta^2 = \frac{1}{\gamma(\gamma - a_{11})},$$

$$u = (a_{11} - \gamma, a_{21}, \ldots, a_{n1})', \qquad \gamma = (a_{11}^2 + a_{21}^2 + \cdots + a_{n1}^2)^{1/2}.$$

It is readily verified that

$$H_1 = H_1' = H_1^{-1}, \qquad H_1 a = \gamma e_1,$$

$$a = \begin{pmatrix} a_{11} \\ a_{21} \\ \vdots \\ a_{n1} \end{pmatrix}, \qquad e_1 = \begin{pmatrix} 1 \\ 0 \\ \vdots \\ 0 \end{pmatrix}.$$

Thus, H_1 is an orthogonal matrix, and, moreover, the elements of the first column of the matrix $A_1 = H_1 A$ located below the principal diagonal are zero.

Further, we find the matrix $A_2 = H_2 A_1$, where H_2 is of the form

$$H_2 = \begin{pmatrix} 1 & 0 \\ 0 & \hat{H}_2 \end{pmatrix},$$

and $\hat{H}_2$ is the matrix of reflections, the same as H_1, but of order $n-1$. Moreover, the elements of the first two columns of the matrix A_2 located below the principal diagonal are zero. Continuing this process, we construct the matrices

$$A_3 = H_3 A_2, \ldots, A_{n-1} = H_{n-1} A_{n-2} = H_{n-1} \ldots H_2 H_1 A = R. \tag{1.14}$$

Here R is the upper triangular matrix. It follows from the equalities (1.14) and the orthogonality of matrices H_i that

$$A = H_1 H_2 \ldots H_{n-1} R = QR, \qquad Q = H_1 H_2 \ldots H_{n-1}.$$

The desired decomposition (1.13) of the matrix A is constructed.

In the second step the sequence of matrices $A^{(k)}$ is constructed according to the formulas

$$A^{(1)} = RQ, \quad A^{(1)} = Q_1 R_1, \quad A^{(2)} = R_1 Q_1, \dots .$$

The matrices $A^{(k)}$ are orthogonally similar to A, since, for example,

$$QA^{(1)}Q^{-1} = QRQQ^{-1} = QR = A.$$

It can be shown that the *sequence* $A^{(k)}$ *converges in form to some Schur matrix* S. Precise definitions and a proof can be found in [14].

To ensure convergence of this process, we proceed as follows. Let a small number ε be given, and as soon as some element $A^{(k)}$ located below the principal diagonal becomes less than ε, it is replaced by zero. This leads to finite duration of the process of reduction to the Schur form, and $A^{(l)} = S$ for some l. Here the orthogonal matrix U that reduces A to S is defined by the relation

$$U = Q \leftrightarrow Q_1 Q_2 \dots Q_{l-1}.$$

The described version of the QR-algorithm has low degree of convergence. To enhance convergence, shifts are introduced in the QR-algorithm. For more details on QR-algorithms and rules of choosing shifts, see [23, 25].

In [25] programs are presented which implement the Jacobi rotational method for symmetric matrices, the QR-algorithm and QL-algorithm, and also the QR-algorithm with a shift.

Example 1.4. Let the matrix A be of the form

$$A = \begin{pmatrix} -10 & -1 & -2 & -3 & -4 \\ -1 & -9 & 1 & -2 & 3 \\ -2 & 1 & -7 & -3 & 5 \\ -3 & -2 & -3 & -12 & 1 \\ -4 & 3 & 5 & 1 & -15 \end{pmatrix}.$$

Its eigenvalues have been computed in [25] to 39 binary digits after the decimal point, using a computer, by means of the QR-algorithm. After 8 iterations the following eigenvalues were obtained: $\lambda_1 = -1.655266620775$, $\lambda_2 = -6.99483783064$, $\lambda_3 = -9.36555492016$, $\lambda_4 = -15.8089207645$, $\lambda_5 = -19.1754202773$. Thus, the matrix A is stable. The errors in determining the eigenvalues are of the order of $40 \cdot 10^{-11}$.

3. The Routh–Hurwitz criterion. In a number of cases transfer functions of the form $W(p) = N(p)/D(p)$, where $N(p)$, $D(p)$ are polynomials, are used to describe linear dynamical SISO-systems. It is necessary and sufficient for a system with transfer function $W(p) = N(p)/D(p)$ to be stable that the roots of the polynomial $D(p)$ lie in the left half-plane. This can be verified by means of the Routh–Hurwitz criterion (see § 5 of Chapter I).

In [7] two programs are presented which can be used to determine whether the roots of the polynomial $D(p)$ lie in the left half-plane or not. A program implementing the Mikhailov criterion as well as a program for determining the stability region in the parameter space for linear systems is also given.

4. The Zubov method of a functional transformation of a matrix. The analysis of stability of the matrix A reduces to either finding all eigenvalues or developing the characteristic equation and using the Routh–Hurwitz criterion. These methods are always labor-consuming and not always numerically stable.

In 1959 V. I. Zubov proposed a method for determining stability that is not connected with eigenvalues. The idea behind the *Zubov method* can be described in the following way. Let all eigenvalues of the matrix A be located in the left half-plane of the complex λ-plane. Consider the mapping of the λ-plane on the ρ-plane according to the formulas

$$\rho = \frac{\lambda+1}{\lambda-1}, \qquad \lambda = \frac{\rho+1}{\rho-1}. \tag{1.15}$$

The linear-fractional transformation (1.15) maps the left half-plane of the λ-plane onto the interior of the unit circle with center at the origin of the complex ρ-plane, and the right half-plane of the λ-plane onto the exterior of the unit circle. The imaginary axis is mapped onto the unit circle. Substituting λ from (1.15) in the equation $\det|A - \lambda I| = 0$, we find that ρ is a root of the equation

$$\begin{aligned} \det\left|A - \frac{\rho+1}{\rho-1} I\right| &= \det|\rho A - A - \rho I - I| \\ &= \det|-A - I - \rho(I - A)| = 0. \end{aligned} \tag{1.16}$$

Multiplying equation (1.16) by $\det|(I-A)^{-1}| \neq 0$, we see that ρ is an eigenvalue of the matrix B:

$$\det|B - \rho I| = 0, \qquad B = (-A - I)(I - A)^{-1}.$$

The matrix B can also be written as follows:

$$B = (-A - I + I - I)(I - A)^{-1} = I - 2(I - A)^{-1}. \tag{1.17}$$

The matrix B is associated with the matrix A under the transformation (1.15). Further, as is known [1, 5, 6], the matrix B all of whose eigenvalues lie inside the unit circle satisfies the relation

$$\lim_{k\to\infty} B^k = 0. \tag{1.18}$$

We can now describe the algorithm for analyzing stability which incorporates the Zubov method.

1°. The matrix B is constructed according to formula (1.17). If $(I - A)^{-1}$ does not exist, then one of the eigenvalues of $I - A$ is zero, and hence one of the eigenvalues of A is equal to 1, i.e., the matrix A is unstable.

2°. The matrices $B^2 = B \cdot B$, $B^4 = B^2 \cdot B^2$, $B^8 = B^4 \cdot B^4$, and so on, are calculated. If for some norm $\|B^{2^k}\| \to 0$ as $k \to \infty$, then the matrix A is stable. Otherwise the matrix A is unstable, i.e., its eigenvalues do not lie in the left half-plane.

Thus, the computation has reduced to finding $(I - A)^{-1}$ and the powers B^{2^k}. Each of these operations requires n^3 multiplications and divisions [5, 6, 21, 24]. The closer to the imaginary axis are the eigenvalues of the matrix A, the less is the degree of convergence of the matrices B^l. If there is at least one eigenvalue of A on the imaginary axis, then $\|B^l\|$ are bounded. But if at least one eigenvalue of A lies in the right half-plane, then $\|B^l\| \to \infty$ as $l \to \infty$.

The method of a functional transformation of a matrix can also be used to determine if all eigenvalues of the matrix A are contained within a prescribed polygon in the complex plane [1].

§ 2. Methods of Solving the Lyapunov Equation

1. Preliminary remarks. In § 7 of Chapter I it has been established that for a stable linear system

$$\dot{x}(t) = Ax(t), \qquad x \in R^n \tag{2.1}$$

there always exists a Lyapunov function of the form

$$V = x'Px \tag{2.2}$$

such that the positive definite symmetric matrix P satisfies the Lyapunov equation

$$A'P + PA = C_1, \tag{2.3}$$

where C_1 is an arbitrary prescribed negative definite symmetric matrix. In a similar way, for a discrete-time system $x_{k+1} = Bx_k$, $x_k \in R^n$, there also exists a quadratic Lyapunov function of the form (2.2), but the matrix P satisfies the equation

$$P - B'PB = C_1, \tag{2.4}$$

where C_1 is also an arbitrary negative definite matrix. Equation (2.3) will be called a *continuous-time Lyapunov equation*, and equation (2.4) a *discrete-time Lyapunov equation.* In this section we consider numerical methods for solving these equations. The construction of a Lyapunov function for linear systems is often very useful for investigating related nonlinear systems.

The matrix P as a symmetric matrix has $n(n+1)/2 \approx n^2/2$ different elements. Therefore, although equation (2.3) can be reduced directly to a system of linear algebraic equations, the order of this system proves to be high and equal to $n^2/2$.

To solve it by the most economical method, the Gauss method, we would need approximately $(n^2/2)^3/3 \approx 0.04n^6$ operations of multiplication and division. Even for $n = 20$ the number of operations becomes very large ($\approx 2.5 \cdot 10^6$). The determination of a Lyapunov function for $n = 20$ by this direct method requires the use of a sufficiently powerful computer; as for $n = 100$ no computer exists that could perform the necessary computation. This circumstance compels us to look for more economical methods and to apply various iterative procedures for solving equations (2.3) and (2.4).

2. The series method. By a direct substitution we can verify that the solution of equation (2.4) can be expressed as a series

$$P = C_1 + B'C_1B + (B')^2C_1B^2 + \dots . \tag{2.5}$$

It can be shown that if the matrix B is stable (i.e., all of its eigenvalues lie inside a unit circle, see § 11 of Chapter I), then the series (2.5) is convergent. It is not very economical to compute the partial sums of the series (2.5) directly. The following method proves to be more effective. Let $P_{\nu(k)}$ be the sum of $2^k - 1$ terms of the series (2.5):

$$P_{\nu(k)} = \sum_{i=1}^{\nu(k)} (B')^{l-1}C_1B^{l-1}, \qquad \nu(k) = 2^k.$$

Then it can be readily verified that the following recurrence relation holds for $P_{\nu(k)}$:

$$P_{\nu(k+1)} = P_{\nu(k)} + (B')^{\nu(k)}P_{\nu(k)}B^{\nu(k)}, \qquad P_{\nu(0)} = C_1,\ B^{\nu(k+1)} = \left[B^{\nu(k)}\right]^2. \tag{2.6}$$

Each computation step requires, according to (2.6), three operations of matrix multiplication. This means that no more than $3n^3$ multiplications are required. The series method has a high degree of convergence, since $k = 6$ corresponds to the sum of $2^6 = 64$ terms of the series (2.5), and $k = 10$ to the sum of 1024 terms of the series. Therefore, we may assume that solving a discrete-time Lyapunov equation will require $20n^3$–$30n^3$ multiplications. This is considerably less compared to direct solution of Lyapunov equation that requires $0.04n^6$ multiplications. In the literature[1] we find the description of a successful application of the series method for solving the Lyapunov equation for $n = 143$.

The subroutine SUM implementing this algorithm has been included into the system ORACLS used by the National Aeronautics and Space Administration (NASA) of the USA for computing linear control systems [2].

Using the series method, we can also solve the continuous-time Lyapunov equation (2.3). It is easily verified that equation (2.3) coincides with equation (2.4) in which

$$B = (\beta I + A)(\beta I - A)^{-1}, \qquad C_1 = -2\beta(\beta I - A')^{-1}C(\beta I - A)^{-1}, \tag{2.7}$$

[1]Larin, V. B., Methods for Solving the Algebraic Riccati Equations, *Izv. Akad. Nauk SSSR Tekhn. Kibernet.*, (1983), no. 2. (Translated into English as *Soviet J. Comput. Systems Sci.*)

where $\beta > 0$ is a specially chosen number. To verify this, we write equation (2.4) with coefficients of the form (2.7) and multiply it on the left by $\beta I - A'$ and on the right by $\beta I - A$. Collecting similar terms gives equation (2.3). The number β in the transformation (2.7) is chosen by one of two methods. The first of these is connected with finding the eigenvalues of the matrix A:

$$\lambda_j(A) = a_j + ib_j, \qquad i^2 = -1,\ |\lambda_1| \leqslant |\lambda_2| \leqslant \dots \leqslant |\lambda_n|,$$

and the number β is defined by the relations

$$\beta^2 = \begin{cases} \gamma = \frac{a_1(a_n^2+b_n^2)-a_n(a_1^2+b_1^2)}{a_n-a_1}, & \text{if } a_n - a_1 \neq 0 \text{ and } \gamma > 0, \\ \frac{1}{n}\sum_{j=1}^{n}(a_j^2+b_j^2)^{1/2}, & \text{if } a_n = a_1 \text{ or } \gamma \leqslant 0. \end{cases} \tag{2.8}$$

In the second method the number β is chosen so that

$$\beta = 2\|A\|_1 = \max_{1\leqslant k\leqslant n} \sum_{j=1}^{n} |a_{jk}|. \tag{2.9}$$

Thus, to solve the continuous-time Lyapunov equation (2.3), we first construct the matrices B and C_1 according to formulas (2.7), where β is defined by either formula (2.8) or (2.9), and then compute the terms of the sequence (2.6). Such a method of solution of the Lyapunov equation and the appropriate subroutine BILIN are described in [2].

The series method can also be applied in the solution of more general equations

$$X = AXC + B, \qquad AX + XC = B,$$

where A, B, and C are matrices having dimensions $n \times n$, $n \times m$, and $m \times m$, respectively.

A deficiency of the series method is that it can be applied only in the case of a stable matrix B. If the matrix B is unstable (i.e., there exists at least one eigenvalue of the matrix B lying on the boundary of or outside the unit circle), then the series (2.5) may be divergent, and the series method does not provide a solution.

3. Method of the matrix sign function. For any nondegenerate matrix A having no pure imaginary eigenvalues, let us define the sequence X_k by the formulas

$$X_{k+1} = \tfrac{1}{2}\left(X_k + X_k^{-1}\right), \qquad X_0 = A,\ k = 0, 1, 2, \dots. \tag{2.10}$$

The sequence X_k is convergent, and its limit denoted by $\operatorname{sgn} A$ is called the *sign function (signum function) of the matrix* A. It has been established by A. A. Abramov [14] that the matrix $B = \operatorname{sgn} A$ has the following property which we give without proof.

Let L_+ and L_- be invariant subspaces of R^n corresponding to eigenvalues of the matrix A with positive and negative imaginary parts, respectively. Here R^n

is a direct sum of L_+ and L_-, i.e., $R^n = L_+ \oplus L_-$. The following equalities are valid:

$$Bx = \begin{cases} x, & \text{for } x \in L_+, \\ -x, & \text{for } x \in L_-. \end{cases} \tag{2.11}$$

But if the matrix A is stable, then it follows from (2.11) that

$$\operatorname{sgn} A = -I. \tag{2.12}$$

The equality (2.12) is the main relation of a certain method for solving the continuous-time Lyapunov equation

$$A'P + PA = C. \tag{2.13}$$

Multiplying equation (2.13) on the left by $(A')^{-1}$ and on the right by A^{-1} leads to the equation

$$PA^{-1} + (A')^{-1}P = (A')^{-1}CA^{-1}. \tag{2.14}$$

It follows from equations (2.13) and (2.14) that the matrix P also satisfies the equation

$$\tfrac{1}{2}(A' + (A')^{-1})P + P\tfrac{1}{2}(A + A^{-1}) = \tfrac{1}{2}(C + (A')^{-1}CA^{-1}).$$

In an analogous way we can verify that P is a solution of the following equations as well:

$$A'_{k+1}P + PA_{k+1} = C_{k+1}, \tag{2.15}$$

where

$$A_{k+1} = \tfrac{1}{2}(A_k + A_k^{-1}), \qquad A_0 = A,\ k = 0, 1, 2, \ldots,$$
$$C_{k+1} = \tfrac{1}{2}(C_k + (A'_k)^{-1}C_kA_k^{-1}), \qquad C_0 = C,\ k = 0, 1, 2, \ldots.$$

Since the matrix A is stable, we find, passing to the limit in (2.15) and taking account of (2.12), that

$$P = -\tfrac{1}{2}\lim_{k\to\infty} C_k. \tag{2.16}$$

Therefore, the value $P = -C_l/2$ may be taken for an approximate value of P, where the number l is sufficiently large. Moreover, if the original matrix $C = C_0$ is negative definite, then the matrix P is positive definite. One step in this method requires an inversion of the matrix A_k and two matrix multiplications for finding the expression $(A'_k)^{-1}C_kA_k^{-1}$, i.e., about $3n^3$ operations of multiplication. If convergence in this method is assumed to be 6–10 steps, then the method requires $20n^3$–$30n^3$ multiplications. However, the degree of convergence in this method depends on how near the spectrum of the stable matrix A is to the imaginary axis. The nearer the spectrum is to the imaginary axis, the less, in general, is the degree of convergence.

4. Application of the QR-algorithm. In the last few years several methods for solving the Lyapunov equation (2.13) based on the QR-algorithm have been proposed [14]. We present one of these called the *BS-algorithm.*[2] First, by means of the QR-algorithm the matrix A is reduced to the Schur form S, where $S = U'AU$. We now multiply the equality (2.13) on the left by U' and on the right by U; moreover, between P and A we insert the product $UU' = E$. Then we obtain the equation

$$U'A'UU'PU + U'PUU'AU = U'CU.$$

Setting $Y = U'PU$, $D = U'CU$, we obtain the equation which is equivalent to equation (2.13):

$$S'Y + YS = D. \tag{2.17}$$

Equation (2.17) is readily solved if we take account of its block-triangular form. Let us write the matrices Y and D in block form that is consistent with the matrix S.

For example, suppose that a 4×4 matrix S^1 is of the following block form:

$$S^1 = \begin{pmatrix} S^1_{11} & \vdots & S^1_{12} & & \vdots & S^1_{13} \\ \cdots & \cdot & \cdots & \cdots & \cdot & \cdots \\ 0 & \vdots & S^1_{22} & & \vdots & S^1_{23} \\ 0 & \vdots & & & \vdots & \\ \cdots & \cdot & \cdots & \cdots & \cdot & \cdots \\ 0 & \vdots & 0 & 0 & \vdots & S^1_{33} \end{pmatrix}, \qquad S^1 = \begin{pmatrix} * & \vdots & * & * & \vdots & * \\ \cdots & \cdot & \cdots & \cdots & \cdot & \cdots \\ 0 & \vdots & * & * & \vdots & * \\ 0 & \vdots & * & * & \vdots & * \\ \cdots & \cdot & \cdots & \cdots & \cdot & \cdots \\ 0 & \vdots & 0 & 0 & \vdots & * \end{pmatrix}.$$

Then the matrix Y^1 is of the form

$$Y^1 = \begin{pmatrix} Y^1_{11} & \vdots & Y^1_{12} & & \vdots & Y^1_{13} \\ \cdots & \cdot & \cdots & \cdots & \cdot & \cdots \\ Y^1_{21} & \vdots & Y^1_{22} & & \vdots & Y^1_{23} \\ & \vdots & & & \vdots & \\ \cdots & \cdot & \cdots & \cdots & \cdot & \cdots \\ Y^1_{31} & \vdots & Y^1_{32} & & \vdots & Y^1_{33} \end{pmatrix}, \qquad Y^1 = \begin{pmatrix} * & \vdots & * & * & \vdots & * \\ \cdots & \cdot & \cdots & \cdots & \cdot & \cdots \\ * & \vdots & * & * & \vdots & * \\ * & \vdots & * & * & \vdots & * \\ \cdots & \cdot & \cdots & \cdots & \cdot & \cdots \\ * & \vdots & * & * & \vdots & * \end{pmatrix}.$$

But if the matrix S^2 is of the following block form:

$$S^2 = \begin{pmatrix} S^2_{11} & \vdots & S^2_{12} & \vdots & S^2_{13} & \\ \cdots & \cdot & \cdots & \cdot & \cdots & \cdots \\ 0 & \vdots & S^2_{22} & \vdots & S^2_{23} & \\ \cdots & \cdot & \cdots & \cdot & \cdots & \cdots \\ 0 & \vdots & 0 & \vdots & S^2_{33} & \\ 0 & \vdots & 0 & \vdots & & \end{pmatrix}, \qquad S^2 = \begin{pmatrix} * & \vdots & * & \vdots & * & * \\ \cdots & \cdot & \cdots & \cdot & \cdots & \cdots \\ 0 & \vdots & * & \vdots & * & * \\ \cdots & \cdot & \cdots & \cdot & \cdots & \cdots \\ 0 & \vdots & 0 & \vdots & * & * \\ 0 & \vdots & 0 & \vdots & * & * \end{pmatrix},$$

[2] Bartels, R. H., and Stewart, G. W., Solution of the Matrix Equation $AX + XB = C$, *Comm. ACM*, (1972), vol. 15, 820–826.

then it is associated with the matrix Y^2 of the form

$$Y^2 = \begin{pmatrix} Y_{11}^2 & \vdots & Y_{12}^2 & \vdots & Y_{13}^2 & \\ \cdots & \cdot & \cdots & \cdot & \cdots & \cdots \\ Y_{21}^2 & \vdots & Y_{22}^2 & \vdots & Y_{23}^2 & \\ \cdots & \cdot & \cdots & \cdot & \cdots & \cdots \\ Y_{31}^2 & \vdots & Y_{32}^2 & \vdots & Y_{33}^2 & \\ & \vdots & & \vdots & & \end{pmatrix}, \qquad Y^2 = \begin{pmatrix} * & \vdots & * & \vdots & * & * \\ \cdots & \cdot & \cdots & \cdot & \cdots & \cdots \\ * & \vdots & * & \vdots & * & * \\ \cdots & \cdot & \cdots & \cdot & \cdots & \cdots \\ * & \vdots & * & \vdots & * & * \\ * & \vdots & * & \vdots & * & * \end{pmatrix}$$

The structures of the matrices D^1 and D^2 are similar to those of the matrices Y^1 and Y^2.

Next, we write equation (2.17) for individual blocks. Since the matrix D cannot have blocks of dimensions greater than 2×2, the system (2.17) disintegrates into a set of linear equations of the first, second, and fourth order. For example, the equations corresponding to the blocks D_{11}^1, D_{12}^1, and D_{22}^1 are equations of the first, second, and fourth order, respectively. It is easy to write formulas for this set of equations, and to solve it consecutively. As a result, we find an intermediate matrix Y. The matrix P is computed further according to the formula

$$P = UYU'.$$

Thus, the BS-algorithm reduces to the following steps:

1°. The matrix A is reduced to the Schur form S: S=U'AU.

2°. The matrix $D = U'CU$ is computed.

3°. Equation (2.17) is solved.

4°. The matrix $P = UYU'$ is computed.

The BS-algorithm can be applied to any matrices A and C, and not just to stable matrices A and symmetric matrices C. This algorithm, as all algorithms based on orthogonal transformations, is highly numerically stable. As estimates indicate, it requires $25n^3$ multiplication and does not depend on the actual arrangement of the spectrum. To implement the BS-algorithm, we would need a memory of $3n^2$ binary words. All this enables the BS-algorithm to be considered one of the best methods for solving the Lyapunov equation (2.13).

5. Construction of stabilizing control. In a number of problems it is required to find optimal control of the form

$$u(t) = -Fx(t) \tag{2.18}$$

such that the system (closed by this control)

$$\dot{x}(t) = Ax(t) + Bu, \qquad x \in R^n,\ u \in R^m, \tag{2.19}$$

is asymptotically stable. The last requirement is equivalent to the following constraints:

$$\operatorname{Re} \lambda_j(A - BF) < 0, \qquad j = 1, \ldots, n. \tag{2.20}$$

The control (2.18) subject to the constraints (2.20) is called the *stabilizing control.*

Construction of stabilizing control reduces to solution of a Lyapunov equation [2].

Let A and B be matrices having the respective dimensions $n \times n$ and $n \times m$. First, we construct a matrix

$$\tilde{A} = -(A + \beta I_n)$$

so that $\operatorname{Re} \lambda_j(\tilde{A}) < 0$. For this purpose it is sufficient to take

$$\beta = s \left[\max_{1 \leqslant j \leqslant n} |\operatorname{Re} \lambda_j(A)| + 0.001 \right], \qquad s > 1,$$

or to determine β by formula (2.9).

Further, consider the Lyapunov equation

$$\tilde{A} Z + Z \tilde{A}' = -2BB' \tag{2.21}$$

and find its solution by one of the methods described above.

If the system (2.19) is completely controllable, then the matrix Z is nondegenerate, and F we may be taken as the matrix

$$F = -B' Z^{-1}. \tag{2.22}$$

Moreover,

$$\operatorname{Re} \lambda_j(A - BF) = -\beta, \qquad j = 1, \ldots, n. \tag{2.23}$$

The system (2.12) can be stabilized even if the pair (A, B) is not completely controllable (see [12] of Part Two). In this case the solution Z of the Lyapunov equation (2.21) is a degenerate matrix, and then F we may be taken as the matrix

$$F = -B' Z^{+}, \tag{2.24}$$

where Z^+ is the pseudo-inverse of the matrix Z. It will be recalled that a matrix Z^+ is called a *pseudo-inverse* of the matrix Z if the following equalities hold:

$$\begin{aligned} ZZ^+Z &= Z, \qquad & Z^+ZZ^+ &= Z^+, \\ (ZZ^+)' &= ZZ^+, \qquad & (Z^+Z)' &= Z^+Z \end{aligned} \tag{2.25}$$

Moreover, it is not necessary that $ZZ^+ = I$.

Thus, to determine the stabilizing control (2.18), it is necessary to construct the matrix $\tilde{A}$, to solve the Lyapunov equation (2.21), find the inverse Z^{-1} or the pseudoinverse Z^+, and compute the feedback matrix F according to formula (2.24).

A procedure for numerical pseudo-inversion has been proposed in [25]. It is based on the singular decomposition of the matrix Z:

$$Z = U\Sigma V', \qquad U'U = V'V = I_n,\ \Sigma = \operatorname{diag}(\sigma_1, \dots, \sigma_n). \tag{2.26}$$

The diagonal elements of the matrix Σ coincide with the nonnegative values of the square roots of the eigenvalues of the matrix $Z'Z$. The numbers $\sigma_1, \dots, \sigma_n$ are called the *singular numbers* of the matrix Z. If the singular decomposition (2.26) of the matrix Z is found, then the pseudo-inverse Z^+ is determined by

$$Z^+ = V\Sigma^+U', \qquad \Sigma^+ = \operatorname{diag}(\sigma_1^+, \dots, \sigma_n^+), \tag{2.27}$$

$$\sigma_i^+ = \begin{cases} (\sigma_i)^{-1}, & \text{for } \sigma_i > 0, \\ 0, & \text{for } \sigma_i = 0. \end{cases}$$

An algorithm which implements the pseudo-inversion of matrices has been published in [25] under the name "algorithm 1.10". The method discussed above used to find stabilizing control has been published in [2] under the name "CSTAB".

6. Computation of the covariance matrix. Consider a system of stochastic equations of the form

$$\dot{x}(t) = Ax(t) + G\eta(t), \tag{2.28}$$

where A is an $n \times n$ matrix, G is an $n \times m$ matrix, $\eta(t)$ is an m-dimensional white-noise process with intensivity matrix Q having dimensions $m \times m$. Let the matrix A be asymptotically stable, i.e., $\operatorname{Re}\lambda_j(A) < 0$, $j = 1, \dots, n$. Then the covariance matrix W of the time-invariant state of the system is the limit of the variance matrices:

$$W = \lim_{t\to\infty} D(t), \qquad D(t) = R(t,t),$$

where $R(t,s)$ is the covariance matrix of the solution $x(t)$. The matrix W satisfies the Lyapunov equation

$$AW + WA' = GQG'. \tag{2.29}$$

The matrix W in the Lyapunov equation (2.29) can be found by one of the methods described above.

§ 3. Controllability and Observability

According to Theorem 2.1 of Chapter VI, the system (1.1) is controllable if the rank of the controllability matrix

$$K = \left(B, AB, \dots, A^{n-1}B\right) \tag{3.1}$$

(where A and B are $n \times n$ and $n \times m$ matrices, respectively) is equal to the order n of the system. Thus, we would need to determine the rank of the matrix K having dimensions $n \times nm$.

Attempts to use the definition of the rank of the matrix K directly as the largest number of linearly independent columns do not yield a good numerical algorithm. Indeed, we would need to take arbitrary n columns of the matrix K and for each resulting matrix K_j compute its determinant. If there exists at least one matrix K_j whose determinant is not zero, then the system is completely controllable; otherwise, it is uncontrollable. This method is too labor-consuming, since there exists

$$\binom{nm}{n} = \frac{nm(nm-1)\dots(n(m-1)+1)}{n!}$$

for different matrices K. The evaluation of each determinant requires about $n^3/3$ arithmetic operations. The number of operations for all determinants is very large. Thus, for example, when using this scheme for $n = 10$, $m = 5$, the computation would require about $0.3 \cdot 10^5$ operations of multiplication and division. Moreover, the evaluation of determinants is not sufficiently numerically stable. From the point of view of both time saving and numerical stability the method for computing the rank based on the singular decomposition of the matrix K is more convenient.

With the help of algorithm 1.10 from [25] the matrix K can be expressed in the form

$$K = U\Sigma V',$$

where U is an $n \times n$ matrix composed of n orthonormal eigenvectors of the matrix KK^+, V is an $nl \times nl$ matrix composed of nl orthonormal eigenvectors of the matrix K^+K. The matrices U and V are orthogonal, i.e., $U'U = I_n$, $V'V = I_{nl}$. The $n \times nl$ matrix Σ is of the form $\Sigma = (S, 0)$, and $S = \operatorname{diag}(\sigma_1, \dots, \sigma_n)$ is a diagonal $n \times n$ matrix having the singular numbers of the matrix K as elements in the diagonal.

If the rank of the matrix K is r, then only r singular numbers are distinct from zero and the other $n - r$ numbers are zero. This assertion lies at the basis of the definition of the rank of the matrix K. Thus, the definition of the rank of the matrix K includes the following steps:

1°. The product KK' is computed with double precision.

2°. The matrix KK' is diagonalized with the help of the QR-algorithm.

3°. The number of nonzero diagonal elements is determined which coincides with the rank of the matrix K.

Example 3.1 [25]. Consider an 8×5 matrix K

$$K = \begin{pmatrix} 22 & 10 & 2 & 3 & 7 \\ 14 & 7 & 10 & 0 & 8 \\ -1 & 13 & -1 & -11 & 3 \\ -3 & -2 & 13 & -2 & 4 \\ 9 & 8 & 1 & -2 & 4 \\ 9 & 1 & -7 & 5 & -1 \\ 2 & -6 & 6 & 5 & 1 \\ 4 & 5 & 0 & -2 & 2 \end{pmatrix}.$$

The singular numbers of this matrix have been found to be the following: $\sigma_1 = \sqrt{1248} = 35.32704346\ldots$, $\sigma_2 = 20$, $\sigma_3 = \sqrt{384} = 19.59591794\ldots$, $\sigma_4 = \sigma_5 = 0$. The process of finding the singular numbers with the help of the QR-algorithm is continued until the off-diagonal elements became less than $46.4 \cdot 10^{-8}$. This requires six iterations of the QR-algorithm. The following values of the diagonal elements of $\tilde{\sigma}$ have been obtained: $\tilde{\sigma}_1 = 0.96 \cdot 10^{-7}$, $\tilde{\sigma}_2 = 19.595916$, $\tilde{\sigma}_3 = 19.999999$, $\tilde{\sigma}_4 = -1.97 \cdot 10^{-7}$, $\tilde{\sigma}_5 = 35.327038$. The maximum error of evaluating $\tilde{\sigma}$ does not exceed $51.8 \cdot 10^{-7}$. Within the limits of computation accuracy $\tilde{\sigma}_1$ and $\tilde{\sigma}_4$ may be taken as zero. Therefore, the rank of the matrix K is three.

§ 4. Linear-Quadratic Time-Invariant Problem of Optimal Stabilization

Consider a time-invariant linear-quadratic problem (LQ-problem) on an infinite interval, having the form

$$\dot{x}(t) = Ax(t) + Bu, \qquad x \in R^n,\ u \in R^m,\ x(0) = x_0, \tag{4.1}$$

with cost functional

$$J(u) = \int_0^\infty [x'(s)N_2x(s) + u'(s)N_3u(s)]\,ds, \qquad N_3 > 0,\ N_2 > 0. \tag{4.2}$$

As was shown in § 3 of Chapter VII, optimal stabilizing control in the problem (4.1), (4.2) is defined by

$$u(x) = -N_3^{-1}B'Px = -Fx. \tag{4.3}$$

The matrix $P > 0$ in formula (4.3) satisfies the matrix algebraic equation

$$A'P + PA + N_2 - PBN_3^{-1}B'P = 0. \tag{4.4}$$

Equation (4.4) is called the *algebraic Riccati equation.*

It should be noted that in practice it is rather simple to realize the P-controller (4.3), since the gain matrix F is a constant. This explains the wide use of control laws of the form (4.3) in engineering applications.

The determination of an F-control (4.3) in the time-invariant optimal stabilization problem (4.1), (4.2) reduces to solving the algebraic Riccati equation (4.4). Let us dwell briefly on some methods used for solving equation (4.4) which are closely connected with matrix calculations.

1. Reduction to a sequence of Lyapunov equations. This method is one of the modifications of the Newton method for solving nonlinear equations by means of successive solutions of linear problems.

It is implemented in the following way. First, a matrix L_0 is chosen so that the spectrum of the matrix $(A + BL_0)$ lies in the left half-plane. The matrix L_0

can, for example, be constructed by the method described in subsection 5 of § 2. Next, we seek solution of a sequence of Lyapunov equations of the form

$$(A' + L_j'B')P_j + P_j(A + BL_j) + N_2 - L_jN_3L_j = 0, \tag{4.5}$$

$$L_{j+1} = -N_3^{-1}B'P_j, \qquad j = 0, 1, 2, \dots . \tag{4.6}$$

It can be shown that:
(a) all matrices $(A + BL_j)$ are stable if the matrix $(A + BL_0)$ is stable;
(b) $P_{j+1} \leqslant P_j$, i.e., the matrix $(P_j - P_{j+1})$ is positive semidefinite;
(c) $\lim_{j\to\infty} P_j = P$.

The degree of convergence of the method (4.5), (4.6), as of any other Newton method, is asymptotically quadratic. However, the degree of convergence of the initial iterations essentially depends on the arrangement of the spectra of the matrices $(A + BL_j)$. The nearer to the imaginary axis are the spectra of the matrices $(A + BL_j)$, the less is the degree of convergence. For more detail on this method, see the paper by V. B. Larin referred to in subsection 2 of § 2.

2. Use of the QR-algorithm. This method is based on the close relation between the matrix Riccati equation (4.4) of dimension n and some linear equation of order $2n$. Consider an equation of order $2n$ of the form

$$\begin{pmatrix} -A & BN_3^{-1}B' \\ N_2 & A' \end{pmatrix} \begin{pmatrix} U_1 \\ U_2 \end{pmatrix} = \begin{pmatrix} U_1 \\ U_2 \end{pmatrix}, \tag{4.7}$$

where U_1 and U_2 are $n \times n$ matrices.

If the matrices U_1 and U_2 are a solution of the linear system (4.7), with the matrix U_1 nondegenerate, then the matrix

$$P = U_2U_1^{-1} \tag{4.8}$$

is a solution of the Riccati equation (4.4). To prove this, we write the equality (4.7) in the form of two equations:

$$-AU_1 + BN_3^{-1}B'U_2 = U_1, \tag{4.9}$$

$$N_2U_1 + A'U_2 = U_2. \tag{4.10}$$

Multiplying equations (4.9) and (4.10) on the right by U_1^{-1} gives

$$-A + BN_3^{-1}B'U_2U_1^{-1} = I_n, \tag{4.11}$$

$$N_2 + A'U_2U_1^{-1} = U_2U_1^{-1}. \tag{4.12}$$

Next, multiplying equation (4.11) on the left by the matrix $U_2U_1^{-1}$, we can write

$$-U_2U_1^{-1}A + U_2U_1^{-1}BN_3^{-1}B'U_2U_1^{-1} = U_2U_1^{-1}. \tag{4.13}$$

Finally, subtracting equation (4.13) from equation (4.12), we find that the matrix $U_2U_1^{-1}$ satisfies the equation

$$A'U_2U_1^{-1} + U_2U_1^{-1}A + N_2 - U_2U_1^{-1}BN_3^{-1}B'U_2U_1^{-1} = 0. \qquad (4.14)$$

If we denote by P the matrix $U_2U_1^{-1}$, we see that equation (4.14) coincides with the Riccati equation (4.4). This completes the proof.

Equation (4.7) can be conveniently solved with the help of the QR-algorithm, which has been considered in subsection 2 of § 1. This can be done in the following steps:

1°. The matrix N_3^{-1} is determined, and the matrix

$$M = \begin{pmatrix} -A & BN_3^{-1}B' \\ N_2 & A' \end{pmatrix}. \qquad (4.15)$$

is constructed.

2°. With the help of the QR-algorithm the matrix M is transformed into the upper Schur form:

$$S = W'MW, \qquad (4.16)$$

where W is an orthogonal matrix, and S is an upper block-triangular matrix.

3°. The Schur form (4.16) is reordered so that the first n elements in the principal diagonal of the new form S_1 are the eigenvalues $\lambda_i(M)$ such that $\operatorname{Re}\lambda_i(M) > 0$, $i = 1, \dots, n$. Such reordering can be effected with the help of some orthogonal matrix V:

$$S_1 = V'SV.$$

4°. The matrix $U = WV$ is found and its $n \times n$ blocks U_{ij} $(i, j = 1, 2)$ are isolated:

$$U = \begin{pmatrix} U_{11} & U_{12} \\ U_{21} & U_{22} \end{pmatrix}.$$

5°. The matrix P is defined by

$$P = U_{21}U_{11}^{-1}.$$

To determine this matrix, using, for example, the Gauss method, we have to solve the matrix equation

$$U'_{11}P = U'_{21}. \qquad (4.17)$$

It follows from the foregoing that the resulting matrix P is a solution of the Riccati equation (4.4).

The application of the QR-algorithm to the matrix M requires $8n^2$ binary words of RAM and about $80n^3$ operations of multiplication. In the Gauss method the inversion of the matrices N_3 and U_{11} requires a number of multiplications of the order of n^3 for each matrix.

Despite such great demands on RAM and large time consumption the use of the QR-algorithm is seemingly the best method for solving the algebraic Riccati equation. Both the QR-algorithm and the Gauss method are numerically stable methods. The application and properties of the method discussed in this subsection are not affected by the arrangement of the spectrum of the matrix M, whereas in the case of the reduction to a sequence of Lyapunov equations this is not so. A more accurate computation [14] shows that the method based on the reduction to a sequence of Lyapunov equations can compete with the QR-algorithm method only if 7 or 8 Lyapunov equations are solved. It should be noted that a procedure for solving the matrix Riccati equation (4.4) based on the QR-algorithm is given in [14].

Example 4.1. Consider the optimal stabilization problem for the system

$$\dot{x}_1 = x_2, \qquad \dot{x}_2 = u,$$
$$J(u) = \int_0^\infty [x_1^2(s) + 2x_2^2(s) + u^2(s)]\, ds \to \inf. \tag{4.18}$$

The matrices entering in the relations (4.1), (4.2) adapted for the problem (4.18) are of the form

$$A = \begin{pmatrix} 0 & 1 \\ 0 & 0 \end{pmatrix}', \qquad B = \begin{pmatrix} 0 \\ 1 \end{pmatrix}, \qquad N_2 = \begin{pmatrix} 1 & 0 \\ 0 & 2 \end{pmatrix}, \qquad N_3 = (1). \tag{4.19}$$

It is easily shown that

$$BN_3^{-1}B' = \begin{pmatrix} 0 & 0 \\ 0 & 1 \end{pmatrix}. \tag{4.20}$$

We apply the QR-algorithm to solve the algebraic Riccati equation (4.4) involving the matrices (4.19), (4.20). To this end, we construct a 4×4 matrix M having the form

$$M = \begin{pmatrix} -A & -BN_3^{-1}B' \\ N_2 & A' \end{pmatrix} = \begin{pmatrix} 0 & -1 & 0 & 0 \\ 0 & 0 & 0 & -1 \\ 1 & 0 & 0 & 0 \\ 0 & 2 & 1 & 0 \end{pmatrix}. \tag{4.21}$$

With the help of the QR-algorithm we find that the matrix

$$U = \begin{pmatrix} 0.5 & -0.1\sqrt{5} & -0.3\sqrt{5} & 0.5 \\ -0.5 & -0.1\sqrt{5} & -0.3\sqrt{5} & -0.5 \\ 0.5 & -0.3\sqrt{5} & 0.1\sqrt{5} & -0.5 \\ -0.5 & -0.3\sqrt{5} & 0.1\sqrt{5} & 0.5 \end{pmatrix} \tag{4.22}$$

reduces the matrix (4.21) to the upper Schur form:

$$S = \begin{pmatrix} 1 & 0 & -1 & 0.5 \\ 0 & 1 & 1 & -1 \\ 0 & 0 & -1 & 0 \\ 0 & 0 & 0 & -1 \end{pmatrix}. \tag{4.23}$$

The operations 3° and 4° of the algorithm need not be carried out, since the first n elements in the principal diagonal of the matrix (4.23) are already positive numbers. In our case, in view of (4.22), equation (4.17) is of the form

$$\begin{pmatrix} 0.5 & -0.5 \\ -0.1\sqrt{5} & -0.1\sqrt{5} \end{pmatrix}\begin{pmatrix} p_{11} & p_{12} \\ p_{21} & p_{22} \end{pmatrix} = \begin{pmatrix} 0.5 & -0.5 \\ -0.3\sqrt{5} & -0.3\sqrt{5} \end{pmatrix}. \tag{4.24}$$

The matrix

$$P = \begin{pmatrix} 2 & 1 \\ 1 & 2 \end{pmatrix}.$$

is a solution of equation (4.24). Next, according to formula (4.3) we determine the optimal stabilizing control

$$u(x) = -N_3^{-1}B'Px = x_1 + 2x_2.$$

3. Landing control of a Boeing-747 airplane[3]**.** A linearized model of the motion of the heavy airplane "Boeing-747" in the landing configuration without the use of the automatic stability enhancement system is of the form

$$\begin{pmatrix} \dot v \\ \dot r \\ \dot p \\ \dot\theta \end{pmatrix} = \begin{pmatrix} -0.089 & -2.19 & 0.328 & 0.319 \\ 0.076 & -0.217 & -0.166 & 0 \\ -0.602 & 0.327 & -0.975 & 0 \\ 0 & 0.150 & 1 & 0 \end{pmatrix}\begin{pmatrix} v \\ r \\ p \\ \theta \end{pmatrix} + \begin{pmatrix} 0 & 0.0327 \\ 0.0264 & -0.151 \\ 0.227 & 0.0636 \\ 0 & 0 \end{pmatrix}\begin{pmatrix} \delta a \\ \delta r \end{pmatrix} + \begin{pmatrix} 0.089 \\ -0.076 \\ 0.602 \\ 0 \end{pmatrix}(vw), \tag{4.25}$$

$$\begin{pmatrix} \dot\psi \\ \dot x \end{pmatrix} = \begin{pmatrix} 0 & 0 \\ 2.21 & 0 \end{pmatrix}\begin{pmatrix} \psi \\ x \end{pmatrix} + \begin{pmatrix} 0 & 1 \\ 1 & 0 \end{pmatrix}\begin{pmatrix} v \\ r \end{pmatrix}. \tag{4.26}$$

Here the following notation for the phase variables are used: v is the sideslip velocity, r is the yaw angular velocity, p is the roll angular velocity, θ is the angle of roll, ψ is the angle of yaw, x is the lateral displacement.

Let $y = (v, r, p, \theta, \psi, x)'$ denote the vector of phase variables (the state vector). The controls in the system (4.25) are: the aileron angle, $\delta\alpha$, and the rudder angle, δr. The disturbance in the system (4.25) is the wind: vw is the wind velocity. The equations (4.25), (4.26) are written in terms of dimensionless quantities. We take 0.01 rad as the unit of angular measure, and 0.305 m/sec as the unit of velocity.

The problem consists in determining F-control of the form

$$\begin{pmatrix} \delta a \\ \delta r \end{pmatrix} = Fy \tag{4.27}$$

[3]Bryson, A. E., Jr., New Concepts in Control Theory, 1959–1984, *J. Guidance Control Dynamics*, (1985), vol. 8, no. 4, 417–425.

which guarantees the execution of coordinated turns of the "Boeing-747" airplane in the landing configuration.

To solve this problem, we can reduce it to an optimal stabilization problem. The cost functional will be taken as

$$J = \int_0^\infty (k_1 |y|^2 + k_2 z^2 + \delta a^2 + \delta r^2)\, dt, \tag{4.28}$$

where z denotes specific lateral force: $z = -0.089v + 0.0327\delta r$. The weight coefficients $k_1 = 1$ and $k_2 = 100$ are chosen so that an increment in the aileron angle of one unit (=0.01 rad) leads to a lateral displacement of 0.305 m (or to a lateral acceleration of $0.0305\,\mathrm{m/sec^2}$).

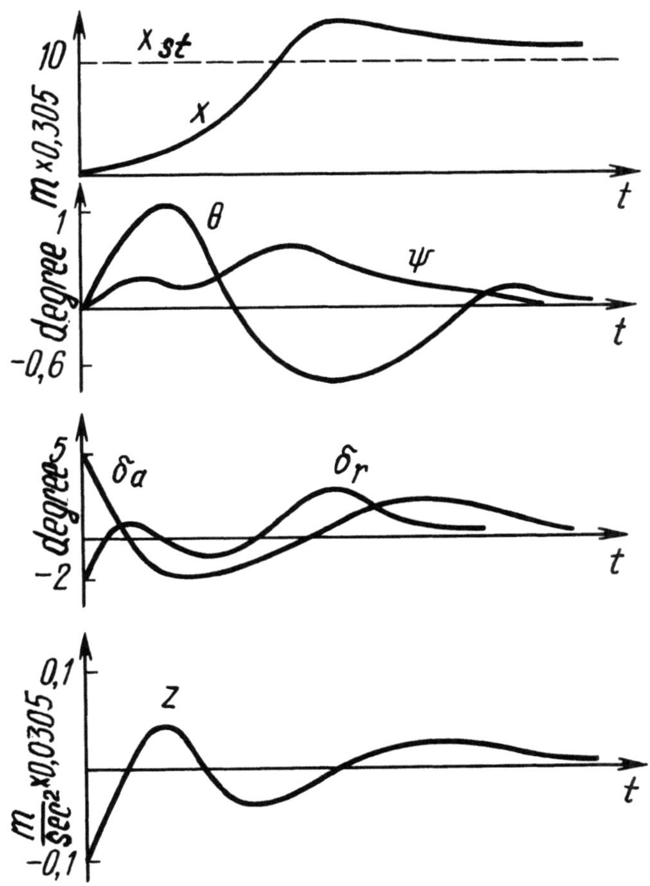

FIG. 4.1. Transients in the control system of the "Boeing-747" airplane executing an S-manoeuvre

A numerical calculation based on the optimal stabilization method has yielded the following control having the form (4.27):

$$\begin{pmatrix} \delta a \\ \delta r \end{pmatrix}' = \begin{pmatrix} -1.947 & -3.59 & -1.421 & -1.672 & -7.29 & -0.859 \\ 1.263 & 6.42 & 0.799 & 1.424 & 6.08 & 0.487 \end{pmatrix} \times \begin{pmatrix} v \\ r \\ p \\ \theta \\ \psi \\ \varphi \end{pmatrix} \tag{4.29}$$

The advantage of such an approach to the synthesis of control is that the motions of the phase variables are well coordinated. Such degree of coordination cannot be achieved by considering isolated control problems for each phase variable and by using traditional methods for synthesizing separate controllers in the pitch, roll, and yaw channels.

Figure 4.1 illustrates the response (computed according to the model (4.25), (4.28), (4.29)) of the "Boeing-747" airplane in the landing configuration to the control signal which must induce a lateral deviation of 3.05. This constitutes a so-called lateral let-down manoeuvre, or an S-manoeuvre. Figure 4.1 shows, in particular, that rudder and aileron controls are well coordinated; this leads to transients in the angles of roll and of yaw having no appreciable overshoot.

Main Results and Formulas of Chapter XIII

Precise methods used to solve the complete eigenvalue problem are not numerically stable and applicable only for $n \leqslant 10$.

Iteration methods: the Jacobi method and the QR-algorithm.

QR-algorithm: $A = QR$, $Q' = Q^{-1}$, R is an upper triangular matrix, $A^{(1)} = RQ$, $A^{(1)} = Q_1R_1$, $A^{(2)} = R_1Q_1$, ..., $A^{(k)} \to S$, S is a Schur matrix.

The Zubov method for the functional transformation of matrices:

$$B = (-A - I)(I - A)^{-1} = I - 2(I - A)^{-1}.$$

If $\lim_{k\to\infty} B^k = 0$, then the matrix A is stable.

The series method for solving the discrete-time Lyapunov equation $P - B'PB = C_1$. The solution is of the form

$$P = \lim_{k\to\infty} P_{\nu(k)}, \quad \text{where } \nu(k) = 2^k,$$
$$P_{\nu(0)} = C_1, \qquad P_{\nu(k+1)} = P_{\nu(k)} + (B')^{\nu(k)} P_{\nu(k)} B^{\nu(k)},$$
$$B^{\nu(k+1)} = [B^{\nu(k)}]^2.$$

The matrix sign function method. The solution P of the Lyapunov equation $A'P + PA = C$ is of the form

$$P = -\tfrac{1}{2} \lim_{k\to\infty} C_k,$$
$$C_{k+1} = \tfrac{1}{2}\big(C_k + (A_k')^{-1} C_k A_k^{-1}\big), \qquad C_0 = C,$$
$$A_{k+1} = \tfrac{1}{2}(A_k + A_k^{-1}), \qquad A_0 = A.$$

The solution of the algebraic Riccati equation

$$A'P + PA + N_2 - PBN_3^{-1}B'P = 0$$

is $P = \lim_{j\to\infty} P_j$, where P_j is the solution of the Lyapunov equations

$$(A' + L_j'B')P_j + P_j(A + BL_j) + N_2 - L_jN_3L_j = 0,$$
$$L_{j+1} = N_3^{-1}B'P_j, \qquad \operatorname{Re}\lambda_j(A + BL_0) < 0.$$

The QR-algorithm for solving the algebraic Riccati equation requires about $80n^3$ multiplications.

CHAPTER XIV

NUMERICAL METHODS FOR THE INVESTIGATION OF NONLINEAR CONTROL SYSTEMS

In this chapter the following topics are treated: analysis of transient processes for nonlinear systems, including those for stiff systems, and numerical methods for solving linear-quadratic problems.

§ 1. Analysis of Transients. The Runge–Kutta Methods

1. On numerical methods of investigating systems. In investigating control systems various problems arise, most of which are not amenable to analytical solution. Most problems, involving, especially, nonlinear systems, can be investigated only on the basis of numerical calculations with the help of a computer. Among such problems are: the problem of determining the output signal of a system, given the input signal and the disturbances; the problem of finding the maximum departure of the output signal from the nominal value, given the type of input signals and disturbances; the modelling of some stochastic characteristics of systems, and so on.

All these problems involving systems described by ordinary differential equations can be reduced to the numerical solution of a Cauchy problem or several Cauchy problems. In this section we are concerned with one-step numerical integration methods for the Cauchy problem. These methods are not only of interest by themselves but they are also incorporated as elements into more complex methods, such as methods employed for solving boundary-value problems for differential equations. Thus, they form a basis for all kinds of investigations of control systems.

Among approximate methods for solving ordinary differential equations we distinguish between approximate analytical, graphical, and numerical methods. Using numerical methods, we obtain a solution in the form of a table. Let us dwell briefly on some of these methods.

2. One-step methods. Consider the following Cauchy problem:

$$\dot{x}(t) = f(t,x), \qquad t_0 \leqslant t \leqslant T,\ x(t_0) = x_0, \tag{1.1}$$

where $x(t) \in R^n$, t_0 and x_0 are prescribed initial values, $f(t,x)$ is a continuous function satisfying the Lipschitz condition

$$|f(t,x) - f(t,y)| \leqslant L|x - y|$$

for all x, y, and t belonging to some region D. Unless otherwise stated, we shall assume that the function $f(t,x)$ is such that the solution $x(t)$ of the problem (1.1) exists and is unique on the whole interval $[t_0, T]$ and has a necessary number of derivatives.

On the interval $[t_0, T]$ we isolate the points $t_0, t_1, \dots, t_N = T$. The aggregate of the points $t_0, t_1, \dots, t_N$ is called a *grid*, and the points t_k are called *nodes*. The quantity $h_k = t_{k+1} - t_k$ $(k = 0, \dots, N-1)$ is called the *step* of the grid at the point t_k. If all $h_k = h = (T - t_0)/N$ are independent of k, then a grid is said to be *uniform*; otherwise, it is *nonuniform*. In numerical methods only the values of the solution $x(t)$ at the nodes t_k of the grid are determined.

At present, a great number (several hundreds) of various numerical methods for solving the Cauchy problem (1.1) are known. There exist several classes of methods: one-step, multistep, predictor-corrector, hybrid, and others.

One-step methods are described by the following formula:

$$x_{n+1} = x_n + h_n \Phi_f(t_n, x_n, h_n), \qquad x_0 = x(t_0). \tag{1.2}$$

Depending on the form of the function Φ, we obtain different methods. Let $\bar{x}(t_{n+1})$ denote the exact solution of the problem (1.1) with initial condition $\bar{x}(t_n) = x_n$.

The quantity $x_{n+1} - \bar{x}_{n+1}$ is called the *local error* of the method (1.2). If $x_{n+1} - \bar{x}_{n+1} = O(h_n^{s+1})$, the method (1.2) is said to provide *approximation of order s*.

The simplest one-step method is *the Euler method*:

$$x_{n+1} = x_n + h_n f(t_n, x_n), \qquad x_0 = x(t_0). \tag{1.3}$$

We expand the exact solution $\bar{x}_{n+1}$ as the following Taylor series:

$$\bar{x}(t_{n+1}) = \bar{x}(t_n) + h_n \bar{x}'(t_n) + \frac{h_n^2}{2} x''(t_n + \theta h_n), \qquad 0 < \theta < 1.$$

Substituting $f(x_n, t_n)$ for $\bar{x}'(t_n)$ according to equation (1.1), we see that in the region, where $|\bar{x}''(t)| \leqslant M$, the following relation holds:

$$x_{n+1} - \bar{x}(t_{n+1}) = \frac{h_n^2}{2} \bar{x}''(t_n + \theta h_n) = O(h_n^2).$$

Thus, the Euler method is a first-order approximation method.

More locally precise methods are the methods that are described by the formulas

$$\begin{gathered} x_{n+1} = x_n + h_n\big(\omega_1 K_1(f) + \cdots + \omega_p K_p(f)\big), \qquad K_1(f) = f(t_n, x_n), \\ K_i(f) = f\left(t_n + \alpha_i h_n, x_n + h_n \sum_{q=1}^{i-1} \beta_{iq} K_q(f)\right). \end{gathered} \tag{1.4}$$

One of the first formulas of the form (1.4) to find practical application, the *classic Runge–Kutta formula*, was derived by C. Runge in 1895 and extended by W. Kutta six years later. Methods based on formulas of the type (1.4) are thus called *Runge–Kutta methods.*

The number p is called the *number of stages* of this formula. To derive Runge–Kutta formulas, we can use the following method. The values of $\bar{x}(t_{n+1})$ and x_{n+1} are expanded as Taylor series, and the coefficients ω_i, α_i, and β_{iq} in formula (1.4) are then chosen so that a maximum number of terms containing like powers of h_n and f coincide.

We now turn to the derivation of formula (1.4) with two stages. We have

$$\begin{aligned}\bar{x}(t_{n+1}) &= \bar{x}(t_n) + h_n\bar{x}'(t_n) + \frac{h_n^2}{2}\bar{x}''(t_n) + O(h_n^3)\\ &= x_n + h_n f(t_n, x_n) + \frac{h_n^2}{2}\left[\frac{\partial f}{\partial t} + \frac{\partial f}{\partial x}f\right] + O(h_n^3).\end{aligned} \tag{1.5}$$

Here the derivatives $\partial f/\partial t$ and $\partial f/\partial x$ are computed at the point (t_n, x_n). On the other hand,

$$\begin{aligned}x_{n+1} &= x_n + h_n[\omega_1 f(t_n, x_n) + \omega_2 f(t_n + \alpha h_n, x_n + h_n\beta f)]\\ &= x_n + h_n\left\{\omega_1 f + \omega_2\left[f + \frac{\partial f}{\partial t}\alpha h_n + \frac{\partial f}{\partial x}h_n\beta f\right]\right\} + O(h_n^3).\end{aligned} \tag{1.6}$$

Comparison of (1.5) and (1.6) yields the following system of equations for the determination of ω_1, ω_2, α and β:

$$\omega_1 + \omega_2 = 1, \qquad \alpha\omega_2 = \frac{1}{2}, \; \beta\omega_2 = \frac{1}{2}. \tag{1.7}$$

The system (1.7) has several solutions, such as: (1) $\omega_1 = 0$, $\omega_2 = 1$, $\alpha = \beta = 1/2$; (2) $\omega_1 = \omega_2 = 1/2$, $\alpha = \beta = 1$. These solutions are associated with two Runge–Kutta formulas

$$\begin{aligned}x_{n+1} = x_n + h_n K_2, \qquad & K_1 = f(t_n, x_n),\\ K_2 = f\left(t_n + \frac{h_n}{2}, x_n + \frac{h_n}{2}K_1\right);&\end{aligned} \tag{1.8}$$

$$\begin{aligned}x_{n+1} = x_n + \frac{h_n}{2}(K_1 + K_2), \qquad & K_1 = f(t_n, x_n),\\ K_2 = f(t_n + h_n, x_n + h_n K_1).&\end{aligned} \tag{1.9}$$

Formulas (1.8) and (1.9) have the same number of stages and also provide the same (second) order of approximation, but these formulas do not yield identical results.

In a similar way, Runge–Kutta formulas of the third, fourth, and higher orders can be derived. It should be noted that each time we can deduce several formulas with the same number of stages providing the same order of approximation, such as the three known formulas—the classic Runge–Kutta formula, the "3/8" formula, and the Gill formula—of the fourth order with four stages. The coefficients of these three formulas are listed in Table 1.1. It is readily seen that the *classic Runge–Kutta formula of the fourth order* is of the form

$$x_{n+1} = x_n + \frac{h_n}{6}(K_1 + 2K_2 + 2K_3 + K_4)$$

$$K_1 = f(t_n, x_n), \qquad K_2 = f\left(t_n + \frac{h_n}{2}, x_n + \frac{h_n}{2}K_1\right),$$

$$K_3 = f\left(t_n + \frac{h_n}{2}, x_n + \frac{h_n}{2}K_2\right), \qquad K_4 = f(t_n + h_n, x_n + h_n K_3);$$

TABLE 1.1

Coefficient	ω_1	ω_2	ω_3	ω_4	α_2	α_3	α_4
Classic Runge–Kutta formula	$\frac{1}{6}$	$\frac{1}{3}$	$\frac{1}{3}$	$\frac{1}{6}$	$\frac{1}{2}$	$\frac{1}{2}$	1
The "3/8" formula	$\frac{1}{8}$	$\frac{3}{8}$	$\frac{3}{8}$	$\frac{1}{8}$	$\frac{1}{3}$	$\frac{2}{3}$	1
The Gill formula	$\frac{1}{6}$	$\frac{1-\frac{1}{\sqrt{2}}}{3}$	$\frac{1+\frac{1}{\sqrt{2}}}{3}$	$\frac{1}{6}$	$\frac{1}{2}$	$\frac{1}{2}$	1

Coefficient	β_{21}	β_{31}	β_{32}	β_{41}	β_{42}	β_{43}
Classic Runge–Kutta formula	$\frac{1}{2}$	0	$\frac{1}{2}$	0	0	1
The "3/8" formula	$\frac{1}{3}$	$-\frac{1}{3}$	1	0	-1	1
The Gill formula	$\frac{1}{2}$	$\frac{\sqrt{2}-1}{2}$	$1 - \frac{1}{\sqrt{2}}$	0	$-\frac{1}{\sqrt{2}}$	$1 + \frac{1}{\sqrt{2}}$

At present Runge–Kutta formulas of still higher orders—6, 8, and 10—are also known. One of the latest formulas of this type to be published is of order 15.

It should be noted that all formulas of the type (1.4) of order higher than four have the following characteristic property. The number of stages of these formulas is always higher than the order. So, the Butcher formula[1] is of order 6 and has 7

[1] Butcher, J. C., On Runge–Kutta Processes of High Order, *J. Austral. Math. Soc.*, (1964), vol. IV, pt. 2, 179–194.

stages, the Curtis formula is of order 7 and has 9 stages, and so on. In view of the above, Runge–Kutta formulas of order 1–4, including the Euler formula (1.3) and the classic Runge–Kutta formula, are widely used in computations.

3. Error estimates for Runge–Kutta methods.. The knowledge of local errors does not make it possible to estimate the error of the approximate solution over the entire interval of integration $[t_0, T]$. In general, small local errors may build up to produce an appreciable total error. In illustration, we shall consider the following example.

Example 1.1. Let the problem

$$\dot{x}(t) = x(t), \qquad x(0) = 1, \tag{1.10}$$

be solved by means of the Euler method with constant step h:

$$x_{n+1} = x_n + h\dot{x}(t_n) = x_n + hx_n = (1+h)x_n. \tag{1.11}$$

At the point $t_n = nh$ we have

$$\bar{x}(t_n) = e^{nh}, \qquad x_n = (1+h)^n = [(1+h)^{1/h}]^{nh}.$$

As is known from the calculus,

$$(1+h)^{1/h} \to e \quad \text{for } h \to 0, \qquad (1+h)^{1/h} < e \quad \text{for } h > 0. \tag{1.12}$$

If the point $t = t_*$ is fixed and $h \to 0$, then by virtue of (1.12) we have

$$x_{n(h)} \to \bar{x}(t_*), \qquad n(h)h = t_*, \qquad h \to 0.$$

In other words, at any fixed point t_* (and on any finite segment $[t_0, t_*]$) the solution obtained by means of the Euler method converges to the exact solution of equation (1.10). Similarly, we can prove that this assertion is true also for an arbitrary system of the form (1.1).

Now let the step h be fixed, but $n \to \infty$. Then

$$\bar{x}(t_n) - x_n = e^{nh} - a^{nh} = e^{nh}\left[1 - \left(\frac{a}{e}\right)^{nh}\right].$$

Here $a = (1+h)^{1/h} < e$; hence, for h sufficiently large we have

$$\bar{x}(t_n) - x_n \geqslant \tfrac{1}{2}e^{nh}.$$

Thus, the total error of the Euler method increases exponentially with n, the exponent of the exponential being the same as that of the solution itself.

In other words, it is impossible to seek the solution of equation (1.10) and of the more general equation (1.1) on large intervals by means of the Euler method, since the error may reach 50% and more.

In an actual calculation the result depends also on computational errors, mostly due to round-off errors.

Example 1.2. Let us consider again the problem (1.10). We shall consider the effect of computational errors under the assumption that computations are performed not according to formula (1.11) but rather according to

$$\tilde{x}_{n+1} = (1+h)\tilde{x}_n + \delta_n. \tag{1.13}$$

Here δ_n is the value of the round-off error at step n. The exact value of δ_n is not known, but we can assume that for all n the inequality is valid:

$$|\delta_n| \leqslant \delta, \qquad n = 0, 1, \ldots \tag{1.14}$$

The values $\tilde{x}_n$ obtained with the help of the Euler method, with regard to computational errors, are expressed by

$$\tilde{x}_n = (1+h)^n + [\delta_{n-1} + (1+h)\delta_{n-2} + \cdots + (1+h)^{n-1}\delta_0].$$

This yields

$$\begin{aligned}|\bar{x}(t_n) - \tilde{x}_n| &\leqslant |\bar{x}(t_n) - (1+h)^n| + |\delta_{n-1} + \cdots + (1+h)^{n-1}\delta_0| \\ &\leqslant e^{t_n}q(h) + \delta[1 + (1+h) + \cdots + (1+h)^{n-1}] \\ &\leqslant e^{t_n}q(h) + \delta\frac{(1+h)^n - 1}{h} \leqslant e^{t_n}\left[q(h) + \frac{\delta}{h}\right],\end{aligned}$$

where $q(h) = 1 - (1+h)^n e^{-nh}$. Since

$$\begin{aligned}q(h) &= q(0) + hq'(\theta h) \\ &= h[-n(1+\theta h)^{n-1}e^{-n\theta h} + (1+\theta h)^n n e^{-n\theta h}],\end{aligned}$$

it follows that $|q(h)| \leqslant Ch$ (C = constant) for all $h > 0$. Finally, we obtain

$$|\bar{x}(t_n) - \tilde{x}_n| \leqslant e^{t_n}\left[Ch + \frac{\delta}{h}\right]. \tag{1.15}$$

Let us analyze the result obtained. First, as distinct from the case of exact computation, the effect of round-off errors is in that the values $\tilde{x}_n$ do not approach $\bar{x}(t_n)$ as $h \to 0$, since $\delta/h \to \infty$ as $h \to 0$. Thus, in actual computations the solution $\tilde{x}_n$ does not approach the exact value $\bar{x}(t_n)$ of the solution of the problem (1.10).

Further, there exists an optimal step h_{opt} for which the error $|\bar{x}(t_n) - \tilde{x}_n|$ attains a minimum. The exact value of h_{opt} is difficult to evaluate, but $h_{\text{opt}} \approx C^{1/2}\delta^{-1/2}$.

A similar estimate is true also for the general problem (1.1) and the general one-step method (1.2). More precisely, *for the general one-step method of order p having the form*

$$\tilde{x}_{n+1} = \tilde{x}_n + h_n\Phi_f(\tilde{x}_n, t_n, h_n) + \delta_n \tag{1.16}$$

the following estimate holds true for the solution $\bar{x}(t_n)$ of the problem (1.1):

$$|\bar{x}(t_n) - \tilde{x}_n| \leqslant C\left[Mh^p + \frac{\delta}{h}\right]e^{L(t_n - t_0)}, \tag{1.17}$$

where C, M, L are constants specified by the functions f and Φ_f.

For the proof of the validity of the estimate (1.17), see the references [4, 5, 13, 19, 22].

Example 1.3 [4]. In illustration of the above results, we shall consider a numerical example. For the problem

$$\dot{x}(t) = t(t+2)x^3(t) + (t+3)x^2(t), \qquad x(0,5) = -1,6, \tag{1.18}$$

we shall try to find a solution at the point $t = 1$ using Runge–Kutta formulas of order 3, 4, and 6 and with different steps. The exact solution of the problem (1.18) is known:

$$x(t) = -2t^{-1}(t+2)^{-1}, \qquad x(1) = -\frac{2}{3}.$$

This allows to determine the total error $\varepsilon = |\bar{x}(1) - \tilde{x}_n|$.

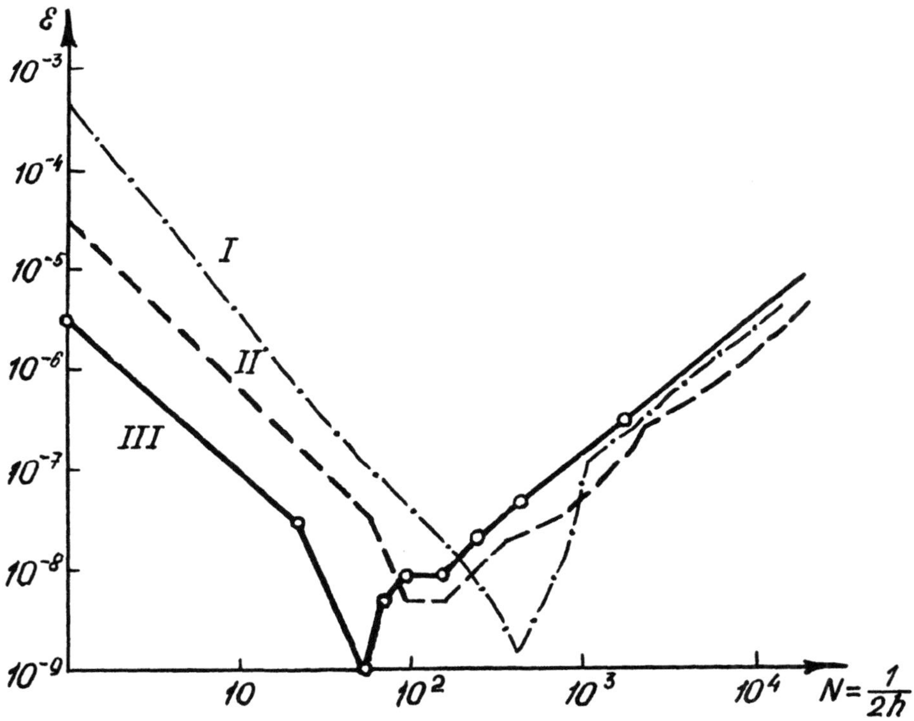

FIG. 1.1. The total error of the solution of equation (1.18) through the use of the Runge–Kutta method:
I, the Runge–Kutta formula of the 3d order; *II*, the Runge–Kutta formula of the 4th order; *III*, the Huta formula of the 6th order

The computation results are illustrated in Figure 1.1, where the logarithmic scale is used for the axes. As is seen from Figure 1.1, for each method (formula) there exist an optimal step h_{opt} and a corresponding division number $N_{\text{opt}} =$

$1/2h_{\text{opt}}$. Thus, for the Runge–Kutta formula of order 3 the optimal division number is $N^3_{\text{opt}} \approx 250$, in the case of four stages it is $N^4_{\text{opt}} \approx 100$, and in the case of six stages it is $N^6_{\text{opt}} \approx 50$. Further, to ensure that the total error of the solution is equal to $\varepsilon = 3 \cdot 10^{-8}$, it is necessary to divide the segment of integration into $N^6_\varepsilon = 20$, $N^4_\varepsilon \approx 65$, and $N^3_\varepsilon = 200$ subintervals. The number of computations Q for evaluating the right side of equation (1.18) in order to achieve the accuracy ε is as follows: $Q^6_\varepsilon = 20 \cdot 8 = 160$, $Q^4_\varepsilon = 65 \cdot 4 = 260$, and $Q^4_\varepsilon = 200 \cdot 3 = 600$.

Therefore, for equation (1.18) the Runge–Kutta method of the sixth order is the most economical. However, for other equations the Runge–Kutta method of different order may prove to be optimal. At present, there is no theory which can furnish an optimal method of solution for a given equation. Only the comparison of numerical results computed according to different methods can specify the best numerical method for a given equation. But if such a comparative investigation (even for a similar problem) is lacking, then, as a rule, the Runge–Kutta method of the fourth order is used.

4. Estimation of solution errors for stable equations. In the general case the estimate (1.17) cannot be improved, and the solution error may increase exponentially as the duration of the interval of integration increases. This is confirmed by the foregoing Example 1.1, where a lower bound on the error was obtained.

However, for narrower classes of equations a non-exponentially increasing estimate can be obtained as is shown in the following example.

Example 1.4. Consider the equation

$$\dot{x}(t) = -\lambda x(t), \qquad x(0) = 1,\ \lambda > 0. \tag{1.19}$$

As distinct from equation (1.10), the solution $x(t) \equiv 0$ of equation (1.19) is asymptotically stable. The exact solution of (1.19) is

$$\bar{x}(t) = \exp(-\lambda t), \qquad \bar{x}(t) \to 0,\ t \to \infty.$$

Using the Euler method, we try to find a solution to equation (1.19) and immediately take account of the computational error. We obtain the following equations, which are similar to (1.13):

$$\begin{aligned} \tilde{x}_{n+1} &= (1 - \lambda h)\tilde{x}_n + \delta_n, \\ \tilde{x}_{n+1} &= (1 - \lambda h)^{n+1} + [\delta_n + (1 - h)\delta_{n-1} + \cdots + (1 - h)^n \delta_0]. \end{aligned} \tag{1.20}$$

If $0 < h < 2\lambda^{-1}$, then the first term of the sum in (1.20) appoaches zero as $n \to \infty$.

Let us show that for each $\varepsilon > 0$ there exists a step h and a computational error level δ such that

$$|\bar{x}(t_n) - \tilde{x}_n| \leqslant \varepsilon, \qquad n = 0, 1, \ldots .$$

Indeed, using the property of an infinite geometric progression, we have

$$|\bar{x}(t_n) - \tilde{x}_n| \leqslant |\bar{x}(t_n) - (1 - \lambda h)^n)| + \delta[1 + \cdots + (1 - \lambda h)^n + \cdots] \leqslant$$
$$\leqslant |\bar{x}(t_n) - (1 - \lambda h)^n| + \frac{\delta}{\lambda h}.$$

In view of the above, we can ensure by choosing a suitably small h_0 that $|\bar{x}(t_n) - (1 - \lambda h)^n| \leqslant \varepsilon/2$ on any finite interval $0 \leqslant n \leqslant N$. Choosing N sufficiently large so that the following inequalities hold true: $|\bar{x}(t_N)| \leqslant \varepsilon/4$, $(1 - \lambda h)^N \leqslant \varepsilon/4$, we conclude that for $h \leqslant h_0$ the following inequality holds uniformly for all n ($0 \leqslant n < \infty$):

$$|\bar{x}(t_n) - (1 - \lambda h)^n| \leqslant \frac{\varepsilon}{2}, \qquad n = 0, 1, \ldots .$$

Let us choose a δ so that $\delta/(\lambda h) \leqslant \varepsilon/2$; then we obtain finally

$$|\bar{x}(t_n) - \tilde{x}_n| \leqslant \varepsilon, \qquad h \leqslant h_0; \qquad 2\delta \leqslant \varepsilon\lambda h, \qquad n = 0, 1, \ldots . \tag{1.21}$$

The estimate (1.21) can be extended to a wider class of systems. Namely, the following assertion holds true. *Let* $f(t, 0) \equiv 0$ *and let the trivial solution of equation* (1.1) *be stable under steady-acting disturbances* (see Definition 1.8 in § 1 of Chapter I). *Further, let the general one-step formula* (1.16) *be used to solve equation* (1.1). *Then for any* $\varepsilon > 0$ *there exist an* h_0 *and a* $\delta > 0$ *such that the following inequalities hold for all* n:

$$|\tilde{x}_n| \leqslant \varepsilon, \qquad n = 0, 1, \ldots, \tag{1.22}$$

if $|\tilde{x}_0| \leqslant \delta$, $h \leqslant h_0$, $|\delta_n| \leqslant \delta h$.

The proof of the validity of the estimate (1.22) is given in [4, 5, 13, 19, 22].

The estimate (1.22) implies the hope that having made a suitable choice of the integration step, the solution, subject to a small error requirement, can be obtained with great accuracy uniformly for all n through the use of the Runge–Kutta method, provided that the original equation was stable under steady-acting disturbances.

5. Standard programs. To analyse transients, standard programs can conveniently be used. Standard programs with an automatic choice of the integration step are most widely applied.

Runge–Kutta methods are implemented by programs with automatic choice of the step, using one of the following approaches: (a) the Runge principle of the *a posteriori* estimation of accuracy; (b) the use of control terms; (c) the use of embedded methods.

The *Runge principle* is based on the comparison of the results obtained by integrating with steps h and $h/2$. It provides a highly reliable estimate of computation accuracy in each step, but requires too many calculations. For example, in the method of the fourth order the right side has to be evaluated 11 times.

Because of this deficiency, this method is presently almost never used in standard programs.

When *control terms* are used, the goal is to estimate a local error, given already-computed values $K_1, K_2, \ldots, K_p$. For example, the control term for the classic Runge–Kutta formula may be taken as

$$\Delta = K_1 - K_2 - K_3 + K_4 \tag{1.23}$$

(for explicit values of K_i, see Table 1.1). Further, we must specify the required computation accuracy ε and the initial step h_0. The automatic choice of the step is implemented in the following way. The quantities K_1, K_2, K_3, K_4, and Δ are evaluated with step h_0 according to formula (1.23). If $\Delta > \varepsilon$, then we assume that $h_0 = h_0/2$ and repeat the computations. If $\varepsilon/32 \leqslant \Delta \leqslant \varepsilon$, then we regard the value of x_1 as computed with accuracy ε, and proceed to the computation of x_2, given that $h_1 = h_0$. But if $\Delta < \varepsilon/32$, then we set $h_1 = 2h_0$. It can be shown that the step stabilizes to the value h_* which assures the stability of the method.

In recent years standard programs which are based on *embedded methods* have found wide application. One of the most widely used programs of this sort implements the *Runge–Kutta-Feldberg formula* [11]. This formula has six stages and is of the fifth order; its coefficients are listed in Table 1.2. The computations are carried out as follows.

TABLE 1.2

α_i	β_{ij}					ω_i	ω_i^*
0						$\frac{16}{35}$	$\frac{25}{216}$
$\frac{1}{4}$	$\frac{1}{4}$					0	0
$\frac{3}{8}$	$\frac{3}{32}$	$\frac{9}{32}$				$\frac{6656}{12825}$	$\frac{1408}{2565}$
$\frac{12}{13}$	$\frac{1932}{2197}$	$-\frac{7200}{2197}$	$\frac{7296}{2197}$			$\frac{28561}{56430}$	$\frac{2197}{4104}$
1	$\frac{439}{216}$	-8	$\frac{3680}{513}$	$-\frac{845}{4104}$		$-\frac{9}{50}$	$-\frac{1}{5}$
$\frac{1}{2}$	$-\frac{8}{27}$	2	$-\frac{3544}{2565}$	$\frac{1859}{4104}$	$-\frac{11}{40}$	$\frac{2}{55}$	0

First, the quantity

$$x_{n+1} = x_n + \sum_{i=1}^{6} \omega_i K_i.$$

is computed. Then the estimate of local accuracy

$$\Delta_{n+1} = \sum_{i=1}^{6} (\omega_i - \omega_i^*) K_i. \tag{1.24}$$

is determined.

Formula (1.24) substitutes for the control term in Runge–Kutta formulas. The value of Δ_{n+1} is compared with the prescribed accuracy ε, and the step is either decreased (by a factor of 10 or less) or increased (by a factor of 5 or less). If the preceding step was a reduced one, then there can be no increase in the value of the subsequent step. A standard FORTRAN RFK45 procedure implementing this method is given in [11].

A large number of various FORTRAN programs developed and thoroughly tested at the Computer Center of Moscow University is given in [3]. Some other FORTRAN programs are also given in [13].

§ 2. Analysis of Transients. Multistep Methods

1. General definitions. In the preceding section we were concerned with one-step methods for solving the Cauchy problem for ordinary differential equations. In addition to one-step methods *multistep* (or *difference*) methods are often used. In this section we deal with the fundamentals of the theory of multistep methods.

For the numerical solution of the Cauchy problem

$$\dot{x}(t) = f(t, x), \qquad t_0 \leqslant t \leqslant T,\ x(t_0) = x_0, \tag{2.1}$$

on the segment $[t_0, T]$ we introduce a uniform grid with step h:

$$t_i = t_0 + ih, \qquad i = 0, 1, \dots, N,\ N = (T - t_0)/h. \tag{2.2}$$

We introduce the notation $x_i = x(t_i)$, $f_i = f(t_i, x_i)$. The relation

$$\sum_{p=0}^{k} \alpha_p x_{i+p} = h \sum_{p=0}^{k} \beta_p f(t_{i+p}, x_{i+p}) \tag{2.3}$$

will be called the *general multistep formula of order k* (or the *k-step formula*). Here α_p and β_p are certain prescribed numbers, $\alpha_k \neq 0$, $|\alpha_0| + |\beta_0| \neq 0$. A particular multistep formula is specified by the particular collection of coefficients α_p, β_p. Formulas of the form (2.3) are often called *difference formulas.* Formula (2.3) is a linear relation between x_i and f_i; hence, formula (2.3) is also called the *linear multistep method.*

To determine the sequence of the values of x_i by means of the k-step method (2.3), it is necessary first to compute k initial values of the form

$$x_0 = x_{0,0},\ x_1 = x_{1,0},\ \dots,\ x_{k-1} = x_{k-1,0}. \tag{2.4}$$

If for $\alpha_k \neq 0$ the coefficient $\beta_k = 0$, then formula (2.3) is called an *explicit* (or *extrapolation formula*). But if $\beta_k \neq 0$, then formula (2.3) is called an *implicit* (or *interpolation formula*). For formula (2.3) to locally approximate the solution of the problem (2.1), it is necessary that the numbers α_p and β_p satisfy certain relations. Expanding the solution of the problem (2.1) as a Taylor series up to power l, we have

$$\bar{x}(t) = q_l(t) + O((t-t_0)^{l+1}), \tag{2.5}$$

where $q_l(t)$ is a polynomial of degree l.

Next, we require that for any polynomial $r(t)$ with the number of stages at most l the following identities hold:

$$\sum_{p=0}^{k} \alpha_p r(t+ph) = \sum_{p=0}^{k} \beta_p r'(t+ph). \tag{2.6}$$

Substituting in (2.6) the powers $1, t, \ldots, t^l$ for $r(t)$ we establish that the following conditions must be satisfied:

$$\begin{gathered}\sum_{p=0}^{k} \alpha_p = 0, \qquad \sum_{p=0}^{k} p\alpha_p = \sum_{p=0}^{k} \beta_p, \\ \cdots\cdots\cdots\cdots\cdots\cdots\cdots\cdots \\ \cdots\cdots\cdots\cdots\cdots\cdots\cdots\cdots \\ \sum_{p=0}^{k} \frac{p^s \alpha_p}{s!} = \sum_{p=0}^{k} \frac{p^{s-1}\beta_p}{(s-1)!}, \qquad s = 2, \ldots, l.\end{gathered} \tag{2.7}$$

Clearly, the identities (2.6) immediately follow from the identities (2.7). In view of the equality (2.5) we find that the local error introduced by replacing the differential equation (2.1) by the difference equation (2.3), subject to the conditions (2.7), is of the order h^{l+1}, i.e.,

$$\bar{x}(t_k) - x_k = O(h^{l+1}).$$

Therefore, the multistep formula (2.3) is said to be *of order* l if its coefficients α_p, β_p satisfy the relations (2.7) up to $s = l$.

2. Particular multistep formulas. We shall give without proof several multistep formulas often used in computations.

The *explicit Adams formulas* can be written as

$$\begin{aligned} x_{k+1} = x_k + h\big(f_k &+ \tfrac{1}{2}\Delta f_{k-1} + \tfrac{5}{12}\Delta^2 f_{k-2} + \tfrac{3}{8}\Delta^3 f_{k-3} + \tfrac{251}{720}\Delta^4 f_{k-4} \\ &+ \tfrac{95}{288}\Delta^5 f_{k-5} + \tfrac{19\,087}{60\,480}\Delta^6 f_{k-6} + \cdots\big). \end{aligned} \tag{2.8}$$

In formula (2.8) we have used the standard notation for finite differences:

$$\begin{aligned} \Delta f_i &= f_{i+1} - f_i, \\ \Delta^2 f_i &= \Delta f_{i+1} - \Delta f_i = f_{i+2} - 2f_{i+1} + f_i, \\ \Delta^3 f_i &= \Delta^2 f_{i+1} - \Delta^2 f_i = f_{i+3} - 3f_{i+2} + 3f_{i+1} - f_i, \ldots \end{aligned}$$

The *implicit Adams formulas* are of the form

$$x_{k+1} = x_k + h\big(f_{k+1} - \tfrac{1}{2}\Delta f_k - \tfrac{1}{12}\Delta^2 f_{k-1} - \tfrac{1}{24}\Delta^3 f_{k-2} - \tfrac{19}{720}\Delta^4 f_{k-3} - \tfrac{3}{160}\Delta^5 f_{k-4} - \tfrac{863}{60\,480}\Delta^6 f_{k-5} - \cdots\big). \tag{2.9}$$

Terminating formula (2.8) or (2.9) after some term, we obtain a computation formula of particular order. For example, keeping in formula (2.8) terms up to $\Delta^2 f_{k-2}$, we get

$$\begin{aligned} x_{k+3} &= x_{k+2} + h\big(f_{k+2} + \tfrac{1}{2}\Delta f_{k+1} + \tfrac{5}{12}\Delta^2 f_k\big) \\ &= x_{k+2} + h\big[f_{k+2} + \tfrac{1}{2}(f_{k+2} - f_{k+1}) + \tfrac{5}{12}(f_{k+2} - 2f_{k+1} + f_k)\big] \\ &= x_{k+2} + h\big[\tfrac{23}{12}f_{k+2} - \tfrac{4}{3}f_{k+1} + \tfrac{5}{12}f_k\big]. \end{aligned} \tag{2.10}$$

It is seen that formula (2.10) is a formula of the third order, in which

$$\alpha_3 = 1, \qquad \alpha_2 = -1, \qquad \alpha_1 = \alpha_0 = 0,$$
$$\beta_3 = 0, \qquad \beta_2 = \frac{23}{12}, \qquad \beta_1 = -\frac{4}{3}, \qquad \beta_0 = \frac{5}{12}.$$

It is easily verified that that these coefficients satisfy the relations

$$\sum_{p=0}^{3}\alpha_p = 0, \qquad \sum_{p=0}^{3}p\alpha_p = \sum_{p=0}^{3}\beta_p, \qquad \sum_{p=0}^{3}\frac{p^2\alpha_p}{2!} = \sum_{p=0}^{3}p\beta_p. \tag{2.11}$$

Hence, formula (2.10) is of the second order. In a similar way, the implicit formula (2.9) of the third order is found to be

$$\begin{aligned} x_{k+3} &= x_{k+2} + h\big[f_{k+3} - \tfrac{1}{2}\Delta f_{k+2} - \tfrac{1}{12}\Delta^2 f_{k+1} - \tfrac{1}{24}\Delta^3 f_k\big] \\ &= x_{k+2} + h\big[f_{k+3} - \tfrac{1}{2}(f_{k+3} - f_{k+2}) - \tfrac{1}{12}(f_{k+3} - 2f_{k+2} + f_{k+1}) \\ &\qquad - \tfrac{1}{24}(f_{k+3} - 3f_{k+2} + 3f_{k+1} - f_k)\big] \\ &= x_{k+2} + h\big[-\tfrac{9}{24}f_{k+3} + \tfrac{19}{24}f_{k+2} - \tfrac{5}{24}f_{k+1} + \tfrac{1}{24}f_k\big], \end{aligned} \tag{2.12}$$

where

$$\alpha_3 = 1, \qquad \alpha_2 = -1, \qquad \alpha_1 = \alpha_0 = 0,$$
$$\beta_3 = \frac{9}{24}, \qquad \beta_2 = \frac{19}{2}4 \qquad \beta_1 = -\frac{5}{24}, \qquad \beta_0 = \frac{1}{24}.$$

Since, in addition to the relations (2.11), the coefficients in formula (2.12) also satisfy the equality

$$\sum_{p=0}^{3}\frac{p^3\alpha_p}{3!} = \sum_{p=0}^{3}\frac{p^2\beta_p}{2!},$$

formula (2.12) is of the third order. For all other formulas having the same order the implicit Adams formula is of order one more than the explicit formula.

For the numerical integration in the problem

$$\ddot{x}(t) = f(t,x) \qquad x(t_0) = x_0,\ \dot{x}(t_0) = x_1 \tag{2.13}$$

the following *Störmer formulas* are used:

the *explicit Störmer formula*

$$\begin{aligned} &x_{k+1} - 2x_k + x_{k-1} \\ &\quad = h^2\big(f_k + \tfrac{1}{12}\Delta^2 f_{k-2} + \tfrac{1}{12}\Delta^3 f_{k-3} + \tfrac{19}{240}\Delta^4 f_{k-4} + \tfrac{3}{40}\Delta^5 f_{k-5} + \cdots\big), \end{aligned} \tag{2.14}$$

the *implicit Störmer formula*

$$\begin{aligned} &x_{k+1} - 2x_k + x_{k-1} \\ &\quad = h^2\big(f_k + \tfrac{1}{12}\Delta^2 f_{k-1} - \tfrac{1}{240}\Delta^4 f_{k-3} - \tfrac{1}{240}\Delta^5 f_{k-4} - \cdots\big). \end{aligned} \tag{2.15}$$

Since

$$\begin{aligned} \Delta^2 f_{k-1} &= \Delta f_k - \Delta f_{k-1} = f_{k+1} - 2f_k + f_{k-1}, \\ \Delta^4 f_{k-3} &= f_{k+1} - 4f_k + 6f_{k-1} - 4f_{k-2} + f_{k-3}, \ldots, \end{aligned}$$

it follows that formula (2.15) is a nonlinear equation in x_{k+1}, which enters explicitly on the left side of equation (2.15) and implicitly on the right side via the function $f_{k+1} = f(t_{k+1}, x_{k+1})$. The Störmer formulas are more convenient to use than to try to reduce the problem (2.13) to a Cauchy problem for the system of two equations of the first order.

To solve the problem (2.1), *predictor-corrector methods* are also often used. Among them the *Milne formulas* are the most popular:

the *Milne predictor formula*

$$x_i^{\text{pred}} = x_{i-4} + \frac{4h}{3}(2f_{i-3} - f_{i-2} + 2f_{i-1}), \qquad f_i^{\text{pred}} = f(t_i, x_i^{\text{pred}}); \tag{2.16}$$

the *Milne corrector formula*

$$x_i^{\text{corr}} = x_{i-2} + \frac{h}{3}(f_{i-2} + 4f_{i-1} + f_i^{\text{pred}}); \tag{2.17}$$

the *Milne control formula*

$$\varepsilon_i^{\text{corr}} = |x_i^{\text{corr}} - \bar{x}(t_i)| \simeq \tfrac{1}{29}|x_i^{\text{corr}} - x_i^{\text{pred}}|.$$

3. Computational scheme for multistep formulas. The computation of a solution according to multistep formulas (2.3) is carried out in the following way.

First, the initial data are determined, i.e., the values $x_1, x_2, \dots, x_{k-1}$. The initial data cannot be determined by the application of the multistep formula (2.3). Any one-step method, such as the Runge–Kutta method of the fourth order, can be used to determine the initial values. Other methods of finding the initial data are known, such as the Krylov method discussed in detail in [6].

Further, to determine subsequent values $x_k, x_{k+1}, \dots$, we can use formula (2.3) directly. However, methods involving explicit and implicit formulas differ essentially. Explicit formulas define certain recurrence relations which are used for computing in succession the values $x_k, x_{k+1}, \dots$. For example, let the explicit Adams formula (2.8) be used:

$$x_{k+3} = x_{k+2} + h\big(f_{k+2} + \tfrac{1}{2}\Delta f_{k+1} + \tfrac{5}{12}\Delta^2 f_k\big). \tag{2.18}$$

The initial data for this formula consists of the points x_0, x_1, x_2. Let us assume that these values are determined somehow. The computational process is illustrated in Table 2.1.

TABLE 2.1

t_i	x_i	f_i	Δf_i	$\Delta^2 f_i$
t_0	x_0	f_0	Δf_0	$\Delta^2 f_0$
t_1	x_1	f_1	Δf_1	$\Delta^2 f_1$
t_2	x_2	f_2	Δf_2	
t_3	x_3	f_3		
t_4	x_4			
$\vdots$	$\vdots$			

First, we compute f_0, f_1, f_2 and then the differences above the broken line in Table 2.1. After that, using formula (2.18), we can obtain x_3, and then compute f_3 and the differences Δf_2 and $\Delta^2 f_1$. This enables us to determine x_4, and so on. The following important characteristic of explicit Adams formulas should be noted. Regardless of the order of the formula, the determination of each subsequent value x_{i+1} requires the computation of only one new value of the function f_i.

An implicit formula of the form (2.9) for each value x_{i+p} is a nonlinear algebraic equation, which can be written as follows:

$$x_{i+p} = \frac{h\beta_p}{\alpha_p} f(t_{i+p}, x_{i+p}) + \varphi_p, \qquad \beta_p \neq 0, \tag{2.19}$$

where φ is a known function of $t_i, \dots, t_{i+p-1}, x_i, \dots, x_{i+p-1}$ and h. To determine x_{i+p}, from equation (2.19), we can use the iteration method. Given a value of $x_{i+p} = x^0_{i+p}$, we obtain in succession $x^1_{i+p}, x^2_{i+p}, \dots$ according to the formula

$$x^k_{i+p} = \frac{h\beta_p}{\alpha_p} f(t_{i+p}, x^{k-1}_{i+p}) + \varphi_p.$$

As soon as the difference $|x^k_{i+p} - x^{k-1}_{i+p}|$ becomes less than a prescribed number, the iteration process is terminated. It can be shown that for a step h sufficiently small (more precisely, for $h \leqslant |\alpha_p/\beta_p|/L$, where L is a Lipschitz constant for the function $f(t,x)$ with respect to x) the iteration method is convergent for any choice of x^0_{i+p}. However, the number of iterations depends on the proximity of x^0_{i+p} to the exact root of equation (2.19). Each iteration step requires one computation of the right side $f(t,x)$.

Although implicit methods possess greater local accuracy as compared to explicit methods, the determination of each value x_{i+p} is more time consuming. Therefore, we cannot state *a priori* that the implicit formulas have an advantage over the explicit ones; likewise, we cannot state the contrary. For some equations it is more advantageous to use an explicit formula, whereas for others an implicit formula is preferable. The issue of selecting a formula for the numerical integration of the problem (2.1) has not been finally settled yet.

Let us now compare multistep formulas with Runge–Kutta one-step formulas. First of all, multistep formulas require a special procedure for determining the initial values; this makes it difficult to change the step in the computational process. Further, if we compare the four-stage explicit Adams formula with the Runge–Kutta formula with the same number of stages, we see that the constant in the estimate of the local error for the Adams formula is almost 1 000 times greater than for the Runge–Kutta formula. This means that, given the same accuracy, the step for the Adams formula must be taken $\sqrt[4]{1\,000} \approx 5.7$ less than for the Runge–Kutta formula. Thus, despite the fact that in each step we need one computation of $f(t,x)$ for the Adams formula and four computations of $f(t,x)$ for the Runge–Kutta formula, the Runge–Kutta method is less time consuming. Since the Runge–Kutta method furnishes an opportunity to effect an automatic change of step, this method is the method that is most widely used in computation work.

It should be noted that in recent years several good programs with variable step and variable order based on the Adams method have been published [3, 8, 13, 19].

4. Error estimation. In using multistep methods, it is important to estimate the error for a large number of steps and not just the local error for one step. The fact is that not all multistep methods are numerically stable.

Example 2.1. To solve the problem

$$\dot{x}(t) = -x, \qquad x(0) = 1, \tag{2.20}$$

we use the formula of the second order

$$x_{n+2} + 2x_{n+1} - 3x_n = h[3f(t_{n+1}, x_{n+1}) + f(t_n, x_n)]. \tag{2.21}$$

The coefficients in formula (2.21) satisfy the relations $1+2-3=0$, $2\cdot 1+2=3+1$, which for this formula coincide with the following general relations:

$$\sum_{p=0}^{2} \alpha_p = 0, \qquad \sum_{p=0}^{2} p\alpha_p = \sum_{p=0}^{2} \beta_p. \tag{2.22}$$

It follows from the relations (2.22) that formula (2.21) has two stages and is of the second order.

Let us consider the step $h = 0.1$, $x_0 = 1$, $x_1 = e^{-0.1} \approx 0.905$. Formula (2.21) for equation (2.20) and $h = 0.1$ becomes

$$x_{n+2} = -2.3x_{n+1} + 3.1x_n, \qquad n \geqslant 2,\ x_0 = 1,\ x_1 = 0.905.$$

TABLE 2.2

n	x_n	n	x_n	n	x_n
0	1.00000	4	2.09685	8	144.17409
1	0.90484	5	−3.39187	9	−466.16637
2	1.01887	6	14.30656	10	1519.12234
3	0.46159	7	−43.40837		

The values of x_n, computed to five decimal places, are given in Table 2.2. It is seen from the table that the module of the solution x_n increases as $n \to \infty$, whereas the exact solution of the problem (2.20) is of the form $\bar{x}(t) = e^{-t} \to 0$ as $t \to \infty$. Thus, formula (2.21) cannot be applied for the numerical solution of the problem (2.20).

The reason for numerical instability of some (but not all) multistep formulas is that the orders of the original differential equation and of the approximating difference equation may differ. Moreover, the difference equation may have additional solutions, which lead to a substantial difference of the solutions of equations (2.1) and (2.3).

It is particularly important to investigate the behavior of the solutions obtained through the use of the multistep formula (2.3) for h small. From the continuous h-dependence of the solutions of (2.3) it follows that for h small the solutions of equation (2.3) lie in the proximity of the solutions of the equation

$$\sum_{p=0}^{k} \alpha_p x_{i+p} = 0. \tag{2.23}$$

If the trivial solution of equation (2.23) is stable in the sense of Lyapunov (Lyapunov-stable), then no additional dangerous solutions of equation (2.3) exist. This implies the following definition[2] [3, 4, 13, 19].

Definition 2.1. The multistep formula (2.3) is said to be *numerically stable* (*Dahlquist-stable*) if the trivial solution of equation (2.23) is Lyapunov-stable.

It will be recalled that for equation (2.23) to be Lyapunov-stable it is necessary and sufficient that the roots of the characteristic equation

$$\sum_{p=0}^{k} \alpha_p \lambda^p = 0 \tag{2.24}$$

lie inside the unit circle and there be no multiple roots on the circle $|\lambda| = 1$ (see § 11 of Chapter I). Therefore, another definition which is equivalent to the preceding one can be given.

Definition 2.2. The multistep formula (2.3) is said to be *numerically stable* (*Dahlquist-stable*) if all the roots of equation (2.24) satisfy the condition $|\lambda| \leqslant 1$, there being no multiple roots on the circle $|\lambda| = 1$.

We shall bring the following assertion without proof [13, 19].

Suppose that the actual computation is carried out according to the formula

$$\sum_{p=0}^{k} \alpha_p \tilde{x}_{i+p} = h \sum_{p=0}^{k} \beta_p f(t_{i+p}, \tilde{x}_{i+p}) + \delta_i, \qquad i = k, k+1, \ldots, \tag{2.25}$$
$$\tilde{x}_0 = x_{0,0} + \theta_0, \ldots, \tilde{x}_{k-1} = x_{k-1,0} + \theta_{k-1}.$$

Here δ_i are some numbers which characterize the computational errors, and θ_i are the errors of the initial values. We assume that

$$|\delta_i| \leqslant \delta, \qquad i = \ldots, k, k+1, k+2, \ldots. \tag{2.26}$$

Theorem 2.1 (G. Dahlquist)[3]. *Suppose that the multistep formula* (2.3) *is of order l and is Dahlquist-stable. Then for $t_0 \leqslant t_i \leqslant t_0 + T$ the following inequality is satisfied*:

$$|\bar{x}(t_i) - \tilde{x}_i| \leqslant e^{C_1 T}\left\{C_2 h^l + C_3 \frac{\delta}{h} + C_4 \max_{0 \leqslant i \leqslant k-1} |\theta_i|\right\}, \tag{2.27}$$

---

[2] Dahlquist, G., Stability and Error Bounds in the Numerical Integration of Ordinary Differential Equations, *Trans. Royal Inst. of Techn.*, Stockholm, (1959), no. 130.

[3] Results similar to those of Dahlquist were obtained one year earlier in the paper Lozinskii, S. M., Error Evaluation for Numerical Integration of Ordinary Differential Equations, *Izv. Vyssh. Uchebn. Zaved. Mat.*, (1958), no. 5, 52-90. (Translated into English as *Soviet Math. (Iz. VUZ)*)

where C_i are some constants depending on the Lipschitz constants of the functions $f(t,x)$, $f'_x(t,x)$.

In the estimate (2.27) the first term describes the error of the method, the second the computational error, and the third the error due to the inaccurate determination of the initial conditions (2.4). Generally speaking, the error increases exponentially as the interval $[t_0, t_0 + T]$ increases. In general, the estimate (2.27) can be characterized similarly to the estimate (1.17). Thus, multistep Dahlquist-stable formulas are similar to one-step formulas errorwise.

To obtain error estimates for a sufficiently large time interval, we must impose more severe requirements on formula (2.3).

Definition 2.3. The multistep formula (2.3) is said to be *strongly Dahlquist-stable* if all the roots λ_i of equation (2.24) except the root $\lambda_1 = 1$ satisfy the condition $|\lambda| < 1$.

It should be noted that in view of the first of the equalities (2.7) equation (2.24) always has the root $\lambda_1 = 1$.

Let $f(t,0) = 0$, and let the trivial solution of equation (2.1) be stable under steady-acting disturbances (see § 1 of Chapter I).

Theorem 2.2. *Suppose that the strongly Dahlquist-stable method* (2.3) *is employed for solving the problem* (2.1) *whose trivial solution is stable under steady-acting disturbances. Then for all $\varepsilon > 0$ there exist $h > 0$ and $\delta > 0$ such that any solution $\tilde{x}_n$ of the difference equation* (2.25) *satisfies the inequality $|\tilde{x}_n| < \varepsilon$, $n = 0, 1, 2\ldots$, if $|\tilde{x}_i| < h\delta$, $i = 0, 1, \ldots, k-1$.*

5. The Butcher formulas[4]. In addition to one-step and multistep methods other approaches combining the merits of both methods are used. In *Butcher formulas*, as in multistep methods, several previous values $x_{n-1}, x_{n-2}, \ldots$ are used to compute the next value x_n and at the same time in each step the right side is computed several times as in Runge–Kutta methods. As an example, let us consider the two-step Butcher formula, according to which three computations of the right side are performed in each step:

$$\begin{aligned}
x_{n-\frac{1}{2}} &= x_{n-2} + \frac{h}{8}(9f_{n-1} + 3f_{n-2}),\\
x_n^1 &= \frac{1}{5}(28x_{n-1} - 23x_{n-2}) + \frac{h}{15}(32f_{n-\frac{1}{2}} - 60f_{n-1} - 26f_{n-2}),\\
x_n &= \frac{1}{31}(32x_{n-1} - x_{n-2}) + \frac{h}{93}(64f_{n-\frac{1}{2}} + 15f_n^1 + 12f_{n-1} - f_{n-2}),\\
&\qquad f_i^k = f(t_i, x_i^k).
\end{aligned} \tag{2.28}$$

The local error of formula (2.28) is equal to $O(h^6)$. In [13, 19, 22] other similar formulas are given and their properties established.

[4]Butcher, J. C., A Multistep Generalization of Runge–Kutta Methods with Four or Five Stages, *J. ACM*, (1967), vol. 14, 84–99.

§ 3. Stiff Systems of Equations

In the investigation of various particular engineering, physical, and biological systems we encounter the so-called "stiffness" phenomenon. Figuratively speaking, this phenomenon consists in the description of a process observed in some time interval by two functions: rapidly varying functions with large derivatives (fast motions) on certain small segments and functions with small derivatives (slow motions) on the remaining part of the interval. Peculiar difficulties in the study of such systems arise if the starting moments for rapid motions are not known beforehand. Stiffness phenomena are especially characteristic of advanced dynamical systems described by systems of differential equations of high order.

One of the first examples of such systems was investigated by A. N. Tikhonov, V. P. Maslov, A. B. Vasilieva, and others. They investigated *systems involving a small parameter at higher derivatives*, or *singular perturbation systems*. Such systems are described by equations of the form

$$\begin{aligned} \mu\dot{y}(t) &= f(t,y,z), \qquad y \in R^m,\ y(t_0) = y_0, \\ \dot{z}(t) &= g(t,y,z), \qquad z \in R^l,\ z(t_0) = z_0. \end{aligned} \tag{3.1}$$

Here $\mu > 0$ is a small number (small parameter).

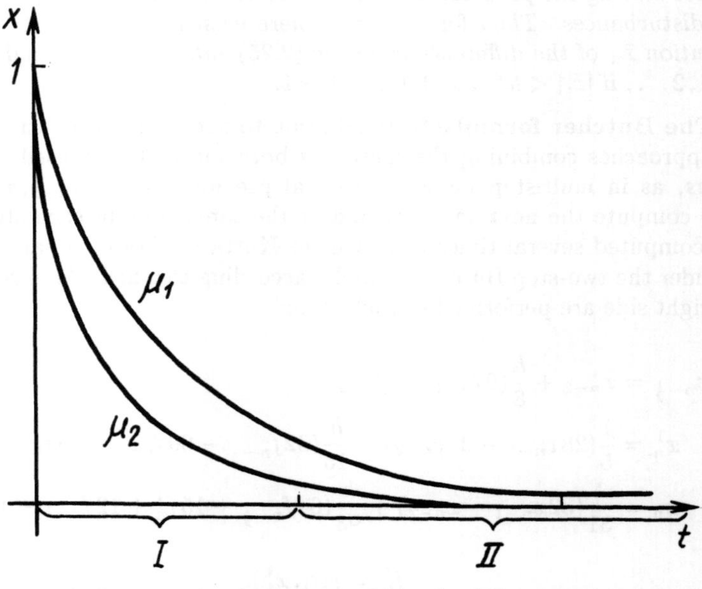

FIG. 3.1. Solutions $x(t,\mu)$ of the problem (3.2) for $\mu_1 > \mu_2$: *I*, the boundary layer region; *II*, the regular solution region

Example 3.1. Let

$$\mu\dot{x}(t) = -ax(t), \qquad x(0) = 1. \tag{3.2}$$

Then

$$x(t,\mu) = e^{-\frac{a}{\mu}t}.$$

For $a > 0$ and for $\mu > 0$ sufficiently small the solution $x(t, \mu)$ is of the form shown in Figure 3.1. As is seen from the figure, the whole time interval $[0, \infty)$ can be divided into two portions. Portion I from $t = 0$ to $t = O(\mu)$, where the solution varies rapidly from 1 to 0 and the derivative is $O(1/\mu)$, is called the *boundary layer region*. Portion II, where t changes from $O(\mu)$ to ∞ and the solution is essentially constant and equal to zero, is called the *regular part* of the solution.

Often we cannot explicitly isolate terms involving a small parameter, but the solution, nevertheless, also consists of portions of fast and of slow motion. This is especially true of systems of differential equations. Such is, for example, the linear time-invariant system

$$\dot{x}(t) = Ax(t), \qquad x \in R^n, \tag{3.3}$$

for which the eigenvalues $\lambda_1, \dots, \lambda_n$ of the matrix A differ widely in magnitude, i.e.,

$$\operatorname{Re}\lambda_1 \ll \operatorname{Re}\lambda_n < 0. \tag{3.4}$$

In this case the component of the solution corresponding to the eigenvalue λ_1 will affect the solution only for those t which are near to the initial value $t = 0$. For large values of t the component corresponding to the eigenvalue λ_n has the greatest effect on the solution of the system (3.3). The numerical integration of systems (3.3), subject to the condition (3.4), is not an easy task as the following example shows.

Example 3.2 [11]. Consider the system

$$\begin{aligned} \dot{x}(t) &= 998x(t) + 1\,998y(t), \\ \dot{y}(t) &= -999x(t) - 1\,999y(t), \\ x(0) &= y(0) = 1. \end{aligned} \tag{3.5}$$

The solution of the problem (3.5) is of the form

$$\begin{aligned} x(t) &= 4\exp(-t) - 3\exp(-1\,000t), \\ y(t) &= -2\exp(-t) + 3\exp(-1\,000t). \end{aligned} \tag{3.6}$$

Figure 3.2 presents a view of the solutions $x(t)$ and $y(t)$. For $t > 0.01$ we practically have $x(t) \approx 4\exp(-t)$, $y(t) \approx -2\exp(-t)$. We shall solve the problem (3.5) by means of the Euler method with step h, using the formulas

$$\begin{aligned} x_{n+1} &= x_n + h(998x_n + 1\,998y_n), \\ y_{n+1} &= y_n + h(-999x_n - 1\,999y_n), \end{aligned}$$

$$x_n = x(nh), \qquad y_n = y(nh), \qquad x_0 = 1, \qquad y_0 = 1.$$

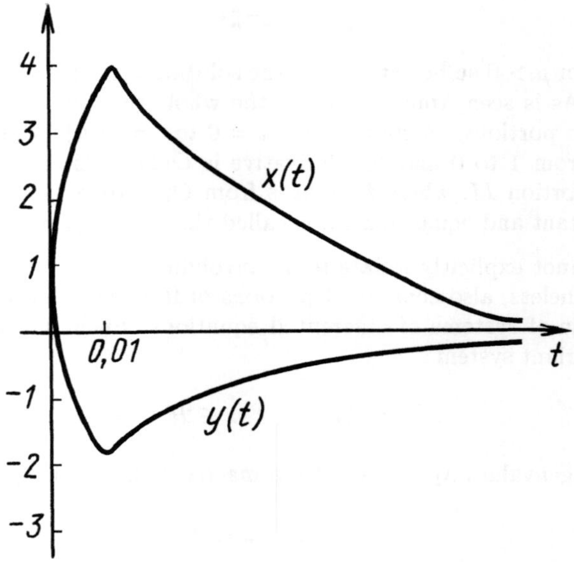

FIG. 3.2. A view of the solution of the stiff system (3.5)

TABLE 3.1

Step h=0.01		
t	$x_n = x(t)$	$y_n = y(t)$
0.05	$1.771508 \cdot 10^5$	$-1.771489 \cdot 10^5$
0.10	$-1.046035 \cdot 10^5$	$1.046035 \cdot 10^5$
0.15	$6.176733 \cdot 10^{14}$	$-6.176733 \cdot 10^{14}$
0.20	$-3.647299 \cdot 10^{19}$	$3.647299 \cdot 10^{19}$
0.25	$2.153693 \cdot 10^{24}$	$-2.153693 \cdot 10^{24}$
0.30	$-1.271734 \cdot 10^{29}$	$1.271734 \cdot 10^{29}$
0.35	$7.509466 \cdot 10^{33}$	$-7.509466 \cdot 10^{33}$
0.40	$-4.434264 \cdot 10^{38}$	$4.434264 \cdot 10^{38}$
0.45	$2.618389 \cdot 10^{43}$	$-2.618389 \cdot 10^{43}$
0.50	$-1.546132 \cdot 10^{48}$	$1.546132 \cdot 10^{48}$

Taking the step $h = 0.01$, we obtain the results listed in Table 3.1. Comparing these results with the exact solutions (3.6), we see that the results obtained are not at all satisfactory. Better results can be obtained by taking $h = 0.001$ (see Table 3.2). But in this case the computation is much more time consuming.

Let us consider in greater detail the difficulties encountered in the solution of the system (3.5). We apply the Euler method with step h to find the solution of

TABLE 3.2

Step h=0.001				
t	$x_n = x(t)$	$y_n = y(t)$	$x_{\text{exact}}(t)$	$y_{\text{exact}}(t)$
0.05	3.804822	−1.902411	3.804917	−1.902458
0.10	3.619168	−1.809584	3.619349	−1.809674
0.15	3.442573	−1.721286	3.442831	−1.721415
0.20	3.274595	−1.637297	3.274923	−1.637461
0.25	3.114813	−1.557406	3.115203	−1.557601
0.30	2.962828	−1.481414	2.936272	−1.481536
0.35	2.818258	−1.409129	2.818752	−1.409376
0.40	2.680743	−1.340371	2.681280	−1.340640
0.45	2.549938	−1.274969	2.550512	−1.275256
0.50	2.425515	−1.212757	2.426122	−1.213061

the equation

$$\dot{x}(t) = -\lambda x(t), \qquad x(0) = 1,\ \lambda > 0. \tag{3.7}$$

The solution of equation (3.7) is of the form

$$\bar{x}(t) = \exp(-\lambda t), \qquad x(t) \to 0,\ t \to \infty.$$

Using the Euler method, we obtain

$$x_{n+1} = x_n - h\lambda x_n = (1 - h\lambda)x_n. \tag{3.8}$$

The sequence (3.8) approaches zero whenever $|1 - h\lambda| < 1$. Hence, to ensure that x_n approaches zero, we must choose h so that

$$0 < h < \frac{2}{\lambda}. \tag{3.9}$$

The condition (3.9) is called the *stability condition for the Euler method* for equation (3.7).

Consider now the system

$$\dot{x}(t) = \Lambda x(t), x \in R^n. \tag{3.10}$$

The matrix Λ is diagonal, the diagonal being composed of the quantities $-\lambda_1, \dots, -\lambda_n$. As above, we see that the Euler method with step h is stable for the system (3.10) whenever $|1-h\lambda_i| < 2, i = 1, \dots, n$, i.e., whenever the step h satisfies the condition

$$0 < h < 2\lambda_{\max}^{-1}, \qquad \lambda_{\max} = \max\{\lambda_1, \dots, \lambda_n\}.$$

The integration interval T for the system (3.10) is specified by the value $\lambda_{\min} = \min\{\lambda_1, \dots, \lambda_n\}$ and is usually equal to $T = (5\text{–}10)\lambda_{\min}^{-1}$.

Suppose that the eigenvalues λ_i in the system (3.10) differ widely in magnitude, for example, $\lambda_{\max}/\lambda_{\min} \simeq 10^4$–$10^5$. Then the integration on the interval $T = 10\lambda_{\min}^{-1}$ with step $h = \lambda_{\max}^{-1}$ would require $T/h \simeq 10^5$–10^6 steps. Such a number of computations is quite large and takes a lot of computer time. But we cannot take a larger step, since this would lead to the infringement of the condition (3.11) for stability, and the solution obtained through the use of the Euler method would differ greatly from the exact solution.

We can thus assume that the system (3.3) is stiff whenever

$$\frac{\max_i \operatorname{Re} \lambda_i(A)}{\min_i \operatorname{Re} \lambda_i(A)} \approx 10^4 - -10^5. \tag{3.11}$$

The condition (3.11) cannot be regarded as the formal definition of a stiff system even for systems of the form (3.3), but it explains the meaning of stiffness.

More acceptable numerical results can be obtained using implicit formulas, such as the implicit Euler formula. For the system

$$\dot{x}(t) = f(t, x), \qquad x \in R^n,\ x(0) = x_0,$$

the *implicit Euler formula* is of the form

$$x_{n+1} = x_n + hf(t, x_{n+1}), \qquad x(0) = x_0. \tag{3.12}$$

Example 3.3. Let us again consider the system (3.7) and find its solution using the implicit Euler formula with step h. Instead of formula (3.8) we now have

$$x_{n+1} = x_n - h\lambda x_{n+1}, \qquad x_{n+1} = \frac{1}{1 + h\lambda} x_n. \tag{3.13}$$

As is seen from this formula, the following condition is satisfied for all $h > 0$:

$$x_n \to 0 \quad \text{as } n \to \infty. \tag{3.14}$$

In view of the foregoing we can give the following definition.

Definition 3.1. A numerical method is said to be *A-stable* if its application to equation (3.7) yields a sequence x_n satisfying the condition (3.14).

Using this definition, we can state the foregoing results thus: the explicit Euler method is not A-stable, but the implicit Euler method is.

As an extension of these results, it has been shown that [4, 8, 19, 22]

(a) no explicit multistep method is A-stable;

(b) Runge–Kutta explicit methods are also not A-stable;

(c) implicit multistep methods with the number of stages greater that two are not A-stable.

In view of this *implicit Runge–Kutta methods* are preferable for integrating stiff systems. We give some particular cases of these formulas obtained by J. R. Cash:[5]

$$n = 2: \qquad x_{n+1} = x_n + \frac{h}{2}(\eta_1 + \eta_2),$$
$$\eta_1 = f(t_{n+1}, x_{n+1}), \qquad \eta_2 = f(t_n, x_{n+1} - h\eta_1);$$

$$n = 3: \qquad x_{n+1} = x_n + \frac{h}{4}(\eta_1 + 3\eta_3),$$
$$\eta_1 = f(t_{n+1}, x_{n+1}), \qquad \eta_2 = f\left(t_{n+1} - \frac{h}{3}, x_{n+1} - \frac{h\eta_1}{3}\right),$$
$$\eta_3 = f\left(t_{n+1} - \frac{2h}{3}, x_{n+1} - \frac{2h\eta_2}{3}\right);$$

$$n = 4: \qquad x_{n+1} = x_n + \frac{h}{6}(\eta_1 + 2\eta_2 + 2\eta_3 + \eta_4),$$
$$\eta_1 = f(t_{n+1}, x_{n+1}), \qquad \eta_2 = f\left(t_{n+1} - \frac{h}{2}, x_{n+1} - \frac{h\eta_1}{2}\right),$$
$$\eta_3 = f\left(t_{n+1} - \frac{h}{2}, x_{n+1} - \frac{h\eta_2}{2}\right), \qquad \eta_4 = f(t_n, x_{n+1} - h\eta_3).$$

In addition to these formulas some other formulas are employed for solving stiff systems [3, 8, 19, 22].

In using implicit formulas, we must solve a system of algebraic equations for x_{n+1} in each step. This is not a simple problem even in the case of the linear system (3.3). Indeed, if the four-stage Cash formula is used to solve the system (3.3), then we obtain the following equation for x_{n+1}:

$$\left(I - hA + \frac{h^2A^2}{2!} - \frac{h^3A^3}{3!} + \frac{h^4A^4}{4!}\right)x_{n+1} = x_n. \tag{3.15}$$

Suppose that the matrix A has real eigenvalues $-\lambda_1 < \cdots < -\lambda_n < 0$ such that the condition number of the matrix A is

$$\operatorname{cond}(A) = \lambda_1/\lambda_n \approx 10^4.$$

Then it can be computed that the condition number of the matrix of the system (3.15) is equal to

$$\operatorname{cond}\left(I - hA + \frac{h^2A^2}{2} - \frac{h^3A^3}{3!} + \frac{h^4A^4}{4!}\right) \approx 10^8\text{–}10^{12}.$$

[5]Cash, J. R., A Class of Implicit Runge–Kutta Methods for the Numerical Integration of Stiff Ordinary Differential Equations, *J. ACM*, (1975), vol. 22, no. 4.

It is hardly worthwhile to numerically integrate systems of linear algebraic equations (3.15) with such a condition number because of the strong effect on the solution of round-off errors (for more details, see [5, 8, 15, 19, 22]).

In view of this difficulty other methods for solving stiff systems have been devised.

For example, for the system (3.3) the following exponential method can be employed. As is known [3, 5, 8, 19], the solution of the system (3.3) can be written as

$$\bar{x}(t) = \exp(At)x_0.$$

Here $\exp(At)$ is a matrix exponential that can be expanded in a power series:

$$\exp(At) = I + At + \frac{A^2t^2}{2!} + \frac{A^3t^3}{3!} + \cdots. \tag{3.16}$$

Suppose it is required to find a solution $\bar{x}(t)$ of the system (3.3) at the moment $t = T$. A number N is chosen so that $T = 2^N h$, where h is a sufficiently small number. For h small, the series (3.16) is fairly rapidly convergent and, to determine $\exp(Ah)$, we can take just a small number of terms in the expansion (3.16). We assume that $B_0 = \exp(Ah)$ is computed by means of the series (3.16) to a necessary degree of accuracy. Further computations can be carried out recursively, using the formulas

$$\begin{aligned} &B_1 = \exp(2Ah) = \exp(Ah)\exp(Ah) = B_0^2, \\ &B_2 = \exp(4Ah) = \exp(2Ah)\exp(2Ah) = B_1^2, \\ &B_3 = \exp(2^3Ah) = B_2^2, \\ &\cdots\cdots\cdots\cdots\cdots\cdots\cdots\cdots\cdots \\ &\cdots\cdots\cdots\cdots\cdots\cdots\cdots\cdots\cdots \\ &B_N = \exp(2^N Ah) = B_{N-1}^2. \end{aligned} \tag{3.17}$$

Computations according to these formulas are not very time consuming. As an example, for $h = 0.01$, $n = 10$ the numerical integration on the interval $T = 2^{10}h \approx 10$ will require a number of arithmetic operations of multiplication and addition of the order of $10n^3 = 10^4$. This number of computations is not very large, and it does not depend on the properties of the matrix A.

Other methods for solving stiff systems are also known. One of the more successful programs for solving stiff systems was developed by C. W. Gear. The Gear method (implemented as the DIFSUB program) employs predictor and corrector formulas of variable order and step.[6] Programs implementing the exponential method (3.17), the implicit Runge–Kutta methods, and the Gear method are given in [3].

As a result of testing various stiff systems, it has been established that the DIFSUB program is only weakly sensitive to the degree of stiffness of the system; it provides the most effective method for solving stiff and nonstiff systems of ordinary differential equations in terms of minimum time consumption, given the prescribed accuracy of the solution.

[6]Gear, C. W., *Numerical Initial Value Problems in Ordinary Differential Equations*, Prentice-Hall, Englewood Cliffs, N. J., (1971).

§ 4. Numerical Methods for the Design of Optimal Control in the Linear-Quadratic Problem

One of the problems that is well studied and often encountered in applications is the linear-quadratic problem (LQ-problem). Let the system under consideration be described by an equation of the form

$$\dot{x}(t) = A(t)x(t) + B(t)u, \qquad 0 \leqslant t \leqslant T,\ x(t) \in R^n,\ u \in R^m, \tag{4.1}$$

with initial condition

$$x(t_0) = x_0. \tag{4.2}$$

It is required to determine an unconstrained F-control $u = u(t,x)$ so as to minimize the quadratic functional

$$J(u) = x'(T)N_1x(t) + \int_{t_0}^{T} [x'(t)N_2(t)x(t) + u'(t)N_3(t)u(t)]\,dt. \tag{4.3}$$

$$N_1 \geqslant 0,\ N_2(t) \geqslant 0,\ N_3(t) > 0.$$

This problem has been investigated in some detail in § 2 of Chapter IV. The corresponding notation and the appropriate results are used in this section.

Of particular interest are stabilization problems (or problems on an infinite time interval) when $T = \infty$.

In this section we are concerned with the following problems:

(a) the determination of an optimal F-control in the problem (4.1)–(4.3);

(b) the stabilization problem and the methods employed for solving the algebraic Riccati equation.

(c) the numerical design of the Kalman filter;

1. Determination of an optimal F-control for time-varying systems. It was shown in § 2 of Chapter IV that the solution of the problem (4.1)–(4.3) can be expressed as

$$u(t,x) = -N_3^{-1}(t)B'(t)P(t)x, \tag{4.4}$$

where the matrix $P(t)$ satisfies the matrix differential Riccati equation

$$\dot{P}(t) + A'(t)P(t) + P(t)A(t) + N_2(t) - P(t)B(t)N_3^{-1}(t)B'(t)P(t) = 0, \tag{4.5}$$

$$t_0 \leqslant t \leqslant T,\ P(T) = N_1.$$

As is known, for given measurable and bounded matrices $A(t)$ and $B(t)$ and a matrix $N_3(t)$ which is uniformly on $[0,T]$ positive definite, equation (4.5) has a unique solution $P(t)$, which is bounded and positive definite, i.e., $CI > P(t) > 0$.

These results lead to the following algorithm for solving the problem (4.1)–(4.3). First, equation (4.5) is integrated backwards (in "reverse time") from $t = T$ to $t = 0$ by means of a numerical method (such as the Euler method, the Runge–Kutta or Adams method). Next, an F-control is determined from formula (4.4). This method has the advantage that the gain matrix

$$F(t) = -N_3^{-1}(t)B'(t)P(t)$$

can be computed just once and used throughout the whole control process. The shortcoming of this method is that for equation (4.5) a sufficiently small integration step must be chosen so that the computed result would have satisfactory accuracy. This shortcoming can be compensated by using integration methods with automatic choice of step (see § 1). Further, since the matrix $P(t)$ must be symmetric, we can integrate not all n^2 equations of the first order making up the system (4.5) but only $n(n+1)/2$ equations for the upper triangular part of the matrix $P(t)$. This almost halves the time consumption and leads automatically to a symmetric matrix $P(t)$.

2. Time-invariant linear-quadratic problem on a finite interval. In the case of a time-invariant LQ-problem (4.1)–(4.3), in which the matrices $A(t)$, $B(t)$, $N_2(t)$, and $N_3(t)$ are constant, the nonlinear equation (4.5) which is to be solved for the matrix $P(t)$ is rather awkward to handle. It is more convenient to consider instead a corresponding linear system of order $2n$, as was done in § 2 of Chapter IV and § 4 of Chapter XIII. The vector composed of phase variables $x(t)$ and adjoint variables $\varphi(t)$ satisfies the linear equation

$$\begin{pmatrix} \dot{x}(t) \\ \dot{\varphi}(t) \end{pmatrix} = \begin{pmatrix} A & -BN_3^{-1}B^{(1)} \\ -N_2 & -A' \end{pmatrix} \begin{pmatrix} x(t) \\ \varphi(t) \end{pmatrix}. \tag{4.6}$$

In this case the solution $P(t)$ of the Riccati equation (4.5) can be expressed by the following recurrence relation

$$\begin{aligned} P(t_{i+1}) = {} & [G_{21}(t_{i+1}, t_i) + G_{22}(t_{i+1}, t_i)P(t_i)] \\ & \times [G_{11}(t_{i+1}, t_i) + G_{12}(t_{i+1}, t_i)P(t_i)]^{-1}, \\ & t_{i+1} - t_i = h = \text{constant}, \quad P(T) = N_1. \end{aligned} \tag{4.7}$$

Here $G_{ij}(t_{i+1}, t_i)$ $(i, j = 1, 2)$ are $n \times n$ blocks of the fundamental matrix $G(t, t_0)$ of the equation (4.6):

$$G(t_{i+1}, t_i) = G(t_{i+1} - t_i) = e^{-Zh}, \qquad Z = \begin{pmatrix} A & -BN_3^{-1}B' \\ -N_2 & -A' \end{pmatrix}.$$

The matrices $G_{ij}(t_{i+1}, t_i)$ can be computed only once, and then the recurrence relation (4.7) be used. To compute $G(t_{i+1}, t_i)$, we expand in a series the matrix exponential $\exp(-Zh)$:

$$G(t_{i+1}, t_i) = I - Zh + \frac{Z^2h^2}{2!} - \frac{Z^3h^3}{3!} + \cdots. \tag{4.8}$$

For h small, the series (4.8) is rapidly convergent. If the real parts of the eigenvalues $\lambda(Z)$ of the matrix Z differ widely in magnitude, for the series (4.8) to converge it is necessary to take very small h. In this case the system (4.6) proves to be stiff and better results can be obtained by using the procedure for computing the matrix exponential described in § 3.

Let us now consider a method for diagonalizing the solution (4.6) which does not involve recursive computations (see [13] of Part Two). To this end, we invoke the following remarkable property of the matrix

$$Z = \begin{pmatrix} A & -BN_3^{-1}B' \\ -N_2 & -A' \end{pmatrix} \tag{4.9}$$

of the system (4.6). If the pair (A, B) is controllable (or at least stabilizable), each eigenvalue $\lambda(Z)$ of the matrix Z is associated with the eigenvalue $-\lambda(Z)$ of the same matrix. Moreover, the matrix Z has no eigenvalues with zero real part. Then the matrix Z can be written as

$$Z = W \begin{pmatrix} \Lambda & 0 \\ 0 & -\Lambda \end{pmatrix} W^{-1}. \tag{4.10}$$

Here Λ is an $n \times n$ diagonal matrix whose diagonal contains eigenvalues $\lambda(Z)$ such that $\operatorname{Re}\lambda(Z) > 0$, the matrix $-\Lambda$ contains eigenvalues $\lambda(Z)$ such that $\operatorname{Re}\lambda(Z) < 0$, and the $2n \times 2n$ matrix W consists of the eigenvectors of the matrix Z. Computing the matrices Λ and W_{ij}, by means of the QR-algorithm, for example, we can write the fundamental matrix $G(s)$ in the form

$$\begin{gathered} G(s) = e^{-\Lambda s} S e^{-\Lambda s}, \\ S = -(W_{21} - P(T)W_{11})^{-1}(W_{22} - P(T)W_{12}). \end{gathered} \tag{4.11}$$

The matrix $e^{-\Lambda s}$is a diagonal matrix whose diagonal is composed of the functions $\exp(-\lambda_i(Z)z)$. In this case the solution $P(t)$ of the Riccati equation (4.5) can be expressed by the formula

$$P(t) = [W_{22} + W_{21}G(T-t)][W_{12} + W_{11}G(T-t)]^{-1}. \tag{4.12}$$

Here t may be arbitrary, and hence formula (4.12), in contrast to (4.7), does not require recursive computations. It has been shown for the controllable pair (A, B) that the solution of the Riccati equation (4.5) tends as $t \to \infty$ to a steady-state value $\hat{P}$, i.e., $\hat{P} = \lim_{t\to\infty} P(t)$. In view of this, it is necessary to introduce a halting rule for computions carried out according to formulas (4.7) or (4.12); this rule would assure that the solution $P(t)$ stabilizes near the equilibrium value $\hat{P}$.

Thus, in the case of a time-invariant LQ-problem the computation of the gain matrix $F(t) = -N_3^{-1}B'P(t)$ reduces to the determination of the eigenvalues and eigenvectors of the matrix Z and the subsequent use of formulas (4.7) or (4.12). This method was implemented in the procedure CNTREG incorporated into the system ORACLS [2].

Example 4.1 [2]. Consider the time-invariant LQ-problem

$$\dot{x}(t) = Ax + Bu, \qquad x \in R^4,\ u \in R^2,$$

$$J(u) = \tfrac{1}{2}x'(20)x(20) + \int_0^{20} [x'(s)x(s) + 100u'(s)u(s)]\,ds$$

with matrices of the form

$$A = \begin{pmatrix} -2.60 & 0.25 & -38.0 & 0.0 \\ -0.075 & -0.27 & 4.40 & 0.0 \\ 0.078 & -0.99 & -0.23 & 0.052 \\ 1.0 & 0.078 & 0.0 & 0.0 \end{pmatrix}, \qquad B = \begin{pmatrix} 17 & 7.0 \\ 0.82 & -3.2 \\ 0.0 & 0.046 \\ 0.0 & 0.0 \end{pmatrix}.$$

TABLE 4.1

Matrix type	Numerical values of 4×4 matrices			
	−1.7233188E−01	5.0257395E−02	2.6582832E−02	−6.2618218E−02
W_{11}	−3.3791761E−02	−1.6344167E−02	9.5183892E−03	−6.9930617E−03
	7.7409397E−03	1.7516130E−06	−8.0834088E−03	3.3935252E−04
	−2.7082403E−01	1.4511446E−02	−1.4186503E−02	−2.2200507E−02
	9.5507170E−02	2.2377640E−02	−2.4483836E−02	1.0441627E−01
W_{12}	7.9262267E−01	3.3965417E−01	−7.8633601E−03	3.4982534E−01
	3.2205187E−01	1.7435213E−01	9.2129547E−01	9.2853245E−01
	3.9327414E−01	−1.4120636E−02	4.7159165E−03	−9.1676080E−03
	−2.4733021E−01	−1.6423055E−01	−8.5528327E−02	6.2134586E−01
W_{21}	4.3326566E−03	1.2843065E−02	3.5983643E−02	3.0404804E−02
	7.5286043E−03	1.7745854E−02	−4.9824656E−03	−2.6988189E−03
	3.8230750E−01	4.5465498E−02	−4.8131275E−02	−2.1922162E−02
	7.2558584E−02	−4.4315091E−02	−1.5780566E−02	8.4269310E−02
W_{22}	5.3390433E−01	−7.4606776E−03	3.2328081E−01	2.3749670E−01
	−4.3207089E−01	8.7411594E−01	−2.9506394E−01	−7.0239243E−01
	5.5697883E−01	−1.2681699E−02	−3.8062689E−02	−8.9888548E−02
	−6.9197483E−01	8.8040174E−02	−7.0617792E−03	−1.0579571E−01
S	5.3014743E−01	1.1451662E−01	−4.2235758E−01	−1.6582063E+00
	1.0292420E+00	−6.8625132E−01	8.0290849E−01	−1.3835163E−01
	−4.1882228E−01	−2.9967598E−01	−4.0340142E−01	1.2407188E+00

The solution of the Riccati equation (4.5) has been found according to formulas (4.7), (4.11), (4.12) for $0 \leqslant t \leqslant 20$. In Table 4.1 the computation results for the matrices W_{ij} and S are listed. In Table 4.2 are recorded the values of the matrix $P(t)$ over the range $t = 20$ to $t = 0$ at intervals of 2. In Table 4.3 the values of the gain matrix $F(t)$ over the same range and at the same intervals are given. As is seen from Tables 4.2 and (4.3), the solution $P(t)$ at $t = 6$ has stabilized to the equilibrium value $\hat{P}$; the matrix $F(t)$ has also stabilized.

The engineering implimentation of the control law with a time-varying gain matrix $F(t)$ is a complicated task. In practice, therefore, controllers with constant

TABLE 4.2

t, sec	Matrix $P(t)$ of dimensions 4×4			
20	5.0000000E−01	0.	0.	0.
	0.	5.0000000E−01	0.	0.
	0.	0.	5.0000000E−01	0.
	0.	0.	0.	5.0000000E−01
18	2.2480611E−01	3.6054050E−01	−1.4217206E+00	3.2023881E−01
	3.6054050E−01	1.0744607E+01	−8.8655022E+00	1.5409135E+00
	−1.4217206E+00	−8.8655022E+00	4.5683890E+01	−2.6235118E+00
	3.2023881E−01	1.5409135E+00	−2.6235118E+00	1.5642209E+00
16	2.3819322E−01	4.0985673E−01	−1.5275098E+00	3.6636522E−01
	4.1302539E−01	1.0991594E+01	−9.0079421E+00	1.7136851E+00
	−1.5275098E+00	−9.0079421E+00	4.9256328E+00	−2.8811826E+00
	3.6636522E−01	1.7136851E+00	−2.9811826E+00	1.7228142E+00
14	2.3924330E−01	4.1302539E−01	−1.5297598E+00	3.6980853E−01
	4.1302539E−01	1.1028965E+01	−9.0076508E+00	1.7278448E+00
	−1.5297598E+00	−9.0076508E+00	4.9271952E+01	−2.9877059E+00
	3.6980853E−01	1.7278448E+00	−2.9877059E+00	1.7346173E+00
12	2.3930520E−01	4.1337855E−01	−1.5297546E+00	3.7003289E−01
	4.1337855E−01	1.1032206E+01	−9.0096823E+00	1.7293204E+00
	−1.5297546E+00	−9.0096823E+00	4.9275911E+01	−2.9880218E+00
	3.7003289E−01	1.7293204E+00	−2.9880218E+00	1.7354617E+00
10	2.3930917E−01	4.1341056E−01	−1.5297727E+00	3.7004881E−01
	4.1341056E−01	1.1032469E+01	−9.0098261E+00	1.7294493E+00
	−1.5297727E+00	−9.0098261E+00	4.9276275E+01	−2.9880973E+00
	3.7004881E−01	1.7294493E+00	−2.9880973E+00	1.7355256E+00
8	2.3930948E−01	4.1341291E−01	−1.5297749E+00	3.7005005E−01
	4.1341291E−01	1.1032489E+01	−9.0098419E+00	1.7294590E+00
	−1.5297749E+00	−9.0098419E+00	4.9276293E+01	−2.9881061E+00
	3.7005005E−01	1.7294590E+00	−2.9881061E+00	1.7355306E+00
6	2.3930915E−01	4.1341309E−01	−1.5297751E+00	3.7005015E−01
	4.1341309E−01	1.1032491E+01	−9.0098433E+00	1.7294597E+00
	−1.5297751E+00	−9.0098433E+00	4.9276294E+01	−2.9881068E+00
	3.7005015E−01	1.7294597E+00	−2.9881068E+00	1.7355310E+00
4	2.3930951E−01	4.1341310E−01	−1.5297751E+00	3.7005015E−01
	4.1341310E−01	1.1032491E+01	−9.0098434E+00	1.7294597E+00
	−1.5297751E+00	−9.0098434E+00	4.9276294E+01	−2.9881068E+00
	3.7005015E−01	1.7294597E+00	−2.9881068E+00	1.7355310E+00

TABLE 4.3

t, sec	Gain matrix of dimensions 2 × 4			
20	−8.5000000E−02	−4.2000000E−02	0.	0.
	−3.5000000E−02	1.6000000E−02	2.3000000E−04	0.
18	4.1173471E−02	1.4939767E−01	−3.1438962E−01	6.7076089E−02
	3.5451406E−03	−3.2266773E−01	−2.0519022E−01	−2.8099331E−02
16	4.3853673E−02	1.5980672E−02	−3.3354180E−01	7.6334306E−02
	2.8554559E−03	−3.2718471E−01	2.0398637E−01	−3.0563703E−02
14	4.4058169E−02	1.6065183E−01	−3.3392190E−01	7.7035778E−02
	2.8265289E−03	−3.2815862E−01	2.0382674E−01	−3.0778783E−02
12	4.4071588E−02	1.6073844E−01	−3.3393768E−01	7.7086018E−02
	2.8195631E−03	−3.2823855E−01	2.0389393E−01	−3.0810439E−02
10	4.4072525E−02	1.6074604E−01	−3.3394193E−01	7.7089781E−02
	2.8188087E−03	−3.2824478E−01	2.0389743E−01	−3.0813714E−02
8	4.4072598E−02	1.6074661E−01	−3.3394244E−01	7.7090072E−02
	2.8187540E−03	−3.2824528E−01	2.0389779E−01	−3.0813714E−02
6	4.4072604E−02	1.6074665E−01	−3.3394248E−01	7.7090094E−02
	2.8187502E−03	−3.2824531E−01	2.0389783E−01	−3.0813729E−02
4	4.4072604E−02	1.6074665E−01	−3.3394148E−01	7.7090096E−02
	2.8187499E−03	−3.2824531E−01	2.0389783E−01	−3.0813730E−02

gain matrices F are used. Such control laws arise when treating LQ-problems on an unbounded interval. For such a constant matrix it is natural to choose the equilibrium state of the matrix $F(t)$. However, the cost functional $J(u)$ associated with this control assumes a larger value (i.e., the performance of the control system deteriorates), and a complicated problem arises of calculating the magnitude of this deterioration in performance. It is easiest to obtain an estimate of this deterioration by the use of numerical methods with the help of the results given in Table 4.3.

3. Optimal stabilization problems and methods for solving the algebraic Riccati equation. In the case of a time-invariant LQ-problem on an infinite time interval or an optimal stabilization problem of the form

$$\begin{gathered} \dot{x}(t) = Ax(t) + Bu, \qquad x(0) = x_0, \\ J(u) = \int_0^\infty [x'(s)N_2x(s) + u'(s)N_3u(s)]\,ds \end{gathered} \tag{4.13}$$

the optimal control is determined by

$$u(x) = -N_3^{-1}B'\hat{P}x = -Fx. \tag{4.14}$$

The proportional regulator (4.14) can be easily implemented, since the gain matrix F in this regulator is constant. The matrix $\hat{P}$ in formula (4.14) satisfies the algebraic equation

$$A'\hat{P} + \hat{P}A + N_2 - \hat{P}BN_3^{-1}B'\hat{P} = 0. \tag{4.15}$$

Thus, construction of an F-control in the optimal stabilization problem reduces to the solution of the algebraic Riccati equation (4.15).

We shall dwell briefly on the stabilization method for solving equation (4.15). As was noted in the preceding subsection,

$$\hat{P} = \lim_{t \to -\infty} P(t). \tag{4.16}$$

Here $P(t)$ is the solution of the differential Riccati equation (4.5). The relation (4.16) lies at the basis of the stabilization method whose algorithm can be described as follows. A constant matrix $N_1 > 0$ is given and then equation (4.5) is solved for $P(t)$ in "reverse time". As soon as the equilibrium value for the matrix $P(t)$ is attained, this value is assumed to be $\hat{P}$. As is illustrated in Example 4.1, the stabilization process evolves rapidly.

Another stabilization method is based on formula (4.12). It will be recalled that for the controllable pair (A, B) the matrix $-\Lambda$ contains quantities $\lambda(Z)$ as elements of the diagonal such that $\operatorname{Re}\lambda(Z) < 0$. Hence, the matrix $G(T-t) \to 0$ as $t \to \infty$, and it follows from formula (4.12) that

$$\hat{P} = W_{22}W_{12}^{-1}. \tag{4.17}$$

Other methods for solving stabilization problems are discussed in § 4 of Chapter XII, where an illustrative example is given.

4. The numerical design of the Kalman filter. Consider the problem of the optimal (in mean-square) estimation of the vector $x(T)$ described by the equation

$$dx(t) = A(t)x(t)\,dt + \sigma(t)\,d\xi(t), \qquad 0 \leqslant t \leqslant T,\ x(0) = x_0, \tag{4.18}$$

given observations y_t, $0 \leqslant t \leqslant T$, of the form

$$dy(t) = Q(t)x(t)\,dt + \sigma_0(t)\,d\xi_0(t), \qquad 0 \leqslant t \leqslant T,\ y(0) = 0. \tag{4.19}$$

Here the same notation is used as for equations (2.1), (2.2) of Chapter XI (§ 2).

As is shown in § 2 of Chapter XI, the solution of the filtering problem (4.18), (4.19) is given the Kalman equations

$$\begin{aligned} \dot{D}(t) = A(t)D(t) + D(t)A'(t) - D(t)Q'(t)(\sigma_0\sigma_0')^{-1}(t)Q(t)D(t) \\ + (\sigma\sigma'), \qquad 0 \leqslant t \leqslant T, \\ D(0) = D_0; \end{aligned} \tag{4.20}$$

$$dm(t) = A(t)m(t)\,dt + D(t)Q'(t)(\sigma_0\sigma_0')^{-1}(t)[dy(t) - Q(t)m(t)\,dt], \quad m(0) = 0. \tag{4.21}$$

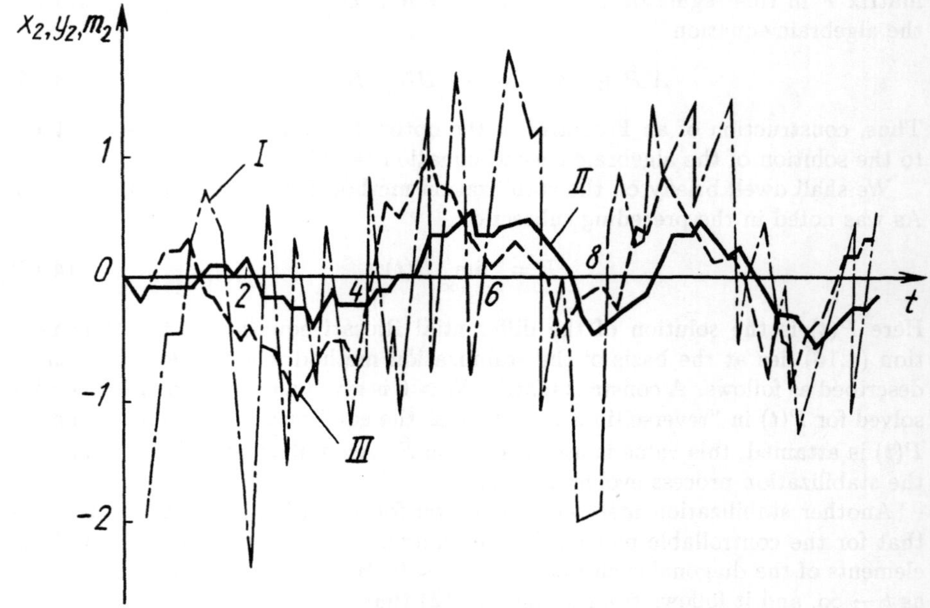

FIG. 4.1. Kalman filtering of the coordinate x_2 :
I, the measured values of the coordinate $y(t)$; II, the optimal estimate $m_2(t)$ of the phase coordinate $x_2(t)$; III, the actual values of the coordinate $x_2(t)$

The equation (4.20) for the covariance matrix is a matrix Riccati equation. Any of the methods described above can be used for its solution. In the process of the numerical solution of equation (4.20) we must verify that the matrix $D(t)$ remains positive definite. It can lose this property due to computational errors and an unfortunate choice of the method of integration of equation (4.20). To overcome this difficulty, we must choose an appropriate method of integration.

To integrate equation (4.21), we must substitute finite differences for the differentials occurring in the equation, and then integrate the resulting finite-difference equation using conventional numerical methods.

Now let us treat an example of the numerical design of a Kalman filter [7]. We shall consider a system of the form

$$\begin{aligned} dx_1 &= x_2\,dt, \\ dx_2 &= (-4x_1 - 0.8x_2)dt + 2\,d\xi(t), t \geqslant 0, \\ &x_1(0) = 0, \qquad x_2(0) = 0. \end{aligned} \tag{4.22}$$

The observations are assumed to have the form

$$dy = dx_2 + d\xi_0(t), \qquad y(0) = 0. \tag{4.23}$$

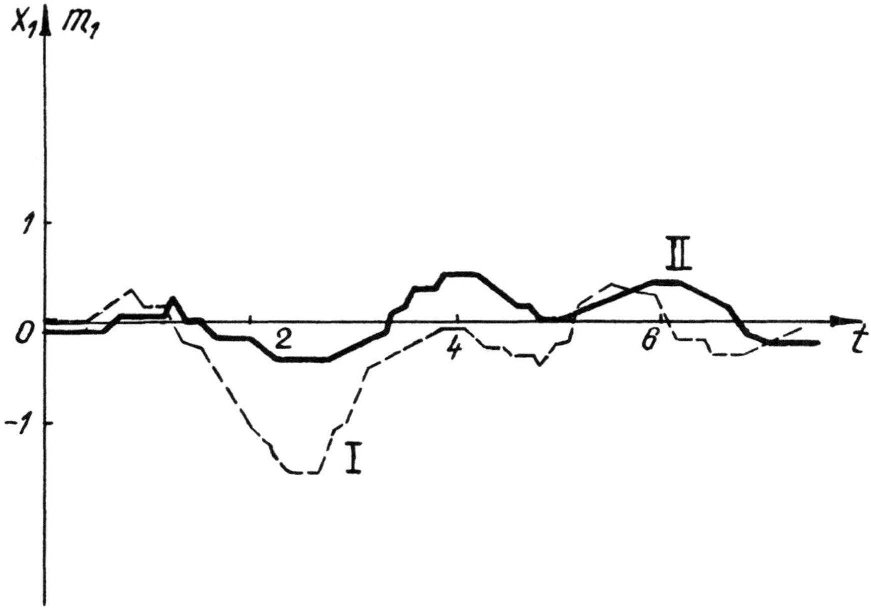

FIG. 4.2. Kalman filtering of the coordinate x_1 :
I, the actual values of the phase coordinate $x_1(t)$; *II*, the optimal estimate $m_1(t)$ of the coordinate $x_1(t)$

Here ξ, ξ_0 denote independent standard Wiener processes. We now have

$$A = \begin{pmatrix} 0 & 1 \\ -4 & -0.8 \end{pmatrix}, \quad \sigma = 2\begin{pmatrix} 0 & 0 \\ 0 & 1 \end{pmatrix}, \quad Q = (\,0 \quad 1\,), \quad \sigma_0 = 1. \tag{4.24}$$

Equations (4.20), (4.21), together with the relations (4.24), describe the optimal Kalman filter for the system (4.22).

Figures 4.1–4.3 illustrate the performance of the obtained Kalman filter. Figure 4.1 shows the measured values of the coordinate $y(t)$, the optimal estimate $m_2(t)$ of the phase coordinate $x_2(t)$, and the actual values of the coordinate $x_2(t)$. Figure 4.2 shows the actual values of the phase coordinate $x_1(t)$ and the optimal estimate $m_1(t)$ of the coordinate $x_1(t)$. As is seen from Figures 4.1 and 4.2, the optimal estimates $m_1(t)$ and $m_2(t)$ follow closely the corresponding phase trajectories $x_1(t)$ and $x_2(t)$. Figure 4.3 depicts the evolution of the elements $D_{ij}(t)$ of the covariance matrix $D(t)$. It is clearly seen that the covariances $D_{ij}(t)$ approach the respective equilibrium states. These steady-state solutions, inserted into the filtering equation (4.21), furnish an optimal time-invariant Kalman filter.

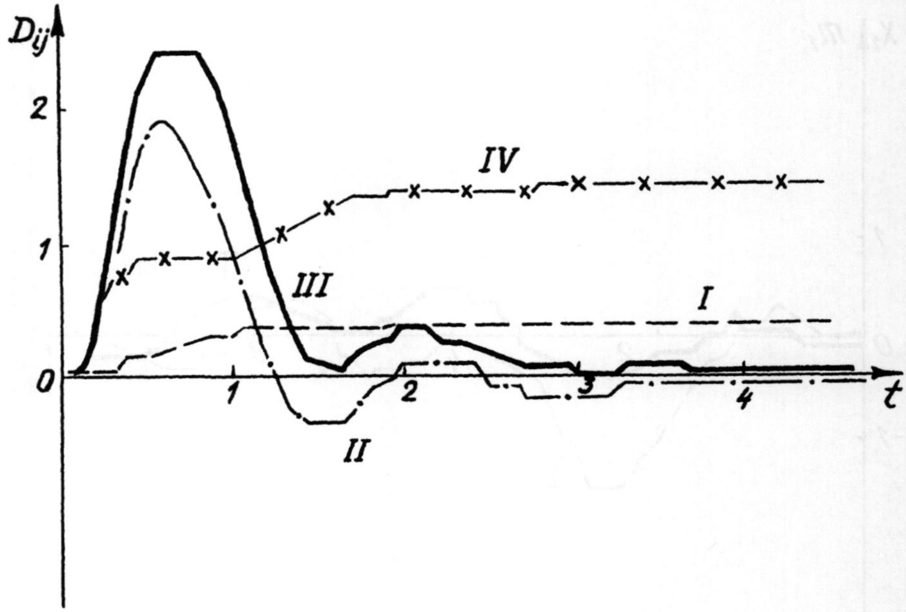

FIG. 4.3. The evolution of the elements of the covariance matrix $D(t)$:
I, the values of $D_{11}(t)$; II, the values of $D_{12}(t)$; III, the values of $D_{21}(t)$; IV, the values of $D_{22}(t)$

Main Results and Formulas of Chapter XIV

One-step methods:

$$x_{n+1} = x + h_n \Phi_f(t_n, x_m, h_n);$$

Euler method

$$x_{n+1} = x_n + h_n f(t_n, x_n);$$

Runge–Kutta formula of the second order

$$x_{n+1} = x_n + h_n K_2, \qquad K_1 = f(t_n, x_n),$$
$$K_2 = f\left(t_n + \frac{h_n}{2}, x_n + \frac{h_n}{2} K_1\right);$$

$$x_{n+1} = x_n + \frac{h_n}{2}(K_1 + K_2), \qquad K_1 = f(t_n, x_n),$$
$$K_2 = f(t_n, h_n, x_n + h_n K_1);$$

classic Runge–Kutta formula of the fourth order

$$x_{n+1} = x_n + \frac{h_n}{6}(K_1 + 2K_2 + 2K_3 + K_4)$$

$$K_1 = f(t_n, x_n), \qquad K_2 = f\left(t_n + \frac{h_n}{2}, x_n + \frac{h_n}{2}K_1\right),$$

$$K_3 = f\left(t_n + \frac{h_n}{2}, x_n + \frac{h_n}{2}K_2\right), \qquad K_4 = f(t_n + h_n, x_n + h_n K_3);$$

error estimate for the general one-step method

$$|x(t_n) - \tilde{x}_n| \leqslant C\left[Mh^p + \frac{\delta}{h}\right]e^{L(t_n - t_0)}.$$

The general k-step formula:

$$\sum_{p=0}^{k} \alpha_p x_{i+p} = h \sum_{p=0}^{k} \beta_p f(t_{i+p}, x_{i+p});$$

the general multistep formula is of order l if

$$\sum_{p=0}^{k} \alpha = 0, \qquad \sum_{p=0}^{k} p\alpha_p = \sum_{p=0}^{k} \beta_p,$$

. . . .

. . . .

$$\sum_{p=0}^{k} \frac{p^s \alpha_p}{s!} = \sum_{p=0}^{k} \frac{p^{s-1}\beta_p}{(s-1)!}, \qquad s = 2, \ldots, l;$$

explicit Adams formulas

$$x_{k+1} = x_k + h\big(f_k + \tfrac{1}{2}\Delta f_{k-1} + \tfrac{5}{12}\Delta^2 f_{k-2} + \tfrac{3}{8}\Delta^3 f_{k-3} + \tfrac{251}{720}\Delta^4 f_{k-4} + \tfrac{95}{288}\Delta^5 f_{k-5} + \tfrac{19\,087}{60\,480}\Delta^6 f_{k-6} + \cdots\big);$$

implicit Adams formulas

$$x_{k+1} = x_k + h\big(f_{k+1} - \tfrac{1}{2}\Delta f_k - \tfrac{1}{12}\Delta^2 f_{k-1} - \tfrac{1}{24}\Delta^3 f_{k-2} - \tfrac{19}{720}\Delta^4 f_{k-3} - \tfrac{3}{160}\Delta^5 f_{k-4} - \tfrac{863}{60\,480}\Delta^6 f_{k-5} - \cdots\big);$$

Milne formulas

$$x_i^{\text{pred}} = x_{i-4} + \frac{4h}{3}(2f_{i-3} - f_{i-2} + 2f_{i-1}),$$

$$x_i^{\text{corr}} = x_{i-2} + \frac{h}{3}(f_{i-2} + 4f_{i-1} + f_i^{\text{pred}}),$$

$$\varepsilon_i^{\text{corr}} = |x_i^{\text{corr}} - \bar{x}(t_i)| \simeq \tfrac{1}{29}|x_i^{\text{corr}} - x_i^{\text{pred}}|;$$

Butcher formulas

$$x_{n-\frac{1}{2}} = x_{n-2} + \frac{h}{8}(9f_{n-1} + 3f_{n-2}),$$

$$x_n^1 = \frac{1}{5}(28x_{n-1} - 23x_{n-2}) + \frac{h}{15}(32f_{n-\frac{1}{2}} - 60f_{n-1} - 26f_{n-2}),$$

$$x_n = \frac{1}{31}(32x_{n-1} - x_{n-2}) + \frac{h}{93}(64f_{n-\frac{1}{2}} + 15f_n^1 + 12f_{n-1} - f_{n-2}),$$

$$f_i^k = f(t_i, x_i^k);$$

accuracy estimate for the multistep, Dahlquist-stable, formula

$$|\bar{x}(t_i) - \tilde{x}_i| \leqslant e^{C_1 T}\left\{C_2 h^l + C_3\frac{\delta}{h} + C_4 \max_{0\leqslant i\leqslant k-1} |\theta_i|\right\};$$

the accuracy estimate for the multistep, strongly Dahlquist-stable, formula employed for solving the problem whose trivial solution is stable under steady-acting disturbances: for all $\varepsilon > 0$ there exist $h > 0$ and $\delta > 0$ such that $|\tilde{x}_n| < \varepsilon$, $n = 0, 1, 2\dots$, if $|\tilde{x}_i| < h\delta$, $i = 0, 1, \dots, k-1$.

For **stiff systems** it is preferable to use implicit Runge–Kutta methods, the Gear method, or the following formulas obtained by J. R. Cash:

$$n = 2: \qquad x_{n+1} = x_n + \frac{h}{2}(\eta_1 + \eta_2),$$

$$\eta_1 = f(t_{n+1}, x_{n+1}), \qquad \eta_2 = f(t_n, x_{n+1} - h\eta_1);$$

$$n = 3: \qquad x_{n+1} = x_n + \frac{h}{4}(\eta_1 + 3\eta_3),$$

$$\eta_1 = f(t_{n+1}, x_{n+1}), \qquad \eta_2 = f\left(t_{n+1} - \frac{h}{3}, x_{n+1} - \frac{h\eta_1}{3}\right),$$

$$\eta_3 = f\left(t_{n+1} - \frac{2h}{3}, x_{n+1} - \frac{2h\eta_2}{3}\right);$$

$$n = 4: \qquad x_{n+1} = x_n + \frac{h}{6}(\eta_1 + 2\eta_2 + 2\eta_3 + \eta_4),$$

$$\eta_1 = f(t_{n+1}, x_{n+1}), \qquad \eta_2 = f\left(t_{n+1} - \frac{h}{2}, x_{n+1} - \frac{h\eta_1}{2}\right),$$

$$\eta_3 = f\left(t_{n+1} - \frac{h}{2}, x_{n+1} - \frac{h\eta_2}{2}\right), \qquad \eta_4 = f(t_n, x_{n+1} - h\eta_3).$$

In addition to these formulas some other formulas are employed for solving stiff systems.

CHAPTER XV

NUMERICAL DESIGN OF OPTIMAL CONTROL SYSTEMS

In this chapter we review some of the methods of reducing optimal control problems to a nonlinear programming problem and a boundary-value problem. Several methods of solving boundary-value problems are described, such as the Newton method, the reduction to a Cauchy problem, the transfer of boundary conditions, and the Abramov method. The Shatrovskii and Fedorenko methods for general optimal control problems are given.

§ 1. On Numerical Methods of Solving Optimal Control Problems

Let us consider the following optimal control problem. Suppose the object under consideration is described by a nonlinear system

$$\begin{gathered} \dot{x}(t) = f(t, x, u), \qquad 0 \leqslant t \leqslant T, \\ x \in R^n,\ u \in U \subset R^m, \\ x(0) = x_0. \end{gathered} \tag{1.1}$$

It is required to determine the P-control $u = u(t)$ or the F-control $u = u(t, x)$ which minimize the cost functional

$$J(u) = \varphi\big(x(T)\big) + \int_0^T F_0\Big(t, x(t), u\big(t, x(t)\big)\Big)\, dt, \tag{1.2}$$

where φ and F_0 are given functions bounded below. Constraints can also be imposed on the phase variables x, such as

$$G\big(x(t), u(t)\big) \leqslant 0, \qquad t \in [0, J]. \tag{1.3}$$

Here $G\colon R^n \times R^m \to R^1$ is a continuously differentiable function. There may be constraints of a more general form. The problem (1.1)–(1.3) encompasses many important cases which are important from an applications point of view.

In this chapter we give a brief exposition of a few methods of numerically determining optimal control in the nonlinear problem (1.1)–(1.3).

At present no universal method of solving this problem exists owing to the difficulties of solving optimal control problems in the nonlinear context.

Numerical solution of optimal control problems is further complicated by a number of diverse reasons. Among these the following are the most important ones:

—the large dimension of the problems to be solved;
—the presence of nondifferentiable cost functionals of the type

$$\max_t F(x(t)), \qquad \max_t F(x(t), u(t)), \qquad \int_0^T |F_0(x(t), u(t))|\, dt;$$

—the presence of complicated control constraints;
—the presence of state (phase) constraints;
—the possibility of several extrema.

In connection with the great variety of problem statements for optimal control and the above-mentioned difficulties of the solution of these problems, there has been devised a great number of diverse methods of numerical solution. By convention these methods are divided into the following groups:

1°. *Reduction of an optimal control problem to a boundary-value problem for ordinary differential equations and subsequent solution of the resulting boundary-value problem through the use of some numerical method.* The specific character of optimal control is reflected, on the whole, in ways of reducing to a boundary-value problem, which are based either on classical variational calculus or on the maximum principle. Although the boundary-value problem can be solved by any of known methods, the importance of solving diverse optimal control problems has led to the creation of new methods for solving boundary-value problems, such as the Abramov method.

2°. *Methods of successive improvements of control, or methods of variations in the control space.* The Shatrovskii method is one of the better known methods; it will be discussed below.

3°. *Methods similar in spirit to the dynamic programming approach of Bellman, or methods of variations in the phase space.* One such method was considered in § 3 of Chapter VII. In this book we shall not go into details of these methods. Their description can be found in [20 and 6 in Part Two].

4°. *Reduction to the problem of the optimization of a function of a finite number of variables subject to variable constraints.*

Two main approaches lie at the basis of most numerical methods: the finite-dimensional approximation of optimal control problems; the linearization of the original continuous or approximating finite-dimensional problem. However, depending on the particular character of approximation and linearization, we obtain diverse methods.

§ 2. Reduction to nonlinear programming

In this section we briefly sketch methods of reducing the optimal control problem (1.1)–(1.3) to a problem in nonlinear programming. To this end, we divide the segment $[0, T]$ into N parts by the points

$$t_j = jh, \qquad h = T/N,\ j = 0, \dots, N. \tag{2.1}$$

Further, we assume that the control $u(t)$ is constant on each subinterval $[t_j, t_{j+1})$:

$$u(t) = u_j, \qquad t \in [t_j, t_{j+1}),\ u_j \in V.$$

Given the control $u = u_0$, we can solve equation (1.1) for $x(t)$, $0 \leqslant t \leqslant t_1$, and check the constraints (1.3). If the constraints (1.3) are met, we evaluate the integral

$$\int_0^{t_1} F_0\Big(s, x(s), u\big(s, x(s)\big)\Big)\, ds.$$

But if the constraints (1.3) are not satisfied, then instead of the control u_0 we take another admissible control and repeat the procedure. Next, we consider the subinterval $[t_1, t_2]$, and so on. Thus, for any admissible control $(u_0, u_1, \dots, u_{N-1})$ we can evaluate the cost functional $J(u) = J(u_0, u_1, \dots, u_{N-1})$. Further, to deduce optimal control, we must solve a problem in nonlinear programming of the form

$$J(u_0, u_1, \dots, u_{N-1}) \to \inf, \tag{2.2}$$
$$u_i \in U,\ i = 0, \dots, N-1, \quad G[x(t), u(t)] \leqslant 0, \quad t \in [0, T].$$

By solving the problem (2.2), we obtain an approximate optimal control.

To obtain a more precise approximation to optimal control, we must increase the value of N. As $N \to \infty$ we may expect convergence of controls thus obtained to the optimal control. The proof of this is not simple, particularly, in view of the fact that the optimal control $u_0(t)$ may be a discontinuous function, with the points of discontinuity not necessarily coinciding with the points of division t_j (2.1) of the segment $[0, T]$. For more details we refer the reader to [20 and 6 of Part Two].

The implementation of this method is very difficult because of the large dimensionality of the problem (2.2). Indeed, the number of variables in the problem (2.2) is mN and steadily increases with N, becoming very large. The numerical determination of the minimum of a function of a large number of variables is a very complicated problem, for which a large number of methods have been devised [5, 15, 20].

Although the reduction to nonlinear programming is often used in practice, there is no reason to regard this approach as a convenient universal method of solving all optimal control problems. In this connection a considerable number of other methods have been devised, some of which will be discussed below.

§ 3. Reduction to a Boundary-Value Problem

We consider this method using the Mayer problem as an example. In the Mayer problem it is required to determine a P-control $u(t) \in U$, for which the functional

$$J(u) = \varphi\big(x(T)\big) \to \inf_{u \in V} \tag{3.1}$$

attains a minimum value on the trajectories of the system (1.1).

1. Boundary-value problem of the maximum principle. It was shown in Chapter V that the optimal control and the optimal trajectory are the solution of the following problem:

$$\begin{gathered}\dot{x}_{\text{opt}}(t) = f(t, x_{\text{opt}}, u_{\text{opt}}), \qquad x_{\text{opt}}(0) = x_0,\\ \dot{\psi}(t) = -H_x\big(t, x_{\text{opt}}(t), u_{\text{opt}}(t), \psi(t)\big), \qquad \psi(T) = -\varphi_x\big(x(T)\big),\\ H\big((t, x_{\text{opt}}(t), u_{\text{opt}}(t), \psi(t)\big) = \max_{u\in U} H\big(t, x_{\text{opt}}(t), u, \psi(t)\big),\\ H(t, x, u, \psi) = \psi' f(t, x, u).\end{gathered} \tag{3.2}$$

Theoretically, we can determine the optimal control

$$u_{\text{opt}} = u_{\text{opt}}(t, x_{\text{opt}}, \psi) \tag{3.3}$$

from the condition for a maximum.

Substituting the expression (3.3) in the first two equations of the system (3.2), we obtain a nonlinear boundary-value problem in closed form with respect to the variables $x(t)$ and $\psi(t)$:

$$\begin{gathered}\dot{x}_{\text{opt}}(t) = f\big(t, x_{\text{opt}}, u_{\text{opt}}(t, x_{\text{opt}}, \psi)\big) = f_1(t, x_{\text{opt}}, \psi),\\ \dot{\psi}(t) = -H_x\big(t, x_{\text{opt}}, u_{\text{opt}}(t, x_{\text{opt}}, \psi)\big) = f_2(t, x_{\text{opt}}, \psi),\\ x_{\text{opt}}(0) = x_0, \qquad \psi(T) = -\varphi_x\big(x_{\text{opt}}(T)\big).\end{gathered} \tag{3.4}$$

The boundary condition for $t = T$ can be written as

$$\psi(T) + \varphi_x\big(x_{\text{opt}}(T)\big) = \Phi\big(\psi(T), x_{\text{opt}}(T)\big) = 0. \tag{3.5}$$

On solving this boundary-value problem, we determine the optimal control with the help of formula (3.3). Thus, the determination of optimal control reduces to the solution of the nonlinear boundary-value problem for a system of ordinary differential equations.

2. The Newton method of solving a boundary-value problem. *The Newton method* is one of the most widely used methods for solving nonlinear boundary-value problems. The scheme of this method is as follows. In an arbitrary way we assign the initial condition $\psi(0) = z$ and solve the Cauchy problem for equations (3.4) with initial condition

$$x(0) = x_0, \qquad \psi(0) = z. \tag{3.6}$$

As a result, we obtain the values $\psi(T, z)$ and $x(T, z)$. Substituting these values in equation (3.5) gives

$$\Phi\big(\psi(T, z), x(T, z)\big) = \Phi_1(z) = 0. \tag{3.7}$$

Thus, the solution of the boundary-value problem (3.4) reduces to finding an initial condition z which is a solution of equation (3.7).

To solve equation (3.7), we can apply the Newton method. To this end, we assign an initial value z_0 and define the successive approximations z_k by the formulas

$$z_k = z_{k-1} - [\Phi_{1z}(z_{k-1})]^{-1}\Phi_1(z_{k-1}), \tag{3.8}$$

where Φ_{1z} is the matrix of derivatives of the function $\Phi_1(z)$ with respect to the vector z.

In the numerical realization of the Newton method considerable difficulties arise. We shall point out some of these.

1°. The Newton method is applied only when the function $\Phi_1(z)$ is differentiable. This can be guaranteed only when the right sides of the boundary-value problem (3.4) are differentiable.

2°. The choice of the initial approximation z_0 is connected mainly with the fact that the convergence of approximations (3.8) can be ensured only when z_0 is sufficiently close to a root of equation (3.7). The region of convergence can be enlarged by replacing the Newton method (3.8) by a modified method of the form

$$z_k = z_{k-1} - \alpha_k[\Phi_{1z}(z_{k-1})]^{-1}\Phi_1(z_{k-1}), \tag{3.9}$$

where the parameter α_k is chosen in a special way.

3°. There may be a nonunique solution of the boundary-value problem (3.4).

4°. The matrix of derivatives Φ_{1z} is computed, in general, by the numerical differentiation of the function $\Phi_1(z)$. The evaluation of each value of the function $\Phi_1(z)$ requires the solution of the Cauchy problem (3.4), (3.6). Moreover, to determine the $n \times n$ matrix $\Phi_{1z}(z)$, it is required to solve at least n^2 Cauchy problems. Although this procedure is somewhat time consuming, any modern computer can handle it within reasonable time limits.

To sum up, although the Newton method is very convenient, its application is accompanied by appreciable difficulties.

3. An example of reducing an optimal control problem to a boundary-value problem. Suppose it is required to control the motion of a simple pendulum with viscous friction at the suspension axis. The control torque must be chosen so that the angular motion of the pendulum $x(t)$ follows as close as possible the prescribed smooth function $W(t)$ and coincides with it at the given finite moment $t = T$.

Mathematically, this problem reduces to the following one:

the equation of motion of the pendulum

$$\ddot{x} + \beta\dot{x} + \omega^2 x = u, \qquad 0 \leqslant t \leqslant T; \tag{3.10}$$

the boundary conditions

$$x(0) = \dot{x}(0) = 0, \qquad x(T) = W(T),\ \dot{x}(T) = \dot{W}(T); \tag{3.11}$$

the cost functional

$$J(u) = \int_0^T [x(t) - W(t)]^2\, dt \to \min_u; \tag{3.12}$$

the control constraint

$$|u(t)| \leqslant 1.$$

Let us reduce this problem to a Mayer problem for the system of the first order. We denote

$$x_1(t) = x(t), \qquad x_2(t) = \dot{x}(t)$$

$$x_3(t) = \int_0^t [x(s) - W(s)]^2 \, ds = \int_0^t [x_1(s) - W(s)]^2 \, ds.$$

Then the problem (3.10)–(3.12) can be rewritten as

$$\begin{aligned} &\dot{x}_1 = x_2, \\ &\dot{x}_2 = -\beta x_2 - \omega^2 x_1 + u, \\ &\dot{x}_3 = \left(x_1(t) - W(t)\right)^2, \\ &x_1(0) = x_2(0) = x_3(0) = 0, \\ &x_1(T) = W(T),\ x_2(T) = \dot{W}(T), \\ &x_3(T) \to \min_u . \end{aligned} \tag{3.13}$$

The Hamilton function for the problem (3.13) is of the form

$$H = \psi_1 x_2 + \psi_2(-\beta \bar{x}_2 - \omega^2 x_1 + u) + \psi_3 (x_1 - W)^2. \tag{3.14}$$

For the auxiliary variables ψ_1, ψ_2, ψ_3 the following equations hold:

$$\begin{aligned} &\dot{\psi}_1 = \omega^2 \psi_2 - 2(x_1 - W)\psi_3, \\ &\dot{\psi}_2 = -\psi_1 + \beta \psi_2, \\ &\dot{\psi}_3 = 0. \end{aligned} \tag{3.15}$$

Let us now determine optimal control $u_{\text{opt}}(t, x, \psi)$ from the condition for a maximum

$$H(u_{\text{opt}}) = \max_{|u| \leqslant 1} H(x_{\text{opt}}, u, \psi).$$

It follows from the expression (3.14) for the Hamiltonian that

$$u_{\text{opt}}(t, x, \psi) = \operatorname{sgn} \psi_2(t). \tag{3.16}$$

The problem (3.13) has five boundary conditions. For a complete boundary-value problem statement we need one more boundary condition. It can be obtained from the transversality conditions and is of the form $\psi_3(T) = -1$.

We can now write the desired boundary-value problem in closed form:

$$\begin{aligned}
\dot{x}_1(t) &= x_2(t),\\
\dot{x}_2(t) &= -\beta x_2(t) - \omega^2 x_1(t) + \operatorname{sgn}\psi_2(t),\\
\dot{x}_3(t) &= \big(x_1(t) - W(t)\big)^2,\\
\dot{\psi}_1(t) &= -2\big(x_1(t) - W(t)\big)\psi_3(t) + \omega^2\psi_2(t),\\
\dot{\psi}_2(t) &= -\psi_1(t) + \beta\psi_2(t),\\
\dot{\psi}_3(t) &= 0,\\
x_1(0) &= 0,\ x_2(0) = 0,\ x_3(0) = 0,\\
x_1(T) &= W(T),\ x_2(T) = \dot{W}(T),\ \psi_3(T) = -1.
\end{aligned} \tag{3.17}$$

Thus, the problem of the optimal control of a linear system of the second order with quadratic criterion and control constraints reduces to a nonlinear boundary-value problem of the sixth order. However, one variable in the problem (3.17) can easily be eliminated, since $\psi_3(t) \equiv -1$. Then we obtain a boundary-value problem of the fifth order with three boundary conditions at one end and two at the other. In this case the right side is nondifferentiable.

A similar situation arises in a large number of diverse optimal control problems. We cannot apply the Newton method for solving such problems, but other handy general methods are not available yet. In this connection the method of reducing to a boundary-value problem cannot be considered a universal method of solving optimal control problems.

However, there exists an important particular case in which the method of reducing to a boundary-value problem proves to be very convenient and highly developed. The next section deals with this case.

§ 4. Solution of Linear Boundary-Value Problems

1. Reduction of an optimal control problem to a linear boundary-value problem. Consider the following optimal control problem. Suppose it is required to transfer the linear system

$$\dot{x} = Ax + Bu, \qquad x \in R^n,\ u \in R^m,\ 0 \leqslant t \leqslant T, \tag{4.1}$$

from a given initial state

$$x(0) = x_0 \tag{4.2}$$

to a given terminal state

$$x(T) = x_1 \tag{4.3}$$

so as to minimize the quadratic functional

$$J(u) = \int_0^T [x'(t)N_2(t)x(t) - u'(t)N_3(t)u(t)]\,dt \to \min_u, \tag{4.4}$$

with the control u being unconstrained.

It should be noted that the problem (4.1)–(4.4) differs from the standard linear-quadratic optimal control problem by involving the boundary condition (4.3), and the determination of optimal control for this problem is, in general, very difficult.

Through the use of the maximum principle the optimal control problem (4.1)–(4.4) can be reduced to the following boundary-value problem:

$$\begin{aligned} \dot{x} &= Ax + \frac{1}{2}BN_3^{-1}(t)B'\psi, \\ \dot{\psi} &= -A'\psi + 2N_2(t)x, \\ x(0) &= x_0,\ x(T) = x_1. \end{aligned} \tag{4.5}$$

Thus, the solution of linear boundary-value problems is of great interest also for determining optimal control.

The importance of developing special methods of solving linear boundary-value problems transcends the mere value of such methods for solving the problem (4.1)–(4.4). Moreover, the techniques for solving linear boundary-value problems can be used for developing certain iteration procedures for solving more general optimal control problems.

2. Reduction of a linear boundary-value problem to Cauchy problems. Let us review some of the general methods of solving linear boundary-value problems. The basic idea behind all these methods is in reducing such problems to a set of Cauchy problems. However, this idea cannot be immediately implemented.

As an example, consider the following boundary-value problem:

$$\begin{aligned} \ddot{x} - p(t) &= f(t), \\ l_1[x] &= \dot{x}(a) - \alpha_0 x(a) = \alpha_1, \\ l_2[x] &= \dot{x}(b) - \beta_0 x(b) = \beta_1. \end{aligned} \tag{4.6}$$

The problem (4.6) with continuous right side is known to have a unique solution, provided that

$$p(t) \geqslant p_0 \geqslant 0. \tag{4.7}$$

We shall use the method of reducing the problem (4.6) to two Cauchy problems. We shall seek a solution of this problem in the form

$$x(t) = Cu(t) + v(t), \qquad C = \text{constant}. \tag{4.8}$$

We require that the functions $u(t)$ and $v(t)$ be solutions of the following Cauchy problems:

$$\begin{cases} \ddot{u}(t) - p(t)u(t) = 0, \\ u(a) = 1,\ \dot{u}(a) = \alpha_0; \end{cases} \tag{4.9}$$

and

$$\begin{cases} \ddot{v}(t) - p(t)v(t) = f(t), \\ v(a) = -\dfrac{\alpha_1}{\alpha_0},\ \dot{v}(a) = 0 & \text{for } \alpha_0 \neq 0, \\ v(a) = 0,\ \dot{v}(a) = \alpha_1 & \text{for } \alpha_0 = 0. \end{cases} \tag{4.10}$$

It is easily verified that for the functions $u(t)$ and $v(t)$ thus defined and for any constant C a function $x(t)$ of the form (4.8) satisfies the relations

$$\ddot{x} - p(t)x = f(t), \qquad l_1[x] = \alpha_1.$$

In general, the second boundary condition will not be satisfied. However, by a suitable choice of the constant C we can ensure that the second boundary condition will be satisfied. Indeed, since the equality

$$l_2[x] = C\dot{u}(b) + \dot{v}(b) - \beta_0[Cu(b) + v(b)] = \beta_1,$$

must hold, we find that

$$C = \frac{\beta_1 - \dot{v}(b) + \beta_0 v(b)}{\dot{u}(b) - \beta_0 u(b)}. \tag{4.11}$$

If $\dot{u}(b) - \beta_0 u(b) \neq 0$, then formula (4.11) uniquely determines the constant C for which a function $x(t)$ of the form (4.8) is a solution of the boundary-value problem (4.6). In this approach the solution of a boundary-value problem reduces to the solution of two Cauchy problems, i.e., to a sufficiently simple procedure.

However, this method cannot be implemented, since *it is not numerically stable.* Actually, the solution $u(t)$ has the following estimate: $u(t) \geqslant u_1(t)$ (see [18] of Part One), where

$$\ddot{u}_1 = p_0 u_1, \qquad \dot{u}_1(a) = 1,\ u_1(a) = \alpha_0 > 0. \tag{4.12}$$

It follows from (4.12) that

$$u(t) \geqslant u_1(t) = C_1 \exp(\sqrt{p_0}\,t) + C_2 \exp(-\sqrt{p_0}\,t),$$

where the constants C_1 and C_2 are determined from the boundary conditions. Thus, the solution $u(t)$ is rapidly increasing (or unstable). In this case, as shown in [4, 15, 19], computational errors are also increasing exponentially.

In [4] the following numerical example is given. The boundary-value problem

$$\frac{d}{dt}[(1+t)\dot{x}] - \gamma x(t) = \pi \cos \pi t - [\gamma + \pi^2(1+t)] \sin \pi t, \tag{4.13}$$
$$x(0) = x(1) = 0,$$

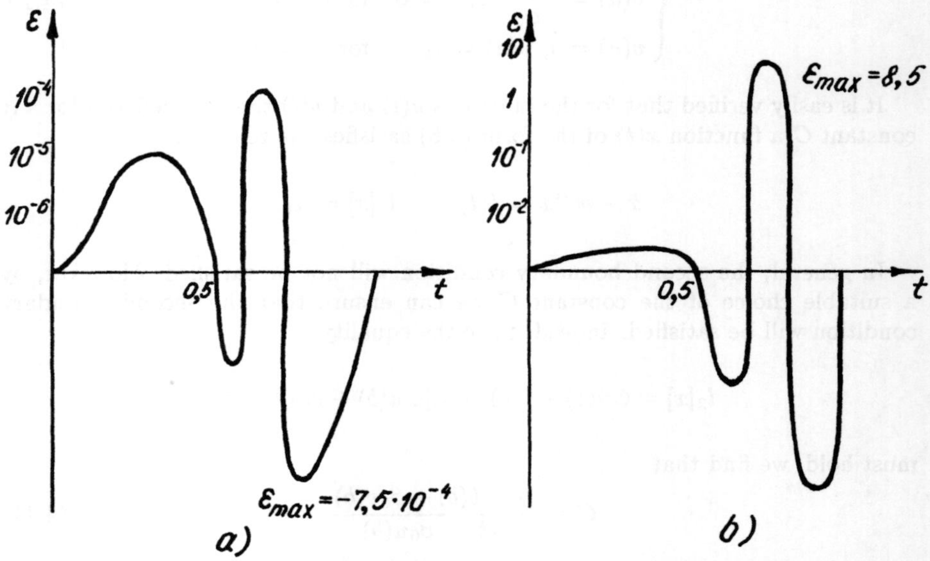

FIG. 4.1. The total error of the solution of the problem (4.13): (a) for $\gamma = 100$; (b) for $\gamma = 500$

for any $\gamma > 0$ has the solution $x(t) = \sin \pi t$. However, the numerical solution according to the above procedure, utilizing the Runge–Kutta formula of the fourth order with step $h = 0.025$ leads to the results shown in Figure 4.1.

Figure 4.1 shows the total error of the solution of the problem (4.13) against time for $\gamma = 100$ and for $\gamma = 500$.

As is seen from this figure, the result for $\gamma = 100$ is quite satisfactory, whereas for $\gamma = 500$ the result is disastrous: the error of the solution is of the order of 1 000% and more.

Some special methods were devised to combat numerical instability in the reduction of a linear boundary-value problem to a Cauchy problem, such as the sweep method for solving problems of the second order and the method of the transfer of boundary conditions for solving general linear boundary-value problems.

3. Method of the transfer of boundary conditions. Let us use a linear boundary-value problem to illustrate this method.

$$\begin{cases} \dot{x}(t) = A(t)x(t) + f(t), & x \in R^n,\ f \in R^n, \\ (l_s, x(t_s)) = \alpha_s, & l_s \in R^n,\ s = 1, \ldots, n. \end{cases} \tag{4.14}$$

In the problem (4.14) the points t_s are prescribed and can either coincide or be distinct; the vectors l_s are also prescribed. Note that any linear boundary condition for equation (4.14) can be written as

$$\big(l_s, x(t_s)\big) = \alpha_s. \tag{4.15}$$

For this purpose, it remains only to choose the vector l_s accordingly.

The idea behind the transfer method is to transfer all boundary conditions to one point t_1, then to determine $x(t_1)$, and to solve the corresponding Cauchy problem.

The boundary conditions are transferred according to the following scheme. Suppose we wish to transfer the condition (4.15) from the point t_s to the point t_1. To this end, we choose two functions $l(t)$ and $\alpha(t)$ which for any solution $x(t)$ of the equation

$$\dot{x}(t) = A(t)x(t) + f(t) \tag{4.16}$$

satisfy the conditions

$$\big(l(t), x(t)\big) = \alpha(t), \qquad l(t_s) = l_s,\ \alpha(t_s) = \alpha_s. \tag{4.17}$$

The relation (4.17) can be regarded as the analytic continuation of the equality (4.15) from the point t_s to the segment $[t_s, t_1]$.

Differentiating the relation (4.17) and taking account of equation (4.14) gives

$$\begin{aligned} \frac{d}{dt}\big(l(t), x(t)\big) &= \big(\dot{l}(t), x\big) + \big(l(t), \dot{x}\big) = \big(\dot{l}(t), x\big) + \big(l(t), Ax\big) + \big(l(t), f(t)\big) \\ &= \big(\dot{l}(t), x\big) + \big(A'l(t), x\big) + \big(l(t), f\big) = \dot{\alpha}(t). \end{aligned} \tag{4.18}$$

The equality (4.18) is identically satisfied with respect to $x(t)$ provided that

$$\dot{l}(t) = -A'(t)l, \qquad l(t_s) = l_s, \tag{4.19}$$

and, moreover,

$$\dot{\alpha}(t) = \big(l(t), f(t)\big), \qquad \alpha(t_s) = \alpha_s. \tag{4.20}$$

Having solved the Cauchy problem (4.19), we determine the function $l(t)$ on the segment $[t_s, t_1]$. Next, we determine $\alpha(t_1)$ by a quadrature in the form

$$\alpha(t_1) = \alpha_s + \int_{t_s}^{t_1} \big(l(s), f(s)\big)\, ds. \tag{4.21}$$

Using the obtained values $l(t_1)$ and $\alpha(t_1)$, we derive the relation

$$\big(l(t_1), x(t_1)\big) = \alpha(t_1). \tag{4.22}$$

The equality (4.22) indicates that the boundary condition (4.14) is transferred from the point t_s to the point t_1.

It should be noted that the functions $l_s(t)$ and $\alpha_s(t)$ are not uniquely determined. The method described is just one of a number of others.

In a similar way, all other boundary values are transferred to the point t_1. Thus, we obtain a system of linear algebraic equations for $x(t_1)$ of the form

$$\begin{aligned} (l_1, x(t_1)) &= \alpha_1, \\ (l_2(t_1), x(t_1)) &= \alpha_2(t_1), \\ &\dots\dots\dots \\ &\dots\dots\dots \\ (l_n(t_1), x(t_1)) &= \alpha_n(t_1). \end{aligned} \tag{4.23}$$

We can find $x(t_1) = \beta$ from the system (4.23) and then $x(t)$ as the solution of a Cauchy problem of the form

$$\dot{x}(t) = A(t)x(t) + f(t), \qquad x(t_1) = \beta. \tag{4.24}$$

It can be easily verified that such a method of the transfer of boundary conditions requires the solution of at most $n-1$ Cauchy problems for linear systems of the nth order of the form (4.19) and $n-1$ quadratures of the form (4.21). Further, it is necessary to solve the system of linear algebraic equations (4.23) and, finally, solve the Cauchy problem (4.24).

If some of the points t_s coincide, then the number of boundary conditions to be transferred diminishes, possibly, to one.

We can thus assume that the working time required for solving the boundary-value problem (4.14) as compared to solving the Cauchy problem for the same equation increases anywhere from twofold to $(n-1)$-fold.

However, the described method of the transfer of boundary conditions does not preclude rapidly increasing solutions, i.e., it is not numerically stable.

4. The Abramov method. In 1961 A. A. Abramov proposed a numerically stable method of the transfer of boundary conditions.[1] He proposed to make use of the fact that the functions $l_s(t)$, $\alpha_s(t)$ are not uniquely defined and to seek functions $g_s(t)$, $\alpha_s(t)$ that satisfy the conditions

$$g_s(t_s) = l_s, \qquad \|g_s(t)\|^2 = (g_s(t), g_s(t)) = \|l_s\|^2, \qquad \alpha_s(t_s) = \alpha_s. \tag{4.25}$$

We shall seek the function $g_s(t)$ in the form

$$g_s(t) = m_s(t)l_s(t). \tag{4.26}$$

Here l_s is a vector function satisfying equation (4.19), and $m(t)$ is a scalar function to be determined. Since

$$\begin{aligned} \dot{g}_s(t) = \dot{m}_s(t)l_s(t) + m_s(t)\dot{l}_s(t) &= -m_s(t)A'(t)l_s(t) \\ + \dot{m}_s(t)\frac{g_s(t)}{m_s(t)} &= -A'(t)g_s(t) + \frac{\dot{m}_s}{m_s}g_s(t), \end{aligned}$$

[1] Abramov, A. A., On the Transfer of Boundary Conditions for Systems of Linear Ordinary Differential Equations, *Zh. Vychisl. Mat. i Mat. Fiz.*, (1961), vol. 1, no. 3. (Translated into English as *U.S.S.R. Comput. Math. and Math. Phys.*)

the ratio $\dot{m}_s/m_s$ can be determined from the equality (4.25). By differentiation we have

$$\frac{d}{dt}(g_s, g_s) = 2(g_s, \dot{g}_s) = 2\left(g_s, -A'(t)\dot{g}_s + \frac{\dot{m}_s}{m_s} g_s\right) = 0;$$

hence

$$\frac{\dot{m}_s}{m_s} = \frac{(g_s, A' g_s)}{(g_s, g_s)}.$$

Finally, we thus obtain the following problem for $g_s(t)$:

$$\begin{gathered} \dot{g}_s(t) = -A'(t) g_s + \frac{(g_s, A' g_s)}{(g_s, g_s)} g_s, \\ g_s(t_s) = l_s. \end{gathered} \tag{4.27}$$

Differentiating the equality $\alpha_s(t) = \big(g_s(t), x(t)\big)$ gives

$$\begin{aligned} \dot{\alpha}_s = (\dot{g}_s, x) + (g_s, \dot{x}_s) &= \left(-A' g_s + \frac{(g_s, A g_s)}{(g_s, g_s)} g_s, x\right) \\ + (g_s, Ax + f) &= \frac{(g_s, A g_s)}{(g_s, g_s)} (g_s, x) + (g_s, f). \end{aligned}$$

Hence, the function $\alpha_s(t)$ is a solution of the problem

$$\begin{gathered} \dot{\alpha}_s = \frac{(g_s, A g_s)}{(g_s, g_s)} \alpha_s + (g_s, f), \\ \alpha_s(t_s) = \alpha_s. \end{gathered} \tag{4.28}$$

Thus, to solve the boundary-value problem (4.14) by the Abramov method, we need to solve from two to $n-1$ nonlinear Cauchy problems in order to determine the vector functions $g_s(t)$ and the same number of linear scalar equations (4.20) in order to determine the functions $\alpha_s(t)$.

All these problems are numerically stable, since the norms of the functions $g_s(t)$ do not increase. Furthermore, if the boundary conditions are transferred so that the resulting problem (4.24) is also stable, then the whole process of solution of a boundary-value problem proves to be numerically stable.

Therefore, by the correct application of the Abramov method the solution of a linear boundary-value problem is numerically stable. Thus, we can assume that there exists an effective numerical method of solution of linear boundary-value problems. For other types of boundary-value problems, such as nonlinear problems or problems with nonlinear boundary conditions, there are, at present, no effective methods of solution.

On the basis of the Abramov method special iteration procedures for solving more general optimal control problems can be devised [16, 20].

§ 5. The Shatrovskii method of successive improvements of control

We present this method for the Mayer problem:

$$\dot{x}(t) = f(t, x, u), \qquad x \in R^n,\ u \in U \subset R^m,\ 0 \leqslant t \leqslant T,$$
$$x(0) = x_0, \tag{5.1}$$
$$J(u) = \varphi(x(T)) \to \min.$$

Here T is a fixed quantity, U is some convex set, and φ is a continuous function bounded below.

As an example of the problem (5.1), we can consider the problem of placing a satellite at a given phase point z at a given time T. The cost functional J may then be taken as a sum of squared deviations:

$$J(u) = \sum_{i=1}^{n} \big(z_i - x_i(T)\big)^2.$$

To solve the problem (5.1), L. I. Shatrovskii proposed a method of successive improvements of control, which can be described as follows.[2]

We first choose an admissible control $\bar{u}(t)$ and determine a corresponding trajectory $\bar{x}(t)$. Next, the cost functional

$$J(\bar{u}) = \varphi\big(\bar{x}(T)\big). \tag{5.2}$$

can be evaluted. The initial approximation $\bar{u}(t)$ is chosen from past experience or from additional information. The choice of the initial approximation affects essentially the degree of convergence of the algorithm.

Further, we seek to determine the control increments $v(t) \in R^m$ that decrease the value of the functional (5.2). The desired improved values of the control and trajectory are denoted by

$$u_1(t) = \bar{u}(t) + v(t), \qquad x_1(t) = \bar{x}(t) + y(t), \tag{5.3}$$

where $y(t)$ is the trajectory increment.

Substituting the expressions (5.3) in the equations of motion (5.1) and linearizing the resulting equations in the neighborhood of the point $(\bar{x}, \bar{u})$ gives

$$\dot{x}(t) = f(t, \bar{x} + y, \bar{u} + v) = f(t, \bar{x}, \bar{u}) + \left\| \frac{\partial f}{\partial x} \right\|_{(\bar{x},\bar{u})} y + \left\| \frac{\partial f}{\partial u} \right\|_{(\bar{x},\bar{u})} v. \tag{5.4}$$

Here the symbols

$$\left\| \frac{\partial f}{\partial x} \right\|_{(\bar{x},\bar{u})} = A(t), \qquad \left\| \frac{\partial f}{\partial u} \right\|_{(\bar{x},\bar{u})} = B(t). \tag{5.5}$$

[2]Shatrovskii, L. I., On a Numerical Method of Solving Optimal Control Problems, *Zh. Vychisl. Mat. i Mat. Fiz.*, (1962), vol. 2, no. 2. (Translated into English as *U.S.S.R. Comput. Math. and Math. Phys.*)

(abbreviated to $A(t)$ and $B(t)$) denote matrices of partial derivatives of f with respect to x_i $(i = 1, \dots, n)$ and u_l $(l = 1, \dots, m)$ calculated at the point $(\bar{x}(t), \bar{u}(t))$.

Note that for known $\bar{u}(t)$ and $\bar{x}(t)$ the matrices A and B can be determined as some known functions of time $A(t)$ and $B(t)$ not depending on $v(t)$ and $y(t)$. Since $\dot{x}_1(t) = \dot{\bar{x}}(t) + \dot{y}(t)$ and $\dot{\bar{x}}(t) = f(t, \bar{x}, \bar{u})$, it follows from (5.4), in view of the abbreviations (5.5), that

$$\dot{y}(t) = A(t)y(t) + B(t)v(t). \tag{5.6}$$

We now evaluate the functional $J(u)$:

$$J(u) = J(\bar{x} + y) = J(\bar{x}) + \big(\operatorname{grad} \varphi(\bar{x}), y\big) + \cdots .$$

The principal linear part δJ of the increment of the functional J is of the form

$$\delta J = \big(\operatorname{grad} \varphi(\bar{x}), y\big) = (C, y). \tag{5.7}$$

The value of $\operatorname{grad} \varphi(\bar{x}) = C$ can be determined, since the values of $\bar{x}$ and $\bar{u}$ are known.

Thus, the problem of improving the control $\bar{u}$ reduces in the linear approximation to the following problem: determine $v(t)$ as the control yielding the minimum of the linear functional

$$\delta J = (C, y) \tag{5.8}$$

on the solutions of the linear system

$$\dot{y}(t) = A(t)y(t) + B(t)v(t), \qquad y(0) = 0. \tag{5.9}$$

It is convenient to use the *method of adjoint equations* for solving the problem(5.8), (5.9). We introduce the vector $p(t)$ which is the solution of the following Cauchy problem:

$$\dot{p}(t) = -A'(t)p(t), \qquad p(t) = -C = -\operatorname{grad} \varphi(\bar{x}). \tag{5.10}$$

The following formula is valid:

$$\delta J = -\big(C, y(T)\big) = -\big(p(T), y(T)\big). \tag{5.11}$$

The expression (5.11) can be computed in a different way. For this purpose we find the derivative of the scalar product $\big(p(t), y(t)\big)$:

$$\begin{aligned} \frac{d}{dt}\big(p(t), y(t)\big) &= (\dot{p}, y) + (p, \dot{y}) = (-A'p, y) + (p, Ay + Bv) \\ &= (p, -Ay + Ay + Bv) = (p, Bv). \end{aligned} \tag{5.12}$$

Since at the initial moment $t = 0$, $\big(p(0), y(0)\big) = 0$, we obtain by integrating equation (5.12):

$$\delta J = -\big(p(T), y(T)\big) = -\int_0^T \frac{d}{dt}\big(p(t), y(t)\big)\,dt = -\int_0^T \big(p(t), Bv(t)\big)\,dt. \tag{5.13}$$

If for each t ($0 \leqslant t \leqslant T$) we now choose $v(t)$ so that $\big(p(t), By(t)\big) > 0$, then $\delta J < 0$, and such a control correction term leads, in the linear approximation, to a decrease in the value of the cost functional $J(u)$. For $v(t)$ we can take the vector $v(t) = kB'(t)p(t)$, where $k > 0$ is a scalar. Further, using the properties of a scalar product, we see that

$$\begin{aligned}\big(p(t), B(t)v(t)\big) &= k\big(p(t), B(t)B'(t)p(t)\big)\\ &= k\big(B'(t)p(t), B'(t)p(t)\big) > 0.\end{aligned}$$

Let us now choose a number k so that

$$u_1(t) = \bar{u}(t) + v(t) = \bar{u}(t) + kB'(t)p(t) \in U. \tag{5.14}$$

If $\bar{u}(t)$ does not lie on the boundary of the region U, then by convexity of U for sufficiently small k we can establish that $u_1(t) \in U$. But if $\bar{u}(t)$ lies on the boundary of the region U, then we also can choose small k so that $u_1(t) \in U$. In the case where there are no such $k > 0$, we can prove that $\bar{u}(t)$ is the optimal control.

To sum up, the algorithm of this method consists of the following steps.

1°. An initial admissible control $\bar{u}(t) \in U$ is chosen and a corresponding trajectory $\bar{x}(t)$ is computed. If the solution $\bar{x}(t)$ is stable under steady-acting disturbances, then it can be determined with a high degree of accuracy by, for example, the Runge–Kutta method.

2°. The matrices $A(t)$ and $B(t)$ are computed according to formula (5.5), along with the vector $C = (\operatorname{grad}\varphi)\big|_{(\bar{x},\bar{u})}$.

3°. The Cauchy problem

$$\dot{p}(t) = -A'(t)p, \qquad p(T) = -C, \tag{5.15}$$

is solved in reverse time from $t = T$ to $t = 0$. Since this problem is solved in reverse time, the function $p(t)$ is determined with a high degree of accuracy, provided that the matrix $A(t)$ is stable.

4°. A number k is chosen so that

$$u_1(t) = \bar{u}(t) + kB'(t)p(t) \in U. \tag{5.16}$$

To this end, we set $k = k_0 > 0$ and verify the condition (5.16). If it is satisfied, then $u_1(t)$ is found. Otherwise, we set $k = k_0/2$, and the computational process is repeated.

5°. Equation (5.3) is solved for $u_1(t)$ and the condition

$$J(u_1) = \varphi\big(x_1(T)\big) < \varphi\big(\bar{x}(T)\big) = J(u_0). \tag{5.17}$$

is verified.

If the condition (5.17) is satisfied, then the control $u_1(t)$ determined from (5.15) leads to a decrease in the value of the functional $J(u)$. Otherwise, we set $u_1(t) =$

$\bar{u}(t) + v(t)/2$, and the condition (5.17) is again verified. But if it is not satisfied again, then we set $u_1(t) = \bar{u}(t) + v(t)/4$, and so on, until the condition (5.17) is satisfied.

6°. The value of the functional $J(u_1)$ is compared with some prescribed admissible value J_{adm}. If

$$J(u_1) \leqslant J_{\text{adm}}, \tag{5.18}$$

then the computational process is terminated; here, $u_1(t)$ is an admissible control such that the functional J assumes a value that is less than the prescribed one. When the condition (5.18) is not satisfied, the obtained value $\bar{u}_1(t)$ is again taken for a new initial control $\bar{u}(t)$, and the process of control improvement is repeated.

The Shatrovskii method is easily converted into an algorithm, and computer programs implementing this method have been created.

It should be noted that this method yields not the optimal control but rather a sufficiently good admissible control. Moreover, the control is obtained in the form of a program control (not an F-control).

The Shatrovskii method can be applied to systems in which certain parameters and the control are to be optimized simultaneously, and also to systems with delay.

§ 6. The Fedorenko method of reducing to a linear programming problem

As in most of other methods discussed above, in the method proposed by R. P. Fedorenko [10] the two basic procedures are also used: the finite-dimensional approximation of the original problem, and linearization. However, the problem resulting in each step from the iteration process is a problem in linear programming.

For simplicity, we examine the Fedorenko method in connection with the problem (5.1):

$$\begin{gathered} \dot{x}(t) = f(t, x, u), \qquad x \in R^n,\ u \in U \subset R^m,\ 0 \leqslant t \leqslant T, \\ x(0) = x_0, \\ J(u) = \varphi\big(x(T)\big) \to \min_u . \end{gathered} \tag{6.1}$$

However, we impose an additional requirement that the trajectory of the problem (6.1) satisfy the following constraints:

$$\int_0^T F_l\big(t, x(t), u(t)\big)\, dt = 0\ (\leqslant 0), \qquad l = 1, \ldots, r. \tag{6.2}$$

As in § 5, we take an admissible initial (reference) control $\bar{u}$ and determine a corresponding reference trajectory $\bar{x}$. Next, we seek a correction to the control $v(t) \in R^m$ which diminishes the value of the functional J, i.e., $J(\bar{u} + v) \leqslant J(\bar{u})$. As in § 5, we establish that $v(t)$ must be chosen so that

$$\delta J = -\int_0^T \big(B'(t)p(t), v(t)\big)\, dt \to \min_v, \qquad \big(\bar{u}(t) + v(t)\big) \in U. \tag{6.3}$$

Let us see how the conditions (6.2) change in passing from $\bar{u}(t)$ to $\bar{u}(t)+v(t)$. We have

$$\int_0^T F_l(t,\bar{x}+y,\bar{u}+v)\,dt = \int_0^T F_l(t,\bar{x},\bar{u})\,dt + \int_0^T (F_{ix}(t,\bar{x},\bar{u}),y(t))\,dt$$
$$+\int_0^T (F_{iu}(t,\bar{x},\bar{u}),v(t))\,dt + O(\|v\|^2), \qquad i=1,\dots,k.$$

Here $y(t)$ is the solution of the problem (5.9); $F_{ix}\in R^n$ and $F_{iu}\in R^m$ are the vectors of derivatives of the scalar functions F_l with respect to x and u, respectively. Denoting

$$F_{ix}(t,\bar{x},\bar{u}) = a_l(t), \qquad F_{iu}(t,\bar{x},\bar{u}) = b_l(t), \qquad \int_0^T F_l(t,\bar{x},\bar{u})\,dt = f_l,$$

we obtain in the linear approximation

$$\int_0^T F_l(t,\bar{x}+y,\bar{u}+v)\,dt = f_l + \int_0^T (a_l(t),y(t))\,dt + \int_0^T (b_l(t),v(t))\,dt. \tag{6.4}$$

We now transform the summand $\int_0^T (a_l(t),y(t))\,dt$. Let us define the function $z(t)$ as the solution of the problem

$$\dot{z}_l(t) + A'(t)z_l(t) = a_l(t), \qquad z_l(T)=0. \tag{6.5}$$

Since

$$\int_0^T \frac{d}{dt}(z_l(t),y(t))\,dt = (z_l(t),y(t))\big|_0^T = 0,$$

then

$$\begin{aligned} 0 &= \int_0^T [(\dot{z}_l,y)+(z_l,\dot{y})]\,dt \\ &= \int_0^T [(-A'z_l+a_l,y)+(z_l,Ay+Bv)]\,dt \\ &= \int_0^T [(a_l,y)+(z_l,Bv)]\,dt. \end{aligned}$$

This yields

$$\int_0^T (a_l,y)\,dt = -\int_0^T (z_l,Bv)\,dt = -\int_0^T (B'z_l,v)\,dt. \tag{6.6}$$

Substituting (6.6) in (6.4) gives in the linear approximation

$$\begin{aligned} \int_0^T F_l(t,\bar{x}+y,\bar{u}+v)\,dt &= f_l + \int_0^T (b_l(t)-B'(t)z_l(t),v(t))\,dt \\ &= f_l + \int_0^T (w_l(t),v(t))\,dt, \end{aligned} \tag{6.7}$$

where $w_l(t)$ denotes the function $w_l(t) = b_l(t) - B'(t)z_l(t)$.

Thus, in the linear approximation the correction $v(t)$ to the admissible initial control $\bar{u}(t)$ can be determined from the problem

$$\begin{gathered} -\int_0^T \big(B'(t)p(t), v(t)\big)\, dt \to \min_v, \\ f_l + \int_0^T \big(w_l(t), v(t)\big)\, dt = 0 \ (\leqslant 0) \qquad \big(\bar{u}(t) + v(t)\big) \in U,\ l = 1, \dots, k. \end{gathered} \tag{6.8}$$

Let us reduce the problem (6.8) to a problemin linear programming. To this end, we divide the segment $[0, T]$ into N subintervals by the points $t_i = (T/N)i$, $i = 0, 1, \dots, N$. Next, the initial control may be taken as

$$\bar{u}(t) = \bar{u}_i = \text{constant} \in U, \qquad t \in [t_i, t_{i+1}],\ i = 0, \dots, N-1.$$

The correction $v(t)$ is also taken as

$$v(t) = v_i = \text{constant}, \qquad t \in [t_i, t_{i+1}],\ i = 0, \dots, N-1.$$

Moreover, each v_i is sought as the expansion:

$$v_i = \sum_{j=1}^{K} s_{ij} e^j, \qquad s_{ij}^- \leqslant s_{ij} \leqslant s_{ij}^+. \tag{6.9}$$

Here e^j is a collection of vectors in R^m, and K is a number which can be less than or greater than m. The numbers s_{ij}^- and s_{ij}^+ are taken so that $\big(\bar{u}(t) + v(t)\big) \in U$. The choice of s_{ij}^- and s_{ij}^+, together with the vectors e^j, is not difficult to make. Substituting the expression (6.9) in the relation (6.8) yields

$$\begin{gathered} -\sum_{j=1}^{K} \sum_{i=0}^{N-1} s_{ij} \int_{t_i}^{t_{i+1}} \big(B'(t)p(t), e^j\big)\, dt \to \min_{s_{ij}}, \\ f_l + \sum_{j=1}^{K} \sum_{i=0}^{N-1} s_{ij} \int_{t_i}^{t_{i+1}} \big(w_l(t), e^j\big)\, dt = 0 \ (\leqslant 0), \\ s_{ij}^- \leqslant s_{ij} \leqslant s_{ij}^+. \end{gathered} \tag{6.10}$$

The problem (6.10) can be rewritten as

$$\begin{gathered} \sum_{j=1}^{K} \sum_{i=0}^{N-1} (h_{ij}, s_{ij}) \to \min_{s_{ij}}, \\ f_l + \sum_{j=1}^{K} \sum_{i=0}^{N-1} (w_{lij}, s_{ij}) = 0 \ (\leqslant 0), \\ s_{ij}^- \leqslant s_{ij} \leqslant s_{ij}^+,\ i = 0, 1, \dots, N-1,\ j = 1, \dots, K,\ l = 1, \dots, k. \end{gathered} \tag{6.11}$$

Here the following notation is used:

$$h_{ij} = -\int_{t_i}^{t_{i+1}} (B'(t)p(t), e^j)\,dt,$$

$$w_{lij} = \int_{t_i}^{t_{i+1}} (w_l(t), e^j)\,dt.$$

Therefore, the problem of constructing a correction $v(t)$ to the control reduces to a problem (6.11) in linear programming in which the variables s_{ij} are the desired quantities. The dimension of the problem (6.11) is equal to NK, and the number of constraints is $2NK + k$. In view of such a large number of constraints, especially, of constraints of inequality type, it is not expedient to use the standard simplex method procedure for solving the problem (6.11). R. P. Fedorenko has devised a special iteration procedure for solving problems in linear programming in connection with problems of the type (6.11).

Thus, to solve an optimal control problem by means of the Fedorenko method, we must state and solve a problem in linear programming of the form (6.11) in each step of the iteration process, determine the correction $v(t)$ to the control and the correction $y(t)$ to the trajectory, and then we must move on to the next iteration step, in which $\bar{u}(t)$ and $\bar{x}(t)$ are replaced by $\bar{u}(t) + v(t)$ and $\bar{x}(t) + y(t)$.

For a fuller treatment of the Fedorenko method and of computational procedures which implement this method, we refer the reader to [10].

Main Results and Formulas of Chapter XV

Reduction of optimal control problems to ones in nonlinear programming leads to problems of large dimension whose solution is not easy.

Reduction to a boundary-value problem is carried out on the basis of the classical calculus of variations or Pontryagin's maximum principle.

Method of the transfer of boundary conditions for the multipoint boundary-value problem

$$\dot{x}(t) = A(t)x(t) + f(t), \qquad x \in R^n,$$
$$\big(l_s, x(t_s)\big) = \alpha_s, \ l_s \in R^n, \ s = 1, \dots, n.$$

is described by the equations

$$\dot{l}_s(t) = -A'(t)l_s(t), \qquad l_s(t_s) = l_s,$$
$$\dot{\alpha}_s(t) = \big(l_s(t), f(t)\big), \qquad \alpha_s(t_s) = \alpha_s.$$

The Abramov method is described by the equations

$$\dot{g}_s(t) = -A'(t)g_s(t) + \frac{\big(g_s(t), A'(t)g_s(t)\big)}{\big(g_s(t), g_s(t)\big)} g_s(t), \qquad g_s(t_s) = l_s.$$

The Shatrovskii method for the problem

$$\dot{x}(t) = f(t, x, u), \qquad x(0) = x_0, \ J(u) = \varphi\big(x(T)\big) \to \min_{u \in U}.$$

is described by the following sequence of operations:

$$(a) \quad A(t) = \left(\frac{\partial f}{\partial x}\right)_{(\bar{x},\bar{u})}, \qquad B(t) = \left(\frac{\partial f}{\partial u}\right)_{(\bar{x},\bar{u})}, \qquad C = (\operatorname{grad} \varphi)\big|_{(}\bar{x}, \bar{u}),$$
$$(b) \quad \dot{p}(t) = -A'(t)p(t), \qquad p(T) = -C,$$
$$(c) \quad u_1(t) = \big(\bar{u}(t) + kB'(t)p(t)\big) \in U,$$
$$(d) \quad \dot{x}_1(t) = f(t, x_1, u_1), \qquad x_1(0) = x_0,$$
$$(e) \quad J(u_1) = \varphi\big(x_1(T)\big) < \varphi\big(\bar{x}(T)\big) = J(\bar{u}).$$

The Fedorenko method reduces an optimal control problem to one in linear programming

$$\sum_{j=1}^{K} \sum_{i=0}^{N-1} s_{ij} \int_{t_i}^{t_{i+1}} \big(B'(t)p(t), e^j\big)\, dt \to \min_{s_{ij}},$$

$$f_l + \sum_{j=1}^{K} \sum_{i=0}^{N-1} s_{ij} \int_{t_i}^{t_{i+1}} \big(w_l(t), e^j\big)\, dt = 0 \ (\leqslant 0), \qquad s_{ij}^- \leqslant s_{ij} \leqslant s_{ij}^+.$$

EXERCISES OF PART FOUR

The exercises that follow are given with a view to obtain computer-assisted solutions. No particular restrictions are placed on the type of the computer, nor on standard programs that are utilized; their choice depends wholly on available resources. The working time required for solving the problems is quite extensive, and so they can be used for undergraduate scientific projects and even degree or diploma work. The models described in the exercises are close to the actual models used in practice, and the numerical data given coincide with data encountered in applications.

1. The following control problem for a submarine is encountered. Test the stability of a matrix $A(p)$ depending on the scalar parameter that varies from 0.5 to 1.

The matrix $A(p)$ is of the form

$$A(p) = \begin{pmatrix} 0 & 850 & 0 & 0 & 0 & 0 & 0 & 0 \\ -850 & -120 & -4100 & 0 & 0 & 0 & 0 & 0 \\ 33 & 0 & -33 & 0 & -700 & 0 & 0 & 0 \\ 0 & 0 & 0 & 0 & 1400 & 0 & 0 & 0 \\ 0 & 0 & 1600 & -450 & -110 & 0 & 0 & 0 \\ 0 & 0 & 0 & 81 & 0 & -1 & 0 & -900 \\ 0 & 0 & 0 & 0 & 0 & 0 & 0 & 110 \\ 0 & 0 & 0 & 0 & 0 & a(8,6) & a(8,7) & a(8,8) \end{pmatrix},$$

where

$$a(8,6) = \frac{12}{2.48 - 1.48p}; \qquad a(8,7) = \frac{1.1 \cdot 0.8065p^2(1.5 - p)}{1 - 0.599968p};$$

$$a(8,8) = \frac{22 \cdot 0.8065p^2(1.5 - p)}{1 - 0.5968p}.$$

(a) Choose an arbitrary value of the parameter p and test the stability of the resulting matrix.

(b) Test the stability of the whole family of matrices $A(p)$ by means of the Kharitonov theorem (see Theorem 5.6 of Chapter I).

2. Investigate numerically the effect of small uncontrollable torque disturbances $M_i(t)$ on the behavior of an axially symmetric spacecraft. The equations of motion of the spacecraft are of the form (see p. 000)

$$\dot{x}_1 = \frac{A-C}{A}x_2x_3 + \frac{M_1(t)}{A},$$
$$\dot{x}_2 = \frac{A-C}{A}x_1x_3 + \frac{M_2(t)}{A},$$
$$\dot{x}_3 = \frac{M_3(t)}{C}, \qquad A > C.$$

Use different models of disturbances, including random ones.

3. *The Marchuk model in immunology.*[1] To describe the process of propagation of a virus infection, G. I. Marchuk proposed the following model:

$$\begin{aligned}
&\dot{V}(t) = \big(h_1 - h_2F(t)\big)V(t),\\
&\dot{C}(t) = \xi(m)h_3F(t-\tau)V(t-\tau) - h_4\big(C(t)-1\big),\\
&\dot{F}(t) = h_5\big(C(t) - F(t)\big) - h_6F(t)V(t),\\
&\dot{m}(t) = h_7V(t) - h_8m(t), \qquad t \geqslant 0,\\
&V(t) = 0, \qquad t < 0,\ V(0) = 10^{-6},\ C(0) = 1,\ m(0) = 0,\\
&F(t) = 1, \qquad t \leqslant 0.
\end{aligned}$$

Here, $V(t)$ is the quantity of viruses of certain type in the organism, $F(t)$ is the number of antibodies, $C(t)$ is the number of plasma cells, $m(t)$ is some characteristic of the weakening of the organism due to illness, and τ is the time lag due to the delay in the reaction of the organism to the virus infection. The function $\xi(m)$ is of the form

$$\xi(m) = \begin{cases} 1, & \text{for } 0 \leqslant m \leqslant 0.1;\\ (1-m)\frac{10}{9}, & \text{for } 0.1 \leqslant m \leqslant 1. \end{cases}$$

This model makes it possible, by varying the coefficients $h_1, h_2, \ldots, h_8$, to model various states of health, such as stable health condition, unstable health condition, acute disease, chronic disease, and so on. Simulate numerically various types and stages of disease by choosing particular values of the coefficients $h_1, h_2, \ldots, h_8$. As an example, take the values $\tau = 0.5$, $h_1 = 2$, $h_2 = 0.8$, $h_3 = 10^4$, $h_4 = 0.5$, $h_5 = 0.17$, $h_6 = 8$, $h_8 = 0.12$. The value of h_7 is to be taken as 10, 100, and 300.

Give an interpretation to the results obtained.

4. *Linear time-optimal problem for the system of the third order* [10, p. 227–233]. The control system under consideration is described by the equations

$$\begin{aligned}
&\dot{x}_1 = x_2,\\
&\dot{x}_2 = x_3,\\
&\dot{x}_3 = u, \qquad |u| \leqslant 1\\
&x_1(0) = x_2(0) = x_3(0) = 0.
\end{aligned}$$

[1] Marchuk, G. I., *Mathematical Models in Immunology*, Nauka, Moscow, (1980).

Determine the control $u(t)$ that minimizes the transfer time T of the system to the terminal state $x_1(T) = 1$, $x_2(T) = 0$, $x_3(T) = 0$. State the boundary-value problem for the maximum principle and solve it.

5. Find the solution of the following boundary-value problems:

$$\begin{aligned}
&(a) \qquad \ddot{x}(t) = x(t) + 10x^3(t) + t^2, \qquad x(0) = x(1) = 0;\\
&(b) \qquad \ddot{x}(t) = x(t) + x^2(t), \qquad x(0) = 0,\ x(1) = 2;\\
&(c) \qquad \ddot{x}(t) = tx^3(t), \qquad x(0) = 0,\ x(1) = 1.
\end{aligned}$$

6. Determine the optimal control of the motion of a two-link manipulator (see p. 000) described by the equations

$$(I_1 + m_2L^2)\ddot{\varphi}_1 = u_1 - u_2, \qquad I_2\ddot{\varphi}_2 = u_2, \qquad I_1 = 50,$$
$$m_2 = 3, \qquad L = 6, \qquad I_2 = 30.$$

The cost functional is to be taken as

$$J(u) = \int_0^\infty (\dot{\varphi}_1^2 + \dot{\varphi}_2^2 + u_1^2 + 10u_2^2)\, dt.$$

7. For the control system of the manipulator determined above, find whether the closed system is stable, and find the eigenvalues of the closed system.

8. *The problem of the vertical ascent of the rocket.*[2] The motion of the rocket is described by the equations

$$\dot{x}_1 = -u, \qquad \dot{x}_2 = x_3, \qquad \dot{x}_3 = -g + \frac{1}{x_1(t)}\left[Vu - Ce^{-\gamma x_3(t)x_3^2(t)}\right],$$
$$0 \leqslant t \leqslant T = 100,\ 0 \leqslant u(t) \leqslant 0.04,\ x_1(0) = 1,\ x_2(0) = x_3(0) = 0.$$

Here, $x_1(t)$ is the variable mass of the rocket, $x_2(t)$ is the altitude, $x_3(t)$ is the vertical velocity of the rocket, V is the constant thrust; g, C, γ are constants that characterize the force of gravity, aerodynamic resistance, and decrease of air density with increased altitude. The numerical values of the constants are as follows: $g = 0.01$, $V = 2.0$, $C = 0.05$, $\gamma = 0.01$.

Deduce the control that maximizes the altitude $x_2(T)$, given the terminal mass of the rocket $x_1(T)$, i.e.,

$$x_2(T) \to \max, \qquad x_1(T) = 0.2.$$

[2]Eneyev, T. M., On the Use of the Gradient Method in Optimal Control Problems, *Kosmicheskiye Issledovaniya*, (1966), vol. 4, no. 5.

See also [10, p. 233–238].

9. For the problem of the control of the landing of a "Boeing-747" aircraft discussed in § 4 (subsection 3) of Chapter XIII consider a linear-quadratic problem in wich the functional (4.28) is replaced by a functional of the form

$$J = \int_0^\infty \left(k_1 |y|^2 + k_2 z^2 + k_3 (\delta a)^2 + k_4 (\delta r)^2\right) dt \to \min .$$

Choose different sets of weights k_1, k_2, k_3, k_4 and for each set determine:

(a) an optimal control;

(b) transient characteristics similar to those given in Figure 4.1 of Chapter XIII;

(c) compare the results obtained with one another and also with those given in § 4 (subsection 3) of Chapter XIII.

10. For the problem of the control of the landing of a "Boeing-747" aircraft described by equations (4.23)–(4.27) of Chapter XIII:

(a) determine an H_∞-optimal static controller;

(b) compute transient characteristics similar to those given in Figure 4.1 of Chapter XIII;

(c) compare the results obtained with those of the preceding exercise.

11. For the system discussed in Example 4.1 of Chapter XIV construct:

(a) an H_∞-optimal static controller;

(b) an H_∞-optimal dynamic controller;

(c) compute transient characteristics for the cases stated above.

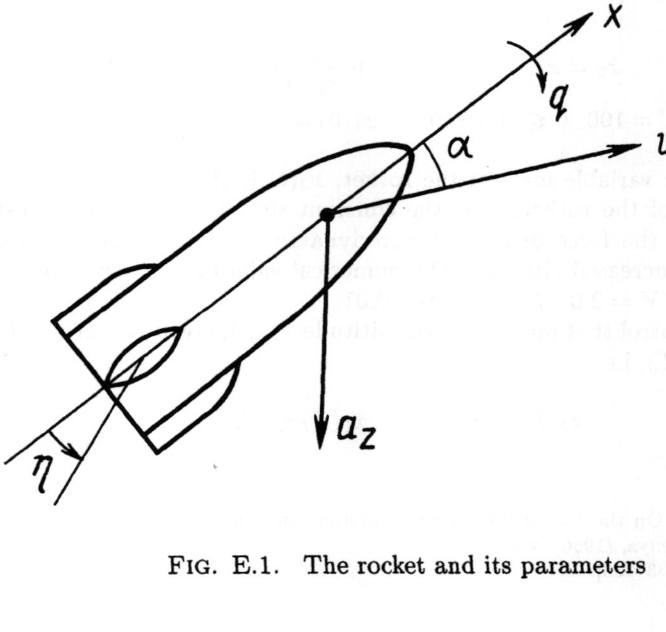

FIG. E.1. The rocket and its parameters

12. *A rocket* regarded as a rigid body can be described by the following equations (see Figure E.1):

$$\dot{x} = Ax + bu, \qquad x \in R^3,\ u \in R^1,$$
$$y = c'x, \qquad y \in R^1,$$
$$e = y - y_m.$$

Here, u (rad) is the elevator deflection control, y (m/sec^2) is the vertical acceleration (a_z in Figure E.1), x_1 (rad/sec) is the velocity of pitch ($\dot{q}$ in Figure E.1), x_3 (rad) is the angle of attack ($\Delta\alpha$ in Figure E.1), x_3 is the elevator angle ($\Delta\eta$ in Figure E.1),, y_m is the desired vertical acceleration, e is the difference between y and y_m.

TABLE E.1

Altitude	0 m				
Mach number	0.8		1.6		2.5
a_{11}	−0.327	−0.391	−0.698	−0.738	−0.886
a_{12}	−63.94	−130.29	−619.27	−651.57	−1068.85
a_{22}	−1.0	−1.42	−2.27	−2.75	−3.38
a_{13}	−155.96	−186.5	−552.9	−604.18	−1004.39
a_{23}	−0.237	−0.337	−0.429	−0.532	−0.582
c_1	0.326	0.35	0.65	0.66	0.79
c_2	−208.5	−272.38	651.11	−913.64	−1926.45
c_3	90.93	75.06	283.44	250.5	402.96

Altitude	1200 m				
Mach number	2.0		3.0		4.0
a_{11}	−1.364	−0.333	−0.337	−0.369	−0.402
a_{12}	−92.82	−163.24	−224.03	−253.71	−277.2
a_{22}	−4.68	−0.666	−0.663	−0.80	−0.884
a_{13}	−128.46	−153.32	−228.72	−249.87	−419.35
a_{23}	−0.087	−0.124	−0.112	−0.135	−0.166
c_1	1.36	0.298	0.319	0.33	0.36
c_2	−184.26	−247.75	−375.75	−500.59	−796.18
c_3	76.43	63.77	117.4	103.76	178.59

It is required to find control of the form $u = u(x_1)$ which satisfies the following requirements:

(a) the overshoot of the output variable y must be less than 10%, and the steady-state error must be within 15%;

(b) the magnitude of the elevator angle, x_3, must not exceed 1/3 rad ($|x_3| \leqslant 1/3$), and its rate of change $\dot{x}_3$ must not exceed 100 rad/sec ($|\dot{x}_3| \leqslant 100$).

The requirements (a) and (b) must be satisfied for 10 prescribed conditions of flight, which depend on the Mach number, the altitude, and the mass of the rocket.

The matrix A and the vectors b and c are of the form

$$A = \begin{pmatrix} a_{11} & a_{12} & a_{13} \\ 1 & a_{22} & a_{23} \\ 0 & 0 & -190 \end{pmatrix}, \qquad b = \begin{pmatrix} 0 \\ 0 \\ 190 \end{pmatrix}, \qquad c = \begin{pmatrix} c_1 \\ c_2 \\ c_3 \end{pmatrix}.$$

Different flight conditions are associated with different values of a_{ij} and c_i. These values are given in Table E.1.

13. For each of the 10 flight conditions of the rocket, verify the controllability and observability of the control system derived in the preceding exercise.

14. *Stabilization of an inverted pendulum system.*[3] Consider an inverted pendulum system of shown in Figure E.2. The linear model of such a system is described by the equations

$$\dot{x}_i = A_i x_i + B_i u_i, \qquad y_i = C_i x_i, \qquad i = 1, 2, 3, 4.$$

The input variables, $y_i = \theta_i$, are the angles of deflection at the pivots; the controls, u_i, are the torques at the pivots. It is required to determine controls that stabilize the angles θ_i near the equilibrium positions $\theta_i = 0$ $(i = 1, 2, 3, 4)$.

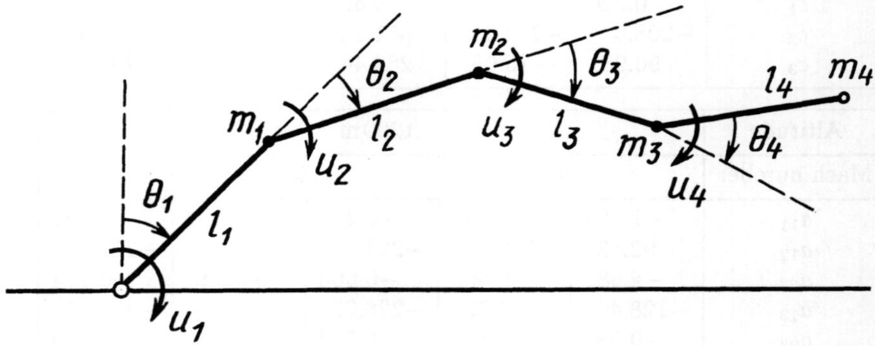

FIG. E.2. A four inverted pendulum system

Investigate the effect of nonmeasurable disturbances on such a system. In the case where all the masses $m_i = 1\,kg$ and all the lengths $l_i = 1\,m$ the numerical values of the matrices A_i, B_i, and C_i $(i = 1, 2, 3, 4)$ are the following ($g = 9.8\,\text{m/sec}^2$ is the acceleration of gravity):

$$A_1 = \begin{pmatrix} 0 & 1 \\ g & 0 \end{pmatrix}, \qquad B_1 = \begin{pmatrix} 0 \\ 1 \end{pmatrix}, \qquad C_1 = (1, 0),$$

[3]Kwakernaak, H., Westdyk, H., Regulability of a Multiple Inverted Pendulum System, *Control Theory Adv. Tech.*, (1985), vol. 1, no. 1, 1–9.

$$A_2 = \begin{pmatrix} 0 & 1 & 0 & 0 \\ g & 0 & -g & 0 \\ 0 & 0 & 0 & 1 \\ -g & 0 & 3g & 0 \end{pmatrix}, \quad B_2 = \begin{pmatrix} 0 & 0 \\ 1 & -2 \\ 0 & 0 \\ -2 & 5 \end{pmatrix}, \quad C_2 = \begin{pmatrix} 1 & 0 & 0 & 0 \\ 0 & 1 & 0 & 0 \end{pmatrix},$$

$$A_3 = \begin{pmatrix} 0 & 0 & 0 & 1 & 0 & 0 \\ 0 & 0 & 0 & 0 & 1 & 0 \\ 0 & 0 & 0 & 0 & 0 & 1 \\ 3g & -2g & -3.6\mathrm{E}-16 & 0 & 0 & 0 \\ -3g & 4g & g & 0 & 0 & 0 \\ 0 & -2 & 2 & 0 & 0 & 0 \end{pmatrix},$$

$$B_3 = \frac{1}{6}\begin{pmatrix} 0 & 0 & 0 \\ 0 & 0 & 0 \\ 0 & 0 & 0 \\ 10 & -16 & 6 \\ -14 & 29 & -21 \\ 4 & -16 & 24 \end{pmatrix}, \quad C_3 = \begin{pmatrix} 1 & 0 & 0 & 0 & 0 & 0 \\ 0 & 1 & 0 & 0 & 0 & 0 \\ 0 & 0 & 1 & 0 & 0 & 0 \end{pmatrix},$$

$$A_4 = \begin{pmatrix} 0 & 0 & 0 & 0 & 1 & 0 & 0 & 0 \\ 0 & 0 & 0 & 0 & 0 & 1 & 0 & 0 \\ 0 & 0 & 0 & 0 & 0 & 0 & 1 & 0 \\ 0 & 0 & 0 & 0 & 0 & 0 & 0 & 1 \\ 4g & -3g & 0 & 0 & 0 & 0 & 0 & 0 \\ -4g & 6g & -2g & 3.63\mathrm{E}-16 & 0 & 0 & 0 & 0 \\ -4.35\mathrm{E}-15 & -3g & 4g & -g & 0 & 0 & 0 & 0 \\ 0 & 0 & -2g & 2g & 0 & 0 & 0 & 0 \end{pmatrix},$$

$$B_4 = \frac{1}{24}\begin{pmatrix} 0 & 0 & 0 & 0 \\ 0 & 0 & 0 & 0 \\ 0 & 0 & 0 & 0 \\ 0 & 0 & 0 & 0 \\ 42 & -66 & 24 & 1.54\mathrm{E}-15 \\ -60 & 124 & -88 & 24 \\ 18 & -74 & 116 & -84 \\ 8.5\mathrm{E}-15 & 16 & -64 & 96 \end{pmatrix},$$

$$C_4 = \begin{pmatrix} 1 & 0 & 0 & 0 & 0 & 0 & 0 & 0 \\ 0 & 1 & 0 & 0 & 0 & 0 & 0 & 0 \\ 0 & 0 & 1 & 0 & 0 & 0 & 0 & 0 \\ 0 & 0 & 0 & 1 & 0 & 0 & 0 & 0 \end{pmatrix}, \qquad \mathrm{E}-i \equiv 10^{-i}.$$

15. Verify the controllability of the four inverted pendulum system described in the preceding exercise.

16. *A binary distillation column*[4] with 8 chambers is depicted in Figure E.3. The linearized model of this system involves three input variables, three output,

[4]Davison, E. J., Control of a Distillation Column with Pressure Variation, *Trans. Inst. Chemical Engineers*, (1967), vol. 45, 228–250.

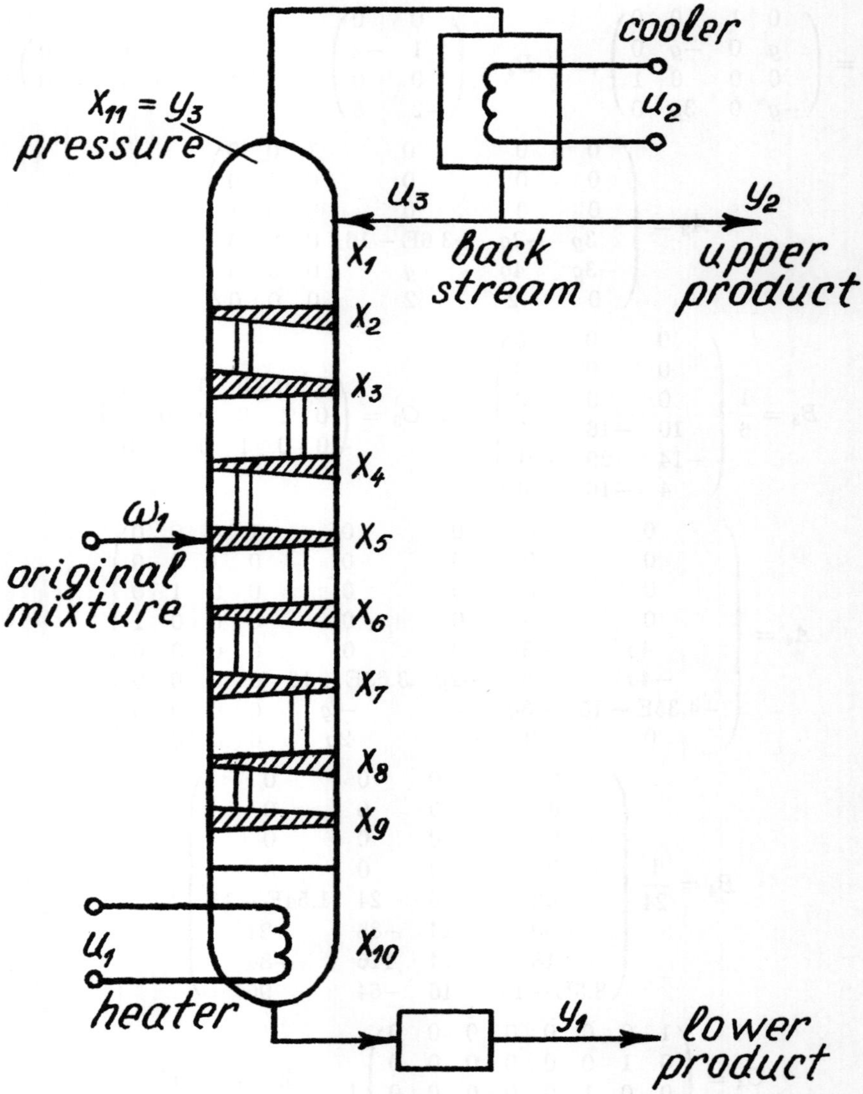

FIG. E.3. A binary distillation column with pressure variation

and one disturbance variable:

$$\dot{x} = Ax + Bu + Fw_1, \qquad y = Cx,$$
$$x \in R^{11},\ y \in R^3,\ u \in R^3,\ w_1 \in R^1.$$

Here, x_1 is the concentration of the light fraction in the cooler, x_i is the concentration of the light fraction in the ith chamber ($i = 2, \ldots, 9$), x_{10} is the concentration of the light fraction in the heater, x_{11} is the pressure, y_1 is the concentration of the light fraction in the lower product, y_2 is the concentration of the light fraction in the upperer product, y_3 is the pressure, u_1 is the temperature of the heater, u_2 is the temperature of the cooler, u_3 is the controlled back stream, w_1 is the deviation of the concentration of the original mixture.

It is required to determine a control of the variations of the output variables y_1, y_2, and y_3 produced by the action of an uncontrollable disturbance w_1, $|w_1(t)| \leqslant 1$, $t \geqslant 0$. The goal of the control is to reduce to zero the deviations y_1, y_2, and y_3. The following constraints are imposed on the control:

$$|u_1(t)| \leqslant 2.5, \qquad |u_2(t)| \leqslant 2.5, \qquad |u_3(t)| \leqslant 0.30, \qquad t \geqslant 0.$$

The matrices A, B, C, and F are of the form

$$A = 10^{-3} \begin{pmatrix} -14 & 4.3 & 0 & 0 & 0 & 0 & 0 & 0 & 0 & 0 & 0 \\ 9.5 & -13.8 & 4.6 & 0 & 0 & 0 & 0 & 0 & 0 & 0 & 0.5 \\ 0 & 9.5 & -14.1 & 6.3 & 0 & 0 & 0 & 0 & 0 & 0 & 0.2 \\ 0 & 0 & 9.5 & -15.8 & 11 & 0 & 0 & 0 & 0 & 0 & 0 \\ 0 & 0 & 0 & 9.5 & -31.2 & 15 & 0 & 0 & 0 & 0 & 0 \\ 0 & 0 & 0 & 0 & 20.2 & -35.2 & 22 & 0 & 0 & 0 & 0 \\ 0 & 0 & 0 & 0 & 0 & 20.2 & -42.2 & 28 & 0 & 0 & 0 \\ 0 & 0 & 0 & 0 & 0 & 0 & 20.2 & -48.2 & 37 & 0 & 0.2 \\ 0 & 0 & 0 & 0 & 0 & 0 & 0 & 20.2 & -57.2 & 42 & 0.5 \\ 0 & 0 & 0 & 0 & 0 & 0 & 0 & 0 & 20.2 & -48.3 & 0.5 \\ 25.5 & 0 & 0 & 0 & 0 & 0 & 0 & 0 & 0 & 25.5 & -18.5 \end{pmatrix},$$

$$B = 10^{-6} \begin{pmatrix} 0 & 0 & 0 \\ 5 & -40 & 2500 \\ 2 & -20 & 5000 \\ 1 & -10 & 5000 \\ 0 & 0 & 5000 \\ 0 & 0 & 5000 \\ -5 & 10 & 5000 \\ -10 & 30 & 3000 \\ -40 & 5 & 2500 \\ -20 & 2 & 2500 \\ 460 & 460 & 0 \end{pmatrix}, \qquad F = \begin{pmatrix} 0 \\ 0 \\ 0 \\ 0 \\ 0.01 \\ 0 \\ 0 \\ 0 \\ 0 \\ 0 \\ 0 \end{pmatrix}, \qquad C = \begin{pmatrix} 0&0&0&0&0&0&0&0&0&0&0 \\ 1&0&0&0&0&0&0&0&0&0&0 \\ 0&0&0&0&0&0&0&0&0&0&1 \end{pmatrix}.$$

17. Verify the controllability and observability of the control system derived in the preceding exercise.

18. *Control of the lateral motion of a bus.*[5] The problem consists in determining a control of the lateral motion of a bus due to several indeterminate causes. The linearized model of the lateral motion of a bus is described by the following equations:

$$\dot{x} = A(q_1, q_2)x + bu, \qquad x \in R^5,\ u \in R^1,$$
$$y = Cx, \qquad y \in R^1.$$

The variables in the model have the following meaning (see Figure E.4): $x_1 = \alpha$ is the slip angle at the center of mass of the bus, $x_2 = \dot{\varepsilon}$ is the drift velocity, $x_3 = d$

[5] Darenberg, W., Automatische Spurführung von Kraftfahrzeugen, *Automobilindustrie*, (1987), 155–159.

is the displacement of the center of mass of the bus from the traffic direction, $x_4 = \varepsilon$ is the drift angle, $x_5 = \beta$ is the turning angle of front wheels, $u = \dot{\beta}$ is the force acting on the hydraulic steering gear, y is the deviation of the front sensor of the bus from the traffic direction, $q_1 = 1/v$ is the reciprocal of the velocity of the bus, $q_2 = \mu/m$ is the ratio of the friction coefficient to the mass of the bus.

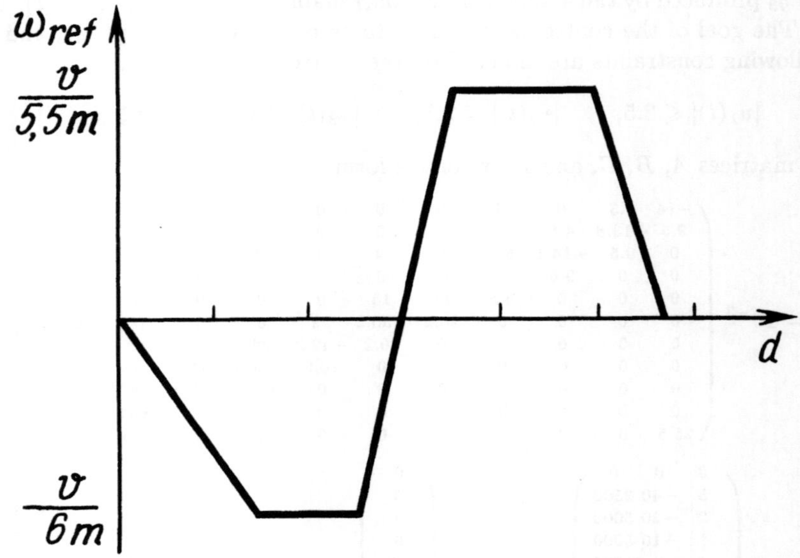

FIG. E.4. The pattern of the lateral motion of the bus:
O, the center of mass of the bus; A, the position of the front sensor

The matrices A, b, and C are of the form

$$A = \begin{pmatrix} a_{11} & a_{12} & 0 & 0 & a_{15} \\ a_{21} & a_{22} & 0 & 0 & a_{25} \\ 1/q_1 & 0 & 0 & 1/q_1 & 0 \\ 0 & 1 & 0 & 0 & 0 \\ 0 & 0 & 0 & 0 & 0 \end{pmatrix}, \qquad b = \begin{pmatrix} 0 \\ 0 \\ 0 \\ 0 \\ 1 \end{pmatrix},$$

$$C = (0 \quad 0 \quad 1 \quad l_s \quad 0),$$

where

$$a_{11} = -2q_1q_2(\delta_F + \delta_R), \qquad a_{12} = -1 - 2q_1^2q_2(\delta_F l_F - \delta_R l_R),$$
$$a_{21} = -2q_2(\delta_F l_F - \delta_R l_R)/c, \qquad a_{22} = -2q_1q_2(\delta_F l_F^2 + \delta_R l_R^2)/c,$$
$$a_{15} = 2q_1q_2\delta_F, \qquad a_{25} = 2q_2\delta_F l_F/c.$$

The bus parameters referred to in the above relations are the following: $l_s =$ 6.12 m is the distance between the front sensor and the center of mass, $l_F = 3.67$ m is the distance between the front axle and the center of mass, $l_R = 1.93$ m is the distance between the rear axle and the center of mass, $\delta_F = 99$ kN/rad is the coefficient of the sideslip force acting on the front wheels, $\delta_R = 235$ kN/rad is the coefficient of the sideslip force acting on the rear wheels, $c = \theta/m = 10.86\,\mathrm{m}^2$, θ is the moment of inertia of the bus about the vertical axis passing through the center of mass, m is the mass of the bus.

Part of the bus parameters is not specified beforehand and may vary, depending on the traffic conditions, within the following limits:

$$0.5 \leqslant \mu \leqslant 1,$$
$$9.55 \leqslant m \leqslant 16 \text{ (tons)},$$
$$1 \leqslant v \leqslant 20 \text{ (m/sec)},$$

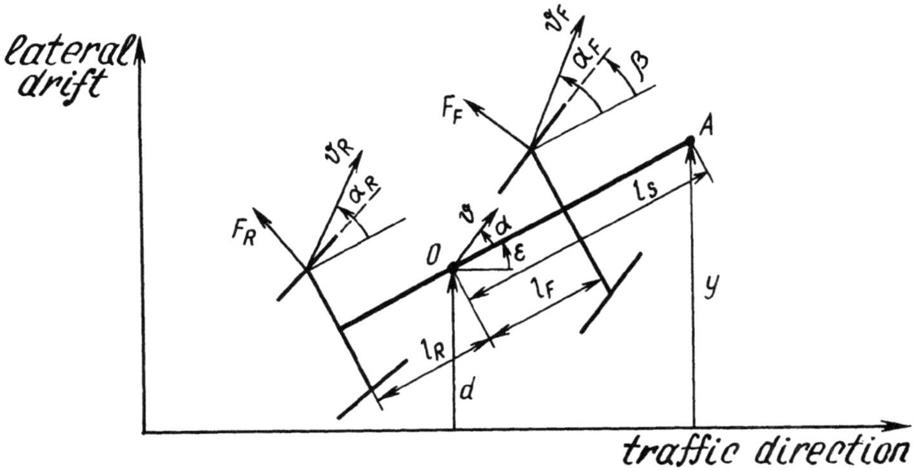

FIG. E.5. The angular velocity ω_{ref} required for stopping at a lay-by: d, the distance (m)

It is required to determine the control that minimizes (in some sense) the error $\varepsilon(t)$ between the actual motion $y(t)$ and the prescribed (reference) motion $y_{\text{ref}}(t)$: $\varepsilon(t) = y(t) - y_{\text{ref}}(t)$.

The motion $y_{\text{ref}}(t)$ can be determined from the equation

$$\ddot{y}_{\text{ref}} = v\omega_{\text{ref}}.$$

We must introduce the following constraints that ensure the comfort of passengers and take account of the bus specifications:

$$|\omega| \leqslant 7.5\,\frac{\text{rad}}{\text{sec}}, \qquad \dot{y} = v\omega,$$

$$|v(\omega - \omega_{\text{ref}})| \leqslant 4\,\frac{\text{m}}{\text{sec}^2},$$

$$|\beta| \leqslant 40, \qquad |\dot{\beta}| \leqslant 23\,\frac{\text{deg}}{\text{sec}}.$$

The cost functional is to be evaluated in three typical situations:

(a) the transfer from the initial state

$$x_1(0) = x_2(0) = x_4(0) = x_5(0) = 0, \qquad x_3(0) = 0.15.$$

to the zero terminal state. This corresponds to a parallel initial displacement of 15 cm from the traffic direction;

(b) the transition from rectilinear motion to turning with constant radius 400 m;

(c) stopping at a narrow lay-by; this is illustrated by the graph of ω_{ref} in Figure E.5.

19. *Spatial controlled motion of a spacecraft in the atmosphere.*[6] Consider a spatial controlled motion in the atmosphere of a spacecraft regarded as a material point. For the phase coordinates we take:

the altitude, y;

the coordinates, x and y, of the projection of the spacecraft on the horizon plane;

the absolute value, v, of the velocity;

the angle, θ, between the velocity vector and the plane (x, z);

the angle, ψ, between the projection of the velocity vector on the (x, z)-plane and the x-axis.

The direction of the x-axis coincides with that of the projection of the initial velocity on the (x, z)-plane. The gravitational force is taken to be constant, and the altitude dependence of the atmospheric density is taken to be exponential. Under these conditions the equations of motion in dimensionless variables are of the form

$$\dot{x} = kv\cos\theta\cos\psi,$$

$$\dot{y} = kv\sin\theta,$$

$$\dot{z} = kv\cos\theta\sin\psi,$$

$$\dot{v} = f_v = -\frac{\sigma}{\mu}\chi v^2 c_x - \sin\theta + \frac{p}{\mu}\left(1 - \frac{\alpha^2}{2}\right)\left(1 - \frac{\beta^2}{2}\right),$$

$$\dot{\theta} = f_\theta = \frac{\sigma}{\mu}\chi v c_y - \frac{\cos\theta}{v} + \frac{p\alpha}{\mu v},$$

$$\dot{\psi} = f_\psi = -\frac{1}{\cos\theta}\left\{\frac{\sigma}{\mu}\chi v c_z + \frac{p\beta}{\mu v}\left(1 - \frac{\alpha^2}{2}\right)\right\}.$$

[6] Chernous'ko, F. L., and Lyubushin, A. A., Method of Successive Approximations for Optimal Control Problems, *Optimal Control Appl. Methods*, (1982), vol. 3, no. 2.

Here the following notation is used:

$$\chi(y) = \exp\left(-\frac{y}{h}\right), \qquad h = \text{constant} > 0,$$

$$p(t) = \begin{cases} p_0, & \text{for } 0 \leqslant t \leqslant \tau, \\ 0, & \text{for } t > \tau, \end{cases}$$

$$\mu(t) = \begin{cases} 1 - \frac{(1-\mu_1)}{\tau}t, & \text{for } 0 \leqslant t \leqslant \tau, \\ \mu_1, & \text{for } t > \tau,\ 0 < \mu_1 \leqslant 1. \end{cases}$$

The aerodynamic coefficients c_x, c_y, c_z are equal to:

$$c_x = c_0 + c_1(c_y^2 + c_z^2), \qquad c_y = a\alpha, \qquad c_z = b\beta.$$

The controlled variables in the problem are the angle of attack $\alpha = \alpha(t)$ and the grazing angle $\beta = \beta(t)$. They are constrained by

$$|\alpha(t)| \leqslant \alpha_{\max}, \qquad |\beta(t)| \leqslant \beta_{\max}.$$

All the parameters of the problem

$$k,\ \alpha_{\max},\ \beta_{\max},\ h,\ \sigma,\ p_0,\ \mu_1,\ \tau,\ a,\ b,\ c_0,\ c_1$$

are positive numbers.

The initial conditions at $t = 0$ have the form

$$x(0) = 0, \quad y(0) = y_0, \quad z(0) = 0, \quad v(0) = v_0, \quad \theta(0) = \theta_0, \quad \psi(0) = 0.$$

The spacecraft trajectory is subject to the following constraints:
with regard to velocity head

$$\omega \equiv \chi(y)\left(\frac{v}{v_m}\right)^2 \leqslant 1, \qquad v_m = \text{constant} > 0;$$

with regard to maximum overload

$$n \equiv \left((f_v + \sin\theta)^2 + (v f_\theta + \cos\theta)^2 + (v f_\psi \cos\theta)^2\right) \leqslant n_m$$
$$n_m = \text{constant} > 0.$$

For the stated problem:

(a) in the (x, z)-plane determine the region of reachability D containing all the points that can be reached by the trajectory of the system starting from the initial state, subject to all the constraints imposed on the control and the phase trajectory. A condition for the termination of the process in our case is the following: $y = 0$.

(b) solve a time-optimal problem with the terminal state

$$x(T) = x_*, \quad z(T) = z_*, \quad y(T) = 0, \qquad T \to \min,$$
$$(x_*, z_*) \in D.$$

(c) solve the maximum distance problem:

$$x(T) \to \max, \qquad y(T) = 0.$$

In the solution of the problems the following values are to be taken:

$$k = 9.2, \quad \alpha_{\max} = \beta_{\max} = 0.3, \quad h = 6.4, \quad \sigma = 10,$$
$$p_0 = 3.2, \quad \mu_1 = 0.5, \quad \tau = 1.14, \quad a = b = 1, \quad c_1 = 1.5,$$
$$c_0 = 0.035, \quad v_0 = 1.5, \quad \theta_0 = 0, \quad y_0 = 10, \quad v_m = 3.$$

The parameters n_m and x_* are to be varied:

$$n_m = 5.5, 6, 12, \infty; \qquad x_* = 300, 200, 50.$$

20. A system of coupled rotating disks with noncollocated sensors and actuators[7] is described by the equation of the eighth order with two neutrally stable poles:

$$\dot{x}(t) = Ax(t) + B_1\omega(t) + B_2u(t),$$
$$z(t) = C_1x(t) + D_{12}u(t),$$
$$y(t) = C_2x(t) + D_{21}\omega(t),$$
$$x \in R^8,\ \omega \in R^2,\ z \in R^2,\ y \in R^1,\ u \in R^1,$$

where

$$A = \begin{pmatrix} -0.161 & 1 & 0 & 0 & 0 & 0 & 0 & 0 \\ -6.004 & 0 & 1 & 0 & 0 & 0 & 0 & 0 \\ -0.5822 & 0 & 0 & 1 & 0 & 0 & 0 & 0 \\ -0.9835 & 0 & 0 & 0 & 1 & 0 & 0 & 0 \\ -0.4073 & 0 & 0 & 0 & 0 & 1 & 0 & 0 \\ -3.982 & 0 & 0 & 0 & 0 & 0 & 1 & 0 \\ 0 & 0 & 0 & 0 & 0 & 0 & 0 & 1 \\ 0 & 0 & 0 & 0 & 0 & 0 & 0 & 0 \end{pmatrix},$$

$$B_2 = \begin{pmatrix} 0 \\ 0 \\ 0.0064 \\ 0.00235 \\ 0.0713 \\ 1.0002 \\ 0.1045 \\ 0.9955 \end{pmatrix}, \qquad \begin{aligned} D_{12} &= \begin{pmatrix} 0 \\ 0 \end{pmatrix}, \\ B_1 &= (\,B_2 \quad 0_{8\times1}\,), \\ C_2 &= (\,1 \quad 0_{1\times7}\,), \\ D_{21} &= (\,0 \quad 1\,), \end{aligned}$$

[7]Cannon, R. H., Jr., and Rosenthal, D. E., Experiments in Control of Flexible Structures with Noncollocated Sensors and Actuators, *J. Guidance Control Dynamics*, (1984), vol. 7, 546–553.

$$C_1 = 10^{-3}\begin{pmatrix} 0 & 0 & 0 & 0 & 0.55 & 11 & 1.32 & 18 \\ 0 & 0 & 0 & 0 & 0 & 0 & 0 & 0 \end{pmatrix}$$

Determine an eighth-order dynamic compensator[8]

$$x_c(t) = A_c x_c(t) + B_c y(t), \qquad u = C_c x_c$$

which satisfies the following design criteria:

(i) the closed-loop system is asymptotically stable;

(ii) the closed-loop transfer function $G(s)$ from $\omega(t)$ to $z(t)$ satisfies the constraint

$$\|G(s)\|_\infty \leqslant \gamma, \qquad \gamma = 0.52.$$

For the theoretical background to this problem, the reader is referred to § 5 of Chapter VI and to Chapter XIII.

[8]Bernstein, D. S., and Haddad, W. H., LOG-Control with an H^∞ Performance Bound: a Riccati Equation Approach, *IEEE Trans. Automat. Control*, (1989), vol. AC-34, no. 3, 293–305.

GENERAL REFERENCES

General References of Part One

[1] Barbashin, E. A., *Introduction to Stability Theory*, Nauka, Moscow, (1967).

[2] Cesari, L., *Asymptotic Behavior and Stability Problems in Ordinary Differential Equations*, Springer-Verlag, Berlin, (1959).

[3] Chetaev, N. G., *Stability of Motion*, Gostechizdat, Moscow, (1955).

[4] Demidovich, B. P., *Lectures on the Mathematical Theory of Stability*, Nauka, Moscow, (1967).

[5] Doob, J. L., *Stochastic Processes*, J. Wiley, New York, (1953).

[6] Gihman, I. I., and Skorohod, A. V., *The Theory of Stochastic Processes*, Springer-Verlag, Berlin, Vol. I (1974), Vol. II (1975).

[7] Hasminskii, R. Z., *Stochastic Stability of Differential Equations*, Sijtjoff and Noordhoff, Alphen, (1980).

[8] Ito, K., and McKean, H. P., *Diffusion Processes and Their Sample Paths*, Springer-Verlag, Berlin, (1965).

[9] Kolmanovskii, V. B., and Nosov, V. R., *Stochastic Stability and Control*, (lecture notes), Moscow University of Electronics and Mathematics, Moscow, (1983).

[10] Kolmanovskii, V. B., and Nosov, V. R., *Stability of Controlled Systems*, (lecture notes), Moscow University of Electronics and Mathematics, Moscow, (1987).

[11] Kolmogorov, A. N., *Foundations of the Theory of Probability*, Nauka, Moscow, (1974).

[12] Krasovskii, N. N., *Stability of Motion*, Stanford University Press, Stanford, (1963).

[13] Lakshmikantham, V., Leela, S., and Martynyuk, A. A., *Stability Analysis of Nonlinear Systems*, Marcel Dekker, New York, (1988).

[14] La Salle, J., and Lefschetz, S., *Stability by Liapunov's Direct Method*, Academic Press, New York–London, (1961).

[15] Lyapunov, A. M., Problème Générale de la Stabilité du Mouvement, *Ann. of Math. Stud.* Princeton University Press, Princeton, (1947).

[16] Malkin, I. G., *Theory of Stability of Motion*, Nauka, Moscow, (1965).

[17] Øksendal, B., *Stochastic Differential Equations*, Springer-Verlag, Berlin, (1992).

[18] Pontryagin, L. S., *Ordinary Differential Equations*, Nauka, Moscow, (1965).

[19] Rumyantsev, V. V., and Oziraner, A. S., *Stability and Partial Stabilization*, Nauka, Moscow, (1987).

[20] Rouche, N., Habets, P., and Laloy, M., *Stability Theory by Liapunov's Direct Method*, Springer-Verlag, New York–Heidelberg–Berlin, (1977).

[21] Stratonovich, R. L., *Conditional Markov Processes and Their Application to the Theory of Optimal Control*, Elsevier, New York, (1968).

[22] Thompson, J. M. T., *Instabilities and Catastrophes in Science and Engineering*, J. Wiley, Chichester–New York, (1982).

[23] Yakubovich, V. A., and Starjinskii, V. M., *Linear Periodic Differential Equations and Their Applications*, Nauka, Moscow, (1972).

General References of Part Two

[1] Alexeyev, V. M., Galeyev, E. M., and Tikhomirov, V. M., *Problem Book in Optimization*, Nauka, Moscow, (1984).

[2] Alexeyev, V. M., Tikhomirov, V. M., and Fomin, S. V., *Optimal Control*, Nauka, Moscow, (1979).

[3] Atans, M., and Falb, P. L., *Optimal Control*, McGraw–Hill, New York, (1966).

[4] Bellman, R., *Dynamic Programming*, Princeton University Press, Princeton, (1957).

[5] Bryson, A. E., and Ho, Y. C., *Applied Optimal Control*, Blaisdell Publ. Co, Toronto–London, (1969).

[6] Chernous'ko, F. L., and Kolmanovskii, V. B., Computational and Approximation Methods of Optimal Control, *J. Sov. Math.*, (1979), vol. 12, pp. 310–353.

[7] Gelfand, I. N., and Fomin, S. V., *Calculus of Variations*, Fizmatgiz, Moscow, (1961).

[8] Girsanov, I. V., *Lectures on the Theory of Extremal Problems*, Moscow University Press, Moscow, (1970).

[9] Krasovskii, A. A., *Analytical Design of Flight Control Systems*, Nauka, Moscow, (1973).

[10] Krasovskii, N. N., *Problems of the Stabilization of Controlled Motion.* Supplement to the book: Malkin, I. G., *Theory of Stability of Motion*, Nauka, Moscow, (1965).

[11] Krasovskii, N. N., *Theory of the Control of Motion*, Nauka, Moscow, (1968).

[12] Kwakernaak, H., and Sivan, R., *Linear Optimal Control Systems*, J. Wiley Interscience, New York–London, (1972).

[13] Krotov, V. F., and Gurman, V. I., *Methods and Problems of Optimal Control*, Nauka, Moscow, (1973).

[14] Lee, E. B., and Marcus, L., *Foundations of Optimal Control Theory*, J. Wiley, New York–London, (1969).

[15] Pontryagin, L. S., Boltyanskii, V. G., Gamkrelidze, R. V., and Mischenko, E. F., *The Mathematical Theory of Optimal Processes*, Interscience, New York, (1962).

[16] Zubov, V. I., *Lectures on Control Theory*, Nauka, Moscow, (1975).

General References of Part Three

[1] Åström, K. J., *Introduction to Stochastic Control Theory*, Academic Press, New York, (1970).

[2] Brammer, K., and Siffling, G., *Kalman–Bucy Filter*, R. Oldenburg Verlag, München–Wien, (1975).

[3] Bensoussan, A., *Stochastic Control by Functional Analysis Methods*, North Holland, Amsterdam, (1982).

[4] Chernous'ko, F. L., and Kolmanovskii, V. B., *Optimal Control under Random Disturbances*, Nauka, Moscow, (1978).

[5] Davis, M. H. A., *Linear Estimation and Stochastic Control*, Chapman and Hall, London, (1977).

[6] Fleming, W. H., and Rishel, R. W., *Deterministic and Stochastic Optimal Control*, Springer-Verlag, Berlin–New York, (1975).

[7] Fleming, W. H., and Soner, H. M., *Controlled Markov Processes and Viscosity Solutions*, Springer-Verlag, Berlin–New York, (1993).

[8] Gihman, I. I., and Skorohod, A. V., *Controlled Random Processes*, Naukova Dumka, Kiev, (1977).

[9] Kolmogorov, A. N., and Fomin, S. V., *Elements of Function Theory and Functional Analysis*, Nauka, Moscow, (1968).

[10] Kolosov, G. E., *Synthesis of Optimal Control Systems under Random Disturbances*, Nauka, Moscow, (1984).

[11] Ladyzhenskaya, O. A., and Uraltseva, N. N., *Linear and Quasilinear Equations of Elliptic Type*, Nauka, Moscow, (1973).

[12] Liptser, R. S., and Shiryayev, A. N., *Statistics of Random Processes*, Springer-Verlag, Berlin, Vol. 1 (1977) and Vol. 2 (1978).

[13] Tikhonov, V. I., and Kharisov, V. N., *Statistical Analysis and Synthesis of Radio Signal Processing Systems*, Radio and Svyaz, Moscow, (1991).

[14] Tikhonov, V. I., and Kulman, N. K., *Nonlinear Filtering and Quasicoherent Reception of Signals*, Sovyetskoye Radio Publishers, Moscow, (1975).

General References of Part Four

[1] *Analysis and Computer-Aided Optimal Design of Control Systems*, (Voronov, A. A., and O'Rourke, I. A., eds.), Nauka, Moscow, (1984).

[2] Armstrong, E. A., *ORACLS: A Design System for Linear Multivariable Control*, Marcel Dekker, New York–Basel, (1980).

[3] Arushanin, O. B., and Zaletkin, S. F., *Numerical Solution of Ordinary Differential Equations Using Fortran*, Moscow University Press, Moscow, (1990).

[4] Babuška, I., Prager, M., and Vitásek, E., *Numerical Processes in Differential Equations*, J. Wiley, Praha, (1966).

[5] Bahvalov, N. S., Zhidkov, N. P., and Kobel'kov, G. M., *Numerical Methods*, Nauka, Moscow, (1987).

[6] Berezin, I. S., and Zhidkov, N. P., *Computational Methods*, Fizmatgiz, Moscow, (1966).

[7] Bolnokin, V. E., and Chinayev, P. I., *Computer-Aided Analysis and Synthesis of Automatic Control Systems. Algorithms and Programs*, Radio and Svyaz', Moscow, (1986).

[8] Dekker, K., and Verwer, J. D., *Stability of Runge–Kutta Methods for Stiff Nonlinear Differential Equations*, North Holland, Amsterdam–New York, (1984).

[9] Faddeev, D. K., and Faddeeva, B. N., *Computational Methods of Linear Algebra*, Fizmatgiz, Moscow–Leningrad, (1963).

[10] Fedorenko, R. P., *Approximate Solution of Optimal Control Problems*, Nauka, Moscow, (1978).

[11] Forsythe, G. E., Malcolm, M. A., and Moler, C. B., *Computer Methods for Mathematical Computation*, Prentice Hall, Englewood Cliffs, N. J., (1977).

[12] Godunov, S. K., *Solving Linear Algebraic Equations*, Nauka, Novosibirsk, (1980).

[13] Hairer, E., Norsett, S. P., and Wanner, G., *Solving Ordinary Differential Equations, I. Nonstiff Problems*, Springer-Verlag, Berlin–Heidelberg, (1987).

[14] Ikramov, H. B., *Numerical Solution of Matrix Equations*, Nauka, Moscow, (1984).

[15] Kalitkin, N. N., *Numerical Methods*, Nauka, Moscow, (1978).

[16] Kolmanovskii, V. B., and Nosov, V. R., *Approximative and Numerical Methods in Control System Design*, (lecture notes), Moscow University of Electronics and Mathematics, Moscow, (1989).

[17] Marchuk, G. I., *Methods of Numerical Mathematics*, Springer-Verlag, New York–Berlin, (1975).

[18] Melsa, J. L., and Jones, S. K., *Computer Programs for Computational Assistance in the Study of Linear Control Theory*, McGraw–Hill, New York, (1973).

[19] *Modern Methods for Ordinary Differential Equations*, (Hall, G., and Watt, J. M. eds.), Clarendon Press, Oxford, (1976).

[20] Moiseev, N. N., *Foundations of the Theory of Optimal Systems*, Nauka, Moscow, (1975).

[21] Nosov, V. R., *Algorithms for Linear Algebra*, (lecture notes), Moscow University of Electronics and Mathematics, Moscow, (1983).

[22] Ortega, J. M., and Poole, W. G., Jr., *An Introduction to Numerical Methods for Differential Equations*, Pitman Publ., (1981).

[23] Voevodin, V. V., *Numerical Foundations of Linear Algebra*, Nauka, Moscow, (1977).

[24] Wilkinson, J. H., *The Algebraic Eigenvalue Problem*, Oxford University Press, London, (1965).

[25] Wilkinson, J. H., and Reinsch, C., *Handbook for Automatic Computation. Linear Algebra*, Springer-Verlag, Berlin, (1973).

SUBJECT INDEX

A

Abramov, method, 630
Adams, formula, 592, 593
Aircraft, 276
Algorithm, for constructing optimal control, 304, 438
  quasioptimal stabilization, 325
Alley, law, 109
Applicability of a quasilinear filter, scalar case, 497
  multidimensional case, 499
Asymptotic, stability, 8
  f-stability, 60
  x-stability, 25
  p-stability, 91
    global, 8
    uniform, 9
Attraction domain, 8
Attractor, 8

B

Barbashin–Krasovskii theorem, 20
Bellman, equation, 303, 399
  function, 302, 399
  theorem, 38, 112
Bernoulli, equation, 160
Binary distillation column, 647
"Boeing-747", landing, 576
Boltz, problem, 129, 156, 186
Boundary layer region, 607
Brownian motion, 77
BS-algorithm, 567
Bus, control of, 649
Butcher, formula, 584, 599

C

Canonical form, of linear time-invariant systems, 258
  of time-varying systems, 260
Case, critical, 47
  noncritical, 47
Cash, formula, 605
Catching of a species, 226
Cauchy–Bunyakovskii inequality, 152
  Cauchy formula, 152
  matrix, 152
Chaplygin, problem, 175
Chattering mode, 359
Chebyshev, inequality, 92
Chetaev, theorem, 15
  sheaf of integrals, 18
Comparison, principle, 23
Complementary slackness, conditions, 198, 271
Construction, of an optimal control, 158, 170, 448
  of observation laws, 513
Control, admissible, 136, 302, 312, 436, 446, 450
  F- (feedback), 133, 303, 394
  optimal, 136, 153, 332, 448
  P- (program), 132
  singular, 351, 353
  stabilizing, 311, 568
  time-optimal, 130, 445
Controllability, of linear time-invariant systems, 252, 570
  of linear time-varying systems, 255
  of a two-link manipulator, 263
  matrix, 252
  set, 249
  with constraints, 255
Controller, P- (proportional), 555
  PD- (proportional-differential), 319
Constraints, control, 132, 394
  isoperimetric, 131, 197
  joint, 135
  trajectory, 131
Covariance matrix, 568

D

Dahlquist, stable multistep formula, 598
strongly stable formula, 599
theorem, 598
Derivative along the solution of an equation, 11
Description, frequency-domain, 281
input-output, 281
state-space, 281
Difference, equation, 64
formula, 591
Distribution, finite-dimensional, 76
function, 74
Drebbel, temperature regulator, 132
Driver cell, 171
Dynamic programming, principle, 301, 398

E

Ecological systems, 219
Electric motor, 137
Elementary event, 74
Equations, for optimal estimate, 476
functional-differential, 8
of evolution,
of a single population, 220
of a system, 128, 169, 393
of second order with periodic coefficients, 40
of two species, 222
stochastic differential, 79
linear, 88, 96
scalar, 98
with delay, 221
Error, control, 283
dynamic, 283
estimates, 565, 596
estimation, 474
of quasioptimal stabilization, 325
Estimate, H_∞-norm, 287
H_2-norm, 288
for Runge–Kutta methods, 585
for multistep methods, 596
for stable equations, 588
of first approximation, 427
of higher order approximation, 429
of zero approximation, 424
Estimation, linear, 468
of Gaussian variables, 469
optimal, 469
Existence theorem, for a minimizing sequence, 337
for an H_∞-suboptimal controller, 288, 290
for optimal control, 344
for optimal stabilizing control, 311
for stochastic equation, 86
using auxiliary functions, 332
Euler, equations, 18, 169, 197, 247
implicit formula, 604
method, 582
stability conditions, 603
"Explorer-1", 19

F

Fedorenko, method, 635
Feedback control system, 133, 280
Filter, discrete, 486
ideal low-pass, 283
nonlinear, 488
quasilinear, 495, 499, 549
quasioptimal, 490
stability, 478
with constant parameters, 480
with degenerate noise in observations, 481
Floquet, multiplier, 39
theorem, 39
Formula, difference, 591
explicit, 591
implicit, 591
numerically stable, 598
predictor-corrector, 594
"3/8", 584
Frequency criteria of stability, 29
Function, auxiliary, 332
negative-semidefinite, 12
positive-definite, 12
transfer, 281
with infinitesimal upper limit, 13
Functional, cost, 129, 396
integral cost, 515
quadratic, 157, 406

G

Gain–frequency characteristic, 30, 284
Gaussian, conditionally, distribution 490
distribution, 75
process, 76
Gear, method, 606

Generator of a Markov process, 87
Generic position, condition, 240
Gill, formula, 584
Goddard's, problem, 145
Goodwin, equation, 111
Grid, 582
Gronwall–Bellman, lemma, 37
Gyrocompass, 118
Gyroscope with spinning vessels, 120
Gyrostat, 444

H

Hamiltonian, 153, 184, 198
Hardy, spaces, 284
Hasminskii, theorem, 93, 102
Hautus, criterion, 261
Heterodyne receiver, 31
H_2, estimate, 288
  space, 285
H_∞, estimate, 287
  optimal control, 286, 383, 644, 654
  problem, 288
  robust stability, 291
  space, 285
Hutchinson, equation, 221
  with diffusion, 222

I

Interpolation, 467, 486
Interval, singular, 353
Iterated logarithm, law, 78
Ito, equation, 79, 81
  formula, 82
  stochastic integral, 79

J

Jacobi, method, 559
Jordan, box, 26
  canonical form, 26

K

Kalman filter, 472
  numerical design, 613
Kelley, conditions, 355
  generalization, 358
Kharitonov, theorem, 35
Kolmogorov, backward equation, 87
  theorem, 76
Kolmogorov–Chapman equation, 400
Kopp–Mayer, conditions, 356
  generalization, 359
Kronecker, product, 46
Krylov, method, 556

L

Lagrange, method of multipliers, 168
  problem, 129
Landing, of a "Boeing-747", 576, 644
  on the Moon, 147, 214
Lappo-Danilevskii, condition, 114
Law, linear-fractional tangent, 179
  pulse-observation, 517
Lax, pair, 116
Liénard, equation, 108
Liénard–Chipart criterion, 29
Leverrier, method, 557
Logarithmic norm, 53
Lotka–Volterra, equation, 223
Luenberger, observer, 273
"Luna-9", 149
Lyapunov, constant, 40
  direct method, 11
    for discrete equations, 64
    for optimal stabilization, 310
    for stochastic systems, 93
  function, 11
    degenerate, 60
    vector, 23
  matrix equation, 43, 563
    continious-time, 563
    discrete-time, 66, 563
    methods of solving, 563
    sequence of, 572
    solvability theorem, 44
  problem, 137
  theorem, first, 12
    on reducibility, 114
    on stability in the first
        approximation, 47
    second, 13
  stability, 4

M

McKendrick, model, 221

Malthus, equation, 220
Marachkov, theorem, 14
Marchuk, model in immunology, 642
Markov, process, 79, 86
  control, 395
Massera, example, 14
  inversion theorem, 16
Material point, damping, 242
  noise control, 528
  pulsed observation, 522
  stochastic problems, 441, 462
Mathematical expectation, 75
  conditional, 74
Mathieu, equation, 41
Matrix, controllability, 252
  covariance, 75, 88, 570
  kinematically similar, 114
  measurement, 266
  monodromy, 39
  observability, 268
  of a system, 292
  pseudo-inverse, 569
  sign function, 565
  transfer, 282
Matrosov, criterion, 22
Mayer, problem, 129, 183, 190
Mayevskii–Krylov, condition, 17
Medecine absorbtion and metabolism, model, 377
Method, A-stable, 604
  embedded, 590
  exponential, 606
  multistep, 591
  of "frozen" coefficients, 36
  one-step, 581
  precise for eigenvalue problem, 555
  series, 564
  successive approximation, 162, 321, 423
Mikhailov, criterion, 30
  hodograph, 30
Milne, formula, 594
MIMO-system, 281, 384
Monod, model, 225
Motion, with regulated friction, 208
Motor shaft, rotation, 189

N

Necessary conditions, for multivariable control, 196
  for optimal control, 153, 157, 169, 184, 191, 197
  for optimality of scalar control, 191
Needle-shaped variations, 185
Newton, method, 622
Node, 582
Nonpositivity conditions, 169, 197
Nuclear reactor, optimal configuration, 206
  optimal reduction of power, 307
  optimization of characteristics, 139
  optimization of power, 140
  optimization of xenon transient process, 143
Number, singular, 570

O

Observer, in stabilization system of an aircraft, 276
  of Luenberger, 273
  of maximum order, 272
  of reduced order, 273
Observability, 268, 570
Olkhovskii, autopilot, 134
Optimal, catching of the species, 226
  combination of control and observation, 529
  control, 132, 153, 168
  estimation of Gaussian variables, 469
  extrapolation, 484
  interpolation, 485
  noise control, 524
Optimization of observation, 509

P

Parametric resonance, 42
Pendulum, 375, 389, 457
  damping, 244
  inverted system, 646
  stochastic control, 437, 457
  with friction, 22, 109, 623
Persidskii, theorem, 13
Phase, vector, 128
  portrait, 6, 8, 223
Point, w-limiting, 20
  switch, 239
Poisson, brackets, 354
  disturbances, 415
Population models, 220

Pontryagin, maximum principle, 184
"Predator–prey" model, control, 225
  equation, 222
  stability, 111
Probability, density, 74
  measure, 74
  space, 74
Problem, boundary-value, 621
    linear, 625
    of the maximum principle, 621
  complete eigenvalue, 555
  fixed-time, 131, 172
  linear-quadratic (LQ-), 157, 308, 320, 529, 607
    stochastic, 406, 409, 415
  regulation, 283
  time-optimal, 130, 227, 239
  tracking, 283
  singularly perturbed, 139
  with isoperimetric constraints, 131, 197
  with variable end points, 131, 183
Process, measurable, 76
  stochastic, 76
  separable, 76
  with independent increment, 77

Q

QR-algorithm, 559, 567, 573
Quetelett–Verhulst, equation, 220

R

Realization, minimal, 282
  problem, 282
Reduction, to a boundary-value problem, 625
  to a Cauchy problem, 626
  to nonlinear programming, 620
  to a sequence of Lyapunov equations, 572
Regulator, 134
Riccati, algebraic equation, 287, 322, 323, 572, 612
  matrix equation, 159, 320, 608
  2-Riccati method, 286, 291
Rigid body, damping, 246, 306
  equations of motion, 18, 128
  stochastic control, 460, 463
  time-optimal control, 439
Robot manipulator, Cyclone-type, 50
  n-link, 315
  single-link, 49
  two-link, 263
Rocket, 130, 177, 376, 643, 645
Routh–Hurwitz, criterion, 29, 561
Rumyantsev, theorem, 25
Runge–Kutta, formula, 582, 584
    error estimate, 585
    implicit, 605
  Feldberg, formula, 590

S

Sample path, 76
Satellite, pitch stability, 99
  stability in probability, 99
  yaw angle stability, 100
Schur, form of a matrix, 559
Schur–Kohn criterion, 66
Sequence, minimizing, 337, 344
Shatrovskii, method, 632
Shell, rotation, 48
  stability of motion, 16
Single-loop systems, 33
Singular perturbation, problem, 139
  system, 600
SISO-system, 280, 383
Sliding mode, 365
  time-optimal, 369
Solution, perturbed, 4
  p-stable, 91
  stable, 4
  unperturbed, 4
  unstable, 6
Space vehicle, axially symmetric, 246
  optimal transfer, 145
  planar transfer, 217
Spacecraft, spatial controlled motion, 652
  optimal control, 145
"Sputnik", 19
Stability, 4
  exponential, 10
  f-, 60
  in mean square, 96
  in probability, 92
  in the first approximation, 46
  partial, 24
  robust, 34, 291
  stochastic, 90
  under steady acting disturbances, 10

uniform, 7
with respect to specified variables, 66
x-, 25
Stabilization, adaptive, 331
asymptotic, 310
exponential, 310
of quasilinear systems, 325
of the ship's heading, 106
of stochastic systems, 417
optimal, 310
quasioptimal, 325
Stationarity, conditions, 169
Stodola, theorem, 28
Störmer, formula, 594
Stratonovich, equation, 81
for conditional probability, 489
integral, 80
System, bilinear, 129
controllable, 252
discrete, 63, 202
evolutionary, xix
invariant, 283
linear, 129
quasilinear, 325
stiff, 600
with an almost constant matrix, 37
with periodic coefficients, 40
Switch points, 239
Switching curves, 213, 216, 232, 244 246, 248, 251, 382, 443, 458

T

Takoma bridge disaster 42
Transfer, from one manifold to another manifold, 196
of a space vehicle, 217
of boundary conditions, 626
Transversality conditions, 169, 197

V

Van-der-Pol, equation, 108
Variable, end points conditions, 169, 197
independent, 76
random, 74
Variance, terminal, 553
Vertical ascent of a rocket, 643

W

Watt, centrifugal governor, 133
Ważewski, inequality, 37
Wedge, 12
Wiener–Hopf equation, 469
Wiener process, 77
Wire reeling, optimal control, 166
control with random disturbances, 416

Y

Yaw angle stability, 99
Yerugin, theorem, 113

Z

Zero, invariant, 291, 385
of a linear system, 292
transfer, 295, 385
Zubov, method, 562

Other *Mathematics and Its Applications* titles of interest:

B.S. Razumikhin: *Physical Models and Equilibrium Methods in Programming and Economics.* 1984, 368 pp. ISBN 90-277-1644-7

N.K. Bose (ed.): *Multidimensional Systems Theory. Progress, Directions and Open Problems in Multidimensional Systems.* 1985, 280 pp. ISBN 90-277-1764-8

J. Szep and F. Forgo: *Introduction to the Theory of Games.* 1985, 412 pp. ISBN 90-277-1404-5

V. Komkov: *Variational Principles of Continuum Mechanics with Engineering Applications.* Volume 1: Critical Points Theory. 1986, 398 pp. ISBN 90-277-2157-2

V. Barbu and Th. Precupanu: *Convexity and Optimization in Banach Spaces.* 1986, 416 pp. ISBN 90-277-1761-3

M. Fliess and M. Hazewinkel (eds.): *Algebraic and Geometric Methods in Nonlinear Control Theory.* 1986, 658 pp. ISBN 90-277-2286-2

P.J.M. van Laarhoven and E.H.L. Aarts: *Simulated Annealing: Theory and Applications.* 1987, 198 pp. ISBN 90-277-2513-6

B.S. Razumikhin: *Classical Principles and Optimization Problems.* 1987, 528 pp. ISBN 90-277-2605-1

S. Rolewicz: *Functional Analysis and Control Theory. Linear Systems.* 1987, 544 pp. ISBN 90-277-2186-6

V. Komkov: *Variational Principles of Continuum Mechanics with Engineering Applications.* Volume 2: Introduction to Optimal Design Theory. 1988, 288 pp. ISBN 90-277-2639-6

A.A. Pervozvanskii and V.G. Gaitsgori: *Theory of Suboptimal Decisions.* Decomposition and Aggregation. 1988, 404 pp. *out of print,* ISBN 90-277-2401-6

J. Mockus: *Bayesian Approach to Global Optimization.* Theory and Applications. 1989, 272 pp. ISBN 0-7923-0115-3

Du Dingzhu and Hu Guoding (eds.): *Combinatorics, Computing and Complexity.* 1989, 248 pp. ISBN 0-7923-0308-3

M. Iri and K. Tanabe: *Mathematical Programming.* Recent Developments and Applications. 1989, 392 pp. ISBN 0-7923-0490-X

A.T. Fomenko: *Variational Principles in Topology.* Multidimensional Minimal Surface Theory. 1990, 388 pp. ISBN 0-7923-0230-3

A.G. Butkovskiy and Yu.I. Samoilenko: *Control of Quantum-Mechanical Processes and Systems.* 1990, 246 pp. ISBN 0-7923-0689-9

A.V. Gheorghe: *Decision Processes in Dynamic Probabilistic Systems.* 1990, 372 pp. ISBN 0-7923-0544-2

Other *Mathematics and Its Applications* titles of interest:

A.G. Butkovskiy: *Phase Portraits of Control Dynamical Systems.* 1991, 180 pp.
ISBN 0-7923-1057-8

A.A. Zhigljavsky: *Theory of the Global Random Search.* 1991, 360 pp.
ISBN 0-7923-1122-1

G. Ruhe: *Algorithmic Aspects of Flows in Networks.* 1991, 220 pp.
ISBN 0-7923-1151-5

S. Walukuwiecz: *Integer Programming.* 1991, 196 pp. ISBN 0-7923-0726-7

M. Kisielewicz: *Differential Inclusions and Optimal Control.* 1991, 320 pp.
ISBN 0-7923-0675-9

J. Klamka: *Controllability of Dynamical Systems.* 1991, 260 pp.
ISBN 0-7923-0822-0

V.N. Fomin: *Discrete Linear Control Systems.* 1991, 302 pp. ISBN 0-7923-1248-1

L. Xiao-Xin: *Absolute Stability of Nonlinear Control Systems.* 1992, 180 pp.
ISBN 0-7923-1988-5

A. Halanay and V. Rasvan: *Applications of Liapunov Methods in Stability.* 1993, 238 pp. ISBN 0-7923-2120-0

D. den Hertog: *Interior Point Approach to Linear, Quadratic and Convex Programming.* 1994, 208 pp. ISBN 0-7923-2734-9

V.S. Tanaev, V.S. Gordon and Y.M. Shafranksy: *Scheduling Theory.* Single-Stage Systems. 1994, 380 pp. ISBN 0-7923-2853-1

V.S. Tanaev, Y.N. Sotskov and V.A. Strusevich: *Scheduling Theory.* Multi-Stage Systems. 1994, 412 pp. ISBN 0-7923-2854-X

L.D. Akulenko: *Problems and Methods of Optimal Control.* 1994, 356 pp.
ISBN 0-7923-2855-8

C. Udrişte: *Convex Functions and Optimization Methods on Riemannian Manifolds.* 1994, 348 pp. ISBN 0-7923-3002-1

B.S. Jensen: *The Dynamic Systems of Basic Economic Growth Models.* 1994, 355 pp. ISBN 0-7923-3091-9

V. Barbu: *Mathematical Methods in Optimization of Differential Systems.* 1994, 259 pp. ISBN 0-7923-3176-1

A. Prékopa: *Stochastic Programming.* 1995, 556 pp. ISBN 0-7923-3482-5

R. Lucchetti and J. Revalski (eds.): *Recent Developments in Well-Posed Variational Problems.* 1995, 266 pp. ISBN 0-7923-3576-7

A. Cheremensky and V. Fomin: *Operator Approach to Linear Control Systems.* 1995, 410 pp. ISBN 0-7923-3765-4

Other *Mathematics and Its Applications* titles of interest:

V.N. Afanas'ev, V.B. Kolmanovskii and V.R. Nosov: *Mathematical Theory of Control Systems Design*. 1996, 681 pp. ISBN 0-7923-3724-7

MIX
Papier aus verantwortungsvollen Quellen
Paper from responsible sources
FSC® C105338

FSC
www.fsc.org

If you have any concerns about our products,
you can contact us on
ProductSafety@springernature.com

In case Publisher is established outside the EU,
the EU authorized representative is:
Springer Nature Customer Service Center GmbH
Europaplatz 3, 69115 Heidelberg, Germany

Printed by Libri Plureos GmbH
in Hamburg, Germany